EXCITATORY AMINO ACIDS AND EPILEPSY

ADVANCES IN EXPERIMENTAL MEDICINE AND BIOLOGY

Editorial Board:

NATHAN BACK, *State University of New York at Buffalo*
NICHOLAS R. DI LUZIO, *Tulane University School of Medicine*
EPHRAIM KATCHALSKI-KATZIR, *The Weizmann Institute of Science*
DAVID KRITCHEVSKY, *Wistar Institute*
ABEL LAJTHA, *Rockland Research Institute*
RODOLFO PAOLETTI, *University of Milan*

Recent Volumes in this Series

Volume 198A
KININS IV, Part A
Edited by Lowell M. Greenbaum and Harry S. Margolius

Volume 198B
KININS IV, Part B
Edited by Lowell M. Greenbaum and Harry S. Margolius

Volume 199
NUTRITIONAL AND TOXICOLOGICAL SIGNIFICANCE OF
ENZYME INHIBITORS IN FOODS
Edited by Mendel Friedman

Volume 200
OXYGEN TRANSPORT TO TISSUE VIII
Edited by Ian S. Longmuir

Volume 201
LIPOPROTEIN DEFICIENCY SYNDROMES
Edited by Aubie Angel and Jiri Frohlich

Volume 202
INFECTIONS IN THE IMMUNOCOMPROMISED HOST
Laboratory Diagnosis and Treatment
Edited by Paul Actor, Alan Evangelista, James Poupard, and Eileen Hinks

Volume 203
EXCITATORY AMINO ACIDS AND EPILEPSY
Edited by Robert Schwarcz and Yehezkel Ben-Ari

Volume 204
NEUROBIOLOGY OF CENTRAL D_1-DOPAMINE RECEPTORS
Edited by George R. Breese and Ian Creese

Volume 205
MOLECULAR AND CELLULAR ASPECTS OF REPRODUCTION
Edited by Dharam S. Dhindsa and Om P. Bahl

A Continuation Order Plan is available for this series. A continuation order will bring delivery of each new volume immediately upon publication. Volumes are billed only upon actual shipment. For further information please contact the publisher.

EXCITATORY AMINO ACIDS AND EPILEPSY

Edited by
Robert Schwarcz
Maryland Psychiatric Research Center
University of Maryland School of Medicine
Baltimore, Maryland

and

Yehezkel Ben-Ari
INSERM-U29
Hôpital de Port-Royal
Paris, France

PLENUM PRESS • NEW YORK AND LONDON

Library of Congress Cataloging in Publication Data

International Symposium on Excitatory Amino Acids and Epilepsy (1st: 1985: Château de Fillerval, France)
Excitatory amino acids and epilepsy.

(Advances in experimental medicine and biology; v. 203)
"Proceedings of the First International Symposium on Excitatory Amino Acids and Epilepsy, held September 2-5, 1985, at Château de Fillerval, France"—T.p. verso.
Includes bibliographies and index.
1. Epilepsy—Congresses. 2. Amino acids—Physiological effect—Congresses. 3. Neurotransmitters—Congresses. I. Schwarcz, Robert. II. Ben-Ari, Yehezkel. III. Title. IV. Series. [DNLM: 1. Amino Acids—pharmacodynamics—congresses. 2. Epilepsy—etiology—congresses. W1 AD559 v.203/WL 385 I6082 1985e]
RC372.5.I53 1985   616.8'53   86-22614
ISBN 978-1-4684-7973-7   ISBN 978-1-4684-7971-3 (eBook)
DOI 10.1007/978-1-4684-7971-3

Proceedings of the First International
Symposium on Excitatory Amino Acids and Epilepsy,
held September 2-5, 1985, at Chateau de Fillerval, France

© 1986 Plenum Press, New York
Softcover reprint of the hardcover 1st edition 1986
A Division of Plenum Publishing Corporation
233 Spring Street, New York, N.Y. 10013

All rights reserved

No part of this book may be reproduced, stored in a retrieval system, or transmitted in any form or by any means, electronic, mechanical, photocopying, microfilming, recording, or otherwise, without written permission from the Publisher

FOREWORD

Human epilepsy is a major public health problem affecting approximately 2 persons per 1000. It is particularly frequent in children where convulsions may lead to brain damage and subsequent seizure activity in adulthood. Temporal lobe epilepsy (synonyms include limbic epilepsy, psychomotor epilepsy and complex partial epilepsy) is the most devastating form of epilepsy in the adult population since: a) it is often extremely resistant to currently available anticonvulsant drugs (i.e., it is more resistant than tonico-clonic or grand mal seizures) and b) it includes loss of consciousness, thereby limiting performance of many normal functions and leaving the individual susceptible to bodily injury. It is also associated with nerve cell loss, in particular in the hippocampus and other structures of the temporal lobes.

In order to promote an appropriate therapy it is essential to understand the etiology of seizures and its relationship to brain damage. Basic research on epilepsy also provides a very useful vehicle to learn about the way the brain functions under normal conditions. For instance, much of our present understanding of the mechanisms of action of GABA and benzodiazepines, control of neuronal activity, etc. has been derived from such studies.

Until recently, basic and applied research in this field has been focused primarily on the role of GABA-mediated inhibition in the prevention of epileptogenesis and the removal of GABA (i.e., disinhibition) as a necessary condition to induce paroxysmal discharge. However, in the past few years considerable evidence has accumulated, which suggests that an enhancement of the activity of excitatory amino acids (EAA) could play a central role in the pathogenesis of the epilepsies. Following the initial observation of Hayashi in 1954, namely that the topical application of glutamic acid produces paroxysmal discharge in animals and man, it is now clear that several endogenous and exogenous EAA can produce epilepsy and brain damage - notably in limbic structures - and that EAA can be used to provide suitable experimental models of human temporal lobe epilepsy.

This book contains the Proceedings of the first symposium devoted to EAA and seizure disorders. The international meeting, held at the Chateau de Fillerval, France, September 1-5, 1985, was attended by approximately 100 participants who provided both basic science and clinical perspectives. While many contributions centered on the hippocampus and other limbic structures, the meeting was truly multidisciplinary in nature: it included sessions on the neurochemistry, pharmacology and physiology of EAA and seizures, the contributions of ion shifts to paroxysmal discharges elicited by EAA, and the mechanistic relationships between epilepsy and brain damage. In addition, sessions were devoted to the anatomy of the limbic system, the problem of the blood-brain barrier with special reference to EAA and EAA-induced seizures, and on the possible contributions of trace metals, notably zinc, to the epileptic actions of EAA. For this volume, the session chairmen have prepared a brief commentary, which has been printed as an appendix following the respective scientific papers in order to provide the reader with an insight into the lively discussions which took place in Fillerval.

The meeting at Fillerval could not have been held without the generous contribution of the Monsanto Co., St. Louis, MO. The symposium was also supported by the Centre National de la Recherche Scientifique (CNRS), the Institute National de la Sante et de la Recherche Medicale (INSERM), and by the following pharmaceutical companies: ASTRA, Ciba-Geigy, Ferrosan, E. Merck, Merck, Sharp and Dohme, Nova, Sanofi, Upjohn, Wander, and the Wellcome Foundation.

We are particularly grateful to P. Wolyniec for her excellent editorial assistance and to G. Charton for his outstanding help with the organization of the symposium. We also thank Drs. W.O. Whetsell, Jr. and J.L. Price for providing the 'official' Fillerval photograph enclosed in this volume. Last but not least, we are highly endebted to Niki, Yasmina, Tamara, Damian and Constance for their patience and understanding.

Baltimore and Paris, June 1986   R. Schwarcz
                                                              Y. Ben-Ari

CONTENTS

SESSION I. THE LIMBIC SYSTEM: NEUROANATOMICAL CONCEPTS RELATING TO EPILEPTIC PHENOMENA

Amygdalohippocampal and Amygdalocortical Projections in
    the Primate Brain   3
  D.G. Amaral

Subcortical Projections from the Amygdaloid Complex 19
  J.L. Price

Cortical and Subcortical Afferents of the Amygdaloid
    Complex 35
  F.T. Russchen

Putative Amino Acid Transmitters in the Amygdala 53
  O.P. Ottersen, B.O. Fischer, E. Rinvik, and
    J. Storm-Mathisen

A Survey of the Anatomy of the Hippocampal Formation,
    With Emphasis on the Septotemporal Organization
    of its Intrinsic and Extrinsic Connections 67
  M.P. Witter

Cytochemical Architecture of the Entorhinal Area 83
  C. Köhler

Session I: Commentary 99
  D.G. Amaral and G.W. Van Hoesen

SESSION II. EPILEPTIC BRAIN TISSUE: NEUROPATHOLOGY AND PHYSIOLOGY IN ANIMALS AND MAN

Neuronal and Glial Pathologies: Morphology and
    Physiology of Human and Monkey Epileptic Foci 105
  A.A. Ward, Jr.

Metabolic, Morphologic and Electrophysiologic
    Profiles of Human Temporal Lobe Foci:
    An Attempt at Correlation 115
  T.L. Babb

Endogenous Excitotoxins as Possible Mediators of
    Ischemic and Hypoglycemic Brain Damage 127
  T. Wieloch

Role of the Substantia Nigra in the Kindling Model of Limbic Epilepsy 139
J.O. McNamara, D.W. Bonhaus, and C. Shin

Long Term Sequelae of Parenteral Administration of Kainic Acid 147
L. Nitecka and E. Tremblay

Electrophysiology of Epileptic Tissue: What Pathologies are Epileptogenic? 157
P.A. Schwartzkroin and J.E. Franck

SESSION III. EXCITATORY AMINO ACIDS AND THE BLOOD-BRAIN BARRIER

Pathophysiological Aspects of Blood-Brain Barrier Permeability in Epileptic Seizures 175
C. Nitsch, G. Goping, and I. Klatzo

Blood-Brain Barrier Permeability to Excitatory Amino Acids 191
J.M. Lefauconnier, Y. Tayarani, and G. Bernard

Limbic Seizures Induced by Systemically Applied Kainic Acid: How Much Kainic Acid Reaches the Brain? 199
M.L. Berger, J.M. Lefauconnier, E. Tremblay, and Y. Ben-Ari

Extravasated Protein as a Cause of Limbic Seizure-Induced Brain Damage: An Evaluation Using Kainic Acid 211
R.E. Ruth

Ultrastructural Analysis of Rat Brain Tissue Following Systemic Kainate Administration 223
H. Lassmann, H. Baran, U. Petsche, K. Kitz, G. Sperk, O. Hornykiewicz, and F. Seitelberger

Session III: Commentary 231
J.M. Lefauconnier and I. Klatzo

SESSION IV. EXCITATORY AMINO ACIDS: RECEPTOR INTERACTIONS

Anatomical Organization of Excitatory Amino Acid Receptors and their Properties 237
C.W. Cotman and D.T. Monaghan

Homocysteic Acid, an Endogenous Agonist of NMDA-Receptor: Release, Neuroactivity, and Localization 253
M. Cuénod, K.Q. Do, P.L. Herrling, W.A. Turski, C. Matute, and P. Streit

Excitatory Amino Acid Pathways in the Brain 263
O.P. Ottersen and J. Storm-Mathisen

Synthesis and Release of Amino Acid Transmitters 285
F. Fonnum, R.H. Paulsen, V.M. Fosse, and B. Engelsen

Na⁺ Fluxes as a Tool to Identify Anticonvulsant
    Antagonists of Neuroexcitation   295
  V.I. Teichberg, M. Beaujean, P. David,
    D. Eisenberg-Tamarin, U. Erez, H. Frenk,
    A. Luini, G. Urca, and O. Goldberg

Involvement of Excitatory Amino Acid Receptors in
    the Mechanisms Underlying Excitotoxic Phenomena   303
  A.C. Foster

Session IV: Commentary   317
  C.W. Cotman and J. Storm-Mathisen

SESSION V. EXCITATORY AMINO ACIDS AND SEIZURES: NEUROCHEMICAL INTERRELATIONSHIPS

Excitatory Amino Acid Antagonists as Novel
    Anticonvulsants   321
  B. Meldrum

The Hyperexcited Brain: Glutamic Acid Release and
    Failure of Inhibition   331
  N.M. van Gelder

Anti-Excitotoxic Actions of Taurine in the Rat
    Hippocampus Studied In Vivo and In Vitro   349
  E.D. French, A. Vezzani, W.O. Whetsell, Jr.,
    and R. Schwarcz

Alterations in Extracellular Amino Acids and Ca²⁺
    Following Excitotoxin Administration and
    During Status Epilepticus   363
  A. Lehmann, H. Hagberg, J.W. Lazarewicz,
    I. Jacobson, and A. Hamberger

Acidic Peptides in Brain: Do They Act at Putative
    Glutamatergic Synapses?   375
  J.T. Coyle, R. Blakely, R. Zaczek, K.J. Koller,
    M. Abreu, L. Ory-Lavollée, R. Fisher,
    J.M.H. ffrench-Mullen, and D.O. Carpenter

Session V: Commentary   385
  J.T. Coyle and N. van Gelder

SESSION VI. MECHANISMS OF EPILEPTOGENESIS

Synaptic Events Underlying Spontaneous and Evoked
    Paroxysmal Discharges in Hippocampal Neurons   391
  D. Johnston, P.A. Rutecki, and F.J. Lebeda

Inward Currents in Cat Neocortical Neurons
    Studied In Vitro   401
  W.E. Crill, P.C. Schwindt, J.A. Flatman,
    C.E. Stafstrom, and W. Spain

Synchronization of Pyramidal Cell Firing by
    Ephaptic Currents in Hippocampus In Situ   413
  K. Krnjević, T. Dalkara, and C. Yim

Excitatory Amino Acids and Regenerative Activity
    in Cultured Neurons 425
  J.F. MacDonald, J.H. Schneiderman, and Z. Miljkovic

Long-Term Alterations in Amino Acid-Induced Ionic
    Conductances in Chronic Epilepsy 439
  R. Pumain, J. Louvel, and I. Kurcewicz

Excitatory Amino Acids and Epilepsy-Induced Changes
    in Extracellular Space Size 449
  U. Heinemann

Session VI: Commentary 461
  K. Krnjević and U. Heinemann

SESSION VII. EXCITATORY AMINO ACIDS: PHYSIOLOGICAL STUDIES

Evidence for the Activation of the N-Methyl-D-Aspartate
    Receptor During Epileptiform Discharge 465
  G.L. King and R. Dingledine

Effects of Kainate on CA1 Hippocampal Neurons Recorded
    In Vitro 475
  E. Cherubini, C. Rovira, M. Gho, and Y. Ben-Ari

Blockade by D-Aminophosphonovalerate or Mg^{2+} of
    Excitatory Amino Acid-Induced Responses on
    Spinal Motoneurons In Vitro 485
  A. Nistri and A.E. King

The Membrane Action of Excitatory Amino Acids on
    Cultured Mouse Spinal Cord Neurons 497
  G.L. Westbrook and M.L. Mayer

A Patch-Clamp Study of Excitatory Amino Acid
    Activated Channels 507
  P. Ascher and L. Nowak

Amino Acid Activated Receptor-Channels at Peripheral
    and Central Synapses 513
  S.G. Cull-Candy

Expression of Vertebrate Amino Acid Receptors in
    Xenopus Oocytes 525
  T.G. Smart, A. Constanti, K. Houamed, G. Bilbe,
    D.A. Brown, E.A. Barnard, and C. VanRenterghem

Session VII: Commentary 539
  D.A. Brown and R. Dingledine

SESSION VIII. METAL IONS AND EPILEPSY

Transition Metal Ions in Epilepsy: An Overview 545
  S.H. Chung, B. Gabrielsson, and D.K. Norris

Zinc-Binding Proteins in the Brain 557
  M. Ebadi and Y. Hama

Neurobehavioral, Neuroendocrine and Neurochemical
   Effects of Zinc Supplementation in Rats 571
   M. Baraldi, P. Zanoli, A. Benelli, M. Sandrini,
   A. Giberti, E. Caselgrandi, G. Tosi, and C. Preti

Excitatory Amino Acids and Divalent Cations in the
   Kindling Model of Epilepsy 587
   J.T. Slevin, E.J. Kasarskis, T.C. Vanaman, and
   M. Zurini

Effect of Zinc on Neuronal Activity in the Rat
   Forebrain 599
   D.M. Wright

Relationship of Glutamic Acid and Zinc to Kindling
   of the Rat Amygdala: Afferent Transmitter
   Systems and Excitability in a Model of Epilepsy 611
   I.L. Crawford

Session VIII: Commentary 625
   S.H. Chung and I.L. Crawford

SESSION IX. SEIZURES AND BRAIN DAMAGE: THE EXCITOTOXIC LINK

Inciting Excitotoxic Cytocide Among Central Neurons 631
   J.W. Olney

Selective and Non-selective Seizure Related Brain
   Damage Produced by Kainic Acid 647
   Y. Ben-Ari, A. Repressa, E. Tremblay, and L. Nitecka

On the Role of Seizure Activity and Endogenous
   Excitatory Amino Acids in Mediating Seizure-
   Associated Hippocampal Damage 659
   R.S. Sloviter

Kainic Acid Seizures and Neuronal Cell Death: Insights
   from Studies of Selective Lesions and Drugs 673
   J.V. Nadler, M.M. Okazaki, M. Gruenthal, B. Ault,
   and D.R. Armstrong

Glutamate and Anoxic Neuronal Death *In Vitro* 687
   S.M. Rothman

Quinolinic Acid: A Pathogen in Seizure Disorders? 697
   R. Schwarcz, C. Speciale, E. Okuno, E.D. French,
   and C. Köhler

Session IX: Commentary 709
   Y. Ben-Ari and R. Schwarcz

Contributors 713

Index 729

SESSION I
THE LIMBIC SYSTEM: NEUROANATOMICAL CONCEPTS RELATING TO EPILEPTIC PHENOMENA

AMYGDALOHIPPOCAMPAL AND AMYGDALOCORTICAL PROJECTIONS IN THE PRIMATE BRAIN

D.G. Amaral

The Salk Institute for Biological Studies and the Clayton
Foundation for Research - California Division
P.O. Box 85800, San Diego, California 92138

INTRODUCTION

Several lines of evidence have supported the view that the amygdaloid complex plays a major role in the initiation and maintenance of temporal lobe or 'psychomotor' seizures. Chronic stereotaxic depth recordings in intractable epileptics, for example, have demonstrated striking correlations between amygdaloid epileptic discharges and the behavioral changes associated with seizures (Wieser, 1983). Electrical stimulation of the amygdaloid complex can also elicit experiences, such as illusions of perception or memory and emotions, such as fear, which are common manifestations of temporal lobe epilepsy (Gloor et al., 1982). Recent experimental models of temporal lobe epilepsy, such as the kindling paradigm (Le Gal La Salle, 1982) and the use of excitatory neurotoxins (Tremblay and Ben-Ari, 1984) have provided additional evidence that the amygdaloid complex and the closely associated hippocampal formation are particularly vulnerable to seizure-producing conditions. In fact, the amygdala and hippocampus are among the most commonly damaged structures in epileptic encephalopathy (Ounsted et al., 1966; Corsellis and Bruton, 1983). Given this context, it would be of obvious importance to determine the extent and organization of the interconnectivity of these two structures as well as their connections with cortical regions which are likely to subserve the perceptual and cognitive functions that are disrupted during seizures. It is only in the last several years, however, that these connections have been systematically studied.

As recently as the late 1960's, the bulk of available evidence was consistent with the inference that the interactions of the amygdaloid complex were largely with olfactory structures or with the hypothalamus (Cowan et al., 1965; Ishikawa et al., 1969). Subsequently, a variety of other amygdaloid connections have been demonstrated. In addition to the substantial connections with the hypothalamus and brain stem, the amygdaloid complex gives rise to prominent projections to the thalamus and neocortex, to the hippocampal formation, to major portions of the striatum and to several cell groups of the basal forebrain including the cholinergic cells of the basal nucleus of Meynert. Many of these recently discovered connections of the amygdaloid complex are reviewed in the chapters by Russchen and Price in this volume. In the present chapter, we will summarize the organization of the connections of the primate amygdaloid complex with the hippocampal formation and the neocortex.

CYTOARCHITECTONIC ORGANIZATION OF THE MONKEY AMYGDALOID COMPLEX

While there is still no generally accepted nomenclature for the primate amygdaloid complex, we have adopted a terminology that closely resembles that employed by Crosby and Humphrey (1941) and colleagues (Lauer, 1945). The major cell groups of the macaque monkey amygdala are illustrated in Fig. 1. The noncortical or deep portion of the amygdala can be divided into the lateral nucleus, which is the largest component of the human amygdala, the basal nucleus, which has magnocellular, parvicellular and paralaminar regions, the accessory basal nucleus, which can also be divided into magnocellular and parvicellular regions, and the central nucleus, which has a medial and lateral subdivision. Not included in this illustration are the anterior amygdaloid area which forms the rostrodorsal border with the substantia innominata and the small, intercalated cell masses located variously below the central nucleus and between the major deep nuclei. The cortical surface of the amygdala comprises a number of distinct regions including the pyriform cortex, the cortical nuclei, the medial nucleus, the periamygdaloid cortex and a transition zone with the hippocampal formation called the amygdalohippocampal area.

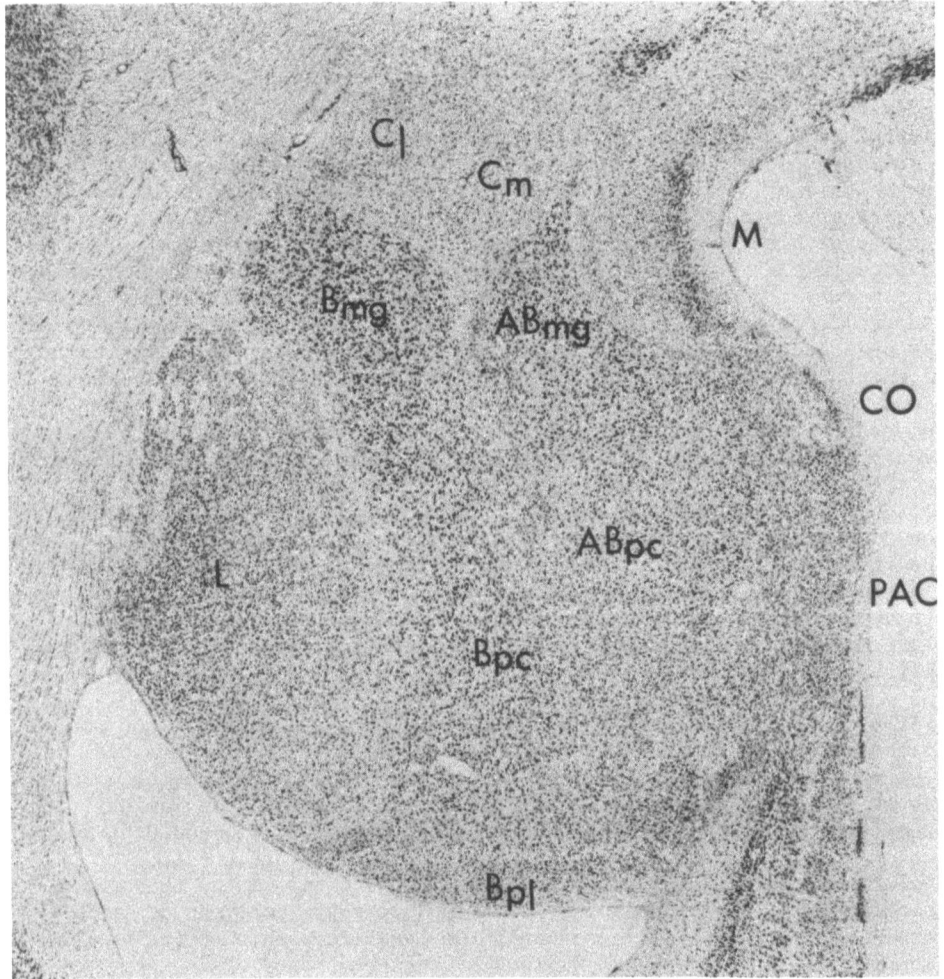

Fig. 1. Nissl-stained coronal section through the amygdaloid complex of the macaque monkey. See text for a description of the major nuclei.

Unfortunately, the nomenclature employed for the amygdala of monkey and man is quite different from that commonly used for nonprimates. For this review we should simply point out that the basolateral nucleus of the nonprimate is homologous with the basal nucleus of the primate. The anterior basolateral nucleus resembles the magnocellular division and the posterior basolateral nucleus is similar, in many respects, to the parvicellular division. The accessory basal nucleus has not generally been defined in nonprimates but would appear to correspond to the medial aspect of the anterior basolateral nucleus. The basomedial nucleus of the rodent and cat, in turn, corresponds in part to the parvicellular division of the accessory basal nucleus.

Early students of the amygdaloid complex often ignored its cytoarchitectonic heterogeneity and considered it to function in a monolithic fashion. As Kaada (1972) noted some time ago, however, there is ample evidence that each of the amygdaloid nuclei functions in a discrete and often antagonistic manner in a variety of behaviors. The findings of recent anatomical studies into the connectivity of each of the nuclei and cortical regions are entirely consistent with this view and have determined that many of the amygdaloid subdivisions give rise to specific and topographically organized projections to the hippocampal formation and neocortex.

AMYGDALOHIPPOCAMPAL PROJECTIONS

Background

Until recently, the amygdaloid complex and the hippocampal formation were thought to be anatomically unrelated despite their close physical association in the temporal lobe. Krettek and Price (1974, 1977) first demonstrated that the rat and cat amygdala have direct projections to the hippocampal formation. In their studies, the lateral nucleus was shown to project heavily to the ventral part of the lateral entorhinal cortex and terminate mainly in layer III. An additional minor projection was noted to the outer and middle thirds of the plexiform layer of the subiculum. Projections arising from the posterior portion of the basolateral nucleus (which, as noted above, is probably homologous to the parvicellular portion of the basal nucleus in the monkey) terminated in the cellular layer and deep one-third of the plexiform layer of the ventral subiculum and in the deep portion of the plexiform layer and superficial aspect of the cellular layer of the parasubiculum; the dorsal subiculum did not appear to receive an amygdaloid projection. The fibers in the plexiform layer of the subiculum were observed to extend slightly into the adjacent molecular layer of the hippocampus. There were also indications that the posterior basolateral nucleus projected to the deep layers of the ventral, lateral entorhinal cortex. The periamygdaloid cortex gave rise to projections to layers I and II of the lateral entorhinal cortex and to the superficial one-third of the plexiform layer of the ventral subiculum.

In the rat, Ottersen (1982) has demonstrated a return projection from the hippocampal formation to the amygdala. The subicular complex was shown to project to the lateral nucleus and the entorhinal cortex projected both to the lateral and basal nuclei. At least a portion of field CA1 of the hippocampus was shown to project to the lateral, basal and cortical nuclei. Similar connections exist in the cat (Witter et al., 1981) and monkey (Rosene and Van Hoesen, 1977; Van Hoesen, 1981). In these animals, the entorhinal cortex projects to both divisions of the basal nucleus and to the lateral and central nuclei. The primate subicular complex projects to the parvicellular division of the basal nucleus and to the periamygdaloid cortex. Physiological studies have demonstrated that stimulation of the rostral hippocampal formation of the monkey modulates neuronal activity in the amygdaloid complex, especially in the parvicellular division of the basal nucleus (Morrison and Poletti, 1980). It should also be noted that depth electrode stimulation and recording in patients with temporal lobe epilepsy demonstrated direct amygdaloid projections to the hippocampal formation (Buser and Bancaud, 1983) though the existence of the return projection was more difficult to establish.

Recent Primate Studies

The data discussed in this section and the subsequent section on amygdalocortical projections are derived from a large series of experiments carried out on mature *Macaca fascicularis* monkeys in which injections of either an anterograde or retrograde tracer had been placed into one of the subdivisions of the amygdaloid complex or one of the fields of the hippocampal formation. Detailed descriptions of the preparative procedures and analysis can be found in Amaral et al. (1983, 1984) and Amaral and Price (1984).

While the general pattern of amygdalohippocampal projections in the monkey is similar to that in the rat and cat, a number of differences have been observed. Notable among these is a more robust projection between the amygdala and the CA1 field of the hippocampus. A summary of the amygdalohippocampal projections is presented in Fig. 4.

The dentate gyrus does not appear to receive an amygdaloid input, though in one case, with a fairly large injection of ^{3}H-amino acids involving the basal and accessory basal nuclei, the extreme rostral portion of the dentate gyrus contained a few labeled fibers in the molecular layer. While caudal levels of field CA3 of the hippocampus do not appear to be innervated by the amygdala, labeled fibers do pass through rostral levels of it en route to field CA1. A projection from the amygdala to the full extent of field CA1 of the hippocampus was not described in the work of Krettek and Price (1974, 1977). However, in experiments in which the ^{3}H-amino acid injection involved the accessory basal nucleus of the amygdala, a dense band of fibers entered the hippocampal formation through the amygdalohippocampal area, swung laterally through the hippocampus and appeared to terminate in the stratum lacunosum-moleculare of field CA1 (Fig. 2). The projection continued in this position through about 80% of the rostrocaudal extent of the hippocampus and at caudal levels extended slightly into the CA1/subiculum border zone.

In order to determine the cells of origin of this projection, an injection of the retrograde tracer WGA-HRP was placed into the rostral hippocampus at the point of entry of the fiber bundle. Consistent with the autoradiographic studies, within the amygdaloid complex most of the retrogradely labeled cells were observed in the magnocellular division of the accessory basal nucleus and to a lesser extent in the periamygdaloid cortex (Fig. 3).

A focus of amygdaloid termination in the hippocampal formation is at the border of field CA1 with the subiculum (Fig. 2). This region can be distinguished both on cytoarchitectonic grounds and on the basis of its histochemical staining pattern; it reacts strongly, for example, for the presence of acetylcholinesterase (Bakst and Amaral, 1984). So distinctive is this area, that several workers have considered it to be a separate division of the hippocampal formation and have labeled it the prosubiculum. Careful examination of this region, however, has led us to the view that it should be considered a border region between the hippo-

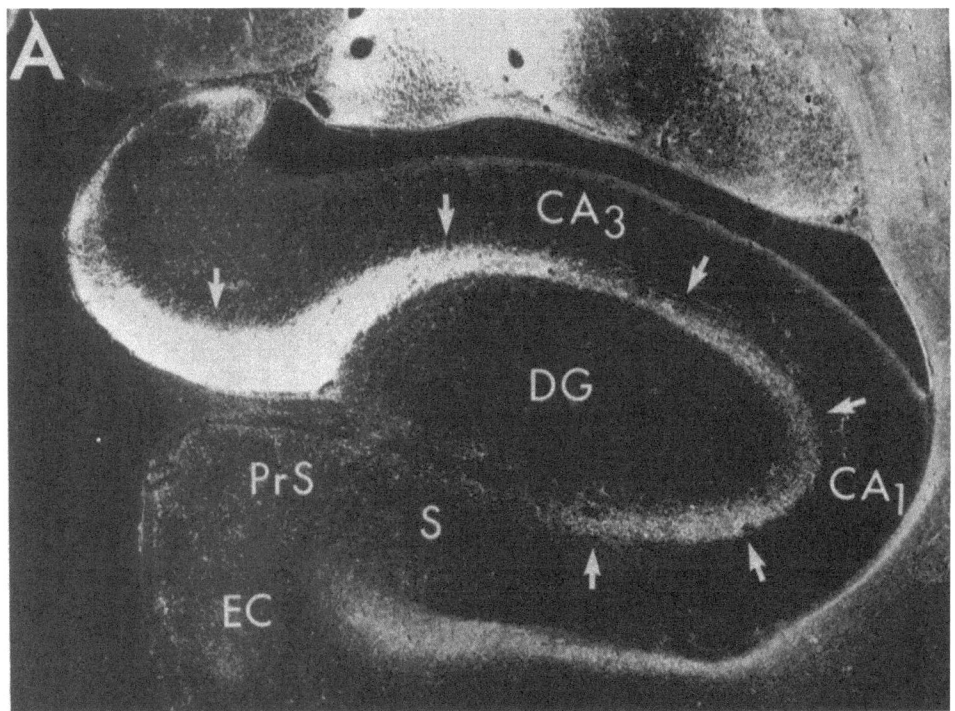

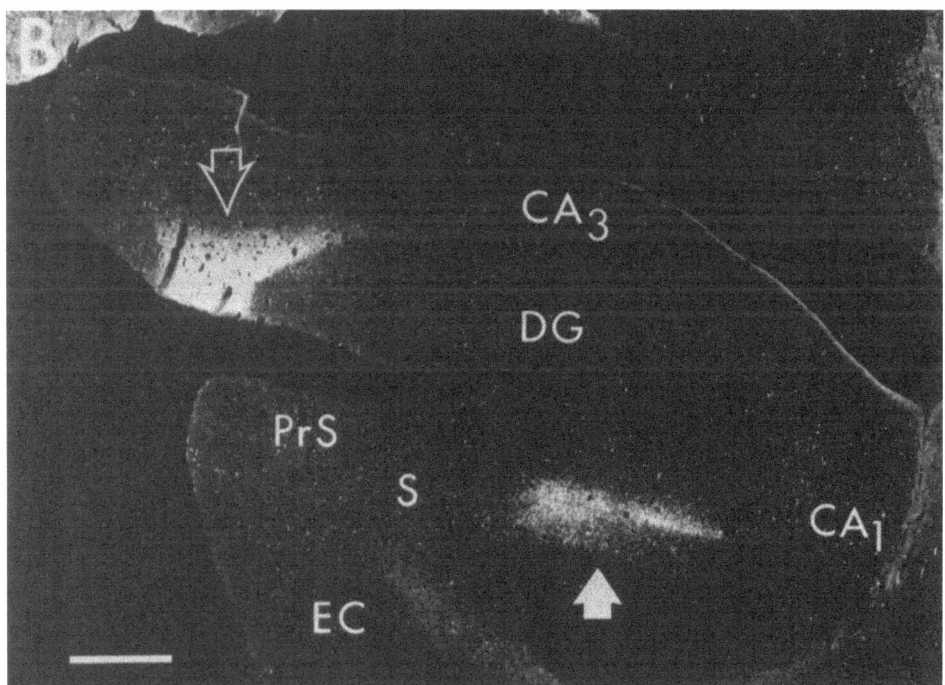

Fig. 2. Darkfield autoradiographs from the rostral hippocampus. In A, a projection originating in the accessory basal nucleus extends laterally through CA3 (arrows) and terminates in CA1. In B, labeled fibers originating in the parvicellular division of the basal nucleus enter the hippocampal formation (open arrows) and terminate exclusively in the CA1/subiculum border zone (filled arrow). Calibration marker = 1 mm.

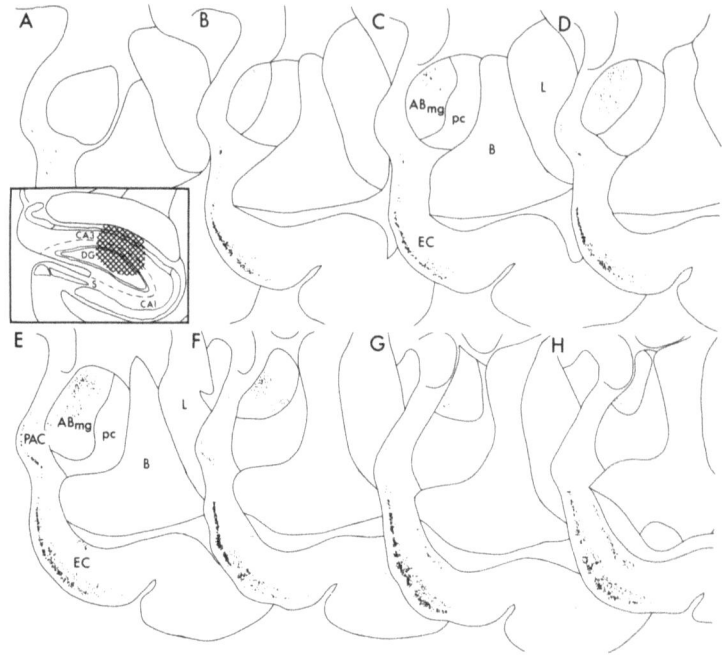

Fig. 3. Computer generated plot of the position of retrogradely labeled cells resulting from an injection (inset) of WGA-HRP into the rostral hippocampal formation. Each dot represents one labeled cell. In addition to the numerous labeled cells in the entorhinal cortex, cells were found in the magnocellular division of the accessory basal nucleus of the amygdala and in the periamygdaloid cortex.

campus and subiculum. In fact, the pyramidal cells of field CA1 overlap those of the subiculum in a mitered fashion throughout this region which is perhaps most descriptively called the CA1/subiculum border zone. In addition to the overlap of the two pyramidal cell layers, this zone is characterized by smaller and more darkly stained CA1 pyramidal cells, by a general increase in cellularity with many of the cells resembling those found in stratum oriens and by a narrow acellular band which separates the two pyramidal cell layers. These characteristics are also found in the same region of the human hippocampal formation.

The amygdaloid projection to the border zone terminates over the superficially located CA1 pyramidal cells and within the overlying molecular layer; there is actually little or no termination over the pyramidal layer of the subiculum though the apical dendrites of subicular cells would extend into the innervated molecular layer. It should be emphasized, however, that the major portion of the subiculum proper does not appear to receive an amygdaloid input. The projection to the border region arises primarily from the parvicellular division of the basal nucleus and to a lesser extent from cells of the periamygdaloid cortex. A minor projection to this region may also originate in the lateral nucleus. While studies in nonprimates suggest that only a portion of the dorsoventral extent of the subiculum receives an amygdaloid input, in the monkey, the amygdaloid projection extends throughout the rostrocaudal extent of the hippocampus with essentially the same terminal density.

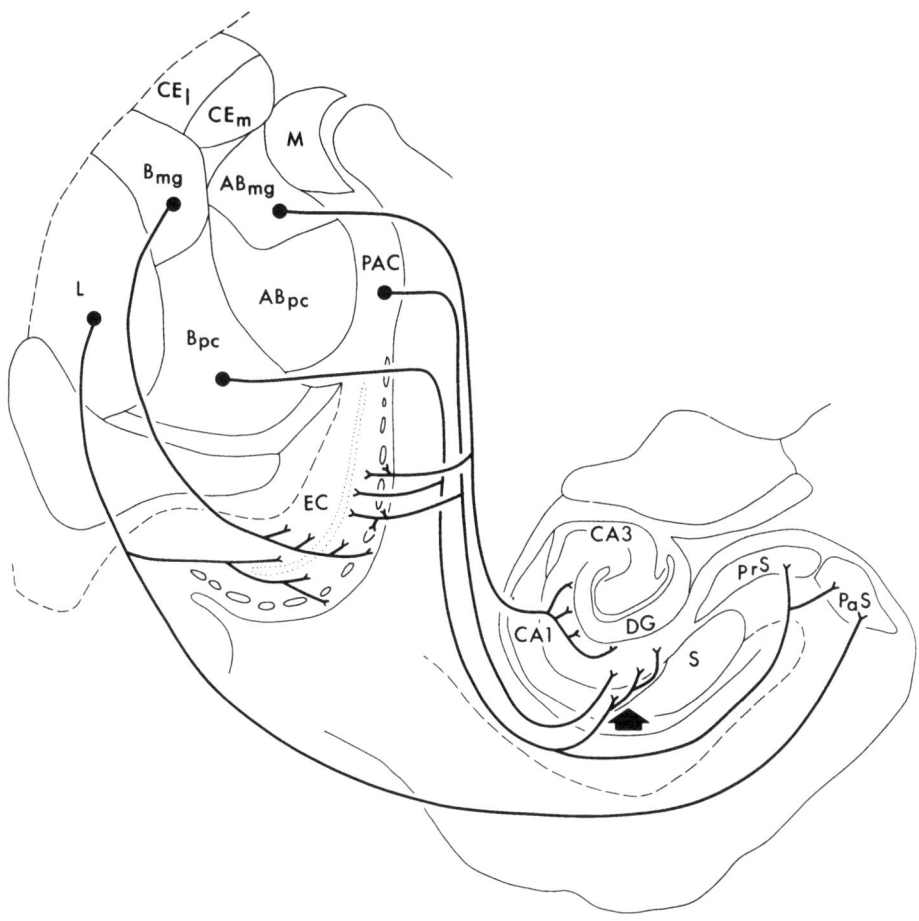

Fig. 4. Summary of amygdaloid projections to the hippocampal formation. Bold arrow indicates the CA1/subiculum border zone.

The projections of the amygdala to field CA1 of the hippocampus and to the border region is particularly interesting in relation to Wilhelm Sommer's original observations on the neuropathology of epilepsy. Sommer (1880) reviewed 90 postmortem macroscopic studies of abnormalities associated with epilepsy and concluded that damage to the hippocampus (which he considered to be a component of the motor system) correlated highly with the illness. He also described microscopic observations of the hippocampal formation from the postmortem analysis of one of his own patients, noting that a fairly restricted region of the hippocampus suffered nearly complete cell loss. Since at that time no nomenclature was available for the subdivisions of the hippocampal formation, he suggested that the hippocampus be thought of as an ellipse with the long axis running transversely and parallel to the horizontal plane. Within the ellipse, the zone of cellular degeneration extended about 60 degrees below the long axis in the ventrolateral quadrant and about 20 degrees above the long axis in the dorsolateral quadrant. This 'sector', i.e., Sommer's sector, correlates most closely with field CA1 of the hippocampus.

The hippocampus and border region are not the only zones in receipt of amygdaloid inputs. The pre- and parasubiculum receive projections from the parvicellular portion of the basal nucleus which terminate both in the plexiform and cellular layers. The parasubiculum receives an additional and somewhat heavier projection from the lateral nucleus.

The entorhinal cortex is perhaps the most heavily innervated component of the monkey hippocampal formation though the laminar organization of the projection is not as clearcut as in the rat and cat. Both anterograde and retrograde studies consistently indicate that the rostral entorhinal cortex, which is closely related to the lateral entorhinal cortex of nonprimates, receives a more prominent projection than caudal levels though at least layer I of the entire rostrocaudal extent of the entorhinal cortex receives some amygdaloid input.

The magnocellular division of the accessory basal nucleus projects to layer I of the entire entorhinal cortex, but rostrally, fibers also terminate in layer III and in the cell free spaces of layer II. The magnocellular division of the basal nucleus projects diffusely to the rostral entorhinal cortex. Terminals are heaviest in layers V and VI, but also extend into layers III and I. The projection continues into the caudal half of the entorhinal cortex but at a much lower density. The parvicellular division of the basal nucleus projects lightly through the rostral half of the entorhinal cortex, primarily deep in layer III. The lateral nucleus projects to layer I of the entire entorhinal cortex. The projection lies deep in layer I and is separated from layer II by a narrow terminal free band. In the rostral entorhinal cortex, the deep portion of layer III is innervated and radially oriented bundles of amygdaloid fibers extend into superficial layer III and through the cell free spaces of layer II. In at least one experimental case with a lateral nucleus injection of ^{3}H-amino acids, only layer VI of the rostral entorhinal cortex was labeled. It is entirely likely that the hippocampally directed projections from each of the major amygdaloid nuclei are topographically organized and that our limited number of experimental cases has, thus far, not resolved these subtleties. Finally, the periamygdaloid cortex also projects to the rostral entorhinal cortex. The termination of this projection is patchy and diffuse, but is located primarily in layer II and in the superficial portion of layer III; layer I is lightly innervated but the deep layers are only traversed by fibers of passage.

AMYGDALOCORTICAL PROJECTIONS

The overall pattern of amygdalocortical connectivity can by summarized with two general statements. First, regardless of the animal studied, the amygdaloid complex appears to project to a greater number of cortical regions than those from which it receives projections. The cortical regions which receive a heavy amygdaloid input, however, invariably reciprocate the projection. Second, of the animals studied, only the primate amygdala is extensively interconnected with the neocortex. Available evidence indicates that in nonprimates such as the rat and cat, the meager cortical input arises from proisocortical and periallocortical regions.

In the rat, the cortical projections originate in the pyriform cortex, which projects to most of the amygdala other than the medial nucleus, in the perirhinal cortex, which projects to the lateral nucleus and in the agranular insula which projects to several amygdaloid nuclei (Veening, 1978; Ottersen, 1982) especially the central nucleus (Saper, 1982). Additional cortical projections arise in the prelimbic and infralimbic regions (areas 32 and 25) which terminate primarily in the central, basal and lateral amygdaloid nuclei (Beckstead, 1979; Ottersen, 1982). A minor projection to the basal nucleus also originates in the anterior cingulate cortex (area 24). Each of these cortical regions receives a reciprocal projection from the amygdaloid complex. These arise mainly from the basolateral nucleus and to a lesser extent from the lateral and cortical nuclei

(Krettek and Price, 1974, 1977; Saper, 1982; Sarter and Markowitsch, 1984). With the use of sensitive retrograde tracers, the rat amygdaloid complex has recently been shown to project to other neocortical regions. The basolateral nucleus, for example, apparently projects to the primary motor and somatosensory cortices (Sripanidkulchai et al., 1984). It is not clear, therefore, whether the phylogenetic differences noted above are due to true species differences in the extent of amygdalocortical interconnectivity or merely reflect the inability of previously employed tracing techniques to demonstrate the connections.

In the cat, the amygdaloid complex also receives projections from the pyriform cortex, the perirhinal cortex and the prelimbic and infralimbic regions of the medial frontal cortex (Russchen, 1982). In addition, there are projections to the lateral, central and basal nuclei from portions of the anterior and posterior sylvian gyri (Druga, 1970; Heath and Jones, 1971; Kamal and Tombol, 1976; Russchen, 1982). The bulk of this projection arises from fields that on cytoarchitectonic grounds are homologous with area 20 in the primate brain.

As in the rat, the cat amygdaloid complex reciprocates its cortical inputs and projects to several cortical regions from which it has not been shown to receive afferents. Retrograde tracer studies have demonstrated that it projects to much of the frontal cortex including motor and premotor regions (Llamas et al., 1977). Macchi et al. (1978) have reported rather widespread amygdaloid projections extending into primary and secondary somatic sensory regions and primary and associational auditory cortex in addition to those to frontal, cingulate, and insular cortices. Attempts at demonstrating amygdaloid projections to the parietal cortex or to primary and associational visual areas of the occipital lobe were unsuccessful.

Based on currently available information, it appears that the extent of amygdalocortical interconnectivity is significantly greater in primates than in the nonprimate species studied. Whitlock and Nauta (1956) first demonstrated that the anterior temporal lobe of the macaque monkey projects directly to the amygdaloid complex and shortly thereafter Nauta (1962) showed that the amygdala projected to the temporal lobe, the insular cortex and at least portions of the orbitofrontal cortex. While these early observations received little attention in the 1960's, several studies beginning in the mid 1970's which employed degeneration techniques (Pandya et al., 1973; Herzog and Van Hoesen, 1976; Leichnetz and Astruc, 1977) provided strong evidence that the monkey amygdaloid complex receives projections from a large number of cortical areas. Subsequently, modern tracing techniques have confirmed and extended the range of cortical areas which project to the amygdaloid complex (Aggleton et al., 1980; Turner et al., 1980; Mufson et al., 1981; Van Hoesen, 1981).

While there is strong evidence for projections from frontal, temporal and insular cortices to the primate amygdaloid complex, the topography of the terminal fields of these projections is still not well understood. In fact, there are two somewhat contradictory assertions related to the amount of convergence of these projections within the amygdaloid complex. Turner et al. (1980) have suggested that there is little, if any, overlap of projections from each of the sensory modalities whereas Van Hoesen (1981) believes that in at least some divisions of the amygdala, and especially in the lateral nucleus, there is evidence for extensive convergence of sensory modalities. Clearly, the resolution of this issue will have important implications for our understanding of the functional circuitry of the amygdaloid complex.

The temporal lobe, or more specifically its anterior half, give rise to the largest component of the cortical projection to the amygdaloid

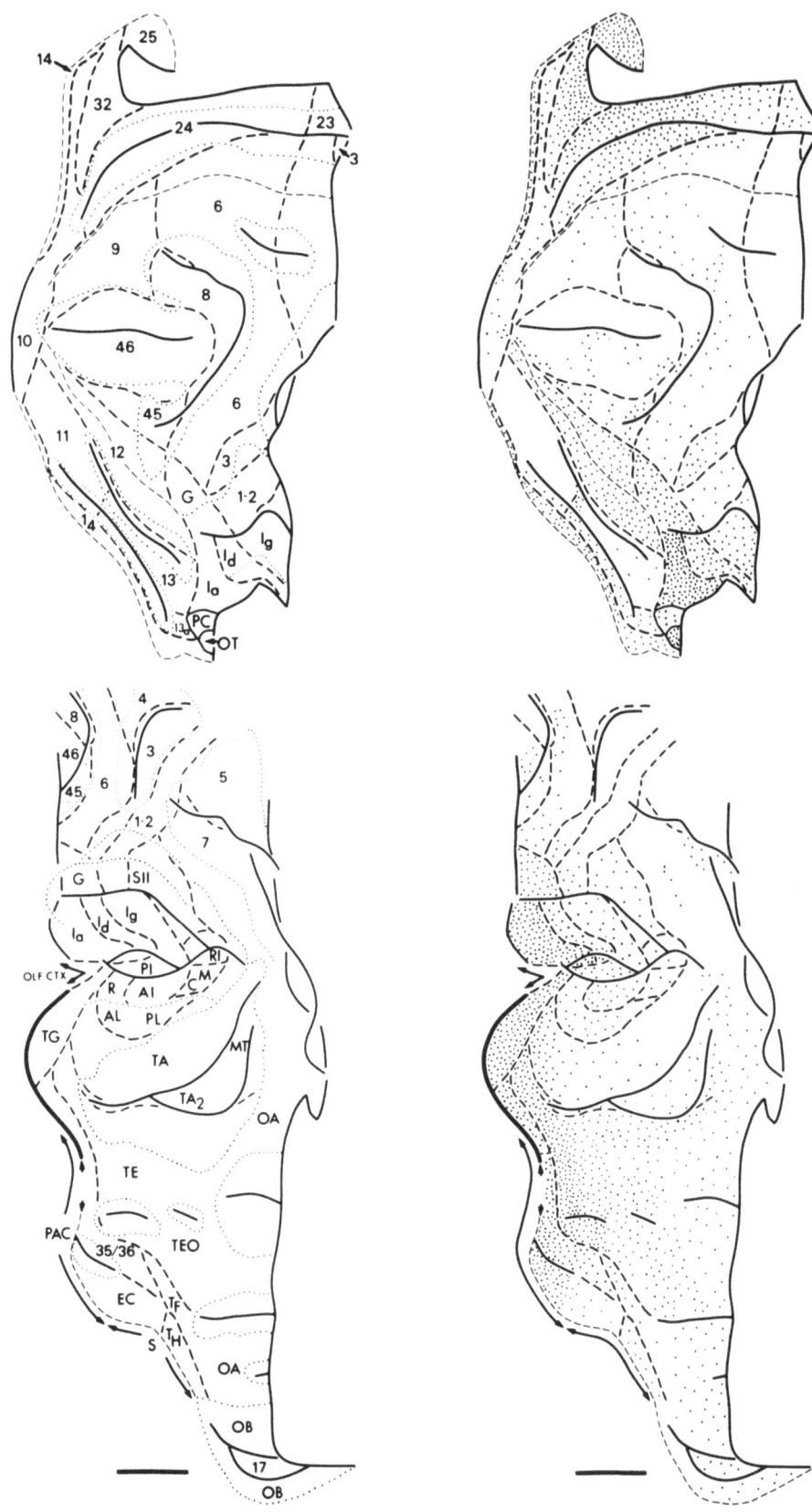

complex. The cortex of the temporal pole (area TG of Bonin and Bailey, area 38 of Brodmann) heavily innervates the amygdala and these fibers terminate mainly in the lateral nucleus and in the accessory basal nucleus. The anterior superior temporal gyrus (area TA), especially the dorsal bank of the superior temporal sulcus, projects primarily to the lateral nucleus. There is no indication of a projection from primary auditory cortex though the closely adjacent auditory associational areas do project to the amygdala. The rostral inferotemporal cortex (area TE), which is generally considered to be visual association cortex, projects to dorsal portions of both the lateral and basal nuclei. The posterior parahippocampal gyrus (areas TF and TH) gives rise to a projection that terminates in the parvicellular division of the basal nucleus and in the accessory basal nucleus.

The primate amygdala is innervated by all regions of the insula though the projections from the rostral agranular and dysgranular divisions are somewhat heavier. The posterior or granular insula projects predominantly to the lateral nucleus and to the lateral division of the central nucleus. The more rostral agranular and dysgranular divisions of the insula project more widely to the amygdaloid complex and contribute fibers to the lateral nucleus, the parvicellular portion of the basal nucleus, the magnocellular division of the basal nucleus, the medial nucleus and the cortical nucleus.

From the frontal cortex, amygdaloid projections arise primarily from orbitofrontal regions including areas 12, 13 and 14 which terminate in the central nucleus, the magnocellular division of the accessory basal nucleus and the lateral nucleus. There is also a projection from the gustatory region of the frontal operculum to the lateral nucleus and at least area 46 of the lateral convexity projects to the magnocellular division of the basal nucleus. On the medial surface, area 32 projects to the magnocellular portion of the accessory basal nucleus and area 25 projects to the magnocellular divisions of both the basal and accessory basal nuclei. The anterior cingulate cortex (area 24) projects to the amygdaloid complex and terminates in the magnocellular division of the basal nucleus. No projections to the amygdaloid complex have been described from the posterior cingulate cortex, from any division of the parietal cortex, from the caudal half of the temporal lobe or from any portion of the occipital cortex.

The literature on amygdalocortical efferents is meager in comparison to that on cortical afferents. However, it is clear that vast portions of frontal, temporal, insular, and occipital cortices receive direct projections from the amygdaloid complex (Jacobson and Trojanowski, 1975; Potter and Nauta, 1979; Mufson et al., 1981; Porrino et al., 1981; Avendano et al., 1983; Amaral and Price, 1984). The parietal cortex, especially along the banks of the intraparietal sulcus, receives an amygdaloid projection which is, nonetheless, rather limited in relation to other cortical areas. It can be summarily stated that the primate amygdala projects most heavily to polysensory cortical regions such as the perirhinal cortex, the temporal polar cortex and the orbitofrontal cortex, but visually related cortices of the temporal lobe also receive very prominent inputs. The auditory cortical regions appear to receive a somewhat weaker input than the visual areas and the somatosensory cortices are only meagerly innervated. While premotor cortex receives a minor projection from the amygdala, primary motor cortex apparently gets no amygdaloid input. Fig. 5 is a summary

Fig. 5. Unfolded cortical maps of the frontal (top) and temporal (bottom) lobes of the macaque monkey. Dots represent the distribution and relative density of the amygdalocortical projections.

of the amygdalocortical projections described by Amaral and Price (1984).

The projections to the frontal lobe arise primarily in the magnocellular divisions of the basal and accessory basal nuclei. The heaviest projections are to medial and orbital regions which include the anterior cingulate cortex (area 24), the prelimbic and infralimbic regions (areas 32 and 25, respectively) area 14, the medial part of area 13 and area 12. Neither the major portion of area 13 nor the rostrally adjacent area 11 receives significant projections. The amygdala sends additional lighter projections to dorsolateral and dorsomedial regions of the frontal lobe including areas 45, 46 and 8 ventral to the principal sulcus and areas 6, 9, and 10 located dorsomedially.

The projections to the insular cortex originate primarily from the lateral nucleus, the parvicellular portion of the basal nucleus, the magnocellular portion of the accessory basal nucleus, the cortical nucleus and the periamygdaloid cortex. The caudal, dysgranular insula receives the bulk of its projections from the lateral nucleus, whereas more anterior regions receive projections from each of the nuclei named above. Fibers which contribute to the insular projection continue around the dorsal limiting sulcus to terminate lightly in the somatosensory fields 3, 1-2 and SII.

All major divisions of the temporal lobe receive an **amygdaloid** input which arises mainly from the magnocellular division of the basal nucleus but at least the temporal polar cortex receives additional projections from the lateral and accessory basal nuclei. The rostral superior temporal gyrus (area TA) receives a fairly heavy innervation which continues more lightly into primary auditory cortex and adjacent auditory association areas though there is little or no projection to caudal levels of the superior temporal gyrus. The entire rostrocaudal extent of the inferotemporal cortex (middle and inferior temporal gyri - area TE and TEO) receives a projection from the amygdala. These projections continue caudally into the visual association fields of the occipital lobe (areas OA and OB or areas 19 and 18) and at least rostral portions of the primary visual cortex (area 17 or OC) are innervated.

The amygdalocortical projections consistently terminate on the border of layers I and II and occasionally in regions which receive a particularly strong input, in layers V and VI. The cortical projections to the amygdala originate from cells in layers II and III and to a lesser extent layer V.

CONCLUSIONS

Recent studies on the connectivity of the primate amygdaloid complex have demonstrated prominent interconnections with the hippocampal formation and with a variety of neocortical regions. These results are consistent with the rapidly emerging view that the amygdala plays a significant role in cognitive functions such as memory formation and associative learning in addition to its more commonly proferred roles in modulating the emotional and visceral responses to environmental stimuli. It is not surprising, therefore, that seizures originating in the amygdaloid complex produce a plethora of behavioral symptoms ranging from visceral sensations to the evocation of emotions to illusions of perception. It is hoped that an understanding of the connectivity of the primate amygdala and a determination of the neurotransmitters that mediate the limbic and cortical projections, may yield some indications for the pharmacological control of temporal lobe epilepsy.

ABBREVIATIONS

ABmg, accessory basal nucleus, magnocellular division
ABpc, accessory basal nucleus, parvicellular division
B, basal nucleus
Bmg, basal nucleus, magnocellular division
Bpc, basal nucleus, parvicellular division
Bpl, basal nucleus, paralaminar division
C_l, CE_l, central nucleus, lateral division
C_m, CE_m, central nucleus, medial division
CA1, CA3, hippocampal fields
CO, cortical nucleus
DG, dentate gyrus
EC, entorhinal cortex
L, lateral nucleus
M, medial nucleus
PAC, periamygdaloid cortex
PrS, presubiculum
PaS, parasubiculum
S, subiculum

ACKNOWLEDGEMENTS

The original work described in this chapter was carried out in collaboration with Drs. J. Price, R. Insausti and W.M. Cowan, and was conducted in part, by the Clayton Foundation for Research - California Division. The author's work is supported by NIH Grant NS-20004.

REFERENCES

Aggleton, J.P., Burton, M.J., and Passingham, R.E., 1980, Cortical and subcortical afferents to the amygdala of the rhesus monkey, Brain Res., 190:347.

Amaral, D.G., Insausti, R., and Cowan, W.M., 1983, Evidence for a direct projection from the superior temporal gyrus to the entorhinal cortex in the monkey, Brain Res., 275:263.

Amaral, D.G., Insausti, R., and Cowan, W.M., 1984, The commissural connections of the monkey hippocampal formation, J. Comp. Neurol., 224:307.

Amaral, D.G., and Price, J.L., 1984, Amygdalo-cortical projections in the monkey (Macaca fascicularis), J. Comp. Neurol., 230:465.

Avendano, C., Price, J.L., and Amaral, D.G., 1983, Evidence for an amygdaloid projection to premotor cortex but not to motor cortex in the monkey, Brain Res., 264:111.

Bakst, I., and Amaral, D.G., 1984, The distribution of acetylcholinesterase in the hippocampal formation of the monkey, J. Comp. Neurol., 225:344.

Beckstead, R.M., 1979, An autoradiographic examination of corticocortical and subcortical projections of the mediodorsal-projection (prefrontal) cortex in the rat, J. Comp. Neurol., 184:43.

Buser, P., and Bancaud, J., 1983, Unilateral connections between amygdala and hippocampus in man: a study of epileptic patients with depth electrodes, Electroencephalogr. Clin. Neurophysiol., 55:1.

Corsellis, J.A.N., and Bruton, C.J., 1983, Neuropathology of status epilepticus in humans, in: Status Epilepticus, A.V. Delgado-Escueta, C.G. Wasterlain, D.M., Treiman, and R.J. Porter, eds., Raven Press, New York, p. 129.

Cowan, W.M., Raisman, G., and Powell, T.P.S., 1965, The connections of the amygdala, J. Neurol. Neurosurg. Psychiat., 28:137.

Crosby, E.C., and Humphrey, T., 1941, Studies of the vertebrate telencephalon. II. The nuclear pattern of the anterior olfactory nucleus, tuberculum olfactorium and the amygdaloid complex in adult man, J. Comp. Neurol., 74:309.

Druga, R., 1970, Neocortical projections to the amygdala (an experimental study with the Nauta method), J. Hirnforsch., 11:467.

Gloor, P., Olivier, A., Quesney, L.F., Andermann, F., and Horowitz, S., 1982, The role of the limbic system in experimental phenomena of temporal lobe epilepsy, Ann. Neurol., 12:129.

Heath, C.J., and Jones, E.G., 1971, The anatomical organization of the suprasylvian gyrus of the cat, Ergebn. Anat. Entwicklgesch., 45:1.

Herzog, A.G., and Van Hoesen, G.W., 1976, Temporal neocortical afferent connections to the amygdala in the rhesus monkey, Brain Res., 115:57.

Ishikawa, I., Kawamura, S., and Tanaka, O., 1969, An experimental study on the efferent connections of the amygdaloid complex in the cat, Acta Med. Okayama, 23:519.

Jacobson, S., and Trojanowski, J.Q., 1975, Amygdaloid projections to prefrontal granular cortex in rhesus monkey demonstrated with horseradish peroxidase, Brain Res., 100:132.

Kaada, B.R., 1972, Stimulation and regional ablation of the amygdaloid cortex with reference to functional representation, in: The Neurobiology of the Amygdala, B.E. Eleftheriou, ed., Plenum Press, New York, p. 205.

Kamal, A.M., and Tombol, T., 1976, Olfactory and temporal projections to the amygdala, Verh. Anat. Ges., 70:283.

Krettek, J.E., and Price, J.L., 1974, Projections from the amygdala to the perirhinal and entorhinal cortices and the subiculum, Brain Res., 71:150.

Krettek, J.E., and Price, J.L., 1977, Projections from the amygdaloid complex and adjacent olfactory structures to the entorhinal cortex and to the subiculum in the rat and cat, J. Comp. Neurol., 172:723.

Lauer, E.W., 1945, The nuclear pattern and fiber connections of certain basal telencephalic centers in the macaque, J. Comp. Neurol., 82:215.

Le Gal La Salle, G., 1982, Amygdaloid organization related to the kindling effect, Electroencephalogr. Clin. Neurophysiol., 36:239.

Leichnetz, G.R., and Astruc, J., 1977, The course of some prefrontal corticofugals to the pallidum, substantia innominata, and amygdaloid complex in monkeys, Exp. Neurol., 54:104.

Llamas, A., Avendano, C., and Reinoso-Suarez, F., 1977, Amygdaloid projections to prefrontal and motor cortex, Science, 195:794.

Macchi, G., Bentivoglio, M., Rossini, P., and Tempesta, E., 1978, The basolateral amygdaloid projections to the neocortex in the cat, Neurosci. Lett., 9:347.

Morrison, F., and Poletti, C.E., 1980, Hippocampal influence on amygdala unit activity in awake squirrel monkeys, Brain Res., 192:353.

Mufson, E.J., Mesulam, M.M., and Pandya, D.N., 1981, Insular interconnections with the amygdala in the rhesus monkey, Neuroscience, 6:1231.

Nauta, W.J.H., 1962, Neural associations of the amygdaloid complex in the monkey, Brain, 85:505.

Ottersen, O.P., 1982, Connections of the amygdala of the rat. IV: corticoamygdaloid and intraamygdaloid connections as studied with axonal transport of horseradish peroxidase, J. Comp. Neurol., 205:30.

Ounsted, C., Linsay, J., and Norman, R., 1966, Biological Factors in Temporal Lobe Epilepsy. Clinics in Developmental Medicine. No. 22, The Spastics Society Medical Education, William Heinemann Medical Books, Ltd., London.

Pandya, K.N., Van Hoesen, G.W., and Domesick, V.B., 1973, A cingulo-amygdaloid projection in the rhesus monkey, Brain Res., 61:369.

Porrino, L.J., Crane, A.M., and Goldman-Rakic, P.S., 1981, Direct and indirect pathways from the amygdala to the frontal lobe in the rhesus monkey, J. Comp. Neurol., 198:121.

Potter, H., and Nauta, W.J.H., 1979, A note on the problem of olfactory associations of the the orbitofrontal cortex in the monkey, Neuroscience, 4:361.

Rosene, D.L., and Van Hoesen, G.W., 1977, Hippocampal efferents reach widespread areas of cerebral cortex and amygdala in the rhesus monkey, Science, 198:315.

Russchen, F.T., 1982, Amygdalopetal projections in the cat. I. Cortical afferent connections. A study with retrograde and anterograde tracing techniques, J. Comp. Neurol., 206:159.

Saper, C.B., 1982, Convergence of autonomic and limbic connections in the insular cortex of the rat, J. Comp. Neurol., 210:163.

Sarter, M., and Markowitsch, H.J., Collateral innervation of the medial and lateral prefrontal cortex by amygdaloid, thalamic, and brain-stem neurons, J. Comp. Neurol., 224:445.

Sommer, W., 1880, Erkrankung des Ammonshorns als aetiologisches Moment der Epilepsie, Arch. Psych. Nervenkrkh., 10:631.

Sripanidkulchai, K., Sripanidkulchai, B., and Wyss, J.M., 1984, The cortical projection of the basolateral amygdaloid nucleus in the rat: a retrograde fluorescent dye study, J. Comp. Neurol., 229:419.

Tremblay, E., and Ben-Ari, Y., 1984, Usefulness of parenteral kainic acid as a model of temporal lobe epilepsy, Rev. Electroencephalogr. Neurophysiol. Clin., 14:241.

Turner, B.H., Mishkin, M., and Knapp, M., 1980, Organization of the amygdalopetal projections from modality-specific cortical association areas in the monkey, J. Comp. Neurol., 191:515.

Van Hoesen, G.W., 1981, The differential distribution, diversity and sprouting of cortical projections to the amygdala in the rhesus monkey, in: The Amygdaloid Complex, Y. Ben-Ari, ed., Elsevier/North-Holland Biomedical Press, Amsterdam, p. 77.

Veening, J.G., 1978, Cortical afferents of the amygdaloid complex in the rat: an HRP study, Neurosci. Lett., 8:191.

Wieser, H.G., 1983, Depth recorded limbic seizures and psychopathology, Neurosci. Biobehav. Rev., 7:427.

Whitlock, D.G., and Nauta, W.J.H., 1956, Subcortical projections from the temporal neocortex in macaca mulatta, J. Comp. Neurol., 106:183.

Witter, M.P., Groenewegen, H.J., and Russchen, F.T., 1981, Entorhinal and perirhinal projections to the striatum, amygdala and claustrum in the cat; a neuroanatomical study using anterograde and retrograde transport techniques, Neurosci. Lett., 7:544.

SUBCORTICAL PROJECTIONS FROM THE AMYGDALOID COMPLEX

J.L. Price

Department of Anatomy and Neurobiology
Washington University School of Medicine
St. Louis, Missouri 63130, USA

INTRODUCTION

It has long been known that the amygdala has substantial subcortical projections to the bed nucleus of the stria terminalis, the hypothalamus and the mediodorsal nucleus of the thalamus (e.g., Nauta, 1961; Cowan et al., 1965; Heimer and Nauta, 1969; De Olmos, 1972). Partially because of this, the function of the amygdaloid complex has been thought to be primarily related to visceral and autonomic mechanisms (e.g., Kaada, 1972). This concept has been somewhat altered by more recent studies. It is now clear that there are also widespread amygdaloid projections to the cerebral cortex (see Amaral, this volume), and that the complex is involved in cognitive as well as visceral functions (e.g., Gloor et al., 1982). However, recent investigations have also emphasized the subcortical projections from the amygdaloid nuclei, and have shown that they are even more complex and extensive than were previously thought. For example, the central amygdaloid nucleus projects not only to the lateral hypothalamus, but also to the full extent of the brainstem and even into the spinal cord (e.g., Hopkins, 1975; Price and Amaral, 1981; Mizuno et al., 1985). At the other end of the neuraxis, the amygdaloid nuclei provide the major telencephalic input to the nucleus basalis and other cell groups in the basal forebrain (Russchen et al., 1985b).

In this review, the subcortical projections of the amygdala will be considered in relation to the part of the brain in which they terminate. The projections to the basal forebrain will be described first, followed by the projections to the thalamus, the hypothalamus and finally the brainstem. The description is based primarily on recent studies using anterograde and retrograde tracers, with special emphasis on recent observations in primates. The experiments were done in collaboration with Drs. D. Amaral, F. Russchen and T. Fuller.

BASAL FOREBRAIN

Fibers from most of the amygdaloid nuclei stream rostromedially through the anterior part of the ventral amygdalofugal pathway to reach the adjacent areas of the basal forebrain. Within this region, the amygdaloid fibers are distributed primarily to the substantia innominata, including the nucleus basalis, as well as to striatal structures, including not only the nucleus accumbens and olfactory tubercle, but also much of the caudate nucleus and putamen (Krettek and Price, 1978a; Kelly et al., 1982; Russchen et al., 1985a,b).

Substantia Innominata

Most of the fibers to the substantia innominata run through and around the cell groups of the magnocellular nucleus basalis and nuclei of the diagonal band (Fig. 1). Although these axons continue to the diencephalon or other parts of the brain, they form synapses-en-passant with the large cells, many of which are cholinergic (Zaborsky et al., 1985; Russchen et al., 1985b). A smaller number of fibers are also distributed to the ventral pallidum, but this projection is much lighter than the projection to the magnocellular nuclei. The projection originates widely within the amygdaloid complex, with the greatest proportion of the fibers arising in the parvicellular basal nucleus, magnocellular accessory basal nucleus and central nucleus (Fig. 2). Surprisingly, the dorsal part of the magnocellular basal amygdaloid nucleus, which receives the heaviest cholinergic projection from the nucleus basalis, provides a very small proportion

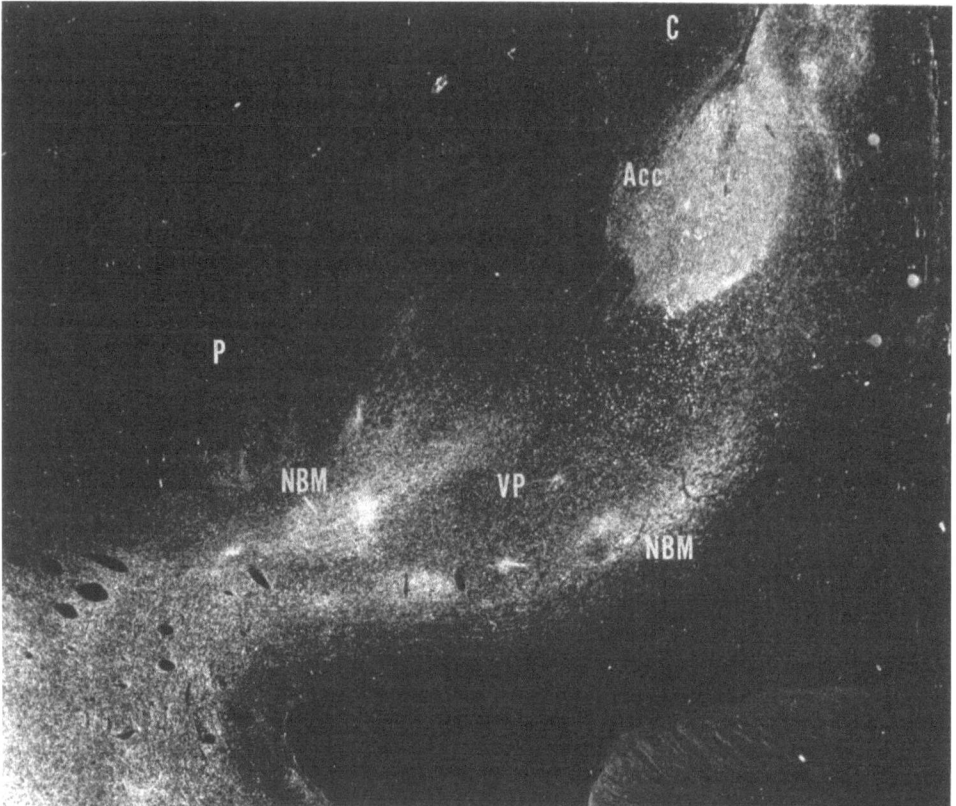

Fig. 1. Darkfield photomicrograph, illustrating the amygdaloid projection to the basal forebrain, labeled by a large injection of ^{3}H-amino acids into the basal and accessory basal amygdaloid nuclei. The bright areas represent concentrations of autoradiographic grains, indicating the presence of radioactively labeled axons. Note the dense termination in the nucleus accumbens, and the cell groups of the nucleus basalis of Meynert. Lighter projections are labeled to the ventral pallidum and portions of the adjacent putamen and caudate nucleus.

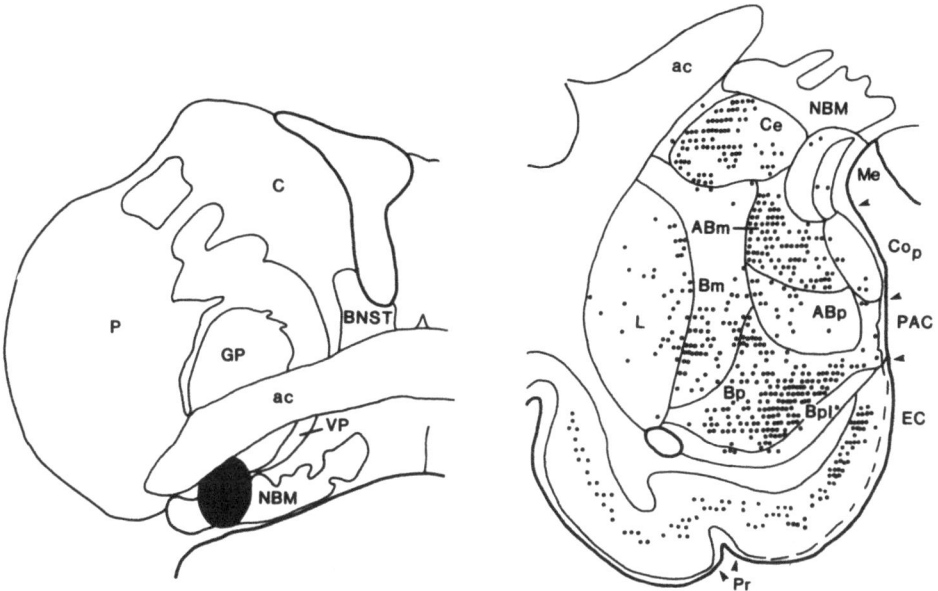

Fig. 2. The distribution of cells in the monkey amygdaloid complex which project to the substantia innominata (including the nucleus basalis and ventral pallidum). The cells were retrogradely labeled by HRP*WGA injected at the site shown in black. Each dot in the amygdala represents one retrogradely labeled cell. Note the concentration of cells in the parvicellular basal nucleus, the magnocellular accessory basal nucleus and the central nucleus, and the absence of cells in the dorsal part of the magnocellular basal nucleus.

of the return projection back to the substantia innominata.

Striatum

Rostral to the main body of the substantia innominata, a very substantial number of amygdaloid fibers continue to the ventral areas associated with the corpus striatum, the nucleus accumbens and the olfactory tubercle (Fig. 1; Krettek and Price, 1978a; Russchen et al., 1985a). In primates the olfactory tubercle is relatively poorly developed, but the projection to the nucleus accumbens is very prominent. As noted above, the ventral pallidal area between the nucleus accumbens and the olfactory tubercle also receives fibers from the amygdala, but this projection is relatively much lighter.

However, the amygdaloid fibers are not limited to the 'ventral striatum'. In both the rat and the monkey, it has been shown that there is a substantial projection to the anteroventral parts of the caudate nucleus and putamen adjacent to the nucleus accumbens, and to the body and tail of the caudate along the stria terminalis. The projection to the tail of the caudate (and even to the adjoining posteroventral edge of the putamen) is especially heavy (Kelly et al., 1982; Russchen et al., 1985a).

The fibers to both the ventral striatal areas and the caudate-putamen arise primarily from the basal and accessory basal nuclei. There is at least a broad topographic organization in the projection; it has been best demonstrated in the rat but also appears to be present in the monkey. The parvicellular part of the basal nucleus (posterior basolateral nucleus in the rat) together with the accessory basal nucleus (basomedial nucleus) project primarily to the medial nucleus accumbens, while the magnocellular basal nucleus (anterior basolateral nucleus in the rat) sends fibers preferentially to more caudolateral portions of the striatum. However, none of the amygdaloid fibers reach the rostral, dorsolateral part of the caudate-putamen which is related to the sensory-motor cortex (Russchen and Price, 1984; Russchen et al., 1985a).

Experiments in the rat using ^{3}H-D-aspartate as a retrograde axonal tracer suggest that the amygdaloid fibers to the basal forebrain use glutamate and/or aspartate as a neurotransmitter. This tracer is taken up by axons which possess a high-affinity uptake mechanism for glutamate and aspartate, and is transported back to the cell body where it may be demonstrated by autoradiography. These cells are presumed to be glutamergic or aspartergic. Injections of ^{3}H-D-aspartate into the substantia innominata and other parts of the basal forebrain label cells in several of the amygdaloid nuclei, especially the nucleus of the lateral olfactory tract, the basolateral (or basal) nucleus and the basomedial (or accessory basal) nucleus (Fig. 3; Fuller et al., 1985).

THALAMUS

The most prominent and best known amygdalo-thalamic projections are

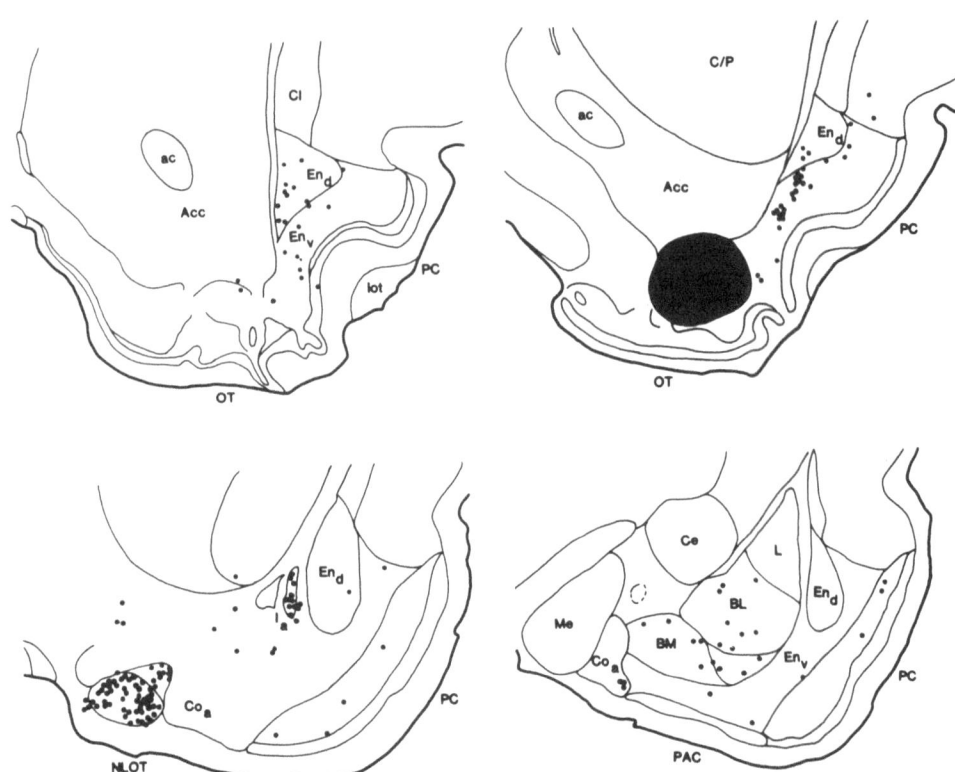

Fig. 3. The distribution of presumptive glutamergic and/or aspartergic cells in the amygdala of the rat, which were labeled from an injection of ^{3}H-D-aspartate into the basal forebrain.

to the mediodorsal nucleus. These are especially interesting because of the interconnections between the mediodorsal nucleus and the areas of the prefrontal cortex which are also directly related to the amygdaloid nuclei. In addition, there are projections from the central amygdaloid nucleus to several intralaminar nuclei on or near to the midline of the thalamus.

Mediodorsal Nucleus

The projection to the mediodorsal nucleus is restricted to the medial, magnocellular part of the nucleus (MDm) (e.g., Krettek and Price, 1977; Porrino et al., 1981). In the monkey, the fibers are distributed within MDm in a complex 'patchy' pattern (Fig. 4). The largest area of termination is in the rostral third of the nucleus; but, in any given experiment demonstrating the projections from a small locus in the amygdala, one or two additional separate areas of termination may be labeled in more caudal parts of MDm. Within each of these, the fibers are concentrated in relatively dense patches, with smaller clusters within the patches (Price, 1981; Aggleton and Mishkin, 1984; Russchen et al., 1986). The significance of this pattern is unknown, but it is possible that patches of fibers from different sources interdigitate with each other.

These fibers to MDm arise from most of the amygdaloid nuclei, with the probable exception of the central and medial nuclei (Fig. 5; Russchen et al., 1986). In experiments with injections of retrograde tracers restricted to MDm, only these nuclei do not show substantial numbers of labeled cells. The cells are scattered throughout the rest of the amygdaloid complex, without any notable concentration in any specific nucleus. Very similar projections to MDm arise from cells scattered throughout the basal forebrain, including the ventral pallidum and other parts of the substantia innominata, the olfactory and entorhinal cortices, the hippocampal formation and even the rostral temporal neocortex. In all of these areas, the cells of origin are large, multipolar neurons, with long, branching dendrites, suggesting that the input to MDm is a very convergent projection that carries integrated information from several sources. Although there is some organization within the projection, with fibers from different sources distributed preferentially to different dorso-ventral or anteroposterior parts of MDm, there may be considerable overlap or interdigitation between the lighter 'patches' of fibers (Aggleton and Mishkin, 1984; Russchen et al., 1986).

Midline Thalamic Nuclei

Although the mediodorsal nucleus does not project back to the amygdaloid nuclei, parts of the intralaminar nuclei along the midline of the thalamus, especially the reuniens nucleus, do send fibers to the amygdala (e.g., see Russchen, this volume). These nuclei also receive a projection from the central amygdaloid nucleus. The fibers are distributed to the full anteroposterior extent of the thalamus, but they are most prominent caudally, at the caudal edge of the massa intermedia (Price and Amaral, 1981). A few fibers from the central nucleus also project further caudally, to the medial edge of the medial pulvinar nucleus.

HYPOTHALAMUS

The amygdaloid fibers to the hypothalamus terminate in several nuclei or areas within both the medial and lateral parts of the hypothalamus. Many of these projections have relatively restricted origins from specific amygdaloid nuclei, and they may terminate in very discrete nuclei in the hypothalamus. The fibers reach the hypothalamus through both the stria

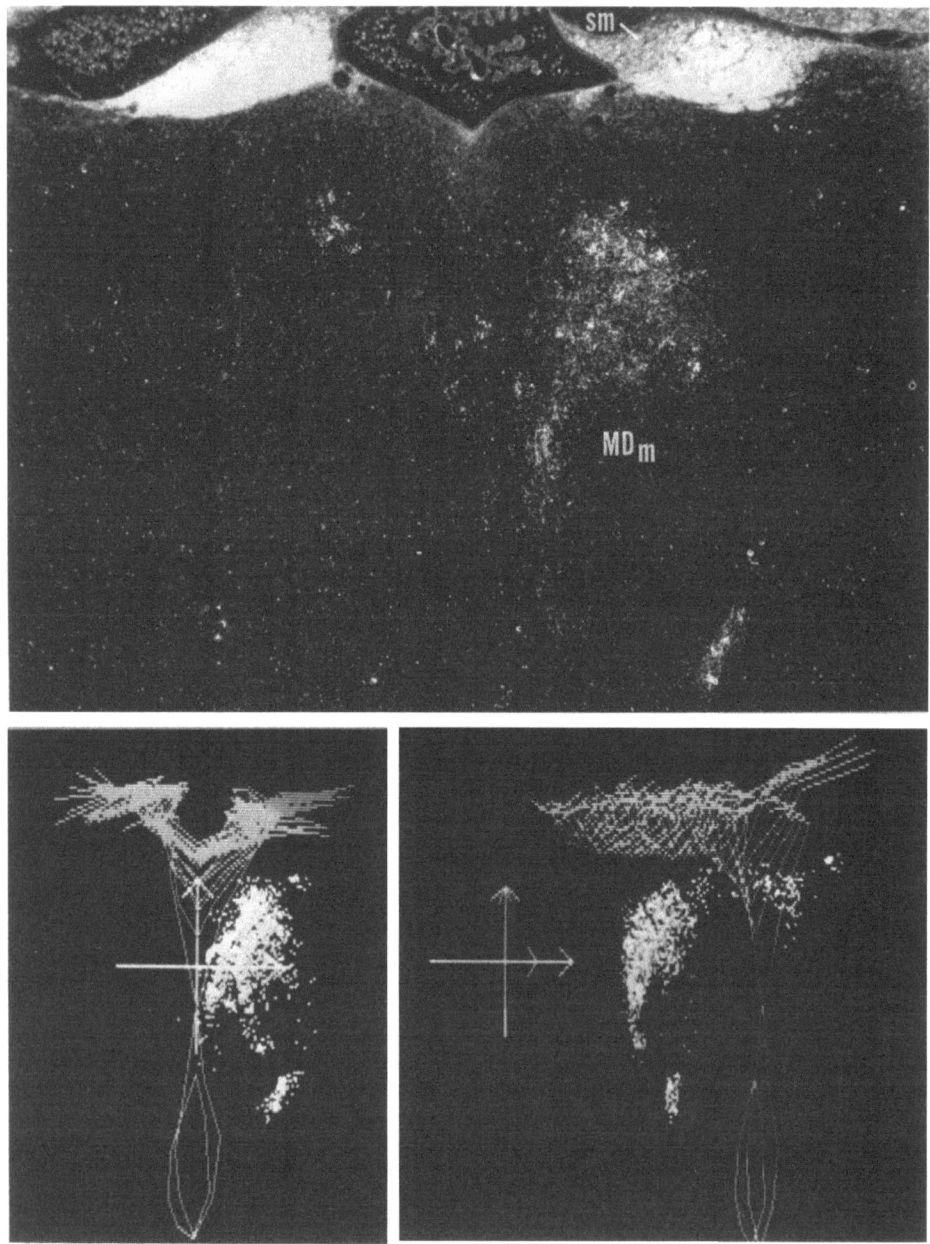

Fig. 4. Above: Darkfield photomicrograph, illustrating the distribution of amygdaloid axons in the medial part of the mediodorsal nucleus, labeled from an injection of ^{3}H-amino acids in the caudal part of the amygdaloid complex. Note the patchy nature of the projection.
Below: Three-dimensional computer reconstruction of the amygdaloid projection in this brain. In addition to the label (represented by white dots), the outline of the third ventricle has been drawn as a reference. The two figures represent an anterior (coronal) view (on the left) and a rotated anterolateral view (on the right).

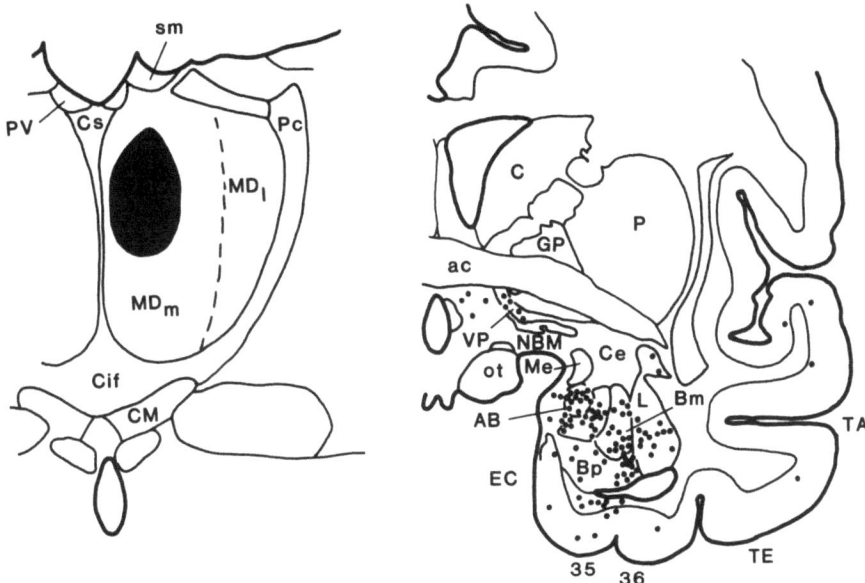

Fig. 5. The distribution of cells in the amygdaloid complex and other parts of the basal forebrain which were retrogradely labeled from an injection of HRP*WGA in the medial part of the mediodorsal nucleus. Note that the cells which project to MDm are scattered throughout all parts of the amygdala (except the central and medial nuclei), as well as in the substantia innominata and the temporal cortex.

terminalis, which curves posteriorly and dorsally around the internal capsule, and the ventral amygdalofugal pathway, which runs between the ventral edge of the internal capsule and the optic tract. In many of the projections, fibers to the same target run through both pathways, and it is likely that they represent the opposite edges of the same pathway which has been 'split' by the internal capsule (e.g., Price and Amaral, 1981).

Bed Nucleus of the Stria Terminalis

Although the bed nucleus of the stria terminalis is not part of the hypothalamus, it forms an intermediate station for many of the projections from the amygdala. Virtually all of the amygdaloid nuclei project through the stria terminalis to the bed nucleus, even if they have little or no direct projection to the hypothalamus (e.g., Krettek and Price, 1978a). The bed nucleus can be divided into medial and lateral parts, which have extensive projections to the medial and lateral hypothalamus as well as the lower parts of the brainstem (e.g., Holstege et al., 1985). The nucleus is in many ways very similar to the central and medial nuclei of the amygdala itself, which also receive fibers from most of the other amygdaloid nuclei and project widely to the hypothalamus (see below). Because of this, it may be suggested that the bed nucleus represents an 'extra-amygdaloid' extension of these nuclei along the stria terminalis.

Preoptic Area/Anterior Hypothalamus

Fibers extend from both the ventral edge of the bed nucleus of the stria terminalis and the ventral amygdalofugal pathway into the junctional region between the anterior hypothalamus and the preoptic area. Most of these fibers appear to arise in the medial and amygdaloid nucleus, with additional fibers from the anterior cortical nucleus and adjacent areas. They are distributed widely within the anterior hypothalamic area, including the regions immediately lateral or dorsal to the magnocellular neurosecretory nuclei (the paraventricular and supraoptic nuclei). Although the amygdaloid fibers do not ramify among the cells of the magnocellular nuclei, they probably contact the dendrites of the neurosecretory cells (Fig. 6; Oldfield and Silverman, 1985).

Ventromedial and Premammillary Nuclei of the Hypothalamus

Further caudally, fibers from the medial nucleus and the adjacent accessory basal amygdaloid nucleus form a dense terminal field in the 'core' of the ventromedial nucleus of the hypothalamus (Fig. 6; Krettek and Price, 1978a). Other fibers from the amygdalo-hippocampal area, together with fibers from the subiculum, terminate in the relatively cell-sparse 'shell' around the nucleus. The fibers from the amygdalo-hippocampal area appear to reach the hypothalamus through the stria terminalis, but many, at least, of the fibers from the medial and accessory basal nuclei run through the ventral pathway (e.g., McBride and Sutin, 1977). Fibers from many of the same amygdaloid nuclei continue behind the ventromedial nucleus to terminate in the premammillary nuclei, although no projection has been identified to the mammillary nuclei themselves.

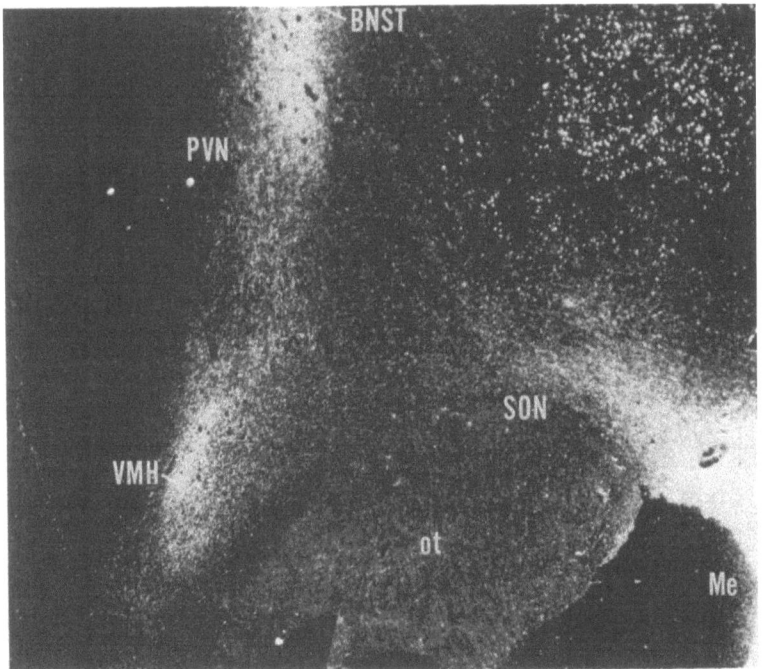

Fig. 6. Darkfield photomicrograph illustrating the projection to the anterior hypothalamus and the 'core' of the ventromedial hypothalamic nucleus which was anterogradely labeled from an injection of ^{3}H-amino acids into the medial amygdaloid nucleus. Note that the labeled axons do not extend around the cells of the supraoptic and paraventricular nuclei, but do run immediately dorsal or lateral to these magnocellular nuclei.

Lateral Hypothalamus

Amygdaloid fibers, primarily from the central nucleus, are distributed to the full extent of the lateral hypothalamic area, from the lateral preoptic area rostrally to the paramammillary nucleus at the caudal edge of the hypothalamus (e.g., Krettek and Price, 1978a; Price and Amaral, 1981). Most of these fibers reach the hypothalamus through the ventral amygdalofugal pathway (Fig. 7), but others run through the stria terminalis; many fibers, in fact, penetrate the internal capsule to run directly between the body of the stria terminalis and the lateral hypothalamus (Price and Amaral, 1981). In the hypothalamus, the fibers run through the medial forebrain bundle and appear to terminate throughout almost all parts of the lateral hypothalamus (Fig. 7).

The only part of the lateral hypothalamus which does not receive fibers from the central nucleus is the lateral tuberal nucleus (Price and Amaral, 1981). However, a very specific projection from the basal amygdaloid nucleus terminates heavily within this small nucleus on the dorsal edge of the optic tract (Fig. 7). These fibers run directly to the nucleus through the ventral pathway; they do not appear to be distributed to any other part of the hypothalamus.

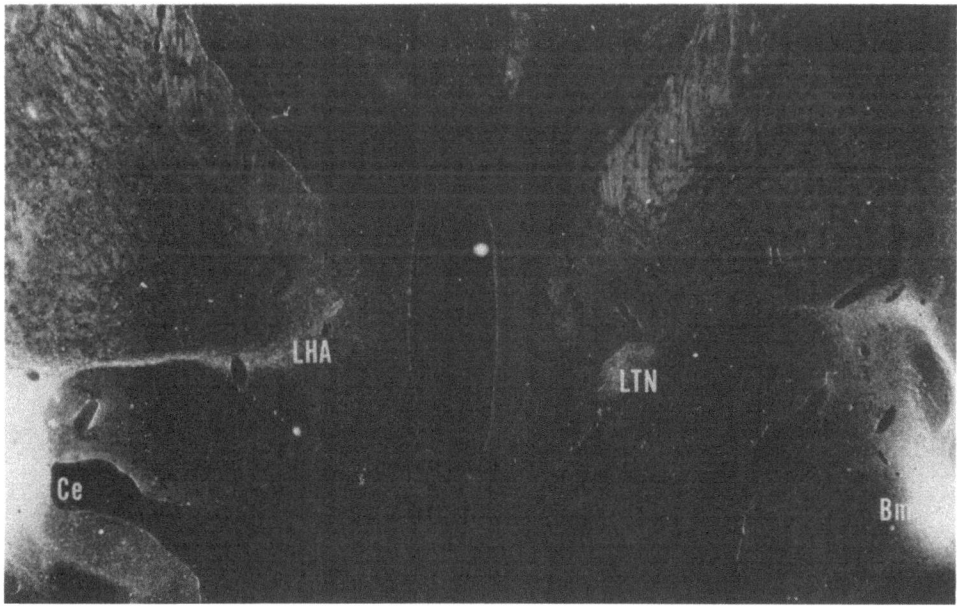

Fig. 7. Darkfield photomicrograph, illustrating fibers to the lateral hypothalamic area, which were labeled from an injection of ^{3}H-amino acids into the central amygdaloid nucleus (on the left), and fibers to the lateral tuberal nucleus, labeled from an injection in the magnocellular basal nucleus (on the right). Note that the two projections are complementary to each other.

BRAINSTEM

Midbrain

Caudal to the hypothalamus, axons from the central nucleus continue from the lateral hypothalamus into the brainstem (e.g., Price and Amaral, 1981). The fibers run over and through the ventral tegmental area and the substantia nigra, especially the A8 and A10 dopaminergic cell groups. The fibers also ramify extensively in the mesencephalic reticular formation and there is a substantial terminal field in ventrolateral and dorsal parts of the periaqueductal grey.

Lower Brainstem

The amygdaloid fibers run caudally through the pontine and medullary reticular formation, with additional areas of termination in the visceral relay nuclei; they appear to arise exclusively from the central amygdaloid nucleus (e.g., Hopkins, 1975; Price and Amaral, 1981; Veening et al., 1984). In the pons, the major area of termination is in the medial and lateral parabrachial nuclei around the superior cerebellar peduncle (Fig. 8). The fibers are not distributed directly around the cells of the locus coeruleus adjacent to the medial parabrachial nucleus, but they may contact the dendrites of the noradrenergic cells. In the medulla, the major projection is to and around the dorsal nucleus of the vagus and the nucleus of the solitary tract (Fig. 8). In the medullary reticular formation, the amygdaloid projection is distributed around all of the catecholaminergic cell groups, although it has not been established whether they make synapses there.

The projection to the vagal nuclei and the reticular formation continues to the spino-medullary junction, and it has recently been shown that some fibers extend still further caudally into at least the cervical spinal cord (Mizuno et al., 1985).

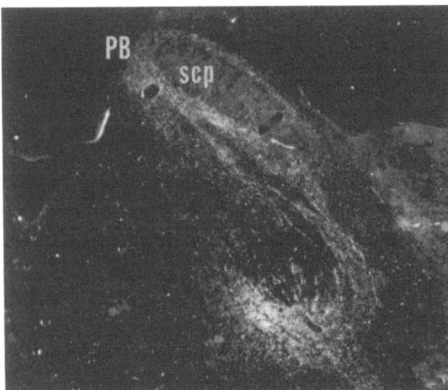 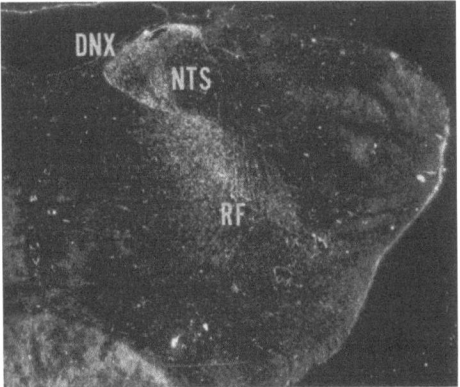

Fig. 8. The amygdaloid projections to the parabrachial nucleus in the pons (on the left), and the dorsal nucleus of the vagus, the nucleus of the solitary tract and the reticular formation of the lower medulla (on the right). These axons were labeled from an injection of ^{3}H-amino acids into the central amygdaloid nucleus.

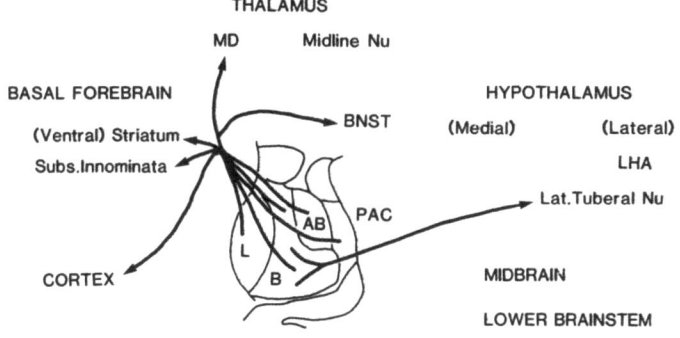

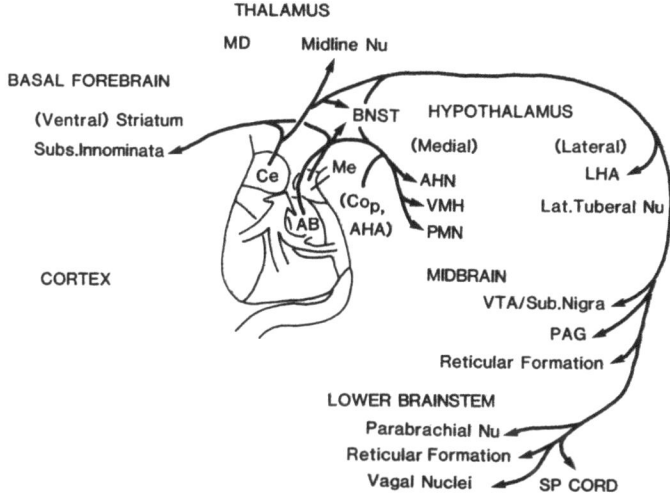

Fig. 9. Summary of the major amygdaloid projections, divided into two systems. The top figure represents the projections of largest amygdaloid nuclei to the cortex, substantia innominata, striatum and mediodorsal thalamic nucleus. The basal nucleus also projects to the lateral tuberal nucleus of the hypothalamus. The lower figure represents the projections of the central and medial nuclei (and adjacent areas, especially the accessory basal nucleus) to the hypothalamus and brainstem. These nuclei also project to the substantia innominata. The two systems are interconnected by intra-amygdaloid fibers to the central and medial nuclei which originate in the other amygdaloid nuclei.

SUMMARY AND FUNCTIONAL CONSIDERATIONS

The amygdaloid complex as a whole projects to a seemingly bewildering array of subcortical targets, in addition to its extensive cortical projections (Amaral, this volume). Although any scheme summarizing these relationships runs the risk of oversimplification, it may be suggested that the projections can be divided into two major systems (Fig. 9). The first is concerned primarily with the cerebral cortex and related structures. Thus, the lateral and basal nuclei, which are the largest amygdaloid nuclei and give rise to a large proportion of the cortical amygdaloid projections (e.g., Porrino et al., 1982; Amaral and Price, 1985), have relatively restricted subcortical projections. These are primarily directed to the substantia innominata, striatum and mediodorsal thalamic nucleus, all of which have major interactions with the cerebral cortex. The second system is more concerned with 'visceral' structures. For example, the central and medial nuclei, which have little if any projection to the cerebral cortex or the mediodorsal thalamic nucleus (e.g., Price and Amaral, 1981), provide the major projections to the structures associated with visceral function in the hypothalamus and brainstem.

The dichotomy is not complete, since the basal accessory nucleus (and probably some of the adjacent areas of the amygdala such as the anterior cortical nucleus and the amygdalo-hippocampal area) project substantially to the cortex and/or the mediodorsal thalamus as well as to the hypothalamus. Similarly, the basal nucleus gives rise to a striking projection to the lateral tuberal nucleus of the hypothalamus and virtually all amygdaloid nuclei project to the substantia innominata. Even more important, there are substantial intra-amygdaloid connections which link the different parts of the complex. In particular, virtually all of the other amygdaloid nuclei project onto the central and medial nuclei, as well as to the bed nucleus of the stria terminalis (e.g., Krettek and Price, 1978b; Price and Amaral, 1981; Aggleton, 1985) so that activity in any part of the amygdaloid complex may be transmitted to the hypothalamus and brainstem.

But, the scheme serves to emphasize the double nature of the amygdaloid complex. On the one hand, amygdaloid nuclei can influence cognitive functions, either by their extensive direct cortical projections or through structures such as the nucleus basalis or the mediodorsal thalamic nucleus. On the other, they can affect visceral function by way of the direct projections to the neuroendocrine centers of the hypothalamus or the afferent and efferent autonomic nuclei of the medulla. On anatomical grounds alone, it might be expected that seizure activity in the amygdala would cause disturbances in such parameters as heart rate and blood pressure while also disrupting more cognitive, sensory-motor functions.

ACKNOWLEDGEMENTS

I would like to thank Mr. H. van Luu and Mr. J. Hayes for drafting and photographic assistance, and Ms. J. Hoffmann for secretarial assistance. This work was supported by NIH research grant NS-09518.

ABBREVIATIONS

ABm, magnocellular accessory basal amygdaloid nucleus
ABp, parvicellular accessory basal amygdaloid nucleus
ac, anterior commissure
Acc, nucleus accumbens
AHA, amygdalo-hippocampal area
AHN, anterior hypothalamic nuclei

BL, basolateral amygdaloid nucleus (rat)
BM, basomedial amygdaloid nucleus (rat)
Bm, magnocellular basal amygdaloid nucleus
Bp, parvicellular basal nucleus
Bpl, paralamellar basal amygdaloid nucleus
BNST, bed nucleus of the stria terminalis
C, caudate nucleus
Ce, central amygdaloid nucleus
Cif, centralis interior thalamic nucleus
Cl, claustrum
CM, centromedian thalamic nucleus
Coa, anterior cortical amygdaloid nucleus
Cop, posterior cortical amygdaloid nucleus
Cs, centralis superior thalamic nucleus
DNX, dorsal nucleus of the vagus
EC, entorhinal cortex
En, endopyriform nucleus
GP, globus pallidus
Ia, anterior intercalated nucleus
L, lateral amygdaloid nucleus
LHA, lateral hypothalamic area
LTN, lateral tuberal nucleus
MD, m and l, mediodorsal thalamic nucleus, medial and lateral parts
Me, medial amygdaloid nucleus
NBM, nucleus basalis of meynert
NTS, nucleus of the solitary tract
ot, optic tract
P, putamen
PAC, periamygdaloid cortex
PAG, periaqueductal grey matter
PB, parabrachial nucleus
PC, pyriform cortex
Pc, paracentral thalamic nucleus
PMN, pre-mammillary nucleus
Pr, perirhinal cortex
PV, periventricular thalamic nucleus
PVN, paraventricular nucleus of the hypothalamus
RF, reticular formation
scp, superior cerebellar peduncle
sm, stria medullaris
SON, supraoptic nucleus of the hypothalamus
VMH, ventromedial hypothalamic nucleus
VP, ventral pallidum
VTA, ventral tegmental area

REFERENCES

Aggleton, J.P., and Mishkin, M., 1984, Projections of the amygdala to the thalamus in the cynomolgus monkey, J. Comp. Neurol., 222:56.
Aggleton, J.P., 1985, A description of intra-amygdaloid connections in old world monkeys, Exp. Brain Res., 57:390.
Amaral, D.G., and Price, J.L., 1985, Amygdalo-cortical projections in the monkey (Macaca fascicularis), J. Comp. Neurol., 230:465.
Cowan, W.M., Raisman, G., and Powell, T.P.S., 1965, The connections of the amygdala, J. Neurol. Neurosurg. Psych., 28:137.
De Olmos, J.S., 1972, The amygdaloid projection field in the rat as studied with the cupric-silver method, in: The Neurobiology of the Amygdala, B.E. Eleftheriou, ed., Plenum Press, New York, p. 295.

Fuller, T.A., Russchen F.T., and Price J.L., 1985, Presumptive glutamergic/aspartergic afferents to the ventral striato-pallidal region, Soc. Neurosci. Abstr., 11:1055.

Gloor, P., Oliver, A., Quesney, L.F., Andermann, F., Horowitz, S., 1982, The role of the limbic system in experiential phenomena of temporal lobe epilepsy, Ann. Neurol., 12:129.

Heimer, L., and Nauta, W.J.H., 1969, The hypothalamic distribution of the stria terminalis in the rat, Brain Res., 182:19.

Holstege, G., Meiners, L., and Tan, K., 1985, Projections of the bed nucleus of the stria terminalis to the mesencephalon, pons, and medulla oblongata in the cat, Exp. Brain Res., 58:379.

Hopkins, D.A., 1975, Amygdalotegmental projections in the rat, cat and monkey, Brain Res., 115:57.

Kaada, B.R., 1972, Stimulation and regional ablation of the amygdaloid complex with reference to functional representations, in: The Neurobiology of the Amygdala, B.E., Eleftheriou, ed., Plenum Press, New York, p. 205.

Kelly, A.E., Domesick, V.B., and Nauta, W.J.H., 1982, The amygdalo-striatal projection in the rat. An anatomical study by anterograde and retrograde tracing methods, Neuroscience, 7:615.

Krettek, J.E., and Price, J.L., 1977, Projections from the amygdaloid complex to the cerebral cortex and thalamus in the rat and cat, J. Comp. Neurol., 171:687.

Krettek, J.E., and Price, J.L., 1978a, Amygdaloid projections to subcortical structures within the basal forebrain and brainstem in the rat and cat, J. Comp. Neurol., 178:225.

Krettek, J.E., and Price, J.L., 1978b, A description of the amygdaloid complex in the rat and cat with observations on intra-amygdaloid axonal connections, J. Comp. Neurol., 178:225.

McBride, R.L., and Sutin, J., 1977, Amygdaloid and pontine projections to the ventromedial nucleus of the hypothalamus, J. Comp. Neurol., 174:377.

Mizuno, N., Takahashi, O., Satoda, T., and Matsushima, R., 1985, Amygdalospinal projections in the macaque monkey, Neurosci. Lett., 53:327.

Nauta, W.J.H., 1961, Fiber degeneration following lesions of the amygdaloid complex in the monkey, J. Anat., 95:515.

Oldfield, B.J., and Silverman, A.J., 1985, A light microscopic HRP study of limbic projections to the vasopressin-containing nuclear groups of the hypothalamus, Brain Res. Bull., 14:143.

Porrino, L.J., Crane, A.M., and Goldman-Rakic, P.S., 1981, Direct and indirect pathways from the amygdala to the frontal lobe in rhesus monkey, J. Comp. Neurol., 198:121.

Price, J.L., 1981, The efferent projections of the amygdaloid complex in the rat, cat and monkey. in: The Amygdaloid Complex, Y. Ben-Ari, ed., Elsevier/North Holland, Amsterdam, p. 121.

Price, J.L., and Amaral, D.G., 1981, An autoradiographic study of the projections of the central nucleus of the monkey amygdala, J. Neurosci., 1:1242.

Russchen, F.T., and Price, J.L., 1984, Amygdalostriatal projections in the rat. Topographical organization and fiber morphology shown using the lectin PHA-L as an anterograde tracer, Neurosci. Lett., 47:15.

Russchen, F.T., Bakst, I., Amaral, D.G., and Price, J.L., 1985a, The amygdalostriatal projections in the monkey. An anterograde tracing study, Brain Res., 329:241.

Russchen, F.T., Amaral, D.G., and Price, J.L., 1985b, The afferent connections of the substantia innominata in the monkey, Macaca fascicularis, J. Comp Neurol., 242:1.

Russchen, F.T., Amaral, D.G., and Price J.L., 1986, Subcortical inputs to the mediodorsal nucleus in the monkey, J. Comp. Neurol., in press.

Veening, J.G., Swanson, L.W., and Sawchenko, P.E., 1984, The organization of projections from the central nucleus of the amygdala to brainstem sites involved in central autonomic regulation: a combined retrograde transport-immunohistochemical study, Brain Res., 303:337.

Zaborszky, L., Leranth, C., and Heimer, L., 1984, Ultrastructural evidence of amygdalofugal axons terminating on cholinergic cells of the rostral forebrain, Neurosci. Lett., 52:219.

CORTICAL AND SUBCORTICAL AFFERENTS OF THE AMYGDALOID COMPLEX

F.T. Russchen

Department of Anatomy
Vrije Universiteit
Amsterdam, The Netherlands

INTRODUCTION

The idea that the amygdaloid complex in some way is involved in the processing of sensory information is closely associated with the fact that this structure plays a central role in the Klüver-Bucy syndrome (Klüver and Bucy, 1939; Weiskrantz, 1956; Rolls and Rolls, 1973). The symptoms of this syndrome all seem to fit the classification discrimination deficits, and it has been suggested that they are due to an inability to associate stimuli with reinforcement (Jones and Mishkin, 1972). The amygdala has therefore been implicated in a variety of functions which relate environmental stimuli to the autonomic and endocrine state of the individual and to past experiences, in order to play a role in producing a meaningful and coordinated response (Kaada, 1972; Gloor, 1978). This suggests an interaction, at the level of the amygdaloid complex, of information from sensory cortical areas with inputs from subcortical areas in the hypothalamus and brainstem, known to be of importance for drive and motivation. At the 1971 symposium on the neurobiology of the amygdaloid complex, Lammers (1972) reviewed the then known sources of afferents to the amygdala. These included the olfactory bulb, the primary olfactory cortex, temporal neocortical areas and the hypothalamus. With the aid of the newly developed anterograde and retrograde transport techniques, in the last decade the knowledge on the connections of the amygdala has greatly expanded. These progresses have been laid down in the proceedings of the INSERM symposium on the amygdaloid complex in 1981. It is now obvious, for example, that the thalamus and the brainstem should be regarded as major sources of inputs at the cost of the quantitative importance of the hypothalamic projections.

The inputs of the amygdala have been studied intensively in the monkey, cat and rat. The present review will summarize and compare the anatomical data available for each of these three species. The sources of inputs will be arranged into three major groups; sensory related, limbic related, and general inputs. The latter group includes aminergic inputs from the brainstem and cholinergic inputs from the basal forebrain. It will be shown that fibers from each group terminate in a distinct set of amygdaloid nuclei (Fig. 4). Furthermore, the arrangement of intraamygdaloid connections suggests that there are different routes through the amygdala for each of these afferent groups so that the major outputs, which also arise from distinct nuclei, probably contain different information.

SENSORY INPUTS

The neocortex has proven to be the major source of afferents that relay sensory information to the amygdaloid complex. Whitlock and Nauta (1956) were the first to demonstrate direct projections from the temporal neocortex to the amygdala in the monkey. Subsequent studies have shown that the neocortical field in the primate that distributes fibers to this structure is much larger and includes parts of the frontal lobe and most of the insula (Jones and Powell, 1970; Pandya et al., 1973; Leichnetz and Astruc, 1975; Leichnetz et al., 1976; Herzog and Van Hoesen, 1976; Leichnetz and Astruc, 1977; Aggleton et al., 1980; Turner et al., 1980; Mufson et al., 1981; Van Hoesen, 1981). Also in the cat neocortical inputs to the amygdala have been found (Druga, 1969; Heath and Jones, 1971; Lescault, 1971; Siegel et al., 1971; Cranford et al., 1976; Russchen, 1982a). In the rat, in contrast, neocortical inputs are very limited or non-existent (Veening, 1978; Ottersen, 1982; Turner and Zimmer, 1984). In all three species, however, meso- and allocortical areas have been described as sources of amygdaloid afferents. They include the perirhinal and insular cortices (M: Herzog and Van Hoesen, 1976; Aggleton et al., 1980; Mufson et al., 1981; Van Hoesen, 1981; C: Russchen, 1982a; Witter and Groenewegen, 1986; R: Veening, 1978; Ottersen, 1982; Turner and Zimmer, 1984), the entorhinal cortex (M: Aggleton et al., 1980; Van Hoesen, 1981; C: Russchen, 1982a; Witter and Groenewegen, 1986; R: Veening, 1978; Yarita et al., 1980; Ottersen, 1982; Luskin and Price, 1983), medial frontal cortical areas (M: Pandya et al., 1973; Aggleton et al., 1980; Van Hoesen, 1981; C: Russchen, 1982a; R: Veening, 1978; Beckstead, 1979; Ottersen, 1982), the pyriform cortex (M: Van Hoesen, 1981; C: Wakefield, 1980; Russchen, 1982a; R: Veening, 1978; Ottersen, 1982; Luskin and Price, 1983) and the hippocampal formation (M: Rosene and Van Hoesen, 1977; Van Hoesen, 1981; C: Russchen, 1982a; R: Veening, 1978; Ottersen, 1982). In Fig. 1, the cortical areas projecting to the amygdala in the monkey, cat and rat are indicated on medial and lateral views of the brains of these animals. The arrows in this figure indicate the direction of the flow of sensory information for the various modalities towards cortical areas which project to the amygdala. Although, at least for the visual, auditory and somatosensory modalities, inputs from primary sensory cortical areas have never been found, cortico-cortical connectivity chains leading from these areas to the amygdala have been demonstrated. These cascades of connections have been studied in most detail in the monkey (e.g., Turner et al., 1980). In this species, visual and auditory connections are comprised of stepwise projections in the inferior and superior temporal gyrus, respectively. In the cat, area 20 and the posterior ectosylvian gyrus (Ep) are neocortical areas that receive visual (e.g., Tusa and Palmer, 1980) and auditory (e.g., Reale and Imig, 1983) inputs, respectively, and project to the amygdala (Fig. 3; Heath and Jones, 1971; Lescault, 1971; Siegel et al., 1971; Russchen, 1982a; Reale and Imig, 1983).

With respect to the somatosensory modality, evidence for a similar connectivity chain recently has become available. In the monkey, projections from SII to portions of the insular cortex, which in turn projects to the amygdala (Mufson et al., 1981), were demonstrated (Friedman et al., 1982). In the cat, the existence of a direct projection from SIV to the amygdaloid complex has been reported (Burton and Kopf, 1984).

It is interesting to note that some degree of retinotopic (Tusa and Palmer, 1980), tonotopic (Reale and Imig, 1983) or somatotopic (Clemo and Stein, 1982) organization has been found in the neocortical visual, auditory and somatosensory areas that distribute fibers to the amygdala. For the auditory modality, the findings of Reale and Imig (1983) suggest that the tonotopic organization is preserved in the projections to the amygdaloid complex.

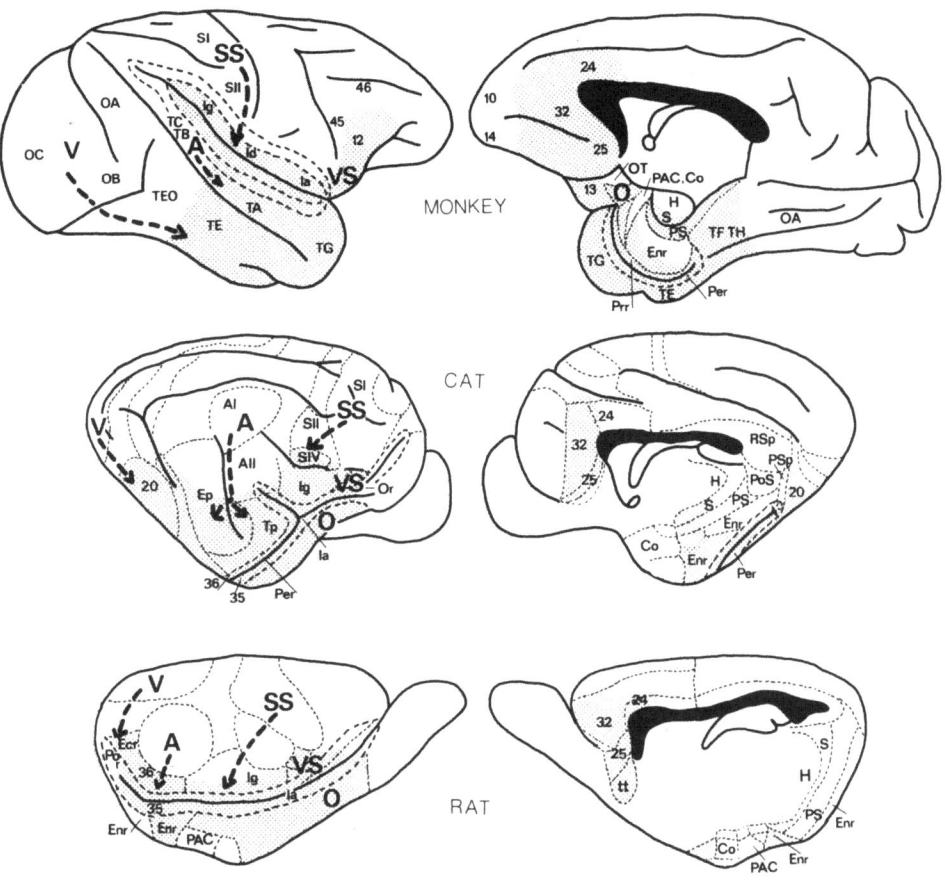

Fig. 1. Medial and lateral views of the brains of a monkey, cat and rat which show the extent of the cortical areas that give rise to corticoamygdaloid projections (shaded). In the medial views, the parahippocampal cortex and the subiculum are depicted as if they have been unfolded. The outlines and parcellations have been modified from figures presented by others (M: Van Hoesen and Pandya, 1975; Van Hoesen, 1981; C: Jimenez-Castellanos and Reinoso-Suarez, 1985; R: Deacon et al., 1983). The letters V, A, SS, VS and O indicate the location of visual, auditory, somatosensory, viscerosensory and olfactory areas, respectively, and the arrows indicate the directions of the information flow. Superficial amygdaloid nuclei have been left unshaded. The delineation of the amygdalopetal cortical fields is still somewhat tentative since retrograde tracing studies in general show a more restricted field of origin than anterograde tracing studies suggest. From anterograde tracing studies, however, it is difficult to determine exactly which portion of a certain area gives rise to the observed projections.

In the monkey and cat, also the mesocortical amygdalopetal areas, especially the perirhinal cortex, receive projections from essentially the same sensory related neocortical areas as the amygdala (M: Jones and Powell, 1970; Van Hoesen and Pandya, 1975; Turner et al., 1980; C: Heath and Jones, 1971; Room and Groenewegen, 1986).

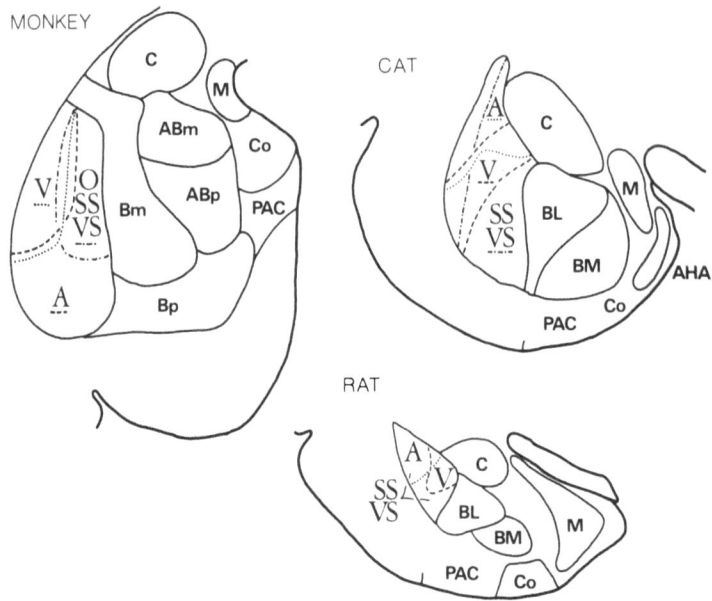

Fig. 2. Frontal sections through the amygdaloid complex of the monkey, cat and rat, showing the delineation of the major nuclei. Most of these nuclei, however, are not homogeneous. In the central nucleus, for example, at least a medial and a lateral subdivision can be distinguished and the basolateral nucleus in the rat and cat is composed of a rostral and a caudal part which most likely correspond to the magnocellular and parvicellular basal nuclei, respectively, of the monkey (Price, 1981). To give some indication of the degree of convergence of information from the various sensory modalities in the lateral nucleus, the major distribution field of corticoamygdaloid fibers for each modality has been indicated with their abbreviations. The outline of each field is marked with different symbols.

ABBREVIATIONS: A, auditory; ABm, magnocellular accessory basal nucleus; ABp, parvicellular accessory basal nucleus; BL, basolateral nucleus; BM, basomedial nucleus; Bm, magnocellular basal nucleus; Bp, parvicellular basal nucleus; C, central nucleus; Co, cortical nucleus; Ecr, ectorhinal cortex; Enr, entorhinal cortex; H, hippocampus; Ig,d,a, granular, dysgranular, agranular, insular cortex; L, lateral nucleus; M, medial nucleus; O, olfactory; Or, orbital cortex; OT, olfactory tubercle; ot, optic tract; P, putamen; PAC, periamygdaloid cortex; Per, perirhinal cortex; Por, postrhinal cortex; PoS, postsubiculum; Prr, prorhinal cortex; PS, pre- parasubiculum; PSp, postsplenial cortex; RSp, retrosplenial cortex; S, subiculum; SS, somatosensory; V, visual; VS, viscerosensory.

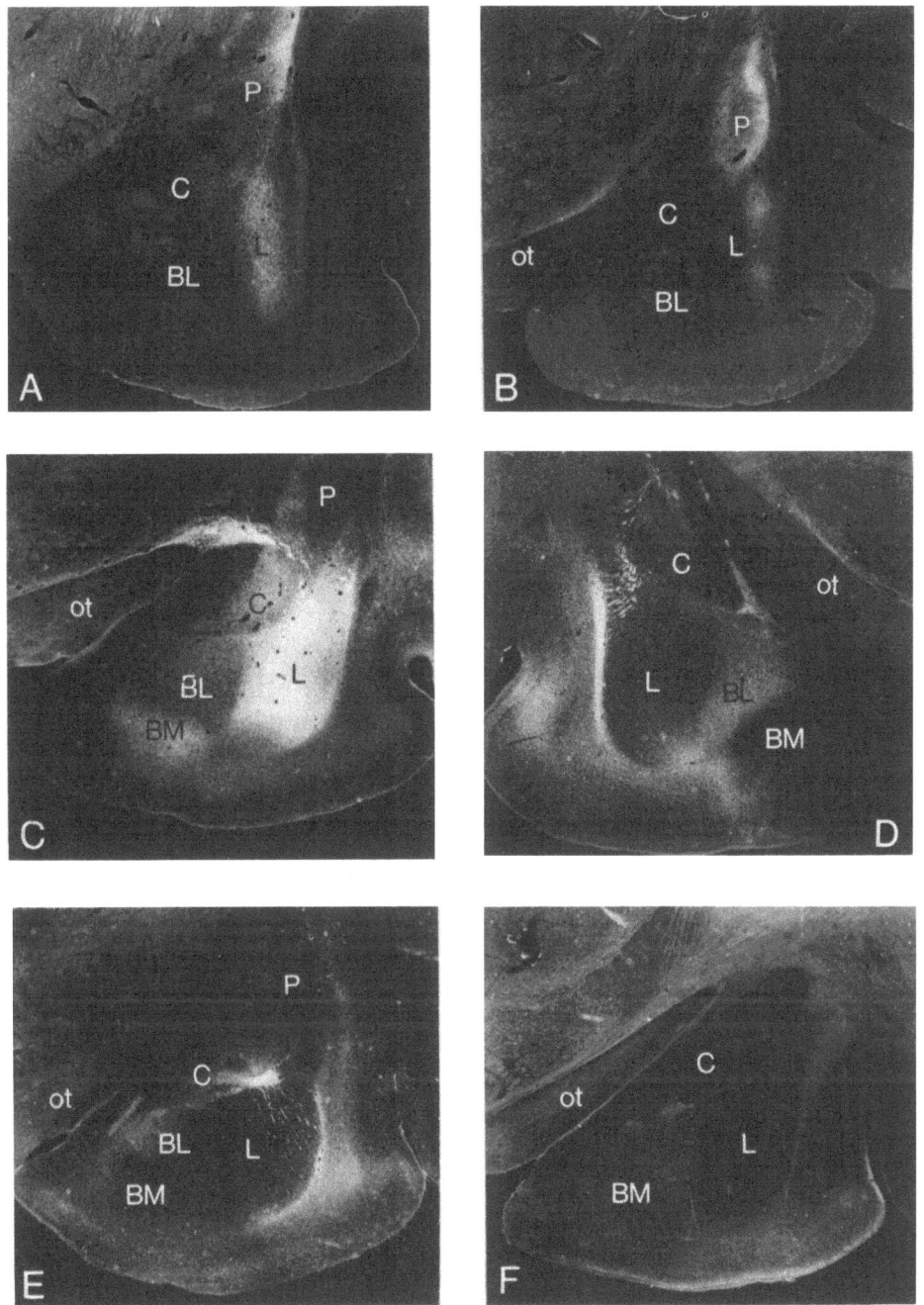

Fig. 3. The distribution of anterogradely transported label in the amygdaloid complex of the cat in six cases in which injections of tritiated amino acids were placed in various cortical areas. The material is from Drs. P. Room and M.P. Witter, and I acknowledge them for the opportunity to use it. The injection sites were placed in: A: visual area 20; B: auditory area Ep; C: area 36 rostrally; D: area 35 and the adjacent part of the entorhinal cortex (DLEA); E: a more medial part of the lateral entorhinal area; F: the agranular insular cortex (AIv, this injection also involves the rostral part of the pyriform cortex).

Since in the rat virtually no neocortical afferents could be demonstrated, in this species the only, or major, sensory related cortical inputs to the amygdaloid complex apparently are comprised of cascades of cortico-cortical connections that reach mesocortical areas in the rhinal sulcus which in turn distribute to the amygdala (Turner and Zimmer, 1980; Ottersen, 1982; Turner and Zimmer, 1984). Thus, unlike in the monkey and cat, in the rat mesocortical areas represent the first stage in the corticocortical sensory processing that has a direct access to the amygdala. With respect to the viscerosensory input, it may well be that the primary sensory cortical area of this modality projects directly to the amygdala. In all three species, the cortical area that receives fibers from the thalamic relay nucleus for visceral afferent information (the ventroposteromedial parvicellular nucleus; VPMpc or VMb; M: Beckstead et al., 1980; Morse et al., 1980; C: Nomura et al., 1980; R: Lasiter et al., 1982; Saper, 1982) has been demonstrated to project to the amygdala (M: Van Hoesen, 1981; C: Yasui et al., 1984; R: Norgren and Grill, 1976; Saper, 1982). In the rat, this cortical area has been shown to receive also more direct inputs from visceral related brainstem nuclei (parabrachial nucleus; PB and the nucleus of the solitary tract; ntr sol; Saper, 1982). Likewise, for the olfactory modality the primary sensory cortex, which receives fibers from the olfactory bulb, has been shown to project directly to the amygdaloid complex (C: Wakefield, 1980; Russchen, 1982a; R: Veening, 1978; Carlsen et al., 1982; Ottersen, 1982; Luskin and Price, 1983). An additional, more indirect route may lead from the primary olfactory cortex to adjacent cortical areas (M: Yarita et al., 1980; R: Luskin and Price, 1983) which in turn project to the amygdala. It may thus be concluded that the cortical olfactory and viscerosensory information flow in all three species seems to have access to the amygdala at an earlier stage of processing than the information conveyed from the visual, auditory and somatosensory modalities.

A similar pattern can be observed when one considers the subcortical structures that are likely to convey sensory information to the amygdaloid complex. Viscerosensory structures of the brainstem have been shown to project directly to the amygdala (M: PB, Aggleton et al., 1980; Mehler, 1980; Norita and Kawamura, 1980; C: PB, Russchen, 1982b; Takeuchi et al., 1982; R: PB and ntr sol., Ricardo and Koh, 1978; Veening, 1978; Saper and Loewy, 1980; Ottersen, 1981). The subcortical structures that are likely candidates for relaying visual, auditory and somatosensory information, in contrast, are located in the posterior thalamus.

In the monkey, the medial pulvinar (Jones and Burton, 1976; Norita and Kawamura, 1980) and in the cat and rat the lateromedial-suprageniculate complex (LM-SG) and the magnocellular portion of the medial geniculate nucleus (C: Heath and Jones, 1971; Graybiel, 1972, 1973; Russchen, 1982b; R: Ottersen and Ben-Ari, 1979) have been reported as sources of amygdaloid afferents. These structures receive projections, for instance, from deep layers of the superior colliculus (coll sup) which are a site of convergence for all three modalities (see, for example, Graybiel and Berson, 1980).

There is some dispute whether the viscerosensory thalamic nucleus (VPMpc) projects to the amygdaloid complex. Mehler and coworkers (Mehler et al., 1981, see also Russchen, 1982b) could not identify projections from VPMpc to the amygdala in the rat, cat or monkey, and instead considered all the cells in the region that were retrogradely filled from injections into the amygdala to be located in the subparafascicular nucleus. Other authors claim that at least part of the projections to the amygdala originate in VPMpc (C: Cechetto et al., 1983; R: Ottersen and Ben-Ari, 1979; Turner and Herkenham, 1981).

DIFFERENTIAL DISTRIBUTION OF AMYGDALOID AFFERENTS

Sensory Inputs; The Lateral and Central Nuclei (Fig. 4A)

A this point it is necessary to mention that the amygdala is not a homogeneous structure, but can be subdivided into several nuclei. In all three species, the gross division that can be made is that into superficial and deep nuclei. In both groups, several nuclei can be distinguished on the basis of cytoarchitectonics and histochemistry. Fig. 2 indicates the parcellations of amygdaloid nuclei for all three species as they are most frequently used. The nomenclature is as proposed by Price (1981).

Whereas in the cat the lateral nucleus, and to a lesser degree the central nucleus, appear to be the only amygdaloid nuclei that receive fibers from neocortical areas associated with visual, auditory (e.g., Figs. 3A, B), somatosensory (Burton and Kopf, 1984) and viscerosensory (Lescault, 1971; Yasui et al., 1984) modalities, in the monkey additional projections reach the dorsal portion of the magnocellular basal nucleus (Turner et al. 1980; Van Hoesen, 1981). For the rat, Turner and Zimmer (1984) showed that perirhinal and insular cortical areas, which receive projections from cortical association areas related to these four modalities, project to the lateral nucleus, but, especially for the viscero- and somatosensory modalities, also to the basolateral and central nuclei. In Fig. 3, it is indicated that projections related to each modality do not distribute diffusely over the lateral nucleus, but rather have their own major termination fields which overlap to some degree (for more detail, see Turner et al., 1980; Turner, 1981; Van Hoesen, 1981; Turner and Zimmer, 1984).

The olfactory bulb projects directly to superficial nuclei of the amygdaloid complex (e.g., Turner et al., 1978), but indirect olfactory inputs do reach deep amygdaloid nuclei including the lateral and central nuclei. It has been shown that the primary olfactory cortex in the cat and rat (C: Wakefield, 1980; Russchen, 1982a; R: Veening, 1978; Ottersen, 1982; Luskin and Price, 1983) and the presumptive olfactory association cortex in the monkey (area 13, e.g., Yarita et al., 1980; Van Hoesen, 1981) distribute fibers to the lateral and central nuclei.

Also the subcortical sources of amygdaloid inputs, that have been indicated above as possible relay centers for sensory information, distribute mainly to the lateral and central nuclei. In Fig. 4A, the distribution of the cortical and subcortical sensory inputs over the amygdaloid complex is shown schematically for the cat. As mentioned above, the situation differs slightly, but probably not essentially, for the two other species. It is obvious from this figure that the basolateral nucleus in the cat remains essentially free from these sensory inputs (compare Turner and Zimmer, 1984 and Van Hoesen, 1981 for the monkey and rat). The basomedial nucleus, however, seems to indirectly receive sensory inputs since it is projected upon by the lateral nucleus. Moreover, like the lateral nucleus, and unlike the basolateral nucleus, it receives a relatively dense projection from area 36 of the perirhinal cortex (Fig. 3, and Witter and Groenewegen, 1986) which is an area of sensory convergence (Heath and Jones, 1971; Room and Groenewegen, 1986).

Limbic Inputs; The Basolateral and Central Nuclei (Fig. 4B)

As discussed before, the distinction between the origins of cortical inputs to the basolateral and lateral nuclei appears to be most definite in the cat. Whereas in this species the basolateral nucleus remains relatively free of neocortical inputs (Fig. 4A), it consitutes, together with the central nucleus, the major termination field of fibers from meso-

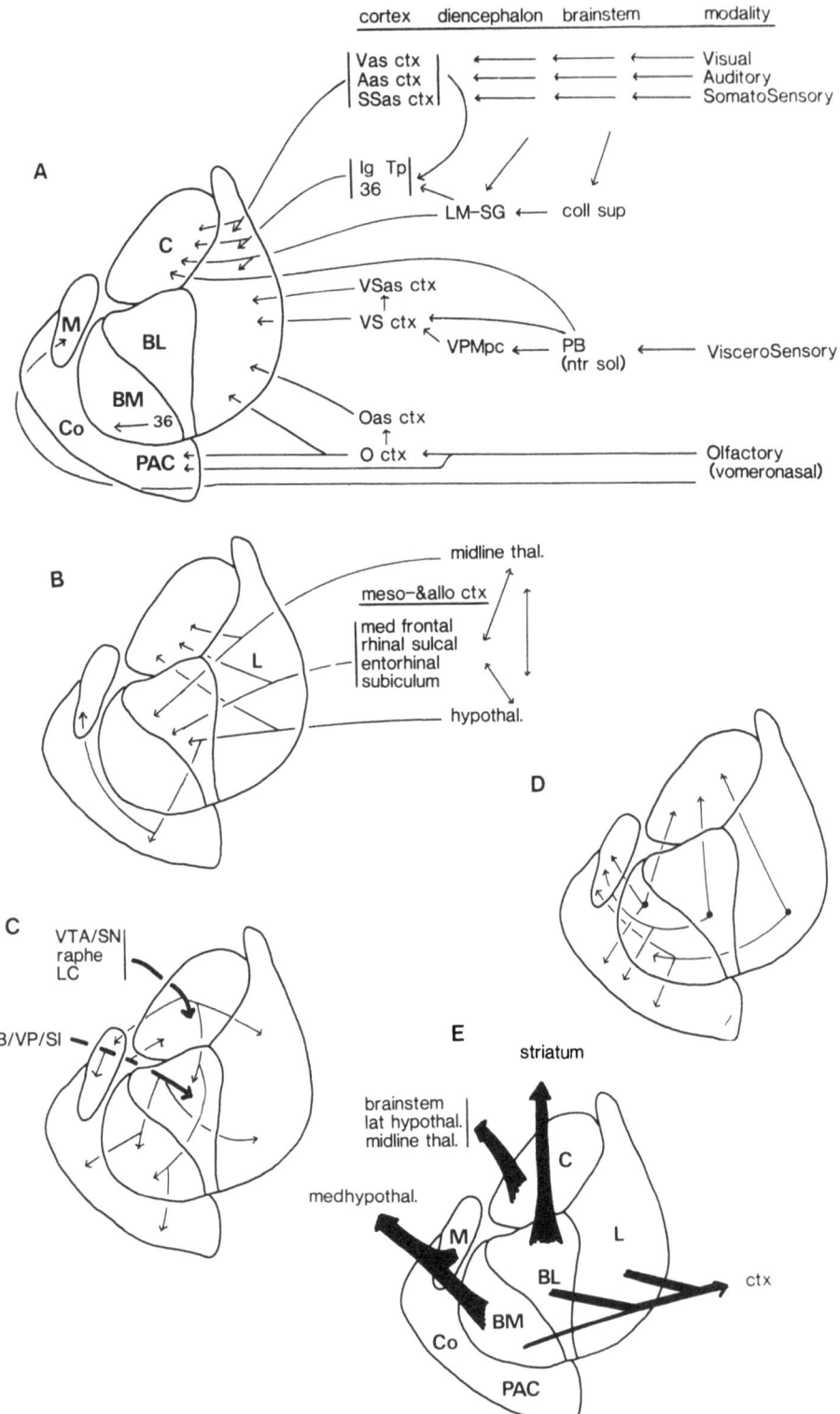

Fig. 4. A schematic representation of the distribution of the three groups of major inputs in the amygdaloid complex (A,B,C), the intraamygdaloid connections (D) and the major outputs (E), in diagrams of the cat's amygdala.

and allocortical areas including parts of the medial frontal cortex, the insular cortex, area 35 of the perirhinal cortex, the lateral entorhinal area and the subiculum (Russchen, 1982a, and Figs. 3, 4B). In the monkey, the probable equivalent of the basolateral nucleus, the basal nucleus (Price, 1981) also constitutes the major termination field of fibers from mesocortical areas. However, in this species the equivalent of the basomedial nucleus, the accessory basal nucleus, also receives fibers from a number of mesocortical areas (see Van Hoesen, 1981 Fig. 12)

Also in the rat, the situation differs from that in the cat, such that mesocortical areas of sensory convergence, which are located in the rhinal sulcus, mainly project to the lateral nucleus (vide supra). In addition, the lateral and central nuclei receive fibers fom mesocortical entorhinal and medial frontal areas and from the allocortical hippocampal formation (Beckstead, 1979; Ottersen, 1982). The basolateral nucleus receives fibers from distinct portions of these latter cortical regions.

There may be a real problem in extrapolating the organization of the cortical amygdaloid inputs as indicated in Figs. 4A and 4B for the cat to the two other species. However, when one compares the distribution of cortical fibers over the lateral, basolateral and basomedial amygdaloid nuclei in the rat, cat and monkey, it appears that a shift has taken place which can be summarized as follows. Whereas in the rat virtually no neocortical input reaches the amygdala, the lateral and basolateral nuclei each receive inputs from distinct portions of the mesocortex. In the cat, neocortical inputs have replaced the mesocortical inputs to the lateral nucleus, so that mesocortical inputs are mainly restricted to the basolateral nucleus. In the monkey, the termination fields of both neocortical and mesocortical projections appear to have expanded, so that neocortical areas not only project to the lateral nucleus but also distribute to the basolateral (basal) nucleus. Mesocortical areas, besides projecting to the basolateral (basal) nucleus, also distribute to the basomedial (accessory basal) nucleus.

Another point worth mentioning is that although the central nucleus on the one hand and the lateral and basomedial nuclei on the other apparently receive projections from the same cortical areas, there is evidence for a difference in the laminar origin of their inputs (Aggleton et al., 1980; Ottersen, 1982; Russchen, 1982a; Cechetto et al., 1983). In most of the amygdalopetal cortical areas, the fibers projecting to the central nucleus arise mainly from layer V whereas those reaching the lateral and basolateral nuclei predominantly originate in layers III and II.

With respect to the subcortical inputs of the basolateral nucleus, in all three species limbic related structures in the thalamus and hypothalamus seem to comprise the major sources. At least in the cat and rat, the basolateral and central nuclei are the major termination sites of fibers arising from midline thalamic nuclei including the paraventricular (PV), reuniens (Re), interanteromedial (IAM), rhomboid (Rh) and centromedial (Cem) nuclei (C: Russchen, 1982b; R: Ottersen and Ben-Ari, 1979; Turner and Herkenham, 1981). Less information is available on the distribution over the amygdaloid nuclei of fibers originating in midline thalamic nuclei in the monkey (Aggleton et al., 1980; Mehler, 1980, Mehler et al., 1981; Norita and Kawamura, 1980). In all three species, hypothalamic inputs arising from the ventromedial (VMH), arcuate (Arc), paraventricular (PVH), premammillary (PM), and supramammillary (SM) nuclei and from the lateral hypothalamic area (LH) most densely project to the central and medial nuclei, and more weakly to the cortical and basolateral nuclei. Few hypothalamic fibers are generally reported to reach the basomedial and lateral nuclei (M: Saper et al., 1978, 1979b; Amaral et al., 1982; C: Saper et al., 1979b; Russchen, 1982b; R: Conrad and Pfaff, 1976a,b; Saper et al., 1976;

Krieger et al., 1979; Ottersen, 1980; Berk and Finkelstein, 1982; Ter Horst et al., 1984). In the monkey, VMH and Arc in addition distribute fibers to the magnocellular accessory basal nucleus (Amaral et al., 1982), which is considered to be equivalent to part of the basomedial nucleus (Price, 1981).

Grouping the mesocortical, midline thalamic and hypothalamic inputs together the way it is presently done gains relevance when one realizes that parts of these structures are interconnected in several ways. Amygdalopetal hypothalamic and midline thalamic nuclei are variably connected with mesocortical areas (Segal, 1977; Beckstead, 1978, 1979; Herkenham, 1978; Wyss et al., 1979; Wyss, 1981; Kita and Oomura, 1981; Amaral et al., 1982; Saper, 1982; Haglund et al., 1984; Saper, 1985; Room and Groenewegen, 1986). Furthermore, the hypothalamic VMH, Arc and LH have been shown to distribute fibers to midline thalamic nuclei including PV and Re (Conrad and Pfaff, 1976a; Saper et al., 1976, 1979a,b; Amaral et al., 1982; Berk and Finkelstein, 1982). In addition, Re has been reported to project diffusely to the hypothalamus (e.g., Herkenham, 1978).

It should be mentioned that the parafascicular nucleus has been indicated by several authors as an additional thalamic source of inputs to the central nucleus (Ottersen and Ben-Ari, 1979; Norita and Kawamura, 1980; Cechetto et al., 1983), a finding which is opposed by others (e.g., Mehler, 1980, Mehler et al., 1981; Russchen, 1982b). This intralaminar thalamic nucleus is known to send a major projection to the striatum (Royce, 1978; Beckstead, 1984).

Two additional thalamic nuclei, the parataenial and peripeduncular nuclei, have been described as sources of amygdaloid afferents. They can, however, not easily be incorporated into one of the presently distinguished groups of inputs to the amygdala. These nuclei have been reported to distribute their fibers predominantly to the lateral and medial nuclei and to a lesser degree also to the central and basomedial nuclei (Jones et al., 1976; Ottersen and Ben-Ari, 1979; Turner and Herkenham, 1981; Russchen, 1982b; Cechetto et al., 1983).

<u>General Inputs (Fig. 4C)</u>

Like most of the telencephalon, the amygdaloid complex receives projections from brainstem aminergic cell groups as well as a cholinergic input from the basal forebrain. These inputs probably can be considered to have some form of general activity modulating influence.

The distribution of brainstem aminergic inputs in the rat has been shown in detail by Fallon (1981). The major termination site of dopaminergic, serotonergic and noradrenergic fibers appears to be the central nucleus, but all other amygdaloid nuclei, the lateral nucleus to the least degree, show fibers with these transmitters. The results of anatomical studies that traced the connections from the presumed sources of these respective inputs, the ventral tegmental area/substantia nigra, the raphe nuclei and the locus coeruleus seem to generally support this distribution pattern (Bobillier et al., 1976; McBride and Sutin, 1976; Jones and Moore, 1977; Azmitia and Segal, 1978; Bowden et al., 1978; Simon et al., 1979; Beckstead et al., 1979; Mehler, 1980; Norita and Kawamura, 1980; Meibach and Katzman, 1981; Ottersen, 1981; Amaral et al., 1982; Russchen, 1982b).

Cholinergic cells are known to be present in the basal forebrain and to distribute fibers to the entire cortex of the telencephalon. Also the amygdala receives such projections which distribute mainly, but not only, to the basolateral nucleus (M: Mesulam et al., 1983; C: Russchen, 1982b; R: Nagai et al., 1982).

It should be kept in mind that the substances indicated in the projections from these brainstem and forebrain structures are not necessarily, and even unlikely, the only ones present. For example, the presence of CCK in amygdalopetal neurons in the ventral tegmentum (Hökfelt et al., 1980), and the existence of GABA-ergic projections from the basal forebrain to telencephalic structures have been shown (e.g., Mesulam et al., 1983; Zaborsky et al., 1986).

Transmitter Candidates in Other Afferent Connections

Recent evidence suggests that cortical (Walker and Fonnum, 1983) thalamic and hypothalamic projections to the amygdala, at least in part, use an excitatory amino acid as a transmitter (own unpublished observations; see also Ottersen, this volume). Furthermore, projections from the bed nucleus of the stria terminalis, the hypothalamus and brainstem may use a wide variety of neuroactive peptides (for review see Shiosaka et al., 1983). Especially the central nucleus is rich in peptide containing fibers (Roberts et al., 1982; Shiosaka et al., 1983), which partly originate from the bed nucleus of the stria terminalis (e.g., Shiosaka et al., 1983). It has been shown, for example, that projections containing neurotensin (Kawakami et al., 1984) and vasoactive intestinal polypeptide (VIP, Marley et al., 1981) originate from the brainstem, and that projections containing beta-endorphin and alfa-MSH arise from the arcuate nucleus (Gray et al., 1984). On the other hand, somatostatin may be mainly intrinsic to the amygdaloid complex (Palkovits et al., 1982).

INTRA-AMYGDALOID CONNECTIONS (FIG. 4D)

Fig. 4D schematically shows the known major connections between the various amygdaloid nuclei (data from Krettek and Price, 1978; Price and Amaral, 1981; Ottersen, 1982; Russchen, 1982b; Aggleton, 1985). An important aspect of the organization of these connections is that the main flow is from lateral to medial and converges on the central and superficial amygdaloid nuclei. Furthermore, the lateral and basolateral nuclei, which receive different extra-amygdaloid inputs, are not interconnected. Also, the basomedial nucleus, which receives fibers from the lateral nucleus, is not directly connected with the basolateral nucleus. This is important since the major output systems of the amygdaloid complex originate in separate nuclei (Fig. 4E). The basolateral nucleus is the major source of projections to the striatum. The lateral nucleus has little or no subcortical output, but projects to the basomedial nucleus, which takes care of the major output to the medial hypothalamus. The central nucleus, on the other hand, is the major source of projections to the lateral hypothalamus and the brainstem. The cortical output of the amygdala arises mainly from the lateral, basolateral and basomedial nuclei. For detailed information on these and other amygdalofugal projections, including the projections to the thalamus, reference is made to the contribution to this volume by Drs. Amaral and Price.

CONCLUSIONS

The first conclusion that can be drawn from the data presented is that not all individual amygdaloid nuclei are to the same degree, or at the same level, involved in the integration of sensory, autonomic and endocrine information. Secondly, there appear to be five major flows of information through the amygdaloid complex.

The first stream seems to represent a more or less strong convergence of relatively direct sensory information from all modalities in the lateral nucleus. The lateral nucleus, in turn, projects to the basomedial nucleus, from where an influence on hypothalamic activities and thus probably the endocrine system is possible.

The second stream involves the basolateral nucleus in which convergence of information from several strongly limbic related structures takes place. The major output of this nucleus is to large portions of the striatum. The striatum has been suggested to be involved in the programming of motor patterns or, more generally, in the formation of procedural memories or habits (Mishkin et al., 1984).

Together, the lateral, basolateral and basomedial nuclei provide the greater part of the amygdaloid output to the cortex, the third stream. This projection may be important in cognitive processes (see Amaral and Price, 1984).

The fourth flow runs via the central nucleus, which apparently is the only amygdaloid nucleus that truly collects all types of input that reach the amygdala. The major output of this nucleus is to the lateral hypothalamus and autonomic centers in the brainstem. Although the central nucleus is not a homogeneous structure, and distinct portions may receive different sets of inputs, this nucleus as a whole apparently provides a waystation for the majority of amygdalopetal structures to exert an influence on autonomic functions.

The fifth flow involves the superficial amygdaloid nuclei. Besides a strong and direct olfactory (and vomeronasal) input, they also receive derivatives of the inputs that reach the other amygdaloid nuclei by way of intra-amygdaloid connections. The superficial nuclei project mainly to the medial hypothalamus.

Kaada (1972), in his review on amygdaloid functional localization, concluded that the corticomedial and central amygdaloid nuclei represent a facilitatory area for feeding and sexual activity, and emotional responses, respectively, whereas the basolateral region exerts a tonic inhibitory influence on these related activities.

From a phylogenetic (Johnston, 1923) and embryologic (Humphrey, 1972; Bayer, 1981) point of view, the central and superficial amygdaloid nuclei have been suggested to be older than the basolateral group of nuclei. The strongest convergence of amygdaloid inputs appears to take place in the oldest group, and together their efferents reach structures involved in autonomic and endocrine functions. The identity of the information flow through the other amygdaloid nuclei seems to support the hypothesis that the basolateral group of nuclei developed to introduce additional functions for the amygdala. One aspect seems to be that the hypothalamic processing by way of these nuclei can be influenced by others than the olfactory modality, which according to Gloor (1978) 'seems to have opened the way which freed animal behavior from the rigidity of simple reflex mechanisms represented at the hypothalamic level'. A second aspect seems to be a role in learning processes that involve the declarative and procedural memory systems or, in Mishkin's terminology, the memories and habits systems, via projections to the cortex and striatum, respectively.

REFERENCES

Aggleton, J.P., Burton, M.J., and Passingham, R.E., 1980, Cortical and subcortical afferents to the amygdala of the rhesus monkey (Macaca mulatta), Brain Res., 190:347.

Aggleton, J.P., 1985, A description of intra-amygdaloid connections in old world monkeys, Exp. Brain Res., 57:390.

Amaral, D.G., Veazey, R.B., and Cowan, W.M., 1982, Some observations on hypothalamo-amygdaloid connections in the monkey, Brain Res., 252:13.

Amaral, D.G., and Price, J.L., 1984, Amygdalo-cortical projections in the monkey (Macaca fascicularis), J. Comp. Neurol., 230:465.

Azmitia, E.C., and Segal, M., 1978, An autoradiographic analysis of the differential ascending projections of the dorsal and median raphe nuclei in the rat, J. Comp. Neurol., 179:641.

Bayer, S.A., 1981, Neurogenesis in the rat amygdala, in: The Amygdaloid Complex, Y. Ben-Ari, ed., Elsevier/North-Holland, p. 19.

Beckstead, R.M., 1978, Afferent connections of the entorhinal area in the rat as demonstrated by retrograde cell-labeling with horseradish peroxidase, Brain Res., 152:249.

Beckstead, R.M., 1979, An autoradiographic examination of corticocortical and subcortical projections of the mediodorsal projection (prefrontal) cortex in the rat, J. Comp. Neurol., 184:43.

Beckstead, R.M., and Domesick, V.B., and Nauta, W.J.H., 1979, Efferent connections of the substantia nigra and ventral tegmental area in the rat, Brain Res., 175:191.

Beckstead, R.M., Domesick, V.B., and Nauta, W.J.H., 1980, The nucleus of the solitary tract in the monkey: projections to the thalamus and brain stem nuclei, J. Comp. Neurol., 190:259.

Beckstead, R.M., 1984, The thalamostriatal projection in the cat, J. Comp. Neurol., 223:313.

Berk, M.L., and Finkelstein, J.A., 1982, Efferent connections of the lateral hypothalamic area of the rat: an autoradiographic investigation, Brain Res. Bull., 8:511.

Bobillier, P., Seguin, S., Petitjean, F., Salvert, D., Touret, M., and Jouvet, M., 1976, The raphe nuclei of the cat brain stem: a topographical atlas of their efferent projections as revealed by autoradiography, Brain Res., 113:449.

Bowden, D.M., German, D.C., and Poynter, W.D., 1978, An autoradiographic, semistereotaxic mapping of major projections from locus-coerulus and adjacent nuclei in Macaca Mulatta, Brain Res., 145:257.

Burton, H., and Kopf, E.M., 1984, Ipsilateral cortical connections from the second and fourth somatic sensory areas in the cat, J. Comp. Neurol., 225:527.

Carlsen, J., De Olmos, J., and Heimer, L., 1982, Tracing of two-neuron pathways in the olfactory system by the aid of transneuronal degeneration: projections to the amygdaloid body and hippocampal formation, J. Comp. Neurol., 208:196.

Cechetto, D.F., Ciriello, J., and Calaresu, F.R., 1983, Afferent connections to cardiovascular sites in the amygdala: a horseradish peroxidase study in the cat, J. Auton. Nerv. Syst., 8:97.

Clemo, H.R., and Stein, B.E., 1982, Somatosensory cortex: a 'new' somatotopic representation, Brain Res., 235:162.

Conrad, L.C.A., and Pfaff, D.W., 1976a, Efferents from medial basal forebrain and hypothalamus in the rat. I. An autoradiographic study of the medial preoptic area, J. Comp. Neurol., 169:185.

Conrad, L.C.A., and Pfaff, D.W., 1976b, Efferents from medial basal forebrain and hypothalamus in the rat. II. An autoradiographic study of the anterior hypothalamus, J. Comp. Neurol., 169:221.

Cranford, J.L., Ladner, S.J., Campbell, C.B.G., and Neff, W.D., 1976, Efferent projections of the insular and temporal neocortex of the cat, Brain Res., 117:195.

Deacon, T.V., Eichenbaum, H., Rosenberg, P., and Eckman, K.W., 1983, Afferent connections of the perirhinal cortex in the rat, J. Comp. Neurol., 220:168.

Druga, R., 1969, Neocortical projections to the amygdala (An experimental study with the Nauta method), J. Hirnforsch., 11:467.

Fallon, J.H., 1981, Histochemical characterization of dopaminergic, noradrenergic and serotonergic projections to the amygdala, in: The Amygdaloid Complex, Y. Ben-Ari, ed., Elsevier/North-Holland, p. 175.

Friedman, D.P., Murray, E.A., and Mishkin, M., 1982, Cortico-limbic pathway for touch: connections via somatosensory cortical fields in the lateral sulcus of the monkey, Soc. Neurosci. Abstr., 8:38.

Gloor, P., 1978, Inputs and outputs of the amygdala: what the amygdala is trying to tell the rest of the brain, in: Limbic Mechanisms, K.E. Livingston and O. Hornykiewicz, eds., Plenum Press, New York, p. 189.

Gray, T.S., Cassell, M.D., and Kiss, J.Z., 1984, Distribution of proopiomelanocortin-derived peptides and enkephalins in the rat central nucleus of of the amygdala, Brain Res., 306:354.

Graybiel, A.M., 1972, Some ascending connections of the pulvinar and nucleus lateralis posterior of the thalamus in the cat, Brain Res., 44:99.

Graybiel, A.M., 1973, The thalamo-cortical projection of the so-called posterior nuclear group: a study with anterograde degeneration methods in the cat, Brain Res., 49:229.

Graybiel, A.M., and Berson, D.M., 1980, Histochemical identification and afferent connections of subdivisions in the lateralis posterior-pulvinar complex and related thalamic nuclei in the cat, Neuroscience., 5:1175.

Haglund, L., Swanson, L.W., and Köhler, C., 1984, The projection of the supramammillary nucleus to the hippocampal formation: an immunohistochemical and anterograde transport study with the lectin PHA-L in the rat, J. Comp. Neurol., 229:171.

Herkenham, M., 1978, The connections of the nucleus reuniens thalami: evidence for a direct thalamo-hippocampal pathway in the rat, J. Comp. Neurol., 177:589.

Heath, C.J., and Jones, E.G., 1971, Anatomical organization of the suprasylvian gyrus of the cat, Ergeb. Anat. Entw. Gesch., 45:1.

Herzog, A.G., and Van Hoesen, G.W., 1976, Temporal neocortical afferent connections to the amygdala in the rhesus monkey, Brain Res., 115:57.

Hökfelt, T., Skirboll, L., Rehfeld, J.F., Goldstein, M., Markey, K., and Dann, O., 1980, A subpopulation of mesencephalic dopamine neurons projecting to limbic areas contains a cholecystokinin-like peptide: evidence from immunohistochemistry combined with retrograde tracing, Neuroscience, 5:2093.

Humphrey, T., 1972, The development of the human amygdaloid complex, in: The Neurobiology of the Amygdala, B.E. Eleftheriou, ed., Plenum Press, New York, p. 21.

Jimenez-Castellanos Jr., J., and Reinoso-Suarez, F., 1985, Topographical organization of the afferent connections of the principal ventromedial thalamic nucleus in the cat, J. Comp. Neurol., 236:297.

Johnston, J.B., 1923, Further contributions to the study of the evolution of the forebrain, J. Comp. Neurol., 35:337.

Jones, B., and Mishkin, M., 1972, Limbic lesions and the problem of stimulus-reinforcement associations, Exp. Neurol., 36:362.

Jones, B.E., and Moore, R.Y., 1977, Ascending projections of the locus coeruleus in the rat. II. Autoradiographic study, Brain Res., 127:23.

Jones, E.G., and Powell, T.P.S., 1970, An anatomical study of converging sensory pathways within the cerebral cortex of the monkey, Brain, 93:793.

Jones, E.G., and Burton, H., 1976, A projection from the medial pulvinar to the amygdala in primates, Brain Res., 104:142.

Jones, E.G., Burton, H., Saper, C.B., and Swanson, L.W., 1976, Midbrain, diencephalic and cortical relationships of the basal nucleus of Meynert and associated structures in primates, J. Comp. Neurol., 167:385.

Kaada, B.R., 1972, Stimulation and regional ablation of the amygdaloid complex with reference to functional representations, in: The Neurobiology of the Amygdala, B.E. Eleftheriou, ed., Plenum Press, New York p. 205.

Kawakami, F., Fukui, K., Okamura, H., Morimoto, N., Yanaihara, N., Nakajima, T., and Ibata, Y., 1984, Influence of ascending noradrenergic fibers on the neurotensin-like immunoreactive perikarya and evidence of direct projection of ascending neurotensin-like immunoreactive fibers in the rat central nucleus of the amygdala, Neurosci. Lett., 51:225.

Kita, H., and Oomura, Y., 1981, Reciprocal connections between the lateral hypothalamus and the frontal cortex in the rat: electrophysiological and anatomical observations, Brain Res., 213:1.

Klüver, H., and Bucy, P.D., 1939, Preliminary analysis of the functions of the temporal lobe in monkeys, Arch. Neurol. Psychiat., 42:979.

Krettek, J.E., and Price, J.L., 1977, Projections from the amygdaloid complex to the cerebral cortex and thalamus in the rat and cat, J. Comp. Neurol., 172:687.

Krettek, J.E., and Price, J.L., 1978, A description of the amygdaloid complex in the rat and cat with observations on intra-amygdaloid axonal connections, J. Comp. Neurol., 178:255.

Krieger, M.S., Conrad, L.C.A., and Pfaff, D.W., 1979, An autoradiographic study of the efferent connections of the ventromedial nucleus of the hypothalamus, J. Comp. Neurol., 183:785.

Lammers, H.J., 1972, The neural connections of the amygdaloid complex in mammals, in: The Neurobiology of the Amygdala, B.E. Eleftheriou, ed., Plenum Press, New York, p. 123.

Lasiter, P.S., Glanszman, D.L., and Mensah, P.A., 1982, Direct connectivity between pontine taste areas and gustatory neocortex in rat, Brain Res., 234:111.

Leichnetz, G.R., and Astruc, J., 1975, Efferent connections of the orbitofrontal cortex in the Marmoset (Saguinus oedipus), Brain Res., 84:169.

Leichnetz, G.R., Povlishock, J.T., and Astruc, J., 1976, A prefrontoamygdaloid projection in the monkey: light and electron microscopic evidence, Neurosci. Lett., 2:261.

Leichnetz, G.R., and Astruc, J., 1977, The course of some prefrontal corticofugals to the pallidum, substantia-innominata and amygdaloid complex in monkeys, Exp. Neurol., 54:104.

Lescault, H., 1971, Some neocortico-amygdaloid connections in the cat, Doctoral thesis, University of Ottawa.

Luskin, M.B., and Price, J.L., 1983, The topographic organization of associational fibers of the olfactory system in the rat, including centrifugal fibers to the olfactory bulb, J. Comp. Neurol., 216:264.

Marley, P.D., Emson, P.C., Hunt, S.P., and Fahrenkrug, J., 1981, A long ascending projection in the rat brain containing vasoactive intestinal polypeptide, Neurosci. Lett., 27:261.

McBride, R.L., and Sutin, J., 1976, Projections of the locus coeruleus and adjacent pontine tegmentum in the cat, J. Comp. Neurol., 165:265.

Mehler, W.R., 1980, Subcortical afferent connections of the amygdala in the monkey, J. Comp. Neurol., 190:733.

Mehler, W.R., Pretorius, J.K., Phelan, K.D., and Mantyh, P.W., 1981, Diencephalic afferent connections of the amygdala in the squirrel monkey with observations and comments on the cat and rat, in: The Amygdaloid Complex, Y. Ben-Ari, ed., Elsevier/North-Holland, p. 105.

Meibach, R.C., and Katzman, R., 1981, Origin, course and termination of dopaminergic substantia nigra neurons projecting to the amygdaloid complex in the cat, Neuroscience, 6:2159.

Mesulam, M.M., Mufson, E.J., Levey, A.I., and Wainer, B.H., 1983, Cholinergic innervation of cortex by the basal forebrain: cytochemistry and cortical connections of the septal area, diagonal band nuclei, nucleus basalis (substantia innominata), and hypothalamus in the rhesus monkey, J. Comp. Neurol., 214:170.

Mishkin, M., Malamut, B., and Bachevalier, J., 1984, Memories and habits: two neural systems, in: Neurobiology of Learning and Memory, G. Lynch, J.L. McGaugh and N.M. Weinberger, eds., Guilford Press, New York, p. 65.

Morse, J.R., Beckstead, R.M., Prichard, T., and Norgren, R., 1980, Ascending gustatory and visceral afferent pathways in the monkey, Neurosc. Abstr., 6:307.

Mufson, E.J., Mesulam, M.M., and Pandya, D.N., 1981, Insular interconnections with the amygdala in the rhesus monkey, Neuroscience, 6:1231.

Nagai, T., Kimura, H., Maeda, T., McGeer, P.L., Peng, F., and McGeer, E.G., 1982, Cholinergic projections from the basal forebrain of rat to the amygdala, J. Neurosci., 2:513.

Nomura, S., Itoh, K., and Mizuno, N., 1980, Topographical arrangement of thalamic neurons projecting to the orbital gyrus in the cat, Exp. Neurol., 67:601.

Norgren, R., and Grill, H.J., 1976, Efferent distribution from the cortical gustatory area in rats, Soc. Neurosci. Abstr., 2:124.

Norita, M., and Kawamura, K., 1980, Subcortical afferents to the monkey amygdala: an HRP study, Brain Res., 190:225.

Ottersen, O.P., and Ben-Ari, Y., 1979, Afferent connections to the amygdaloid complex of the rat and cat: I. Projections from the thalamus, J. Comp. Neurol., 187:401.

Ottersen, O.P., 1980, Afferent connections to the amygdaloid complex of the rat and cat: II. Afferents from the hypothalamus and the basal telencephalon, J. Comp. Neurol., 194:267.

Ottersen, O.P., 1981, Afferent connections to the amygdaloid complex of the rat. III. Afferents from the lower brain stem, J.Comp. Neurol., 202:335.

Ottersen, O.P., 1982, Connections of the amygdala of the rat. IV. Cortico-amygdaloid and intraamygdaloid connections as studied with axonal transport of horseradish peroxidase, J. Comp. Neurol., 205:30.

Palkovits, M., Tapia-Arancibia, L., Kordon, C., and Epelbaum, J., 1982, Somatostatin connections between the hypothalamus and the limbic system of the rat brain, Brain Res., 250:223.

Pandya, D.N., Van Hoesen, G.W., and Domesick, V.B., 1973, A cinguloamygdaloid projection in the rhesus monkey, Brain Res., 61:369.

Price, J.L., 1981, Toward a consistent terminology for the amygdaloid complex, in: The Amygdaloid Complex, Y. Ben-Ari, ed., Elsevier/North-Holland, p. 13.

Price, J.L., and Amaral, D.G., 1981, An autoradiographic study of the projections of the central nucleus of the monkey amygdala, J. Neurosci., 11:1242.

Reale, R.A., and Imig, T.J., 1983, Auditory cortical field projections to the basal ganglia of the cat, Neuroscience, 8:67.

Ricardo, J.A., and Koh, E.T., 1978, Anatomical evidence of direct projections from the nucleus of the solitary tract to the hypothalamus, amygdala, and other forebrain structures in the rat, Brain Res., 153:1.

Roberts, G.W., Woodhams, P.L., Polak, J.M., and Crow, T.J., 1982, Distribution of neuropeptides in the limbic system of the rat: the amygdaloid complex, Neuroscience, 7:99.

Rolls, E.T., and Rolls, B.J., 1973, Altered food preferences after lesions in the basolateral region of the amygdala in the rat, J. Comp. Physiol. Psychol., 83:248.

Room, P., and Groenewegen, H.J., 1986, Connections of the parahippocampal cortex in the cat. I. Cortical afferents and II. Subcortical afferents, J. Comp. Neurol., in press.

Rosene, D.L., and Van Hoesen, G.W., 1977, Hippocampal efferents reach widespread areas of cerebral cortex and amygdala in the rhesus monkey, Science, 198:315.

Royce, G.J., 1978, Cells of origin of subcortical afferents to the caudate nucleus: a horseradish peroxidase study in the cat, Brain Res., 153:465.

Russchen, F.T., 1982a, Amygdalopetal projections in the cat. I. Cortical afferent connections. A study with retrograde and anterograde tracing techniques, J. Comp. Neurol., 206:159.

Russchen, F.T., 1982b, Amygdalopetal projections in the cat. II. Subcortical afferent connections. A study with retrograde tracing techniques, J. Comp. Neurol., 207:157.

Saper, C.B., Swanson, L.W., and Cowan, W.M., 1976, The efferent connections of the ventromedial nucleus of the hypothalamus of the rat, J. Comp. Neurol., 169:409.

Saper, C.B., Swanson, L.W., and Cowan, W.M., 1978, The efferent connections of the anterior hypothalamic area of the rat, cat and monkey, J. Comp. Neurol., 182:575.

Saper, C.B., Swanson, L.W., and Cowan, W.M., 1979a, An autoradiographic study of the efferent connections of the lateral hypothalamic area in the rat, J. Comp. Neurol., 183:689.

Saper, C.B., Swanson, L.W., and Cowan, W.M., 1979b, Some efferent connections of the rostral hypothalamus in the squirrel monkey (saimiri sciureus) and cat, J. Comp. Neurol., 184:205.

Saper, C.B., and Loewy, A., 1980, Efferent connections of the parabrachial nucleus in the rat, Brain Res., 197:291.

Saper, C.B., 1982, Convergence of autonomic and limbic connections in the insular cortex of the rat, J. Comp. Neurol., 210:163.

Saper, C.B., 1985, Organization of cerebral cortical afferent systems in the rat. II. Hypothalamocortical projections, J. Comp. Neurol., 237:21.

Segal, M., 1977, Afferents to the entorhinal cortex of the rat studied by the method of retrograde transport of horseradish peroxidase, Exp. Neurol., 57:750.

Shiosaka, S., Sakanaka, M., Inagaki, S., Senba, E., Hara, Y., Takatsuki, K., Takagi, H., Kawai, Y., and Tohyama, M., 1983, Putative neurotransmitters in the amygdaloid complex with special reference to peptidergic pathways, in: Chemical Neuroanatomy, P,C. Emson, ed., Raven Press, New York, p. 359.

Siegel, A., Sasso, L., and Tassoni, J.P., 1971, Fiber connections of the temporal lobe with the corpus striatum and related structures in the cat, Exp. Neurol., 33:130.

Simon, H., Le Moal, M., and Calas, A., 1979, Efferents and afferents of the ventral tegmental-A10 region studied after local injection of ^{3}H-leucine and horseradish peroxidase, Brain Res., 178:17.

Takeuchi, Y., McLean, J.H., and Hopkins, D.A., 1982, Reciprocal connections between the amygdala and parabrachial nuclei: ultrastructural demonstration by degeneration and axonal transport of horseradish peroxidase in the cat, Brain Res., 239:583.

Ter Horst, G.J., Groenewegen, H.J., Karst, H., and Luiten, P.G.M., 1984, Phaseolus vulgaris leuco-agglutinin immunohistochemistry. A comparison between autoradiographic and lectin tracing of neuronal efferents, Brain Res., 307:379.

Turner, B.H., Cupta, K.C., and Mishkin, M., 1978, The locus and cytoarchitecture of the projection areas of the olfactory bulb in macaca mulatta, J. Comp. Neurol., 177:381.

Turner, B., Mishkin, M., and Knapp, M.,1980, Organization of the amygdalopetal projections from modality-specific cortical association areas in the monkey, J. Comp. Neurol., 191:515.

Turner, B., and Zimmer, J., 1980, Connections between the cerebral cortex and amygdala in the rat, Soc. Neurosci. Abstr., 6:113.

Turner, B.H., 1981, The cortical sequence and terminal distribution of sensory related afferents to the amygdaloid complex of the rat and monkey, in: The Amygdaloid Complex, Y. Ben-Ari, ed., Elsevier/North-Holland, p. 51.

Turner, B., and Herkenham, M., 1981, An autoradiographic study of thalamo-amygdaloid connections in the rat, Anat. Rec., 199:260A.

Turner, B.H., and Zimmer, J., 1984, The architecture and some of the interconnections of the rat's amygdala and lateral periallocortex, J. Comp. Neurol., 227:540.

Tusa, R.J., and Palmer, L.A., 1980, Retinotopic organization of areas 20 and 21 in the cat, J. Comp. Neurol., 193:147.

Van Hoesen, G.W., and Pandya, D.N., 1975, Some connections of the entorhinal (area 28) and perirhinal (area 35) cortices of the rhesus monkey. I. Temporal lobe afferents, Brain Res., 95:1.

Van Hoesen, G.W., 1981, The differential distribution, diversity and sprouting of cortical projections to the amygdala in the rhesus monkey, in: The Amygdaloid Complex, Y. Ben-Ari, ed., Elsevier/North-Holland, p. 77.

Veening, J.G., 1978, Cortical afferents of the amygdaloid complex in the rat: an HRP study, Neurosci. Lett., 8:191.

Wakefield, C., 1980, The topographical organization and laminar origin of some cortico-amygdaloid connections, Neurosci. Lett., 20:21.

Walker, J.E., and Fonnum, F., 1983, Regional cortical glutamergic and aspartergic projections to the amygdala and thalamus of the rat, Brain Res., 267:371.

Weiskrantz, L., 1956, Behavioral changes associated with ablation of the amygdaloid complex in monkeys, J. Comp. and Physiol. Psychol., 49:381.

Whitlock, D.G., and Nauta, W.J.H., 1956, Subcortical projections from the temporal neocortex in Macaca mulatta, J. Comp. Neurol., 106:183.

Witter, M.P., and Groenewegen, H.J., 1986, Connections of the parahippocampal cortex in the cat. III. Cortical and thalamic efferents and IV. Subcortical efferents, J. Comp. Neurol., in press.

Wyss, J.M., Swanson, L.W., and Cowan, W.M., 1979, A study of subcortical afferents to the hippocampal formation in the rat, Neuroscience, 4:463.

Wyss, J.M., 1981, An autoradiographic study of the efferent connections of the entorhinal cortex in the rat, J. Comp. Neurol., 199:495.

Yarita, H., Iino, M., Tanabe, T., Koguere, S., and Takagi, S.F., 1980, A transthalamic olfactory pathway to orbito-frontal cortex in the monkey, J. Neurophysiol., 43:69.

Yasui, Y., Itoh, K., and Mizuno, N., 1984, Projections from the parvocellular part of the posteromedial ventral nucleus of the thalamus to the lateral amygdaloid nucleus in the cat, Brain Res., 292:151.

Zaborszky, L., Carlsen, J., Brashear, H.R., and Heimer, L., 1986, Cholinergic and GABA-ergic projections to the olfactory bulb in the rat, J. Comp. Neurol., in press.

PUTATIVE AMINO ACID TRANSMITTERS IN THE AMYGDALA

O.P. Ottersen, B.O. Fischer, E. Rinvik and
J. Storm-Mathisen

Anatomical Institute, University of Oslo
Karl Johans gate 47
N-0162 Oslo 1
Norway

INTRODUCTION

The amygdala is involved in temporal lobe seizures, in man as well as in animal models (refs. in Ben-Ari, 1981), and is among the brain structures from which epileptiform seizures can be most effectively elicited by repeated electrical stimulation (Racine, 1981) or by topical injections of neuroexcitants (Ben-Ari et al., 1980; Tremblay et al., 1983). A better understanding of how the amygdala participates in these epileptic phenomena requires, among other things, more insight into its transmitter mechanisms. The amino acid transmitters gamma-aminobutyrate (GABA), glutamate (Glu), and aspartate (Asp) are of particular interest in this respect since it is assumed that they play decisive roles in the pathogenesis and sequelae of epilepsy (Meldrum, 1984). It is now time for increased efforts to unravel the amino acid transmitter mechanisms in the amygdala. First, the efferent and afferent fiber systems of the various amygdaloid subdivisions have recently been mapped in considerable detail (refs. in Ben-Ari, 1981), facilitating the undertaking and interpretation of experiments aimed at tracing transmitter-specific pathways. Second, new methods have been introduced to supplement those based on autoradiography of in vitro high affinity uptake of radiolabeled amino acids or on immunocytochemistry of glutamic acid decarboxylase (GAD), which have hitherto been the predominant histological techniques in amino acid transmitter research (for references see Ottersen and Storm-Mathisen, 1984a). Thus, evidence has accumulated that D-[^{3}H]Asp and [^{3}H]GABA, injected in vivo, are selectively transported in axons of Glu/Asp-ergic and GABA-ergic neurons, respectively, leading to labeling of the parent cell bodies (Streit, 1980). Further, we have developed immunocytochemical techniques for the demonstration of GABA, Glu, and Asp, using antisera raised against the amino acids conjugated to protein by glutaraldehyde (Storm-Mathisen et al., 1983, 1986; Ottersen and Storm-Mathisen, 1984a,b, 1985; Storm-Mathisen and Ottersen, 1986). Antisera have also been raised against conjugates of taurine (Tau; Madsen et al., 1985; Ottersen et al., 1985), which is a more equivocal transmitter candidate than GABA, Glu, and Asp. In the present paper we report data obtained with amino acid immunocytochemistry and autoradiography of amino acid uptake and transport. Some of the results have been published elsewhere (Fischer et al., 1982; Ottersen and Storm-Mathisen, 1984a; Storm-Mathisen and Ottersen, 1986).

MATERIALS AND METHODS

Immunocytochemistry

Antisera were raised in rabbits by immunizing with GABA, L-Glu, L-Asp, or Tau conjugated to bovine serum albumin (BSA) by distilled glutaraldehyde. The crude sera were absorbed in solid phase with glutaraldehyde-treated BSA, and BSA-glutaraldehyde conjugates of one or more amino acids depending on the test results. After purification, the antisera showed a high degree of selectivity toward their respective antigens (Ottersen and Storm-Mathisen, 1984b; Storm-Mathisen et al., 1986). Our novel test system (Ottersen and Storm-Mathisen, 1984b) permitted simultaneous processing of test antigens and sections (Figs. 2,4), increasing the reliability of the test results. The specific nature of the antisera was corroborated by solid phase absorption experiments with the homologous antigens. Interestingly, the amino acids against which the serum was raised, when added to the serum at concentrations of 50 mM and above, inhibited staining of the corresponding test spots and of tissue sections. This shows that the antibodies are able to interact with the amino acids as such. However, it is the reactivity with the test spots of fixation products that is relevant for the immunocytochemical use of the antibodies. The antisera were applied to free-floating Vibratome or frozen sections (8-30 µm) obtained from mice, rats, guinea pigs, cats, or baboons (Papio papio) that had been perfusion fixed though the heart or abdominal aorta with 5% glutaraldehyde diluted from 25% in 0.1 M phosphate buffer, pH 7.4, preceded by a flush of phosphate buffered 2% dextran. Three of the rats received a unilateral intraamygdaloid injection of colchicine (30 µg in 3 µl sodium phosphate buffer, pH 7.4) 24 hours prior to perfusion. Bound antibodies were visualized by the peroxidase-antiperoxidase technique of Sternberger (1979).

Autoradiography of D-[^{3}H]Asp and [^{3}H]GABA Uptake In Vitro

Transverse slices (200 µm) of brains, rapidly removed from mice or rats after stunning and decapitation, were cut in the cold on a Sorvall tissue chopper and immediately transferred to Krebs' solution at 0°C. The slices were incubated for five minutes in Krebs' medium at 25°C and then for 10-15 minutes in fresh medium containing either D-[^{3}H]Asp (2 µM, 13800 Ci/mol; New England Nuclear) or [^{3}H]GABA (0.5 or 1 µM, 65000 Ci/mol; Amersham; Taxt and Storm-Mathisen, 1984). After three rinses (two minutes each) in the vehicle solution, the slices were fixed for one hour in 5% phosphate-buffered glutaraldehyde. Some of the fixed slices were transferred to 30% sucrose in phosphate buffer and resectioned at 25 µm on a freezing microtome. The sections and slices were treated with 1 M ethanolamine and ethanols, mounted from 70% ethanol on glass slides, allowed to dry, and covered with Ilford L4 or Kodak NTB2 emulsion. The autoradiograms were developed in Kodak D19 (4 minutes, 18°C) after one or two weeks.

Axonal D-[^{3}H]Asp Transport

D-[^{3}H]Asp (100 nl, 25 µCi, 16 mM) was stereotaxically injected by pressure over 20 minutes in the amygdala of five rats using a glass micropipette (see Ottersen et al., 1983, for details). A horizontal approach (Ottersen and Ben-Ari, 1979) was used in order to avoid contamination of the caudatoputamen. The rats were reanesthetized after 12-24 hours and perfused as for immunocytochemistry. Frozen sections (30 µm) were covered with Ilford L4 emulsion and exposed for 15 weeks before development in Kodak D19B. The distribution of labeled neurons was mapped using an X-Y plotter, and compared with the distribution of neurons labeled after horseradish peroxidase (HRP) injections of similar size and location (Ottersen and Ben-Ari, 1979; Ottersen, 1980, 1981, 1982).

RESULTS AND COMMENTS

The nomenclature used in the following is based on that agreed upon by the participants of the 1981 INSERM symposium on the Amygdaloid Body (Price, 1981). Unless stated otherwise, the results described are from mice or rats, which were essentially similar.

GABA

The amygdala exhibited a highly differentiated pattern of GABA-like immunoreactivity (GABA-LI) which, unlike the neuropeptide distributions (Roberts et al., 1982), largely adhered to cytoarchitectonical borders. The central (AC) and medial (Am) nuclei (Figs. 1, 2), the anterior amygdaloid area (AAA), and the intraamygdaloid portion of the bed nucleus of stria terminalis were stained at the same high intensity as the hypothalamus, and were only slightly paler than the globus pallidus, which is the forebrain structure richest in GABA-LI (Fig. 1). A much lower level of immunoreactivity, comparable to that of the neocortex, was found in the basolateral (BL), lateral (AL), and cortical nuclei. Thus, it seems that the major dividing line with respect to GABA-LI corresponds to the phylogenetical partition (Stephan, 1975); the phylogenetically older centromedial part (including the AAA) being richer in GABA-LI than the more recently differentiated corticobasolateral part. The basomedial nucleus displayed an intermediate level of immunoreactivity. The present results agree with biochemical data on GABA contents in microdissected amygdaloid nuclei (Ben-Ari et al., 1976).

Differences in staining intensity occurred within the individual nuclei. Layer I (the superficial plexiform layer) and II of the nucleus of the lateral olfactory tract (NTOL) were far richer in neuropil staining than layer III, and the Am was less intensely stained ventrally than dorsally (Fig. 1A). Further, the staining intensity decreased on moving from the medial subdivision of AC into the intermediate subdivision (Fig. 1A). At rostral levels a laterally situated part of AC with an even lower GABA-LI could be identified; this part apparently corresponds to the lateral capsular subdivision (McDonald, 1982). The neuropil of the intercalated cell islands were invariably intensely stained; this was also true for the small cell nests embedded within layer III of the periamygdaloid cortex (Krettek and Price, 1978b). The plexiform layer of the anterior cortical nucleus and periamygdaloid cortex displayed less immunoreactivity in the deep sublamina (IB; Heimer, 1978) than more superficially (sublamina IA), whereas the plexiform layer of the posterior cortical nucleus was rich in GABA-LI throughout (Fig. 2A).

Most of the neuropil staining for GABA-LI resided in bouton-like dots, many of which appeared to contact unstained, or, to a lesser extent, stained neuronal cell bodies. GABA-LI positive axons were also easily distinguished; they occurred in all nuclei but were particularly concentrated in the stria terminalis (Fig. 2).

The distribution of GABA-LI positive perikarya (Fig. 1) did not match the pattern of neuropil staining. Thus, the AL, BL, and cortical nuclei showed a high density of labeled cells; these were spread throughout the nuclei, were generally smaller than the unstained ones, and seemed to belong to the non-pyramidal cell types (Millhouse and DeOlmos, 1983). The morphology of the GABA-LI positive cells in the BL fitted the description of neurons in this site labeled with another GABA antiserum (McDonald, 1985). A much lower concentration of positive cells occurred in the AC and Am; this could be considerably enhanced by colchicine pretreatment, but remained lower than the concentration in the BL and AL of untreated animals (Fig. 1). The number of GABA-LI positive cells in the latter

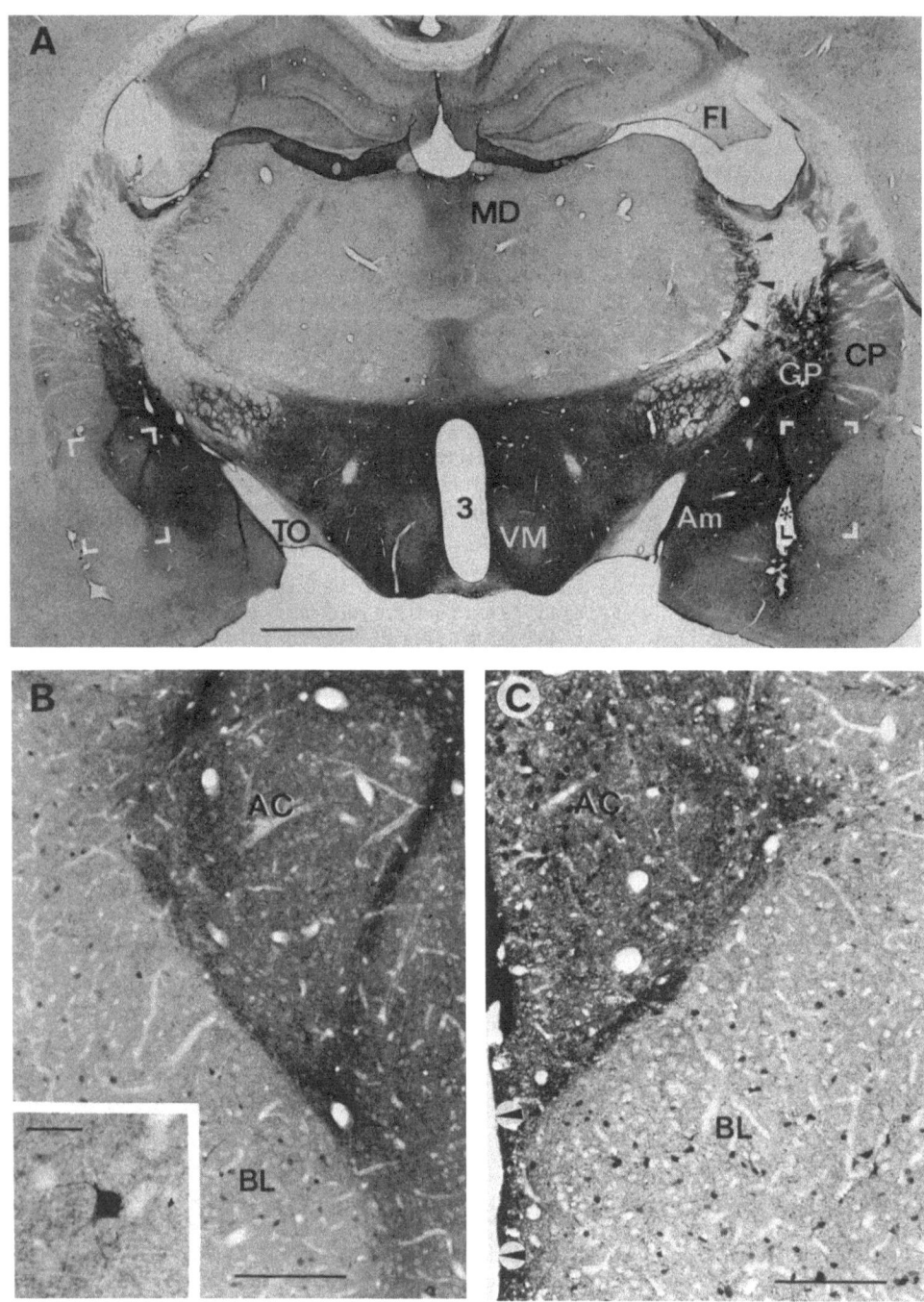

nuclei was only marginally increased by colchicine. These results raise the question whether the GABA-LI rich centromedial amygdala receives an extrinsic GABA innervation in addition to that provided by local neurons. Since lesion studies do not support the existence of a major GABA input from sources outside the amygdala (Le Gal La Salle et al., 1978), it is tempting to speculate that some of the GABA in the AC and Am may depend on connections originating in the BL, AL, or cortical nuclei (Ottersen, 1982). Also, electrophysiological studies suggest the existence of inhibitory internuclear connections within the amygdala (Le Gal La Salle, 1976). However, the case for a contribution of GABA-containing fibers from the BL is weakened by the observation of Smith and Millhouse (1985) that only the pyramidal (i.e., presumably non-GABAergic) cells of BL send axons to AC.

Our finding of numerous GABA-LI positive fibers in the stria terminalis is in agreement with previous data suggesting that this pathway is richer in GAD than many other fiber systems, such as the optic tract, anterior commissure, corpus callosum, and fimbria (Ben-Ari et al., 1976). Lesion experiments indicate that the stria terminalis conveys GAD-containing fibers to its bed nucleus and that it also contributes a small contingent of such fibers to the AC (Le Gal La Salle et al., 1978).

Cat and baboon showed, like rodents, a higher GABA-LI in the centromedial part than in the corticobasolateral part of the amygdala. The morphology (inset, Fig. 1B) and distribution of labeled perikarya also seemed to be largely similar across species.

Autoradiograms of slices incubated with [^{3}H]GABA indicated a higher uptake intensity in the centromedial amygdala, including the AAA, than in the BL and AL (Fig. 3A), thus matching the immunocytochemical data. The NTOL seemed to be relatively poorer in GABA uptake sites than in GABA-LI (cf. Ottersen and Storm-Mathisen, 1984a).

Taurine

In the corticobasolateral subdivision of the amygdala, Tau-like immunoreactivity was found in most of the neurons but only in a small proportion of the glial cells. Conversely, glial cell labeling predominated over neuronal labeling in the medial and central nuclei, except laterally in the central nucleus where a substantial number of immunopositive neurons were detected.

Fig. 1. Photomicrographs showing GABA-like immunoreactivity in the forebrain. A: Low power view from rat COL 7. This animal received an injection of colchicine in the right amygdala 24 hours prior to perfusion (asterisk indicates cannula track). Note increased staining intensity of the amygdala, basal ganglia, and reticular thalamic nucleus (arrowheads) on the injected side. Hooks indicate areas enlarged in B and C. B,C: Labeling of cell bodies and neuropil is more intense on the injected side (C) than contralaterally (B). Stained perikarya in the central nucleus are only seen on the injected side. Arrowheads indicate cannula track. Inset in B: Immunoreactive multipolar neuron in the basolateral nucleus of a baboon (case B1). Transverse sections, antiserum 26 diluted 1:250. Bar = 1 mm (A); 200 μm (B,C); 25 μm (inset in B). Abbreviations, see separate list. Modified from Storm-Mathisen and Ottersen, 1985.

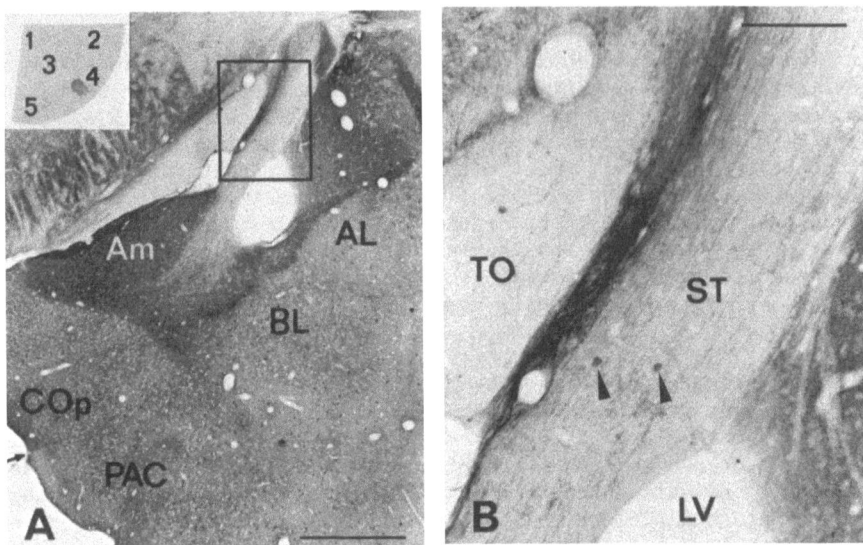

Fig. 2. Photomicrographs showing GABA-like immunoreactivity in the caudal amygdala of rat. A: Note intense staining of the medial nucleus. Arrow indicates the border between the peri-amygdaloid cortex and posterior cortical nucleus; frame shows area enlarged in B. Inset in A: Test filter (radius 6 mm) processed together with the sections. The antiserum reacts strongly with the GABA conjugates (spot 4), but insignificantly with conjugates of aspartate (1), glutamate (2), or taurine (5). Center spot (3) contains glutaraldehyde-treated brain macromolecules (no amino acid added). B: The antiserum stains numerous fibers and some neurons (arrowheads) in the stria terminalis. Transverse section, antiserum 26 diluted 1:250. Bar = 0.5 mm (A); 100 µm (B).

Glutamate and Aspartate

Our results so far are consistent with the notion that GABA occurs exclusively in neurons using GABA as a transmitter (the hippocampal mossy fiber system may be an exception; see Storm-Mathisen and Ottersen, 1986). In contrast, much of the free Glu and Asp in the brain serve metabolic purposes unrelated to transmitter function (Fonnum, 1984), implying that the presence of Glu-LI or Asp-LI within a neuron is difficult to interpret in terms of transmitter identity (see Ottersen and Storm-Mathisen, 1985). Corresponding to the situation in most brain areas, Glu-LI and Asp-LI were found in a vast majority of the neurons in the amygdala (Fig. 4). Glu-LI was, however, higher in the perikarya of the corticobasolateral amygdala (Fig. 4B) than in those of the centromedial part. In agreement with previous biochemical data, the mean levels of Glu-LI and Asp-LI did not differ significantly among amygdaloid nuclei (Ben-Ari et al., 1976), but there was a pronounced difference in staining intensity between the amygdala and the globus pallidus (Fig. 4A). The latter structure is among the brain structures poorest in Glu-LI (Ottersen and Storm-Mathisen, 1984b).

To assess the density of putative Glu/Asp-ergic terminals in the various parts of the amygdaloid complex, we incubated brain slices in D-[^{3}H]Asp under conditions favoring selective uptake in such terminals (Taxt and Storm-Mathisen, 1984). The BL, AL, and layer I and II of the NTOL showed the highest uptake activities, corresponding to the level

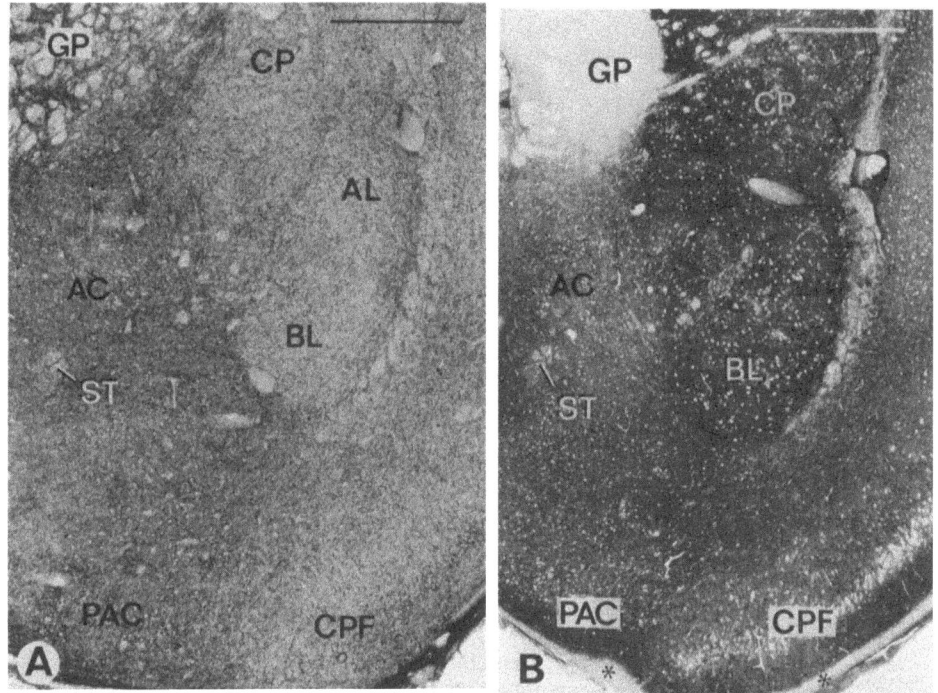

Fig. 3. Photomicrographs showing [^{3}H]GABA (A) and D-[^{3}H]aspartate (B) uptake in the amygdala of the mouse. The two uptake patterns exhibit a striking complementarity. Note in B virtual absence of labeling in the globus pallidus, indicating a very low concentration of aspartate uptake sites, and intense labeling of the caudatoputamen and the basolateral and lateral amygdaloid nuclei. Asterisks in B indicate the weakly stained sublamina IA of the plexiform layer. Transverse sections. Bar = 0.5 mm.

in the caudatoputamen (Fig. 3B). The staining intensity in the rest of the amygdala including layer III of the NTOL was comparable to that of the neocortex. Sublamina IA in the plexiform layer of the prepyriform and periamygdaloid cortices displayed a much weaker labeling than the inner sublamina (IB); this lightly stained zone disappeared at the transition toward neocortex and corresponds to the termination zone of the lateral olfactory tract fibers (Heimer, 1978).

As the next step we attempted to trace the sources of the Glu/Asp input to the amygdala by use of axonal D-[^{3}H]Asp transport (Streit, 1980). The injections were centered in the BL/AL complex which, according to the uptake data presented above, should contain the greatest density of Glu/Asp-ergic terminals. Fig. 5A shows a representative injection site (case D9) with most of the injected label confined to the BL and AL. As after HRP injections in this site (Ottersen, 1982), there was an abrupt decline in labeling intensity at the nuclear borders, probably reflecting reduced penetrability due to fiber bundles and laminar organization of dendrites (see, e.g., McDonald, 1984). Retrogradely labeled neurons were found bilaterally, with an ipsilateral predominance, in layer III of the NTOL (Figs. 5C, D and Fig. 6), and ipsilaterally, in the dorsal subdivision of the lateral entorhinal area (DLEA; mostly in layer IV) and in the prepyriform cortex (mostly in layer III; Fig. 6). Smaller numbers of stained cells occurred in the deep layers of the posterior agranular insular area (AIp; not illustrated) and the ventral subdivision of the lateral entorhinal

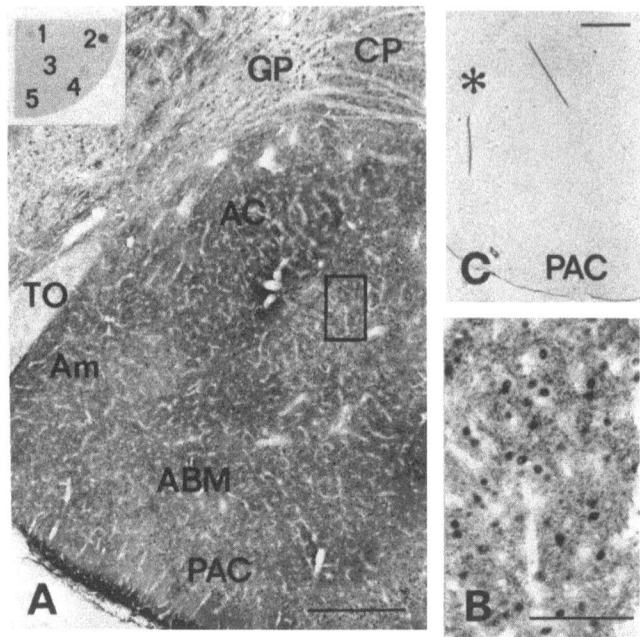

Fig. 4. Photomicrographs showing sections of rat amygdala treated with glutamate antiserum 13 (A,B) and a preimmune serum (C). A: The glutamate-like immunoreactivity is rather homogeneously distributed among the amygdaloid nuclei. Frame shows area enlarged in B. Inset in A: Test filter incubated together with the sections. Note selective staining of the glutamate conjugates (code as in Fig. 2A). B: Intensely stained neurons in the basolateral nucleus. C: Note absence of staining with the preimmune serum. Asterisk indicates optic tract. Transverse sections. Bar = 400 µm (A,C); 100 µm (B).

area (Fig. 6G). The group of labeled neurons in the subiculum extended into the adjacent part of CA1, where the marked cells were found deep in the stratum pyramidale (Fig. 6F). Only a few subcortical structures exhibited noteworthy retrograde labeling; apart from the Am, which contained a modest number of marked neurons, these were midline or medial thalamic nuclei. High densities of stained neurons occurred in the paraventricular nucleus (particularly in its caudal portion; Fig. 6D) and interanteromedial nucleus (Fig. 5B), but the labeling defied cytoarchitectonical borders and invaded adjacent structures including the parataenial, central medial, and rhomboid nuclei.

From the injection site, stained axons could be traced via the stria terminalis to its bed nucleus and the anterior commissure, and thence into the contralateral stria (Fig. 6A, B). Stained fibers could also be followed into the ventral amygdalofugal pathway and, dorsally, into the external capsule and along the border between the caudatoputamen and globus pallidus (Fig. 6C).

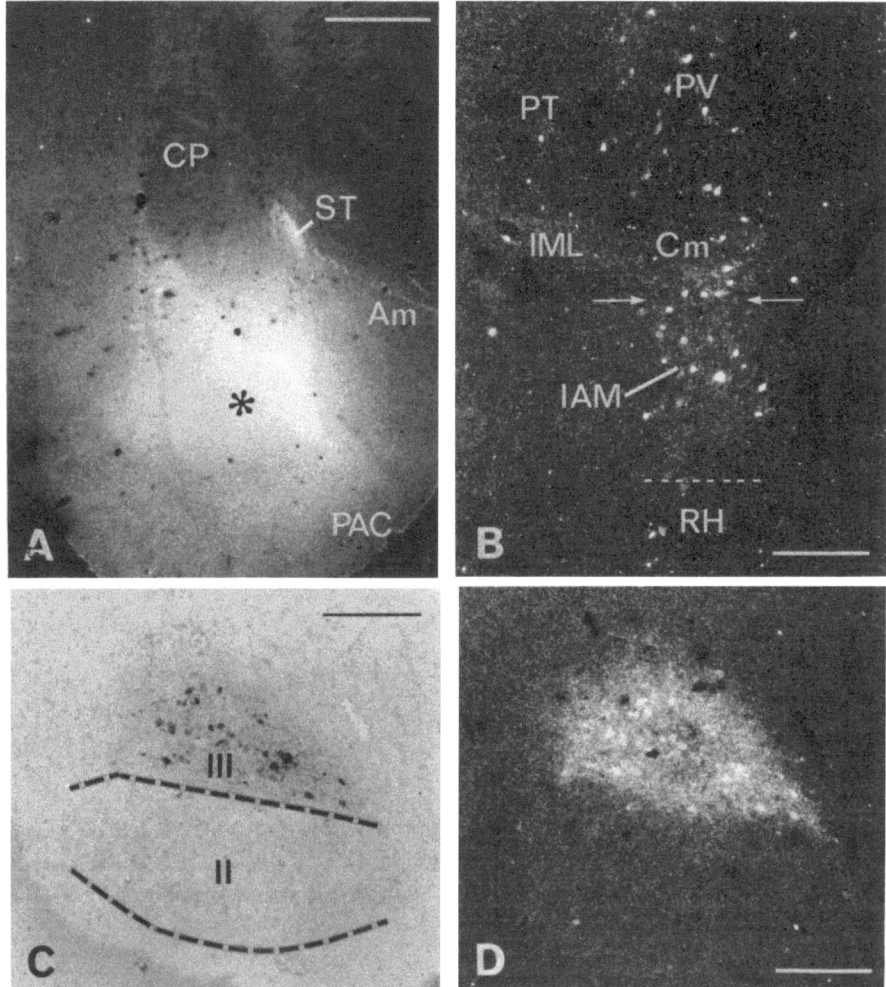

Fig. 5. Photomicrographs showing labeled neurons and fibers after an injection of D-[^{3}H]aspartate in the amygdala (asterisk in A indicates center of injection site in the basolateral nucleus). In A, note heavy labeling of the stria terminalis. B: Retrogradely labeled neurons in midline thalamic nuclei at a level adjacent to that represented in Fig. 6B. Dashed line and arrows indicate nuclear borders. C,D: Bright field (C) and dark field (D) view of the same section, showing intensely labeled perikarya and neuropil in layer III of the nucleus of the lateral olfactory tract. There is also a modest concentration of radioactivity in layer II. Transverse sections. Bar = 700 μm (A); 200 μm (B-D).

The other cases with injections involving BL and AL showed a labeling that was similar to but less extensive than that reported here for case D9.

All of the structures that displayed retrograde labeling after D-[^{3}H]Asp injections also did so after similarly placed HRP injections (Ottersen and Ben-Ari, 1979; Ottersen, 1982; unpublished results). In addition, HRP-filled neurons occurred in several sites showing no evidence of retro-

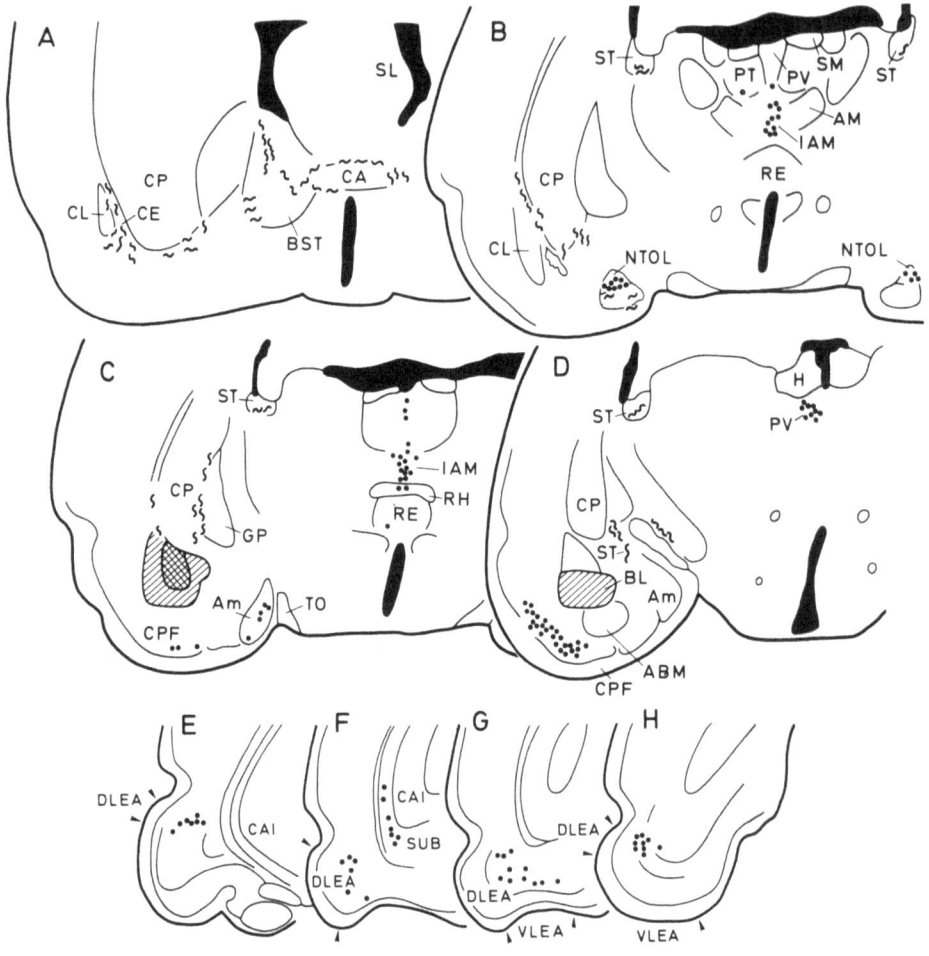

Fig. 6. Distribution of labeled fibers and neurons after an injection of D-[^{3}H]aspartate centered in the basolateral amygdaloid nucleus (same case as in Fig. 5). Crosshatching and hatching indicate central and peripheral zones of the injection site. In the central zone the labeling obscures all tissue details when viewed in dark field (Fig. 5A). Each dot represents one labeled neuron; wavy lines indicate labeled nerve fibers. Note that most of the marked cells are found in cortical or midline thalamic structures. Transverse sections, A is most frontal. Abbreviations, see separate list.

grade D-[^{3}H]Asp transport; these sites included various hypothalamic nuclei, the ventral pallidum, and the dorsal raphe nucleus (Ottersen, 1980, 1981). Our results thus suggest that the ability to transport D-[^{3}H]Asp is a property shared by some, but not all, of the afferents of the BL and AL. However, when comparing two different tracers, differences in labeling patterns must be interpreted with caution since the extent of the effective uptake area cannot be determined precisely. It should also be pointed out that nerve pathways thought to use an excitatory amino acid as transmitter differ in their uptake activity (e.g., Taxt and Storm-Mathisen, 1984) and that not all such pathways are easily labeled with D-[^{3}H]Asp (Baughman and Gilbert, 1981; Wiklund et al., 1982).

The present data are consistent with a transmitter role for Glu or Asp in a proportion of the corticoamygdaloid fibers and strengthen the notion that these amino acids are important transmitters in corticofugal projections in general (Fonnum, 1984; Ottersen and Storm-Mathisen, this volume). Interestingly, in AIp and DLEA, where HRP positive cells occurred in deep as well as in superficial layers (Ottersen, 1982), mainly the deep layers contained cells sustaining D-[^{3}H]Asp transport, suggesting that the corticoamygdaloid fibers do not constitute a uniform population in terms of transmitter identity. Evidence for a corticoamygdaloid Glu/Asp-ergic input was also provided by Walker and Fonnum (1983), who found that the amygdala lost up to 72% of its D-[^{3}H]Asp uptake capacity after various cortical lesions. Neurons in the 'nonspecific' thalamic nuclei have previously been shown to transport D-[^{3}H]Asp retrogradely from the striatum (Streit, 1980) and neocortex (Ottersen et al., 1983).

The fibers that were labeled after intraamygdaloid D-[^{3}H]Asp injections followed the course of anatomically recognized amygdalofugal pathways (Krettek and Price, 1977, 1978a). Anterograde transport of D-[^{3}H]Asp may occur subsequent to uptake by axon collaterals within the injection site (Storm-Mathisen, 1982; Wiklund et al., 1982). The possibility should thus be considered that the BL and AL not only receive but also emit a substantial number of Glu/Asp-ergic fibers. This issue is now being explored by analysis of retrograde D-[^{3}H]Asp transport following injections in the various targets of amygdalofugal fibers.

CONCLUSIONS

The present observations indicate that the amygdala contains a dense GABA circuitry, and that the concentration of GABA-containing terminals is particularly high in the centromedial subdivision. The close correspondence between the patterns of GABA-LI and GABA uptake supports the assumption that most of the GABA in the brain belongs to a transmitter pool. No such correspondence was found between Glu-LI or Asp-LI and D-[^{3}H]Asp uptake activity, suggesting that in immunocytochemical preparations the transmitter pools of Glu and Asp are masked by the metabolic pools. In contrast to the situation for GABA, much of the putative Glu/Asp-ergic input seems to originate outside the amygdala, most notably in restricted parts of cortex and thalamus. However, this study has also provided evidence for intraamygdaloid Glu/Asp projections, such as between NTOL and BL/AL. A challenge for future research is to resolve how the different amino-acid transmitter systems in the amygdala cooperate with each other and with other transmitter systems, under physiological as well as under pathological conditions.

ACKNOWLEDGEMENTS

We are grateful to Dr. Brian Meldrum for supplying us with the Senegalese baboons. This study was supported by the Norwegian Research Council for Science and the Humanities.

ABBREVIATIONS

AAA, anterior amygdaloid area
ABM, basomedial amygdaloid nucleus
AC, central amygdaloid nucleus
AIp, posterior agranular insular area
AM, anteromedial thalamic nucleus
Am, medial amygdaloid nucleus

Asp, aspartate
BL, basolateral amygdaloid nucleus
BSA, bovine serum albumin
BST, bed nucleus of the stria terminalis
CA, commissura anterior
CA1, field of Ammon's horn
CE, capsula externa
CL, claustrum
Cm, central medial thalamic nucleus
COp, posterior cortical amygdaloid nucleus
CP, caudatoputamen
CPF, prepyriform cortex
DLEA, dorsal subdivision of the lateral entorhinal area
FI, fimbria
GABA, gamma-aminobutyrate
GAD, glutamic acid decarboxylase
Glu, glutamate
GP, globus pallidus
H, habenula
HRP, horseradish peroxidase
IAM, interanteromedial thalamic nucleus
IML, internal medullary lamina
-LI, -like immunoreactivity (e.g., GABA-LI)
LV, lateral ventricle
MD, mediodorsal thalamic nucleus
NTOL, nucleus of the lateral olfactory tract
PAC, periamygdaloid cortex
PT, parataenial thalamic nucleus
PV, paraventricular thalamic nucleus
RE, nucleus reuniens
RH, nucleus rhomboideus
SL, lateral septal area
SM, stria medullaris
ST, stria terminalis
SUB, subiculum
Tau, taurine
TO, tractus opticus
VLEA, ventral subdivision of the lateral entorhinal area
VM, ventromedial hypothalamic nucleus
3, third ventricle

REFERENCES

Baughman, R.W., and Gilbert, C.D., 1981, Aspartate and glutamate as possible neurotransmitters in the visual cortex, J. Neurosci., 1:427.
Ben-Ari, Y., Kanazawa, I., and Zigmond, R.E., 1976, Regional distribution of glutamate decarboxylase and GABA within the amygdaloid complex and stria terminalis system of the rat, J. Neurochem., 26:1279.
Ben-Ari, Y., Tremblay, E., and Ottersen, O.P., 1980, Injections of kainic acid into the amygdaloid complex of the rat: an electrographic, clinical and histological study in relation to the pathology of epilepsy, Neuroscience, 5:515.
Ben-Ari, Y., 1981, The Amygdaloid Complex, Elsevier/North-Holland Biomedical Press, Amsterdam.
Fischer, B.O., Ottersen, O.P., and Storm-Mathisen, J., 1982, Labelling of amygdalopetal and amygdalofugal projections after intra-amygdaloid injections of tritiated D-aspartate, Neuroscience, 7(Suppl.):S69.
Fonnum, F., 1984, Glutamate: a neurotransmitter in mammalian brain, J. Neurochem., 42:1.

Heimer, L., 1978, The olfactory cortex and the ventral striatum, in: Limbic Mechanisms, K.E. Livingston and O. Hornykiewicz, eds., Plenum Press, New York, p. 95.

Krettek, J.E., and Price, J.L., 1977, Projections from the amygdaloid complex to the cerebral cortex and thalamus in the rat and cat, J. Comp. Neurol., 172:687.

Krettek, J.E., and Price, J.L., 1978a, Amygdaloid projections to subcortical structures within the basal forebrain and brainstem in the rat and cat, J. Comp. Neurol., 178:225.

Krettek, J.E., and Price, J.L., 1978b, A description of the amygdaloid complex in the rat and cat with observations on intra-amygdaloid axonal connections, J. Comp. Neurol., 178:255.

Le Gal La Salle, G., 1976, Antidromic identification of amygdaloid multipolar neurons, Brain Res., 118:479.

Le Gal La Salle, G., Paxinos, G., Emson, P., and Ben-Ari, Y., 1978, Neurochemical mapping of GABAergic systems in the amygdaloid complex and bed nucleus of the stria terminalis, Brain Res., 155:397.

Madsen, S., Ottersen, O.P., and Storm-Mathisen, J., 1985, Immunocytochemical visualization of taurine: neuronal localization in the rat cerebellum, Neurosci. Lett., 60:255.

McDonald, A.J., 1982, Cytoarchitecture of the central amygdaloid nucleus of the rat, J. Comp. Neurol., 208:401.

McDonald, A.J., 1984, Neuronal organization of the lateral and basolateral amygdaloid nuclei in the rat, J. Comp. Neurol., 222:589.

McDonald, A.J., 1985, Immunohistochemical identification of γ-aminobutyric acid-containig neurons in the rat basolateral amygdala, Neurosci. Lett., 53:203.

Meldrum, B., 1984, Amino acid neurotransmitters and new approaches to anticonvulsant drug action, Epilepsia, 25(Suppl.2):S140.

Millhouse, O.E, and DeOlmos, J., 1983, Neuronal configurations in lateral and basolateral amygdala, Neuroscience, 10:1269.

Ottersen, O.P., and Ben-Ari, Y., 1979, Afferent connections to the amygdaloid complex of the rat and cat. I. Projections from the thalamus, J. Comp. Neurol., 187:401.

Ottersen, O.P., 1980, Afferent connections to the amygdaloid complex of the rat and cat: II. Afferents from the hypothalamus and the basal telencephalon, J. Comp. Neurol., 194:267.

Ottersen, O.P., 1981, Afferent connections to the amygdaloid complex of the rat with some observations in the cat. III. Afferents from the lower brain stem, J. Comp. Neurol., 202:335.

Ottersen, O.P., 1982, Connections of the amygdala of the rat. IV: Corticoamygdaloid and intraamygdaloid connections as studied with axonal transport of horseradish peroxidase, J. Comp. Neurol., 205:30.

Ottersen, O.P., Fischer, B.O., and Storm-Mathisen, J., 1983, Retrograde transport of D-[^{3}H]aspartate in thalamocortical neurones, Neurosci. Lett., 42:19.

Ottersen, O.P., and Storm-Mathisen, J., 1984a, Neurons containing or accumulating transmitter amino acids, in: Handbook of Chemical Neuroanatomy, A. Björklund, T. Hökfelt, and M.J. Kuhar, eds., Elsevier/North-Holland, Amsterdam, p. 141.

Ottersen, O.P., and Storm-Mathisen, J., 1984b, Glutamate- and GABA-containing neurons in the mouse and rat brain, as demonstrated with a new immunocytochemical technique, J. Comp. Neurol., 229:374.

Ottersen, O.P., Madsen, S., Meldrum, B.S., and Storm-Mathisen, J., 1985, Taurine in the hippocampal formation of the Senegalese baboon, Papio papio: an immunocytochemical study with an antiserum against conjugated taurine, Exp. Brain Res., 59:457.

Ottersen, O.P., and Storm-Mathisen, J., 1985, Different neuronal localization of aspartate-like and glutamate-like immunoreactivities in the hippocampus of rat, guinea pig, and Senegalese baboon (Papio papio), with a note on the distribution of GABA, Neuroscience, 16:589.

Price, J.L., 1981, Toward a consistent terminology for the amygdaloid complex, in: The Amygdaloid Complex, Y. Ben-Ari, ed., Elsevier/North-Holland Biomedical Press, Amsterdam. p. 13.

Racine, R.J., 1981, Kindling: a model of amygdaloid epileptogenesis, in: The Amygdaloid Complex, Y. Ben-Ari, ed., Elsevier/North-Holland Biomedical Press, Amsterdam, p. 431.

Roberts, G.W., Woodhams, P.L., Polak, J.M., and Crow, T.J., 1982, Distribution of neuropeptides in the limbic system of the rat: the amygdaloid complex, Neuroscience, 7:99.

Smith, B.S., and Millhouse, O.E., 1985, The connections between the basolateral and central amygdaloid nuclei, Neurosci. Lett., 56:307.

Stephan, H., 1975, Allocortex, Springer-Verlag, Berlin.

Sternberger, L.A., 1979, Immunocytochemistry, Second edition, John Wiley, New York.

Storm-Mathisen, J., 1982, Amino acid compartments in hippocampus: an autoradiographic approach, in: Neurotransmitter Interaction and Compartmentation, H.F. Bradford, ed., Plenum Press, New York, p. 395.

Storm-Mathisen, J., Leknes, A.K., Bore, A.T., Vaaland, J.L., Edminson, P., Haug, F.-M.Š., and Ottersen, O.P., 1983, First visualization of glutamate and GABA in neurones by immunocytochemistry, Nature, 301:517.

Storm-Mathisen, J., and Ottersen, O.P., 1986, Antibodies against amino acid transmitters, in: Neurohistochemistry Today, P. Panula, H. Päivärinta, and S. Soinila, eds., Alan R. Liss, New York, in press.

Storm-Mathisen, J., Ottersen, O.P., and Fu-long, T., 1986, Antibodies for the localization of excitatory amino acids, in: Excitatory Amino Acids, P.J. Roberts, J. Storm-Mathisen, and H.F. Bradford, eds., MacMillan, London, in press.

Streit, P., 1980, Selective retrograde labeling indicating the transmitter of neuronal pathways, J. Comp. Neurol., 191:429.

Taxt, T., and Storm-Mathisen, J., 1984, Uptake of D-aspartate and L-glutamate in excitatory axon terminals in hippocampus: autoradiographic and biochemical comparison with γ-aminobutyrate and other amino acids in normal rats and in rats with lesions, Neuroscience, 11:79.

Tremblay, E., Ottersen, O.P., Rovira, C., and Ben-Ari, Y., 1983, Intraamygdaloid injections of kainic acid: regional metabolic changes and their relation to the pathological alterations, Neuroscience, 8:299.

Walker, J.E., and Fonnum, F., 1983, Regional cortical glutamergic and aspartergic projections to the amygdala and thalamus of the rat, Brain Res., 267:371.

Wiklund, L., Toggenburger, G., and Cuénod, M., 1982, Aspartate: possible neurotransmitter in cerebellar climbing fibers, Science, 216:78.

A SURVEY OF THE ANATOMY OF THE HIPPOCAMPAL FORMATION, WITH EMPHASIS ON

THE SEPTOTEMPORAL ORGANIZATION OF ITS INTRINSIC AND EXTRINSIC CONNECTIONS

M.P. Witter
Department of Anatomy
Vrije Universiteit
Amsterdam, The Netherlands

INTRODUCTION

The hippocampal formation (HF) is composed of the hippocampus proper, i.e., the cornu Ammonis (CA) and the fascia dentata (FD), and the subiculum (Sub). Despite a rather extensive knowledge of its extrinsic and intrinsic connections, the functions of the HF are still an enigma. The suggestion has been made that the hippocampus contains a map of the external spatial environment, but also that it plays a more general role in memory and learning processes (O'Keefe and Nadel, 1978; Olton et al., 1982; Olton, 1983; Squire, 1983). A most relevant notion as regards the subject of this symposium is that the HF has long been implicated in temporal lobe epilepsy. Already in 1880, Sommer reported that many epileptic patients showed extensive loss of neurons, in particular in field CA1 of the CA. Recently, the anatomy of the hippocampal circuitry in relation to the occurrence of seizures has attracted special attention (Somogyi et al., 1983a, b). Although it appears that the basic circuitry is similar in all parts of the hippocampus, the present analysis of the topographical organization of the major intrinsic and extrinsic connections suggests that different parts along the septotemporal axis of the HF are connected with a different set of extra-hippocampal structures (Ruth et al., 1982; Roberts et al., 1984; Witter and Groenewegen, 1984; Van Groen and Lopes da Silva, 1985).

ANATOMY OF THE HIPPOCAMPAL FORMATION

The position of the HF in a certain species depends on the degree of the development of the temporal lobe. In the rat, for example, the HF extends virtually from the septum dorsally to the caudal part of the amygdala ventrally, forming a cashew nut-shaped structure, with the septal pole as its main part. In the monkey, on the other hand, the HF is located temporally in the floor and medial wall of the inferior horn of the lateral ventricle. In the cat, the HF takes an intermediate position between the rat and monkey, and has a small part close to the septum, but its main part lies temporally. For descriptive purposes, the HF of all three species will be divided in a dorsal or septal part, located closest to the septum, and a ventral or temporal part in the temporal lobe, whereas the intermediate portion will be referred to as the posterior or splenial part.

In all mammals, in the HF two C-shaped interlocking principal neuronal cell layers can be recognized: the granular cell layer of the FD, and the pyramidal cell layer of the CA and adjacent Sub (Fig. 1). The granular cells of FD project to the CA, whereas the pyramidal cells of the CA and the Sub are considered to be output neurons. In addition to these principal cells, all cortical fields contain several types of interneurons. In the center of the FD, its hilus can be distinguished by the presence of large polymorph neurons. In the present report, these cells will be considered to belong to the FD, thus omitting the designation of field CA4 as described by Lorente de Nó (1934; Blackstad, 1956; Amaral, 1978). Field CA3 consists of large pyramidal cells and is continuous with CA1, which contains smaller pyramidal neurons. Lorente de Nó distinguished a separate area CA2 between CA3 and CA1. Most later researchers did not identify a field CA2, as it appears to be difficult to delineate this field in most species. However, recent reports paid special attention to this transition area as the possible site of generation of rhythmical cell bursts (Miles et al., 1984). Field CA1 is continuous with the Sub which constitutes the transition to the parahippocampal cortex. The basic neuronal assembly of the FD, the CA and the Sub is very similar (Figs. 1, 2): one sheet of neurons of which the apical dendrites extend into a molecular layer. The majority of the fibers that reach these fields cross the dendrites at roughly right angles, and make synapses <u>en passage</u> on a restricted segment of the dendritic trees. Underneath the cellular layer, a fiber zone is present which in the FD contains the axons of the granular cells, i.e., the mossy fibers, and in the CA fields and Sub,

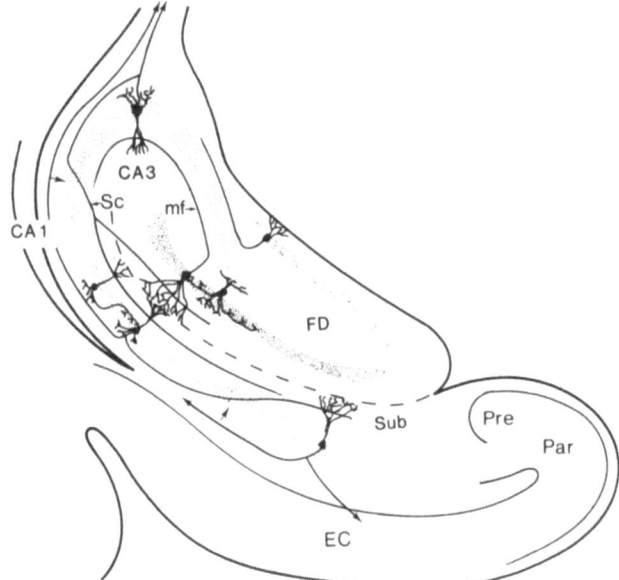

Fig. 1. The main hippocampal constituents, i.e., the fascia dentata (FD), cornu Ammonis: fields CA_3 and CA_1, and the subiculum (Sub), and the sequence of intrinsic connections from FD to Sub outlined in a schematic drawing of a coronal section through the temporal hippocampal formation of the cat. Abbreviations: EC, entorhinal cortex; mf, mossy fiber; Par, parasubiculum; Pre, presubiculum; Sc, Schaffer collateral.

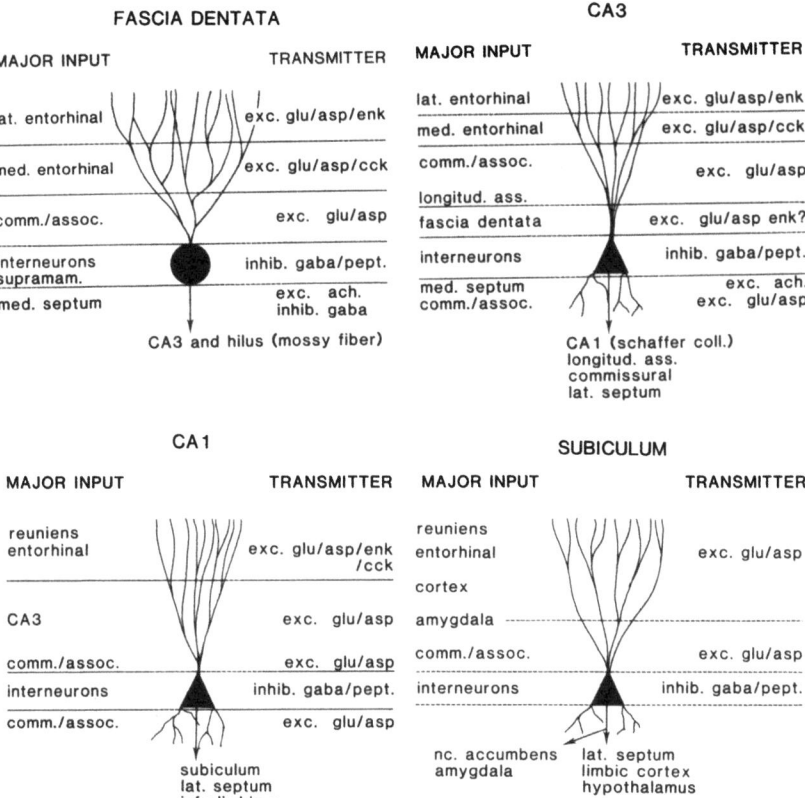

Fig. 2. Schematic representation of the major input/output relations of the hippocampal subfields with special emphasis on the terminal organization along the apical dendrites and the putative transmitters of these pathways. Inputs that terminate diffusely along the dendrites have not been depicted. See text for further details.

the basal dendrites and the axons of the pyramidal cells of these fields.

INTRINSIC CONNECTIONS

Lamellar Organization

The HF contains a sequence of almost completely unidirectional connections. These connections are preferentially organized in narrow lamellae which are oriented more or less perpendicularly to the longitudinal septo-temporal axis of the hippocampus (Fig. 2; Andersen et al., 1971). The granule cells, located in one lamella, distribute their mossy fibers to the entire extent of CA3 in the same lamella. The majority of the mossy fibers terminate on the apical dendrites in the stratum lucidum, close to the somata of the CA3 pyramidal neurons (Figs. 1,2). A relatively small infrapyramidal bundle, which arises mainly from the outer blade of the granule cell layer, distributes to the stratum oriens underneath the CA3 pyramidal neurons (Blackstad et al., 1970; Swanson et al., 1978). The pyramidal cells of CA3, in turn, give rise to axons with at least two branches, one of which leaves the hippocampus via the fimbria and fornix, whereas the other branch forms the so-called Schaffer collateral which reaches the CA1 region. The Schaffer collaterals synapse in the stratum radiatum on distal portions of the apical dendrites of the CA1 pyramidal neurons (Figs. 1, 2). Finally, also the CA1 neurons issue an axon, of which one branch contributes to the hippocampal output via the alveus and fornix/fimbria, whereas collaterals reach the Sub. These latter fibers terminate in the deep half of the molecular layer and in the pyramidal cell layer of the Sub (Figs. 1, 2; Swanson et al., 1978). The putative neurotransmitters of these, and a number of other hippocampal connections, have been indicated in Fig. 2 (Gall et al., 1981; Walaas, 1983; Fredens et al., 1984; Gall, 1984; Roberts et al., 1984; Taxt and Storm-Mathisen, 1984).

Associational and Commissural Connections

In addition to the sequence of connections which are contained within the lamellae, various associational and commissural connections have been described, some of which distribute over extensive lengths of the septotemporal axis. Whereas CA3 and the FD provide the major associational and commissural projections, the Sub gives rise to only minor associational projections (Berger et al., 1980; Amaral et al., 1984; Schwerdtfeger, 1984; Köhler, 1985), and CA1 does not contribute significantly to either system (Swanson et al., 1978; Amaral et al., 1984). Fibers from CA3 reach the hilus of the FD and the Sub in lamellae close to that of the cells of origin (Swanson and Cowan, 1977; Swanson et al., 1978). Furthermore, CA3, and in particular the area which probably corresponds to CA2, give rise to the so-called longitudinal association system (Lorente de Nó, 1934; Swanson et al., 1978). These longitudinal association fibers travel in the stratum radiatum, and to a lesser degree in the stratum oriens, and reach extensive portions of the CA along its septotemporal axis. These fibers terminate in the stratum radiatum close to the somata of the CA3 pyramids (Fig. 2; Swanson et al., 1978). Neurons in CA3 also give rise to an extensive commissural projection system, which within the contralateral hippocampus follows a largely similar course as the ipsilateral CA3 projection. These commissural CA3 fibers also terminate upon the same segments of the dendrites of the target cells as described for the ipsilateral pathways. However, the ipsilateral association fibers reach a wider range of the septotemporal axis than the commissural fibers (Gottlieb and Cowan, 1973; Swanson et al., 1978). In the rat, the longitudinal association fibers, the Schaffer collaterals and the commissural fibers, originating in CA3, appear to arise from the same neurons (Laurberg, 1979; Swanson et al., 1980; Laurberg and Sørensen, 1981; Swanson et al., 1981).

Neurons in the hilus project to the inner one-third of the molecular layer of the FD (Fricke and Cowan, 1978; Swanson et al., 1978). Collaterals of these axons reach the homotopic portion of the FD of the contralateral side (Fricke and Cowan, 1978). In contrast to the CA3 projections, the ipsi- and contralateral terminal fields of hilar cell projections are equally extensive as regards their distribution along the septotemporal axis.

Although the overall organization patterns of associational and commissural pathways are quite similar in the rat and monkey, some marked differences exist with respect to the septotemporal extent of the termination fields. In the rat, all parts of the HF along its septotemporal extent contribute to both the commissural and associational connections. However, fibers which arise in the septal and splenial parts do not distribute to the temporal pole, and fibers with their origin in the temporal pole do not reach the splenial and septal portions of the HF (Fricke and Cowan, 1978; Swanson et al., 1978; Köhler, 1985). In contrast, recent data reveal that in the monkey marked differences exist between the commissural and the associational fiber systems with respect to their septotemporal origin and distribution (Amaral et al., 1984; Demeter et al., 1985). Whereas the associational system in the monkey is similarly organized as in the rat, the commissural fibers in this species arise only from the temporal and uncal parts of the HF. This restricted commissural projection system exhibits a similar distribution and termination pattern in CA1 and CA3 as described for the rat (Amaral et al., 1984).

Interneurons

The main cell layers and the plexiform strata of all fields of the HF contain interneurons, or local circuit neurons (Fig. 1). In a recent review the morphological, histochemical and physiological features of these neurons have been described (Buzsaki, 1984). A wide variety of neuronal cell types are present, most of which are presumably GABA-ergic (Ribak et al., 1978; Frotscher et al., 1984). Furthermore, many peptide-containing neurons have been observed in the hippocampus. Interestingly, these peptidergic neurons for the greater part belong to the class of interneurons, and exhibit morphological features similar to those of the GABA-containing neurons (Köhler, 1983; Leranth and Frotscher, 1983; Somogyi et al., 1983b; Buzsaki, 1984; Roberts et al., 1984; Nunzi et al., 1985). Interneurons are supposed to form part of feedback inhibitory circuits and to subserve a crucial role in feed-forward inhibition (Ashwood et al., 1984; Buzsaki, 1984; Schwartzkroin and Kunkel, 1985). Dendrites of interneurons not only distribute to the principal cell layers, but also reach the terminal fields of the commissural and entorhinal inputs, which synapse directly on the dendrites of the interneurons (Frotscher and Zimmer, 1983; Frotscher et al., 1984; Taxt and Storm-Mathisen, 1984; Bakst et al., 1985; Schwartzkroin and Kunkel, 1985; Vincent et al., 1985). The principal neurons also project to these interneurons (Knowles and Schwartzkroin, 1981), which, in turn, distribute their axons over relatively long distances along the septotemporal axis of the HF, and make synaptic contacts on the somata, the proximal part of the dendrites, and the initial segment of the axons of several hundred principal neurons (Buzsaki, 1984; Somogyi et al., 1983a, b). Therefore, the interneurons may exert a direct inhibitory control over the activity of a large number of principal cells. This notion may be of relevance to the understanding of the processes which underlie the generation of spontaneous cell bursts in the hippocampus (Struble et al., 1978; Somogyi et al., 1983a, b). Furthermore, the interneurons, by way of the relatively widespread septotemporal distribution of their axons, may interconnect the different lamellae which are present in the HF.

EXTRINSIC CONNECTIONS OF THE HIPPOCAMPAL FORMATION

Connections With the Entorhinal Cortex

Entorhinal-hippocampal projections. The entorhinal cortex is considered to be a site of convergence of fibers from functionally different cortical areas, which represent all sensory modalities (Van Hoesen, 1982; Witter et al., 1986). Entorhinal-hippocampal fibers arise mainly in layers II and III, whereas a much smaller contribution comes from the deeper layers (Steward and Scoville, 1976; Schwerdtfeger, 1984; Witter and Groenewegen 1984; Köhler, 1985). In the rat and the cat, the projection from the entorhinal cortex distributes along a proximodistal gradient over the dendrites of the principal cells of the FD and CA3: afferents from the medial entorhinal area (MEA) terminate in the middle one-third of the molecular layer, whereas those from the lateral entorhinal area (LEA) reach the outer one-third (Fig. 2; Steward, 1976; Wyss, 1981; Witter and Groenewegen 1984). The two components of the perforant pathway also exhibit differences with respect to their neuroactive substances and electrophysiological characteristics (McNaughton, 1980; Fredens et al., 1984). Based on the organization of the extrinsic and intrinsic connections of the entorhinal cortex, we suggested that the lateral and medial entorhinal areas each have their unique role with respect to hippocampal functioning (Witter et al., 1986).

Entorhinal fibers also terminate in the molecular layer of CA1 and the Sub and exhibit a less clear proximodistal gradient in the terminal fields than exists in their projections to the FD and CA3 (Steward, 1976; Wyss, 1981; Van Groen and Lopes da Silva, 1986; Van Groen et al., 1986). The projections to the Sub arise predominantly from MEA. In the monkey, projections from the entorhinal cortex to the most temporal portions of the HF are organized as described for the rat and the cat, whereas at more septal levels the FD is the main recipient of entorhinal fibers (Van Hoesen and Pandya, 1975; Schwerdtfeger, 1984).

In both the rat and the cat, relatively small areas of the entorhinal cortex project to widespread parts of the HF along its septotemporal axis. Therefore, the lamellar organization as outlined for the sequence of projections from the FD to the Sub via the CA appears not to fit with the organization of the entorhinal inputs (Wyss, 1981; Witter and Groenewegen, 1984).

In all three species mentioned, the entorhinal cortex gives rise to additional projections to the contralateral HF. This crossed pathway reaches the septal part of the FD in the rat, the cat, and the monkey (Steward, 1976; Wyss, 1981; Amaral et al., 1984; own observations). In the cat and the monkey, also more temporal levels of the HF, in particular the Sub, are in receipt of crossed entorhinal fibers which, as is the case for the ipsilateral pathway, originate for the most part in MEA (Amaral et al., 1984; Van Groen et al., 1986).

<u>Hippocampal-entorhinal projections</u>. In the rat, the deep layers of the entorhinal cortex receive a small projection from the temporal one-third of CA3 (Hjorth-Simonsen, 1971; Swanson and Cowan, 1977) and from the entire septotemporal extent of field CA1. Also the Sub projects to layer IV of MEA, and this projection extends into the perirhinal cortex (Swanson and Cowan, 1977; Sørensen and Shipley, 1979; Swanson et al., 1981). Likewise, in the cat and the monkey, projections from the various hippocampal subfields to the deep layers of the medial entorhinal cortex have been observed (Rosene and Van Hoesen, 1977; Schwerdtfeger, 1984; Van Groen et al., 1986). However, in the cat the subiculo-entorhinal pathway predominantly reaches superficial layers of the entorhinal cortex (Van Groen et al., 1986). In the rat, this pathway appears to be strictly ipsilateral, whereas in the other two animals there is also a crossed pathway. In the cat, the entire septotemporal extent of the Sub contributes to this projection, whereas in the monkey only the temporal pole gives rise to this crossed connection (Amaral et al., 1984; Demeter et al., 1985).

<u>Septotemporal organization of hippocampal-entorhinal connections</u>.
In the rat, the projections from the entorhinal cortex to the HF are topographically organized such that a septotemporal axis in the hippocampus is related either to a dorsoventral axis (Hjorth-Simonsen and Jeune, 1972; Wyss, 1981) or to a posterodorsolateral to anteroventromedial axis (Ruth et al., 1982; Pohle and Ott, 1984). In our studies in the cat, we observed a comparable organization such that the lateral parts of both LEA and MEA project preferentially to septal parts of the HF, whereas progressively more medial parts of LEA and MEA project to more temporal levels (Fig. 3; Witter and Groenewegen, 1984; Van Groen and Witter, 1985; Van Groen and

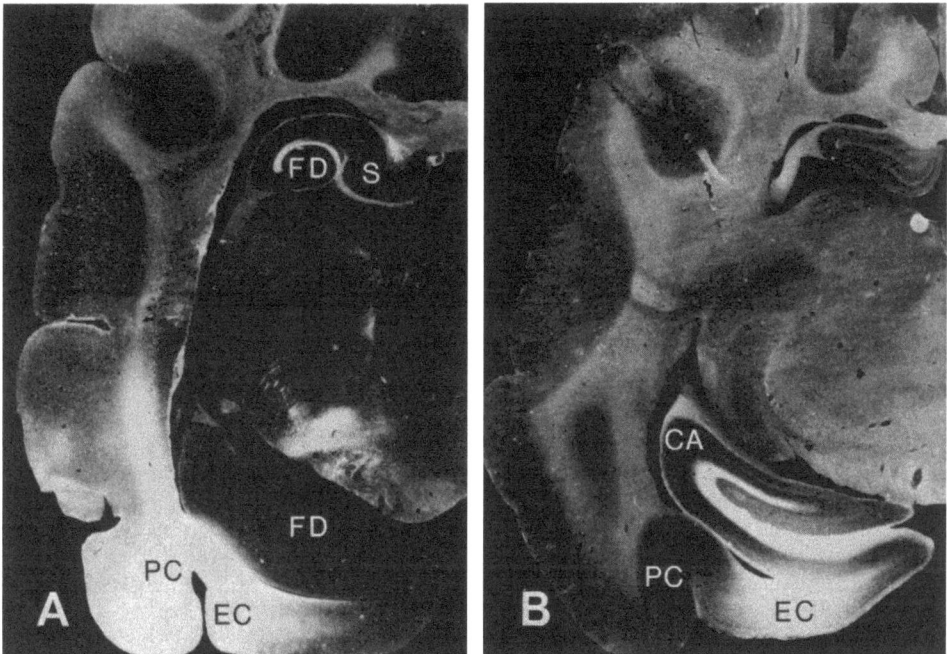

Fig. 3. Darkfield photomicrographs of two coronal sections of the brain of the cat, illustrating the septotemporal organization of the entorhinal-hippocampal pathway. Following an injection laterally in LEA, labeling is present in the septal part of the hippocampal formation (A), whereas a more medially located injection in LEA results in labeling restricted to the temporal part. Note that heavy labeling is present in FD, CA3 and Sub (B). A similar septotemporal organization is observed following injections in MEA, although in these cases the terminal fields on the dendrites are shifted towards a more proximal position (cf. Witter and Groenewegen, 1984).

Lopes da Silva, 1986; Van Groen et al., 1986). The crossed entorhinal-subicular pathway exhibits a similar septo-temporal organization (Van Groen et al., 1986). Furthermore, in the rat and the cat, the reciprocal connections from CA1 and the Sub are organized also according to a septotemporal axis of the HF (Swanson and Cowan, 1977; Van Groen and Lopes da Silva, 1986; Van Groen et al., 1986). Furthermore, in the rat the temporal part of CA3 issues fibers to the entorhinal cortex (Swanson and Cowan, 1977).

In conclusion, in the rat and the cat all reciprocal connections between subfields of the HF and the entorhinal cortex appear to be topographically organized according to the same septotemporal principle. There are no detailed data on the topographical organization of these connections in the monkey, but there appear to exist some connectional differences with respect to these connections between the temporal and more septal parts of the HF (Van Hoesen and Pandya, 1975).

Connections With Other Cortical Areas

In general, it can be stated that the HF and in particular the Sub is reciprocally connected with widespread parts of the 'limbic cortex', including the perirhinal cortex, presubiculum, parasubiculum, postsubiculum, retrosplenial and cingulate areas, and the infralimbic and prelimbic areas

(Leichnetz and Astruc, 1975, 1976; Swanson and Cowan, 1977; Van Hoesen et al., 1979; Swanson, 1981; Kosel et al., 1983; Schwerdtfeger, 1984; Witter and Groenewegen, 1984; Köhler, 1985; Room et al., 1985). The cortical afferent projections terminate for the most part in the molecular layer of the hippocampal subfields, although in case of projections reaching the CA termination has also been observed in the strata oriens and pyramidale. In comparing these connections in the rat and the cat with those in the monkey, in the latter species there appears to be a shift towards more direct neocortical connections of the Sub, in particular with prefrontal and temporal cortical areas. This is accompanied by a connectional shift of the cortical connections within the HF such that in the monkey CA1 becomes more involved in these extrinsic cortical connections (Rosene and Van Hoesen, 1977; Van Hoesen et al., 1979; Schwerdtfeger, 1984).

Subcortical Connections of the Hippocampal Formation

Connections with the septal complex. The septohippocampal pathway appears to be the most extensively studied subcortical afferent system to the HF. The projections arise mainly from the medial septal nucleus and the nucleus of the diagonal band, in particular from its vertical limb (R: Segal and Landis, 1974; Swanson and Cowan, 1979; Wyss et al., 1979a; C: Siegel and Tassoni, 1971b; Krayniak et al., 1980; M: Amaral and Cowan, 1980). The septal fibers terminate predominantly in the infragranular and granular layers of the FD, the stratum oriens of CA3, and the deep part of the molecular layer of the Sub. Only moderate to weak terminal fields have been observed in the stratum radiatum of CA3 and the most inner zone of the molecular layer of the FD (R: Swanson and Cowan, 1979; Crutcher et al., 1981; Chandler and Crutcher, 1983). It further appears that CA1 does not receive septal inputs (see however Monmaur and Thomson, 1983).

All fields of the CA and the Sub send projections back to the septum, terminating in largely overlapping areas in the lateral septal nucleus (R: Swanson and Cowan, 1977, 1979; C: Siegel and Tassoni, 1971a; own observations; M: Rosene and Van Hoesen, 1977; Krayniak et al., 1979; Schwerdtfeger, 1984). In the rat, hippocampal neurons have been identified which project to the lateral septum and also contribute collaterals to the associational and commissural connections within the HF as well as to the entorhinal cortex (Swanson et al., 1980, 1981).

The reciprocal connections between the septal area and the HF in the rat and the cat show a clear septotemporal organization. The septal part of the HF receives fibers which originate medially in the medial septal nucleus and the most ventromedial part of the nucleus of the diagonal band, and in turn distributes fibers to the dorsomedial part of the lateral septum. Temporal hippocampal levels, in contrast, are projected upon by the more lateral parts of the medial septal nucleus and dorsolateral parts of the vertical limb of the diagonal band, and issue fibers to the ventrolateral part of the lateral septum (Siegel and Tassoni, 1971a, b; Siegel et al., 1974; Swanson and Cowan, 1977, 1979; Krayniak et al., 1980; own observations). In the monkey, no detailed data with respect to the septotemporal organization of the septum-hippocampal connections are available (Amaral and Cowan, 1980; Schwerdtfeger, 1984).

Connections with the amygdaloid complex and nucleus accumbens. In the rat, the cat, and the monkey it has been reported that the amygdala, in particular the basolateral nucleus, anterior amygdaloid area, and the posterior cortical nucleus or periamygdaloid cortex, projects to the molecular layer of the temporal Sub (Krettek and Price, 1977; Wyss et al., 1979a; Amaral and Cowan, 1980; Schwerdtfeger 1984). In turn, the temporal Sub is the site of origin of projections to the basolateral nucleus (Rosene and Van Hoesen, 1977; Swanson and Cowan, 1977; Russchen, 1982; Schwerdtfeger, 1984; see also the contributions of Amaral and Russchen).

The Sub also projects to the nucleus accumbens. Although, in the rat, this projection is described to arise from the temporal Sub, our own results indicate that also the septal part of the Sub contributes to this projection (Newman and Winans, 1980; Kelley and Domesick, 1982; Schwerdtfeger, 1984). As already described for the cat, this projection is topographically organized such that the septal part of the Sub projects to the ventrolateral nucleus accumbens, whereas progressively more temporal parts project to more dorsomedial parts of this nucleus (Groenewegen et al., 1981, 1982; Lopes da Silva, 1984; Lopes da Silva et al., 1985).

Connections with the diencephalon. The HF receives substantial, bilateral projections from the hypothalamus, in particular from the supramammillary nucleus and the lateral hypothalamic area (Segal and Landis, 1974; Pasquier and Reinoso-Suarez, 1976, 1978; Wyss et al., 1979a, b; Amaral and Cowan, 1980; Stanfield et al., 1980; Dent et al., 1983; Haglund et al., 1984; Köhler et al., 1984). In the rat, fibers from the supramammillary nucleus terminate predominantly in the outer part of the granular cell layer and the directly adjacent part of the molecular layer, whereas weaker projections reach the strata lacunosum/moleculare, pyramidale and oriens of CA3 (Wyss et al., 1979b; Stanfield et al., 1980; Dent et al., 1983; Haglund et al., 1984). In contrast, in the monkey, projections to the CA are restricted to CA2 (Veazey et al., 1982; see also in the rat, Segal, 1979). Projections from the lateral hypothalamus terminate diffusely in all subfields of the HF (Köhler et al., 1984). Although in the cat these hypothalamic afferents are to some extent topographically organized along the septotemporal axis of the HF (Pasquier and Reinoso-Suarez, 1976, 1978), in the rat they are reported to distribute equally along the longitudinal hippocampal axis. In all species studied, the Sub send projections back to the hypothalamus (Rosene and Van Hoesen, 1977; Swanson and Cowan, 1977; Krayniak et al., 1979; Haglund et al., 1984; Köhler et al., 1984; Schwerdtfeger, 1984). In the rat, the septal part of the Sub projects preferentially to the mammillary body, whereas fibers from the temporal part mainly reach the ventromedial hypothalamus (Swanson and Cowan, 1977). In the cat, the projections from the Sub to the mammillary bodies are topographically organized such that septal levels project dorsally and more temporal levels project ventromedially in the mammillary bodies (own observations).

With respect to connections of the HF with the thalamus, it is well documented that the nucleus reuniens projects to the molecular layer of the Sub and CA1 (Herkenham, 1978; Amaral and Cowan, 1980). In the rat, these projections show a weak septotemporal differentiation (Herkenham, 1978). As yet, there is no agreement with respect to whether the anterior thalamus is connected with the Sub or with the presubiculum and the parasubiculum (Meibach and Siegel, 1977a,b; Rosene and Van Hoesen, 1977; Krayniak et al., 1979; Amaral and Cowan, 1980; Stanfield et al., 1980; Schwerdtfeger, 1984).

Projections from the brainstem. In several species it has been demonstrated that the raphe nuclei, in particular the medial raphe nucleus, and the locus coeruleus project to the HF (Segal and Landis, 1974; Pasquier and Reinoso-Suarez, 1977; Azmitia and Segal, 1978; Pasquier and Reinoso-Suarez, 1978; Wyss et al., 1979a; Amaral and Cowan, 1980; Schwerdtfeger, 1984). In the rat, the projections from the raphe nuclei terminate bilaterally in the molecular and infragranular layers of the FD. Raphe neurons do not project to both hippocampi, but they project by way of collaterals to the HF and the entorhinal cortex, and to the HF and the septum (Köhler and Steinbusch, 1982; Crunelli and Segal, 1985). Although no septotemporal

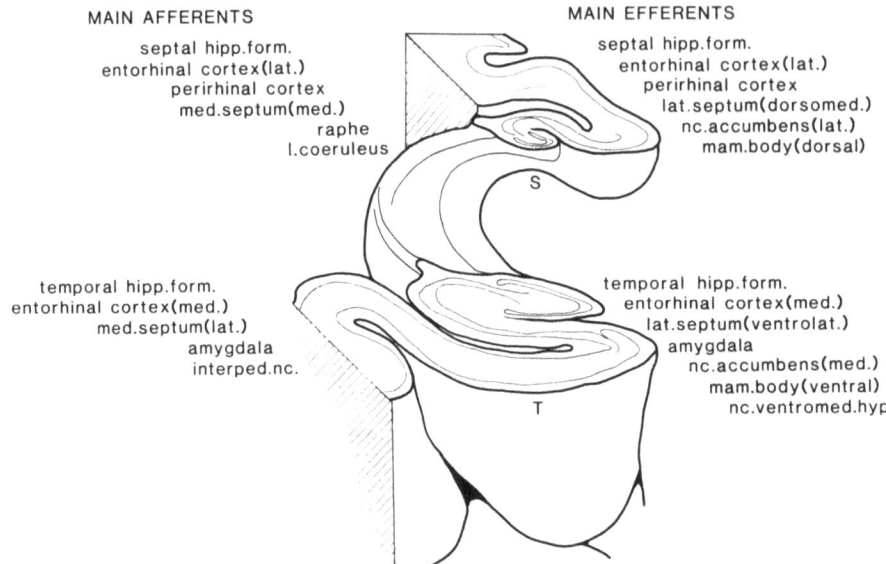

Fig. 4. Diagram of the hippocampal formation (HF) of the cat, summarizing the differential extrinsic and intrinsic connections of its septal (S) and temporal (T) parts. Note that data from predominantly the rat and the cat are taken together, and that the gradual change in site of origin or termination of these connections, with the splenial part of the HF as an intermediate between S and T, is omitted.

organization of the raphe-hippocampal projections has been described, there is a septo-temporal differential distribution of serotonin (Gage and Thompson, 1980; Roberts et al., 1984). Furthermore, only the temporal part of the HF receives additional input from the interpeduncular nucleus (Groenewegen et al., 1983). In the cat, the projections from the raphe nuclei are to some extent topographically organized along the longitudinal axis of the HF (Pasquier and Reinoso-Suarez, 1978). The projections from the locus coeruleus terminate for the most part in the septal portion of the HF, particularly in the hilus and in all layers of the CA (Pasquier and Reinoso-Suarez, 1978; Wyss et al., 1979a; see also Frotscher and Zimmer, 1983; Roberts et al., 1984).

CONCLUDING REMARKS

Although our knowledge of the intrinsic organization of the HF is mainly based on studies in the rat, a comparison with the less detailed data available for the monkey leads to the conclusion that the intrinsic organization is largely similar in both species. Furthermore, both the subcortical efferent and afferent connections constitute a conservative connectional system which is present in all species studied. In contrast, the cortical connections appear to vary more between different species, such that in the monkey with a higher developed neocortex, the connections of the Sub and CA1 with the neocortex are intensified (cf. Schwerdtfeger, 1984).

It is of interest that area CA2 has recently been put forward as the hippocampal structure which is crucial for the generation of spontaneous bursts of cellular activity (Miles et al., 1984), a feature which seems to be important in relation to the occurrence of hippocampal seizures.

Unfortunately, since CA2 is not generally considered a separate entity, no detailed data concerning its connections are available. As yet, it seems that next to the observation that CA2 contributes heavily to the longitudinal association pathway, it is characterized, at least in the monkey and possibly also in the rat, by a specific input from the supramammillary region.

The organization of intrinsic and extrinsic connections of the HF all point to a septotemporal differentiation within this structure (Fig. 4). In both the rat and the monkey, the organization of the commissural and associational connections indicate a difference between the temporal one-third and the remaining septal and splenial two-thirds. Furthermore, in the rat and the cat it is clear that the two major inputs to the HF, i.e., the projections from the entorhinal cortex and the medial septum, are topographically organized along the septotemporal axis of the hippocampus. The reciprocal pathways are similarly arranged. Furthermore, several of the hypothalamic and brainstem connections exhibit to some extent a septotemporal differentiation in their termination. In particular, the organization of the entorhinal-hippocampal pathway taken in conjunction with the arrangement of entorhinal afferents has led to the suggestion that the septal part of the HF mainly receives sensory related cortical information whereas the temporal part is involved in the processing of more visceral related information (Witter and Groenewegen, 1984; Room and Groenewegen, 1986; Witter et al., 1986; cf. Amaral et al., 1984; Demeter et al., 1985). The functional differentiation within the HF may be of relevance with respect to the comparison of experimental data. For example, in lesion experiments used for the study of hippocampal functions in non-primates, for the most part animals are used in which the septal part of the HF is lesioned. In contrast, studies in primates rely heavily on animals which received temporal hippocampal lesions. Comparisons between these two sets of data have to be regarded with caution, since they probably reflect the characteristics of functionally different parts of the HF.

REFERENCES

Amaral, D.G., 1978, A Golgi study of cell types in the hilar region of the hippocampus in the rat, J. Comp. Neurol., 182:851.
Amaral, D.G., and Cowan, W.M., 1980, Subcortical afferents to the hippocampal formation in the monkey, J. Comp. Neurol., 189:573.
Amaral, D.G., Insausti, R., and Cowan, W.M., 1984, The commissural connections of the monkey hippocampal formation, J. Comp. Neurol., 224:307.
Andersen, P., Bliss, T.V.P., and Skrede, K.K., 1971, Lamellar organization of hippocampal excitatory pathways, Exp. Brain Res., 13:222.
Ashwood, T.J., Lancaster, B., and Wheal, H.V., 1984, In vivo and in vitro studies on putative interneurons in the rat hippocampus: possible mediators of feed-forward inhibition, Brain Res., 293:279.
Azmitia, E.C., and Segal, M., 1978, An autoradiographic analysis of the differential ascending projections of the dorsal and the median raphe nuclei in the rat, J. Comp. Neurol., 179:641.
Bakst, I., Morrison, J.H., and Amaral, D.G., 1985, The distribution of somatostatin-like immunoreactivity in the monkey hippocampal formation, J. Comp. Neurol., 236:423.
Berger, T.W., Swanson, G.W., Milner, T.A., Lynch, G.S., and Thompson, R.F., 1980, Reciprocal anatomical connections between hippocampus and subiculum in the rabbit: evidence for subicular innervation of regio superior, Brain Res., 183:265.
Blackstad, T.W., 1956, Commissural connections of the hippocampal region in the rat, with special reference to their mode of termination, J. Comp. Neurol., 105:417.

Blackstad, T.W., Brink, K., Hem, J., and Jeune, B., 1970, Distribution of hippocampal mossy fibers in the rat. An experimental study with silver impregnation methods, J. Comp. Neurol., 138:433.

Buzsaki, G., 1984, Feed-forward inhibition in the hippocampal formation, Progress in Neurobiol., 22:131.

Chandler, J.P., and Crutcher, H.A., 1983, The septohippocampal projection in the rat: an electronmicroscopic horseradish peroxidase study, Neuroscience, 10:685.

Crunelli, V., and Segal, M., 1985, An electrophysiological study of neurones in the rat median raphe and their projections to septum and hippocampus, Neuroscience, 15:47.

Crutcher, H.A., Madison, R., and Davis, J.N., 1981, A study of the rat septohippocampal pathway using anterograde transport of horseradish peroxidase, Neuroscience, 6:1961.

Demeter, S., Rosene, D.L., and Van Hoesen, G.W., 1985, Interhemispheric pathways of the hippocampal formation, presubiculum and entorhinal and posterior parahippocampal cortices in the rhesus monkey: the structure and organization of the hippocampal commissures, J. Comp. Neurol., 233:30.

Dent, J.A., Galvin, N.J., Stanfield, B.B., and Cowan, W.M., 1983, The mode of termination of the hypothalamic projection to the dentate gyrus: an EM autoradiographic study, Brain Res., 258:1.

Fredens, K., Stengaard-Pedersen, K., and Larsson, L.I., 1984, Localization of enkephalin and cholecystokinin immunoreactivities in the perforant path terminal fields of the rat hippocampal formation, Brain Res., 304:255.

Fricke, R., and Cowan, W.M., 1978, Autoradiographic study of the commissural and ipsilateral hippocampal-dentate projections in the adult rat, J. Comp. Neurol., 181:253.

Frotscher, M., and Zimmer, J., 1983, Commissural fibers terminate on non-pyramidal neurons in the guinea pig hippocampus. A combined Golgi/EM degeneration study, Brain Res., 265:289.

Frotscher, M., Leranth, Cs., Lübbers, K., and Oertel, W.H., 1984, Commissural afferents innervate glutamate decarboxylase immunoreactive non-pyramidal neurons in the guinea pig hippocampus, Neurosci. Lett., 46:137.

Gage, F.H., and Thompson, R.G., 1980, Differential distribution of norepinephrine and serotonin along the dorso-ventral axis of the hippocampal formation, Brain Res. Bull., 5:771.

Gall, C., Brecha, N., Karten, H.J., and Chang, K.-J., 1981, Localization of enkephalin-like immunoreactivity to identified axonal and neuronal populations of the rat hippocampus, J. Comp. Neurol., 198:335.

Gall, C., 1984, The distribution of cholecystokinin-like immunoreactivity in the hippocampal formation of the guinea pig: localization in the mossy fibers, Brain Res., 306:73.

Gottlieb, D.I., and Cowan, W.M., 1973, Autoradiographic studies of the commissural and ipsilateral association connections of the hippocampus and dentate gyrus of the rat. I. The commissural connections, J. Comp. Neurol., 149:393.

Groenewegen, H.J., Arnolds, D.E.A.T., and Lopes da Silva, F.H., 1981, Afferent connections of the nucleus accumbens in the cat, with special emphasis on the projections from the hippocampal region. An anatomical and electrophysiological study, in: The Neurobiology of the Nucleus Accumbens, R.B. Chronister, and J.F. De France, eds., Hear Institute for Electrophysiological Res., Brunswick, p. 41.

Groenewegen, H.J., Room, P., Witter, M.P., and Lohman, A.H.M., 1982, Cortical afferents of the nucleus accumbens in the cat, studied with anterograde and retrograde transport techniques, Neuroscience, 7:977.

Groenewegen, H.J., Haber, S.N., and Nauta, W.J.H., 1983, Structure and efferent connections of the interpeduncular nucleus in the rat. An immunohistochemical and neuroanatomical tracer study, Neurosci. Lett. Suppl., 14:145.

Haglund, L., Swanson, L.W., and Köhler, C., 1984, The projection of the supramammillary nucleus to the hippocampal formation: an immunohistochemical and anterograde transport study with the lectin PHA-L in the rat, J. Comp. Neurol., 229:171.

Herkenham, M., 1978, The connections of the nucleus reuniens thalami: evidence for a direct thalamo-hippocampal pathway in the rat, J. Comp. Neurol., 177:589.

Hjorth-Simonsen, A., 1971, Hippocampal efferents to the ipsilateral entorhinal area: an experimental study in the rat, J. Comp. Neurol., 142:417.

Hjorth-Simonsen, A., and Jeune, B., 1972, Origin and termination of the hippocampal perforant path in the rat studied by silver impregnation, J. Comp. Neurol., 144:215.

Kelley, A.E., and Domesick, V.B., 1982, The distribution of the projection from the hippocampal formation to the nucleus accumbens in the rat: an anterograde- and retrograde-horseradish peroxidase study, Neuroscience, 7:2321.

Knowles, W.D., and Schwartzkroin, P.A., 1981, Local circuit synaptic interactions in hippocampal brain slices, J. Neurosci., 1:318.

Köhler, C., and Steinbusch, H., 1982, Identification of serotonin and non-serotonin-containing neurons of the mid-brain raphe projecting to the entorhinal area and the hippocampal formation. A combined immunohistochemical and fluorescent retrograde tracing study in the rat brain, Neuroscience, 7:951.

Köhler, C., 1983, A morphological analysis of vasoactive intestinal polypeptide (VIP)-like immunoreactive neurons in the area dentata of the rat brain, J. Comp. Neurol., 221:247.

Köhler, C., Haglund, L., and Swanson, L.W., 1984, A diffuse α-MSH-immunoreactive projection to the hippocampus and spinal cord from individual neurons in the lateral hypothalamic area and zona incerta, J. Comp. Neurol., 223:501.

Köhler, C., 1985, Intrinsic projections of the retrohippocampal region in the rat brain. I. The subicular complex, J. Comp. Neurol., 236:504.

Kosel, K.C., Van Hoesen, G.W., and Rosene, D.L., 1983, A direct projection from the perirhinal cortex (area 35) to the subiculum in the rat, Brain Res., 269:347.

Krayniak, P.F., Siegel, A., Meibach, R.C., Fruchtman, D., and Scrimenti, M., 1979, Origin of the fornix system in the squirrel monkey, Brain Res., 160:401.

Krayniak, P.F., Weiner, S., and Siegel, A., 1980, An analysis of the efferent connections of the septal area in the cat, Brain Res., 189:15.

Krettek, J.E., and Price, J.L., 1977, Projections from the amygdaloid complex and adjacent olfactory structures to the entorhinal cortex and to the subiculum in the rat and cat, J. Comp. Neurol., 172:723.

Laurberg, S., 1979, Commissural and intrinsic connections of the rat hippocampus, J. Comp. Neurol., 184:685.

Laurberg, S., and Sørensen, K.E., 1981, Associational and commissural collaterals of neurons in the hippocampal formation (hilus fascia dentatae and subfield CA3), Brain Res., 212:287.

Leichnetz, G.R., and Astruc, J., 1975, Efferent connections of the orbitofrontal cortex in the marmoset (Saguinus oedipus), Brain Res., 84:169.

Leichnetz, G.R., and Astruc, J., 1976, The efferent projections of the medial prefrontal cortex in the squirrel monkey (Saimiri sciureus), Brain Res., 109:455.

Leranth, Cs., and Frotscher, M., 1983, Commissural afferents to the rat hippocampus terminate on vasoactive intestinal polypeptide-like immunoreactive non-pyramidal neurons. An EM immunocytochemical degeneration study, Brain Res., 276:357.

Lopes da Silva, F.H., Arnolds, D.E.A.T., and Neijt, H.C., 1984, A functional link between the limbic cortex and ventral striatum: physiology of the subiculum accumbens pathway, Exp. Brain Res., 55:205.

Lopes da Silva, F.H., Groenewegen, H.J., Holsheimer, J., Room, P., Witter, M.P., Van Groen, Th., and Wadman, W.J., 1985, The hippocampus as a set of partially overlapping segments with a topographically organized system of inputs and outputs: the entorhinal cortex as a sensory gate, the medial septum as a gain-setting system and the ventral striatum as a motor interface, in: Electrical Activity of the Archicortex, G. Buzsaki and C.H. VanderWolf, eds., Akademiai Kiado, Budapest, p. 83.

Lorente de Nó, R., 1934, Studies on the structure of the cerebral cortex. II. Continuation of the study of the ammonic system, Psychol. Neurol., 46:113.

McNaughton, B.L., 1980, Evidence for two physiologically distinct perforant pathways to the fascia dentata, Brain Res., 199:1.

Meibach, R.C., and Siegel, A., 1977a, Efferent connections of the hippocampal formation in the rat, Brain Res., 124:197.

Meibach, R.C., and Siegel, A., 1977b, Thalamic projections of the hippocampal formation: evidence for an alternate pathway involving the internal capsule, Brain Res., 134:1.

Miles, R., Wong, R.K.S., and Traub, R.D., 1984, Synchronized afterdischarges in the hippocampus: contribution of local synaptic interactions, Neuroscience, 12:1179.

Monmaur, P., and Thomson, M.A., 1983, Topographic organization of septal cells innervating the dorsal hippocampal formation of the rat: special reference to both the CA1 and dentata theta generators, Exp. Neurol., 82:366.

Newman, R., and Winans, S.S., 1980, An experimental study of the ventral striatum of the golden hamster. I. Neuronal connections of the nucleus accumbens, J. Comp. Neurol., 191:167.

Nunzi, M.G., Gorio, A., Milan, F., Freund, T.F., Somogyi, P., and Smith, A.D., 1985, Cholecystokinin-immunoreactive cells form symmetrical synaptic contacts with pyramidal and nonpyramidal neurons in the hippocampus, J. Comp. Neurol., 237:485.

O'Keefe, J., and Nadel, L., 1978, The Hippocampus as a Cognitive Map, Clarendon Press, Oxford.

Olton, D.S., Walker, J.A., and Wolf, W.A., 1982, A disconnection analysis of hippocampal function, Brain Res., 233:241.

Olton, D.S., 1983, Memory functions and the hippocampus, in: Neurobiology of the Hippocampus, W. Seifert, ed., Academic Press, London, p. 335.

Pasquier, D.A., and Reinoso-Suarez, F., 1976, Direct projections from hypothalamus to hippocampus in the rat demonstrated by retrograde transport of horseradish peroxidase, Brain Res., 108:165.

Pasquier, D.A., and Reinoso-Suarez, F., 1977, Differential efferent connections of the brain stem to the hippocampus in the cat, Brain Res., 120:540.

Pasquier, D.A., and Reinoso-Suarez, F., 1978, The topographic organization of hypothalamic and brainstem projections to the hippocampus, Brain Res. Bull., 3:373.

Pohle, W., and Ott, T., 1984, Localization of entorhinal cortex neurons projecting to the dorsal hippocampal formation. A stereotaxic tool in three dimensions, J. Hirnforsch., 25:661.

Ribak, C.E., Vaughn, J.E., and Saito, K., 1978, Immunocytochemical localization of glutamic acid decarboxylase in neuronal somata following colchicine inhibition of axonal transport, Brain Res., 140:315.

Roberts, G.W., Woodhams, P.L., Polak, J.M., and Crow, T.J., 1984, Distribution of neuropeptides in the limbic system of the rat: the hippocampus, Neuroscience, 11:35.

Room, P., Russchen, F.T., Groenewegen, H.J., and Lohman, A.H.M., 1985, Efferent connections of the prelimbic (area 32) and the infralimbic (area 25) cortices. An anterograde tracing study in the cat, J. Comp. Neurol., 242:40.

Room, P., and Groenewegen, H.J., 1986, Connections of the parahippocampal cortex in the cat. I. Cortical afferents, J. Comp. Neurol., in press.

Rosene, D.L., and Van Hoesen, G.W., 1977, Hippocampal efferents reach widespread areas of cerebral cortex and amygdala in the rhesus monkey, Science, 198:315.

Russchen, F.T., 1982, Amygdalopetal projections in the cat. I. Cortical afferent connections. A study with retrograde and anterograde tracing techniques, J. Comp. Neurol., 206:159.

Ruth, R.E., Collier, T.J., and Routtenberg, A., 1982, Topography between the entorhinal cortex and the dentate septotemporal axis in rats: I. Medial and intermediate entorhinal projecting cells, J. Comp. Neurol., 209:69.

Schwartzkroin, P.A., and Kunkel, D.D., 1985, Morphology of identified interneurons in the CA1 region of guinea pig hippocampus, J. Comp. Neurol., 232:205.

Schwerdtfeger, W.K., 1984, Structure and fiber connections of the hippocampus. A comparative study, Adv. Anat. Embryol. Cell. Biol., 83:1.

Segal, M., and Landis, S., 1974, Afferents to the hippocampus of the rat studied with the method of retrograde transport of horseradish peroxidase, Brain Res., 78:1.

Segal, M., 1979, A potent inhibitory monosynaptic hypothalamo-hippocampal connection, Brain Res., 162:137.

Siegel, A., and Tassoni, J.P., 1971a, Differential efferent projections from the ventral and dorsal hippocampus of the cat, Brain Behav. Evol., 4:185.

Siegel, A., and Tassoni, J.P., 1971b, Differential efferent projections of the lateral and medial septal nuclei to the hippocampus in the cat, Brain Behav. Evol., 4:201.

Siegel, A., Edinger, H., and Ohgami, S., 1974, The topographical organization of the hippocampal projection to the septal area: a comparative neuroanatomical analysis in the gerbil, rat, rabbit, and cat, J. Comp. Neurol., 157:359.

Somogyi, P., Nunzi, M.G., Gorio, A., and Smith, A.D., 1983a, A new type of specific interneuron in the monkey hippocampus forming synapses exclusively with the axon initial segments of pyramidal cells, Brain Res., 259:137.

Somogyi, P., Smith, A.D., Nunzi, M.G., Gorio, A., Takagi, H., and Wu, J.Y., 1983b, Glutamate decarboxylase immunoreactivity in the hippocampus of the cat, J. Neurosci., 3:1450.

Sørensen, K.E., and Shipley, M.T., 1979, Projections from the subiculum to the deep layers of the ipsilateral presubicular and entorhinal cortices in the guinea pig, J. Comp. Neurol., 188:313.

Squire, L.R., 1983, The hippocampus and the neuropsychology of memory, in: The Neurobiology of the Hippocampus, W. Seifert, ed., Academic Press, London, p. 491.

Stanfield, B.B., Wyss, J.M., and Cowan, W.M., 1980, The projection of the supramammillary region upon the dentate gyrus in normal and reeler mice, Brain Res., 198:196.

Steward, O., 1976, Topographic organization of the projections from the entorhinal area to the hippocampal formation of the rat, J. Comp. Neurol., 167:285.

Steward, O., and Scoville, S.A., 1976, Cells of origin of entorhinal cortical afferents to the hippocampus and fascia dentata of the rat, J. Comp. Neurol., 169:347.

Struble, R.G., Desmond, N.L., and Levy, W.B., 1978, Anatomical evidence for interlamellar inhibition in the fascia dentata, Brain Res., 152:580.

Swanson, L.W., and Cowan, W.M., 1977, An autoradiographic study of the organzition of the efferent connections of the hippocampal formation in the rat, J. Comp. Neurol., 172:49.

Swanson, L.W., Wyss, J.M., and Cowan, W.M., 1978, An autoradiographic study of the organization of intrahippocampal association pathways in the rat, J. Comp. Neurol., 181:681.

Swanson, L.W., and Cowan, W.M., 1979, The connections of the septal region in the rat, J. Comp. Neurol., 186:621.

Swanson, L.W., Sawchenko, P.E., and Cowan, W.M., 1980, Evidence that the commissural, associational and septal projections of the regio inferior of the hippocampus arise from the same neurons, Brain Res., 197:207.

Swanson, L.W., 1981, A direct projection from Ammon's horn to prefrontal cortex in the rat, Brain Res., 217:150.

Swanson, L.W., Sawchenko, P.E., and Cowan, W.M., 1981, Evidence for collateral projections by neurons in Ammon's horn, the dentate gyrus and the subiculum. A multiple retrograde labeling study in the rat, J. Neurosci., 1:548.

Taxt, T., and Storm-Mathisen, J., 1984, Uptake of D-aspartate and L-glutamate in excitatory axon terminals in hippocampus: autoradiographic and biochemical comparison with γ-aminobutyrate and other amino acids in normal rats and in rats with lesion, Neuroscience, 11:79.

Van Groen, Th., and Lopes da Silva, F.H., 1985, Septotemporal distribution of entorhinal projections to the hippocampus in the cat: electrophysiological evidence, J. Comp. Neurol., 238:1.

Van Groen, Th., and Witter, M.P., 1985, Electrophysiological and tracing study of the septotemporal distribution of entorhinal projections to the hippocampus in the cat, in: Electrical Activity of Archicortex, G. Buzsaki and C.H. VanderWolf, eds., Academiai Kiado, Budapest, p. 107.

Van Groen, Th., and Lopes da Silva, F.H., 1986, The organization of the reciprocal connections between the subiculum and the entorhinal cortex in the cat. II. An electrophysiological study, J. Comp. Neurol., in press.

Van Groen, Th., van Haren, F., Witter, M.P., and Groenewegen, H.J., 1986, The organization of the reciprocal connections between the subiculum and the entorhinal cortex in the cat. I. A neuroanatomical tracing study, J. Comp. Neurol., in press.

Van Hoesen, G.W., and Pandya, D.N., 1975, Some connections of the entorhinal (area 28) and perirhinal (area 35) cortices of the rhesus monkey. III. Efferent connections, Brain Res., 95:39.

Van Hoesen, G.W., Rosene, D.L., and Mesulam, M.-M., 1979, Subicular input from temporal cortex in the rhesus monkey, Science, 205:608.

Van Hoesen, G.W., 1982, The parahippocampal gyrus. New observations regarding its cortical connections in the monkey, TINS, 5:345.

Veazey, R.B., Amaral, D.G., and Cowan, W.M., 1982, The morphology and connections of the posterior hypothalamus in the cynomolgus monkey (Macaca fascicularis). II. Efferent connections, J. Comp. Neurol., 207:135.

Vincent, S.R., McIntosh, C.H.S., Buchan, A.M.J., and Brown, J.C., 1985, Central somatostatin systems revealed with monoclonal antibodies, J. Comp. Neurol., 238:169.

Walaas, I., 1983, The hippocampus, in: Chemical Neuroanatomy, P.C. Emson, ed., Raven Press, New York, p. 337.

Witter, M.P., and Groenewegen, H.J., 1984, Laminar origin and septotemporal distribution of entorhinal and perirhinal projections to the hippocampus in the cat, J. Comp. Neurol., 224:371.

Witter, M.P., Room, P., Groenewegen, H.J., and Lohman, A.H.M., 1986, Connections of the parahippocampal cortex in the rat. V. Intrinsic connections; comments on input/output connections with the hippocampus, J. Comp. Neurol., in press.

Wyss, J.M., Swanson, L.W., and Cowan, W.M., 1979a, A study of subcortical afferents to the hippocampal formation in the rat, Neuroscience, 4:463.

Wyss, J.M., Swanson, L.W., and Cowan, W.M., 1979b, Evidence for an input to the molecular layer and the stratum granulosum of the dentata gyrus from the supramammillary region of the hypothalamus, Anat. Embryol., 156:165.

Wyss, J.M., 1981, An autoradiographic study of the efferent connections of the entorhinal cortex in the rat, J. Comp. Neurol., 199:495.

CYTOCHEMICAL ARCHITECTURE OF THE ENTORHINAL AREA

Ch. Köhler

Astra Läkemedel AB
Department of Neuropharmacology
Södertälje, Sweden

INTRODUCTION

The hippocampal region[1] consists of serially arranged cortical fields which become increasingly simplified in their basic laminar structure as one proceeds from the entorhinal area (EA) to the area dentata (Swanson, 1982b; Fig. 1). The EA is the largest retrohippocampal subfield and it gives rise to one of the most prominent association pathways within the entire hippocampal region. This pathway, the perforant path, originates primarily from stellate and pyramidal cells in layers II and III of the medial and the lateral EA, and terminates in the dentate gyrus where it synapses onto the granular cell dendrites in the outer two-thirds of the molecular layer (Blackstad, 1958; Hjorth-Simonsen, 1971; Steward and Scoville, 1976). The medial EA, in turn, receives a topographically organized input from the subfields of the subicular complex: the deep layers of the EA are innervated by the subiculum proper and the outer three layers by the para- and presubiculum, respectively (Shipley, 1975; Köhler, 1985). It is through these latter projections that thalamic as well as some commissural inputs are relayed to the EA, and the association pathways from the subicular complex to the medial EA represent important routes through which afferents of extra-hippocampal origin may affect neurotransmission along the perforant path. The EA thus occupies a nodal position within the hippocampal region where inputs from a number of cortical and subcortical brain areas converge and several of these may be critical for maintaining normal transmission along the perforant path. Since the perforant path exerts a powerful excitatory drive upon the intrinsic hippocampal trisynaptic circuit (Andersen et al., 1966, 1971), unraveling the neural circuits directly involved in the regulation of activity along the perforant path may be of critical importance in understanding the development of some forms of hippocampal seizure disorders. At present, there exists considerable support for a role of an acidic amino acid (e.g., glutamate, aspartate) as an excitatory transmitter at the perforant path to granular cell synapse (Storm-Mathisen, 1977; Andersen et al., 1982; Streit, 1984). In agreement with this hypothesis, the cells giving rise to the perforant path could be retrogradely labeled after intrahippocampal injections of [^{3}H]-D-aspartate

---

[1] The term hippocampal region is used as a collective name for the retrohippocampal region (entorhinal area and subicular complex), the Ammon's horn and the area dentata.

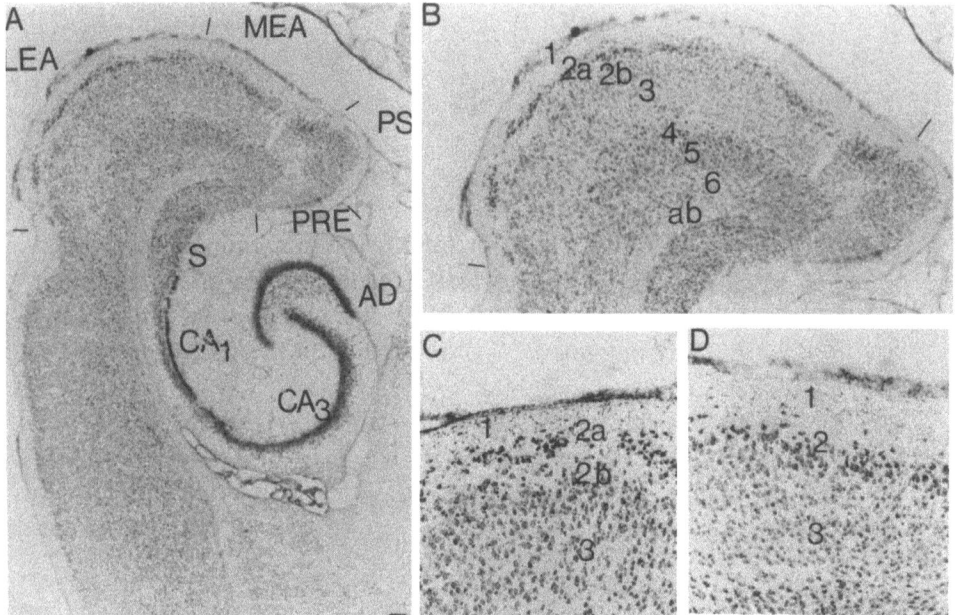

Fig. 1. Low power photomicrographs showing the hippocampal region (A) and the entorhinal area (B) in a horizontal section through the rat brain. Details of the cortical lamina in the medial and lateral EA are shown in D and C, respectively. Bar = 50 µm.

(Fig. 2), a method claimed (Streit, 1984) to selectively label pathways utilizing excitatory amino acids as transmitters. Importantly, however, recent studies suggest that not all perforant path fibers release the same neurotransmitter(s) as it has been shown that the lateral perforant path contains the opioid peptide leucine-enkephalin and that stimulation of the lateral perforant path results in the release of this peptide (Gall et al., 1981; Charkin et al., 1985). Yet other studies (Fredens et al., 1984) have shown that hippocampal afferents running in the temporo-ammonic tract contain CCK-8-immunoreactivity, but so far the release of this peptide after stimulation of these fibers has not been demonstrated. Apart from the studies on the perforant path, little is still known about the chemical identities of the other major efferents of the EA. Despite the fact that our knowledge about the neurotransmitters used by the efferent, afferent and intrinsic pathways of the hippocampal region is still rudimentary, important progress has been made in recent years. In the present chapter, the principal organization of several histochemically identified neurons and terminal of the EA will be summarized, with the aim of providing a comprehensive picture of chemical messengers contained in some entorhinal interneurons and afferent pathways.

CHEMICAL MESSENGERS IN EA INTERNEURONS

GABA Containing Cells

Neurons immunoreactive (-i) for the inhibitory (Andersen et al., 1982) neurotransmitter gamma-aminobutyric acid (GABA) or its synthesizing enzyme, glutamic acid decarboxylase (GAD), are scattered throughout all six layers of the medial and lateral EA (Fig. 3). These cells show large individual differences in size, they are morphologically heterogeneous and the large number of immunohistochemically stained cells suggests that

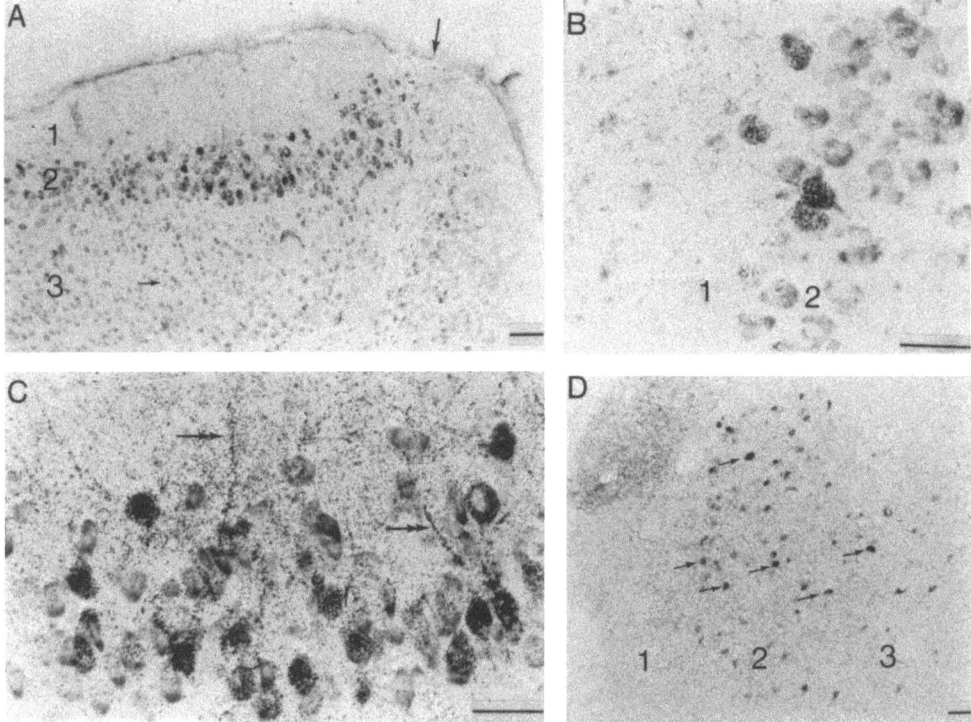

Fig. 2. Cells in layer II of the medial EA retrogradely labeled after injections of ^{3}H-D-aspartate into the dentate gyrus of the dorsal hippocampus. B and C show details of labeled cells in the lateral and medial EA, respectively. Cells in layer II and III are retrogradely labeled in presubiculum after ^{3}H-D-aspartate injections into the medial EA. Bar = 50 μm.

GAD and GABA are contained within most of the different classes of interneurons described previously within the EA using the Golgi-method (Ramon y Cajal, 1911; Lorente de Nó, 1934). Many GABA-i cells are situated in the second and third layer of the medial and the lateral EA (Fig. 3) and they may belong to a group of interneurons having horizontal axons or extensively arborizing axons forming dense terminal plexuses close to the parent cell bodies (Lorente de Nó, 1934; Jones and Hendry, 1984). All layers of the EA are rich in GABA- or GAD-i terminals (Fig. 3). In all layers, the GABA preterminal processes are seen as immunoreactive punctate structures, which probably represent GABA terminals. Frequently, GABA and GAD-i puncta are situated close to cell bodies forming pericellular baskets which outline the profile of the innervated cell. Layer II is particularly rich in such terminals, which are found around the somata of unstained and, occasionally, around GAD or GABA-i cell bodies. A closer examination of the GABA and GAD-i terminals in layer II shows that the palisades of immunoreactive terminals around unstained cells represent actual symmetric axo-somatic synapses formed between the GABA terminals and stellate, pyramidal as well as smaller cells of unknown identity (Köhler et al., 1985a). In layers I and III, many of the GAD-i terminals form symmetric axo-dendritic synapses and direct axo-somatic contacts are present here as well (Köhler et al., 1985a). In layers IIa and b of the lateral EA, the GABA-i terminals are present around a large number of unstained somata. Since a vast majority of the layer II cells are stellate cells giving rise to the perforant path, it follows that the majority of the perforant path neurons receive a GABAergic input, which presumably is inhibitory in nature.

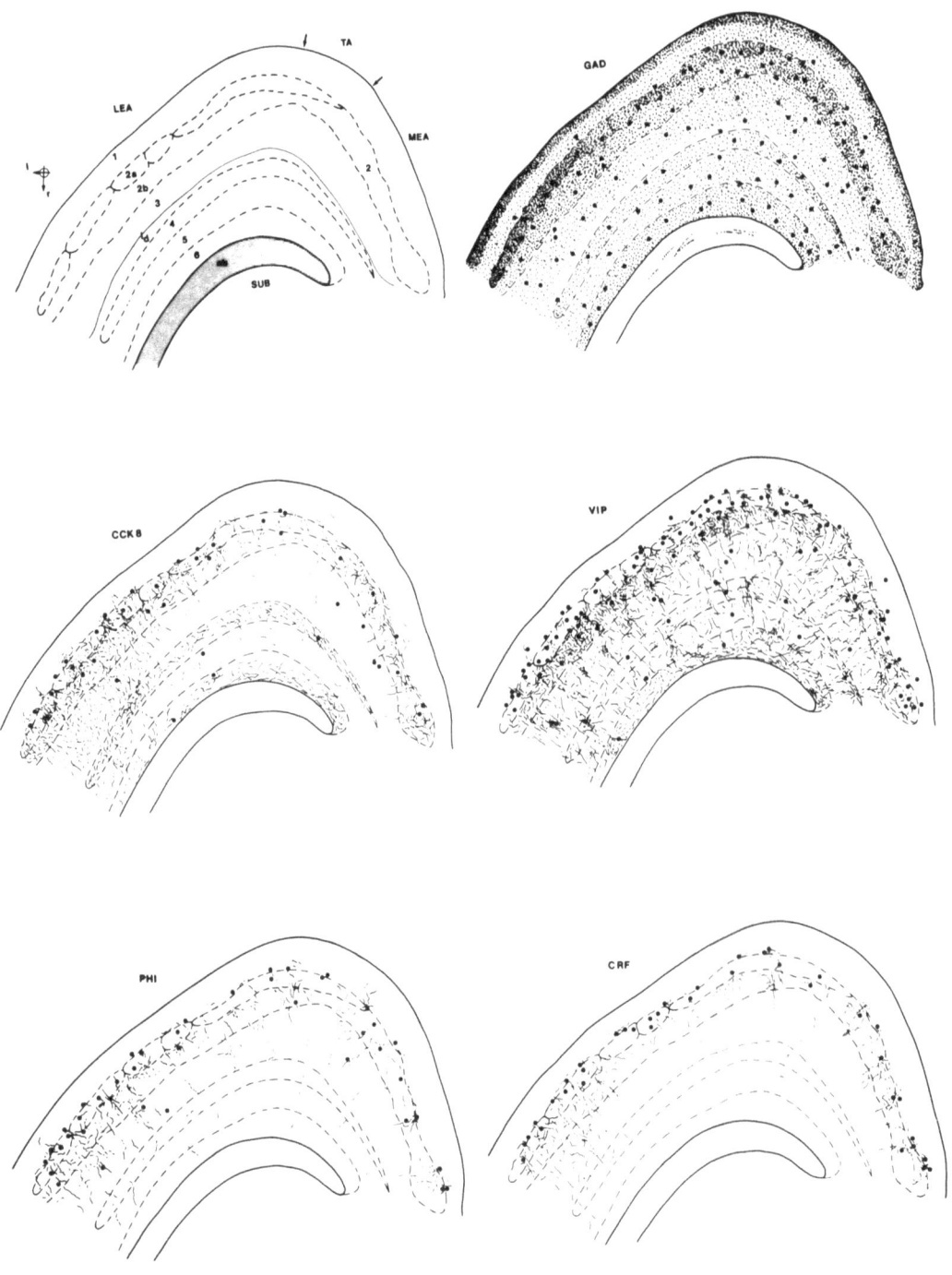

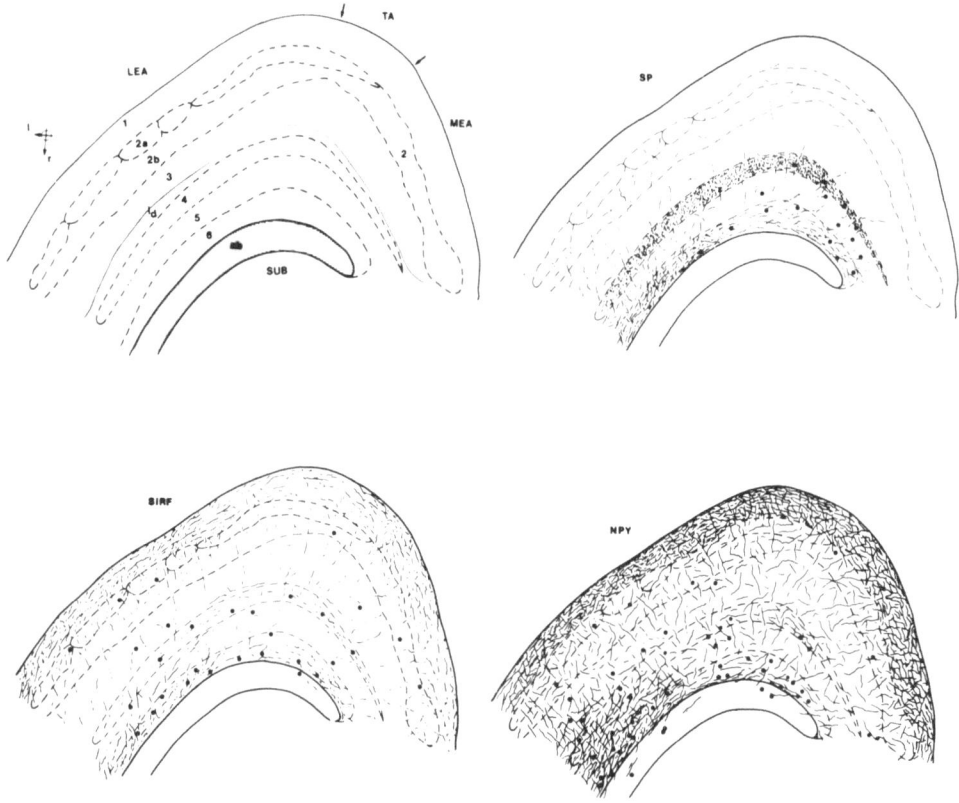

Fig. 3. Line drawings showing, in schematical form, the distribution of cells and preterminal processes immunoreactive for different chemical messengers in the EA.

The exact origin of this GABAergic input to the perforant path cells is not known but most likely it derives from local interneurons, as has been suggested to be the case for the GABA-i axo-somatic terminals found in the hippocampal formation (Ribak et al., 1978; Somogyi et al., 1983b; Roberts et al., 1984) and the neocortex (Houser et al., 1984). In the Ammon's horn, the basket cells situated within, or adjacent to, the pyramidal cell layer give rise to a dense GABAergic innervation of the somata, the proximal dendrites and the axon-initial segment of the pyramidal cells (Somogyi et al., 1983b, 1985). It is quite likely that by a principally similar organization, GABA cells in layers II and III of the EA are the origins of the axo-somatic inputs to the stellate and pyramidal cells of these layers. However, GABA cells in the deep layers may also contribute to the innervation of the superficial layers as has been shown to be the case in the monkey striate cortex (Somogyi et al., 1983a).

The high density of GAD- and GABA-i terminals as well as GABA receptors (Köhler, in preparation) in layer II suggests that GABA mediated inhibition of perforant path neurons is an important mechanism whereby control of firing along this pathway may occur. While the density of GABA terminals is highest in layer II, all other layers are rich in GAD and GABA-i terminals as well. These may represent GABAergic axo-somatic, axo-dendritic and axo-axonic synapses as well as fibers of passage deriving from local interneurons, short association projection neurons and neurons located outside the hippocampal region (see below).

CCK-8 Containing Cells

The gut peptide, cholecystokinin (CCK) is present in cells of the EA as CCK-octapeptide. The CCK-8 neurons are far fewer in number than the GABA-i neurons. The majority of CCK-8-i cells are situated in the outer three layers and their number increases at successively more ventral and lateral parts of the EA (Köhler and Chan-Palay, 1982). In the medial EA, CCK-8 is present predominantly in a group of medium to large size multipolar and polymorphic cells while in the lateral EA primarily small spheroid or bipolar, bitufted cells contain CCK-8-i (Fig. 3). In general, the CCK-8-i cells present in the EA are smaller than those found in the Ammon's horn and this is particularly true for the lateral EA. However, CCK-8-i cells are likely to serve principally similar functions in the different subfields of the hippocampal region and thus, most of the hippocampal cells containing CCK-8-i are probably local interneurons, since studies involving transections of the major hippocampal afferents have failed to reduce hippocampal CCK-8 levels (Handelman et al., 1981). Furthermore, the position of the CCK-8 cells does not correspond to known origins of hippocampal efferents. The morphology and the location of some CCK-8 positive cells in the Ammon's horn (Greenwood et al., 1982) suggest that they are identical to the different types of basket cells situated within, and adjacent to, the stratum pyramidale. Similarly, CCK-8-i cells are situated within layers II and III of both the medial and lateral EA (Fig. 3). Within layer II, a dense network of CCK-8 positive varicose axons form pericellular plexuses around somata of unstained cells. In some cases the axons can be followed from the CCK-8-i cells to their termination around other layer II cells. This arrangement of pericellular baskets is quite similar to that seen for cortical basket cells and it resembles the patterns seen for some GABA and CCK-8 terminals in the hippocampus and other parts of the cortex. Interestingly, the pericellular innervation of layer II cells in the medial EA usually involves clusters of CCK-8-i terminals innervating groups of layer II somata separated by cells apparently lacking CCK-8 innervation.

The reason(s) for this arrangement of the CCK-8-i terminals in the medial EA is unknown at present but it could indicate that subpopulations of perforant path cells in layer II are under the influence of CCK-8 containing interneurons. If this is indeed the case, it could imply differential control of individual perforant neurons at the level of their origin in the EA. The pericellular CCK-8 innervations present in layer II (and to some extent layer IV) resemble that of the GABA terminals in this layer and it is possible that both CCK-8 and GABA are co-localized within the same interneurons. This is not unlikely since such co-localization has been shown to occur in other cortical areas (Hendry et al., 1984), and ultrastructural studies of rat hippocampal CCK-8-i cells show morphological features of these cells similar to those of GABA-i cells (Peters et al., 1983). Furthermore, the medium size cells immunoreactive for CCK-8 in the medial EA resemble some GAD positive cells also present here. The presence of these cells in layers II and III, together with their general morphological characteristics, could indicate that they are basket cells, utilizing both GABA and CCK-8. The relatively small number of CCK-8 stained cells in this area indicate, however, that only a smaller group of GABA cells may utilize CCK-8 as well.

Most of the CCK-8 stained terminals are present in layers II, IV and VI (Fig. 3). In the ventral part of the lateral EA, however, all layers receive a relatively homogeneous innervation by CCK-8-i axons, and at the most ventral levels CCK-8 axons enter the EA from the pyriform cortex and the periamygdaloid area, forming a dense network throughout all cortical layers. In layer IV, like in layer II, the CCK-8-i terminals form what appear to be axo-somatic contacts with unstained somata. These

CCK-8 innervated cells may be local interneurons situated in this layer and/or pyramidal cells projecting to the hippocampus or to various extra-hippocampal regions (Köhler and Eriksson, 1981). The CCK-8-i terminals may, thus, innervate a restricted group of these projection cells, while other cells in this layer may not be under control by neurons containing CCK-8.

It appears from these studies that most of the CCK-8 axons in the EA derive from local interneurons. While this is probably true for most of the middle and dorsal parts of the EA, the intrinsic CCK-8 system is supplemented at ventral levels by a CCK-8-i input originating in the ventral tegmental area and the substantia nigra (Köhler, unpublished observations).

VIP and PHI(27) Containing Cells

The 28 amino acid residue polypeptide vasoactive intestinal peptide (VIP) is found in cells present within all hippocampal subfields, including the EA (Köhler and Chan-Palay, 1983b). Most of the VIP-i cell bodies are present in layer II and the superficial part of layer III but VIP-i cells can be found in other layers as well. The number of VIP-i cells is largest in the lateral EA, especially at ventral levels (Figs. 3,4). The typical VIP stained cell of the EA is small with dendrites emerging from polar regions of the soma, either as single dendrites or bitufts. Compared to the typical bipolar cells containing VIP-i in other cortical areas (e.g., visual, cingulate cortex; Lorén et al., 1976; McDonald et al., 1982), the EA VIP-i cells have shorter dendrites which often span only a few layers and, furthermore, the VIP-i cells of the EA show larger morphological differences compared to the cells in, for example, the visual cortex. Thus, small and medium size multipolar cells staining for VIP are not uncommon in the EA. Most of the VIP-i cells, however, closely resemble the type of bipolar neuron shown in Golgi studies (Peters and Saint Marie, 1984) to have restricted axonal plexuses, innervating translaminar stripes of cortex, and it has been proposed that VIP cells are involved in the control of neural activity in vertically oriented cortical columns. This may be the case in the EA as well, but due to the high density of terminals innervating all layers, except layer I, it is difficult to determine how the terminals in different layers are associated with the VIP cells of this cortex. Numerous VIP-i cells in the lateral EA projecting into layer IIb or layer III are situated in layer IIa and the axons of these cells form dense terminal plexuses among stellate and pyramidal cells in layers II and III. Some VIP-i cells appear to have more restricted projections which terminate primarily among the stellate cells of layer IIa. Still other cells, situated at the border between layer II and I, appear to send axons towards the pial surface of the brain.

Within the area dentata of the rat, VIP-i is present in certain types of basket cells (Köhler, 1983). Some of these innervate long segments of the granular cell layer while other VIP-i cells have axons innervating the underlying hilar region (Köhler, 1983).

Similar arrangements seem to exist in the more differentiated EA, where the VIP-i cells in layers IIa and IIb apparently innervate the perforant path neurons or send axons which innervate the deeper layers of the cortex. Importantly, however, while we still do not know the targets of these axons, recent ultrastructural studies carried out in the hippocampus have shown that VIP-i terminals form symmetric synapses with pyramidal cells in the pyramidal cell layer and with dendrites in the radial layer (Leranth et al., 1984).

It is generally believed that all VIP positive cells in the neocortex are local circuit interneurons and the same has been proposed to be the

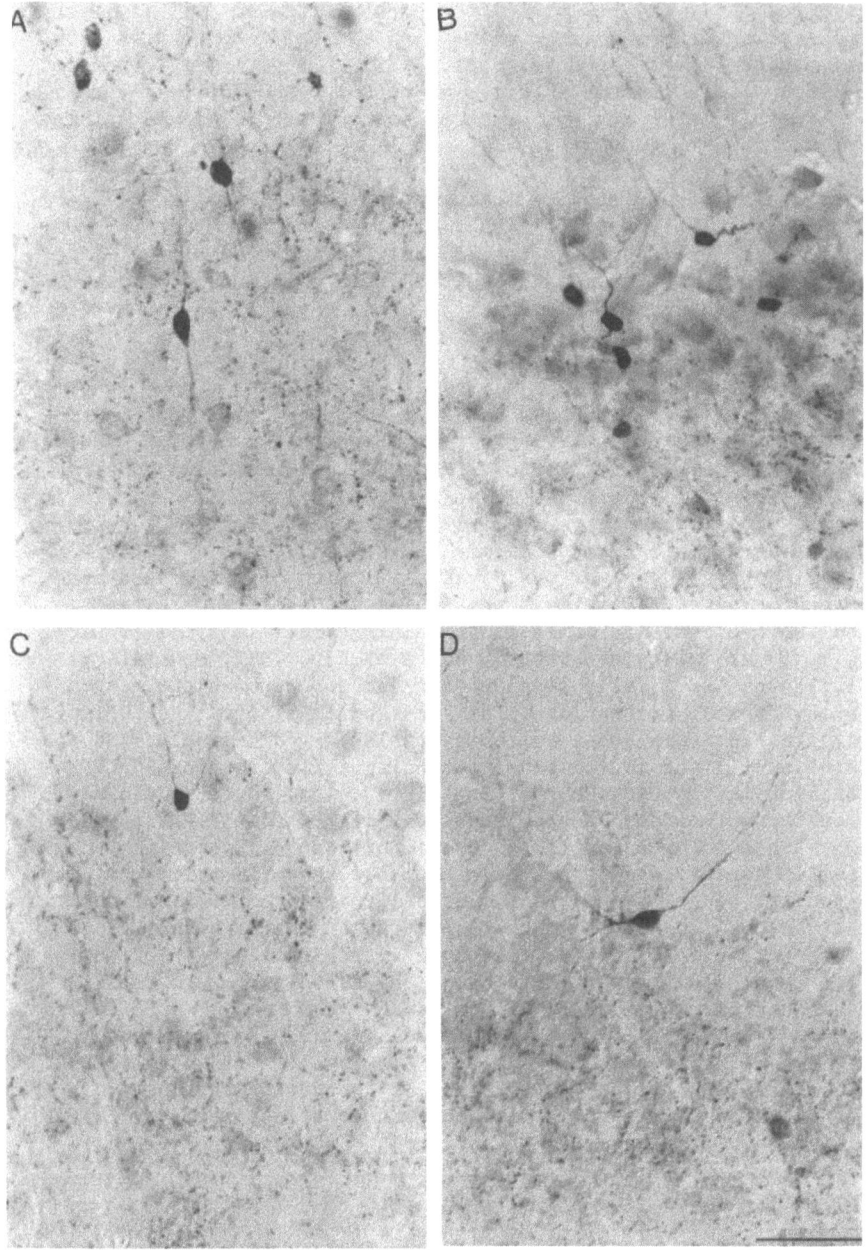

Fig. 4. Photomicrographs showing cells and terminals in the outer layers of the lateral EA immunoreactive for CCK-8 (A), VIP (B), PHI(27) (C), and CRF (D). Bar = 50 μm.

case in the hippocampal formation as well: after transection of hippocampal afferent fibers no significant reduction in VIP levels occurs (Leranth et al., 1984). Although the same type of experiment has not been performed in the EA, it is likely that the situation is the same here as in the Ammon's horn. It should be pointed out, however, that a small VIP-i projection to the EA from the supramammillary nucleus in the rat may exist (Haglund et al., 1984).

A 27 amino-acid residue peptide with an amino-acid sequence partly similar to that of VIP, but containing histidine at its NH_2-terminal and isoleucine amide at its COOH-terminal, PHI(27) (Tatemoto and Mutt, 1981), has recently been detected in brain (Hökfelt et al., 1982). Nucleotide sequencing indicates that this peptide may be part of the VIP precursor molecule (Itoh et al., 1983). Immunohistochemical studies suggest that VIP and PHI(27) may not always be co-localized in the same cells, possibly due to differences in post-translational processing of the two peptides in different neurons. The EA contains neurons immunoreactive for PHI(27) and, as expected, the morphology and the general distribution of the PHI(27) cells are similar to cells containing VIP-i. Thus, the PHI(27) cells are small and bipolar and they are found primarily in the outer three layers of the EA. Like the VIP-i cells, the PHI(27)-i neurons predominate in the lateral and ventral parts of the EA. However, in consecutive sections cut through the EA, there are far more VIP than PHI(27) cells (Figs. 3 and 4) and preterminal processes. It remains to be determined if this reflects differences in post-translational processing, turn-over rates of the two peptides or if technical factors such as binding capacity of the antibodies used in our experiments can explain these observations. It is, however, possible that only some VIP-i cells actually release both peptides and that there may exist differences among the bipolar cells of the EA with respect to the degree to which they utilize both PHI(27) and VIP as transmitters. It remains an important finding, however, that PHI(27)-i is present in the EA. If PHI(27) is released, this peptide may act upon EA cells, including the perforant path neurons, in layers II and III of the medial and lateral EA.

CRF Containing Cells

Neurons immunoreactive for corticotropin releasing factor (CRF) have previously been localized in several extra-hypothalamic brain areas, including the neocortex and the hippocampal region of the rat brain (Swanson et al., 1983). Using antibody against the rat sequence of CRF (Köhler and Swanson, in preparation), CRF-i can be detected in the EA (Fig. 3). The CRF-i is found primarily in small bipolar or round cells, which closely resemble cells also stained with VIP and PHI(27) antisera (see above). The CRF-i cells are situated in layers I and II and only rarely does CRF-i appear in cells in deep layers. This is in contrast to the VIP-i cells which are far more abundant than those containing CRF and which are present in all cortical layers. Like VIP and PHI(27), however, successively more CRF positive cells are present at ventral levels, especially in the lateral EA (Figs. 3 and 4).

While the total number of CRF-i cells is more or less the same in the pyriform cortex and the lateral EA, the medial EA harbors significantly fewer CRF cells than any of the other cortical areas examined (e.g., the perirhinal, visual and cingulate cortex).

The CRF-i preterminal processes show the same general distribution as the VIP-i cells, albeit the density of the CRF innervation is more modest. This is probably due to the significantly fewer CRF neurons present in the EA. Given the morphological similarities and the laminar distribution of CRF-i and VIP-i or PHI(27)-i cells, it is quite possible that these peptides may be co-localized within the same cells.

SRIF Containing Cells

The neurons in the medial and lateral EA which stain with antibody to growth-hormone release inhibiting factor (SRIF; somatostatin) are situated primarily in layers IV through VI. The more superficial layers contain few scattered SRIF-i cells which increase in number at lateral and ventral

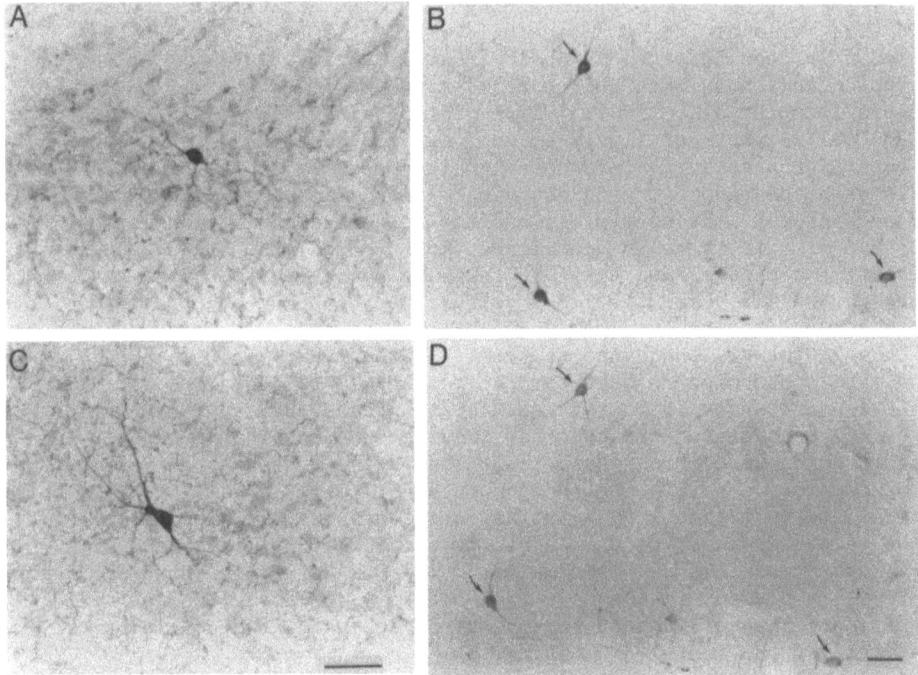

Fig. 5. Photomicrographs showing NPY-i cells in the superficial (A) and deep (B) layers of the EA. Co-localization of NPY-i and SRIF-i is shown in C and D. The sections were first incubated in NPY antiserum and after elution of the binding sites restained with antibody to somatostatin 28. Bar = 50 μm.

levels. The SRIF-i cells are multipolar and fusiform with the fusiform cells dominating in the deep layers. While most of the multipolar cells are found in layers IV-VI of the lateral EA, layers II and III may harbor such cells. SRIF-i fibers are present in all layers of the EA with a predominance in layer I, where numerous horizontally oriented fibers are present. In addition, layers II and IV contain axons immunoreactive for SRIF. Similar to the NPY innervation (see below), most of the SRIF-i fibers apparently do not form axonal plexuses around somata of entorhinal cells, as seen for GABA, CCK-8 and VIP-i cells. The SRIF- and NPY-i neurons show morphological similarities and their similar laminar distribution suggests that the same cell-type may harbor these peptides. In other cortical areas, SRIF- and NPY-i have been shown to be co-localized within some cells (Chronwall et al., 1984; Hendry et al., 1984). Experiments employing double immunostaining have shown that, indeed, many cells in the EA contain immunoreactivity for both NPY and SRIF (Fig. 5). Thus, a group of EA cells having short association projections within this area may use NPY and SRIF as transmitters. Studies carried out in the cortex of the monkey (Hendry et al., 1984) suggested that most, if not all, SRIF-i cells also contain GAD-i and these may, thus, be GABA containing as well. In light of the large number of GAD and GABA-cells within the EA, it is quite possible that most SRIF (and NPY) positive cells utilize GABA as well.

Substance P Containing Cells

While the hippocampal region has generally been considered poor in substance P (SP)-i, recent studies (Gall and Selawaski, 1984; Davies and Köhler, 1985) have shown specific staining of a group of small round or spheroid and some multipolar cells in the deep layers of the EA. These cells are restricted to the ventral parts of the EA and their position and general morphology suggest that they are different from the other cells described above. These cells are also likely to be local interneurons and they may contribute some of the SP-i terminals present in layers IV-VI. As can be seen in Fig. 3, SP-i terminals predominate in layer IV of the medial and lateral EA but at more ventral levels some innervation is present in all the other layers, including layer II (Fig. 3). It is unlikely that all SP innervation of the EA is of intrinsic origin since SP-i terminals can be followed as they enter the EA through a ventral and a lateral route. At present, the origin of this projection is unknown in the rat.

Neuropeptide Y Containing Cells

Neuropeptide Y (NPY), a 36 amino acid peptide of the pancreatic polypeptide family (Tatemoto, 1982), is present in large quantities in several cortical areas (Allen et al., 1983; Emson and Hunt, 1984), including the EA (Köhler et al., 1985b). In the EA, NPY-i is present in medium to large sized multipolar or fusiform cells situated primarily within the deep (IV-VI) layers. Compared to other cortical areas, the EA is relatively poor in NPY-i cell bodies, although the number of immunostained cells increases at ventral levels. In both the medial and lateral EA, the NPY-i cells in the deep layers are of medium size and most of these are fusiform or oval in shape. In the lateral EA, the medium to large size multipolar and fusiform cells dominate. No pyramidal or stellate cells contain NPY-i. While most NPY-i cells are situated in the deep layers, some NPY-i cells are present in layers I-III, especially in the lateral EA. In some of these cells, the axons could be traced into layer II where varicose terminals could be seen branching among layer II cells. These NPY-i cells are of medium size and multipolar and may belong to a group of basket cells which innervate neurons in layer II.

In contrast to the relatively few NPY-i bodies present in the EA, this cortical field is exceptionally rich in NPY-i preterminal processes. These form a dense network which distributes relatively homogeneously throughout all cortical layers, with a slightly higher density in layers I and IV. The NPY-i terminals usually do not form the types of pericellular plexuses around somata in the EA as seen for the other EA interneurons. However, the large number and the spatial distribution of the NPY-i fibers indicate that virtually every neuron of the EA is in close vicinity of a NPY-i axon (Fig. 3).

Some of the NPY-i axons in the EA derive from intrinsic cells (see above). In addition, cells in other brain areas contribute NPY-i afferents to the EA. By using retrograde fluorescence tracing in combination with immunohistochemistry, several sources of the NPY innervation have been established (Köhler et al., in preparation). Thus, the NPY-i cells send short association projections from the deep to the superficial layers of the EA, where some axons may synapse while others continue to innervate the adjacent fields of the hippocampal region. In addition, NPY-i cells in the perirhinal cortex, the pyriform cortex, the endopyriform nucleus, the lateral and basolateral nuclei of the amygdala and the locus ceruleus project to the EA.

CHEMICAL MESSENGERS IN EA AFFERENTS

The present chapter has focused primarily upon the chemical identities of EA interneurons. However, considerable work has been carried out in attempts to identify the neurotransmitter candidates of various EA afferents. While some progress has been made with regard to identifying the chemical messengers associated with projections from the brainstem, the hypothalamus and the septum, the chemical identities of the cortical and thalamic afferents remain unknown. There is evidence to suggest (see Streit, 1984), however, that excitatory amino acids such as glutamate and aspartate may be utilized by some of these latter projections but presently available neuroanatomical tracing techniques have not provided conclusive evidence to prove this hypothesis.

One of the best studied afferent inputs to the EA originates in the medial septum and the diagonal band of Broca (dbB; Alonso and Köhler, 1984). This projection terminates primarily in layers II and IV of the medial and lateral EA, and several studies suggest that this projection may be partly cholinergic (Mesulam and Van Hoesen, 1976; Alonso and Köhler, 1984). In the rat brain, a considerable number of septo-entorhinal afferents originate from GAD-i cells situated in the MS and dbB (Köhler and Chan-Palay, 1983a; Köhler et al., 1984a), albeit the exact termination of these axons remains unknown. A second source of presumed GABA containing afferents to the EA are the GAD positive neurons in the tuberomammillary nucleus of the posterior hypothalamus (Köhler et al., 1985c). Like the septal GABA projection to the EA, the termination of the hypothalamic GABA input is still unknown, but it is unlikely that these inputs contribute to the pericellular baskets around layer II cells since they are not removed by cutting major afferents to the EA (Köhler, unpublished observations).

Other hypothalamic afferents which diffusely innervate the EA include a projection from the zona incerta and the lateral hypothalamus containing alpha-melanocyte-stimulating hormone (αMSH)-i material (Köhler et al., 1984b) as well as a projection from the supramammillary region (Haglund et al., 1984). So far, the transmitter candidate(s) of this latter projection has not been identified in the rat brain. In the guinea pig, on the other hand, SP-i has been claimed to be contained within the supramammillary afferents to the area dentata (Gall and Selawaski, 1984) and it is possible that in this species the SP projection reaches the EA as well. In the rat, however, SP cells in the supramammillary region are not retrogradely labeled from the medial EA (Haglund et al., 1984). The EA receives dopamine (DA), 5-hydroxytryptamine (5-HT; serotonin) and noradrenalin (NA) containing afferents from the substantia nigra/ventral tegmental area, the raphe nuclei and the locus ceruleus, respectively. Of these, the DA and the 5-HT innervations have been extensively studied in the rat. The DA innervation originates in the lateral part of the ventral tegmental nucleus of Tsai and from DA cells in the reticular formation (Swanson, 1982a; Björklund and Lindvall, 1984). It terminates as well restricted islands in the ventral EA (Björklund and Lindvall, 1984), while the more dorsal parts of the EA are sparsely innervated by DA fibers (Björklund and Lindvall, 1984). The cells in the ventral tegmental area which contribute to the EA innervation also contain CCK-8 as shown by retrogradely fluorescent tracing (Köhler, unpublished observations). The dorsal and median raphe nuclei provide a dense network of 5-HT fibers which diffusely innervate all cortical layers with little variation in density (Köhler et al., 1981; Köhler and Steinbusch, 1982). The only exception is a remarkably dense plexus of 5-HT terminals in a thin dorsoventral segment of layer III in the lateral EA (Köhler et al., 1981). Lastly, the NA input from the locus ceruleus to the EA is, like the 5-HT innervation of this area, diffusely distributed throughout all layers with a slight dominance in layers I and II (Swanson and Hartman, 1975; Moore and Card, 1984).

CONCLUSIONS

The inhibitory neurotransmitter GABA and several neuropeptides have been localized to interneurons in the EA by immunohistochemistry. While GABA may be utilized by the majority of EA interneurons, neuropeptides such as CCK-8, VIP, PHI(27), CRF, SP, NPY and SRIF are present in a more restricted population of cells. GABA, as well as all peptides, is present in cell bodies and terminals in layer II and III of the EA. In these layers, they may make synaptic contacts with stellate and pyramidal cells projecting to the area dentata. Thus, the neural activity along the perforant path can be regulated at the level of its origin in the EA through interneurons utilizing a variety of chemical messengers. Normal transmission along the perforant path, and consequently along the intrinsic trisynaptic hippocampal circuit, may critically depend upon the normal operation of GABA and peptide containing EA interneurons together with extra-entorhinal afferents innervating the perforant path cells in layers II and III.

ABBREVIATIONS USED IN FIGS. 1 AND 3

ab, angular bundle
AD, area dentata
CCK-8, cholecystokinin octapeptide
CRF, corticotropin releasing factor
CA1, CA3, subfields of Ammon's horn
GAD, glutamic acid decarboxylase
LEA, MEA, lateral and medial entorhinal area
NPY, neuropeptide Y
PHI, porcine peptide histidine isoleucine amide
PRE, presubiculum
PS, parasubiculum
SP, substance P
SRIF, somatostatin
SUB, S, subiculum
T, transitional area
VIP, vasoactive intestinal polypeptide

REFERENCES

Allen, S., Adrian, T.E., Allen, J.M., Tatemoto, K., Crow, T.J., Bloom, S.R., and Polak, J.M., 1983, Neuropeptide Y distribution in the rat brain, Science, 221:877.
Alonso, A., and Köhler, Ch., 1984, A study of the reciprocal connections between the septum and the entorhinal area using anterograde and retrograde axonal transport methods in the rat brain, J. Comp. Neurol., 225:327.
Andersen, P., Holmquist, B., and Voorhoeve, P.E., 1966, Entorhinal activation of dentate granule cells, Acta Physiol. Scand., 66:448.
Andersen, P., Bliss, T.V.P., and Skrede, K.K., 1971, Lamellar organization of hippocampal excitatory pathways, Exp. Brain Res., 13:222.
Andersen, P., Bie, B., and Ganes, T., 1982, Distribution of GABA sensitive areas on hippocampal pyramidal cells, Exp. Brain Res., 45:357.
Blackstad, T.W., 1958, On the termination of some afferents to the hippocampus and fascia dentata. An experimental study in the rat, Acta Anat. 35:202.
Björklund, A., and Lindvall, O., 1984, Dopamine-containing systems in the CNS, in: Handbook of Chemical Neuroanatomy, Vol. 2, Part 1, Classi-

cal Transmitters in the CNS, A. Björklund and T. Hökfelt, eds., Elsevier, Amsterdam, p. 55.

Charkin, Ch., Shoemaker, W.J., McGinty, J.F., Bayon, A., and Bloom, F.E., 1985, Characterization of the prodynorphin and proenkephalin neuropeptide systems in rat hippocampus, J. Neurosci., 5:808.

Chronwell, B.M., Chase, T.N., O'Donohue, T., 1984, Co-existence of neuropeptide Y and somatostatin in human cortical and rat hypothalamic neurons, Neurosci. Lett., 52:213.

Davies, S., and Köhler, Ch., 1985, The substance P innervation of the hippocampus in the rat, Anat. Embryol., 173:45.

Emson, P.C., and Hunt, S.P., 1984, Peptide-containing neurons of the cerebral cortex, in: Cerebral Cortex, Vol. 2, Functional Properties of Cortical Cells, E.G. Jones and A. Peters, eds., Raven Press, New York, p. 145.

Fredens, K., Stengaard-Pedersen, K., and Larsson, L.-I., 1984, Localization of enkephalin and cholecystokinin immunoreactivities in the perforant path terminal fields in the rat hippocampal formation, Brain Res., 304:255.

Gall, C., Brecha, N., Karten, H.J., and Chang, K.-I., 1981, Localization of enkephalin-like immunoreactivity to identified axonal and neuronal populations of the rat hippocampus, J. Comp. Neurol., 198:335.

Gall, C., and Selawaski, 1984, Supramammillary efferents to guinea pig hippocampus contain substance P-like immunoreactivity, Neurosci. Lett., 51:171.

Greenwood, R.S., Godar, S., Reaves, T.A., and Haywood, J.N., 1982, Cholecystokinin in hippocampal pathways, J. Comp. Neurol., 198:335.

Haglund, L., Swanson, L.W., and Köhler, Ch., 1984, The projection of the supramammillary nucleus to the hippocampal formation: an immunohistochemical and anterograde transport study with the lectin PHA-L in the rat, J. Comp. Neurol., 229:171.

Handelman, G.E., Meyer, D.U., Beinfeld, M.C., and Oertel, W.H., 1981, CCK-containing terminals in the hippocampus are derived from intrinsic neurons. An immunohistochemical and radioimmunological study, Brain Res., 224:181.

Hendry, S.H.C., Jones, E.G., DeFelipe, J., Schmechel, D., Brandon, and Emson, P.C., 1984, Neuropeptide containing neurons of the cerebral cortex are also GABAergic, Proc. Natl. Acad. Sci. USA, 81:6526.

Hjorth-Simonsen, A., 1971, Hippocampal efferents to the ipsilateral entorhinal area: an experimental study in the rat, J. Comp. Neurol., 142:417.

Hökfelt, T., Fahrenkrug, J., Tatemoto, K., and Mutt, V., 1982, PHI, a VIP-like peptide, is present in the rat median eminence, Acta Physiol. Scand., 78:6603.

Houser, C.R., Vaughn, J.E., Hendry, S.H.C., Jones, E.G., and Peters, A., 1984, GABA neurons in the cerebral cortex, in: Cerebral Cortex, Vol. 2, Functional Properties of Cortical Cells, E.G. Jones and A. Peters, eds., Raven Press, New York, p. 63.

Itoh, N., Obata, K.-I., Yanaihara, N., and Okamoto, H., 1983, Human preprovasoactive intestinal polypeptide contains a novel PHI-27-like peptide, PHM-27, Nature, 304:547.

Jones, E.G., and Hendry, S.H.C., 1984, Basket cells, in: Cerebral Cortex, Vol. 1, Cellular Components of the Cerebral Cortex, A. Peters and E.G. Jones, eds., Raven Press, New York. p. 309.

Köhler, Ch., Chan-Palay, V., and Steinbusch, H., 1981, The distribution and orientation of serotonin fibers in the entorhinal and other retrohippocampal areas. An immunohistochemical study with anti-serotonin antibodies in the rat's brain, Anat. Embryol., 161:237.

Köhler, Ch., and Eriksson, L.C., 1981, Efferent projections from the entorhinal area to the basal forebrain and frontal cortex originates in layer IV, Soc. Neurosci. Abstr., 8:420.

Köhler, Ch., and Chan-Palay, V., 1982, The distribution of cholecystokinin-like immunoreactive neurons and nerve terminals in the retrohippocampal region in the rat and guinea pig, J. Comp. Neurol., 210:136.

Köhler, Ch., and Steinbusch, H., 1982, Identification of serotonin and non-serotonin containing neurons of the mid-brain raphe projecting to the entorhinal area and the hippocampal formation. A combined immunohistochemical and fluorescent retrograde tracing study in the rat brain, Neuroscience, 7:951.

Köhler, Ch., 1983, A morphological analysis of vasoactive intestinal polypeptide (VIP)-like immunoreactive neurons in the area dentata of the rat brain, J. Comp. Neurol., 221:247.

Köhler, Ch., and Chan-Palay, V., 1983a, Distribution of gamma-aminobutyric acid containing neurons and terminals in the septal area. An immunohistochemical study using antibodies to glutamic acid decarboxylase in the rat brain, Anat. Embryol., 167:53.

Köhler, Ch., and Chan-Palay, V., 1983b, Somatostatin and vasoactive intestinal polypeptide-like immunoreactive cells and terminals in the retrohippocampal region of the rat brain, Anat. Embryol., 167:151.

Köhler, Ch., Chan-Palay, V., and Wu, J.-Y., 1984a, Septal neurons containing glutamic acid decarboxylase immunoreactivity project to the hippocampal region in the rat brain, Anat. Embryol., 169:41.

Köhler, Ch., Haglund, L., and Swanson, L.W., 1984b, A diffuse αMSH-immunoreactive projection to the hippocampus and spinal cord from individual neurons in the lateral hypothalamic area and zona incerta, J. Comp. Neurol., 223:501.

Köhler, Ch., 1985, Intrinsic connections of the retrohippocampal region in the rat's brain. I. The subicular complex, J. Comp. Neurol., 236:504.

Köhler, Ch., Chan-Palay, V., and Wu, J.-Y., 1985a, Neurons and terminals in the retrohippocampal region in the rat's brain identified by anti-γ-aminobutyric acid and anti-glutamic acid decarboxylase immunocytochemistry, Anat. Embryol., 173:35.

Köhler, Ch. Eriksson, L., Davies, S., and Chan-Palay, V., 1985b, Neuropeptide Y innervation of the hippocampal region in the rat and monkey brain, J. Comp. Neurol., in press.

Köhler, Ch., Haglund, L., Swanson, L.W., and Wu, J.-Y., 1985c, The cytoarchitecture, histochemistry and projections of the tubero-mammillary nucleus in the rat brain, Neuroscience, 16:85.

Leranth, C.S., Frotscher, M., Tömbold, T., and Palkovits, M., 1984, Ultrastructure and synaptic connections of vasoactive intestinal polypeptide-like immunoreactive non-pyramidal neurons and axon terminals in the rat hippocampus, Neuroscience, 12:531.

Lorén, I., Emson, P.C., Fahrenkrug, J., Björklund, A., Alumets, J., Håkansson, R., and Sundler, F., 1976, Distribution of vasoactive intestinal polypeptide in the rat and mouse brain, Neuroscience, 4:1935.

Lorente de Nó, R., 1934, Studies on the structure of the cerebral cortex, II. Continuation of the study of the Ammonic system, J. Psychol. Neurol., 46:113.

McDonald, J.F., Parnavelas, J.G., Karamanlidis, A.N., and Brecha, N., 1982, The morphology and distribution of peptide-containing neurons in the adult and developing visual cortex of the rat. II. Vasoactive intestinal polypeptide, J. Neurocytol., 11:825.

Mesulam, M.M., and Van Hoesen, G.W., 1976, Acetylcholinesterase-rich projections from the basal forebrain of the rhesus monkey to the neocortex, Brain Res., 109:152.

Moore, R.Y., and Card, P.J., 1984, Noradrenaline-containing neuron systems, in: Handbook of Chemical Neuroanatomy, Vol. 2, Part 1, Classical Transmitters in the CNS, A. Björklund and T. Hökfelt, eds., Elsevier, Amsterdam, p. 123.

Peters, A., Miller, M., and Kimmerer, L.M., 1983, Cholecystokinin-like immunoreactive neurons in rat cerebral cortex, Neurosci., 8:431.

Peters, A., and Saint Marie, R.L., 1984, Smooth and sparsely spinous nonpyramidal cells forming local axonal plexuses, in: Cerebral Cortex, Vol. 1, Cellular Components of the Cerebral Cortex, A. Peters and E.G. Jones, eds., Raven Press, New York, p. 419.

Ramon y Cajal, S.R., 1911, Histologie du Systeme Nerveux de l'homme et des vertebrates, Vol. II, A. Maloine, Paris.

Ribak, C.E., Vaughn, J.E., and Saito, K., 1978, Immunocytochemical localization of glutamic acid decarboxylase in neuronal somata following colchicine inhibition of axonal transport, Brain Res., 140:315.

Roberts, G.W., Woodhams, P.L., Polak, J.M., and Crow, T.J., 1984, Distribution of neuropeptides in the limbic system of the rat: the hippocampus, Neuroscience, 11:35.

Seress, L., and Ribak, C.E., 1983, GABAergic cells in the dentate gyrus appear to be local circuit and projection neurons, Exp. Brain. Res., 50:173.

Shipley, M.T., 1975, The topographical and laminar organization of the presubiculum's projection to the ipsi- and contralateral entorhinal cortex in the guinea pig, J. Comp. Neurol., 160:127.

Somogyi, P., Cowey, A., Kisvarday, Z.F., Freund, T.F., and Szentagothai, J., 1983a, Retrograde transport of ^{3}H-GABA reveals specific interlaminar connections in the striate cortex of the monkey, Proc. Natl. Acad. Sci. USA, 80:2385.

Somogyi, P., Smith, A.D., Nunzi, M.G., Gorio, A., Takagi, H., and Wu, J.-Y., 1983b, Glutamate decarboxylase immunoreactivity in the hippocampus of the cat. Distribution of immunoreactive terminals with special reference to the axon initial segment of pyramidal neurons, J. Neurosci., 3:1450.

Somogyi, P., Freund, T.F., Hodgson, A.J., Somogyi, J., Beroukas, D., and Chubb, I.W., 1985, Identified axo-axonic cells are immunoreactive for GABA in the hippocampus and visual cortex of the cat, Brain Res., 332:143.

Steward, O., and Scoville, S.A., 1976, Cells of origin of entorhinal cortical afferents to the hippocampus and fascia dentata of the rat, J. Comp. Neurol., 169:347.

Storm-Mathisen, J., 1977, Glutamic acid and excitatory nerve endings: reduction of glutamic acid uptake after axotomy, Brain Res., 120:379.

Streit, P., 1984, Glutamate and aspartate as transmitter candidates for systems of the cerebral cortex, in: Cerebral Cortex, Vol. 2, Functional Properties of Cortical Cells, E.G. Jones and A. Peters, eds., Raven Press, New York, p. 119.

Swanson, L.W., and Hartman, B., 1975, The central adrenergic system: an immunofluorescent study of the localization of cell bodies and their efferent connections in the rat utilizing dopamine-β-hydroxylase as a marker, J. Comp. Neurol., 163:467.

Swanson, L.W., 1982a, The projections of the ventral tegmental area and adjacent regions: a combined fluorescent retrograde tracer and immunofluorescence study in the rat, Brain Res. Bull., 9:321.

Swanson, L.W., 1982b, The anatomy of the septo-hippocampal pathway, in: Alzheimers Disease. A Report of Progress. Ageing, Vol. 19, S. Corkin, et al., eds., Raven Press, New York.

Swanson, L.W., Sawchenko, P.E., Rivier, J., and Vale, W.W., 1983, Organization of ovine corticotropin-releasing factor immunoreactive cells and fibers in the rat brain: an immunohistochemical study, Neuroendocrinology, 36:165.

Tatemoto, K., and Mutt, V., 1981, Isolation and characterization of the intestinal peptide porcine PHI (PHI-27), a new member of the glucagon secretion family, Proc. Natl. Acad. Sci. USA, 78:6603.

Tatemoto, K., 1982, Neuropeptide Y: complete amino acid sequence of the brain peptide, Proc. Natl. Acad. Sci. USA, 79:5485.

Wyss, J.M., Swanson, L.W., and Cowan, W.M., 1979, A study of subcortical afferents to the hippocampal formation in the rat, Neuroscience, 4:463.

COMMENTARY - THE LIMBIC SYSTEM: NEUROANATOMICAL CONCEPTS RELATING TO EPILEPTIC PHENOMENA

D.G. Amaral and G.W. Van Hoesen

The session on 'Limbic System Neuroanatomy as it Relates to Epileptic Phenomena' covered a range of topics relating primarily to limbic structures of the temporal lobe in mammals. It focused on the organization of the intrinsic and extrinsic connections of the amygdaloid complex and hippocampal formation. These are topics that neuroanatomists can take some pride in since many advances have been made here in recent years. These have served to greatly alter the classical, and somewhat simplistic, view that these structures are involved solely in olfactory sensation or emotional expression. All of the speakers have made substantive contributions to this progress and to the changes in opinion that have accrued.

The decision to concentrate on the temporal lobe was not made arbitrarily or without consideration for the fact that seizure activity can occur at many locations along the neuraxis. However, temporal lobe epilepsy is a commonly encountered form of seizure disorder and for this reason has been the focus of epilepsy research for many decades. Moreover, much of the work on excitatory amino acids has been carried out in the temporal lobe structures since a variety of them and their receptors are found in these regions. Innumerable other physiological, clinical and therapeutic facts likewise support this decision. For example: (1) It is well-known that temporal lobe limbic structures are particularly vulnerable to neurotoxins, antimetabolites, anoxia and ischemia, head trauma and intracranial viral infections. These all have as potential sequelae seizure disorders. (2) It is also well-known that certain temporal lobe limbic structures have unusually low thresholds for seizures, and that these can be triggered by sensory stimuli of various modalities. (3) The occasional intractability of temporal lobe or partial complex seizures to antiseizure medication is also well-recognized by clinicians and ictal and interictal changes in behavior exacerbate the problems of an already debilitating illness. (4) The neurosurgical treatment of intractable temporal lobe epilepsy by lobectomy, and the increased availability of this tissue to the pathologist and basic scientist for various types of study, requires as a back-drop a detailed study of normal temporal lobe anatomy in experimental animals and man. (5) Finally, many phenomena associated with temporal lobe epilepsy afford an opportunity to better understand brain-behavior relationships, and it is not unreasonable to believe that knowledge here will contribute ultimately to a better understanding of other diseases that affect the temporal lobe. Some of the major observations presented in our session, and our comment on their relevance to seizure disorders are summarized below.

While seizure-like activity occurs spontaneously both *in vivo* and in cells in tissue culture and slice preparations, stimulus elicited seizures likewise occur and have been capitalized on in animals as models for epilepsy. These observations suggest that seizure-prone structures, like the amygdala and hippocampus, receive input from parts of the brain that normally play a role in monitoring and interpreting the external and internal environments. There is now abundant anatomical evidence to support the contentions that both the amygdala and hippocampal formation receive inputs from all sensory modalities.

In the case of the amygdala, cortical afferents arise both from unimodal association cortices and from several polysensory or multimodal cortical areas. Cortically derived sensory input is not distributed uniformly to all of the various amygdaloid nuclei, but rather is distributed to only certain of the nuclei or even to subregions of each nucleus. While there is evidence for segregation of the various modalities in some amygdaloid nuclei, it is equally clear that there is also substantial sensory convergence in some regions. Moreover, the abundant intrinsic connections between various nuclei suggest substantial intra-amygdaloid synthesis. In contrast, the hippocampus appears to preferentially receive more highly processed, multimodal sensory information. Neuroanatomical evidence suggests that its sensory input is likely to be derived from neocortical regions, such as the parahippocampal gyrus, which are known to be areas of sensory convergence. Thus, the two structures both receive sensory input, but the directness and presumed level of integration varies substantially. Pharmacologically specific pathways containing acetylcholine, serotonin, dopamine and norepinephrine are well-known inputs of the amygdaloid complex and hippocampal formation.

The behavioral consequences of temporal lobe seizures are often unpredictable. On the one extreme they can lead to tonic-clonic activity if they generalize, whereas in others the involvement of motor structures may be minimal and involve only automatisms. In some cases, behavioral changes may include aggressiveness or other emotional activities, perceptual alterations and a variable memory lapse. A common observation is the experience of visceral sensations with strong autonomic overtones. Endocrine changes are also known to occur but their genesis is understood poorly. Such observations suggest that temporal lobe structures involved in seizures must have efferent projections to those parts of the cortex and thalamus involved in higher cognitive processes, such as perception and memory, and to cortical and subcortical areas that are linked to motor, autonomic and endocrine effectors. The neuroanatomical evidence is strongly suggestive that such structural linkages are present, especially for the amygdaloid complex. The amygdala projects to several basal forebrain and brainstem areas that may be related to motor behavior. There are also projections to various parts of the cingulate cortex that project strongly to the premotor and supplementary motor cortices, direct projections to the nucleus accumbens, which has relays to locomotor centers in the midbrain, and direct projections to the medial reticular formation and the basal nucleus of Meynert. The former may influence descending reticulospinal neurons, whereas the latter, via strong input to the somatomotor cortices, no doubt affects corticospinal neurons.

Neuroanatomical findings supporting the existence of projections from the hippocampal formation and amygdala to the cortex have revealed that these are sizeable and widespread. In the case of the amygdala, the cortical efferents are to well-known sensory-specific and multi-modal association cortices. Those from the hippocampal formation project to adjacent polysensory regions, and through these exert potentially widespread influences on nearly all areas of association cortex. Seizure disorders that activate these pathways may be correlated with the behavioral changes in cognition and memory that accompany limbic seizures.

Finally, recent neuroanatomical observations further strengthen the classical concept of a strong linkage between temporal limbic structures and the hypothalamus, and add important new details. Such connections may play a major role in modulating endocrine and other regulatory mechanisms vested in the ventral diencephalon. Direct projections from the amygdala to autonomic centers in the brainstem, along with indirect projections via the hypothalamus, provide an interesting structural correlate for the autonomic changes that both herald and accompany many temporal lobe seizures.

The identification of putative transmitters and neuromodulators in experimentally defined pathways is a major goal of systems neuroanatomy, and one that holds substantial promise for understanding the normal and pathological functioning of brain structures. While only some of the recent achievements in this important area of chemical neuroanatomy were surveyed in the session, its potential as a progressive step was clearly manifest. For example, recent autoradiographic studies strongly indicate that excitatory amino acids are localized in projection neurons that provide important sources of input to both the hippocampal formation and the amygdala. In addition, it is clear that local circuit neurons in both the amygdala and the hippocampus contain one or more of a variety of identified neuropeptides. It would be expected that these play a role in modulating the flow of information through these structures and ultimately in the output of projection neurons.

In conclusion, this session generated substantial information regarding the cortical and subcortical neural systems that arise from or interact with temporal lobe limbic structures. These structures have been linked with the pathogenesis of temporal lobe epilepsy and are now known to contain several forms of excitatory amino acids and their receptors. Little is known, however, about the chemical mediator and physiological correlates of many of these projections and this must be viewed as a major goal of future research efforts.

SESSION II
EPILEPTIC BRAIN TISSUE: NEUROPATHOLOGY AND PHYSIOLOGY IN ANIMALS AND MAN

NEURONAL AND GLIAL PATHOLOGIES: MORPHOLOGY AND PHYSIOLOGY OF HUMAN AND MONKEY EPILEPTIC FOCI

A.A. Ward, Jr.

Department of Neurological Surgery
University of Washington School of Medicine
Seattle, Washington 98195 U.S.A.

Focal epilepsy in the human is characterized by the occurrence of chronic, spontaneous seizures. The animal model that best approximates the histological characteristics, natural history and electrophysiological properties of the human condition is the chronic monkey model in which the cortical focus is induced by the subpial injection of alumina gel (Lockard and Ward, 1980). The seizures become clinically apparent many weeks after the cortical scarring has stabilized. They are of focal onset, occur spontaneously at random intervals and, once their pattern is established, their frequency is relatively constant and the epilepsy appears to persist indefinitely.

The behavior of neurons in a cortical epileptic focus can be monitored with an extracellular microelectrode. Extracellular recording from cortical neurons in man was first reported in 1955 (Ward and Thomas, 1955). But, even with subsequent refinements in technique, our observations are necessarily limited because of restricted access to synaptic inputs in the human and because of an already long operation under local anesthesia which limits the time which can be devoted to such observations. The behavior of neurons in the focus of awake, undrugged monkeys, on the other hand, can be studied for long periods of time and their behavior can be manipulated by a variety of experimental physiological and behavioral synaptic inputs.

The neurons in such a focus in the monkey show a spectrum of abnormalities ranging from neurons having normal physiological properties through a sequence to those that are highly epileptic and which appear to serve as pacemakers to the focus. Under such circumstances, it appears that the pacemaker neurons are responsible for the focal cortical epileptogenicity. They appear to periodically induce clinical seizures by recruiting surrounding neurons into epileptic activity, thereby producing the necessary critical mass in the circuitry to sustain a propagating discharge.

The physiological properties of neurons in the focus have been extensively described in the past (Wyler and Ward, 1980). In summary, such neurons in both monkey and man fire autonomously in high frequency bursts. Under waking conditions, normal precentral neurons in monkey rarely fire with interspike intervals less than 5 msec. There are strongly epileptic or pacemaker neurons in the focus, on the other hand, which fire almost exclusively in bursts in which the interspike intervals are less than 5 msec (typically 2 msec). Their activity is stereotyped and not easily

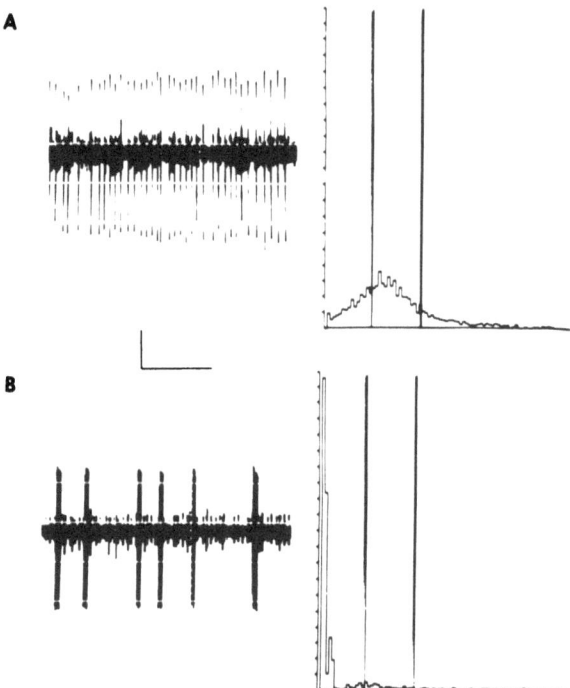

Fig. 1. Action potential and inter-spike interval histograms from a normal neuron (A) and an epileptic pacemaker neuron (B). Histograms are from 3 minutes of spontaneous activity. Calibration: 100 μV and 200 msec; cursors on histograms at 30 and 60 msec. (From Wyler et al., 1980).

modified by any synaptic inputs. The distinction between normal and pacemaker neuronal behavior is shown in Fig. 1. Above we see a sample of activity from a normal pyramidal tract neuron firing at a modestly high, tonic rate. The interspike interval histogram to the right is typical for such cells and shows only a few intervals less than 5 msec. This cell is illustrated because it is firing at a high rate for a normal cell but even so less than 1% of its interspike intervals are less than 5 msec (cursors at 30 and 60 msec).

In contrast, the lower records show an example of a highly epileptic or pacemaker pyramidal tract neuron in a monkey focus. The slow sweep speed demonstrates bursts characteristic of such epileptic neurons. This cell fires only in such bursts. The bursts are composed of interspike intervals less than 5 msec and the interspike interval histogram to the right shows a sharp peak at intervals between 2 and 4 msec. The dramatic difference between the behavior of normal and epileptic neurons is obvious.

Neurons have also been described in the focus that fire with a peculiarly structured burst pattern, a long first interval (LFI; Calvin et al., 1968). This pattern is characterized by an initial action potential followed by an interval that is terminated by an extremely stereotyped burst or 'afterburst' (Fig. 2). Since the first interspike interval is much longer than those of the afterburst, it was named the LFI burst. This long first

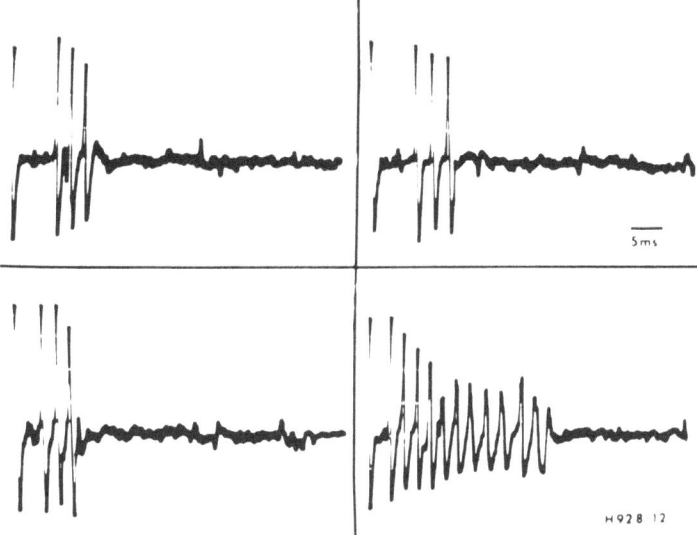

Fig. 2. Examples of structured epileptic burst activity from human epileptic foci. (From Calvin et al., 1973).

interval is often extremely stable, sometimes having a variability of less than 2% of its mean - a stability of timing that is unusual in biological systems. This pattern of neuronal firing is seen in the monkey only in epileptic foci; within the focus, it is seen only in cells identified as pyramidal tract neurons. Such LFI bursts have also been recorded (Fig. 2) in man (Calvin et al., 1973) and are indistinguishable from those recorded in the monkey.

A spectrum of weakly epileptic neurons are widely distributed within the focus and their activity lies between that of physiologically normal neurons and the highly epileptic or pacemaker cells. Their behavior may, at times, be indistinguishable from normal neurons; but, with altered synaptic input, their activity may become progressively epileptic and the majority of their firing may occur in high frequency bursts. If, indeed, the epileptogenicity of the focus is maintained by pacemaker neurons, it would be predicted that the density of such neurons within any one focus would correlate with the frequency of clinical seizures (Wyler et al., 1978). A plot based on the activity of 1,617 neurons recorded from foci in 12 epileptic animals is presented in Fig. 3. Here the ratio of pacemaker neurons in the focus is plotted against the mean daily seizure frequency for each monkey. At the far left is a monkey who was having one seizure about every 10 days and only 1% of the neurons in its focus were highly epileptic. At the top of the curve are two animals showing epilepsia partialis continua with almost continuous jerking of the fingers of the contralateral hand. Here pacemaker neurons constituted 88% and 99%, respectively, of cells in the focus. In the animal having slightly less than one seizure per day, 24% of the neurons were pacemaker cells. This is a log plot and the correlation between this derived index of epileptogenicity and seizure frequency is highly significant ($p = 0.001$). This logarithmic relationship shows the power such pacemaker neurons have on recruitment of normal cells into abnormal activity.

In both the epileptic monkey and man, such neurons are embedded in a dense astrogliosis (Fig. 5). The neurons themselves also exhibit morpho-

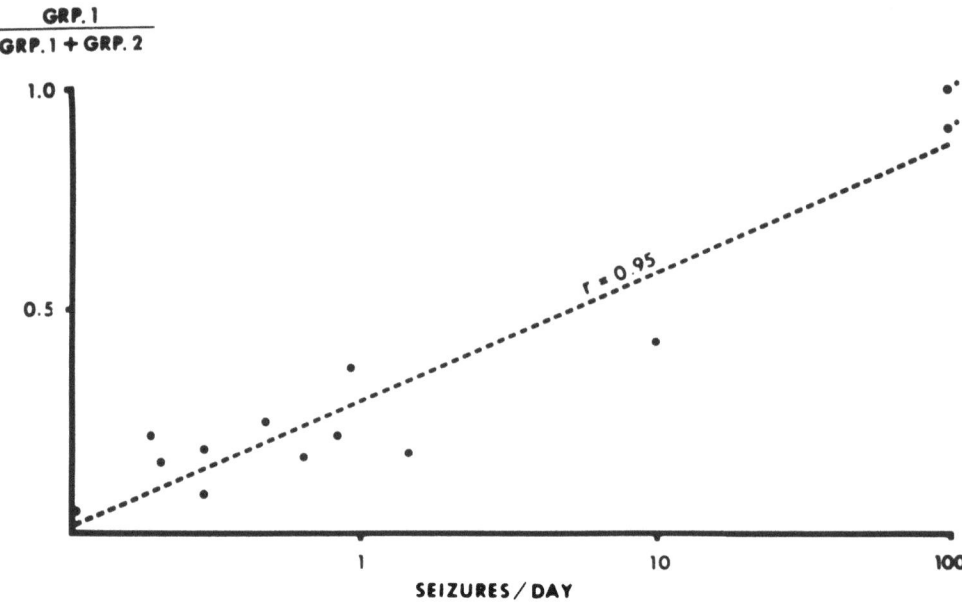

Fig. 3. Logarithmic plot of ratio of pacemaker neurons to all epileptic neurons against seizure frequency for 13 epileptic monkeys. (From Wyler et al., 1978).

logical changes (Westrum et al., 1964) characterized by a reduction or loss of dendritic spines and a reduction of dendritic branching. In Fig. 4, the appearance of neurons from both normal cortex and the epileptic focus is compared. Similar changes have been reported for neurons in human foci (Brown, 1973). There is a consistent loss of GABAergic synaptic endings in the focus of monkey (Ribak et al., 1979; Ribak, 1986). Finally, there is structural evidence that the extracellular space in the focus is reduced (Harris and Jenkins, 1975; Lewis et al., 1977).

High frequency neuronal firing can be predicted to raise the $[K^+]_o$ very significantly. Thus, it is not surprising that increases in $[K^+]_o$ are associated with propagating seizures in monkey neocortex (Sypert and Ward, 1974). From the observation that such increases are significantly less than those predicted by the neuronal firing rate (Lux et al., 1986), it must be assumed that K^+ clearance is unusually effective. Since the extracellular space is apparently reduced in the chronic focus and, in addition, it should shrink even further during seizures (Lux et al., 1986), other mechanisms must be operating. Glotzner (1973) has shown that glia in the chronic monkey epileptic focus are more efficient than normal glia in spatial buffering of potassium. This would be consistent with the observation by Harris (1975) that there is ultrastructural evidence of increased cell to cell coupling in the focus as well as morphological evidence of increased metabolic activity in such glia. Metabolically, astrocytes have been reported to have a greater dehydrogenase activity in foci (Brotchi, 1978). Finally, not only are glia in the focus more efficient in maintaining K^+ homeostasis, but there are many more of them. It may be no accident that gliosis is perhaps the most prominent structural characteristic of epileptic foci in both experimental models and in man. The proposal that the gliosis is an adaptive response to a high K^+ environment would be consistent with the hypothesis that, in the monkey model, the gliosis is maximal in layers 4 and 5 where $[K^+]_o$ is known to be maximal in seizures.

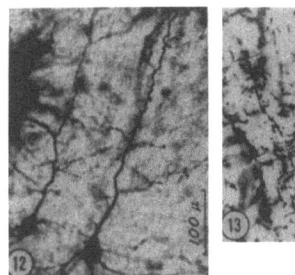
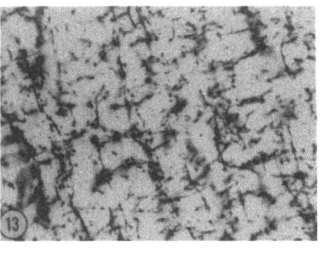

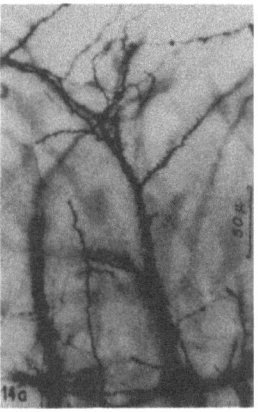
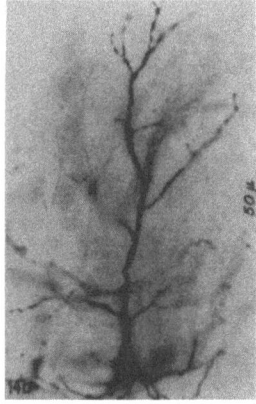

Fig. 4. Morphology of neurons (Golgi impregnation) in normal (14a) and epileptic (B) cortex. 12 demonstrates distortion of dendritic tree in epileptic neurons. 13 documents axonal degeneration (reduced silver impregnation). (From Westrum et al., 1964).

As Lux et al. (1986) have pointed out, epileptic activity is generally associated with considerable alterations in the extracellular ionic constitution as well as with reductions in the extracellular space. Not only are seizures associated with increases in $[K^+]_o$, but also decreases in $[Ca^{2+}]_o$ and both of these support the development of seizure activity. Such factors may well play a role in the recruitment of nearby weakly epileptic cells and ultimately trigger bursting activity in normal neurons. The development of local synchrony of epileptic discharges is poorly understood but such alterations in the ionic environment may well be playing a role.

A second possibly relevant factor may be the reduced extracellular space in the focus. This would tend to not only augment the ionic changes, but it would significantly increase the local concentrations of possible excitatory amino acid transmitters which may be released in association with the high frequency discharges. Finally, the reduced extracellular space might possibly decrease the distances between adjacent neuronal membranes, thereby facilitating a role for ephaptic mechanisms in the development of such synchrony. The field potentials during the onset and propagation of a seizure are of significant magnitude and such spatial factors would augment their role in neuronal synchronization.

In early studies of the morphology of the chronic focus in the monkey,

Fig. 5. Gliosis in cortex excised from epileptic focus in human (Cajal gold chloride sublimate impregnation).

it was noted that, in addition to the alterations in neuronal morphology shown in the Golgi impregnations, reduced silver techniques demonstrated widespread axonal degeneration (Westrum et al., 1964). Subsequent studies (Harris, 1972, 1980) confirmed such axonal degeneration and, furthermore, showed that such degeneration in the focus was specifically related to the occurrence of clinical seizures in the animal. If the chronic, recurrent, clinical seizures were suppressed by suitable medication with an anticonvulsant drug (diphenyl hydantoin) and the seizure-free state maintained a suitable period of weeks so that the degeneration products in cortex could be transported away, such degeneration was no longer visualized. In addition, in those animals having spontaneous seizures, the magnitude of neuronal degeneration appeared to be roughly proportional to the seizure frequency. The conclusion to be drawn is that neurons die during spontaneous seizures. Such degeneration within the focus has also been confirmed in foci in man (Brown, 1973). This observation is of major clinical importance since it signifies that clinical seizures are not innocuous in patients and that efforts to suppress seizures by either medication or surgical excision of the focus should be accomplished early in the course of the disease.

The mechanisms by which spontaneous propagating seizures induce neuronal death is not clear. Recognizing the well established observation that Ca^{2+} appears to move from the extracellular space into neurons during seizure activity, it had been proposed that the $[Ca^{2+}]_i$ might reach toxic levels. However, it has been recently shown that neuronal death induced by certain excitatory amino acids may be the consequence of other mechanisms. Rothman (1985) has recently reported that the toxicity of glutamate, N-methyl-D-aspartate, and kainate was unaffected when calcium was deleted and tetrodotoxin added to balanced salt solution bathing the neurons. He proposes that amino acid neurotoxicity may be based on rather straightforward mechanisms. Chloride is passively distributed, for the most part, across neuronal membrane. Depolarization increases the intracellular concentration, which cannot be compensated by outward anion flux since the bulk of intracellular anions are impermeable. Thus, more cations are drawn into the cell with the concomitant increase in intracellular osmolarity. This leads to water entry and eventually the cell membrane lyses. Any technique which induces significant cell depolarization will involve this sequence. Hence similar findings follow exposure of cells to high external potassium concentrations and veratridine. Glutamate release may contribute to the brain damage in neurons during major epileptic activity.

Such neuronal damage might play a role in the genesis of the gliosis which characterizes the epileptic focus. It is known that neuronal damage or transection of nerve fibers in adult mammalian brain evokes the early appearance of reactive microglia. It has been proposed that such cells may be one source of interleukin-1 in the brain. It is known that the peptide hormone interleukin-1 is a potent mitogen for astroglia but not for oligodendroglia. Thus, Givlian and Lachman (1985) have proposed that interleukin-1 may promote the formation of astroglial scars by astroglia in areas of brain sustaining neuronal damage. Coupling this proposal with the observation that dense gliosis in the experimental focus in the monkey occurs in the middle cortical layers and the evidence of neuronal damage in the focus, one might speculate that it is the neuronal depolarization during seizures and the subsequent cell death that results in gliosis rather than the local accumulation of extracellular potassium. If that is the case, then blocking the initial seizures in a neuronal pool might block the development of a chronic focus. There are clues that patients sustaining significant head injury will not develop post-traumatic epilepsy if they are placed on prophylactic anticonvulsant drugs shortly after trauma and thereby prevented from having seizures for 6-12 months after trauma.

Unfortunately, it is probable that more than one factor is operating in the development of the gliosis which characterizes the focus. The gliosis which follows neuronal death never approaches that seen in Fig. 5. However, it presumably can be induced by the local liberation of interleukin-1. This would be consistent with the dense gliosis which follows brain wounds. Of course, such dense gliosis can occur without the induction of seizures.

Finally, although the environmental effects which have been discussed may well be most significant to understanding the mechanisms of spread of seizures, they do not explain the mechanisms that are intrinsic to those pacemaker neurons which are responsible for the epileptogenicity of the focus. A variety of mechanisms may be proposed. a) There is both physiological and morphological evidence that neurons in the epileptic focus are partially deafferented. Deafferentation induced by lesions can induce autonomous neuronal firing (Loeser and Howe, 1980) and epileptiform discharges can occur some weeks after kainate lesions in hippocampus (Franck and Schwartzkroin, 1983). b) As Schwartzkroin and Wyler (1980) have proposed, some of the morphological changes seen in neurons in the focus might result in conductance shifts which, in turn, give the dendrites a more active role in burst generation (e.g., higher density of calcium channels and closer approximation to initial segment trigger zones).
c) Other factors may include distortion of dendrites by the gliosis in which they are embedded (Ward, 1978) since it appears that dendrites are mechanosensitive.

Finally, it has been pointed out that the epileptic focus is characterized by a loss of GABAergic endings in the focus of the chronic monkey model (Ribak and Harris, 1979). Ribak (1986) has also shown that this is related to the demonstrated 20-30% loss of small GABAergic somata (basket and chandelier cells) in epileptic cortex. Bakay and Harris (1981) reported that CSF over a monkey hemisphere containing an epileptic focus showed a significantly decreased GABA concentration. The epileptic cortex itself also demonstrated markedly reduced GABA receptor binding and diminished tissue GABA concentration and GAD activity. Similar reduction of GABA-mediated synaptic transmission has also been described in human epileptic foci (Lloyd et al., 1981).

Thus, the current data indicate that the autonomous burst activity of neurons in the epileptic focus of the monkey model and man may be related to the reduction of inhibitory synaptic control over pyramidal neurons. At the current level of understanding, it would appear that the factors involved in sustaining the chronic epileptic focus are multiple. However, hypotheses are now available so that refinements of models of mechanisms underlying the epileptic process should proceed more rapidly.

REFERENCES

Bakay, R.A.E., and Harris, A.B., 1981, Neurotransmitter, receptor and biochemical changes in monkey cortical epileptic foci, Brain Res., 206:387.

Brotchi, J., 1978, The activated astrocyte - a histochemical approach to the epileptic focus, in: Dynamic Properties of Glia Cells, E. Schoffeniels, G. Franck, D.B. Towers, and L. Hertz, eds., Pergamon, Oxford, p. 429.

Brown, W.J., 1973, Structural substrates of seizure foci in human temporal lobe, in: Epilepsy, its Phenomena in Man, M.A.B. Brazier, ed., Academic Press, New York, p. 337.

Calvin, W.H., Sypert, G.W., and Ward, A.A. Jr., 1968, Structured timing patterns within bursts from epileptic neurons in undrugged monkey cortex, Exp. Neurol., 21:535.

Calvin, W.H., Ojemann, G.A., and Ward, A.A. Jr., 1973, Human cortical neurons in epileptogenic foci. Comparison of interictal firing patterns to those of 'epileptic neurons' in monkeys, EEG Clin. Neurophysiol., 34:337.

Franck, J.E., and Schwartzkroin, P.A., 1983, Do kainate-lesioned hippocampi become epileptogenic? Brain Res., 329:309.

Givlian, D., and Lachman, L.B., 1985, Interleukin-1 stimulation of astroglial proliferation after brain injury, Science, 228:497.

Glotzner, F.L., 1973, Membrane properties of neuroglia in epileptogenic gliosis, Brain Res., 55:159.

Harris, A.B., 1972, Degeneration in experimental epileptic foci, Arch. Neurol., 26:434.

Harris, A.B., 1975, Cortical neuroglia in experimental epilepsy, Exp. Neurol., 49:691.

Harris, A.B., and Jenkins, D.P., 1975, Intercellular space in epileptic brain, Soc. Neurosci. Abstr., 1:719.

Harris, A.B., 1980, Structural and chemical changes in experimental epileptic foci, in: Epilepsy: A Window to Brain Mechanisms, J.S. Lockard and A.A. Ward, Jr., eds., Raven Press, New York, p. 149.

Lewis, D.V., Matsuga, N., Schuette, W.H., and Van Buren, J., 1977, Potassium clearance and reactive gliosis in the alumina gel lesion, Epilepsia, 18:499.

Lockard, J.S., and Ward, A.A. Jr., 1980, Epilepsy: A Window to Brain Mechanisms, Raven Press, New York.

Lloyd, K.G., Munari, C., Bossi, L., Bancaud, J., Talairach, J., and Morselli, P.L., 1981, Biochemical evidence for the alterations of GABA-mediated synaptic transmission in human epileptic foci, in: Neurotransmitters, Seizures and Epilepsy, P.L. Morselli, K.G. Lloyd, W. Löscher, B. Meldrum, and E.H., Reynolds, eds., Raven Press, New York, p. 331.

Loeser, J.D., and Howe, J.F., 1980, Deafferentation and neuronal injury, in: Epilepsy: A Window to Brain Mechanisms, J.S. Lockard and A.A. Ward Jr., eds., Raven Press, New York, p. 123.

Lux, H.D., Heinemann, U., and Dietzel, I., 1986, Ionic changes and alterations in the size of the extracellular space during epileptic activity, in: Basic Mechanisms of the Epilepsies: Molecular and Cellular Approaches, A.V. Delgado-Escueta, A.A. Ward Jr., D.M. Woodbury, and R.J. Porter, eds., Raven Press, New York, in press.

Ribak, C.E. Harris, A.B., Vaughn, J.E., and Roberts, E., 1979, Inhibitory, GABAergic nerve terminals decrease at sites of focal epilepsy, Science, 205:211.

Ribak, C.E., 1986, Contemporary methods in neurocytology and their application to the study of epilepsy, in: Basic Mechanisms of the Epilepsies: Molecular and Cellular Approaches, A.V. Delgado-Escueta, A.A. Ward Jr., D.M. Woodbury, and R.J. Porter, eds., Raven Press, New York, in press.

Rothman, S.M., 1985, The neurotoxicity of excitatory amino acids is produced by passive chloride influx, J. Neurosci., 5:1483.

Sypert, G.W., and Ward, A.A. Jr., 1974, Changes in extracellular potassium activity during neocortical propagated seizures, Exp. Neurol., 45:19.

Schwartzkroin, P.A., and Wyler, A.R., 1980, Mechanisms underlying epileptiform burst discharge, Ann. Neurol., 7:95.

Ward, A.A. Jr., and Thomas, L.B., 1955, The electrical activity of single units of the cerebral cortex of man, EEG Clin. Neurophysiol., 7:135.

Ward, A.A. Jr., 1978, Glia and epilepsy, in: Dynamic Properties of Glial Cells, F. Schoffeniels, G. Franck, and L. Hertz, eds., Pergamon Press, Oxford, p. 413.

Westrum, L.E., White, L.E. Jr., and Ward, A.A. Jr., 1964, Morphology of the experimental epileptic focus, J. Neurosurg., 21:1033.

Wyler, A.R., Burchiel, K.J., and Ward, A.A. Jr., 1978, Chronic epileptic foci in monkeys: correlation between seizure frequency and proportion of pacemaker epileptic neurons, Epilepsia, 19:475.

Wyler, A.R., and Ward, A.A. Jr., 1980, Epileptic neurons, in: Epilepsy: A Window to Brain Mechanisms, J.L. Lockard and A.A. Ward Jr., eds., Raven Press, New York, p. 51.

METABOLIC, MORPHOLOGIC AND ELECTROPHYSIOLOGIC PROFILES OF HUMAN TEMPORAL LOBE FOCI: AN ATTEMPT AT CORRELATION

T.L. Babb

Department of Neurology
UCLA School of Medicine
Los Angeles, California

HISTORY

In attempting to understand the relationship between brain damage and recurrent seizures, Hughlings Jackson, as early as 1872, pioneered the concept of hyperexcitable tissue adjacent to frontal cortex tumors or infarcts (Jackson, 1931a). He adopted the same explanation to partial complex seizures in which there was an intellectual aura, only slight movements and a 'dreamy state' defect in consciousness (Jackson, 1931b). These partial seizures were localized to the temporal lobe when the necropsy reports identified a temporo-sphenoidal tumor in one epileptic (Jackson and Beevor, 1890) and a softening of the uncinate gyrus in another (Jackson and Coleman, 1898). These cases confirmed earlier correlations of partial seizures and temporal lobe pathologies (Sanders, 1874; Hamilton, 1882; Anderson, 1887). Hence, the Jacksonian concept of hyperexcitable tissue adjacent to neocortical damage was accepted; however, that concept was not extended to the mesial temporal lobe (archicortex), where damage was well-established in epileptics.

The earliest report of hippocampal damage in epilepsy was published by Bouchet and Cazauvieilh (1825) when they described visible or palpable changes of Ammon's horn in postmortem analysis of epileptics classified as having suffered from alienation epilepsy and epileptic aura. However, this report, as well as later reports of hippocampal damage (e.g., Pfleger, 1880; Alzheimer, 1898; Spielmeyer, 1927), viewed the hippocampal damage as a <u>consequence</u> of the epileptic attacks, and the notion that hyperexcitable hippocampal formation could <u>cause</u> seizures when the hippocampus was apparently dead was not accepted. In fact, Spielmeyer's (1927) arguments that hippocampal atrophy was caused by seizure-induced ischemia was further promoted by Scholz (1933) who called the ischemic mechanism 'ictal damage'.

Throughout this period of interest in Ammon's horn, one major paper was misunderstood or ignored by those advocating the view that the hippocampal damage was caused by seizures. In 1880, Sommer (1880) summarized all the available autopsies on epileptics and proposed that the hippocampus had a very low threshold for damage, and that this damage in turn could cause epileptic attacks. This concept, like Jackson's hyperexcitable cortical cells, did not define the physiologic mechanisms for hippocampal seizures; however, it persuaded Bratz (1899) to agree that epileptic attacks were the result of hippocampal damage. Nevertheless, the role of the hippocampus in _generating_ partial seizures became ignored, and such seizures were classified as idiopathic. Ironically, Sommer's excellent cytologic descriptions were well-received, even by those believing the damage was ischemic, and his name was used to define Sommer's sector, the region of prosubiculum that is most damaged in epilepsy.

In 1935, Stauder (1935) reported Ammon's horn damage in 33 of 36 postmortem exams of patients who were classified as temporal lobe epileptics. Stauder's view was that this percentage was so much higher than the percentage of hippocampal damage in other epileptics that the damage must _cause_ the temporal lobe seizures. Shortly thereafter, electroencephalography (EEG) was used to diagnose partial complex seizures. In 1941, Jasper and Kershman (1941) localized sharp waves and a 6 Hz rhythm to the temporal lobe during psychomotor seizures, and prophetically they deduced that '... the temporal lobe and subjacent structures, probably in the archipallium, are the regions primarily involved. This is in accord with the electrographic localization, which so frequently seems to be deep to the temporal lobe (e.g., the hippocampus) near the midline....' (Jasper and Kershman, 1941, p. 932).

Despite this EEG localization in 1941, which suggested that epileptic foci reside in the hippocampus, for 10 years Jasper carried out intra-operative EEG analyses for guiding the surgical excision of temporal lobe foci, and in only 2 of 51 cases was the hippocampus removed (Penfield and Flanigan, 1950). Again, the concept of damaged hippocampus as hyperexcitable was ignored in favor of surgical removal of visible neocortical cysts, tumors or _neocortex_ exhibiting EEG 'spikes'. One interesting finding was that while many of these neocortical excisions cured the focal seizures, as a _group_ only 53% were fully cured. This suggested strongly that the remaining mesial temporal structures (hippocampus) were epileptogenic. In 1952, Penfield and Baldwin (1952) presented their birth-injury model of incisural sclerosis which proposed that the eventual Ammon's horn sclerosis led to seizure generation and that surgical excision must include the hippocampus.

Surgeries for temporal lobe epilepsy which included removal of the hippocampus have led to cures of 68% (Green and Sheetz, 1964), 86% (Falconer et al., 1955) and 80% with large resections but only 50% with small resections (Van Buren et al., 1975). Hence, these surgical reports indicate that typical temporal lobe seizures are generated by epileptic tissue in the hippocampal formation, despite the fact that there is severe loss of neurons in Ammon's horn. More recent studies have made the association between Ammon's horn sclerosis and epileptogenic hippocampus even more conclusive. Falconer and Cavanagh (1959) reported that with clearly-defined temporal lobe seizures, e.g., attacks preceded by an aura, hippocampal sclerosis was present in 90% of the resected temporal lobes. Falconer et al. (1964) further reported that surgical benefit may be as high as 92% when the resected temporal lobe exhibited clear mesial temporal sclerosis. Finally, Babb et al. (1984a, b) demonstrated that when hippocampal seizures were initiated from _only_ the anterior hippocampus, _only_ the anterior hippocampus was sclerotic while the posterior hippocampal neuron densities were not significantly lower than those of non-epileptic controls (Babb et al., 1984b). In addition, when patients were cured by anterior temporal lobectomy (including the hippocampus) there was _more_ damage anteriorly than posteriorly. However, when the hippocampal damage was equally severe back to the posterior line of resection, there was seizure reduction but not cure, indicating remnant _posterior_ hippocampal damage was linked to posterior epileptogenic tissue (Babb et al., 1984a). In conclusion, it appears that damage to the hippocampal formation leads to synaptic reorganization that is epileptogenic. Two questions that remain to be answered are: what altered circuitry can be definitively linked to epileptic tissue, and what physiologic mechanisms generate <u>typical</u> and <u>chronic</u> focal seizures that we can now classify as partial complex seizures originating from mesial temporal lobe?

PATHOLOGY OF TEMPORAL LOBE EPILEPSY

There are numerous types of tissue abnormalities found in temporal lobes resected for relief of psychomotor seizures. Falconer (1970) estimated that approximately 20% of lobes had hamartomas, a generic term which he said included congenital malformations such as glial malformations, tuberose sclerosis, small capillary angiomas and even dermoid cysts. In our experience with over 120 temporal lobes, about 35% have significant abnormalities such as astrocytomas, meningiomas, arachnoid cysts, epidermoid inclusions, Schwannoma, gangliogliomas, oligodendrogliomas, vascular anomalies and heterotopia. The most interesting of these is the heterotopia, which are displaced cells that may be found clustered in temporal lobe white matter or even as deep as the subiculum. Most of the abnormalities are large enough to be considered a focus of damage around which epileptic tissue develops. Accordingly, the hippocampus usually shows little or no cell loss with such extra-hippocampal lesions (Babb et al., 1984a). However, with some clusters of heterotopia, hippocampal sclerosis is commonly found. This suggests that there may be a congenital malformation responsible for both the Ammon's horn sclerosis and the heterotopia elsewhere in the same temporal lobe.

Fig. 1 is a schematic drawing of the normal cytology found in a transverse section of the human hippocampal formation extending from the infolded fascia dentata (FD) around to the medially-situated presubiculum (PRE), which has a five to six-layered cortical arrangement. The major afferent and efferent fiber tracts (ALVEUS and PERFORANT PATH) are shown; however,

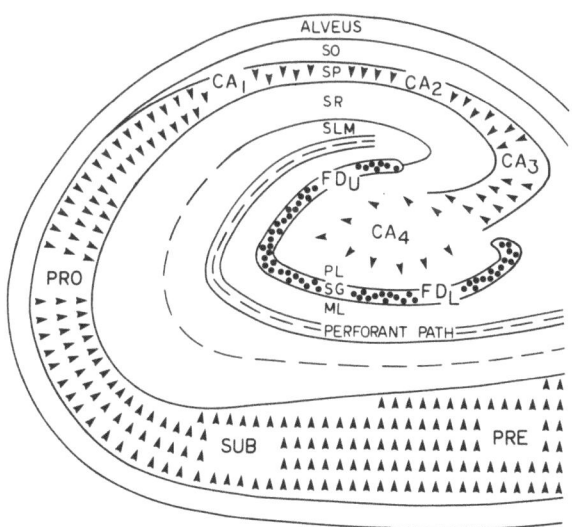

Fig. 1. Schematic of the normal human hippocampus with granule cells (SG) and pyramidal cells (SP) drawn to demonstrate the cytoarchitecture of different Ammon's horn (CA) fields and the subicular complex (PRO, SUB, PRE).

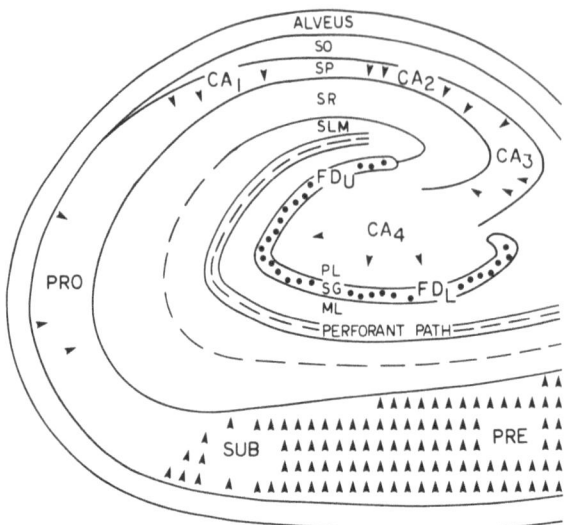

Fig. 2. Schematic drawing of hippocampal sclerosis typically found in resected temporal lobes of psychomotor epileptics. Note greatest loss in Sommer's sector (CA1 and PRO), less loss in SUB and no loss in presubiculum (PRE).

not shown are the dendrites that extend from each cell body (SG, SP) through to the molecular layer (ML, SLM, and above dashed line in subicular complex).

Fig. 2 is a schematic drawing that represents typical Ammon's horn or hippocampal sclerosis. As has been previously reported, Sommer's sector (CA1, PRO) is the most damaged, and there is also significant neuron loss in all other fields of Ammon's horn and fascia dentata. There is often, though not always, some sparing of CA2 neurons. The most interesting finding is that in the subicular complex there is <u>statistically</u> no significant loss in subiculum or presubiculum. This may be of critical physiological importance because this multi-layered subicular cortex may represent hyperexcitable cortex adjacent to the damaged hippocampus or it may represent an important excitatory relay for the hippocampal discharges to propagate to other cortical and sub-cortical areas. For example, it is known that subicular regions project to entorhinal cortex, thalamus, septum (Swanson and Cowan, 1977), cingulate (Finch et al., 1984) and accumbens (Lopes da Silva et al., 1984). These regions in turn have excitatory projections to various subcortical and neocortical areas that would generate the 'visceral' (hypothalamus) and 'cognitive' (temporal gyri) disturbances typical of psychomotor seizures.

MICROANATOMY OF EPILEPTIC HIPPOCAMPUS

One of the most difficult problems to resolve in attempting to define the morphologic basis of epileptic tissue is the fact that the microscopic evidence and some physiologic evidence suggests that the hippocampus is <u>less-excitable</u>, not hyperexcitable. In addition to the previously described cell loss, more detailed studies of hippocampus have revealed degenerated post-synaptic dendritic branches and spines (Scheibel and Scheibel, 1973;

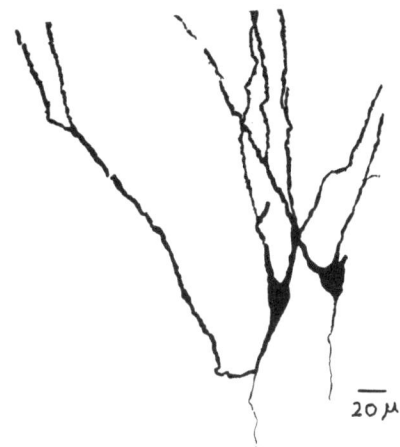

Fig. 3. Golgi-imbedded granule cells from sclerotic, epileptic hippocampus. Note the sparse dendritic branches and lack of spines, both suggesting loss of excitatory efficacy.

Babb and Brown, 1986). For example, Fig. 3 shows the marked loss of complex apical and basilar dendrites in sclerotic fascia dentata as well as loss of spines which are normally abundant in undamaged hippocampus (Fig. 4).

Ultrastructural studies of the epileptic, sclerotic hippocampus have shown that both axosomatic (putative inhibitory) and axodendritic (putative excitatory) synapses may be intact or may be occasionally 'blocked' by glial fibrils in all fields of Ammon's horn. However, as expected from light microscopy of hematoxylin-eosin stains of cells and glia, the presubiculum is free of such glial sheaths (Babb and Brown, 1986). Furthermore, as expected, such glial fibrils near synapses have not been apparent in the few cases examined where the patients had typical psychomotor seizures but no microscopic evidence of cell loss or gliosis. The ultrastructure of such neuropil appeared to show more intact synapses (excitatory and inhibitory) than the sclerotic hippocampus; however, more quantitative studies are needed. In conclusion, all the microscopic studies of neurons, dendrites and synapses suggest that sclerotic hippocampus is relatively inexcitable, a conclusion supported by evidence of increased electrical after-discharge thresholds (Cherlow et al., 1977), reduced pharmacologic activation and reduced glucose metabolism (Engel et al., 1981).

By contrast to this notion of inexcitability, there is strong evidence that despite these synaptic 'deficiencies' the sclerotic hippocampus is 'spontaneously hyperexcitable' (SEEG seizure onset), especially near maximum hippocampal sclerosis (Babb et al., 1984b). Furthermore, microelectrode recordings of viable neurons in sclerotic Ammon's horn often have anomalous firing patterns (e.g., synchronized and/or sustained bursts) which represent hyperexcitability or neuronal epileptogenesis (Babb and Crandall, 1976).

Fig. 5 is normal burst firing from two adjacent neurons in a normal, non-sclerotic anterior hippocampus for comparison with burst patterns often recorded from two neurons in sclerotic, epileptic anterior hippocampus. The normal recordings are typical of hippocampal pyramidal cell burst patterns in lower animals, and we routinely record them contralateral

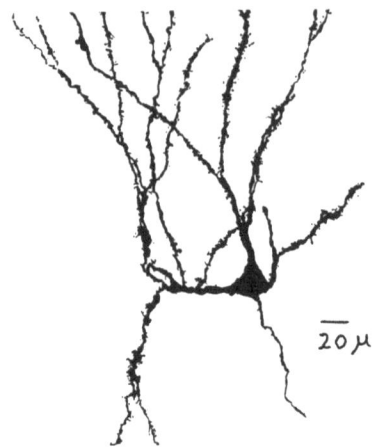

Fig. 4. Golgi-embedded granule cell from temporal lobe epileptic without hippocampal sclerosis. Note the multiple-branching of apical and basilar dendrites, both of which have normal appearing spines, suggesting normal excitatory functions.

to the unilateral epileptic focus or in hippocampi subsequently found to have little or no hippocampal sclerosis (e.g., in extrahippocampal pathologies). The synchronous sustained bursts of epileptic neurons in Fig. 5 do not occur often; however, they may be essential as a synchronizing output that initiates hippocampal seizures. Such synchronization may result from a 'release of inhibition', especially since the hippocampus is known to have such powerful recurrent inhibition. Rebound excitation has been reported extensively as an intrinsic mechanism of firing of hippocampal pyramids (e.g., 'rebound excitation' in Spencer and Kandel, 1961, p. 154 and 'late excitation and repetitive firing' in Spencer and Kandel, 1962, p. 96).

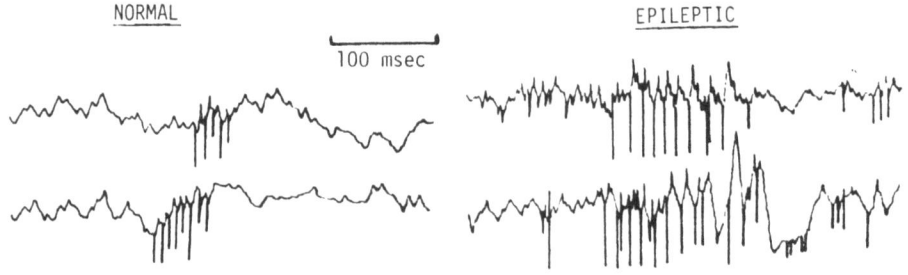

Fig. 5. Examples of spontaneous firing patterns of Ammon's horn pyramidal cells in normal and epileptic hippocampus. In the left column, (normal) pyramidal firing may have bursts with short durations and short interspike intervals. By contrast, in epileptic hippocampus (right column) bursts have similar short interspike intervals but sustained and synchronous firing, perhaps due to release of widespread inhibition.

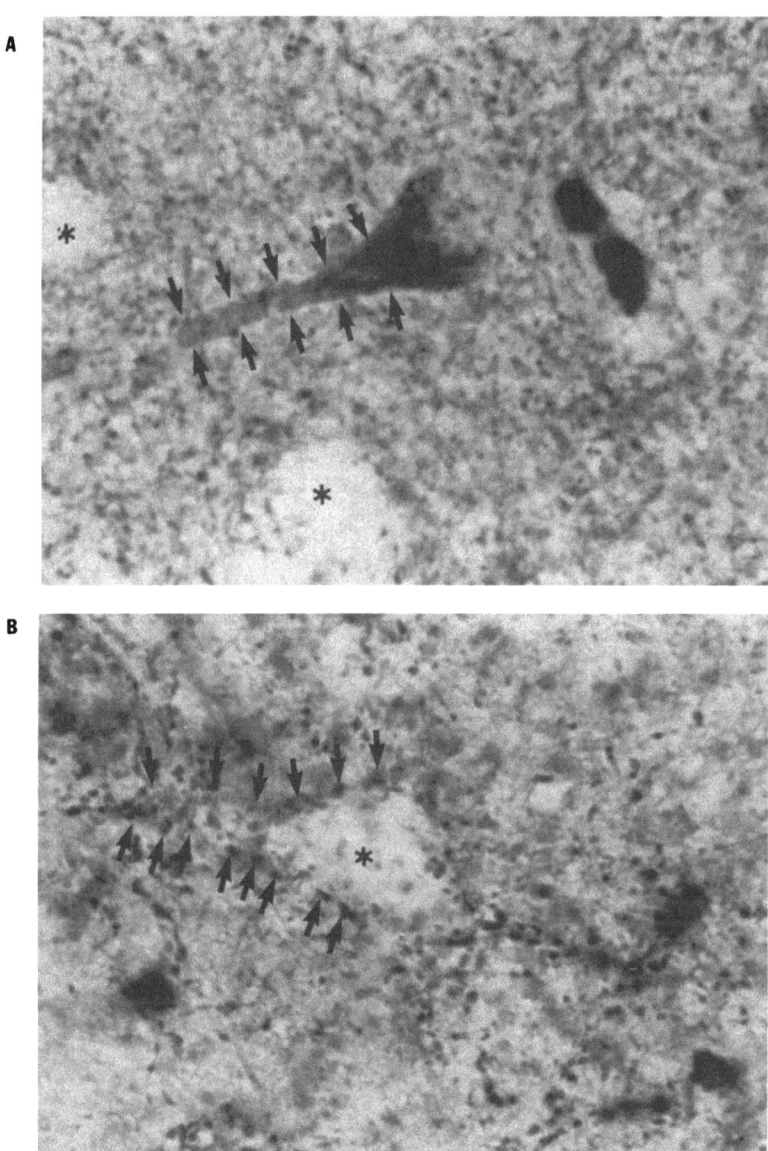

Fig. 6. Prosubiculum. In A there is a darkly-stained inhibitory neuron with an extensive process (small arrow) located in the sclerotic prosubiculum. In B, the inhibitory terminals (GAD puncta) that surround a remnant pyramidal cell (asterisk) extend arround its soma and along its proximal dendrite (small arrows). Hence, sclerotic epileptic tissue has intact inhibitory neurons and terminals on the remnant cells.

In conclusion, although morphologic studies of sclerotic hippocampus
suggest that the neurons there may be less-excitable, the evidence of
hyperexcitability indicates that there are synaptic mechanisms in sclerotic
hippocampus that initiate seizures. We have proposed a model where synchronized bursts from remaining hippocampal neurons project to the cytologically-
preserved presubiculum to excite this large population of neurons which,
in turn, propagate the seizure to numerous other regions. Because there
is both physiological and immunocytochemical evidence (see below) that
inhibitory interneurons function normally or even acutely 'over-inhibit'
pyramids in sclerotic hippocampus, we propose that synchronized sustained
bursts (see Fig. 5) represent one mechanism for seizure genesis.

GABA-ERGIC INHIBITION IN EPILEPTIC HIPPOCAMPUS

Although there is evidence that loss of inhibitory circuits, loss
of releasable GABA, or GABA receptor blockers may induce seizure-like
discharges (for reviews see Fariello et al., 1984 and Bartholini et al.,
1985); nevertheless, in human epileptic hippocampus there is only a mild
or _no loss_ of inhibitory interneurons in sclerotic hippocampus (Babb and
Brown, 1986). For example, compared to _losses_ of principal neurons that
exceed 70% of normal in CA1-prosubiculum, GABA-ergic inhibitory interneurons,
as detected by glutamic acid decarboxylase (GAD) immunocytochemistry,
are _not_ decreased in CA1-prosubiculum (Sommer's sector), the major efferent
to pre-subiculum (Babb, 1985). Figs. 6A and B are photomicrographs of
the presence of typical GAD-positive inhibitory interneurons (A) and terminals (B) in sclerotic, epileptic prosubiculum. In this case, despite
over 70% loss of pyramidal cells, the remaining 30% of the output neurons
have normal, effective recurrent inhibition. Because of the relatively
high _ratio_ of inhibitory interneurons to principal output cells, it is
conceivable that there is a profound inhibition that synchronizes these
remnant cells and results in phase-locked rebound-excitation. Such a
mechanism could be sufficient and may be necessary for triggering a seizure
when there are so few output cells remaining in sclerotic hippocampus.

Another synaptic mechanism that _may_ occur is 'sprouting' or reinnerva-
tion by intra-hippocampal _excitatory_ circuits. We have no evidence yet
of increased 'recurrent-excitation' in sclerotic hippocampus; however,
we have begun to routinely label the major intra-hippocampal pathways
(e.g., mossy fibers, Schaffer collaterals) using the Timm's sulphide procedure. In freshly-cut transverse epileptic hippocampus it is possible
to identify the alveus, the perforant path fibers to middle and outer
molecular layer of the fascia dentata, the CA3/CA4-derived fibers to the
inner molecular layer, the mossy fibers from granule cells to CA3 and
the Schaffer collaterals from CA3 to CA1 stratum radiatum. Although such
enhanced recurrent excitation by mossy fibers on inner molecular layer
has been demonstrated following kainic acid lesions of CA3/CA4 in rats
(Frotscher and Zimmer, 1983; Tauck and Nadler, 1985), we have not yet
observed such a pattern in sclerotic human hippocampus with equivalent
loss of CA3/CA4 neurons.

SUMMARY

Our current understanding of focal seizures strongly suggests a model
of damage and associated synaptic reorganization that leads to periodic
'spontaneous' hyperexcitation and/or sustained firing that generates seizures. Our working model of a human epileptic focus assumes that there
will often be _inexcitability_ near the damage (hippocampus proper); however,

the anomalous circuitry will occasionally lead to <u>hyperexcitability</u> whenever these anomalous (epileptic) circuits are activated <u>synchronously</u> in sufficient numbers to propagate discharges to normal tissue (e.g., the presubiculum) which would be normally excitable.

The key role of <u>rebound excitation</u> following prolonged inhibition in the hippocampus is strongly supported by both physiologic and GAD immunologic results. Our ability to directly test the level of excitability of hippocampal and presubicular neurons <u>in vivo</u>, followed by microanatomical studies of the same resected tissue, will allow us to test our model and revise it as our results become more complete. Also, our findings will be important for relating synaptic mechanisms of seizure genesis to those demonstrated in various experimental models of focal hippocampal seizures.

ACKNOWLEDGEMENTS

This work was possible through the collaborative efforts of 1) Drs. P. Crandall and L. Cahan, who implanted recording electrodes and removed the temporal lobes en bloc, 2) J. Pretorius, L. Reed and M. Hall, who provided expert histology, 3) Drs. C. Wilson and M. Isokawa-Akesson, who performed various physiologic tests and analyses, and 4) all the staff of the Clinical Neurophysiology Project (NIH Grant NS02808). This work would not have been possible without the expert microdissection and microscopic interpretations of temporal lobe pathology made by my close colleague Dr. W.J. Brown. Special thanks are extended to K. Kelly for typing this paper.

REFERENCES

Alzheimer, A., 1898, Ein Beitrag zur pathologischen Anatomie der Epilepsie, <u>Monatsch. Psych. Neurol.</u>, 4:345.

Anderson, J., 1887, On sensory epilepsy: a case of basal tumor, affecting the left temporosphenoidal lobe, and giving rise to paroxysmal taste-sensation and dreamy state, <u>Brain</u>, 9:385.

Babb, T.L., and Crandall, P.H., 1976, Epileptogenesis of human limbic neurons in psychomotor epileptics, <u>Electroenceph. Clin. Neurophysiol.</u>, 40:225.

Babb, T.L., Brown, W.J., Pretorius, J., Davenport, C., Lieb, J.P., and Crandall, P.H., 1984a, Temporal lobe volumetric cell densities in temporal lobe epilepsy, <u>Epilepsia</u>, 25:729.

Babb, T.L., Lieb, J.P., Brown, W.J., Pretorius, J., and Crandall, P.H., 1984b, Distribution of pyramidal cell density and hyperexcitability in the epileptic human hippocampal formation, <u>Epilepsia</u>, 25:721.

Babb, T.L., 1985, GABA-mediated inhibition in the Ammon's horn and presubiculum in human temporal lobe epilepsy: GAD immunocytochemistry, <u>in</u>: Workshop on Neurotransmitters in Epilepsy, G. Nistico, K.G. Lloyd, R. Fariello and P.L. Morselli, eds., Raven Press, New York, in press.

Babb, T.L., and Brown, W.J., 1986, Neuronal, dendritic and vascular profiles of human temporal lobe epilepsy correlated with cellular physiology <u>in vivo</u>, <u>in</u>: Basic Mechanisms of the Epilepsies, A.V. Delgado-Escueta, A.A. Ward, Jr., and D.M. Woodbury, eds., Raven Press, New York, in press.

Bartholini, G., Bossi, L., Lloyd, K.G., and Morselli, P.L., eds., 1985, <u>Epilepsy and GABA Receptor Agonists</u>, L.E.R.S. Monograph Series 3, Raven Press, New York.

Bouchet et Cazauvieilh, Y., 1825, De l'epilepsie consideree dans ses rapports avec l'alienation mentale. Recherche sur la nature et le siege de ces deux Malades; memoire quia remporte le prix au concours establi par M. Esquirol, <u>Arch. Gen. Med.</u>, 9:510.

Bratz, E., 1889, E. Ammonshornbefunde der Epileptischen Krämpfe, Arch. Psychiat. Nervenkrkh., 31:820.

Cherlow, D., Dymond, A.M., Crandall, P.H., Walter, R.D. and Serafetinides, E.A., 1977, Evoked response and after-discharge thresholds to electrical stimulation in temporal lobe epileptics, Arch. Neurol., 34:527.

Engel, J., Rausch, R., Lieb. F., Kuhl, D., and Crandall, P.H., 1981, Correlation of criteria used for localizing the epileptic focus in patients considered for surgical therapy of epilepsy, Ann. Neurol., 9:215.

Falconer, M.A., Hill, D., Meyer, A., Mitchell, W., and Pond, D.A., 1955, Treatment of temporal-lobe epilepsy by temporal lobectomy. A survey of findings and results, Lancet, 1:827.

Falconer, M.A., and Cavanagh, J.B., 1959, Clinicopathological considerations of temporal lobe epilepsy due to small focal lesions, Brain, 82:483.

Falconer, M.A., Serafetinides, E.A., and Corsellis, J.A.N., 1964, Etiology and pathogenesis of temporal lobe epilepsy, Arch. Neurol., 10:233.

Falconer, M.A., 1970, The pathological substrate of temporal lobe epilepsy, Guy's Hospital Report, 119:47.

Fariello, R.G., Morselli, P.L., Lloyd, K.G., Quesney, L.F., and Engel, Jr., J.E., eds., 1984, Neurotransmitters, Seizures and Epilepsy II, Raven Press, New York.

Finch, D.M., Derian, E.L., and Babb, T.L., 1984, Excitatory projection of the rat subicular complex to the cingulate cortex and synaptic integration with thalamic afferents, Brain Res., 301:25.

Frotscher, M., and Zimmer, J., 1983, Lesion-induced mossy fibers to the molecular layer of the rat fascia dentata: identification of postsynaptic granule cells by the Golgi-E.M. technique, J. Comp. Neurol., 215:299.

Green, J.R., and Scheetz, D.G., 1964, Surgery of epileptogenic lesions of the temporal lobe, Arch. Neurol., 10:135.

Hamilton, A.M., 1882, On cortical sensory discharging lesions (sensory epilepsy), New York Medical Journal and Obstetrical Review, 36:575.

Jackson, J.H., and Beever, C.E., 1890, Case of tumor of the right temporosphenoidal lobe bearing on the localization of the sense of smell and on the interpretation of a particular variety of epilepsy, Brain, 12:346.

Jackson, J.H., 1931a, On the anatomical and physiological localization of movements in the brain, in: Selected Writings of John Hughling Jackson, J. Taylor, ed., Hodder and Stoughton, Ltd., London, p. 37.

Jackson, J.H., 1931b, On a particular variety of epilepsy ('intellectual aura'), one case with symptoms of organic brain disease, in: Selected Writings of John Hughling Jackson, J. Taylor, ed., Hodder and Stoughton, Ltd., London, p. 385.

Jackson, J.H., and Coleman, W.S., 1931, Case of epilepsy with tasting movements and 'dreamy state' - very small patch of softening in the left uncinate gyrus, (Brain, 1898, 21:580), in: Selected Writings of John Hughling Jackson, J. Taylor, ed., Hodder and Stoughton, Ltd., London, p. 458.

Jasper, H.H., and Kershman, J., 1941, Electroencephalographic classification of the epilepsies, Arch. Neurol. Psychiat., 45:903.

Lopes da Silva, F.H., Arnolds, D.E.A.T., and Neijt, H.C., 1984, A functional link between the limbic cortex and ventral striatum: physiology of the subiculum accumbens pathway, Exp. Brain Res., 55:205.

Penfield, W., and Baldwin, M., 1952, Temporal lobe seizures and the technique of subtotal temporal lobectomy, Ann. Surg., 136:625.

Pfleger, L., 1880, Beobachtungen über Schrumpfung und Sklerose des Ammonshorns bei Epilepsie, Allg. Z. Psych., 36:359.

Sanders, W., 1874, Epileptische Anfälle mit Geruchs-Empindungen bei Zerstörung des linken tractus olfactorius durch einen Tumor, Arch. Psych. Nervenkrkh., 4:234.

Scheibel, M.E., and Scheibel, A.B., 1973, Hippocampal pathology in temporal lobe epilepsy. A Golgi survey, in: Epilepsy: Its Phenomena in Man, M.A.B. Brazier, ed., Academic Press, New York, p. 311.

Scholz, W., 1933, Über die Entstehung des Hirnbefundes bei der Epilepsie, Z. ges. Neurol. Psych., 145:471.

Sommer, W., 1880, Erkrankung des Ammonshorns als aetiologisches Moment der Epilepsie, Arch. Psych. Nervenkrkh., 10:631.

Spencer, W.A., and Kandel, E.R., 1961, Hippocampal neuron response to selective activation of recurrent collaterals of hippocampofugal axons, Exper. Neurol., 4:149.

Spielmeyer, W., 1927, Die Pathogenese des epileptischen Krampfes, Z. ges. Neurol. Psych., 109:501.

Stauder, K.H., 1935, Epilepsie und Schläfenlappen, Arch. Psych. Nervenkrkh., 104:181.

Swanson, L.W., and Cowan, W.M., 1977, An autoradiographic study of the organization of the efferent connections of the hippocampal formation in the rat, J. Comp. Neurol., 172:49.

Tauck, D.L., and Nadler, J.V., 1985, Evidence of functional mossy fiber sprouting in hippocampal formation of kainic acid-treated rats, J. Neurosci., 5:1016.

Van Buren, J.M., Ajmone-Marsan, C., Mutsuga, N., and Sadowsky, D., 1975, Surgery of temporal lobe epilepsy, in: Neurosurgical Management of the Epilepsies, D.P. Purpura, J.K. Penry, and R.D. Walter, eds., Raven Press, New York, p. 155.

ENDOGENOUS EXCITOTOXINS AS POSSIBLE MEDIATORS OF ISCHEMIC AND HYPOGLYCEMIC

BRAIN DAMAGE

T. Wieloch

Laboratory for Experimental Brain Research
University Hospital, University of Lund
S-221 85 Lund, Sweden

INTRODUCTION

Hypoglycemia and cerebral ischemia are conditions associated with a vast number of complicated clinical situations (Marks, 1972; Graham, 1985). If severe, both hypoglycemia and cerebral ischemia inevitably lead to irreversible neuronal damage (Brierley, 1976). Similarly as in epilepsy, the damage incurred following hypoglycemia and ischemia is distributed to specific vulnerable areas of the brain. However, the distribution of the damage differs between these three conditions (Siesjö and Wieloch, 1986), suggesting that at least in some aspects different pathophysiological mechanisms prevail. Although this selective neuronal damage has been known for several decades (Spielmeyer, 1925), its pathophysiological and molecular mechanisms are largely elusive.

Epilepsy, hypoglycemia and ischemia have many acute metabolic changes in common. The insults lead to membrane depolarization and excessive transmitter release, disturbances in energy metabolism and activation of catabolic processes (Siesjö and Wieloch, 1985a). During recent years, it has become evident that excitatory amino acids (EAA) may play an important role in the pathogenesis of ischemic and hypoglycemic brain damage (for ref. see Wieloch, 1985a). The EAA or related excitotoxins could thus be envisaged to be the common pathogenetic factors in epilepsy, hypoglycemia and cerebral ischemia, and the differences in the distribution of damage could be explained by differences in the site of release, receptor activation, or secondary intracellular reactions such as acidosis. With the advent of small animal models allowing long-term recovery following the insults (Pulsinelli and Brierley, 1979; Auer et al., 1984b; Smith et al., 1984), in vivo studies of these conditions have been possible to accomplish. Such models are necessary since the acute morphological changes following the insults do not entirely reflect the density and distribution of the final neuronal damage observed at the end of longer recovery periods.

This article describes some aspects of the pathophysiology and neuropathology of hypoglycemic and ischemic brain damage in rat models allowing long-term recovery. It discusses the distribution of the damage in relation to the known EAA systems and addresses the question whether excitotoxins (ET) such as the EAA's are involved in the pathogenesis of ischemic and hypoglycemic encephalopathies.

HYPOGLYCEMIA

Neurochemical Alterations

When blood glucose levels reach concentrations below 1 µmol/g following insulin administration, a gradual slowing of the EEG is noted (Lewis et al., 1974). After a period with burst-supression, the EEG becomes isoelectric (Fig. 1). The onset of isoelectricity is associated with an increase in extracellular potassium levels and decrease in calcium levels, signifying a depolarization of cellular plasma membranes (Harris et al., 1984; Fig. 1). The energy stores of the brain are rapidly decreased, and cellular metabolism is shifted towards catabolism of endogenous substrates (Siesjö and Agardh, 1983). Most prominent is the shift of the aspartate aminotransferase equilibrium leading to an increase in the aspartate/glutamate ratio. Thus, in control brains, the levels of glutamate are approximately 11 $\mu mol \cdot g^{-1}$ and aspartate 3 $\mu mol \cdot g^{-1}$, while following 30 minutes of hypoglycemia the levels have shifted to 7 and 3 $\mu mol \cdot g^{-1}$ for aspartate and glutamate, respectively (Siesjö and Agardh, 1983; Wieloch et al., 1985b). Furthermore, an increase in the levels of free fatty acids, in particular arachidonic acid, is noted which suggests that an activation of phospholipases, possibly mediated by receptor activation, takes place (Siesjö and Agardh, 1983; Wieloch et al., 1984).

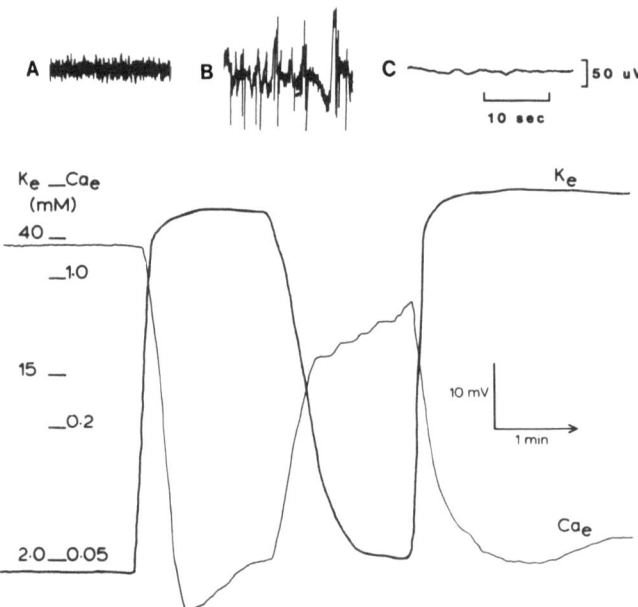

Fig. 1. Top: Changes in the electroencephalogram of an insulin injected rat. A) Control. B) Burst-suppression pattern. C) Isoelectricity. (from Tossman et al., 1985). Bottom: The extracellular cortical activities of calcium (Ca_e) and potassium (K_e) as measured by an ion-selective microelectrode around the onset of isoelectricity during insulin-induced hypoglycemia (from Harris et al., 1984).

The decrease in extracellular levels of calcium at the onset of iso-electricity probably reflects an influx of the ion into intracellular stores. Consequently, a release of transmitter substances can be expected to occur. Using an in vivo dialysis technique, we have demonstrated a release of amino acid transmitters (Tossman et al., 1985; Wieloch et al., 1985a). Thus, the extracellular levels of aspartate increased 15-fold and glutamate 9-fold (Fig. 2). This finding has several implications. It demonstrates that the extracellular levels of EAA's indeed are elevated during hypoglycemia. Moreover, the earlier observed increase in whole tissue aspartate/ glutamate ratio, mainly reflecting the metabolic pool, is also valid for the transmitter pools. Furthermore, it contrasts the changes observed during ischemia, where a relatively larger increase in glutamate is noted (see below).

Distribution of Neuronal Damage

Fig. 3c shows the distribution of neuronal necrosis in the rat brain following 30 minutes of hypoglycemic coma and one week of recovery. The

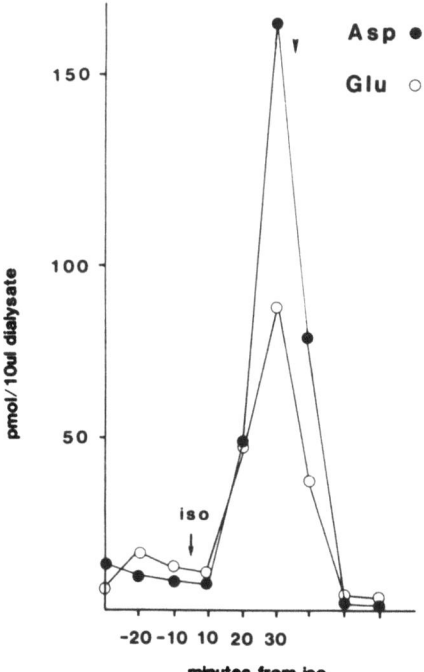

Fig. 2. Changes in the extracellular levels of aspartate and glutamate as measured in an in vivo microdialysis experiment. ISO denotes onset of isoelectricity; GLU: start of infusion of glucose (from Wieloch et al., 1985a).

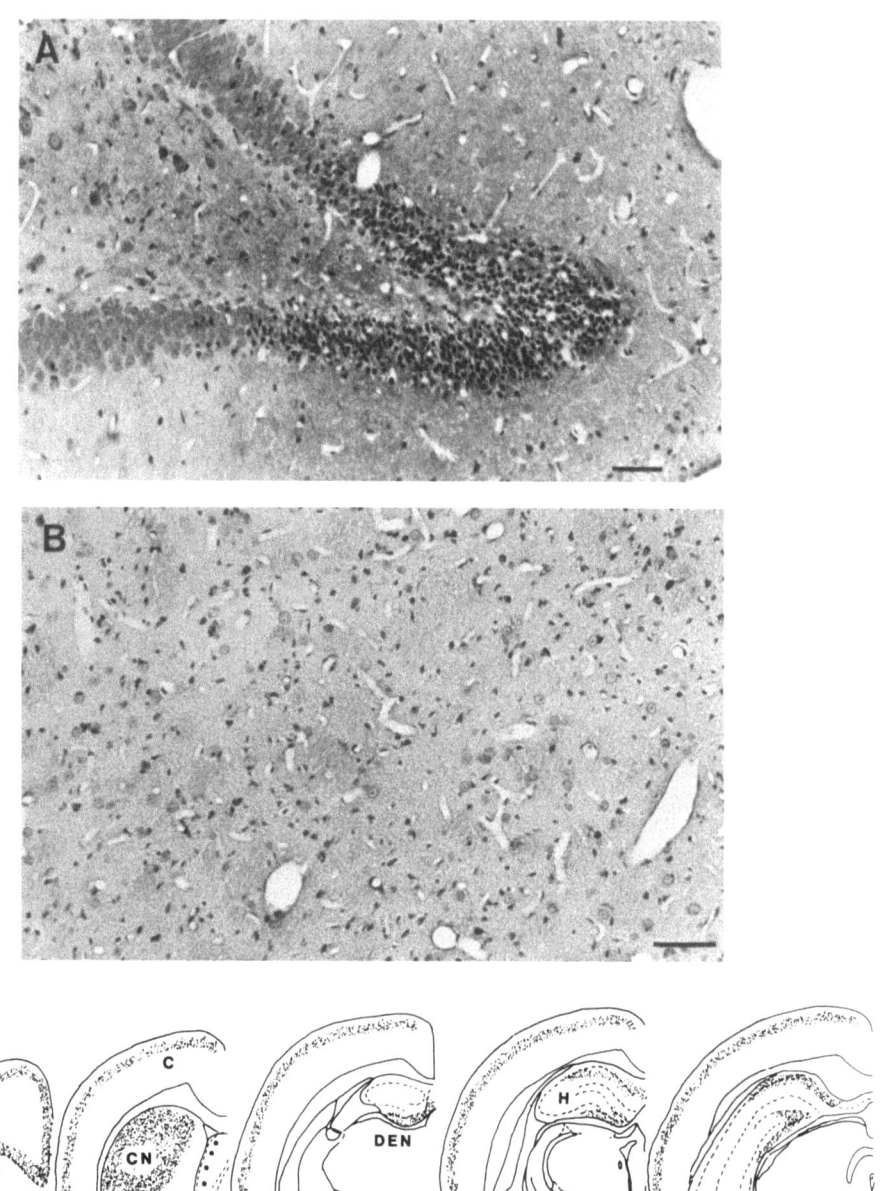

Fig. 3. Neuronal necrosis in the rat brain following 30 minutes of hypoglycemic coma and one week of recovery. A) Photomicrograph of the dentate gyrus. Necrotic granule cells (arrows) are noted at the tip of the dentate gyrus and in the hilus. Bar = 200 µm. B) The dorsal striatum. Approximately 70% of the total neuronal population are necrotic. Bar = 200 µm. C) The distribution of the damage. C: cortex layers 2-3; CN: caudate nucleus; H: hippocampus; DEN: dentate gyrus.

areas most prominently affected are the superficial layers of the cortex, the granule cells at the crest and preferentially the ventral blade of the dentate gyrus (Fig. 3a), and the hippocampal pyramidal cells close to ventricular spaces. Although neuronal damage is confined to areas with high EAA receptor density, there is no obvious correlation between the distribution of the damage and the neuroanatomy of excitatory afferents. It has been suggested that a toxin circulating in the CSF could be responsible for this peculiar distribution of the damage (Auer et al., 1984a).

A second feature of hypoglycemic brain damage is that observed in the striatum (Fig. 3b). This is particularly interesting since the damage is observed first following glucose administration, suggesting that detrimental posthypoglycemic events prevail, such as posthypoglycemic seizure activity (Wilkinson and Prockop, 1976).

Is Hypoglycemic Brain Damage Mediated by an Excitotoxin?

Several pieces of evidence support the notion that hypoglycemic brain damage is mediated by an excitotoxin. First, the damage is not incurred if the EEG pattern has not become isoelectric (Auer et al., 1984b), i.e., there is a prerequisite for a membrane depolarization to occur. Since the membranes are depolarized and the energy levels seriously curtailed, the uptake mechanisms of EAAs are seriously impeded. The extracellular levels of aspartate and glutamate are increased and EAAs can exert their toxic action. Furthermore, the distribution of the damage following hypoglycemia is similar to that observed following intraventricular injections of aspartate or glutamate (Sloviter and Dempster, 1985, see also Sloviter, this volume). In addition, an electronmicroscopical study of the outer molecular layer of the dentate gyrus reveals extensive dendritic swellings and mitochondrial damage while axonal elements are preserved (Auer et al, 1985; Fig. 4). This dendro-somatic axon-sparing lesion is a feature characteristic for excitotoxin-induced brain damage (Olney, 1978; see Olney, this volume).

The caudate nucleus receives a major, presumably glutamatergic, excitatory input from the neocortex (Fonnum, 1984). Since the damage to the striatum is located in the dorsal and lateral aspects of the structure, a correlation of the damage to the cortico-striatal pathways is evident. Thus, ablation of the neocortex overlying the striatum leads to a 10% decrease of striatal glutamate content. If hypoglycemia is induced in lesioned animals, no damage is observed in an area subjacent to the lesion (Wieloch et al., 1985b), demonstrating the importance of synaptic events for the development of hypoglycemic damage. Similarly, injections of a NMDA-antagonist, AP7, into the striatum decreases the damage on the injected hemisphere as compared to the contralateral hemisphere and both hemispheres of saline injected animals (Wieloch, 1985b; Fig. 5).

This demonstrates the involvement of EAA receptors, in particular NMDA receptors, in the induction of the neuronal damage. Thus, there is ample evidence indicating that the mediators of hypoglycemia-induced brain damage are excitatory amino acids or related compounds.

CEREBRAL ISCHEMIA

Neurochemical Alterations

Following complete cessation of blood flow to the brain, electrical activity is abolished within five to ten seconds, and membrane ionic gradients are equilibrated one to two minutes following induction of ischemia. ATP and PCr levels rapidly decrease and within five minutes 90%

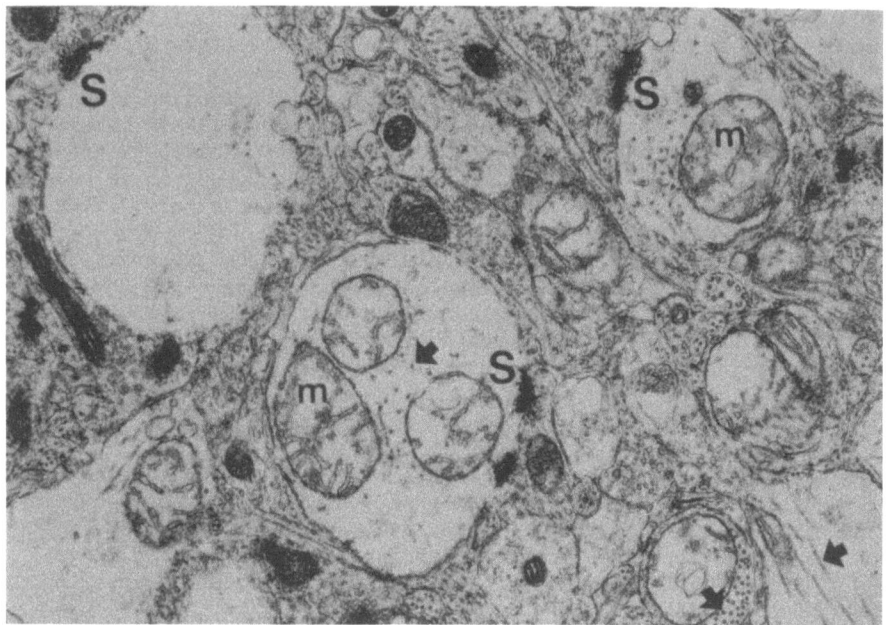

Fig. 4. Electronmicrograph of a portion of the molecular layer of the dentate gyrus following ten minutes of isoelectricity. Note the swollen dendrites with damaged mitochondria, while axonal endings with normal or contracted mitochondria are preserved (from Auer et al., 1985).

of the ATP stores are depleted and catabolic reactions such as lipolysis are activated (Siesjö and Wieloch, 1985b). During the first ten minutes following ischemia, EAA levels are rapidly increased. Thus, glutamate levels increase eight-fold and aspartate two-fold during ischemia and decrease to normal within 20 minutes following recirculation (Benveniste et al., 1984).

Distribution of Neuronal Damage

Similar as in hypoglycemia, ischemia leads to selective neuronal damage (Fig. 6). Following relatively short periods of ischemia, the damage is confined to areas such as the CA_1, CA_4 regions of the hippocampus, subiculum, entorhinal cortex, the lateral septum and olfactory tubercle (Blomqvist and Wieloch, 1985). If ischemia is prolonged, neocortical layers 3 and 4 and the dorsal and lateral caudate nucleus are also affected (Smith et al., 1984). Another feature of ischemic neuronal damage is the delayed neuronal death observed in the CA1 region of the hippocampus. Thus, during two to three days following ischemia the neurons appear normal under the light microscope. On the third or fourth day following ischemia a massive degeneration of cell bodies takes place (Kirino, 1982). Thus, although the acute reactions during ischemia or immediately following ischemia are determinants for the development of ischemic neuronal damage, neuronal degeneration is not manifested until several days following the insult.

Do Excitotoxins Play a Role in Ischemic Brain Damage?

The distribution of damage following short periods of cerebral ischemia coincides with the distribution of EAA receptors in several brain areas (Monaghan et al., 1983). In particular, the distribution of NMDA and

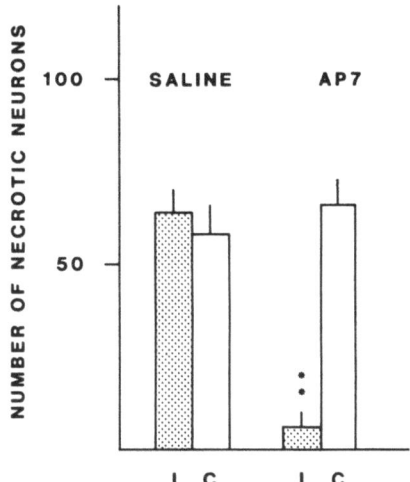

Fig. 5. The number of necrotic neurons in an area of the striatum injected with 40 µg of AP7 (2-amino-7-phosphonoheptanoic acid) and saline, respectively, prior to induction of 30 minutes of hypoglycemic coma. Total number of neurons = 111 ± 4. I: ipsilateral to the injection; C: contralateral to the injection. Values are means ± S.E.M. * $p < 0.01$ (from Wieloch, 1985b).

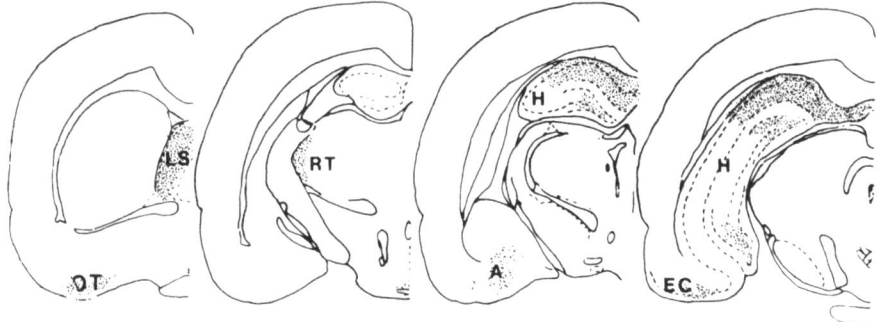

Fig. 6. The distribution of ischemic brain damage in the rat following six minutes of cardiac arrest. H: hippocampus, OT: olfactory tubercle, A: amygdala, ENT CX: entorhinal cortex, LS: lateral septum, RT: reticular nucleus of the thalamus (from Blomqvist and Wieloch, 1985).

quisqualate receptors in hippocampus, septum and cerebral cortex, correlate with ischemic brain damage. Moreover, the distribution of the damage has similarities to that observed following sustained perforant path stimulation (Sloviter and Damiano, 1981). The effects of cortical ablations leading to transections of the main excitatory inputs to the hippocampus on neuronal damage is shown in Fig. 7. In control ischemic animals, approximately 50% of all CA1 neurons were necrotic following six minutes of cardiac arrest and one week of recovery. Transection of the septal input of the hippocampus did not affect the damage. However, cortical ablations leading to lesion of the perforant path fibers decreased the neuronal damage by 50%, suggesting that synaptic events are indeed important for the development of the ischemic damage (Wieloch et al., 1985c).

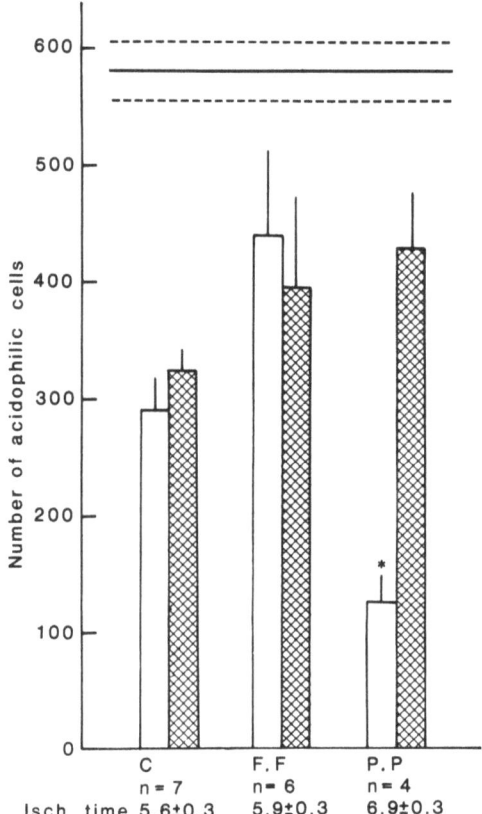

Fig. 7. Number of necrotic neurons in the CA1 region of the hippocampus in control rats (C), rats with fornix lesions (FF), and rats with lesions of the perforant path (PP), subjected to ischemia and one week survival. White and cross hatched bars denote the two different hemispheres. Solid line is the total number of neurons in the counted area. Values are means ± S.E.M. * $p < 0.05$ (from Wieloch et al., 1985c).

The noradrenergic locus coeruleus (LC) system is a principal inhibitory and seizure depressant system in the brain. Lesions of the LC lead to an almost complete degeneration of the CA1 neurons following a six minute ischemic insult, compared to the ischemic control situation where less than 50% neuronal necrosis is seen (Blomqvist et al., 1985; Fig. 8). This demonstrates that inhibitory transmitters and modulators such as noradrenaline, and possibly also adenosine and GABA, modulate the ischemic damage by modulating the deleterious effects of excessive glutamate receptor activation.

Glutamate toxicity has been reported in several *in vitro* studies using pyramidal neuronal cultures exposed to anoxia (Rothman, 1984, 1985; see also Rothman, this volume). Furthermore, the acute morphological changes observed following 30 minutes cerebral ischemia could be ameliorated by an intracerebral injection of a NMDA-receptor antagonist, AP7 (Simon et al., 1984). Thus, evidence prevails suggesting EAAs could be important factors in the development of ischemic brain damage. Still, mechanisms underlying the delayed neuronal death have not been unequivocally proven to be an EAA mediated process.

Two major hypotheses on the mechanisms of ischemic-anoxic cell death have been proposed. Based on experiments performed *in vitro* in hippocampal neuronal cultures, one hypothesis favors the induction of damage by an

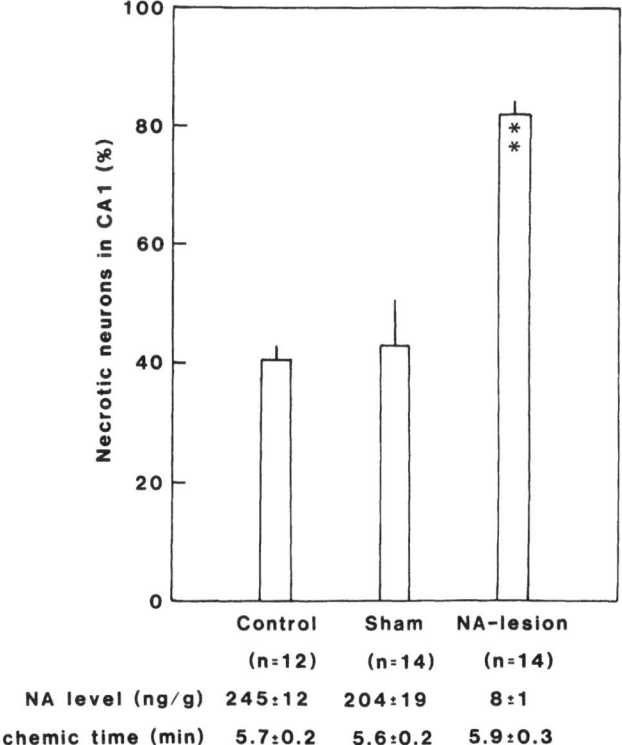

Fig. 8. Neuronal necrosis as percent of the total counted population of neurons in the CA1 region of the hippocampus in control, sham operated and noradrenaline depleted rats, following five to six minutes of cerebral ischemia (from Blomqvist et al., 1985).

EAA-stimulated passive influx of chloride and water into the neurons leading to osmolysis and neuronal degeneration during the anoxic period (Rothman, 1984, 1985). However, pharmacological interventions in the postischemic period have proved to be protective against neuronal damage (Deshpande and Wieloch, 1985). Furthermore, neuronal necrosis is not seen until two to three days following the insult, suggesting that postischemic events are important for the development of the damage (Kirino, 1982). The second hypothesis proposes that an aberrant postischemic calcium homeostasis (Siesjö and Wieloch, 1985a) is the detrimental event inducing lethal catabolic reactions (Schlapfer and Zimmerman, 1985), disturbances in protein synthesis mechanisms (Bodsch et al., 1985), and other regulatory mechanisms of cell function such as protein phosphorylation (Chin et al., 1985), eventually leading to cell death. Still though an unequivocal demonstration of calcium accumulation prior to cell death has not been reported.

Although there is ample circumstantial evidence for an involvement of EAAs in hypoglycemic and ischemic neuronal damage, the differences observed in the distribution of the damage between the two conditions warrant an explanation. At least four major important differences can be speculated upon. First, in hypoglycemia 30-40% of the ATP levels are preserved, which may suppress the deleterious catabolic reactions that are fully activated in the ischemic situation. Second, since there is no bulk CSF flow during ischemia, the transport of a potent toxin via this route cannot be accomplished. Thus, while in hypoglycemia the ET would act mainly along the CSF routes, except perhaps in the case of the caudate nucleus, in ischemia, as in epilepsy, it acts mainly at the site of release. Third, apart from higher extracellular aspartate concentrations during hypoglycemia, quinolinic acid and other breakdown products can be formed, differing in neurotoxic action. Fourth, in ischemia a massive accumulation of lactic acid with a subsequent deleterious decrease in pH takes place, while in hypoglycemia negligible amounts are formed (Siesjö, 1985).

REFERENCES

Auer, R.N., Wieloch, T., Olsson, Y., and Siesjö, B.K., 1984a, The distribution of hypoglycemic brain damage, Acta Neuropathol., 64:177.
Auer, R.N., Olsson, Y., and Siesjö, B.K., 1984b, Hypoglycemic brain injury in the rat, Diabetes, 33:1090.
Auer, R., Kalimo, H., Olsson, Y., and Wieloch, T., 1985, The dentate gyrus in hypoglycemia: pathology implicating excitotoxin-mediated neuronal necrosis, Acta Neuropathol., 67:279.
Benveniste, H., Drejer, J., Schousboe, A., and Diemer, N.H., 1984, Elevation of the extracellular concentrations of glutamate and aspartate in rat hippocampus during transient cerebral ischemia monitored by intracerebral microdialysis, J. Neurochem., 43:1369.
Blomqvist, P., Lindvall, O., and Wieloch, T., 1985, Lesions of the locus coeruleus system aggravate ischemic damage in the rat brain, Neurosci. Lett., 58:353.
Blomqvist, P., and Wieloch, T., 1985, Ischemic brain damage in rats following cardiac arrest using a long-term recovery model, J. Cereb. Blood Flow Metabol., 5:420.
Bodsch, W., Takahashi, K., Barbier, A., Grosse Ophoff, B., and Hossmann, K.-A., 1985, Cerebral protein synthesis and ischemia, Prog. Brain Res., 63:197.
Brierley, J.B., 1976, Cerebral hypoxia, in: Greenfield's Neuropathology 3rd Edition, W. Blackwood and J.A.N. Corsellis, eds., Edward Arnold Publishers, Ltd., London, p. 43.

Chin, J.H., Buckholz, T.M., and DeLorenzo, R.J., 1985, Calmodulin and protein phosphorylation: implications in brain ischemia, Prog. Brain Res., Vol. 63, in press.

Deshpande, J., and Wieloch, T., 1985, Amelioration of ischemic brain damage by postischemic treatment with flunarizine, Neurol. Res., 7:27.

Fonnum, F., 1984, Glutamate: a neurotransmitter in mammalian brain, J. Neurochem., 42:1.

Graham, D.I., 1985, The pathology of brain ischemia and possibilities for therapeutic intervention, Br. J. Anesth., 57:3.

Harris, R.J., Wieloch, T., Symon, L., and Siesjö, B.K., 1984, Cerebral extracellular calcium activity in severe hypoglycemia: relation to extracellular potassium and energy state, J. Cereb. Blood Flow Metabol., 4:187.

Kirino, T., 1982, Delayed neuronal death in the gerbil hippocampus following ischemia, Brain Res., 239:57.

Lewis, L.D., Ljunggren, B., Ratcheson, R.A., and Siesjö, B.K., 1974, Cerebral energy state in insulin-induced hypoglycemia, related to blood glucose and to EEG, J. Neurochem., 23:673.

Marks, V., 1972, Spontaneous hypoglycemia, Br. Med. J., 1:430.

Monaghan, D.T., Holets, V.R., Toy, D.W., and Cotman, C.W., 1983, Anatomical distributions of four pharmacologically distinct ^{3}H-L-glutamate binding sites, Nature, 306:176.

Olney, J., 1978, Neurotoxicity of excitatory amino acids, in: Kainic Acid as a Tool in Neurobiology, E.G. McGeer, J.W. Olney, and P.L. McGeer, eds., Raven Press, New York, p. 95.

Pulsinelli, W.A., and Brierley, J.B., 1979, A new model of bilateral hemispheric ischemia in the unanesthetized rat, Stroke, 10:267.

Rothman, S., 1984, Synaptic release of excitatory amino acid neurotransmitter mediates anoxic neuronal death, J. Neurosci., 4:1884.

Rothman, S.M., 1985, The neurotoxicity of excitatory amino acids is produced by passive chloride influx, J. Neurosci., 5:1483.

Schlapfer, W.W., and Zimmerman, U.-J.P., 1985, Mechanisms underlying the neuronal response to ischemic injury. Calcium-activated proteolysis of neurofilaments, Prog. Brain Res., Vol. 63, in press.

Siesjö, B.K., and Agardh, C.-D., 1983, Hypoglycemia, in: Handbook of Neurochemistry, 2nd Edition, Vol. 3 Metabolism in the Nervous System, A. Lajtha, ed., Plenum Press, New York, p. 353.

Siesjö, B.K., 1985, Acid-base homeostasis in the brain: physioloby, chemistry and neurochemical pathology, Prog. Brain Res., Vol. 63, in press.

Siesjö, B.K., and Wieloch, T., 1985a, Brain injury: neurochemical aspects, in: Central Nervous System Trauma Status Report, D.P. Becker, J.T. Povlishock, eds., William Byrd Press, Inc., Richmond, p. 513.

Siesjö, B.K., and Wieloch, T., 1985b, Cerebral metabolism in ischemia: neurochemical basis for therapy, Br. J. Anaesth., 57:47.

Siesjö, B.K., and Wieloch, T., 1986, Epileptic brain damage, pathophysiology and neurochemical pathology, in: Basic Mechanisms of Epilepsies, A.V. Delgado-Escueta and W. Ward, eds., Raven Press, New York, in press.

Simon, R.P., Griffith, T., Evans, M.C., Swan, J.H., and Meldrum, B.S., 1984, Calcium overload in selectively vulnerable neurons of the hippocampus during and after ischemia: an electron microscopy study in the rat, J. Cereb. Blood Flow Metabol., 4:350.

Sloviter, R.S., and Damiano, B.P., 1981, Sustained electrical stimulation of the perforant path duplicates kainate-induced electrophysiological effects and hippocampal damage in rats, Neurosci. Lett., 24:279.

Sloviter, R.S., and Dempster, D.W., 1985, 'Epileptic' brain damage is replicated qualitatively in the rat hippocampus by central injection of glutamate or aspartate but not by GABA or acetylcholine, Brain Res. Bull., 15:39.

Smith, M.-L., Auer, R.N., and Siesjö, B.K., 1984, The density and distribution of ischemic brain injury in the rat following 2-10 min. of forebrain ischemia, Acta Neuropathol., 64:319.

Spielmeyer, W., 1925, Zur Pathogenese örtlich elektiver Gehirnveränderungen, Z. Ges. Neurol. Psychiatr., 99:756.

Tossman, U., Wieloch, T., and Ungerstedt, U., 1985, GABA and taurine release in the striatum of the rat during hypoglycemic coma, studied by microdialysis, Neurosci. Lett., 62:231.

Wieloch, T., Harris, R., Symon, L., and Siesjö, B.K., 1984, Influence of severe hypoglycemia on brain extracellular calcium and potassium activities, energy charge, and phospholipid metabolism, J. Neurochem., 43:160.

Wieloch, T., 1985a, Neurochemical correlates to selective neuronal vulnerability, Prog. Brain Res., 63:69.

Wieloch, T., 1985b, Hypoglycemia-induced neuronal damage prevented by an N-methyl-d-aspartate antagonist, Science, 230:681.

Wieloch, T., Auer, R.N., Westerberg, E., Tossman, U., Ungerstedt, U., and Engelsen, B., 1985a, Hypoglycemic brain damage is mediated by excitotoxins, in: Excitatory Amino Acids, P. Roberts, J. Storm-Mathisen, and H.F. Bradford, eds., MacMillan Press, London, in press.

Wieloch, T., Engelsen, B., Westerberg, E., and Auer, R., 1985b, Lesions of the glutamatergic cortico-striatal projections in the rat ameliorate hypoglycemic brain damage in the striatum, Neurosci. Lett., 58:25.

Wieloch, T., Lindvall, O., Blomqvist, P., and Gage, F.H., 1985, Evidence for amelioration of ischemic neuronal damage in the hippocampal formation by lesions of the perforant path, Neurol. Res., 7:24.

Wilkinson, D.S., and Prockop. L.D., 1976, Hypoglycemia effects on the central nervous system, in: Handbook of Clinical Neurology, Vol. 27, Elsevier North-Holland Publishing Company, Amsterdam, p. 53.

ROLE OF THE SUBSTANTIA NIGRA IN THE KINDLING MODEL OF LIMBIC EPILEPSY

J.O. McNamara, D.W. Bonhaus, and C. Shin

Departments of Medicine (Neurology) and Pharmacology
Duke University Medical Center and Epilepsy Research
Laboratory
Veterans Administration Medical Center
Durham, North Carolina 27705

INTRODUCTION

The kindling model of epilepsy was discovered relatively recently (Goddard et al., 1969), yet it is already the most extensively studied animal model of epilepsy. Despite extensive study, the mechanisms responsible for kindling are unknown. Our intent in this chapter is: 1) to identify some difficulties confronting investigations of the mechanisms; 2) to describe our studies of the substantia nigra in kindling; and 3) to consider how these studies of the substantia nigra may contribute to understanding the mechanisms of kindling.

KINDLING MODEL

The term 'kindling' refers to a phenomenon in which repeated administration of an initially subconvulsive electrical stimulus results in progressive intensification of seizure activity, culminating in a generalized clonic motor seizure. In rats stimulated in the amygdala, the initial stimulus often elicits focal electrical seizure activity (afterdischarge recorded on the electroencephalogram) without overt behavioral seizure activity. Subsequent stimulations induce the development of seizures, generally evolving through the following classes as described by Racine (1972): 1) facial clonus; 2) head nodding; 3) forelimb clonus; 4) rearing; and 5) rearing and falling. The behavior observed in Classes 1 and 2 is similar to that found in human complex partial or limbic seizures; the behavior in later classes is comparable to limbic seizures spreading to secondarily generalized clonic motor seizures. Hereafter, the term limbic seizure will be used to refer to behavior in which the rat is immobilized with or without associated facial clonus or head nodding; these behaviors are termed limbic seizures when correlated with afterdischarge recorded from an electrode in the limbic system. The term motor seizure will be used to refer to the clonic or tonic activity of the extremities during a Class 4 or 5 seizure.

Once the enhanced sensitivity to electrical stimulus has developed as evidenced by a Class 5 seizure, the animal is said to be 'kindled'. This effect is long lasting and perhaps permanent, since kindled animals left unstimulated for as long as twelve months will respond to one of the first two electrical stimuli with a Class 5 seizure (Goddard et al., 1969). Thus the term 'kindling' implies that the change is permanent.

Kindling can be induced by electrical stimulation of many, but not all, sites in the brain. The amygdala is the structure most commonly used because relatively few stimulations are required to induce kindling; other limbic structures often used include the entorhinal cortex, hippocampus, and pyriform cortex. Kindled seizures have not been elicited with electrodes placed in superior colliculus, reticular formation, or cerebellum (Goddard et al., 1969).

A key question confronting every model is whether its mechanisms are identical to those of the disorder under investigation. Definitive resolution of this question awaits delineation of the mechanisms of both kindling and human epilepsy. The similarities between kindled and human complex partial seizures in their behavioral and electroencephalographic features and in their response to anticonvulsants support the validity of the kindling model (see McNamara et al., 1985, for consideration of this issue).

Investigations over the past two decades have further characterized the phenomenon of kindling but have yet to delineate the underlying mechanisms. Thus we currently lack a definitive understanding of this phenomenon in cellular or molecular terms (McNamara et al., 1985). Application of techniques from multiple scientific disciplines will be required to understand this phenomenon, yet application of these techniques requires knowing where in the brain to look for the critical changes. It is therefore our view that identifying the anatomic network[1] underlying kindling is a necessary first step to elucidating the basic mechanisms.

THE NETWORK OF THE KINDLING MODEL

The idea underlying our approach is that only a tiny fraction of brain neurons need to be modified in order for kindling to develop. Moreover, the alterations responsible for kindling likely reside in multiple sites within a network anatomically remote from the site of stimulation. If this idea is correct, then identifying the key nuclei in such a network would provide an anatomic framework for studies seeking the cellular and molecular mechanisms.

Identification of the structures involved in a kindling network is a complex and ambitious task. One approach is to identify structures activated (i.e., exhibiting afterdischarge in electroencephalographic recordings or altered 2-deoxyglucose uptake) during kindled seizures. The idea is that the structures responsible for initiation and propagation of a kindled seizure may provide clues to the structures underlying kindling itself. Unfortunately, the poor spatial resolution of electroencephalographic recording methods and the poor temporal resolution of the 2-deoxyglucose autoradiographic methods severely limit the information obtainable with these methods. Nonetheless, studies employing these methods have provided a catalog of limbic, basal ganglia, and brain stem structures which appear to be activated during kindled seizures (Racine, 1972; Wada and Sato, 1974; Wada et al., 1975; Wada and Osawa, 1976; Engel et al., 1978; Blackwood et al., 1981; Collins et al., 1983). Indeed the brain regions implicated in kindled seizures by electrophysiologic studies in cats and monkeys

---

[1] The term network will refer to a discrete constellation of brain nuclei interrelated by monosynaptic or polysynaptic connections.

are similar to the areas exhibiting altered 2-deoxyglucose uptake in kindled rats. Whether these structures are directly, or even indirectly, responsible for the behavioral manifestations of kindled seizures is unknown. Activation of only some critical structures may be responsible for the behavior of the seizure and others may be activated as a consequence; in this instance activation of non-essential structures could be merely an epiphenomenon with respect to the seizure. Interventional studies such as lesion experiments or pharmacologic manipulations are required to distinguish these possibilities.

EVIDENCE IMPLICATING SUBSTANTIA NIGRA AS A CRITICAL STRUCTURE IN LIMBIC SEIZURES IN THE KINDLING MODEL

Evidence for the involvement of the substantia nigra in limbic seizures was discovered serendipitously during investigations seeking the anatomic structures responsible for clonic motor seizures in kindled rats. Observations by other investigators suggested that the substantia nigra might be involved in the propagation of motor seizures in the kindling model. Iadarola and Gale (1982) found the substantia nigra to be a key site of GABA mediated anticonvulsant action against seizures induced by electroshock or chemoconvulsants; they found that microinjection of muscimol, an agonist of the inhibitory neurotransmitter GABA, into the substantia nigra decreased susceptibility to motor seizures in these models. Myslobodsky et al. (1979) had previously found that systemic administration of γ-vinyl γ-aminobutyric acid (γ-vinyl GABA; GVG), a GABA transaminase inhibitor, suppressed the clonic motor component of kindled seizures. We therefore hypothesized that GABAmimetic drugs applied to the substantia nigra might suppress clonic motor seizures in the kindling model. If correct, the substantia nigra would then be implicated as a brain stem structure serving a key role in the propagation of clonic motor seizures in the kindling model.

To test this idea, we examined the effects of microinjected drugs on clonic motor and limbic seizures in kindled rats. The duration of clonic motor seizures was determined by timing the movements of the extremities. The duration of limbic seizures was determined by measuring afterdischarge recorded on the electroencephalogram from electrodes placed in limbic structures.

We found that microinjection of the GABA agonist, muscimol, into the area of the substantia nigra bilaterally markedly suppressed clonic motor seizures in animals kindled from amygdala, olfactory structures or lateral entorhinal cortex (McNamara et al., 1984). Microinjection of saline did not suppress the seizures. The suppressant effect of muscimol 1) dissipated after several hours and was dependent on dose; 2) was due to an elevation of the seizure threshold, since typical seizures could be elicited by electrical current far exceeding the threshold; and 3) exhibited spatial specificity since muscimol injections 1-2 mm dorsal to the substantia nigra or into neocortex did not suppress the seizures.

Destruction of brain stem structures was subsequently produced by microinjection of the neurotoxin, N-methyl-D,L-aspartate. As predicted, seizures were dramatically suppressed in animals with bilateral destruction of the substantia nigra but not in animals in which the substantia nigra was spared.

In addition to these effects on clonic motor seizures, muscimol microinjections into the regions of the substantia nigra or destruction of the substantia nigra also suppressed limbic seizures. This finding was unexpected. The suppression of limbic seizures was apparent from both behavioral and electroencephalographic observations. The duration of afterdischarge recorded from electrodes in the limbic system was reduced by approximately 85% in comparison to the duration in vehicle injected controls or animals in which brain stem lesions spared the substantia nigra (McNamara et al., 1984).

Thus, two pieces of evidence implicate the substantia nigra as a pivotal site responsible for suppression of kindled motor and limbic seizures: 1) the spatial selectivity of the microinjected GABA receptor agonist, muscimol, together with the fact that the substantia nigra is the principal site of GABA receptors in ventral midbrain; and 2) the correlation between seizure suppression and neurotoxin-mediated destruction of the substantia nigra.

Our results together with those of others suggest that reduction of neuronal activity within the substantia nigra underlies the seizure-suppressant effect. One, frank destruction of substantia nigra, undoubtedly eliminates its output and this has anticonvulsant effects. Two, microinjection of muscimol almost certainly reduces the output from substantia nigra. Iontophoresis of muscimol suppresses nigral unit activity (Waszczak et al., 1980). Although microinjection of large amounts (50 ng) of muscimol could differ from iontophoretic application in its effects on nigral output, behavioral observations and single unit recordings in structures receiving nigral projections indicate that microinjected muscimol does reduce nigral output (Hikosaka and Wurtz, 1985). Three, the effects of microinjection of other compounds into substantia nigra further suggest that decreasing nigral excitability has anticonvulsant effects. Thus nigral injection of antagonists of the putative excitatory transmitter, substance P, suppresses electroshock seizures (Gale, 1985). Intranigral injection of clonazepam and GVG, drugs which presumably enhance the inhibitory effects of GABA, also suppress seizures (Iadarola and Gale, 1982; King and McNamara, 1984; Le Gal La Salle et al., 1983; McNamara et al., 1983, 1984).

ELECTROPHYSIOLOGIC EVIDENCE THAT SUBSTANTIA NIGRA NEURONS ARE ACTIVATED DURING KINDLED SEIZURES

The anticonvulsant actions of the various nigral manipulations cited in the preceding paragraph implicate the substantia nigra in the generation of seizures. The role of the substantia nigra almost certainly involves propagation, not initiation, since kindled seizures presumably initiate in limbic structures undergoing stimulation. The mechanism by which the substantia nigra promotes seizure propagation is unknown.

One mechanism by which the substantia nigra could promote seizure propagation would involve a tonic net seizure facilitatory action on other pathways through which the seizure is transmitted. Alternatively the substantia nigra itself may transmit seizure activity from rostral sites of origin to target structures. To begin to investigate these possibilities, we recorded single unit activity in the substantia nigra through the course of electrically induced seizures in both naive and kindled rats. We reasoned that demonstrating the absence of a clear change in nigral unit firing rate or pattern during a kindled seizure would exclude the second possibility.

In collaboration with Dr. Judy Walters at the National Institutes of Health, we performed extracellular recordings in substantia nigra in both naive and kindled rats in a paralyzed, ventilated condition (Bonhaus et al., 1986). The principal finding was that nigral units of kindled animals exhibited a dramatic change in firing pattern during the seizure. The pattern consisted of the cells firing in bursts of action potentials; moreover, the bursts were time-locked to components of afterdischarge recorded in the electroencephalogram from electrodes in either amygdala or caudate. The likelihood that the nigral units in pars reticulata would fire in bursts during the afterdischarge was far greater in kindled (19 of 22 units, 88%) than in naive (3 or 15 units, 20%) (p < .05) rats. A similar trend was observed for the dopamine cells recorded.

The latency from the time of onset of the afterdischarge to the time at which the activity of the cell became locked to a component of the afterdischarge was variable. The burst firing of nigral units began 0 to 30 seconds after initiation of the afterdischarge in different animals. However, once nigral units began firing in bursts, this pattern always continued to the end of the afterdischarge.

To determine whether the burst firing occurred in neurons throughout the brain stem, units were recorded in additional kindled animals during afterdischarge in the paralyzed ventilated condition. Only one of fifteen cells recorded in the pontine reticular formation caudal to the substantia nigra fired in bursts during afterdischarge. Only four of twenty-three locus ceruleus neurons fired in bursts during afterdischarge. Thus from the limited number of structures studied to date, burst firing during afterdischarge was found in the majority of neurons only in the substantia nigra. It is clear that burst firing of neurons during afterdischarge is not a general phenomenon throughout the brain stem.

Identifying this change in firing patterns suggests that the seizure facilitatory action of the substantia nigra is mediated by directly transmitting seizure information from rostral sites of origin to target structures. Importantly, this does not exclude the possibility of a tonic seizure facilitatory action on target structures which themselves transmit the activity. Understanding the temporal relationship between the nigral burst firing and behavioral features of the seizures was not possible in the paralyzed ventilated preparations; such correlative observations could aid in defining the role of the nigral burst firing in the behavioral expression of the seizure. Likewise examining the anticonvulsant effects of a compound which eliminates nigral burst firing without modifying the tonic firing rate could further aid in delineating the mechanism of the substantia nigra's seizure facilitory action.

EVIDENCE IMPLICATING SUBSTANTIA NIGRA IN THE DEVELOPMENT OF KINDLING

The evidence implicating the substantia nigra in propagation of kindled seizures led us to hypothesize that the substantia nigra also contributed to the development of kindling. To address this question, we took advantage of the relatively prolonged course of action of intranigral GVG. This compound presumably produces its anticonvulsant effects through inhibition of GABA transaminase, an enzyme which degrades GABA. Previous studies demonstrated that intranigral GVG produced anticonvulsant effects between 16 and 72 hours following injection (Iadarola and Gale, 1982; McNamara et al., 1984). We therefore examined the rate of kindling development produced by amygdala stimulation at 90 minute intervals commencing 16 hours following microinjection of GVG into the substantia nigra bilaterally (Shin et al., 1985). We found that intranigral GVG produced a marked (70%) suppression of the rate of kindling development in comparison to vehicle injected controls (p < .05). This effect exhibited spatial specificity, since injection of GVG 1-2 mm dorsal to the substantia nigra did not modify the rate of kindling development. We interpret these results to indicate that suppressing neuronal excitability within the substantia nigra retards kindling development.

Preliminary results of studies of destruction of the substantia nigra by an excitotoxin or by thermocoagulative lesions suggest that elimination of the nigra does not retard the rate of kindling development (Shin et al., 1985). One explanation for the discrepant results of the pharmacologic and lesion experiments is that an intact substantia nigra is part of the network of kindling development and GVG, by presumably suppressing nigral excitability, retards kindling. However, in the absence of the nigra, kindling may proceed through an alternate network. Kindling through this alternate network may proceed as efficiently as through the network which exists in the unlesioned brain. Once kindling is established through the intact substantia nigra, the substantia nigra becomes a critical structure in seizure propagation, thereby accounting for the anticonvulsant effects of both nigral lesions and intranigral GVG in kindled animals.

IMPLICATIONS OF NIGRAL STUDIES FOR UNDERSTANDING KINDLING

A particularly striking finding of the present work is that focal application of a GABA agonist to the substantia nigra can suppress limbic seizure activity including electrical afterdischarge at the site of stimulation. Thus the substantia nigra is not simply regulating the expression of clonic motor seizures generated by caudal structures. Rather this nucleus in the midbrain is altering the intrinsic neuronal excitability of multiple sites in the cerebral hemispheres including lateral entorhinal cortex, amygdala, and olfactory structures.

Precisely which neuronal constituents within the substantia nigra are responsible for this action is not clear. The dopaminergic neurons are not likely to be involved, since neither dopamine receptor agonists nor antagonists suppress kindled seizures (Babington and Wedeking, 1973). The much greater sensitivity to muscimol of neurons in pars reticulata than in pars compacta (Waszczak and Walters, 1980) supports pars reticulata as the key site. The pars reticulata also seems more likely in view of its extensive, often bilateral and collateralized, projections to numerous thalamic and brainstem nuclei (Gerfen et al., 1982; Parent et al., 1983); these connections render it capable of modulating neuronal activity diffusely throughout the brain.

This regulatory influence is almost certainly mediated through polysynaptic pathways, since no direct connections either to or from the substantia nigra on the one hand and olfactory structures and lateral entorhinal cortex on the other have been identified (Gerfen et al., 1982). Whether the interactions between amygdala and the substantia nigra are mediated through polysynaptic connections is less clear because reciprocal connections do exist between the central nucleus of the amygdala and pars lateralis of the substantia nigra (Bunney and Aghajanian, 1976; Loughlin and Fallon, 1983).

The inhibitory effects of intranigral GVG on kindling development strengthen the possibility that this structure is part of the network underlying kindling itself in the intact (i.e., unlesioned) brain. Understanding the connections mediating the interactions between the substantia nigra and limbic structures may provide a clue to other key sites in the network. Likewise, understanding the mechanism of the substantia nigra's seizure facilitating action may shed light on how the network functions during the evolution of kindling and in the kindled condition.

ACKNOWLEDGEMENTS

This work was supported by VA Medical Research funds and a postdoctoral fellowship award NS 07614 and grant NS 17771 from the National Institutes of Health. The authors thank Mrs. Eloise Pittman for her assistance in preparation of the manuscript.

REFERENCES

Babington, R.G., and Wedeking, P.W., 1973, The pharmacology of seizures induced by sensitization with low intensity brain stimulation, Pharmacol. Biochem. Behav., 1:461.

Bonhaus, D.W., Walters, J.R., and McNamara, J.O., 1986, Activation of substantia nigra neurons: role in the propagation of seizures in kindled rats, J. Neurosci., in press.

Blackwood, D.H.R., Kapoor, V., and Martin, M.J., 1981, Regional changes in cerebral glucose utilization associated with amygdaloid kindling and electroshock in the rat, Brain Res., 224:204.

Bunney, B.S., and Aghajanian, G.K., 1976, The precise localization of nigral afferents in the rat as determined by a retrograde tracing technique, Brain Res., 117:423.

Collins, R.C., Tearse, R.G., and Lothman, E.W., 1983, Functional anatomy of limbic seizures: focal discharges from medial entorhinal cortex in rat, Brain Res., 280:25.

Engel, J., Jr., Wolfson, L., and Brown, L., 1978, Anatomical correlates of electrical and behavioral events related to amygdaloid kindling, Ann. Neurol., 3:538.

Gale, K., 1985, Mechanisms of seizure control mediated by gamma aminobutyric acid: role of the substantia nigra, Fed. Proc., 44:2414.

Gerfen, C.R., Staines, W.A., Arbuthnott, G.W., and Fibiger, H.C., 1982, Crossed connections of the substantia nigra in the rat, J. Comp. Neurol., 207:283.

Goddard, G.V., McIntyre, D.C., and Leech, C.K., 1969, A permanent change in brain function resulting from daily electrical stimulation, Exp. Neurol., 25:295.

Hikosaka, O., and Wurtz, R.H., 1985, Modification of saccadic eye movements by GABA-related substances. II. Effects of muscimol in monkey substantia nigra pars reticulata, J. Neurophys., 53:292.

Iadarola, M.J., and Gale, K., 1982, Substantia nigra: site of anticonvulsant activity mediated by γ-aminobutyric acid, Science, 218:1237.

King, P.H., and McNamara, J.O., 1984, Microinjection of clonazepam into substantia nigra suppressed kindled seizures, Soc. Neurosci. Abstr., 10:343.

Le Gal La Salle, G., Kajima, M., and Feldblum, S., 1983, Abortive amygdaloid kindled seizures following microinjection of γ-vinyl-GABA in the vicinity of substantia nigra in rats, Neurosci. Lett., 36:69.

Loughlin, S.E., and Fallon, J.H., 1983, Dopaminergic and non-dopaminergic projections to amygdala from substantia nigra and ventral tegmental area, Brain Res., 262:334.

McNamara, J.O., Rigsbee, L.C., and Galloway, M.T., 1983, Evidence that substantia nigra is crucial to neural network of kindled seizures, Eur. J. Pharmacol., 86:485.

McNamara, J.O., Galloway, M.T., Rigsbee, L.C., and Shin, C., 1984, Evidence implicating substantia nigra in regulation of kindled seizure threshold, J. Neurosci., 4:2410.

McNamara, J.O., Bonhaus, D., Crain, B.J., Gellman, R.J., Giacchino, J.L., and Shin, C., 1985, The kindling model of epilepsy: a critical review, CRC Crit. Rev. Neurobiol., 1:341.

Myslobodsky, M.S., Ackerman, R.F., and Engel, J., 1979, Effects of γ-acetylenic GABA and γ-vinyl GABA on metrazol-activated and kindled seizures, Pharmacol. Biochem. Behav., 11:265.

Parent, A., Mackey, A., Smith, Y., and Boucher, R., 1983, The output organization of the substantia nigra in primate as revealed by a retrograde double labeling method, Brain Res. Bull., 10:529.

Racine, R.J., 1972, Modification of seizure activity by electrical stimulation. II. Motor seizure, Electroencephalogr. Clin. Neurophysiol., 32:281.

Shin, C., Silver, J.M., Bonhaus, D.W., and McNamara, J.O., 1985, The role of substantia nigra in the development of kindling, Soc. Neurosci. Abstr., 11:40.

Wada, J.A., and Sato, M., 1974, Generalized convulsive seizures induced by daily electrical stimulation of the amygdala in cats: correlative electrographic and behavioral features, Neurology, 24:565.

Wada, J.A., Sato, M., and McCaughran, J.A., Jr., 1975, Cortical electrographic correlates of convulsive seizure development induced by daily electrical stimulation of the amygdala in rats and cats, Folia Psychiatr. Neurol. Japonica, 29:329.

Wada, J.A., and Osawa, T., 1976, Spontaneous recurrent seizure state induced by daily electric amygdaloid stimulation in Senegalese baboons (Papio papio), Neurology, 26:273.

Waszczak, B.L., Eng, N., and Walters, J.R., 1980, Effects of muscimol and picrotoxin on single unit activity of substantia nigra neurons, Brain Res., 188:185.

LONG TERM SEQUELAE OF PARENTERAL ADMINISTRATION OF KAINIC ACID

L. Nitecka[1,2] and E. Tremblay[1]

LPN, CNRS, Gif-sur-Yvette, F-91190 France
[1]Present Address: U29 INSERM, 123 Bd de Port-Royal
75014 Paris, France
[2]On leave from the Department of Anatomy of the Medical
Academy of Gdansk, Poland

INTRODUCTION

Neuronal loss and the development of gliotic scars in various brain regions, in particular in limbic structures, following administration of kainic acid (KA) has been extensively described in the literature (Olney et al. 1979; Ben-Ari et al., 1980; Nadler et al., 1980; Schwob et al., 1980; Ben-Ari et al., 1981; Lothman and Collins, 1981; French et al., 1982; Nitecka et al., 1984; Ben-Ari, 1985). The pattern of brain damage is reminiscent of that seen in human temporal lobe epileptics (Falconer et al., 1964; Margerison and Corsellis, 1966; Brown, 1973; Scheibel and Scheibel, 1973; Scheibel et al., 1974; Dam, 1980).

These and other observations (e.g., Cavalheiro et al., 1982; Franck and Schwartzkroin, 1983; Lancaster et al., 1983; Franck, 1984; Tremblay and Ben-Ari, 1984; Ben-Ari, 1985) suggest that administration of KA is a good model of temporal lobe epilepsy. Human epilepsy is a long lasting disorder with a complex relationship between pathological processes and seizures. It is therefore essential to determine in detail the long term sequelae produced in animal models notably in respect to the occurrence of spontaneous seizures and late pathological processes.

This report will focus on the long term pathological changes occuring in animals showing spontaneous limbic motor seizures (LMS). The pathological alteration in animals which did not display overt epileptic behavior during two to eight weeks after kainate injections will also be described in short. A comparison of both patterns of brain lesions may contribute to the understanding of the process of epileptogenesis.

EXPERIMENTAL PROCEDURE

The present study is based on the histological examination of eight rat brains after systemic administration of KA (9 mg/kg). The behavioral changes were observed directly after injections and occasionally during the survival time. After various survival time periods (one to eight weeks), the animals were deeply anesthetized with equithesin (Jensen Salisbury, 4 ml/kg i.p.), and brains fixed by intracardial perfusion with

10% formalin or a mixture of formalin, 40% acetic acid and methanol. Paraffin sections (10 μm) were cut in frontal plane throughout the entire brain and stained with cresyl violet. In one case (case 1455), the location of every pathological change was depicted in camera lucida drawings (Fig. 1).

RESULTS

The pattern of the degeneration will be described with special reference to cases which have developed spontaneous limbic motor seizures following external stimuli (e.g., by handling the animals). Not all rats pretreated with KA show overt spontaneous LMS. However, electrographic studies indicate that systemic KA produces paroxysmal discharges in the hippocampus and amygdala as long as (at least) one month after injection (Tremblay and Ben-Ari, 1984). However, a detailed analysis and comparison of the patterns of the neuropathological sequelae in cases with evident spontaneous LMS and the cases which failed to develop seizures have shown notable differences in the distribution of the pathological alterations. A detailed description of the sequelae will be made in one case (1455) with spontaneous limbic motor seizures.

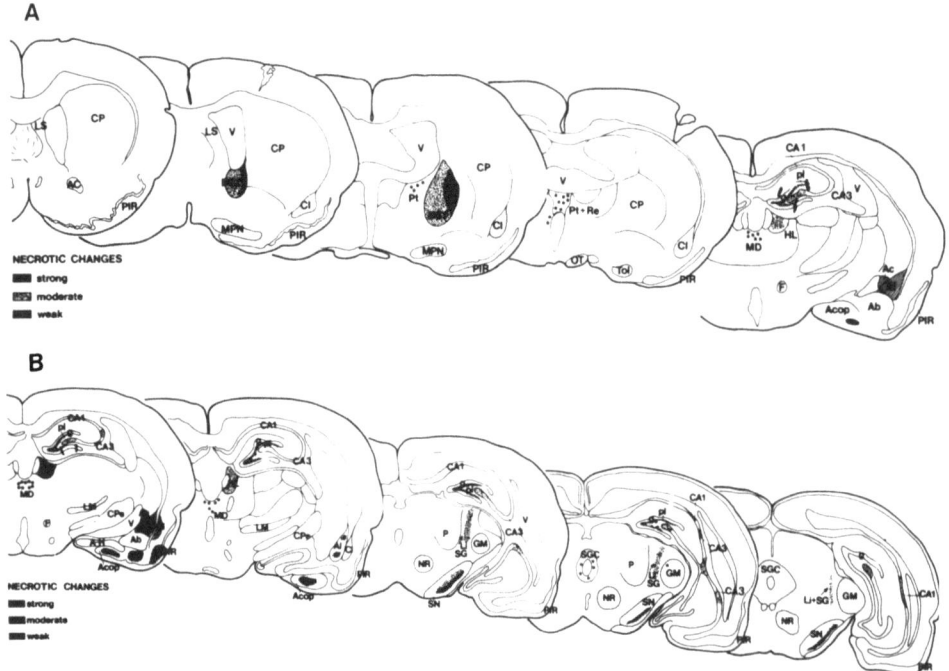

Fig. 1. Diagrams to illustrate the regional distribution of pathological changes in case 1455 (spontaneous motor seizures were observed in this case; survival time: 8 weeks). The location of the degenerating dark neurons (asterisks), the microinfarcts in dentate gyrus (irregular dark areas) and the necrotic neuronal changes followed by proliferation of the reactive nicroglia seen in serial sections are depicted in the camera drawings. Strong neuronal necrosis corresponds to regions of almost complete neuronal loss and associated gliosis; moderate necrotic changes correspond to partial neuronal loss and the weak necrosis indicates scattered neuronal loss. Cresyl violet stained paraffin sections. See abbreviations list.

Case 1455

The seizure pattern produced by KA was similar to that described earlier (Ben-Ari, 1981; Nitecka et al., 1984). Eight days after the injections, spontaneous seizures were evoked by handling the animal. During the eight weeks survival time, the animal displayed aggressive behavior; it was sacrificed and perfused rapidly after a fully developed spontaneous LMS.

A distinct enlargement of the lateral ventricles was found, accompanied by changes in shape and size of the thalamus and septum. The thalamus was especially diminished in size and flattened in the ventrodorsal direction (Fig. 1). It seems to indicate the compensatory character of the observed hydrocephalus.

The most prominent histopathologicl alterations were observed in hippocampal formation, pyriform cortex, amygdaloid nuclei, thalamic nuclei, and septum. The alterations included shrinkage of tissue, progressive degeneration of neurons, necrosis associated with neuronal cell loss and proliferation of glial elements and abnormalities of capillary vessels (Figs. 2, 3).

Hippocampal formation. Prominent alterations were observed in the dentate polymorph layer and the hilar zone. A significant loss of neurons and moderate proliferation of micro- and astroglia were found. Degenerating neurons were also observed (Fig. 2A). Enlarged capillary vessels showing hypertrophic endothelium were seen in all layers of the dentate gyrus. Usually small coagulative scars (indicating the microinfarcts) were found in association with the enlarged capillaries; they were most conspicuous in the granular layer (Fig. 2A). Degenerating granule cells were sporadically noted. Neuronal cell loss and proliferation of glial elements (microglia and astroglia) were detected in CA3 subfield. The alterations never involved very large areas. The pathological alterations in CA1 subfield were weakly expressed and restricted to a very small area, where the loss of neurons and proliferation of the reactive microglia were found (Fig. 2A, B).

Moreover, a few degenerating neurons undergoing phagocytosis were seen in the dentate molecular layer as well as in hippocampal strata radiatum and oriens. The continuous presence of degenerating neurons undergoing phagocytosis is worth stressing. It may indicate that the pathological process is in progress. Pathological changes did not occur throughout the full extent of hippocampal formation. They involved only restricted areas in anteroposterior direction and were located asymmetrically in both hippocampal and dentate gyri.

Pyriform cortex. Sclerotic scars were found posterior in layers II and III (Fig. 2C). In layer II, partial loss of neurons and extensive hyperplasia of microglia were noted; in layer III, the loss of neurons was almost complete and the development of microglia very extensive.

Amygdaloid complex. Areas with extensive gliosis involving hyperplasia of microglia and complete loss of neurons were detected in the posterior region of the amygdaloid cortical nucleus (Fig. 2D). The lateral nucleus and lateral part of the central nucleus showed moderate loss of neurons and weakly expressed glial changes; the nuclei seemed to be shrunken and smaller than in control cases. These changes were subtle and difficult to interpret due to the complex cytoarchitecture of the amygdaloid region.

Thalamic nuclei. The medial dorsal parataenial and reuniens nuclei were significantly diminished. They contained almost exclusively shrunken

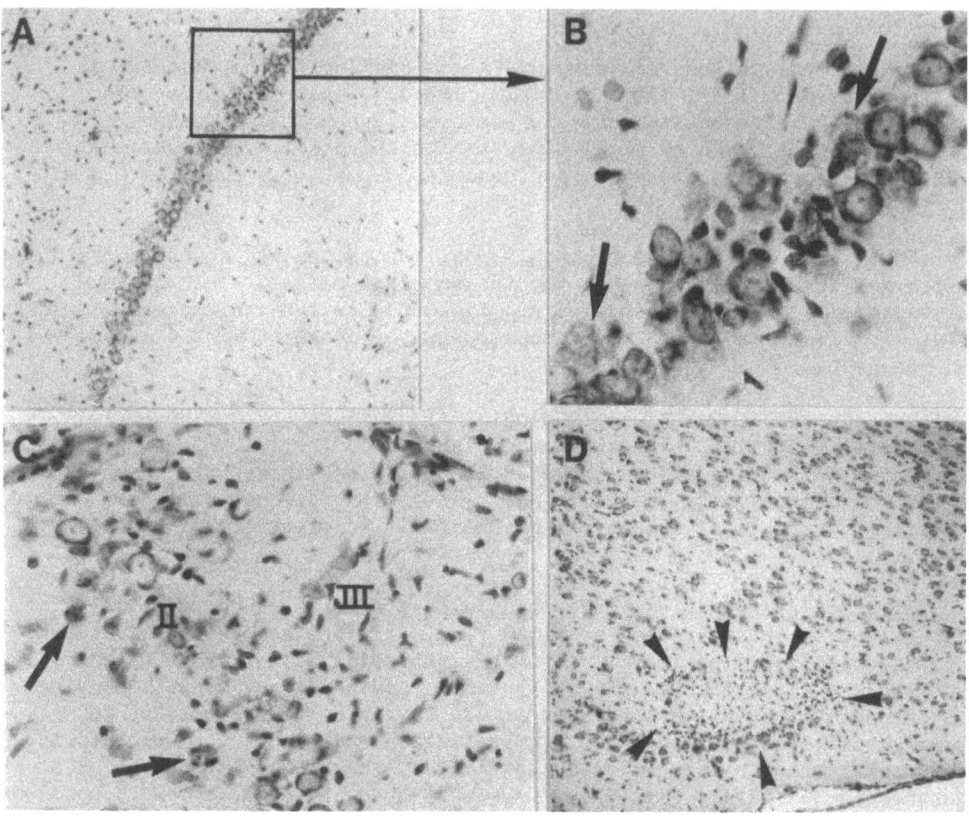

Fig. 2. Photomicrographs to illustrate the type of neuronal changes and capillary abnormalities in case 1455 (see Fig. 1). A: Coagulative necrosis in granular layer of fascia dentata (small arrows); note the hypertrophy of the endothelial cells of the capillary (thick arrow). Degenerating cells and the reactive microglia (arrow heads) in plexiform layer are present; note the significant loss of cells (x396). B: Significant loss of neurons and presence of a few necrotic cells (thick arrow) in the hilar pyramidal cells region; dark degenerating cell (arrow with double head), reactive microglia (small arrow) and astroglia (arrowheads) are present (x396). C-E: neuronal loss in CA3; note the presence of atrophic cells (arrows) and microglia (arrowheads); C. (x88.5), D. (x396), E. (x990). Cresyl violet stain paraffin sections.

and deeply cresyl violet stained neurons. A conspicuous loss of neurons and hyperplasia of microglia were noted in nuclei of the thalamic posterior region; the most prominent changes were found in suprageniculate-limitans complex (Fig. 1).

Neocortex. In comparison to control cases, the frontoparietal cortex was relatively thinner; the changes there did not involve any particular layer. Vessels with strongly hypertrophic endothelial cells (intensely stained with cresyl violet) were particularly numerous in that part of the cortex. Usually they ran perpendicularly to the cortical surface. No evidence of pathological alterations of neurons and glial cells in the area adjacent to the vessels was noted.

Small gliotic scars were observed in the substantia nigra on both sides.

Cases Without Overt Spontaneous LMS

Although in these cases the overt spontaneous LMS were not revealed in response to handling of animals, it is difficult to exclude completely the possibility that they occurred when the animals were not observed. However, the pattern of the distribution of neuronal damage in these animals was different. Thus, clearcut pathological alterations were observed only in the hippocampal formation and, in striking contrast to cases showing the spontaneous LMS, the neurons of dentate polymorph layer did not undergo conspicuous degeneration. In contrast, the atrophy of CA1 field was present and the loss of cells and proliferation of glia in CA3 were observed.

DISCUSSION

The conclusions which can be drawn from the present study can be summarized as follows:

1) In the hippocampal formation, particularly prominent pathological alterations are noted in the fascia dentata with long survival periods. This includes coagulative scars in the granular layer and neuronal loss notably in the adjacent polymorph zone where a large proportion of presumably GABAergic interneurons are destroyed. In a preliminary study using antibodies directed against GABA, we have observed in two cases pathological alterations in GABAergic neurons in this zone; systematic study is in progress (see also Ben-Ari, this volume). The progressive character of the pathological changes (degenerating cells were even observed after eight weeks survival time) is in keeping with the dynamic changes also reported by Franck (1984).

2) The damage in the posterior thalamic region deserves particular emphasis considering the importance of external stimuli in eliciting the 'spontaneous seizures'. The nuclei of the posterior thalamic region are connected with the lateral, central and medial nuclei of the amygdala (Jones and Burton, 1976; Nitecka, 1979) which play a central role in the development of the limbic motor seizures (Ben-Ari et al., 1981).

Long Term Experimental Pathological Alterations and Possible Relevance to Human Temporal Lobe Epilepsy

The data on human epilepsy underline the importance of the pathological changes in the hippocampal formation and amygdala in the etiology of temporal lobe epilepsy. The classical model of damage in the hippocampal formation involved extensive loss of cells and proliferation of glia in the CA1 region, moderate damage of the CA3 and CA4 areas and little damage to neurons of the CA2 region and fascia dentata. In addition, mild neuronal destruction was described in the amygdaloid body, thalamus and some cortical fields (Falconer et al., 1964). More recent studies of Margerison and Corsellis (1966) indicate a significant loss of neurons in endfolium (hilar region of the hippocampal formation). These authors underline that 'cell loss and gliosis in the endfolium with complete sparing of the Sommer's sector and dentate gyrus are in fact not uncommonly found'. Later, quantitative studies of neuronal cell loss in human epileptics led Dam (1980) to the conclusion that 'the loss of neurons was most marked in the endfolium, field H3 and fascia dentata'. Furthermore, these studies revealed a correlation between neuronal loss and duration of the epilepsy and frequency of seizures (Dam, 1980). Using the Golgi method, Scheibel and Scheibel (1973) and Scheibel et al. (1979) observed in hippocampi surgically resected

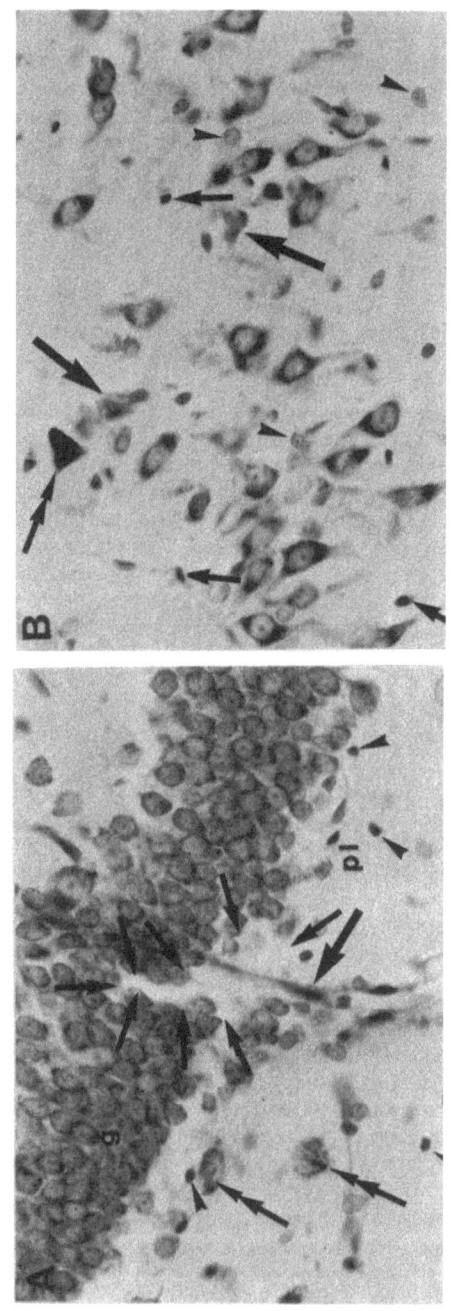

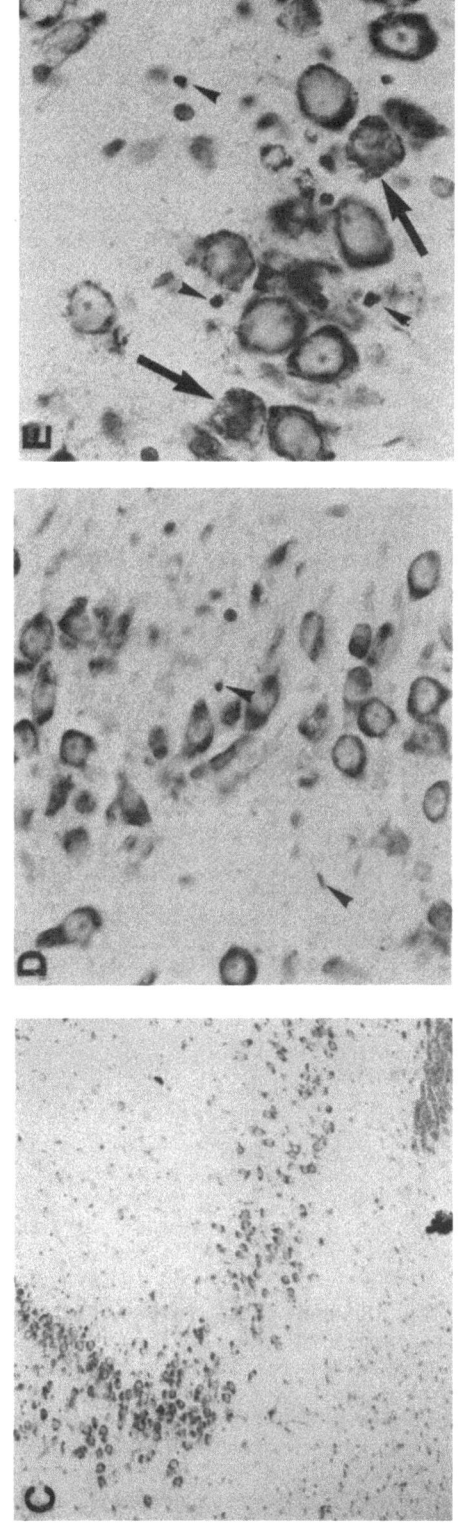

Fig. 3. Photomicrographs to illustrate the pathological alterations in case 1455. A: Patchy development of reactive microglia and neuronal alterations in CA1 (x88.5). The rectangle which shows pyramidal neurons surrounded by reactive microglia is enlarged in B (x396). The dying cells are present (arrows; x396). C: Gliotic scar in pyriform cortex. Note the significant proliferation of reactive microglia and the presence of a few pathologically changed cells in layer II (arrows; x396). D: Gliotic scar (arrowheads) in the posterior part of the cortical nucleus of amygdala (x88.5). Cresyl violet stain. Paraffin sections.

from epileptic brains the pathological abnormalities in the dendrites of the pyramidal cells and the dentate granule cells. Moreover, they indicated that the most prominent gliosis involved the zone of polymorph cells located deeply to the fascia dentata (Scheibel and Scheibel, 1973). Characteristically, very few 'basket cells' in the polymorph layer were impregnated with Golgi method in epileptic brains (Scheibel and Scheibel, 1973). These observations are therefore comparable with the pattern of damage seen with long-term survival after parenteral administration of KA. The further quantitative studies of the abnormalities in the granular layer and their mossy fibers would be of great importance in particular in order to take into account synaptic rearrangement. The plasticity of this region is well documented and there is now good evidence to suggest that it could play a role in epileptogenesis (Nadler et al., 1983).

ABBREVIATIONS USED IN FIGURES

Ab, basal lateral nucleus of the amygdala
Ac, central nucleus of the amygdala
AC, anterior commissure
Acop, posterior cortical nucleus of the amygdala
A-H, transitional amygdalo-hippocampal area
Al, lateral nucleus of the amygdala
BST, bed nucleus of the stria terminalis
CA1-3, fields of the Ammon's horn according to Lorente de Nó
CGS, periaqueductal grey substance
Cl, claustrum
CP, caudate-putamen
CPe, cerebral penduncle
F, fornix
g, granular layer of the dentate gyrus
GM, medial geniculate body
Hl, lateral habenular nucleus
Li+SG, limitaus and suprageniculate thalamic nuclei
LM, medial lemniscus
LS, lateral septal nucleus
MD, medial dorsal thalamic nucleus
MPN, magnocellular preoptic nucleus
NR, red nucleus
OT, optic tract
P, pretectal area
PIR, pyriform cortex
pl, polymorph layer of dentate gyrus
Pt, parataenial thalamic nucleus
Re, reuniens thalamic nucleus
Tol, nucleus of the lateral olfactory tract
V, ventricle

REFERENCES

Ben-Ari, Y., Tremblay, E., and Ottersen, O.P., 1980, Injections of kainic acid into the amygdaloid complex of the rat: an electrographic, clinical and histological study in relation to the pathology of epilepsy, Neuroscience, 5:515.
Ben-Ari, Y., Tremblay, E., Riche, D., Ghilini, G., and Naquet, R., 1981, Electrographic, clinical and pathological alterations following systemic administration of kainic acid, bicuculline and pentetrazol: metabolic mapping using the deoxyglucose method with special reference to the pathology of epilepsy, Neuroscience, 6:1361.

Ben-Ari, Y., 1985, Limbic seizure and brain damage produced by kainic acid: mechanism and relevance to human temporal lobe epilepsy, Neuroscience, 14:375.

Brown, W.J., 1973, Structural substrate of seizures foci in the human temporal lobe, in: Epilepsy: Its Phenomenon in Man, M.A.B. Brazier, ed., Academic Press, New York, p. 339.

Cavalheiro, E.A., Riche, D.A., and Le Gal La Salle, G., 1982, Long-term effects of intrahippocampal kainic acid injection in rats: a method for inducing spontaneous recurrent seizures, Electroencephal. Clin. Neurophys., 53:581.

Dam, A.M., 1980, Epilepsy and neuronal loss in the hippocampus, Epilepsia, 21:617.

Falconer, M.A., Seragetinides, E.A., and Corsellis, J.A.N., 1964, Etiology and pathogenesis of temporal lobe epilepsy, Arch. Neurol., 10:233.

Franck, J.E., and Schwartzkroin, P.A., 1983, Kainate lesioned hippocampi become epileptogenic, Soc. Neurosci. Abstr., 9:908.

Franck, J.E., 1984, Dynamic alterations in hippocampal morphology following intra-ventricular kainic acid, Acta Neuropathol., 62:242.

Jones, E.G., and Burton, H., 1976, A projection from the medial pulvinar to the amygdala in primates, Brain Res., 104:142.

Lancaster, B., Wheal, H.V., and Ashwood, T.J., 1983, Hippocampal electrophysiology after kainic acid treatment: a chronic model of focal epilepsy, Soc. Neurosci. Abstr., 9:908.

Lothman, E.W., and Collins, R., 1981, Kainic acid induced limbic seizures: metabolic, behavioral, electroencephalographic and neuropathological correlates, Brain Res., 218:299.

Margerison, J.H., and Corsellis, J.A.N., 1966, Epilepsy and the temporal lobe. A clinical, electroencephalographic and neuropathological study of the brain in epilepsy with particular reference to the temporal lobes, Brain, 89:499.

Nadler, V.J., Tauck, D.L., Evenson, D.A., and Davis, J.N., 1983, Synaptic rearrangements in the kainic acid model of Ammon's horn sclerosis, in: Excitotoxins, K. Fuxe, P. Roberts, and R. Schwarcz, eds., MacMillan Press, London, p. 256.

Nitecka, L., 1979, Connections of the posterior thalamus with the amygdaloid body of the rat, Acta Neurobiol. Exp., 39:49.

Nitecka, L., Tremblay, E., Charton, G., Bouillot, J.P., Berger, M., and Ben-Ari, Y., 1984, Maturation of kainic acid seizure-brain damage syndrome in the rat. II. Histopathological sequelae, Neuroscience, 13:1073.

Olney, J.W., Fuller, T., and De Gubareff, T., 1979, Acute dendrotoxic changes in hippocampus of kainate treated rats, Brain Res., 176:91.

Scheibel, M.E., and Scheibel, A.B., 1973, Hippocampal pathology in temporal lobe epilepsy. A Golgi survey, in: Epilepsy: Its Phenomenon in Man, M.A.B Brazier, ed., Academic Press, New York, p. 311.

Scheibel, M.E., Grandall, P.H., and Scheibel, A.B., 1974, The hippocampal-dentate complex in temporal lobe epilepsy. A Golgi study, Epilepsia, 15:15.

Schwob, J.E., Fuller, T., Price, J.L., and Olney, J.W., 1980, Widespread patterns of neuronal damage following systemic or intracerebral injections of kainic acid: a histological study, Neuroscience, 5:991.

Tremblay, E., and Ben-Ari, Y., 1984, Usefulness of parenteral kainic acid as a model of temporal lobe epilepsy, Rev. EEG Neurophysiol., 14:241.

ELECTROPHYSIOLOGY OF EPILEPTIC TISSUE: WHAT PATHOLOGIES ARE EPILEPTOGENIC?

P.A. Schwartzkroin[1,2] and J.E. Franck[1]

Departments of [1]Neurological Surgery and [2]Physiology and
Biophysics
University of Washington
Seattle, Washington

INTRODUCTION

What defines the abnormality of epileptic tissue? What constitutes an epileptic focus? Experimental studies have been carried out for many years to try and answer these questins, and a large number of alterations have been proposed as basic to generation of epileptiform activity (Jasper et al., 1969; Lockard and Ward, 1980; Schwartzkroin and Wheal, 1984; Delgado-Escueta et al., 1986). Yet, our real information about the abnormalities intrinsic to the epileptic condition remains limited. In studying potential causes of abnormal cell electrical activity, we have become familiar with many of the normal properties of neocortical and hippocampal neurons, but have not been able to define the essential nature of the epileptic defect. Morphological, pharmacological, biochemical, and electrophysiological features have been suggested which could be involved in generation of epileptiform activity (Westrum et al., 1964; Heinemann et al., 1977; Woodbury and Kemp, 1977; French and Siggins, 1980; Schwartzkroin and Wyler, 1980; Lux and Heinemann, 1982; Ribak et al., 1982; Dudek et al., 1983; Grisar et al., 1983; Scheibel et al., 1983; Schwartzkroin, 1983; DeLorenzo, 1984; Johnston and Brown, 1984; Somjen, 1984). For most of these, it remains unclear whether the abnormality is present before, and is responsible for, the development of epileptiform activity, or whether these abnormalities are produced as a result of abnormal electrical activity in the brain. It is this issue of defining causal, rather than correlative features of epileptogenesis that has been so difficult in epilepsy research. The drive for such causality has led to the establishment of a host of animal model systems.

MODEL SYSTEMS

Because it has proved impractical to study the human epileptic cortex or hippocampus in experimental detail, many of the implicated abnormalities have been 'discovered' in animal models of the epilepsies (Purpura et al., 1972). These animal model studies have generated a wide range of possible mechanisms leading to abnormal neuronal activities, but have yet to show how any one of these possibilities is critical to the development of human epileptic activity. Just as in human epileptic brain, there are varied abnormal phenomenologies correlated with, or resulting from,

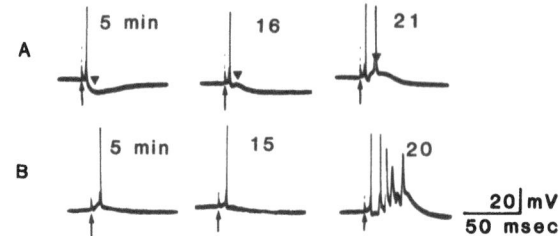

Fig. 1. Reduction of IPSP and development of burst discharge in two hippocampal pyramidal cells (guinea pig) bathed in medium containing 2 mM penicillin (in vitro slice preparation). Numbers above each trace show interval from time penicillin was added to the solution in the reservoir bottle. Note the loss of a hyperpolarization following the action potential and the growth of a prolonged depolarization. Arrows show stimulus delivery.

seizure discharges. However, many of these models leave us uncertain about the cause-effect relationship of treatment (i.e., that procedure used to generate the model) and resulting epileptiform activities. What is it about a given treatment that causes epileptiform activity?

There is currently widespread experimental support for the hypothesis that cell hyperexcitability and epileptogenesis result from a loss or decrease in the degree of inhibition in cortical structures. A variety of studies have shown that epileptiform events can be produced experimentally by blocking inhibition produced by gamma-aminobutyric acid (GABA) in cortex or hippocampus using such drugs as bicuculline, picrotoxin, or penicillin (Ajmone-Marsan, 1969; Prince, 1978; Dingledine and Gjerstad, 1980; Schwartzkroin and Prince, 1980). With such agents applied to brain tissue, one can measure the blockade of the inhibitory conductances normally associated with inhibitory postsynaptic potentials (IPSPs) (Dingledine and Langmoen, 1980), and carefully examine the development of hyperexcitable and synchronized activity in the tissue (Fig. 1). Unquestionably, such blockade of inhibition results in epileptiform events, either locally - when the GABA blocker is applied to a circumscribed and discrete region of brain, or generally - when the GABA blocker is infused systemically such that the entire brain is exposed (Gloor, 1984). It now appears, especially in light of the data from these latter studies, that inhibition need not be entirely abolished. In fact, in the generalized model of systemic penicillin application, little alteration in the IPSP may be measured. That finding suggests that even subtle generalized decrements in the degree of inhibition in neural tissue may lead to epileptic-like events.

From such experimental results has emerged the hypothesis that brain trauma, or even genetic make-up, may result in a loss of inhibition, leading to the development of epilepsy in human patients. Such a loss may stem from a number of deficits. For example, inhibition in hippocampus and neocortex is mediated largely via interneuronal cell types which use GABA as their neurotransmitter; a decrease in number or loss of the GABA-containing interneurons might well have devastating effects on the control of excitability. In immunocytochemical studies of monkey alumina gel foci, it has been shown that there is a reduction in the number of GABA-containing

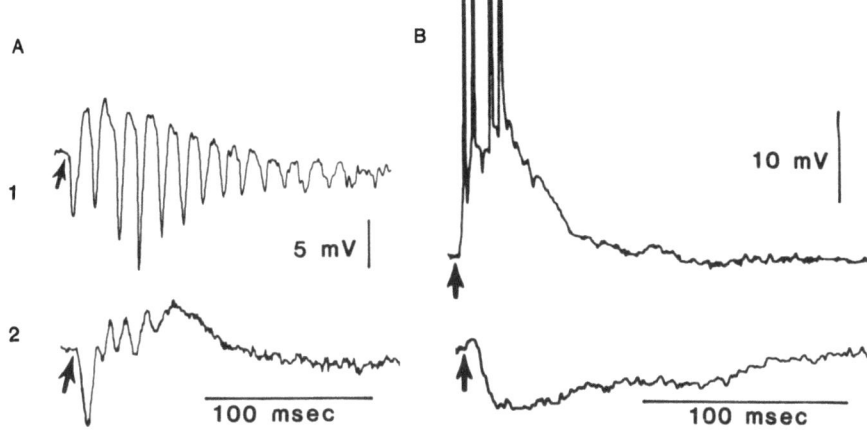

Fig. 2. Short- and long-term effects of intraventricular kainate injections on excitability of the CA1 region of hippocampus (in vitro slice preparation from rat). A. Field potential (1) and intracellular (2) responses to stimulation (at arrows) in tissue prepared 2-4 weeks after kainate injection. At both population and cellular levels, stimulation evoked prolonged burst discharges; no IPSP was apparent in the intracellular record. B. Field potential (1) and intracellular (2) responses at 2-4 months after kainate injection. The population response appeared less hyperexcitable, and intracellular recordings often revealed effective IPSPs.

terminals (Ribak et al., 1982). This change apparently occurs before the electrical discharges typical of the alumina focus occur, and therefore seems temporally appropriate as a causal contributor to the epileptiform activity. It is unlikely that the loss of GABAergic interneuron profiles is entirely attributable to the effects of seizure activity on these interneurons. Biochemical studies carried out on alumina foci have further shown that there is a loss of GABA receptors in epileptic tissue (Bakay and Harris, 1981), indicating an additional alteration that could aggravate the loss of inhibitory interneurons.

In the kainic acid model, too, there is evidence that inhibition, as reflected in inhibitory postsynaptic potentials, is reduced (Wheal et al., 1984; Franck and Schwartzkroin, 1985; Fig. 2A). Since this reduction occurs over a time period during which the tissue is hyperexcitable, it has been proposed that loss of inhibition may account for the kainate-induced hyperexcibabililty. However, in chronically-studied animals which have been treated with kainate, the hyperexcitability (manifest in behavioral seizures) gradually wanes (Cavalheiro et al., 1982). During the course of this behavioral return to normality, the frequency and strength of IPSPs also gradually return toward normal in the tissue (Franck and Schwartzkroin, 1985; Fig. 2B). There is recent evidence to suggest that inhibitory interneurons may be particularly sensitive to the effects of kainate, and may be preferentially damaged by kainate treatment (Nitecka et al., 1984). However, since inhibitory potentials are restored to the affected hippocampal tissue, one must postulate (at the very least) that surviving interneurons in the tissue retain the plasticity to develop new inhibitory connections, perhaps by growth of additional axon collaterals.

The transient nature of decreased inhibitory PSPs in kainate-induced epileptiform discharges is only one reason to question its primacy in a hypothesis for epileptogenesis. Some prior event seems necessary to account for the hypothesized destruction (or reduced efficacy) of the

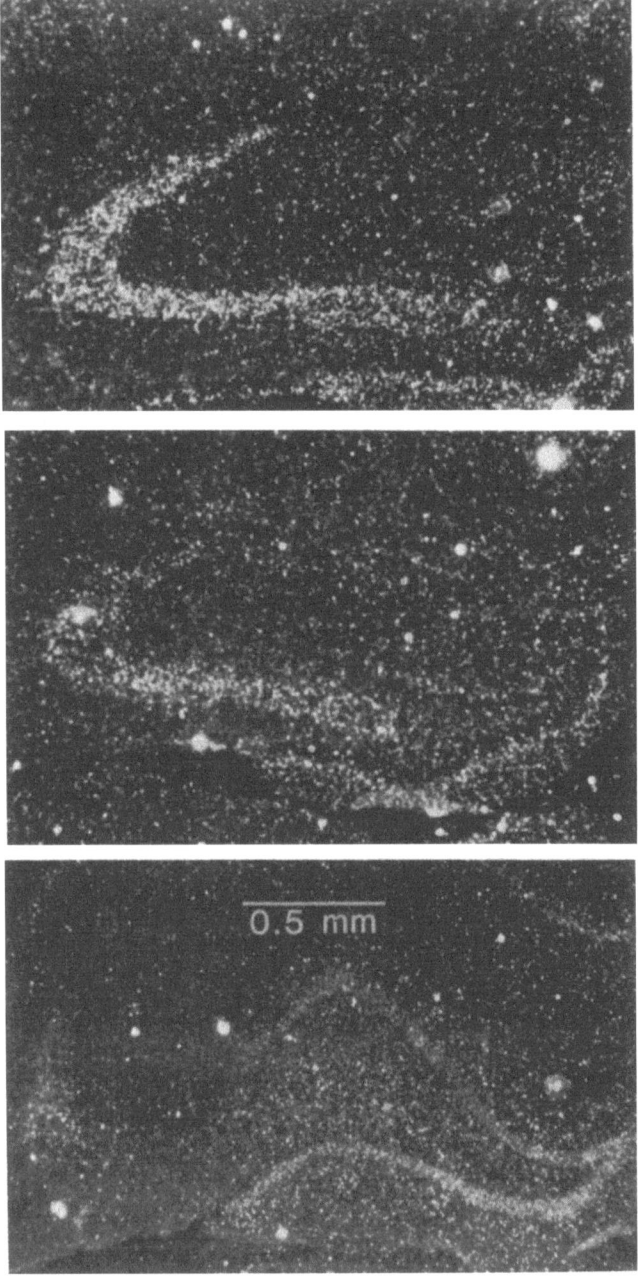

Fig. 3. ^{3}H-Kainate autoradiographic profiles of normal rat hippocampus (top) and kainate-lesioned hippocampus (i.c.v. injection) at two weeks (middle) and two months (bottom) after injection. The prominent white band in the normal tissue reflects kainate receptors associated with the mossy fiber/CA3 synapse. These receptors gradually disappeared following the lesion of CA3 pyramidal cells by kainic acid.

inhibitory interneuron population. Experimental evidence indicates that the CA3 pyramidal cell population is particularly sensitive to kainic acid (Nadler et al., 1980). These cells (or the mossy fiber terminals impinging on them) have large numbers of kainate-preferring receptors (Berger and Ben-Ari, 1983; Fig. 3), and are destroyed when kainate is injected locally. Are interneurons also especially sensitive to the kainic acid itself, or are they rather sensitive to the hyperexcitability induced by the kainate in the CA3 population? The CA1 pyramidal cells in hippocampus display the latter type of sensitivity, and may be damaged secondarily by trauma-related events which increase excitability, disrupt the cellular local environment, and create stressful energy/metabolic demands (Sperk et al., 1983). Systemic injections of kainate, which induce generalized convulsive effects, produce the same kind of CA1 degeneration as found often in animals which are oxygen deprived; indeed, CA1 is far more vulnerable to hypoxic/ischemic damage than is the kainate-sensitive CA3 region (Karnushina et al., 1983). Interneurons have been traditionally thought to exhibit a similar vulnerability to traumatic stimulation and/or hypoxic/ischemic insult.

The results of these kainate studies lead one to postulate a possible primary role for an excitatory neurotransmitter, at least in some forms of epileptiform acitivity. Indeed, even kindling studies have provided some evidence consistent with such an idea. The kindling model has been widely used to study basic mechanisms of epileptogenesis, with many of these studies focused on the demonstration of changes of receptors in the affected brain regions (e.g., Valdes et al., 1982). Recent kindling reports indicate that there is, at least transiently, an up-regulation of glutamate binding in the dentate region of hippocampus (Savage et al., 1984). This change, seen at one day following the last kindling trial, disappears over time, and is not seen at a one month interval. Thus, like the change in IPSPs in the kainate model, it is unclear what the significance of such increased glutamate binding might be for the tissue. Unlike the reversion to a normal state in the kainate model, however, the kindled animal continues to display burst firing discharge and stimulus-evoked kindled events, even in the absence of increased glutamate binding. It is certainly possible, for both the transient changes in inhibition and in glutamate receptor number, that an initial alteration is sufficient to trigger the epileptic process. Once accomplished, the tissue may readjust in an attempt to 'compensate' for the resultant increased excitability.

Other morphological changes, with longer-lasting visibility, have been demonstrated in epileptic tissue. Epileptic foci in human patients are often characterized by some degree of gliosis (Ward, 1978). It is still not clear, however, whether the gliosis is a primary or a secondary event in the development of the epileptic focus. Certainly, gliosis may occur in tissue which is not epileptic - and epileptiform activity may be generated in the absence of gliosis. The increase in extracellular potassium which is recorded from epileptic foci during electrical discharges (Heinemann et al., 1977) may be seen as a stimulus for the gliosis; that is, the tissue 'needs' more glial cells to sponge up the extracellular potassium. Alternatively, the extracellular potassium increases may be viewed as a result of a glial incapacity to absorb the release of potassium from active neurons (Pollen and Trachtenberg, 1970). The increased gliosis may also provide mechanical barriers against normal diffusion of transmitter substances or potassium, such that they accumulate to abnormally high concentrations in the extracellular space. Diffusional barriers may, in this way, contribute to, or even be a cause of, increases in excitability of the epileptic focus. Decreases in extracellular space are common in epileptic foci, and are clearly related to mechanisms of increased excitation. Changes in extracellular space may have critical consequences not only for normal ion and transmitter diffusion (Lux et al., 1986) but also for the effectiveness of ephaptic coupling (Knowles et al., 1985).

The potential for positive feedback 'circuits' within an epileptic focus is enormous, since the consequence of each causal event may act as a secondary cause to increase cell excitability. For example, the gliotic lesion in a focus appears to be associated, too, with changes in the morphology of the neurons within the focus (Scheibel et al., 1983). Several reports have now described long-term changes in neuronal morphology; dying cell and fiber profiles are found in virtually all epileptic foci. The best characterized changes, as seen in animal models (such as the chronic alumina focus) as well as in human foci, occur in the dendritic structure of pyramidal cells in neocortex or hippocampus (Westrum et al., 1964; Scheibel et al., 1983). These cells lose their spines and fine dendritic processes, and may take on abnormal dendritic tree shapes. Such changes in cellular morphology not only have obvious consequences for the synaptic input to these cells, but also for the mechanisms by which transmitted information is integrated electrotonically within the cell (Turner and Schwartzkroin, 1984). It is the excitatory input to these cells that primarily impinges on dendritic spines and fine dendritic processes; thus, loss of the spines and fine dendrites might lead to a preferential loss of excitatory input. How could that loss make functional sense in a 'focus' showing increased excitability? One possible explanation is that when cells lose their spines, the excitatory afferents make contact directly on the dendritic stalk. The nature of EPSPs mediated by such input could be dramatically changed; electrotonic transfer of the EPSP current to the level of the action potential trigger zone would be more effective in dendritically-shortened neurons. In addition, the deafferentation process may elicit receptor supersensitivity (Sharpless, 1969; Bird and Aghajanian, 1975), so that remaining afferents, which do make contact, have larger excitatory effects on the neuron. Structural changes in the cell may also give rise to new (and multiple) spike trigger zones.

Such conjectures are natural consequences of the rather dramatic changes seen in the morphologic profiles of epileptic cortex. Many studies of experimental models, however, suggest that no morphologic change need be produced to initiate epileptiform discharges. No significant changes have been identified in foci produced by application of such drugs as penicillin or bicuculline. Similarly, in the chronic kindling model, no morphologic alterations are visible during the time that epileptiform activity develops. Although some changes in cell morphology may occur as a result of long-term epileptiform activity in these chronic animals — <u>consequences</u> of hyperactivity — the initial epileptiform insult may not be a morphological one. Indeed, it has yet to be shown that the morphologically-abnormal cells of a focus are identical to the cells which are hyperexcitable.

HUMAN RECORDINGS

If these changes in epileptic brain tissue are relevant to the development of human epilepsies, one would expect that studies of human epileptic foci should reveal some evidence of the proposed mechanistic phenomena (reduced inhibition, enhanced excitatory input, etc.). Relatively little work has been done with microelectrophysiological tools, however, to define the nature of the human epileptic focus. Both technical and ethical issues have limited our progress. It has recently become possible, however, to obtain human 'epileptic' tissue for electrophysiological analysis. Thus, it is now — at least in theory — possible to test those hypotheses on human epileptic brain which have been developed from animal model experiments. Tissue from lateral aspects of the cortex (primarily from temporal lobe) and from mesial aspects of the temporal lobe (primarily hippocampus and surrounding parahippocampal gyrus) have been obtained and studied (Schwartzkroin and Prince, 1976; Prince and Wong, 1981; Schwartzkroin et al., 1983; Schwartzkroin and Knowles, 1984) using the in vitro slice preparation. This technique has been widely used to study the hippocampus and neocortex of animal models (Dingledine, 1984).

In our studies, tissue has been provided by the neurosurgeons in the Department of Neurological Surgery, University of Washington - Dr. G.A. Ojemann and Dr. A.R. Wyler (now at the University of Tennessee). In conjunction with the Epilepsy Center at the University of Washington, the Department of Neurological Surgery maintains a program for resecting epileptic foci (primarily temporal lobe) which are intractable to treatment with anti-epileptic medications. The surgeries are normally carried out under local anesthetic so that the surgeon can obtain detailed and accurate information, both corticographic and behavioral, about the tissue to be removed. An extensive electrocorticographic survey is done before the resection to explore both lateral temporal cortical areas, and mesial regions (with 'strip' electrodes inserted under the temporal lobe). With such corticographic techniques, combined with Brevital activation tests (when needed) and complementary localization procedures (e.g., CAT scans, sodium amytal tests), data regarding the localization of the focus are generated. Defining the focus is obviously important for the surgeon - to facilitate removal of only that part of the brain which is epileptic. For experimental purposes, it is also critical to know whether a particular piece of tissue is epileptic - by clinically relevant standards - or is relatively normal. Based primarily on electrocorticography, we can define our samples from lateral or mesial temporal lobe as being electrographically 'hot' or relatively normal. Such diagnoses can be supported by later histopathological analysis of the tissue.

Tissue slices are cut from 600 to 800 microns in thickness, and maintained in an in vitro recording chamber according to the same methods we have used to maintain slices from animal hippocampus (Schwartzkroin, 1981; Fig. 4). Micropipettes are introduced into the tissue to obtain intracellular recordings, and stimulating electrodes are placed in appropriate positions on the tissue to activate fiber bundles. Using intracellular microelectrode techniques, we have analyzed the intrinsic cell properties of neurons from lateral and mesial temporal lobe structures, both from

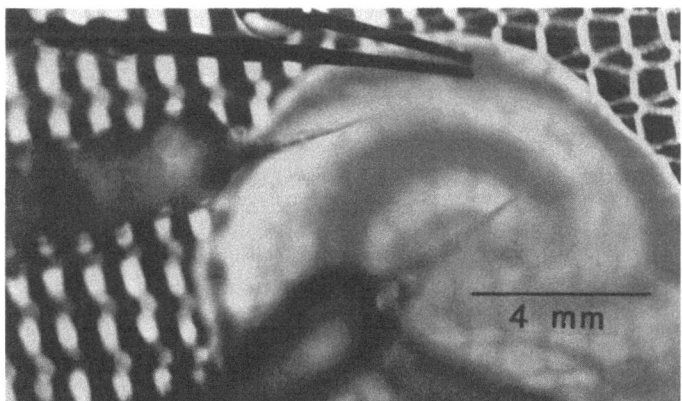

Fig. 4. Photograph of hippocampal slice made from monkey. Two microelectrodes have been introduced into the tissue to record spontaneous, synchronous activities. A bipolar stimulating electrode has been positioned on the alveus.

electrographically very active and from electrographically relatively normal sites. In addition, we have assessed the synaptic efficacies of inputs to these cells, and studied the spontaneous synaptic activity. Finally, we have also been able to intracellularly stain a number of these neuronal elements with the fluorescent dye, Lucifer Yellow (Stewart, 1978). The morphological data derived from such an injection can then be combined with the electrophysiological results of recording from that cell, allowing us to correlate the anatomy and physiology of a particular neuron.

The results of these studies have been at the same time disappointing, confusing, and intriguing. We have found little evidence of spontaneous epileptiform activity in any of the slices surveyed. Thus, even in tissue from electrographically very active epileptic regions of the brain, there is little sign of spontaneous cell bursting activity, or of hyperexcitability manifest in population events. With stimulation of afferent inputs to these cells, the burst discharge pattern that is often associated with epileptic activity is not evident in these slices.

Since the obvious signs of epileptogenicity in these brain slices are absent, it is possible that the process of tissue sampling and slicing may disrupt the critical epileptogenic processes. If, however, there is a localized physical abnormality which is responsible for the development of the epileptic activity, then at some level of analysis, the specific cell features that correlate with occurrence of seizure activity in vivo

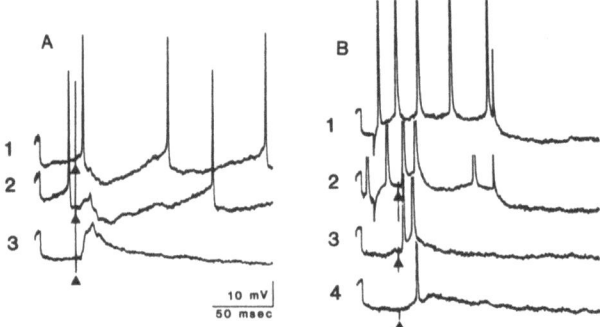

Fig. 5. Intracellular recordings from human 'epileptic' tissue maintained in vitro. A. Records from a neocortical neuron. Stimulation of the tissue (arrow) evoked an EPSP-IPSP sequence which was clearly visible as the cell's resting potential was manipulated (from hyperpolarized in 3 to depolarized in 1). B. Records from a hippocampal pyramidal cell. Current injection into the cell evoked repetitive spiking (1). Stimulation of the tissue (arrows) at weak and strong intensity elicited one and two action potentials, respectively (4 and 3), followed by a hyperpolarization. Interposing the stimulus during a current-elicited spike train (2) shows the inhibitory effect of the hyperpolarization (cf. 1 and 2).

should become evident. Studying the intrinsic membrane properties of cells in slices of epileptic tissues, we have found that input resistance and time constant measures fall within the range of 'normal' cells in cortical tissues. This assessment, of course, is somewhat suspect, since we have no 'normal' human tissue on which to base 'normal' values. If comparisons are made with pyramidal cells recorded from cortex of experimental animals, however, the cells recorded in the human tissue do not appear to be abnormal. Further, if our slices are divided into groups composed of tissue that was electrographically epileptic and tissue which appeared relatively normal, no differences are apparent in the intrinsic cell properties in these slices.

In all tissue, too, afferent stimulation produced relatively normal appearing synaptic sequences (Fig. 5A). Excitatory postsynaptic potentials followed by inhibitory postsynaptic potentials were commonly evoked, even in tissue which was reputedly epileptogenic. There was, however, a tendency for the IPSP to be less obvious or less frequent in tissues which were electrographically epileptic. Given that the process of slice preparation itself may affect the degree of inhibition in a slice, this distinction between epileptic versus relatively normal tissue must be interpreted with great care. In some cells, particularly those from neocortical slices, increasing stimulus intensity often led to burst discharges in response to the stimulation. These bursts were not 'epileptiform' in the sense of the paroxysmal depolarization shift (PDS; Prince, 1968). The PDS is an all-or-none event which is easily triggered at low stimulus intensity (Fig. 6A). The bursts occasionally seen in neocortical neurons from our

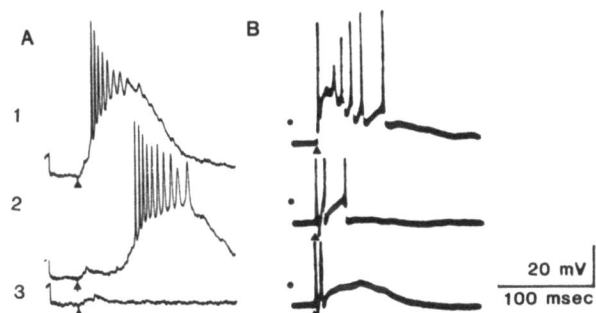

Fig. 6. Burst discharges in human cortical slices. A. PDS-like bursts evoked by tissue stimulation. At moderate stimulus intensities, the all-or-none burst was triggered at short latency (1) (stimulation at arrows). With very low intensities (2 and 3), the burst was only sometimes elicited. The low intensity stimulation produced a small, short-latency EPSP (3); when the PDS did occur (2), it was triggered at long and variable latencies. B. More common, graded burst discharge in response to stimulation (arrows). At moderate stimulus intensity, only one or two action potentials were triggered by the stimulus (3). Increasing the stimulus intensity (3 to 1) led to a longer burst, with more action potentials.

human brain slices were produced, however, in a graded fashion (Fig. 6B), as a response to growing EPSP amplitude. Particularly striking was the tendency for bursting of this nature to be more apparent in tissue taken from relatively normal regions of neocortex, as opposed to those areas that were electrographically epileptogenic. In recordings from hippocampal cells, the normal synaptic sequences were consistently elicited by stimulation (Fig. 5B); bursting was rare, even with rather intense stimuli. In our studies from dozens of slices, representing tissue from over 30 patients, we have recorded from several hundred neurons. In only one preparation have we been able to detect cells which discharged in a typical PDS fashion (Fig. 6A). This activity, paradoxically, was recorded from a slice of neocortex which had been 'diagnosed' as electrographically relatively normal.

The Lucifer Yellow intracellular staining of electrophysiologically characterized neurons provided us with little information about the possible correlation between the physiology and morphology of 'epileptic' neurons. Lucifer Yellow staining revealed primarily normal cortical neurons, with intact dendritic trees, fine dendritic processes and dendritic spines (Fig. 7). This result is consistent with the lack of abnormal electrophysiology of these cells, but tells us little about the morphology of epileptic

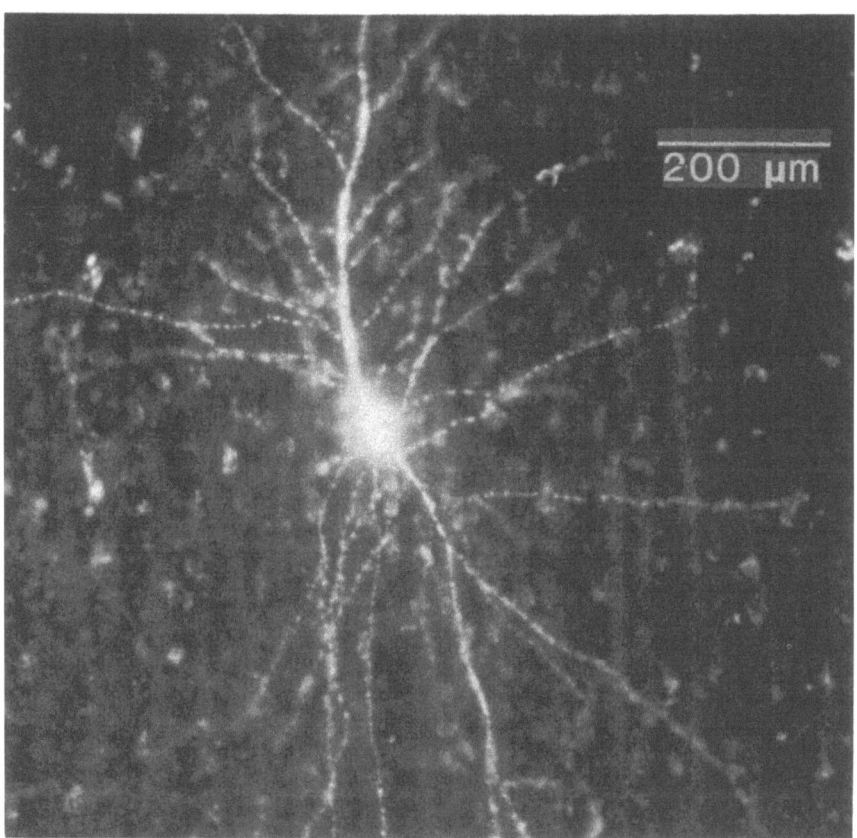

Fig. 7. Neocortical neuron from human 'epileptic' tissue. This cell was intracellularly impaled, electrophysiologically characterized, and injected with the fluorescent dye, Lucifer Yellow. Many fine dendritic processes and dendritic spines can be resolved.

burst-firing neurons. Differentiation of our electrophysiologically-defined neurons on the basis of whether they discharged in bursts or not provided no further insights; there was no apparent morphological correlate. Both bursting and non-bursting neurons had normal dendrites with intact fine processes and spines typical of these cells. Although subtle alterations in the morphology of these neurons cannot be ruled out on the basis of our light microscopy, such differences would not correspond to the rather gross abnormalities previously described for cells in the epileptic focus.

One set of experimental data has proved to be rather provoking. In recordings from tissue of mesial origin (hippocampus), spontaneous rhythmic synaptic events were found in a high proportion of slices (Fig. 8). These rhythmic events were often composed of both depolarizing and hyperpolarizing components, and could sometimes result in cell discharge when the depolarizing component reached spike-initiation threshold. To determine whether these spontaneous events were occurring synchronously within the cell population of the slice, simultaneous intracellular recordings were

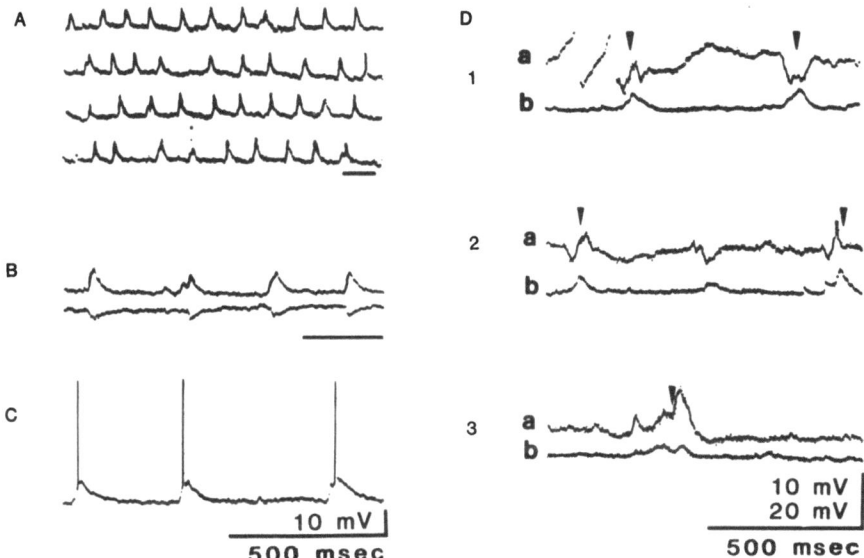

Fig. 8. Intracellular recordings from a slice of human mesial temporal tissue. A. Continuous record of spontaneously-occurring, rhythmic events which were evident in a high proportion of these slices. B. These spontaneous events sometimes triggered action potentials. C. Simultaneous intracellular recording from two neurons showed synchronous occurrence of these PSP-like events throughout the slice. Note that the waveform/polarity of the events could differ from cell to cell. D. Simultaneous recording from two neurons while manipulating membrane potential of one of them (cell 1 of pair) with intracellularly injected current. Cell 1 is shown at high gain, while cell 2 is shown for reference at lower gain. At depolarized potentials, the spontaneous events (arrowhead) in cell 1 were hyperpolarizing (1); near resting potential (2), the events were biphasic; and at a hyperpolarized level, the events were depolarizing. Reversal potential for the initial component of these spontaneous events was similar to the reversal potential of the early component of stimulus-elicited IPSPs.

made from two cells within the slice. The distance between recorded cells was varied from fifty microns to three millimeters. In all cases where spontaneously rhythmic activity was observed, the activity occurred synchronously (or with a fixed phase relation) in the recorded pair; all the cells in the tissue appeared to be involved in its generation. Although this activity did not result in grossly epileptic phenomena, it appeared to us that such activity must reflect a circuitry which might well be involved in mediation of epileptiform synchronous cell firing. For that reason, we have further analyzed these events, both in human tissue, and in hippocampus taken from experimental animals. In the human tissue, we found these events could be manipulated as true synaptic events, with determinable reversal potentials. The apparent PSPs tended to be primarily hyperpolarizing, with reversal potentials similar to that seen for the IPSP (as measured when afferent stimulation was used to evoke IPSPs). Both the IPSP produced by the stimulation, and the spontaneously occurring synchronous events, proved to be sensitive to bicuculline, a GABA-blocking agent (Schwartzkroin and Prince, 1980). In characterizing this activity, the proposed 'epileptogenic' event in this human tissue, we thus showed it to be primarily inhibitory. This surprising finding was further aggravated by our observation that these events were often seen in slices taken from normal monkey hippocampus, and even in slices made routinely from our guinea pig preparations. Again, the events appeared to be truly synaptic, rhythmic and synchronous within the population, sensitive to bicuculline, and primarily inhibitory. Clearly, these events are not epileptiform in nature, and perhaps simply reflect the intrinsic circuitry of the hippocampus.

What can we now say about the various model-generated hypotheses given the data generated from studies of human tissue? Can these in vitro studies of human brain actually shed light on the mechanisms underlying epileptogenesis? Can we get any closer to the primary pathology or pathologies that might be responsible for the genesis of epileptic activity? Our studies of the human epileptic tissue have left us with more questions than answers. We were not able to obtain data that would support hypotheses generated by animal experimentation. For example, there appeared to be only weak indication that the inhibitory circuitry of these 'epileptic' slices was abnormal or had been disrupted. Indeed, in many cases, the IPSP circuitry appeared extremely robust. We were unable to find burst firing neurons, and did not record from cells that were morphologically abnormal. No signs of intrinsic cell abnormalities (i.e., resting membrane properties) were found. In fact, on the basis of the recordings from the tissue we obtained, it would be impossible to maintain that the tissue was from an epileptic region of the brain.

Thus arises the question, What is an epileptic focus? Is it, in fact, a definable region of the brain which is determined by specific structural properties - properties that perhaps need not be manifest in abnormal electrical activity? The question remains unanswered. There are as yet no data which directly correlate the abnormal electrophysiological burst firing which we associate with epileptic neurons with abnormal morphological features (such as truncated dendrites). Are the morphologically abnormal cells seen in epileptic foci electrically active, or is the electrophysiology of the focus produced by morphologically normal cells (perhaps at the border of a focus)? Is inhibition in the tissue actually reduced and therefore responsible for the hyperexcitability of the tissue? Is the reduction of inhibition subtle, only quantitatively definable? Is the observation of relatively normal EPSPs, and of relatively normal burst-firing in response to afferent stimulation consistent with the idea of abnormal transmitter physiology (e.g., changes in receptor characteristics) in the epileptic focus?

Although we can try to answer many of these questions using animal models, where the treatment to the tissue is known and relatively discrete, it is not clear how well we can answer these questions for human epileptic activities. From animal studies of kindling (McNamara et al., 1984), as well as our own experiments here on epileptic slices, it seems likely that the epileptogenicity of the tissue is conferred upon it by a distributed circuit which is disrupted when brain slices are isolated. If this is the case, if epileptic activity is the result of a distributed abnormality rather than a focal one, the concept of the epileptic focus must be revised. This conclusion may be relevant, even if the electrographic abnormalities appear focally confined and removal of the 'focus' eliminates seizure activity. Such a situation would obviously make it much more difficult for us to know where to look for abnormalities in such features as transmitter sensitivities, receptor numbers, or binding sensitivities. Nor is it clear that the regions of the brain which are morphologically abnormal are those that are primarily involved in generating the abnormal electrophysiology. Clearly, brain regions which have suffered from the onslaught of the epileptic barrage may 'die' and become quiescent. Thus, there may be regions of abnormal brain which do not generate high levels of electrical activity. Such 'inactive' foci are common among experimental models, where morphological damage (for example, intense gliosis) may be produced without concurrent epileptic activity. Experimental foci — freeze lesions, cobalt and alumina implants — are known to 'burn out'; cells in the region die and/or are no longer capable of generating discharges.

It is important to realize that although we may propose a whole host of mechanisms that may be responsible for hyperexcitability in epileptic foci, it is essential that we be able to show how such mechanisms are causally involved in generation of the abnormality in human brain. This is by no means an easy task, and will continue to challenge epilepsy researchers for some time to come.

REFERENCES

Ajmone-Marsan, C., 1969, Acute effects of topical epileptogenic agents, in: Basic Mechanisms of the Epilepsies, H.H. Jasper, A.A. Ward, Jr., and A. Pope, eds., Little, Brown and Company, Boston, p. 299.

Bakay, R.A.E., and Harris, A.B., 1981, Neurotransmitter, receptor and biochemical changes in monkey cortical epileptic foci, Brain Res., 206:387.

Berger, M., and Ben-Ari, Y., 1983, Autoradiographic visualization of ^{3}H-kainic acid receptor subtypes in the rat hippocampus, Neurosci. Lett., 39:237.

Bird, S.J., and Aghajanian, G.K., 1975, Denervation supersensitivity in the cholinergic septo-hippocampal pathway: a microiontophoretic study, Brain Res., 100:355.

Cavalheiro, E.A., Riche, D.A., and Le Gal La Salle, G., 1982, Long-term effects of intrahippocampal kainic acid injection in rats: a method for inducing spontaneous recurrent seizures, Electroenceph. Clin. Neurophysiol., 53:581.

Delgado-Escueta, A.V., Ward, A.A., Jr., Woodbury, D.M., and Porter, R., eds., 1986, Basic Mechanisms of the Epilepsies, Raven Press, New York, in press.

DeLorenzo, R.J., 1984, Calmodulin systems in neuronal excitability: a molecular approach to epilepsy, Annals Neurol., 16(Suppl):S104.

Dingledine, R., and Gjerstad, L., 1980, Reduced inhibition during epileptiform activity in the in vitro hippocampal slice, J. Physiol., 305:297.

Dingledine, R., and Langmoen, I.A., 1980, Conductance changes and inhibitory actions of hippocampal recurrent IPSPs, Brain Res., 185:277.

Dingledine, R., ed., 1984, *Brain Slices*, Plenum Press, New York.

Dudek, F.E., Andrew, R.D., MacVicar, B.A., Snow, R.W., and Taylor, C.P., 1983, Recent evidence for and possible significance of gap junctions and electrotonic synapses in the mammalian brain, in: *Basic Mechanisms of Neuronal Hyperexcitability*, H.H. Jasper and N.M. VanGelder, eds., Alan R. Liss, New York, p. 31.

Franck, J.E., and Schwartzkroin, P.A., 1985, Do kainate-lesioned hippocampi become epileptogenic? *Brain Res.*, 329:309.

French, E.D., and Siggins, G.P., 1980, An iontophoretic survey of opioid peptide actions in the rat limbic system: in search of opiate epileptogenic mechanisms, *Reg. Pept.*, 1:127.

Gloor, P., 1984, Electrophysiology of generalized epilepsy, in: *Electrophysiology of Epilepsy*, P.A. Schwartzkroin and H.V. Wheal, eds., Academic Press, New York, p. 107.

Grisar, T., Franck, G., and Delgado-Escueta, A.V., 1983, Glial contribution to seizures: K^+ activation of (Na^+, K^+)-ATPase in bulk isolated glial cells and synaptosomes of epileptogenic cortex, *Brain Res.*, 261:75.

Heinemann, U., Lux, H.D., and Gutnick, M.J., 1977, Extracellular free calcium and potassium during paroxysmal activity in the cerebral cortex of the cat, *Exp. Brain Res.*, 27:237.

Jasper, H.H., Ward, A.A., Jr., and Pope, A., eds., 1969, *Basic Mechanisms of the Epilepsies*, Little, Brown and Company, Boston.

Johnston, D., and Brown, T.H., 1984, Mechanisms of neuronal burst generation, in: *Electrophysiology of Epilepsy*, P.A. Schwartzkroin and H.V. Wheal, eds., Academic Press, London, p. 277.

Karnushina, I., Susuki, R., Padgett, W., and Daly, J.W., 1983, Degeneration of CA1 neurons in hippocampus after ischemia in mongolian gerbils: cyclic AMP systems, *Brain Res.*, 268:87.

Knowles, W.D., Traub, R.D., Wong, R.K.S., and Miles, R., 1985, Properties of neural networks: experimentation and modeling of the epileptic hippocampal slice, *TINS*, 8:73.

Lockard, J.S., and Ward, A.A., Jr., eds., 1980, *Epilepsy: A Window to Brain Mechanisms*, Raven Press, New York.

Lux, H.D., and Heinemann, U., 1982, Consequences of calcium electrogenesis for the generation of paroxysmal depolarisation shift, in: *Epilepsy and the Motor System*, E.J. Speckmann and H. Elger, eds., Urban and Schwarzenberg, Munich, p. 101.

Lux, H.D., Heinemann, U., and Dietzel, I., 1986, Ionic changes and alterations in the size of the extracellular space during epileptic activity, in: *Basic Mechanisms of the Epilepsies*, A.V. Delgado-Escueta, A.A. Ward, Jr., D.M. Woodbury, and R. Potter, eds., Raven Press, in press.

McNamara, J.O., Galloway, M.T., Rigsbee, L.C., and Shin, C., 1984, Evidence implicating substantia nigra in regulation of kindled seizure threshold, *J. Neurosci.*, 4:2410.

Nadler, J.V., Perry, B.W., Gentry, C., and Cotman, C.W., 1980, Degeneration of hippocampal CA3 pyramidal cells induced by intraventricular kainic acid, *J. Comp. Neurol.*, 192:333.

Nitecka, L., Tremblay, E., Charton, G., Bouillot, J.P., Berger, M.L., and Ben-Ari, Y., 1984, Maturation of kainic acid seizure-brain damage syndrome in the rat. II. Histopathological sequelae, *Neuroscience*, 13:1073.

Pollen, D.A., and Trachtenberg, M.C., 1970, Neuroglia: gliosis and focal epilepsy, *Science*, 167:1252.

Prince, D.A., 1968, The depolarization shift in 'epileptic' neurons, *Exp. Neurol.*, 21:467.

Prince, D.A., 1978, Neurophysiology of epilepsy, *Ann. Rev. Neurosci.*, 1:395.

Prince, D.A., and Wong, R.K.S., 1981, Human epileptic neurons studied in vitro, *Brain Res.*, 210:323.

Purpura, D.P., Penry, J.K., Tower, D., Woodbury, D.M., and Walter, R., eds., 1972, *Experimental Models of Epilepsy*, Raven Press, New York.

Ribak, C.E., Bradburne, R.M., and Harris, A.B., 1982, A preferential loss of GABAergic symmetric synapses in epileptic foci: a quantitative ultrastructural analysis of monkey neocortex, J. Neurosci., 2:1725.

Savage, D.D., Werling, L.L., Nadler, J.V., and McNamara, J.O., 1984, Selective and reversible increase in the number of quisqualate-sensitive glutamate binding sites on hippocampal synaptic membranes after angular bundle kindling, Brain Res., 307:332.

Scheibel, A.B., Paul, L., and Fried, I., 1983, Some structural substrates of the epileptic state, in: Basic Mechanisms of Neuronal Hyperexcitability, H.H. Jasper and N.M. VanGelder, eds., Alan R. Liss, New York, p. 109.

Schwartzkroin, P.A., and Prince D.A., 1976, Microphysiology of human cerebral cortex studied in vitro, Brain Res., 115:497.

Schwartzkroin, P.A., and Prince, D.A., 1980, Changes in excitatory and inhibitory synaptic potentials leading to epileptogenic activity, Brain Res., 183:61.

Schwartzkroin, P.A., and Wyler, A.R., 1980, Mechanisms underlying epileptiform burst discharge, Ann. Neurol., 7:95.

Schwartzkroin, P.A., 1981, To slice or not to slice, in: Electrophysiology of Isolated Mammalian CNS Preparations, G.A. Kerkut and H.V. Wheal, eds., Academic Press, London, p. 15.

Schwartzkroin, P.A., 1983, Local circuit considerations and intrinsic neuronal properties involved in hyperexcitability and cell synchronization, in: Basic Mechanisms of Neuronal Hyperexcitability, H.H. Jasper and N.M. VanGelder, eds., Alan R. Liss, New York, p. 75.

Schwartzkroin, P.A., Turner, D.A. Knowles, W.D., and Wyler, A.R., 1983, Studies of human and monkey 'epileptic' neocortex in the in vitro slice preparation, Ann. Neurol., 13:249.

Schwartzkroin, P.A., and Knowles, W.D., 1984, Intracellular study of human epileptic cortex; in vitro maintenance of epileptiform activity?, Science, 223:709.

Schwartzkroin, P.A., and Wheal, H.V., eds., 1984, Electrophysiology of Epilepsy, Academic Press, London.

Sharpless, S.K., 1969, Isolated and deafferented neurons: disuse supersensitivity, in: Basic Mechanisms of the Epilepsies, H.H. Jasper, A.A. Ward, Jr., and A. Pope, eds., Little, Brown and Company, Boston, p. 329.

Somjen, G.G., 1984, Interstitial ion concentration and the role of neuroglia in seizures, in: Electrophysiology of Epilepsy, P.A. Schwartzkroin and H.V. Wheal, eds., Academic Press, London, p. 303.

Sperk, G., Lassmann, H., Baran, H., Kish, S.J., Seitelberger, F., and Hornykiewicz, O., 1983, Kainic acid induced seizures: neurochemical and histopathological changes, Neuroscience, 10:1301.

Stewart, W.W., 1978, Functional connections between cells as revealed by dye-coupling with a highly fluorescent naphthalimide tracer, Cell, 14:744.

Turner, D.A., and Schwartzkroin, P.A., 1984, Passive electronic structure and dendritic properties of hippocampal neurons, in: Brain Slices, R. Dingledine, ed., Plenum Press, New York, p. 25.

Valdes, F., Dashieff, R.M., Birmingham, F., Crutcher, K., and McNamara, J.O., 1982, Benzodiazepine receptor increases following repeated seizures: evidence for localization to dentate granule cells, Proc. Natl. Acad. Sci. USA, 79:193.

Ward, A.A., Jr., 1978, Glia and epilepsy, in: Dynamic Properties of Glial Cells, F. Schoffeniels, G. Franck, L. Hertz, and D.B. Tower, eds., Pergamon Press, Oxford, p. 413.

Westrum, L.E., White, L.E., Jr., and Ward, A.A., Jr., 1964, Morphology of the experimental epileptic focus, J. Neurosurg., 21:1033.

Wheal, H.V., Ashwood, T., and Lancaster, B., 1984, A comparative in vitro study of the kainic acid lesioned and bicuculline treated hippocampus: chronic and acute models of focal epilepsy, in: Electrophysiology of Epilepsy, P.A. Schwartzkroin and H.V. Wheal, eds., Academic Press, London, p. 173.

Woodbury, D.M., and Kemp, J.W., 1977, Basic mechanisms of seizures: neurophysiological and biochemical etiology, in: *Psychopathology and Brain Dysfunction*, C. Shagass, S. Gershon, and A.J. Friedhof, eds., Raven Press, New York, p. 149.

SESSION III
EXCITATORY AMINO ACIDS AND THE BLOOD-BRAIN BARRIER

PATHOPHYSIOLOGICAL ASPECTS OF BLOOD-BRAIN BARRIER PERMEABILITY IN EPILEPTIC SEIZURES

C. Nitsch, G. Goping and I. Klatzo

Laboratory of Neuropathology and Neuroanatomical Sciences
NINCDS, NIH, Bethesda, Maryland 20205, and (C.N.) Anatomical
Institute, University of Munich, Pettenkoferstr. 11
D-8000 München 2, F.R.G.

SUMMARY

Blood-brain barrier (BBB) permeability to macromolecules was assessed during seizures induced by pentylenetetrazole, bicuculline, methoxypyridoxine, methionine sulfoximine, and kainic acid. It was observed that each convulsant induced a specific pattern of regional BBB opening. This was, however, only the case when systemic blood pressure (BP) rose with seizure onset. The analysis of regional cerebral blood flow revealed that a high increase in flow in rabbits with BP rise is related to the normal flow at rest in the single brain region, but not to BBB permeability. In rabbits without BP increase, regional flow increase was low but well modulated and is possibly a better indicator for neuronal activity. The ultrastructural analysis showed that macromolecular transport over the cerebrovascular endothelium is by pinocytosis, an neurotransmitter controlled process. It is suggested that seizure-induced regional BBB opening is determined by two factors: release of neurotransmitters due to the process of autoregulation during peripheral pressure increase, and change in local neurotransmitter milieu due to the action of the convulsant and/or the seizure activity.

THE PRINCIPLE OF THE BLOOD-BRAIN BARRIER

These are exactly 100 years since Paul Ehrlich coined the term blood-brain barrier (BBB), based on the observation that dyes injected intravenously into laboratory animals, although staining other organs, did not penetrate in the brain (Ehrlich, 1885). Fifty years ago, Hugo Spatz demonstrated that the barrier is localized at the luminal surface of the intracerebral vessels (Spatz, 1934). This view was generally accepted only when Brightman and Reese (1969) proved on the ultrastructural level that a belt of interendothelial tight junctions (TJ) effectively restricts the passive transport of macromolecules from the vessel lumen to the brain parenchym and, vice versa, from the nervous tissue to the blood stream. Spatz (1934) also claimed that the problem of BBB is only part of the basic problem of vascular permeability in general ('Die Permeabilität im Gehirn ist nur gradweise von der der übrigen Organe verschieden'...sie 'ist ein Teil des großen Problems der Durchlässigkeit der Gefäßwände überhaupt.'). In fact, interendothelial TJ are also present in continuous capillaries of peripheral tissue (e.g., Simionescu, 1983); the main structural difference is the paucity of pinocytotic vesicles in endothelial cells of cerebral vessels (Wolff, 1963; Westergaard, 1980). Capillary pinocytosis can be mediated by the cyclic AMP system (Wagner et al., 1972; Sato et al., 1974) and, actually, brain vessels do possess adenylate cyclase (Joó et al., 1975; Szumanska et al., 1984). Thus, it may be argued that it is not so much a lack of capacity but some as yet unknown inhibitory factor(s) which prevent the macromolecular transport in the cerebrovascular system. Stated in other terms, do neurons and glia (i.e., cells of ectodermal origin) have the capacity to control inherent qualities of the endothelium (cells of mesodermal origin)?

One possible approach to elucidate the mechanisms controlling the transport over the BBB would be to alter the permeability in an experimentally controlled design. Events causing a readily reversible BBB opening such as hypertension (Johansson et al., 1970; Nagy et al., 1979; Hardebo, 1980) or epileptic seizures (Lorenzo et al., 1972; Petito et al., 1977; Westergaard et al., 1978; Nitsch and Klatzo, 1983) look promising in this respect.

REGIONAL DISTRIBUTION OF SEIZURE-EVOKED BBB OPENING

There exist numerous possibilities to induce seizures accompanied by BBB opening, yet, not in all studies published care was taken to control concomitant factors which, per se, might result in a BBB dysfunction. Besides hypertension, these are hypercapnia (Cutler and Barlow, 1966), hypoxia (Dux et al., 1984), and severe metabolic alterations such as lactacidosis (Paljarvi et al., 1983) and increased ammonia level (Laursen and Westergaard, 1977).

We investigated in rabbits the BBB opening by recording the extravasation into brain tissue of Evans blue (EB) given intravenously before seizure onset. Several convulsant drugs with different action mechanisms resulting in varying clinical pictures were tested. As a rule, BBB opening takes place at the beginning of generalized seizure activity and is reversible within five minutes (probably earlier, but we did not check this; Nitsch and Klatzo, 1983).

Pentylenetetrazole

Pentylenetetrazole (PTZ) is widely used in screening tests for potential anticonvulsants. It is thought to interfere with intracellular Ca^{2+}-storage (Sugaya et al., 1982) but recent studies suggest also a benzodiazepine receptor blocking property (Chweh et al., 1983). A single rapid intravenous administration of 50 mg/kg PTZ resulted in immediate generalized tonic-clonic convulsions which lasted with intermittent pauses for 5 to 15 minutes. After a short postconvulsive stuporous period, the animal recovered to full preexperimental activity. BBB opening was present in all rabbits with full developed convulsions. It was conspicuous in cerebellum and basal periventricular brain regions (Table 1). Detailed pictures have been published previously (Nitsch and Klatzo, 1983).

Bicuculline

Bicuculline (BC) is an effective convulsant with a clinical picture similar to PTZ. Its action mechanism is well established: it blocks GABA receptors and, therefore, prevents GABAergic transneuronal inhibition (Curtis et al., 1971). A dose of 0.4 mg/kg of +-bicuculline intravenously induced generalized tonic-clonic convulsions which ceased after about 20 minutes with full recovery of the rabbit. Only rarely, a transition

Table 1. Regional Distribution of Seizure-Induced BBB Opening

Region	PTZ	BC	MP	MSO	KA
Neocortex	-	-	++	+	+++ *
Entorhinal cortex	+	++	-	+	-
Basal forebrain	++	++	-	-	+
Preoptic area	++++	+++	+	+	-
Septum	+++	++	++	-	++
Amygdala	-	+	-	-	++
Caudate nucleus	-	-	-	-	-
Putamen	-	+	-	-	-
Pallidum	++	++++	-	-	-
Thalamus	++	++	+++	++	++ *
Hypothalamus	++++	++++	++	+	++
Corpus mamillare	-	-	-	++++	-
Hippocampus	-	++	+++	-	+
Midbrain tegmentum	+++	+++	++	++	+
Superior colliculus	-	-	-	-	-
Inferior colliculus	++	+	-	+	-
Medulla oblongata	+++	+++	++	+	-
Cerebellar cortex	++++	++	+	+	-

-: no EB extravasation, +: rare staining, ++: staining observed in up to half of the animals tested, +++: frequent EB staining, ++++: regular and very dense EB staining. *: asymmetric BBB opening.

occurred to prolonged status epilepticus. With respect to BBB opening, the distinct selective bilateral staining of the pallidum was remarkable (Fig. 1a). It was often accompanied by EB extravasation into preoptic area (Fig. 1a) and hypothalamus (Fig. 1b). In thalamus, single nuclei were stained in a strictly symmetric fashion (Fig. 1b). In some cases, also the basotemporal parts of the hippocampus showed BBB opening (Fig. 1b). For the posterior brain areas, see summary given in Table 1.

Methoxypyridoxine

Methoxypyridoxine (MP) induces, in a dosage of 100 mg/kg, generalized seizures after a delay of 15 minutes in freely moving rabbits, up to 25 minutes in immobilized ones. As a vitamin B6 antagonist, MP blocks the activity of glutamate decarboxylase (Nitsch, 1980), the biosynthetic enzyme

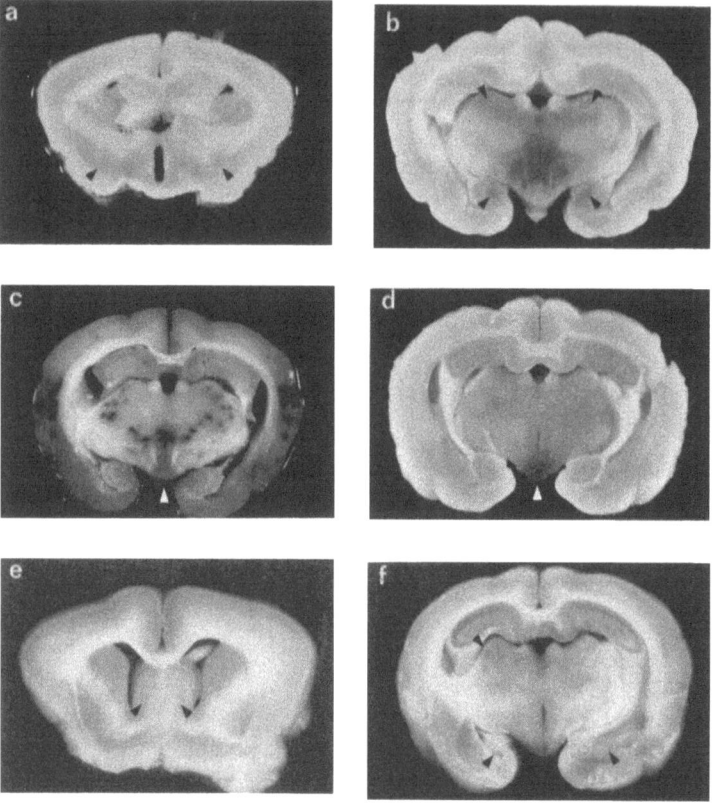

Fig. 1. Seizure induced BBB opening to Evans blue-albumin complex. a) BC seizure, bilateral BBB opening in pallidum and preoptic area. b) BC seizure, symmetric EB leakage in single nuclei of the thalamus and in temporal part of hippocampus. c) MSO seizure, EB given before seizure onset stains the corpus mamillare, HRP given 30 minutes later cannot enter corpus mamillare but leaks into thalamus and hypothalamus and in temporal neocortex. d) Selective MSO induced EB staining of corpus mamillare. e) KA seizure, bilateral BBB opening in septum. f) KA seizure, BBB opening in basotemporal part of hippocampus and unilateral leakage in laterodorsal thalamus.

of the inhibitory transmitter GABA. Thus, GABA content decreases preictally in brain regions with a high GABA turnover (Nitsch and Okada, 1976), leading to generalized seizures which proceed toward a fatal status epilepticus. BBB opening during the early seizure periods occurred only in half of the rabbits tested. It was mainly found in the temporal half of the hippocampus and in thalamus (Table 1). Although, as with BC, an impairment of GABA transmission is responsible for the seizures, in pallidum BBB remained intact during MP induced convulsions (see also Nitsch and Klatzo, 1983).

Methionine Sulfoximine

Methionine sulfoximine (MSO) is an irreversible inhibitor of glutamine synthetase (Lamar and Sellinger, 1965), so that the level of the excitatory substances, glutamate and ammonium, increases. With 350 mg/kg MSO, intermittent paroxysmal myoclonic fits started after two hours. Transition into the fatal status epilepticus took place after three hours. Quite outstandingly, BBB opening was confined in many cases to the mamillary bodies (Fig. 1d), a region affected by none of the other treatments. Sometimes, the thalamus showed EB extravasation in a spotlike symmetric fashion. Also, midbrain areas were occasionally involved (Table 1).

With MSO, due to the attenuated progression of the convulsions, it was possible to test BBB permeability at various time intervals. EB given before the first seizure event stained the mamillary bodies selectively. Horseradish peroxidase (HRP) injected 30 minutes later could not enter the mamillary bodies, but was present in thalamic and basocortical regions (Fig. 1c). This disparity in regional distribution of BBB opening suggests that during seizure development alternating factors control in a regionally selective manner the transport over the BBB.

Kainic Acid

Kainic acid (KA) must be given to rabbits in a relatively high dosage (35 mg/kg) in order to induce generalized seizures by systemic application. The animals died during the second or third generalized convulsion. EB leakage was comparable to that described in the meantime in rats (Zucker et al., 1983; Ruth, 1984). It mainly involved limbic regions such as septum (Fig. 1e) and hippocampus, where only the basotemporal part was affected (Fig. 1f). In contrast to all the other convulsants used, KA elicited asymmetric BBB leakage in thalamus (Fig. 1f) and in occipital cortex (Table 1).

Influence of Systemic Blood Pressure Increase on BBB Opening

The main result of the experiments described above is the selective, strictly bilateral BBB opening in anatomically circumscribed brain regions, in areas which can be regarded as functional entities. Since the convulsants were given systemically and no toxic action on mesodermal cells is known, but on the other hand, the convulsants interfere with neurotransmitters distributed highly selectively in the various neuronal populations, the conclusion may be drawn that BBB is controlled by neurotransmitters. A disequilibrium in their level, induced by the convulsant and/or the seizure, could bring about an opening of the BBB solely in selected regions.

However, as already shown by Petito et al. (1977) and Westergaard et al. (1978), and again demonstrated by us (Nitsch and Klatzo, 1983), BBB opening occurs only in those animals in which the systemic blood pressure (BP) increases with seizure onset. Still, seizure-induced BBB opening differs in several aspects from hypertension-induced BBB leakage:
1. velocity and amplitude of BP increase must be much higher to cause a hypertension-induced BBB opening than is the case with PTZ, BC and MP (Nitsch and Klatzo, 1983) and
2. the localization of hypertension-induced BBB leakage differs, i.e., it is found in an asymmetric fashion in neocortical areas (Blomstrand et al., 1975) and never in deep brain regions. (This may provide the explanation for the asymmetric BBB opening in occipital cortex of KA treated rabbits. Systemic KA induces a preictal rise in BP up to 180 mm Hg, suggesting that pure hypertension induced BBB leakage is intermingled with seizure evoked BBB opening.)

Nevertheless, the BP involvement in the induction of BBB opening caused us to investigate their relations to regional cerebral blood flow (rCBF) increase.

rCBF INCREASE AT SEIZURE ONSET

rCBF was estimated using the ^{3}H-nicotine method of Ohno et al. (1979), as applied to rabbits by Suzuki et al. (1984). In short, ^{3}H-nicotine was infused intravenously over 40 seconds and the presence of the fully extracted tracer detected in freshly dissected discrete brain areas by liquid scintillation. The single rCBF values were then calculated in relation to the integral of the arterial blood concentration of the tracer. rCBF was determined in controls, during acute adrenaline-induced hypertension (Suzuki et al., 1984) and at the onset of seizures evoked by BC, MP or KA (Nitsch et al., 1985). (MSO was not included, since the protracted interval between application of the convulsant and seizure onset would necessitate an overly long immobilization of the experimental animal).

Bicuculline

BC seizures were accompanied with a rise of BP in five out of the six rabbits tested. In these animals, CBF increased three- to five-fold. No relation was detectable between relative CBF increase and presence or absence of BBB opening (Nitsch et al., 1985). A data analysis on the basis of absolute increases in flow over their respective control values provided some more insight. Whereas again no relation between CBF increase and BBB leakage was evident, it was, on the other hand, obvious that a

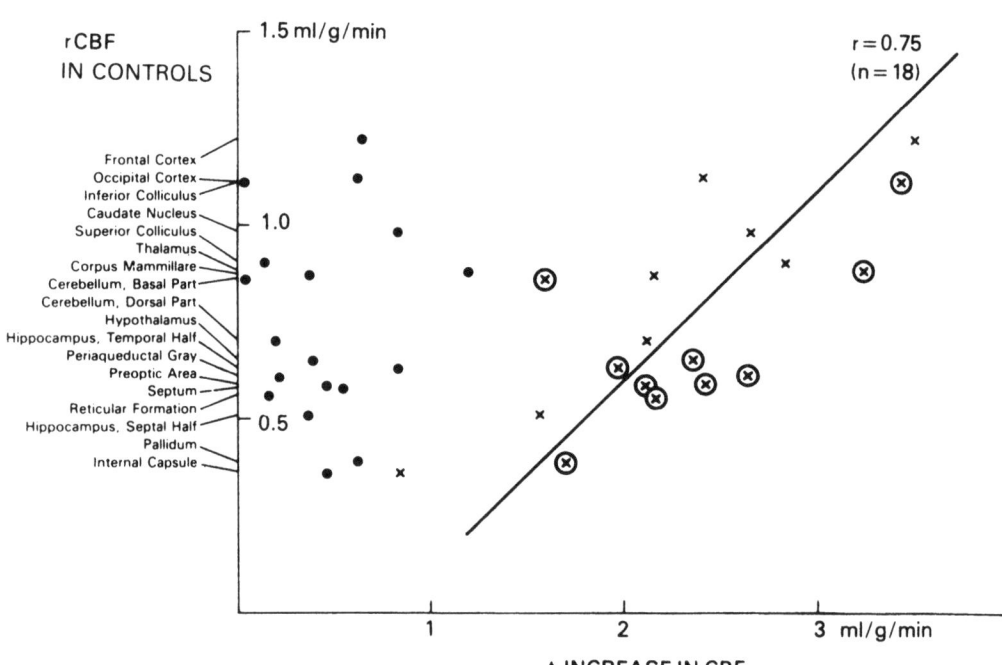

Fig. 2. Dependence of absolute increase in rCBF and respective control CBF is given in the five animals with BP increase at seizure onset. Their values are indicated by crosses. The regression line is drawn in and the regression coefficient for these two parameters given. Values from regions in which BBB opening was observed are encircled. The black dots are the CBF data from the one animal with unchanged BP.

positive relation existed between control CBF and seizure-induced CBF (Fig. 2). This suggests that the capacity to increase CBF is determined by the flow at rest.

In the one animal without a BP rise at seizure onset, CBF increased nevertheless, although only up to 200%. A relation between control and experimental values was absent in this single case.

Methoxypyridoxine

MP seizures are less often accompanied by BP increase, a fact which explains the absence of BBB leakage in about half of the animals tested. In the present series, four rabbits belonged to the group with unchanged BP. This allowed a more reliable evaluation of rCBF not influenced by peripheral pressure changes than in the case of the single BC treated animal described above. rCBF remained unaltered in inferior colliculus, it doubled in most brain regions and reached 260% of control in thalamus and 370% in pallidum (Nitsch et al., 1985). In absolute terms, rCBF increase ranged between 0 and 1.4 ml/g/hr (Fig. 3). These are values well above those detected in hypertension induced CBF in areas exhibiting BBB leakage (Suzuki et al., 1984).

In the six MP treated rabbits with increased BP at seizure onset,

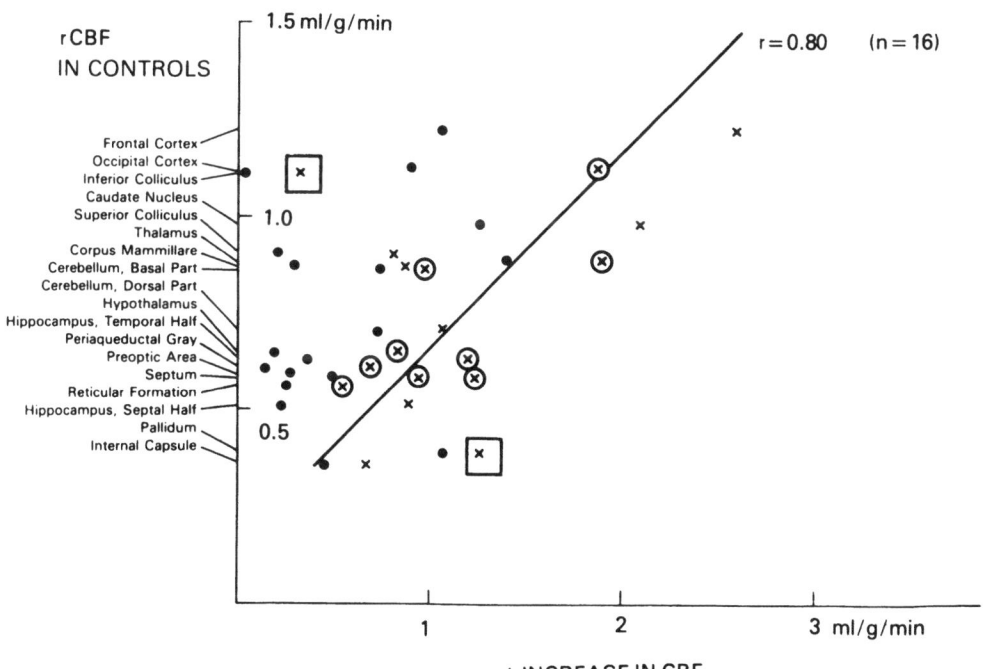

Fig. 3. Relation between absolute increase in rCBF at seizure onset and control CBF for four animals with unchanged BP (black dots) and six animals with BP increase (crosses). Regression coefficient was calculated using values of 16 regions from the animals with increased BP. The crosses surrounded by a square indicate the two areas not included, inferior colliculus and pallidum, and the encircled data are from regions exhibiting BBB opening.

rCBF rose two- to three-fold, i.e., somewhat less than in the case of
BC seizures (Nitsch et al., 1985). Nevertheless, a good relation existed
for most brain areas between normal CBF and CBF rise due to MP seizures
(Fig. 3). Two areas did not fit into the general pattern: the inferior
colliculus with an insignificant change and the pallidum with too high
an increase. Still, in pallidum, as in other areas with substantial CBF
elevations, BBB remained unpermeable for macromolecules, and no interaction
seemed to exist between CBF increase and BBB opening.

Kainic Acid

Intravenously applied KA induces a BP increase before seizure onset
(Nitsch et al., 1985). Yet, overall rise in CBF was lower at the beginning
of seizure activity as compared to BC and MP seizures. CBF was unaltered
from normal in inferior colliculus and both parts of the cerebellum.
It was, however, relatively high in regions belonging, in the broad sense,
to the limbic system. These were the mamillary bodies, hypothalamus,
preoptic area, septum and hippocampus. Of these, only hypothalamus, septum
and temporal part of the hippocampus showed BBB leakage. When these regions
and the inferior colliculus were omitted from the calculation, a significant
positive relation between original CBF and KA-induced CBF increase was
revealed for the remaining brain areas (Fig. 4).

Conclusion

There does not seem to exist any relationship between degree of CBF

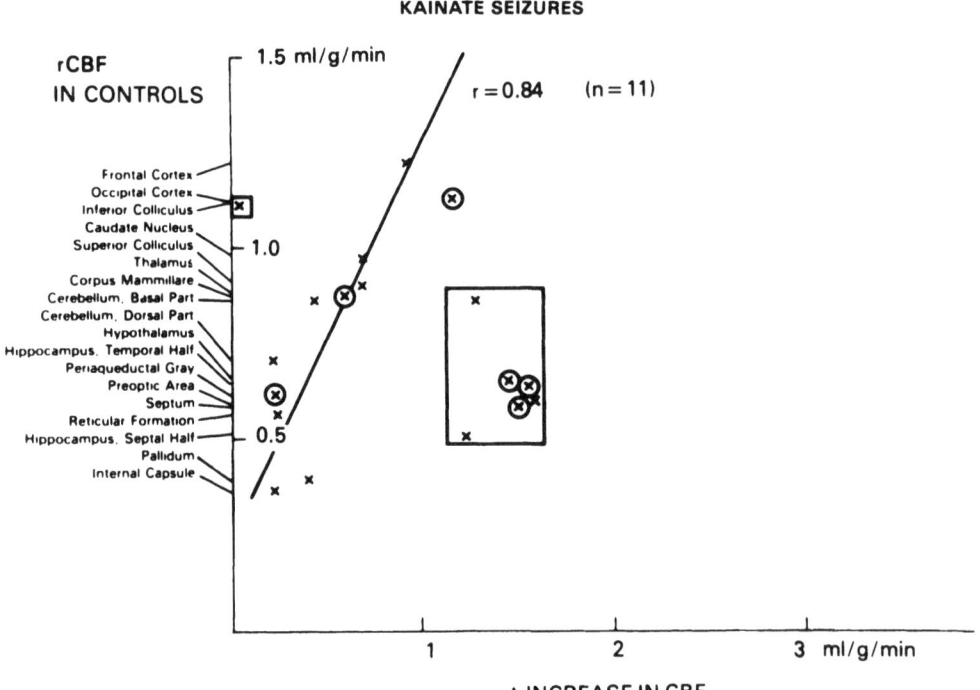

Fig. 4. Comparison between control CBF and absolute increase in flow
from seven KA treated rabbits. Values from inferior colliculus
and limbic brain areas are surrounded by a square and were omitted
from the calculation of the regression.

increase and macromolecular transport over the BBB during seizures. In the contrary, amount of CBF increase was determined by the preexisting flow rate. Only if seizure activity was outstandingly intense in a few regions, as under KA in areas belonging to the limbic system (Lothman and Collins, 1981), was the increase higher than could be expected. It would be worthwhile to investigate in greater detail rCBF during seizures with unchanged BP. These data may be a better correlate of regional neuronal activity than rCBF values merged with peripheral changes in flow.

On the other hand, a connection between CBF increase and hypertension induced BBB leakage was clearly established (Suzuki et al., 1984; Abdul-Rahman et al., 1979). Obviously, a seizure induced tripling of CBF is well tolerated by the BBB, whereas a doubling of CBF during hypertension brings about BBB leakage. CBF is independent from systemic BP at pressures between 60 mm Hg and 160 mm Hg (Pollay and Roberts, 1980). This autoregulation is achieved by vasodilation at increasing peripheral pressure and vasoconstriction at decreasing pressure. These processes are under the control of neurotransmitters (McCulloch and Edvinssson, 1984). Some of the transmitters liberated during the process of autoregulation are also involved in the control of transendothelial pinocytosis. Histamine is a vasodilatator (McCulloch and Edvinsson, 1984), is linked to adenylate cyclase (Karnushina et al., 1980) and stimulates transendothelial pinocytosis (Dux and Joó, 1982). Also, acetylcholine dilates brain vessels and induces an increased transport of macromolecules over BBB (Domer et al., 1983). Adrenaline and noradrenaline act at the α_2-receptor as vasoconstrictors (McCulloch and Edvinsson, 1984) and may regulate adenylate cyclase activity (Karnushina et. al., 1982). Serotonin, a modulator of preexisting vessel tone (McCulloch and Edvinsson, 1984), is known to induce transendothelial pinocytosis (Westergaard, 1980).

Yet, before definitely claiming that BBB permeability is determined by the neurotransmitter milieu around the endothelium, it must be verified whether, under our experimental conditions, macromolecules take the transendothelial route on their way to the neuropil.

TRANSENDOTHELIAL OR INTERENDOTHELIAL ROUTE?

There are still some discrepancies in the literature on whether macromolecules leaking over the BBB take the transendothelial passage, via pinocytosis, or the interendothelial one, via disruption of TJ. Although most papers, both on epilepsy (Petito et al., 1977; Westergaard et al., 1978) and on hypertension (Westergaard et al., 1977), favor the transendothelial route, the claim of Brightman et al. (1970) and Rapoport (1976) is difficult to reject in that often massive accumulation of tracer in the neuropil is hard to explain by the energy-consuming process of pinocytosis alone.

For our detailed regional ultrastructural investigations we choose BC. Preliminary experiments were also carried out with MSO and KA. As electron dense tracer for the transport of macromolecules, HRP (Type II, Sigma, 100 mg/kg) was injected intravenously, followed by BC 10 minutes later. After 10 to 15 minutes of seizure activity, rabbits were perfused in deep Nembutal anesthesia with saline and aldehyde fixatives. Slices of the brain were submitted to the peroxidase reaction, using the enzymatic oxidation of glucose with glucose oxidase as H_2O_2 donor and diaminobenzidine as indicator. Selected areas showing the brown reaction product were processed for electron microscopy.

In brain areas showing tracer extravasation, HRP was present in the basement membranes of arterioles and, more or less diffuse, in surrounding

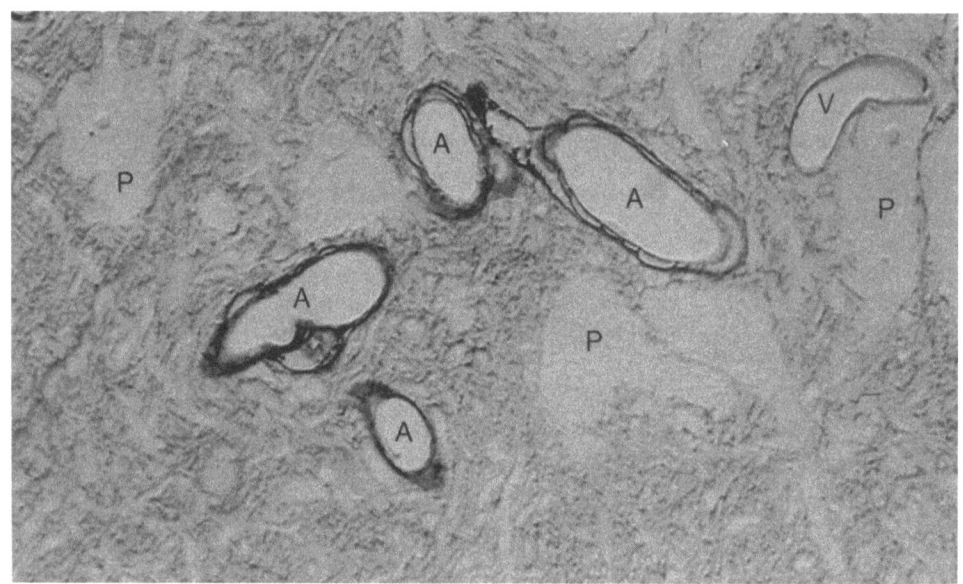

Fig. 5. Semithin section from the hypothalamus of a BC-treated rabbit. Dense HRP reaction product is present in the basement membranes of arterioles (A) and is surrounding smooth muscle cells. Basement membranes of venules (V) are only partially stained. Tracer is found spotlike in the neuropil. Neuronal perikarya (P), however, are always unstained.

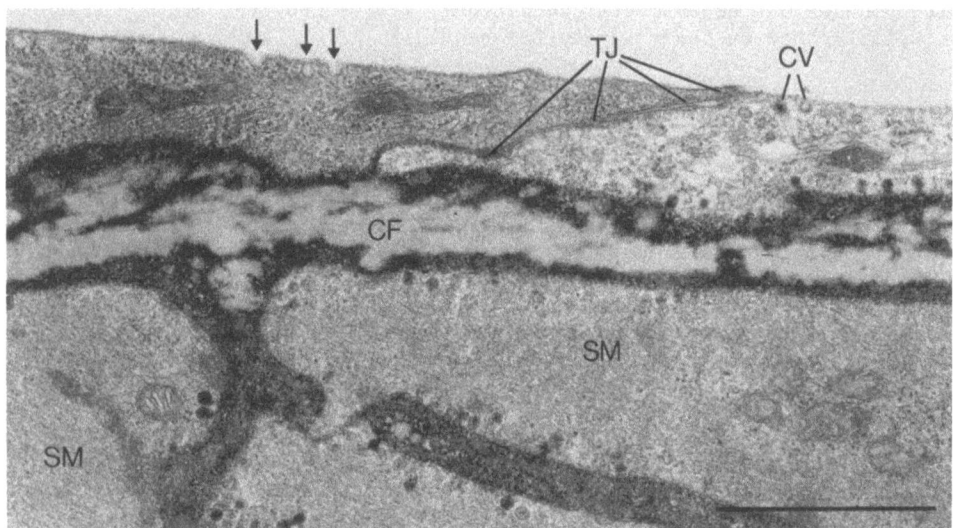

Fig. 6. Wall of an arteriole in the midbrain periaqueductal gray of a BC-treated rabbit. The tight junctions (TJ) are intact. Empty vesicles (arrow) are present at the luminal surface of an endothelial cell. In the neighboring cell with light cytoplasm, coated vesicles (CV) are involved in HRP transport. At the abluminal site, many tracer-filled vesicles open to the basement membrane with its collagen fibrils (CF). Caveolae of smooth muscle cells (SM) contain reaction product.

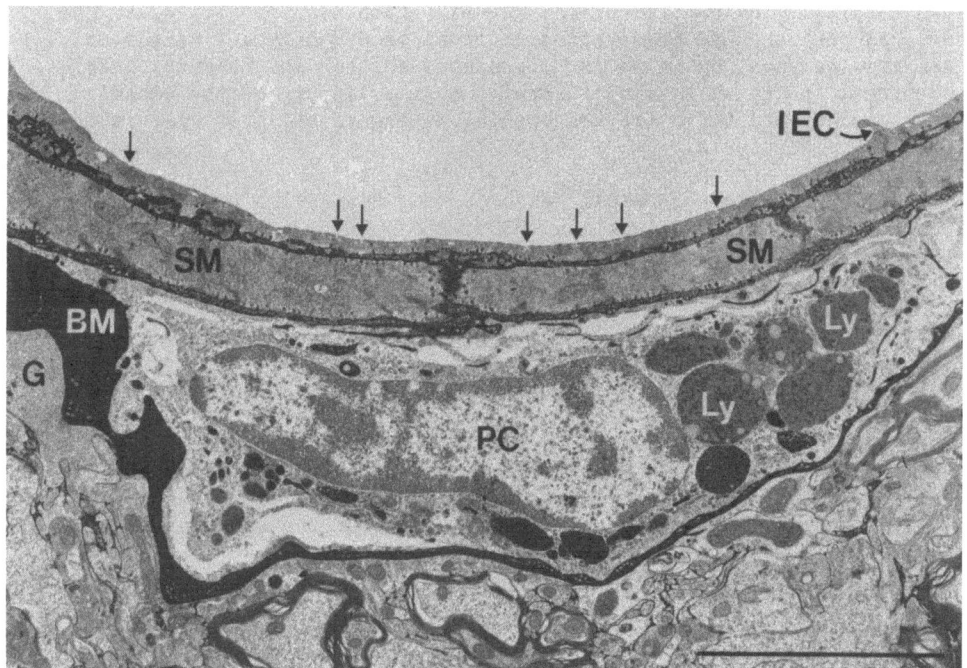

Fig. 7. Arteriole in gray matter of medulla oblongata after 10 minutes of BC induced seizure activity. At the luminal side of the endothelium, numerous empty pinocytotic vesicles (arrow) are present and at the abluminal side even more tracer-filled vesicles. The network of basement membranes (BM) is filled with HRP. The pericyte (PC) exhibits peroxidase activity in some lysosomes (Ly) and in vacuoles. Tracer is leaking between glial cells (G) into the neuropil.

neuropil (Fig. 5.). Uptake into neurons was never observed at the light microscopic level. At the ultrastructural level, it became evident that the tracer was using pinocytotic vesicles to traverse the endothelial cells (Figs. 6, 7). Some of the transport vesicles exhibited clearly the features of coated vesicles (Fig. 6). The TJ between endothelial cells remained intact (Fig. 6, 7). In few cases, in hippocampus, hypothalamus and midbrain periaqueductal gray, tracer was found in between two TJ, but there it was possible to explain this by intraendothelial vesicles ejecting their content into the interendothelial cleft (Nitsch et al., 1986). In most brain areas, arterioles were the vascular segment exhibiting HRP transport. Only in thalamus and hippocampus, capillaries were involved as well. In thalamus, hypothalamus and septum, pinocytosis was also found in venules (see Nitsch et al., 1986).

The pinocytotic vesicles ejected their content into the space of the vascular basement membranes where it surrounded collagen fibrils (Figs. 6, 7). Reaction product was present in caveolae of smooth muscle cells. Uptake in pericytes was conspicuous (Fig. 7). These cells are known to take up HRP injected intracerebrally (Mato et al., 1980), but they are obviously not an effective barrier for further spread of HRP. The tracer leaked from the glial basement membrane between the perivascular glial endfeet into the surrounding neuropil (Fig. 7).

In the case of MSO, tracer transport was via pinocytosis alone (unpublished result). In the case of KA, the situation was complicated by the occasional perivascular hemorrhages in areas of BBB leakage, namely in cortex and thalamus, but also in hippocampus (Nitsch and Hubauer, 1986). In contrast, in septum no indication for a gross injury of the vessel wall was found, and there the BBB opening during KA seizures was via transendothelial pinocytosis.

Thus, under mild reversible seizures (BC) and during prolonged seizures before entering status epilepticus (MSO), BBB opening is the result of an energy-consuming, second-messenger regulated process.

SYNOPSIS

The exact localization of BBB opening is probably determined by two factors:
1. release of substances which regulate the vascular tone during an event which increases BP within the limits of autoregulation, and
2. the neurotransmitter microenvironment in the vicinity of these vessels, which evolves as a result of the action of the convulsant and/or the seizure.

The milieu present at seizure onset of different genesis will be quite different in various functional entities and thus, might antagonize or might enhance the adenylate cyclase stimulating effect of the vessel tone regulating transmitters.

The present experiments lend support to the hypothesis that neurons (and glia in its role for neurotransmitter inactivation by uptake) participate in the control of vascular transport. This, of course, opens a spectrum of new aspects. It may even be questioned whether BBB opening is a pathological, a negative event, as generally assumed (see Rapoport, 1976; Bradbury, 1979). There might well exist circumstances under which an unrestricted diffusion is desirable, of nurturing substances from blood to nervous tissue and of toxic metabolites in the other direction. Further experiments will show whether this situation is present during epileptic seizures.

REFERENCES

Abdul-Rahman, A., Dahlgren, N., Johansson, B.B., and Siesjö, B.K., 1979, Increase in local cerebral blood flow induced by circulating adrenaline - Involvement of blood-brain barrier dysfunction, Acta Physiol. Scand. 107:227.
Blomstrand, C., Johansson, B., and Rosengren, B., 1975, Blood-brain barrier lesions in acute hypertension in rabbits after unilateral X-ray exposure of brain, Acta Neuropathol., 31:97.
Bradbury, M., 1979, The Concept of a Blood-Brain Barrier, John Wiley and Sons, Chichester.
Brightman, M.W., and Reese, T.S., 1969, Junctions between intimately apposed cell membranes in the vertebrate brain, J. Cell Biol., 40:648.
Brightman, M.W., Klatzo, I. Olsson, Y., and Reese, T.S., 1970, The bloodbrain barrier to proteins under normal and pathological conditions, J. Neurol. Sci., 10:215.
Chweh, A.Y., Swinyard, E.A., and Wolf, H.H., 1983, Pentylenetetrazole may discriminate between different types of benzodiazepine receptors, J. Neurochem., 41:830.
Curtis, D.R., Duggan, A.W., Felix, D., and Johnston, G.A.R., 1971, Bicuculline, an antagonist of GABA and synaptic inhibition in the spinal cord of the cat, Brain Res., 32:69.

Cutler, R.W.P., and Barlow, C.F., 1966, The effect of hypercapnia on brain permeability to protein, Arch. Neurol., 14:54.

Domer, F.R., Boertje, S.B., Bing, E.G., and Reddix, I., 1983, Histamine- and acetylcholine-induced changes in the permeability of the blood-brain barrier of normotensive and spontaneously hypertensive rats, Neuropharmacology, 22:615.

Dux, E., and Joó, F., 1982, Effects of histamine on brain capillaries, Exp. Brain Res., 47:252.

Dux, E., Temesvari, P., Joó, F., Adam, G., Clementi, F., Dux, L., Hideg, J., and Hossmann, K.-A., 1984, The blood-brain barrier in hypoxia: ultrastructural aspects and adenylate cyclase activity of brain capillaries, Neuroscience, 12:951.

Ehrlich, P., 1885, Das Sauerstoffbedürfnis des Organismus. Eine farbenanalytische Studie, cited from: Paul Ehrlich - Gesammelte Arbeiten, F. Himmelweit, ed., Springer, Berlin, 1956.

Hardebo, J.E., 1980, A time study in rat on the opening and reclosure of the blood-brain barrier after hypertensive or hypertonic insult, Exp. Neurol., 70:155.

Johansson, B., Li, C.L., Olsson, Y., and Klatzo, I., 1970, The effect of acute arterial hypertension on the blood-brain barrier to protein tracers, Acta Neuropathol., 16:117.

Joó, F., Rakonczay, Z., and Wollemann, M., 1975, cAMP-mediated regulation of the permeability in the brain capillaries, Experientia, 31:582.

Karnushina, I.L., Palacios, J.M., Barbin, G., Dux, E., Joó, F., and Schwartz, J.C., 1980, Studies on capillary-rich fraction isolated from brain: histaminic components and characterization of the histamine receptors linked to adenylate cyclase, J. Neurochem., 34:1201.

Karnushina, I.L., Spatz, M., Bembry, J., 1982, Cerebral endothelial cell culture. I. The presence of α_2- and β_2-adrenergic receptors linked to adenylate cyclase activity, Life Sci., 33:849.

Lamar, Jr., C., and Sellinger, O.Z., 1965, The inhibition in vivo of cerebral glutamine synthetase and glutamine transferase by the convulsant methionine sulfoximine, Biochem. Pharmacol., 14:489.

Laursen, H., and Westergaard, E., 1977, Enhanced permeability to horseradish peroxidase across cerebral vessels in the rat after portocaval anastomosis, Neuropath. Appl. Neurobiol., 3:29.

Lorenzo, A.V., Shirahige, I., Liang, M., and Barlow, C.F., 1972, Temporary alteration of cerebrovascular permeability to plasma protein during drug-induced seizures, Am. J. Physiol., 223:268.

Lothman, E.W., and Collins, R.C., 1981, Kainic acid induced limbic seizures - metabolic, behavioral, electroencephalographic and neuropathological correlates, Brain Res., 218:299.

Mato, M., Ookawara, S., and Kurihara, K., 1980, Uptake of exogenous substances and marked infoldings of the fluorescent granular pericyte in cerebral fine vessels, Am. J. Anat., 157:329.

McCulloch, J., and Edvinsson, E., 1984, Cerebrovascular smooth muscle reactivity: a critical appraisal of in vitro and in situ techniques, J. Cerebr. Blood Flow Metab., 4:129.

Nagy, Z., Mathieson, G., and Hüttner, I., 1979, Blood-brain barrier opening to horseradish peroxidase in acute arterial hypertension, Acta Neuropath. 48:45.

Nitsch, C., and Okada, Y., 1976, Differential decrease of GABA in the substantia nigra and other discrete regions of the rabbit brain during the preictal period of methoxypyridoxine-induced seizures, Brain Res., 105:173.

Nitsch, C., 1980, Regulation of GABA metabolism in discrete rabbit brain regions under methoxypyridoxine - Regional differences in cofactor saturation and the preictal activation of glutamate decarboxylase activity, J. Neurochem., 34:822.

Nitsch, C., and Klatzo, I., 1983, Regional patterns of blood-brain barrier breakdown during epileptiform seizures induced by various convulsive agents, J. Neurol. Sci., 59:305.

Nitsch, C., Suzuki, R., Fujiwara, K., and Klatzo, I., 1985, Incongruence of cerebral blood flow increase and blood-brain barrier opening in rabbits at the onset of seizures induced by bicuculline, methoxypyridoxine, and kainic acid, J. Neurol. Sci., 67:67.

Nitsch, C., Goping, G., Laursen, H., and Klatzo, I., 1986, The blood-brain barrier to horseradish peroxidase at the onset of bicuculline-induced seizures in hypothalamus, pallidum, hippocampus and other selected regions of the rabbit, Acta Neuropathol., 69:1.

Nitsch, C., and Hubauer, H., 1986, Distant blood-brain barrier opening in subfields of the rat hippocampus after intrastriatal injections of kainic acid but not ibotenic acid, Neurosci. Lett., 64:53.

Ohno, K., Pettigrew, K.D., and Rapoport, S.I., 1979, Local cerebral blood flow in the conscious rat as measured with ^{14}C-antipyrine, ^{14}C-iodoantipyrine and ^{3}H-nicotine, Stroke, 10:62.

Paljärvi, L., Rehncrona, S, Söderfeldt, B., Olsson, Y., and Kalimo, H., 1983, Brain lactic acidosis and ischemic cell damage: quantitative ultrastructural changes in capillaries of rat cerebral cortex, Acta Neuropathol., 60:232.

Petito, C.K., Schaefer, J.A., and Plum F., 1977, Ultrastructural characteristics of the brain and blood-brain barrier in experimental seizures, Brain Res., 127:251.

Pollay, M., and Roberts, P.A., 1980, Blood-brain barrier: a definition of normal and altered function, Neurosurgery, 6:675.

Rapoport, S.I., 1976, Blood-Brain Barrier in Physiology and Medicine, Raven Press, New York.

Ruth, R.E., 1984, Increased cerebrovascular permeability to protein during systemic kainic acid seizure, Epilepsia, 25:259.

Sato, T., Garcia-Bunuel, R., and Brandes, D., 1974, Ultrastructural cytochemical localization of adenylate cyclase in the rat nephron, Lab. Invest., 30:222.

Simionescu, N., 1983, Cellular aspects of transcapillary exchange, Physiol. Rev., 63:1536.

Spatz, H., 1934, Die Bedeutung der vitalen Färbüng für die Lehre vom Stoffaustausch zwischen dem Zentralnervensystem und dem übrigen Körper, Arch. Psychiat. Nervenkr., 101:267.

Sugaya, E., Onozuka, M., Furuichi, H., Sugaya, A., and Tsuda, T., 1982, Intracellular calcium and bursting activity, in: Physiology and Pharmacology of Epileptogenic Phenomena, M.R. Klee, H.D. Lux, and E.J. Speckmann, eds., Raven Press, New York, p. 325.

Suzuki, R., Nitsch, C., Fujiwara, K., and Klatzo, I., 1984, Regional changes in cerebral blood flow and blood brain barrier permeability during epileptiform seizures and in acute hypertension in rabbits, J. Cerebr. Blood Flow Metab., 4:96.

Szumanska, G., Palkama, A., Lehtosalo, J.I., and Uusitalo, H., 1984, Adenylate cyclase in microvessels of the rat brain, Acta Neuropathol., 62:219.

Wagner, R.C., Krainer, P., Barnett, R.J., and Bitensky, M.W., 1972, Biochemical characterization and cytochemical localization of a catecholamine adenylate cyclase in isolated capillary endothelium, Proc. Natl. Acad. Sci. USA, 69:3176.

Westergaard, E., van Deurs, B., and Brandsted, H.E., 1977, Increased vesicular transfer of horseradish peroxidase across cerebral endothelium, evoked by acute hypertension, Acta Neuropathol., 37:141.

Westergaard, E., Hertz, M.M., and Bolwig, T.G., 1978, Increased permeability to horseradish peroxidase across cerebral vessels, evoked by electrically induced seizures in rat, Acta Neuropathol., 41:73.

Westergaard, E., 1980, Ultrastructural permeability properties of cerebral microvasculature under normal and experimental conditions after application of tracers, Adv. Neurol., 28:55.

Wolff, J. 1963, Beiträge zur Ultrastruktur der Kapillaren in der normalen Großhirnrinde, <u>Z. Zellforsch.</u>, 60:409.

Zucker, D.K., Wooten, G.F., and Lothman, E.W., 1983, Blood-brain barrier changes with kainic acid-induced limbic seizures, <u>Exp. Neurol.</u>, 79:422.

BLOOD-BRAIN BARRIER PERMEABILITY TO EXCITATORY AMINO ACIDS

J.M. Lefauconnier, Y. Tayarani and G. Bernard

INSERM U 26 (Dir: J.M. Bourre)
200 Rue du Faubourg St Denis 75010 Paris, France

INTRODUCTION

The blood-brain barrier is situated at the layer of endothelial cells of brain capillaries (Rapoport, 1976; Bradbury, 1979). These cells are connected together by tight junctions, which prevent significant paracellular diffusion, and they contain very few transfer vesicles. As a consequence of these characteristics, a blood-borne solute must successively pass through the two membranes (luminal and antiluminal) and the cytoplasm of the endothelial cells before reaching the brain extracellular space. Molecules pass through the membranes by passive diffusion as a function of the lipid solubility of their undissociated form and of the extent of their dissociation at physiological pH (only the undissociated form is able to traverse the membranes). Molecules like amino acids, which are dissociated at normal pH, are thus nearly excluded from blood-brain transport unless they are transported by a specific carrier mechanism. These specific carriers are not all symmetrically distributed on the two membranes. Carriers on the luminal membrane seem to function mainly as exchange systems, while some carriers on the antiluminal membrane can produce active transport (Goldstein and Betz, 1983). An additional regulation of the passage of solutes can be exerted by cytoplasmic enzymes which can catabolize metabolites which have entered the endothelial cytoplasm (Bertler et al., 1966).

Most neurotransmitters in the blood are prevented from reaching the brain either by their polarity and the absence of carrier on the luminal membrane, or because they are degraded inside the endothelial cells. In contrast, some of them, such as monoamines, are actively taken up into the wall of microvessels from the antiluminal side (Hardebo and Owman, 1980). Such an active high affinity uptake into isolated brain capillaries has been recently described for glutamate (Hutchinson et al., 1985).

However, as far as anionic amino acid neurotransmitters and analogs are concerned, the restriction of passage does not seem to be complete. Oldendorf and Szabo (1976) have described a blood-brain transport system of low capacity for glutamate and aspartate. In addition, systemically injected kainic acid is able to produce limbic seizures (Ben-Ari, 1985). It seemed thus of interest to try to quantitate blood-brain transport of two selected anionic amino acids: glutamate and kainate. In addition, in vitro experiments were performed to test the effects of some other anionic amino acid on glutamate high affinity uptake of brain capillaries and the possiblity of the presence of acidic amino acid receptors on the membranes of brain capillary endothelial cells, as several receptors to neurotransmitters have been already described on these membranes (Joó, 1985).

METHODS

Radioactive Products

The isotopically labeled compounds studied were L-(U-^{14}C) glutamic acid (>250 mCi/mmol), L-(1-^{14}C) glutamic acid (50-60 mCi/mmol), L-(G-^{3}H) glutamic acid (20-40 Ci/mmol), (G-^{3}H) kainic acid (2-10 Ci/mmol), and (U-^{14}C) sucrose (>350 mCi/mmol). They were obtained from Amersham Ltd. (Amersham, England).

Blood-Brain Transport of Amino Acids

Blood-brain transport of amino acids was determined according to the methods described by Ohno et al. (1978) and Blasberg et al. (1983). Two-month-old male and female Sprague-Dawley rats were anesthetized with ether, place on a heating pad and warmed if necessary with a heating lamp to obtain a rectal temperature between 36 and 37°C. A saphenous vein was catheterized and 200 μl of heparin containing saline (200 IU heparin/ml) was injected. The catheter was then connected to a syringe containing the radiotracer (5 μCi/rat of ^{14}C-labeled glutamate or 15 μCi/rat of ^{3}H-labeled kainic acid) in 200 or 300 μl of saline. A femoral artery was also catheterized and blood withdrawn at a constant rate with a pump Infu 362 (Datex, Switzerland). Five seconds after the beginning of the blood collection, the radioactive tracer was rapidly injected. In experiments lasting more than three minutes, further sampling was discontinuous. The last arterial sample was collected just before decapitation. Blood was collected from the severed neck, the brain was removed from the skull and put on ice in a Petri dish. The meninges were carefully removed and the brain was dissected into the following regions: olfactory bulbs, parietal cortex, cingulate cortex, pyriform cortex, striatum, hippocampus, thalamus, midbrain, pons-medulla, cerebellum and cervical spinal cord. The choroid plexuses, hypophysis and a small piece of muscle were also sampled. Each region was placed in a preweighed scintillation vial. The vial and region was then reweighed. The region was digested in a tissue solubilizer (Soluene, Packard) in a water bath at 50°C. After cooling, a scintillation mixture containing 10 ml toluene, 40 mg 2,5-diphenyl-oxazole and 1 mg dimethyl 1,4-di(2-5-phenyloxazolyl)benzene was added. The following day, the vials were counted in an Intertechnique SL 3000 scintillation spectrophotometer. Counts per minutes were converted to disintegrations per minute by using external standardization and predetermined efficiency curves that allowed correction for quenching.

Calculations

The blood-brain influx rate of a substance is equal to the product of its concentration in plasma by a transfer constant Ki. The transfer constants were calculated differently for glutamate and kainate.

The transfer constant for glutamate should be determined over a very short period of time in order to avoid the systemic metabolism of the injected radioactive amino acid. This was thus performed by decapitating the rats 40 seconds after the injection of either L-(U-^{14}C) or L-(1-^{14}C) glutamate. The following equation was used (Ohno et al., 1978):

$$K_i = \frac{Q_m(T) - Q_{vasc}}{\int_0^T C_a(t)\, dt} \quad (1)$$

Where $Q_m(T)$ is the measured tracer content in brain at the time of sacrifice T, C_a the concentration of tracer in arterial plasma. The integral $\int_0^T C_a(t)\, dt$ is also equal to $\overline{C_a} \times T$, with $\overline{C_a}$ the mean concentration of tracer in arterial plasma from time 0 to time T. Q_{vasc} is the radioactivity remaining in blood vasculature or in compartments equilibrated with plasma at the time of decapitation. It was calculated as follows:

$$Q_{vasc} = C_a(T) \times V(T) \quad (2)$$

Where $C_a(T)$ is the concentration of radioactivity in plasma at the time of sacrifice, $V(T)$ is the plasma volume plus the brain volume whose radioactive concentration is equilibrated with that of plasma at the time of sacrifice. The values used for $V(T)$ in the experiments with glutamate were those obtained for kainate in the following experiments. No correction was made for radioactivity present in the blood cells which remained negligible 40 seconds after injection.

<u>The transfer constant for kainate</u>, an exogenous substance which is not rapidly metabolized, was determined according to the method described by Patlak et al. (1983). This method has the advantage that it does not require a separate determination of vascular radioactivity. The equation used is a modification of equation (1) in which all terms are divided by $C_a(T)$, the radioactivity of the last arterial plasma sample.

$$\frac{Q_m(T)}{C_a(T)} = \frac{\int_0^T C_a\, dt}{C_a(T)} + V_i \quad (3)$$

V_i is the brain space equilibrating rapidly with plasma. It can differ from $V(T)$ as even a rapid equilibration may not have been complete 40 seconds after the injection. When various experimental times are chosen for a sufficient number of animals, the plot of $Q_m/C_a(T)$ versus $\int_0^T C_a(t)\, dt$ gives V_i as the ordinate intercept and K_i as the slope of the graph. In these experiments, rats were thus sacrificed at various times between 40 seconds and 30 minutes.

<u>Uptake of Amino Acids into Isolated Rat Brain Microvessels</u>

Microvessels were isolated from brains of two month old rats according to a slight modification of the method described by Goldstein et al. (1975). 200 µl of microvessels suspension containing about 150 µg of proteins were preincubated at 37°C for 20 minutes. The incubation buffer consisted of 140 mM NaCl, 4 mM KCl, 3.3 mM $CaCl_2$, 1.2 mM $MgCl_2$, 10 mM HEPES (pH 7.4), 5 mM glucose and 1% serum albumin and was oxygen-saturated. The uptake was initiated by addition of 50 µl aliquot of buffer containing 1 µCi of tritiated glutamate or kainate, 0.2 µCi ^{14}C-labeled sucrose and unlabeled carrier at a concentration of 10 µM. Unlabeled competitors at a concentration of 1 mM were added when needed. In studies done in the absence of sodium, sodium chloride was replaced by choline chloride. The uptake was stopped one minute later by addition of cold sodium chloride followed by filtration on a 43 µ nylon mesh. The sieves with the retained microvessels were then sonicated for two minutes in 1 ml water. An aliquot of the sonicate was used for protein determination according to the method of Lowry et al. (1951) after an overnight digestion in 0.1 N NaOH. Another

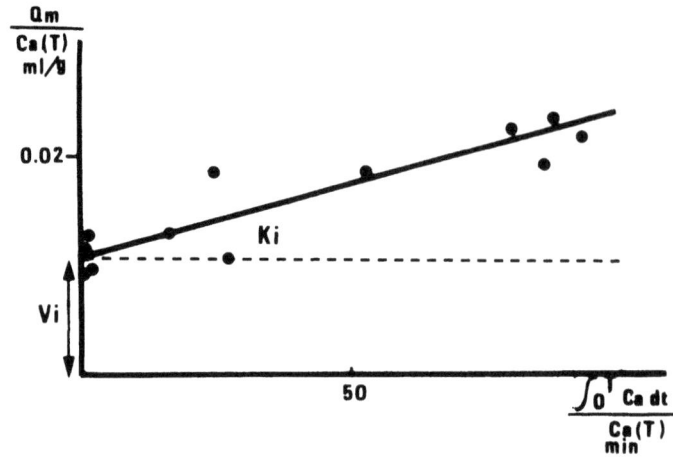

Fig. 1. Graphical analysis of blood-brain transport of kainate in hippocampus. Abscissa: plasma arterial radioactivity integral/ final plasma radioactivity. Ordinate: hippocampus radioactivity/ final plasma radioactivity.

aliquot was digested in Soluene for counting of total radioactivity (i.e., transported radioactivity + radioactivity possibly bound to membranes). The rest of the sonicate was centrifuged and radioactivity in the supernatant counted. The difference between total radioactivity and that found in the supernatant represents membrane-bound radioactivity. The uptake was expressed as dpm/min/µg of proteins or as distribution ratio, using an amount of intracellular water of 2.28 ± 0.56 µl/mg of proteins as determined in previous experiments.

RESULTS

Transfer Constant of Kainate

Fig. 1 shows the graph obtained by applying equation (3) to the results of transport of kainate in hippocampus. The solid line represents the best linear fit by a least square analysis of the data points. The ordinate intercept Vi (brain space equilibrating rapidly with plasma) was equal to 11.1 ± 0.9 ml/g x 10^3 and the slope (transfer constant) was equal to 0.13 ± 0.03 ml/min/g x 10^3. Table 1 lists these two parameters for selected brain regions. It can be seen that both vary from one region to another.

Transfer Constant of Glutamate

The transfer constant of glutamate was calculated using equations (1) and (2). It was supposed, as will be discussed later, that V(T) which appears in equation (2), was equal to the volume of distribution of kainate during the same interval of time. This volume was calculated as the ratio of radioactive content of kainate in brain to its radioactive concentration in plasma at the time of decapitation. Table 2 shows the values of the transfer constant when the experiment was performed with L-(U-^{14}C) glutamate. The results were very similar when L-(1-^{14}C) glutamate was used, indicating that decarboxylation of glutamate was not very active during the experimental time. The concentration of glutamate was also enzymatically determined in the plasma of experimental animals. This allowed the calculation of

Table 1. Transfer Constant (K) of Kainate and Brain Space Equilibrating Rapidly With Plasma (Vi) in Some Brain Regions

Region	Vi	Ki
Olfactory bulbs	36.1 ± 8.4	0.37 ± 0.17
Hypothalamus	17.2 ± 2.3	0.29 ± 0.04
Striatum	7.7 ± 1.8	0.15 ± 0.03
Parietal cortex	13.4 ± 2.4	0.21 ± 0.05
Hippocampus	11.1 ± 0.9	0.13 ± 0.02
Piriform cortex	13.2 ± 1.3	0.15 ± 0.03
Thalamus	11.2 ± 1.0	0.15 ± 0.02
Midbrain	16.3 ± 1.9	0.12 ± 0.04
Pons-medulla	20.4 ± 1.6	0.18 ± 0.01
Cerebellum	24.7 ± 3.1	0.11 ± 0.06

Values are mean ± SD, expressed as ml/g × 10^3 for Vi and as ml/min/g × 10^3 for Ki, n = 13

blood-brain influx of glutamate which was in hippocampus was 0.7 ± 0.2 nmol/min/g.

Uptake of Amino Acids into Brain Microvessels

This uptake was linear for about five minutes. It was 75% sodium-dependent and a ratio tissue/medium of more than 10 was obtained in five minutes. When the uptake of L-(1-^{14}C) glutamate was measured in the same experiment as that of tritiated glutamate, the ratio ^{14}C/^{3}H was only slightly lower in microvessels than in medium, indicating that decarboxylation was only a minor route of glutamate metabolism during the experimental time. Incorporation of glutamate into proteins was studied on a one hour period of time and shown to be 2.5% of total incorporated radioactivity. No binding of glutamate to the membranes could be observed in these experiments.

Table 2. Transfer Constant of Glutamate in Some Brain Regions

Region	Ki
Olfactory bulbs	13.8 ± 2.2
Hypothalamus	19.9 ± 7.2
Striatum	5.3 ± 1.7
Parietal cortex	5.3 ± 0.8
Cingulate cortex	11.7 ± 5.3
Piriform cortex	9.7 ± 2.5
Hippocampus	5.7 ± 1.5
Thalamus	6.4 ± 1.5
Midbrain	6.7 ± 1.6
Pons-medulla	12.1 ± 3.5
Cerebellum	6.3 ± 2.2
Cervical spinal cord	19.6 ± 7.6
Choroid plexus	232 ± 75

Values are mean ± SD expressed as ml/min/g × 10^3
Number of animals: 4

Table 3. Inhibition of L-glutamate Uptake into Brain Microvessels

Inhibitors	% of unhibited rate
None	100
L-Glutamate	1
D-Glutamate	58
L-Aspartate	28
D-Aspartate	43
L-Cysteate	23
L-Cysteine Sulfinate	10
Kainate	120
n-Methyl-D-aspartate	106
Quisqualate	35
L-cystine	109

Glutamate concentration: 10 µM
Inhibitors concentration: 1 mM

Table 3 shows the effect of some competitors on glutamate uptake. Radioactive kainate was not taken up by microvessels nor was there any binding to the membranes.

DISCUSSION

The study of blood-brain transport of anionic amino acids is hampered by three difficulties: 1) the radioactivity measured in brain is the sum of the radioactivity which has effectively crossed the blood-brain barrier and of the radioactivity which remains in plasma or in compartments which rapidly equilibrate with plasma (Patlak et al., 1983). As these amino acids have a very low permeability through the blood-brain barrier, at any experimental time, a large part of the measured radioactivity is situated in these compartments whose accurate determination is of prime importance; 2) since glutamic and aspartic acid are present in plasma, competition can exist for the hypothetical carrier between exogenous amino acids at tracer concentration and endogenous amino acids through the blood-brain barrier; 3) the study of transport of endogenous amino acids is greatly complicated by the rapid metabolism of the radioactive tracer.

The first problem can be solved by using the method of Patlak et al. (1983) for determination of the transfer constant. It is noteworthy that the brain space rapidly equilibrating with plasma obtained by this method is superior to plasma volume, but very similar to the apparent volume of distribution of kainate 40 seconds after injection. This probably means a very rapid (<40 seconds) equilibration of radioactivity between this space and plasma. In addition, this volume is very similar to that obtained for sucrose in other experiments (in preparation), suggesting that this space is extracellular. In fact, it was observed in experiments on isolated capillaries that kainate does not enter the endothelial cells. The transfer constant of kainate is even lower than that of sucrose which is an extracellular substance with a low permeability through the blood-brain barrier (Ohno et al., 1978). The possibility of competition with plasma amino acids seems unlikely for two reasons: 1) as mentioned earlier, no carrier transport for kainate exists in isolated brain microvessels; 2) in two experiments, cold kainate (2 mg/kg body weight) was coinjected with radioactive kainate and the results obtained did not deviate from those obtained with a tracer concentration. This transfer constant for kainate varied from one region to another without any correlation with the area in which the first seizure activity appears after systemic injection of kainate. Since seizure activity in hippocampus is due to a direct effect of kainate on brain cells (Ben-Ari, 1985), it seemed important to estimate the concentration of kainate in susceptible areas after injection. Kainate concentration 30 minutes after a 2 mg/kg i.v. injection was about 200 nM in hippocampus, which is lower than the concentration found in olfactory bulbs (570 nM) or parietal cortex (320 nM).

The transfer constant for glutamate was twenty times more than that for kainate. However, two reservations can be made concerning these results: 1) injected radioactive glutamate is very rapidly metabolized (Lajtha et al., 1959). For this reason we chose a short experimental time and checked by thin layer chromatography that radioactivity present in the plasma of animals at the time of decapitation was recovered on the glutamic acid spot. However, we cannot exclude the possibility that a radioactive metabolite in minute quantity, with a higher permeability through the blood-brain barrier could be responsible in part for the transported radioactivity. Thus, the measure constant should be considered as the upper limit of the real constant; 2) the choice of kainate as a marker for V(T) may not be appropriate. The reason for this choice is that kainate and glutamate have probably similar physico-chemical characteristics and that kainate may thus have the same distribution as the fraction of glutamate which has not crossed the endothelial cell membrane and is in equilibrium with plasma glutamate. This transfer constant for glutamate is lower than that of most other amino acids, but higher than that of other hydrophilic molecules, with similar molecular weight and charge but without carrier-mediated transport (i.e., kainate). This is in accordance with the results of Oldendorf and Szabo (1976), who have described a blood-brain saturable transport for aspartate and glutamate.

Our results on isolated brain capillaries are in agreement with those of Hutchinson et al. (1985), who have shown the presence of a high affinity concentrative uptake of glutamate into these capillaries. These authors suggested that this system would move, _in vivo_, glutamate out of the brain. This does not favor the possibility for a ready exchange of radioactive glutamate between plasma and brain. The hypothesis can be made that the exchange takes place between plasma and endothelial cell through the luminal membrane. In this case the measured constant would not really be a blood-brain transfer constant.

The study of the effect of competitors on glutamate uptake into isolated brain microvessels shows that it is strongly inhibited by other anionic amino acids and not by cystine. The characteristics of this transport are similar to those of the uptake system observed in brain cells (Christensen and Makowske, 1983; Foster and Fagg, 1984). In contrast, agonists of acidic amino acid receptors had no effect on uptake with the exception of quisqualate. This suggests that with the exception of the uptake system, no receptor exists for acidic amino acids on the membrane of the endothelial cells.

Finally, in spite of the fact that glutamate decarboxylase is present in brain microvessels endothelial cells (Mrsulja and Djuricić, 1980), there does not seem to exist an enzymatic barrier to glutamate. In fact, _in vivo_ influx as well as _in vitro_ uptake do not differ significantly when uniformly-labeled or $(1-^{14}C)$-labeled glutamate are used for the experiments. Furthermore, the addition of amino oxyacetate, known to inhibit glutamate decarboxylase, to the uptake medium did not increase the capillary radioactivity.

ACKNOWLEDGEMENTS

This study was supported by the Institut National de la Santé et de la Recherche Médicale (INSERM). We thank Dr. Ben-Ari for helpful discussion.

REFERENCES

Ben-Ari, Y., 1985, Limbic seizures and brain damage produced by kainic acid: mechanisms and relevance to human temporal lobe epilepsy, Neuroscience, 14:375.

Bertler, A., Falck, B., Owman, C., and Rosengren, E., 1966, The localization of monoaminergic blood-brain barrier mechanisms, Pharmacol. Rev., 18:369.

Blasberg, R.G., Fenstermacher, J.D., and Patlak, C.S., 1983, Transport of α-aminoisobutyric acid across brain capillary and cellular membranes, J. Cereb. Blood Flow Metabol., 3:8.

Bradbury, M., 1979, The Concept of a Blood-Brain Barrier, Wiley and Sons, London.

Christensen, H.N., and Makowske, M., 1983, Recognition chemistry of anionic amino acids for hepatocyte transport and for neurotransmittory action compared, Life Sci., 33:2255.

Foster, A.C., and Fagg, G.E., 1984, Acidic amino acid binding sites in mammalian neuronal membranes: their characteristics and relationship to synaptic receptors, Brain Res. Rev., 7:103.

Goldstein, G.W., Wolinski, J.S., Csejtey, J., and Diamond, I., 1975, Isolation of metabolically active capillaries from the brain, J. Neurochem., 25:715.

Goldstein, G.W., and Betz, A.L., 1983, Recent advances in understanding brain capillary function, Ann. Neurol., 14:389.

Hardebo, J.E., and Owman, C., 1980, Barrier mechanisms for neurotransmitter monoamines and their precursors at the blood-brain interface, Ann. Neurol., 8:1.

Hutchinson, H.T., Eisenberg, H.M., and Haber, B., 1985, High-affinity transport of glutamate in rat brain microvessels, Exp. Neurol., 87:260.

Joó, F., 1985, The blood-brain barrier in vitro: ten years of research on microvessels isolated from the brain, Neurochem. Int., 7:1.

Lajtha, A., Berl, S., and Waelsch, H., 1959, Amino acid and protein metabolism of the brain - IV - The metabolism of glutamic acid, J. Neurochem., 3:322.

Lowry, O.H., Rosebrough, N.J., Farr, A.L., and Randall, R.J., 1951, Protein measurement with the Folin phenol reagent, J. Biol. Chem., 193:265.

Mrsulja, B.B., and Djuricić, B.M., 1980, Biochemical characteristics of cerebral capillaries, in: The Cerebral Microvasculature, H.M. Eisenberg and R.L. Suddith, eds., Plenum Press, New York, p. 29.

Ohno, K., Pettigrew, K.D., and Rapoport, S.I., 1978, Lower limit of cerebrovascular permeability to non electrolytes in the conscious rat, Am. J. Physiol., 235:H299.

Oldendorf, W.H., and Szabo, J., 1976, Amino acid assignment to one of three blood-brain amino acid carriers, Am. J. Physiol., 230:94.

Patlak, C.S., Blasberg, R.G., and Fenstermacher, J.D., 1983, Graphical evaluation of blood-to-brain transfer constants from multiple-time uptake data, J. Cereb. Blood Flow Metabol., 3:1.

Rapoport, S.I., 1976, Blood-Brain Barrier in Physiology and Medicine, Raven Press, New York.

LIMBIC SEIZURES INDUCED BY SYSTEMICALLY APPLIED KAINIC ACID: HOW MUCH KAINIC ACID REACHES THE BRAIN?

M.L. Berger[1], J.-M. Lefauconnier[2], E. Tremblay[1], and Y. Ben-Ari[1]

LPN, CNRS, Gif-sur-Yvette, F-91190 France
[1]and INSERM-U29, 123 Bd de Port-Royal
Hôpital de Port-Royal, 75014 Paris, France
[2]INSERM Unité de Toxicologie Experimentale, Hospital Fernand Widal, F-75475 Paris, France

Systemic or intracerebral injection of the neurotoxin kainic acid (KA) in rats induces a limbic seizure and brain damage syndrome, which has been proposed as an experimental animal model for human temporal lobe epilepsy (Nadler, 1981; Ben-Ari, 1985). The electrographic, neurochemical, metabolic and histopathological consequences involve preferentially limbic structures such as the hippocampal formation, the septum, the amygdaloid complex, the cingulate cortex, the claustrum, and several thalamic nuclei (Ben-Ari et al., 1980; Schwob et al., 1980; Ben-Ari et al., 1981; Lothman and Collins, 1981). If the neurotoxin is injected systemically (i.e., intravenously, intraperitoneally, or subcutaneously), doses from 8 to 12 mg/kg are required to induce a full seizure response (Ben-Ari et al., 1981; Lothman and Collins, 1981; Sperk et al., 1984). Upon central application, doses of 1 µg or less are sufficient (Schwarcz et al., 1978; Ben-Ari et al., 1980; Nadler et al., 1980). This suggests a central site of action and a poor permeability of the blood-brain barrier (BBB) for the neurotoxin. From in vitro studies, it is known that KA can exert a variety of effects on central nervous tissue. In the hippocampal slice preparation, it initiates epileptiform discharges at concentrations down to 20 nM, and produces irreversible depolarization above 1 µM (Robinson and Deadwyler, 1981; Westbrook and Lothman, 1983). In cerebellar slices, 5 µM KA elevates cGMP levels (Garthwaite, 1982) and induces first necrotic changes (Garthwaite and Wilkin, 1982). Sodium permeability (Luini et al., 1981) is stimulated by 10-100 µM KA. To provoke the release of [^{3}H]aspartic acid from cerebellar slices, 10 µM KA (Potashner and Gerard, 1983) or at least 500 µM KA (Ferkany and Coyle, 1983) are necessary. 100 µM KA or more are required to elevate cAMP levels in cerebellar slices (Schmidt et al., 1976). Finally, specific membrane binding sites have been described for the neurotoxin in neuronal tissue, with affinity constants ranging from 1 to 200 nM (London and Coyle, 1979; Berger et al., 1984; Berger et al., in press). It remains to be clarified which of these neurochemical effects are involved in the seizure syndrome induced by KA. In this connection, an estimate of the actual brain tissue concentration of KA after systemic application of a convulsive dose would be of great interest.

EXPERIMENTAL PROCEDURES

When this study was undertaken, we investigated at the same time the ontogenesis of the KA seizure-brain damage syndrome (Ben-Ari et al., 1984; Berger et al., 1984; Nitecka et al., 1984; Tremblay et al., 1984). According to our results, we have chosen two developmental stages of our experimental animals, which can be regarded as immature (10-13 days) or completely mature (30 days), respectively, in terms of clinical signs, metabolic changes in the brain, histopathological outcome and distribution of high affinity KA binding sites.

Brain Uptake After a Single Pass

Blood-brain transport of KA and glutamic acid was studied in 10 day and 30 day old male wistar rats, using Oldendorf's method (Oldendorf, 1970, 1981) with modifications described for immature rats (Lefauconnier and Trouvé, 1983). The method is based on the uptake of the test substance in brain tissue during a single passage in a 'bolus' produced by rapid injection into the carotid circulation. A diffusible co-injected substance is used as an internal reference. The ratio of the test to the reference substance in brain tissue, expressed in % of this ratio in the injection solution, was introduced by Oldendorf as the 'brain uptake index' (BUI; Oldendorf, 1970).

Rats with an age of 10 days and 30 days were anesthetized with ether and the right brachial artery rapidly exposed and cannulated under microscopic control with a 30 gauge needle. 0.12 ml (10 day old rats) or 0.20 ml (30 day old rats) of a buffered solution (10 mM HEPES, 145 mM Na^+, 5 mM K^+, 3 mM Ca^{2+}, 153 mM Cl^-, 5 mM glucose) containing 50 µCi/ml of tritiated KA (2 Ci/mmole, Amersham) and 0.5 µCi/ml of [^{14}C]iodoantipyrine (54 mCi/mmole, Amersham) were rapidly injected in a retrograde direction in the brachial artery. This resulted in a clear bolus in the right common artery in the orthograde direction. 10 day old rats were decapitated 10 seconds, 30 day old rats 7 seconds, after the injection. The brains were rapidly removed from the skull and frozen in isopentane/dry ice. Twenty-four hours later, the right hemispheres were dissected by free hand at $-10°C$. Regions were taken from frontal slices cut with a razor blade. From the first slice, which included the main part of the striatum, we cut with a scalpel the fronto-parietal and cingulate cortices, the septum, and the striatum itself. From the second slice, which included the whole hypothalamus, we took the hippocampus, the thalamus, the amygdala, and the pyriform lobe. A third slice included the pedunculus and furnished the mesencephalon. Tissues were weighed in tightly closed miniature scintillation vials (biovials, Beckman) and solubilized in 0.5 ml protosol (NEN) at $55°C$ overnight. After addition of 60 µl concentrated acetic acid to avoid luminescence and 3 ml toluol scintillator (aquasol-2, NEN), samples were counted in a liquid scintillation counter (LKB) with a double label program and automatic quench correction, calibrated with solubilized tissue quenched standards. BUIs for KA were obtained as ratio of ^3H KA to ^{14}C iodoantipyrine counts, expressed in % of this ratio in the injection solution.

Using the same experimental procedure, we also determined the BUIs of [^3H]glutamic acid (22 Ci/mmole, Amersham, diluted to 2 Ci/mmol with unlabeled glutamic acid, Sigma) in several brain regions of 10 day and 30 day old rats. Finally, we directly compared the apparent uptake of [^3H] KA to that of [^{14}C] saccharose (specific activity 540 mCi/mmole). Saccharose is known to have a very low permeability at the BBB (Ohno et al., 1978) and after a single pass of the bolus can be considered as exclusively intravascular. We injected the same volumes as described above. The concentration of radioactive saccharose was 10 µCi/ml, that of KA again 50 µCi/ml.

Brain Uptake After Longer Delays

In a series of preexperiments, we used altogether 33 rats (10-13 days old) to validate our method. They received 20 μCi [^{3}H] KA (2 Ci/mmole) and/or 4 μCi [^{14}C] saccharose i.p. in 100 μl phosphate buffered saline and were killed 5-180 minutes later by decapitation or by extensive perfusion through the heart under equithesin anesthesia. Heparine was injected together with the anesthetic and also added to the phosphate buffered saline perfusate. Blood samples were withdrawn upon sacrifice, weighed, and counted with blood calibrated quench correction. The absolute amount of KA injected was approximately 0.1 mg/kg and did not induce any abnormal behavioral signs. In three of these rats, we added sufficient unlabeled KA to reach a dose of 3 mg/kg. These rats exhibited tonico-clonic generalized convulsions (Tremblay et al., 1984).

In another experiment, we injected i.v. KA and saccharose together into 30 day old rats. Under ether anesthesia, they were cannulated in the femoral vein. The needle was connected with a short tube to a 1 ml syringe. Local anesthesia was applied to the wound, and the incision closed around the tube. When the animals regained consciousness, we injected 100 μl phosphate buffered saline with heparine, and thereafter 20 μCi [^{3}H] KA and 3 μCi [^{14}C] saccharose together with 0.43 mg unlabeled KA in 0.1 ml phosphate buffered saline; the resulting KA dose was 4 mg/kg. The rats developed mild limbic seizures, manifestated by wet dog shakes, staring, mastication, facial clonus, and forelimb clonus (Tremblay et al., 1984). One hour after the injection, the rats were anesthetized with equithesin and perfused as described above.

Brains were removed from the skull and dissected fresh on an ice cold glass dish. We dissected the pyriform cortex, the amygdala, the cingulate cortex, the striatum, the hippocampal CA3 region, and the remaining part of the hippocampal formation, termed 'DG-part'. Details of the dissection procedure will be published elsewhere (Berger et al., in press).

RESULTS

Brain Uptake After a Single Pass

In 30 day old rats, the BUI for [^{3}H] KA as determined by a modification (Lefauconnier and Trouvé, 1983) of Oldendorf's method (Oldendorf, 1970) varied between 1.16% (thalamus) and 3.56% (pons; Table 1). In 10 day old rats, the range was from 2.63% (thalamus) to 8.2% (cingulate cortex). At the same age, we also determined the BUI for [^{3}H] glutamic acid. Taking into account all regions analyzed, the BUI of [^{3}H] glutamic acid was significantly higher than of [^{3}H] KA, by 61% in 10 day old, and by 22% in 30 day old rats ($p < 0.05$, paired Student's t-test).

Simultaneous injection of [^{3}H] KA and [^{14}C] saccharose yielded a surprising result, inasmuch as ^{3}H/^{14}C ratios in brain tissue were lower than in the injection solution, by 10% in 10 day old rats, but by 40-50% in 30 day old rats (Table 2). Since we have seen even higher differences with longer delays after systemic injection (see below), we suspected pharmacokinetic complications arising from radiotracer recirculation (see also Fig. 1). Therefore, we repeated the experiment with 30 day old rats, co-injecting [^{3}H] KA and [^{14}C] saccharose into the carotid, which can be cannulated at that age. With this modification, we obtained in brain tissue ^{3}H/^{14}C ratios which were only 10-20% lower than those in the injection solution (Table 2). There were no detectable regional variations in the

Table 1. Apparent BUI for [^{3}H] KA and [^{3}H] Glutamic Acid in Various Brain Regions in 10 and 30 Day Old Rats

	BUI - KA (%)		BUI - glu (%)	
	10 d	30 d	10 d	30 d
am	3.43 + 1.67	2.22 + 0.78	9.14 + 4.6	1.81 + 0.75
pir	2.66 + 0.21	1.77 + 0.34	3.57 + 0.49	2.14 + 0.50
par	3.13 + 0.86	1.81 + 0.19	3.07 + 0.57	1.90 + 0.26
cing	8.2 + 5.4	3.12 + 1.45	5.48 + 2.41	3.00 + 0.88
str	3.96 (2)	1.29 + 0.05	3.19 + 0.98	1.66 + 0.27
hpc	4.80 (2)	1.68 + 0.76	11.9 + 4.5	2.78 + 0.95
spt	3.31 (2)	1.58 + 0.17	6.18 + 2.73	2.25 + 0.91
thal	2.63 + 1.05	1.16 + 0.18	4.29 + 0.23	1.63 + 0.37
mes	3.29 + 0.88	1.41 + 0.16	3.18 + 0.06	1.69 + 0.25
pons	4.34 + 1.14	3.56 + 1.35	4.85 + 1.15	2.51 + 0.29
	(n = 3)	(n = 3)	(n = 3)	(n = 4)

Means + SD; n indicated at the bottom of each column; some values are only duplicates. am, amygdala; pir, piriform cortex; par, fronto-parietal cortex; cing, cingulate cortex; str, striatum; hpc, hippocampus; spt, septum; thal, thalamus; mes, mesencephalon.

ratio ^{3}H/^{14}C (Table 2), indicating that both radiotracers distributed identically.

We also measured the ^{3}H/^{14}C ratio in two brain regions (fronto-parietal cortex and hippocampus) after orthograde carotid co-injection of [^{3}H] glutamic acid (1.5 µM) and [^{14}C] saccharose into 30 day old rats. This was the only occasion where we found in brain tissue higher ^{3}H/^{14}C ratios than in the injection solution (by 70% in both regions, data not shown).

Table 2. ^{3}H/^{14}C Ratio in Various Brain Regions of 10 Day and 30 Day Old Rats, Some Seconds After Intra-arterial Co-injection of [^{3}H] KA and [^{14}C] Saccharose, as Related to this Ratio in the Injectate

	KA/sacch. (tissue) / KA/sacch. (injected)		
	10 d a)	30 d a)	30 d b)
am	0.94 + 0.13	0.55 + 0.06	0.89 + 0.10
pir	0.93 + 0.14	0.59 + 0.08	0.88 + 0.14
par	0.86 + 0.10	0.56 + 0.04	0.77 + 0.06
cing	0.91 + 0.16	0.53 + 0.02	0.83 + 0.07
str	0.88 + 0.18	0.58 + 0.06	0.91 + 0.16
hpc	0.98 + 0.13	0.55 (2)	0.85 + 0.02
spt	0.92 + 0.17	0.46 + 0.11	n.d.
thal	0.97 + 0.10	0.57 + 0.02	0.90 + 0.08
mes	0.94 + 0.11	0.52 (2)	n.d.
pons	0.93 + 0.10	0.62 (2)	n.d.
	(n = 4)	(n = 3)	(n = 4)

Means + SD; n indicated at the bottom of each column; some values are only duplicates. Abbreviations as in Table 1.

a) Retrograde injection into the right brachial artery; b) orthograde injection into the carotid; n.d., not determined.

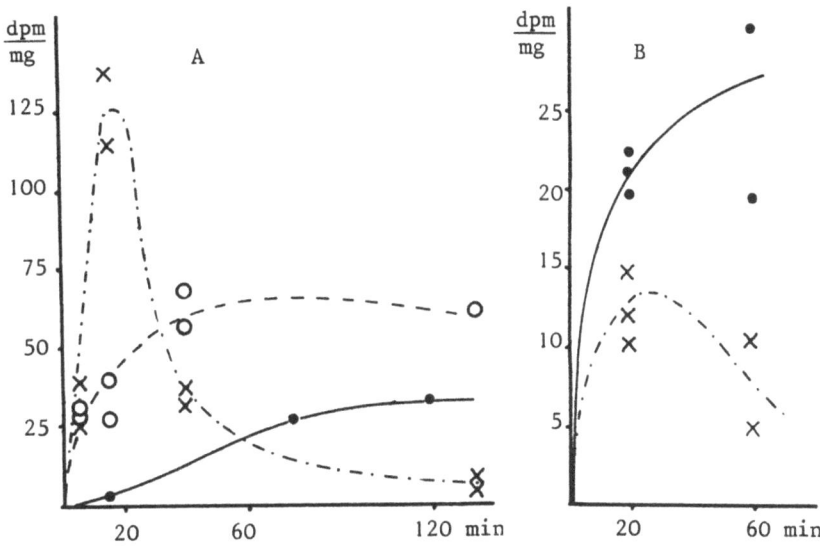

Fig. 1. Mean brain tissue concentrations of ^{3}H-label (dpm/mg wet tissue) several delays after i.p. injection of [^{3}H] KA into 10-12 day old rats. In A, the specific activity of [^{3}H] KA was 2 Ci/mmole (1 dpm = 0.23 fmole; total amount injected 20 μCi 0.11 mg KA/kg), in B a convulsive dose of KA was added (3 mg/kg; 1 dpm = 32 fmole; total amount injected 4 μCi). Filled circles: perfused brains; open circles: unperfused brains; crosses: blood (scale 1/50). Raising the KA dose from 0.11 to 3 mg/kg elevated the ^{3}H-content of the brain 100-fold (7 to 800 fmole/mg). This might indicate binding of KA to vascular components.

Brain Uptake After Longer Delays

Extensive perfusion 15 minutes after i.p. injection of [^{3}H] KA to 10-11 day old rats removed a higher proportion of radioactivity from the brain, whereas after one hour or later, more than 50% remained in the tissue (Fig. 1A). After approximately equimolar injection of [^{14}C] saccharose, the results were similar, with the difference that the amount of radioactivity not removable by perfusion increased earlier (Fig. 2). After injection of [^{3}H] KA together with a convulsive dose of unlabeled KA (3 mg/kg), the appearance of ^{3}H label in the brain was faster, with plateau levels reached already 20 minutes after the injection (Fig. 1B). At that time point, the rat pups developed first seizure signs (trembling, scratching; compare Tremblay et al., 1984). The actual mean brain concentration can be calculated at that stage as 670 fmole/mg wet tissue (Table 3).

Fig. 3 shows the regional distribution of ^{3}H and ^{14}C label in the perfused brain of 12-13 day old rats, one hour after i.p. co-injection of [^{3}H] KA and [^{14}C] saccharose in comparable doses of approximately 0.1 mg/kg. The cingulate cortex accumulated most of both radiotracers, the hippocampal CA3 region showed some preference for [^{3}H] KA. The tissue concentration of KA in the CA3 subpart of the hippocampus after perfusion was calculated as 8.2 ± 0.9 fmole/mg wet tissue (0.07 mg KA/kg i.p.). There were no significant regional variations in the amount of radioactivity removable by perfusion (empty parts of the columns). The ^{3}H/^{14}C ratio in the injection solution was 5.4, in blood samples taken upon sacrifice between 4.3 and 5.4, in unperfused brains between 3.1 and 4.1, and in perfused brains between 2.7 and 3.6.

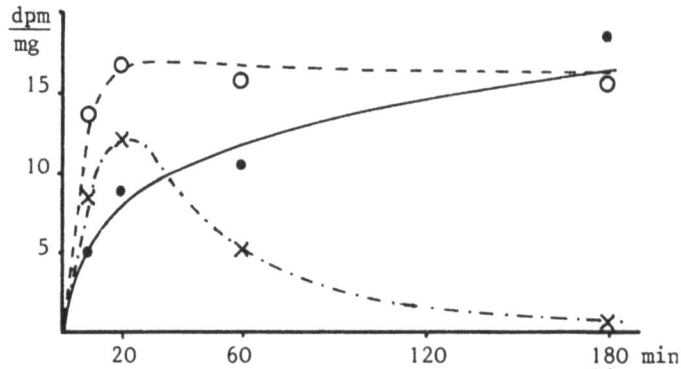

Fig. 2. Mean brain tissue concentrations of ^{14}C-label (dpm/mg wet tissue) several days after i.p. injection of 4 µCi [^{14}C] saccharose (0.13 mg/kg). With longer delays, ^{14}C-label was cleared from the blood and perfusion did no longer reduce the brain's ^{14}C-content.

The results from 30 day old rats injected i.v. with a mixture of [^{3}H] KA, [^{14}C] saccharose, and unlabeled KA are summarized in Table 4 (values normalized by whole brain averages). The dose of KA injected was 4 mg/kg (i.v.) and induced mild limbic seizures within one hour. After one hour, above average accumulation of both radiotracers was found in the septum, the cingulate cortex, and in the pons. In the hippocampal CA3 region, the amygdala, and in the pyriform lobe only, [^{3}H] KA but not [^{14}C] saccharose, was better retained than in the rest of the brain. As a matter of fact, we observed in the CA3 region the lowest ^{14}C counts per mg wet tissue (Table 4). Of 12 different brain regions analyzed, 6 were part of or related to the limbic system, and all of these but one accumulated more [^{3}H] KA than the remaining 6 non-limbic ones. The notable exception was that part of the hippocampal formation, which did not contain the CA3 region ('DG-part'). After perfusion, the concentration of KA in limbic structures of 30 day old rats with KA-induced limbic seizures was around 250 fmole/mg wet tissue (Table 3). The ratio ^{3}H/^{14}C in the injection solution was 8.0, in blood samples withdrawn after opening of the heart between 5.4 and 7.2, and in the perfused brains between 1.5 and 3.2.

Table 3. Tissue Concentrations of KA in fmole/mg Wet Tissue after Systemic Injection of a Convulsive Dose into 10-11 Day and 30 Day Old Rats (Perfused Brains)

	10-11 d (20 min after 3 mg/kg i.p.)			30 d (60 min after 4 mg/kg i.v.)	
am + pir	640 $\pm$ 70	(3)	am	245 $\pm$ 80	(5)
hpc	700 $\pm$ 25	(3)	CA3	250 $\pm$ 50	(5)
whole brain	670 $\pm$ 45	(3)	whole brain	215 $\pm$ 55	(5)

am, amygdala; pir, piriform cortex; hpc, hippocampus; CA3, 1/4 of the hpc enriched in the CA3 region. Number of rats in parentheses; means $\pm$ SD.

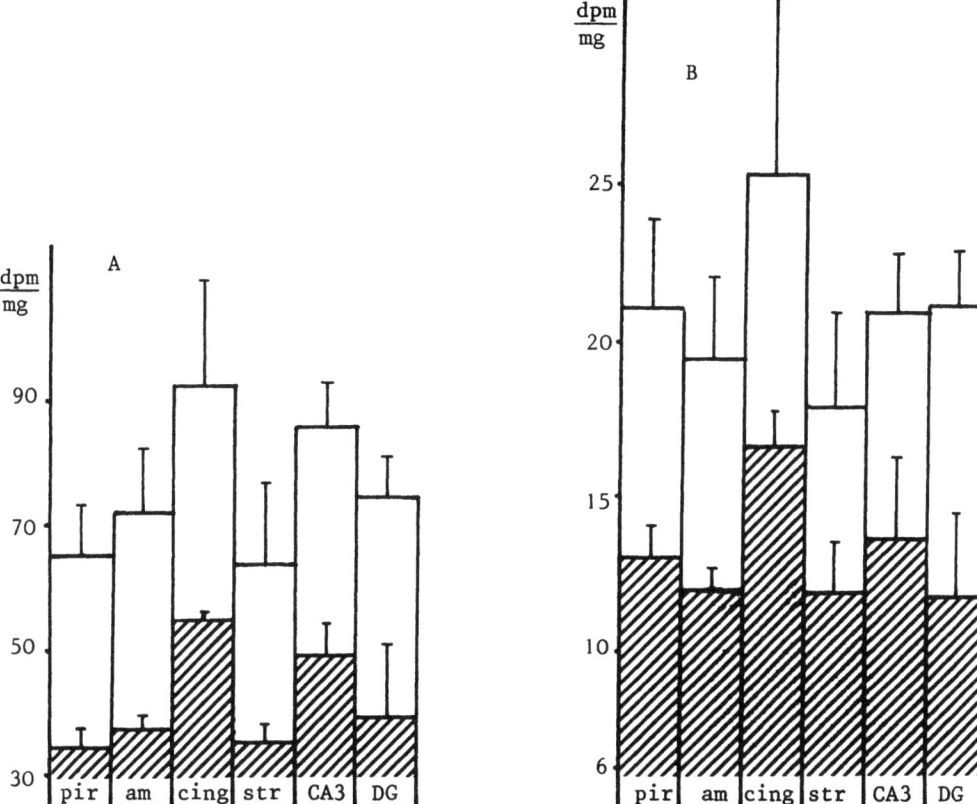

Fig. 3. Tissue concentrations of ^{3}H and ^{14}C (dpm/mg wet tissue) in various brain regions one hour after i.p. co-injection of 20 μCi [^{3}H] KA (0.06 mg/kg; A) and 4 μCi [^{14}C] saccharose (0.13 mg/kg; B) into 12-13 day old rats. Shaded columns: perfused brains (3 rats); open columns: unperfused brains (5 rats). The ordinates in A and B have different scales, according to the different amounts of ^{3}H and ^{14}C injected, and do not start with zero. pir, pyriform cortex; am, amygdala; cing, cingulate cortex; str, striatum; CA3 and DG, CA3 part (=1/4) and remaining part (=3/4) of the hippocampal formation, respectively. Note that the open parts of the columns are all of comparable height, and that regional variations are mainly due to the fraction of radioactivity that was not removed by perfusion.

DISCUSSION

Does KA Have a Measurable Brain Uptake Index?

Oldendorf's method has been extensively used to measure the permeability of the BBB, including recently a series of amino acids in developing rats (Lefauconnier and Trouvé, 1983). For KA, the present study demonstrates very low values, and the direct comparison between KA and poorly diffusible saccharose suggests that KA in the time range of seconds behaved like an intravascular tracer. Thus, the permeability of the BBB to KA seems to be below the detection limit of Oldendorf's method. In that respect, KA differs from glutamic acid. The permeability of the BBB to glutamic acid is low, but measurable by Oldendorf's method (Oldendorf and Szabo, 1976).

Table 4. [^{3}H] KA and [^{14}C] Saccharose as Ratios to Whole Brain Averages, One Hour After 4 mg/kg KA and 18 µg/kg Saccharose i.v. (30 Day Old Rat, Mild Seizures, Perfused Brains)

[^{3}H] KA

	I		II		III
spt	2.12 ± 0.32**	par	1.00 ± 0.12	cbl	0.92 ± 0.10
cing	1.22 ± 0.25	thal	0.98 ± 0.11	mes	0.88 ± 0.11
CA3	1.19 ± 0.18			str	0.86 ± 0.17
pir	1.14 ± 0.12			DG	0.84 ± 0.08
am	1.13 ± 0.15				
pons	1.12 ± 0.13				

[^{14}C] Sacch.

	I		II		III
spt	1.32 ± 0.17*	pir	0.98 ± 0.11	str	0.93 ± 0.04
cing	1.14 ± 0.12*	cbl	0.97 ± 0.08	DG	0.90 ± 0.03*
par	1.06 ± 0.05*	am	0.94 ± 0.06	CA3	0.86 ± 0.08*
thal	1.05 ± 0.04*				
pons	1.06 ± 0.10				
mes	1.01 ± 0.06				

[^{3}H] KA/[^{14}C] Sacch.

	I		II		III
spt	1.62 ± 0.24*	pir	1.17 ± 0.18	par	0.96 ± 0.15
CA3	1.38 ± 0.16*	cing	1.10 ± 0.29	cbl	0.95 ± 0.13
am	1.20 ± 0.17	pons	1.07 ± 0.13	thal	0.94 ± 0.09
				DG	0.93 ± 0.10
				str	0.92 ± 0.17
				mes	0.88 ± 0.13

n = 5; means ± SD; each value in columns I is significantly different (p < 0.05) from each value in columns III (vice versa); * p < 0.01 instead of 0.05; ** significantly different from every other value (p < 0.01). cbl, cerebellum; for other abbreviations see Tables 1 and 3.

The highest apparent BUI values for [^{3}H] KA were observed in the cingulate cortex, the septum, and the hippocampus. However, these high values occurred with inconsistancy and were not accompanied by a high [^{3}H] KA/[^{14}C] saccharose ratio. An explanation could be that in these cases meninges or small parts of choroid plexus have been dissected together with the frozen structures to varying extents.

Slow Entry of KA Into the Brain

In the second part of our study we demonstrated that the appearance of systemically injected [^{3}H] KA in the brain is a gradual process. In 10 day old rats, KA reached plateau values in perfused brain tissue approximately with seizure onset (i.e., 20 minutes). The slower penetration of a subconvulsive dose was most likely due to a higher relative proportion of [^{3}H] KA transitorily bound by components of blood and extraneuronal tissue. In nearly adult rats (30 days), we measured at seizure onset lower levels of KA in perfused brain tissue (Table 3). Interestingly, the brains of 30 day old rats contain KA binding sites with a higher affinity for the toxin than brains of 10 day old rats (Ben-Ari et al., 1984; Berger et al., 1984). In 30 day old rats, the highest [^{3}H] KA levels not removable by perfusion were detected in the septum, the cingulate cortex, and the hippocampus (CA3 part). Among these structures, the CA3 region occupies a special position since it is the only one with a high level of slow dissociation rate KA binding sites (Monaghan and Cotman, 1982; Berger and Ben-Ari, 1983; Unnerstall and Wamsley, 1983) which seem to mediate the convulsive action of KA (Ben-Ari et al., 1984; Berger et al., 1984). In agreement, metabolic studies have shown that the CA3 region is one of the first to get activated after systemic KA injection (Ben-Ari et al., 1981; Lothman and Collins, 1981)

The inert control tracer [^{14}C] saccharose, which crossed the BBB to a similarly poor extent, distributed over various brain structures in similar proportions as [^{3}H] KA. Thus, regions as the septum and the cingulate cortex appear to be more readily accessible for poorly penetrating substances than others, probably independent of the molecular structure. On the other hand, chemical structure seems to be more important in the CA3 subpart of the hippocampus, which after perfusion was rich in [^{3}H] KA, but poor in [^{14}C] saccharose. It has been shown that during KA induced seizures, the fall in [^{14}C] saccharose concentration in the blood (Fig. 2) will allow for the efflux of ^{14}C tracer from brain tissue, especially in regions of increased metabolism and blood flow, but [^{3}H] KA, once bound to high affinity sites, will dissociate only very slowly.

Recently, a regional increase in the cerebrovascular permeability during systemic KA seizures has been described (Ruth, 1984). However, the seizures induced in our experiments with 30 day old rats were mild and in a stage of onset (compare Zucker et al., 1983), and a possible seizure-induced extravasation should have been seen with both radiotracers, not with only one of them, and also in other regions as, for example, the thalamus (Ruth, 1984).

The Problem of Tissue Content and Active Concentration of KA

Although at this point of our discussion it seems clear that KA one hour after systemic application of doses inducing mild seizures was present inside the brain, and although we can even indicate its amounts in fmole per mg wet tissue (=nmole per kg), there remains the problem that we have obtained the overall molality of KA in a heterogeneous tissue, which can be far away from the molarity in an ideal test tube system. Several questions remain open. What fraction of brain KA exists in a free form, and where? How accessible are, for example, synaptic clefts? It must be taken into consideration, that the diffusible space in tissue is surely smaller than tissue volume (especially for a polar substance which is not taken up by cells). On the other hand, total binding of [^{3}H] KA will strongly reduce its free concentration, and a high proportion of [^{3}H] KA in the tissue can be expected to be in bound form (largely to unspecific sites). When 16 μm slices (rat forebrain, frontal sections) are incubated in the presence of 20 nM [^{3}H] KA, total binding is about 40 fmole/mg wet tissue (Berger, unpublished observation). This seems not too far away from 215 fmole/mg wet tissue (Table 3). Thus, in the slice the 'concentration' of bound [^{3}H] KA doubles that of free [^{3}H] KA in the bath. One could argue that also our 215 fmole/mg value more likely indicates a free KA concentration below than above 215 nM. However, the situation _in vivo_ can be completely different from that in an _in vitro_ slice.

Conclusion

Relying on our results we think that free brain KA concentrations above 1 μM in the brain during systemic KA seizures can be excluded with some certainty. Evident implications from this upper limit are, that none of the hitherto described molecular actions of KA on brain tissue _in vitro_ (cGMP formation, Garthwaite, 1982; stimulation of Na$^+$ fluxes, Luini et al., 1981; release of acidic amino acids, Ferkany and Coyle, 1983; Potashner and Gerard, 1983) can be a serious candidate for a molecular explanation of KA's convulsive and neurotoxic effects after systemic injection. All these effects might be secondary consequences of seizure activity. However, KA concentrations slightly below 1 μM should be sufficient to saturate specific KA binding sites.

ACKNOWLEDGEMENTS

We thank Dr. E. Cherubini for his criticism of the manuscript, G. Charton for excellent technical assistance, and W. Krivanek for typing. M.L.B. received financial support from 'La Foundation pour la Recherche Medicale'.

REFERENCES

Ben-Ari, Y., Tremblay, E., and Ottersen, O.P., 1980, Injections of kainic acid into the amygdaloid complex of the rat: an electrographic, clinical and histological study in relation to the pathology of epilepsy, Neuroscience, 5:515.

Ben-Ari, Y., Tremblay, E., Riche, D., Ghilini, G., and Naquet, R., 1981, Electrographic, clinical and pathological alterations following systemic administration of kainic acid, bicuculline and pentetrazole: metabolic mapping using the deoxyglucose method with special reference to the pathology of epilepsy, Neuroscience, 6:1361.

Ben-Ari, Y., Tremblay, E., Berger, M., and Nitecka, L., 1984, Kainic acid seizure syndrome and binding sites in developing rats, Developm. Brain Res., 14:284.

Ben-Ari, Y., 1985, Limbic seizure and brain damage syndrome produced by kainic acid: mechanisms and relevance to human temporal lobe epilepsy, Neuroscience, 14:375.

Berger, M., and Ben-Ari, Y., 1983, Autoradiographic visualization of [^{3}H] kainic acid receptor subtypes in the rat hippocampus, Neurosci. Lett., 39:237.

Berger, M.L., Tremblay, E., Nitecka, L., and Ben-Ari, Y., 1984, Maturation of kainic acid seizure-brain damage syndrome. III. Postnatal development of kainic acid binding sites in the limbic system, Neuroscience, 13:1095.

Berger, M.L., Charton, G., and Ben-Ari, Y., The effect of seizures induced by intra-amygdaloid kainic acid on kainic acid binding sites in rat hippocampus and amygdala, J. Neurochem., in press.

Ferkany, J.W., and Coyle, J.T., 1983, Kainic acid selectively stimulates the release of endogenous excitatory acidic amino acids, J. Pharmacol. Exp. Ther., 225:399.

Garthwaite, J., 1982, Excitatory amino acid receptors and guanosine 3', 5'-cyclic monophosphate in incubated slices of immature and adult rat cerebellum, Neuroscience, 7:2491.

Garthwaite, J., and Wilkin, G.P., 1982, Kainic acid receptors and neurotoxicity in adult and immature rat cerebellar slices, Neuroscience, 7:2499.

Lefauconnier, J.-M., and Trouvé, R., 1983, Developmental changes in the pattern of amino acid transport at the blood-brain barrier in rats, Developm. Brain Res., 6:175.

London, E.D., and Coyle, J.T., 1979, Specific binding of [^{3}H]kainic acid to receptor sites in rat brain, Molec. Pharmacol., 15:492.

Lothman, E.W., and Collins, R.C., 1981, Kainic acid-induced limbic motor seizures: metabolic, behavioral, electroencephalographic and neuropathological correlates, Brain Res., 218:299.

Luini, A., Goldberg, O., and Teichberg, V.I., 1981, Distinct pharmacological properties of excitatory amino acid receptors in the rat striatum: study by Na^+ efflux assay, Proc. Natl. Acad. Sci. USA, 78:3250.

Monaghan, D.T., and Cotman, C.W., 1982, The distribution of [^{3}H] kainic acid binding sites in rat CNS as determined by autoradiography, Brain Res., 252:91.

Nadler, J.V., Perry, B.W., Gentry, G., and Cotman, C.W., 1980, Degeneration of hippocampal CA3 pyramidal cells induced by intraventricular kainic acid, J. Comp. Neurol., 192:333.

Nadler, J.V., 1981, Kainic acid as a tool for the study of temporal lobe epilepsy, Life Sci., 29:2031.

Nitecka, L., Tremblay, E., Charton, G., Bouillot, J.P., Berger, M.L., and Ben-Ari, Y., 1984, Maturation of kainic acid seizure-brain damage syndrome in the rat. II. Histopathological sequelae, Neuroscience, 13:1073.

Ohno, K., Pettigrew, K.D., and Rapoport, S.J., 1978, Lower limits of cerebrovascular permeability to nonelectrolytes in the conscious rat, Am. J. Physiol., 235:H299.

Oldendorf, W.H., 1970, Measurement of brain uptake of radiolabelled substances using a tritiated water internal substance, Brain Res., 24:372.

Oldendorf, W.H., and Szabo, J., 1976, Amino acid assignment to one of three blood-brain barrier amino acid carriers, Am. J. Physiol., 230:94.

Oldendorf, W.H., 1981, Clearance of radiolabelled substances by brain after arterial injection using a diffusible internal standard, in: Research Methods in Neurochemistry, Vol. 5, N. Marks and R. Rodnight, eds., Plenum Press, New York, p. 91.

Pinard, E., Tremblay, E., Ben-Ari, Y., and Seylaz, J., 1984, Blood flow compensates oxygen demand in the vulnerable CA3 region of the hippocampus during kainate-induced seizures, Neuroscience, 13:1039.

Potashner, S.J., and Gerard, D., 1983, Kainate-enhanced release of D-[^{3}H]-aspartate from cerebral cortex and striatum: reversal by baclofen and pentobarbital, J. Neurochem., 40:1548.

Robinson, J.H., and Deadwyler, S.A., 1981, Kainic acid produces depolarization of CA3 pyramidal cells in the in vitro hippocampal slice, Brain Res., 221:117.

Ruth, R.E., 1984, Increased cerebrovascular permeability to protein during systemic kainic acid seizures, Epilepsia, 25:259.

Schmidt, M.J., Ryan, J.J., and Mollay, B.B., 1976, Effects of kainic acid, a cyclic analogue of glutamic acid, on cyclic nucleotide accumulation in slices of rat cerebellum, Brain Res., 112:113.

Sperk, G., Lassmann, H., Baran, H., Kish, S.J., Seitelberger, F., and Hornykiewicz, O., 1984, Kainic acid-induced seizures: neurochemical and histopathological changes, Neuroscience, 10:1301.

Schwarcz, R., Zaczek, R., and Coyle, J.T., 1978, Microinjection of kainic acid into the rat hippocampus, Eur. J. Pharmacol., 50:209.

Schwob, E., Fuller, T., Price, J.L., and Olney, J.W., 1980, Widespread patterns of neuronal damage following systemic or intracerebral injection of kainic acid: a histological study, Neuroscience, 5:991.

Tremblay, E., Nitecka, L., Berger, M., and Ben-Ari, Y., 1984, Maturation of kainic acid seizure-brain damage syndrome in the rat. I. Motor, electrographic, and metabolic observations, Neuroscience, 13:1051.

Unnerstall, J.R., and Wamsley, J.K., 1983, Autoradiographic localization of high-affinity [^{3}H] kainic acid binding sites in the rat forebrain, Eur. J. Pharmacol., 86:361.

Westbrook, G.L., and Lothman, E.W., 1983, Cellular and synaptic basis of kainic acid-induced hippocampal epileptiform activity, Brain Res., 273:97.

Zucker, D.K., Wooten, G.F., and Lothman, E.W., 1983, Blood-brain barrier changes with kainic acid-induced limbic seizures, Exp. Neurol., 79:422.

EXTRAVASATED PROTEIN AS A CAUSE OF LIMBIC SEIZURE-INDUCED BRAIN DAMAGE:

AN EVALUATION USING KAINIC ACID

R.E. Ruth

Institute for the Study of Developmental Disabilities
and Committee on Neuroscience
The University of Illinois at Chicago
Chicago, Illinois 60608

INTRODUCTION

Tonicoclonic seizure-induced alterations in cerebrovascular permeability (CVP) to protein (CVP-p) have been recognized for several years (see Klatzo, 1983). Although the mechanisms of this permeability change are not completely understood, free arachidonic acid (Bazan, 1970; Rodriguez de Turco et al., 1983; Siesjö et al., 1983) has been implicated (Chan et al., 1983). Several other characteristics of the change are known. The morphologic basis for increased CVP-p appears to be the pinocytotic vesicle, sometimes coalescing to form transendothelial channels which effectively shunt endothelial tight junctions (Westergaard, 1980). Alterations in CVP-p also depend partly on the increased arterial blood pressure accompanying seizures (Johansson and Nilsson, 1977). However, as Nitsch and Klatzo (1983) pointed out, the degree of hypertension attained during an attack would not be, in the absence of the seizure, of sufficient magnitude to provoke increases in CVP-p. Thus, seizure-induced disturbances in CVP-p probably involve an interaction between global systemic factors and more localized molecular phenomena occurring in the microenvironment of recruited neurons. Adding to the complexity of CVP alterations, there appears to be a pathoclisis whereby certain regions are more prone to alterations in CVP than are other recruited regions.

Recently, several laboratories have demonstrated increased CVP to both small and large molecules (Nitsch and Klatzo, 1983; Zucker et al., 1983; Lassmann et al., 1984; Ruth, 1984) during limbic motor seizures caused by the systemic administration of kainic acid (Olney et al., 1974). Electrophysiologically, metabolically and neuropathologically, kainate-induced seizures in rats resemble temporal lobe seizures and status epilepticus in humans (Ben-Ari et al., 1981; Nadler, 1981). It seems likely that alterations in CVP produced by kainate seizures in the rat would also occur in psychomotor epileptics. In this regard, Mihály and Bozoky (1984) demonstrated serum albumin-like immunoreactivity in astrocytes from a temporal lobe epileptic; astrocytes participate in the clearance of extravasated protein (Baker et al., 1971).

The present chapter provides a brief overview of recent experiments with kainic acid in our laboratory. At issue here is the consequence of increased CVP-p on neuronal viability during limbic seizures. Because increased CVP-p is closely associated with vasogenic edema (Klatzo, 1967; Chan and Fishman, 1984), its occurrence represents a possible contributor to seizure-induced neuronal death. To date, there is little direct evidence either to support or to refute a role for CVP-p in the genesis of such damage. Indeed, it has been reasonably suggested that blood-brain barrier injury during kainate seizures may be a consequence, rather than a cause, of neuronal deterioration (Zucker et al., 1983). If, in fact, increased CVP-p does contribute to neuron death in a particular region, then at least two fundamental conditions must hold: (i) extravasation of endogenous protein must occur in the region; and (ii) increased CVP-p must precede the onset of irreversible neuronal deterioration. We analyzed these conditions in two structures that are virtually destroyed by kainic acid: the amygdala and the hippocampus.

EXTRAVASATION OF ENDOGENOUS PROTEIN

Serum albumin-like immunoreactivity was detected in the brains of rats seizing from systemic kainic acid administration (Lassmann et al., 1984). In rabbits, kainate induced the penetration of Evans Blue into the neocortex, but in only 50% of the cases examined (Nitsch and Klatzo, 1983). Since this dye binds readily to serum albumin, it may be presumed that endogenous serum proteins are extravasated, albeit inconsistently. Using rats and horseradish peroxidase (HRP) as the exogenous tracer, we demonstrated a consistent, highly localized extravasation of HRP into several regions of the rat forebrain (Ruth, 1984). These latter experiments were therefore repeated (using antibodies to endogenous rat serum proteins and immunocytochemical methodolgy) to determine if increased CVP-p occurred for endogenous protein during kainate seizures.

Adult female albino rats were injected with kainate (9 mg/kg, i.p.) to induce limbic seizures. Approximately 120 minutes after injection, animals were anesthetized with pentobarbital and perfused with 1% paraformaldehyde/0.5% glutaraldehyde in phosphate buffer (pH = 7.4). Coronal or sagittal sections (50 μm, 1-in-2 series) were incubated with preimmune rabbit serum, followed by purified biotinylated rabbit anti-rat immunoglobulin G (anti-IgG; Vector Labs.) at a dilution of 1:700. After washing, the sections were exposed to an avidin-biotinylated HRP complex (Vector Labs.) and then washed again. After processing for the blue HRP reaction product (Mesulam, 1976), the tissue was analyzed under a Zeiss Photomicroscope I. Vehicle-injected (control) and kainate-injected animals were routinely processed together to ensure identical conditions throughout each procedure.

The location of rat IgG-like immunoreactivity (IgG-R) in the brains of control and seizing animals was, with a few exceptions, qualitatively similar to the distribution of extravasated HRP in our previous work. For the present, it is important to note that weak to nonexistent IgG-R occurred within the amygdaloid complex or neocortex of seizing animals. On the other hand, IgG-R was always observed within the hippocampal formation (see below).

Control reactions indicated that positive staining was probably due to the presence of extravasated IgG. Biotinylated antibody blocked with excess purified rat IgG (Sigma), as well as frank deletion of the antibody from an otherwise identical reaction protocol, produced no staining. Additionally, in control animals lower dilutions of biotinylated antibody produced staining everywhere (especially in myelin). But as the antibody dilution was increased toward 1:700, the brain was virtually unreactive except for the pineal gland, area postrema, dorsal solitary nucleus, basomedial hypothalamus and choroid plexus. Because they lack a structural barrier (Rapoport, 1976), these latter regions would be expected to contain endogenous serum protein in the extracellular space.

SURVIVAL TIME OF SEIZING AMYGDALOID AND HIPPOCAMPAL NEURONS

In order for CVP-p to participate in the destruction of seizing neurons, barrier injury should occur prior to irreversible neuronal injury. Therefore, knowing precisely when during the ongoing seizure process a given set of neurons is irreversibly injured was critical for this analysis. The neuropathologic 'endpoint' of sublethal kainate-induced status epilepticus is well-known: massive death in the amygdala, hippocampus and several other regions of the limbic system and thalamus (Schwob et al., 1980; Ben-Ari et al., 1981). However, no information was availabe on when during the seizure process itself the actual transition between reversible and irreversible injury occurred (see Fuller and Olney, 1981).

A simple technique was devised to permit a light-microscopic approximation of the time-sequence of neuronal death during kainate seizures. This approach has yielded consistent results and a time-line of neuronal death, along which regions of the rat forebrain are differentially distributed. The essence of the method was threefold : (i) permit seizures to occur and then 'quench (halt)' the seizure process pharmacologically; (ii) introduce a seizure-free survival period of sufficient duration for irreversibly injured neurons to degenerate to the point where they are detectable with the light-microscope (at the same time, reversibly injured neurons would recover); and (iii) apply histopathologic methods to assess the location and degree of irreversible injury extant around the time of seizure quenching.

Fifty-one rats injected systemically with kainate (as before) were closely observed throughout the post-injection period. Their behaviors were rated according to the following ordinal stages: 1 (staring or any normal behaviors); 2 (wet-dog shakes); 3 (automatisms and stereotypies); 4 (rearing with forelimb clonus); 5 (circling); 6 (stereotyped escape). In most animals injected with kainate, the onset of status epilepticus was preceded by several episodes of behaviorally 'discrete' limbic motor seizures. That is, animals in stage 1 (punctuated sporadically by stage 2 behaviors) rather suddenly would enter stage 3 for 30-90 seconds, followed by stage 4 behavior for up to 150 seconds, followed by a return to stage 1 for several minutes before the entire progression repeated itself. This sequence was termed a discrete limbic motor seizure episode (Ben-Ari et al., 1981). Finally (after 4-8 such episodes), a point in the seizure process was reached where stage 4 attacks did not subside to stage 1 or 2. The onset of behavioral status epilepticus was operationally defined as the time of occurrence of the first stage 4 event which was followed by five consecutive minutes of exclusive stage 3 or greater behaviors. Animals sustaining five consecutive minutes of such incessant motor seizure rarely returned to stage 1-2 until spontaneous remission several hours later. By timing the seizure process according to the animal's own progression through the process (rather than simply according to time elapsed since injection), a reduction in the variability of brain damage across animals appears to occur. Undoubtedly it is the good correlation between these behavioral stages and the underlying electrical seizure patterns (Lothman and Collins, 1981) which permits their utility as time markers of the ongoing seizure process.

Prior to (following four to six discrete limbic seizures) or after (timed in minutes of status) the onset of behaviorally defined status epilepticus, rats were injected with pentobarbital (40 mg/kg, i.p.); control animals were yoked to the seizing rats. Within four minutes of barbiturate administration, all behavioral indices of seizure were quenched and the animals became anesthetized for at least three hours. Upon full recovery from anesthesia, rats which sustained up to 45 minutes of status epilepticus prior to quenching appeared remarkably healthy. Although brief and sporadic periods of stage 1-3 behaviors were occasionally observed in some animals, no stage 4 or greater behaviors were observed. Forty-eight hours after the barbiturate injection, animals were reanesthetized (pentobarbital, 50 mg/kg, i.p.) and perfused with formalin. Brains were processed for histopathologic examination using Fink-Heimer (1967) and Nissl methods as previously described (Ruth, 1982).

Sufficient cases have been examined thus far (n = 18) to permit several preliminary suggestions and an approximation of the time relationship between increased CVP-p and neuronal death in the amygdala and hippocampus. Neither region contained significant neuronal losses in any of the cases (n = 8) where seizures were quenched prior to the onset of status epilepticus. In the amygdala, fewer than 5% of the neurons were argyrophilic and pyknotic, despite up to six severe discrete seizures (up to 115 minutes after kainate injection). Less than 1% of hippocampal neurons were irreversibly injured prior to the onset of status epilepticus (see also Sperk et al., 1983). This overall picture did not deteriorate further after five minutes of status (n = 2). However, between 5 and 30 minutes after status epilepticus began (n = 4 at 15 minutes; 2 at 30 minutes), there was a dramatic increase in presumed cell death within several amygdaloid nuclei (Fig. 1A) and within area TR (Haug, 1976) of the anteroventral entorhinal cortex (Fig. 1B). Damage within the amygdala ranged from 20% to 45% of the neurons within a given nucleus.

Somewhat unexpectedly in view of current thinking that the hippocampus is the most vulnerable brain area to kainate-induced seizures, no significant hippocampal cell death was observed in cases quenched at or prior to 15 minutes of status epilepticus. After 30 minutes of status, one to three small patches of pyramidal cell argyrophilia were found in dorsal field CA3 in both cases (Fig. 2A). Field CA3 damage was not noticeably increased after 45 minutes of status (n = 2), but now a moderate amount of damage was readily apparent within dorsal field CA1 (Fig. 2B). No damage was found in the ventral hippocampus at any of the durations examined. This relatively mild hippocampal damage after 45 minutes of status epilepticus may be compared with cases permitted to seize incessantly for hours until spontaneous remission, i.e., near the neuropathologic endpoint of kainate seizures. At that time, virtually all pyramidal (excluding field CA2) and hilar neurons were killed throughout the dorsoventral length of the hippocampus.

Several preliminary suggestions are possible from these data. First, it appears that events associated with behavioral status epilepticus, rather than with discrete seizures per se, represent the condition which compromises neuronal viability in the hippocampus; the relationship could actually be reversed in the amygdala. This notion might explain why French et al. (1982) did not find significant hippocampal neuron loss despite intense electrical seizures from low doses of locally applied kainate; the injections reportedly produced only occassional stage 4 (mostly stages 1-3) behavior and no status epilepticus as operationally defined above.

Second, the collapse of hippocampal neurons may proceed in a dorsal-to-ventral direction. If this observation is supported in future work, it will also be difficult to reconcile with the idea that discrete seizure activity per se kills hippocampal pyramidal cells. The ventral hippocampal formation is the first region to manifest electrical and metabolic indices of systemic kainate seizures (Collins et al., 1980; Ben-Ari et al., 1981), and would be expected to succumb earlier than the more slowly recruited

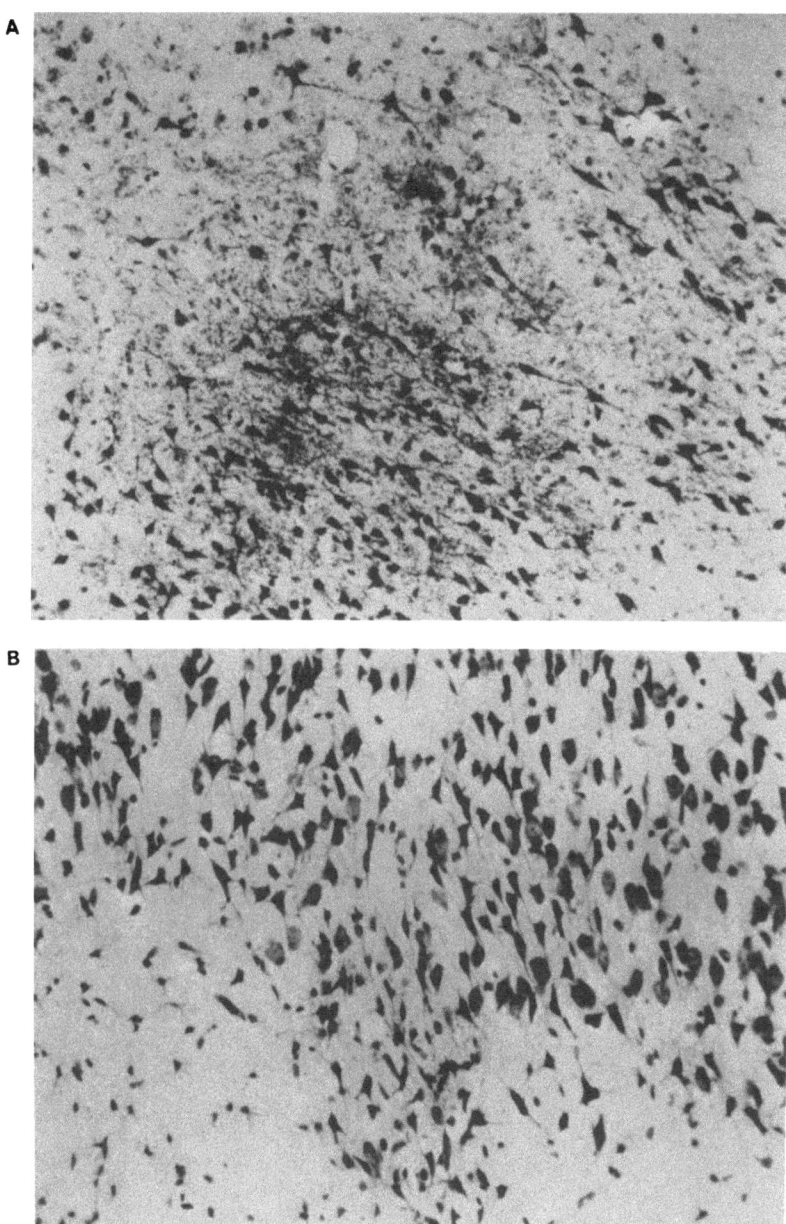

Fig. 1. Early neuronal losses during kainate-induced seizure.
A: Necrotic cells in the cortical and basolateral
(upper right) amygdaloid nuclei after 30 minutes of
status epilepticus; Fink-Heimer stain. B: Profound
damage within the multicellular 'islands' of entorhinal area TR after 15 minutes of status (thionin stain).
Note the few surviving neurons (ovoid-shaped with distinct nuclei and nucleoli) in this section.

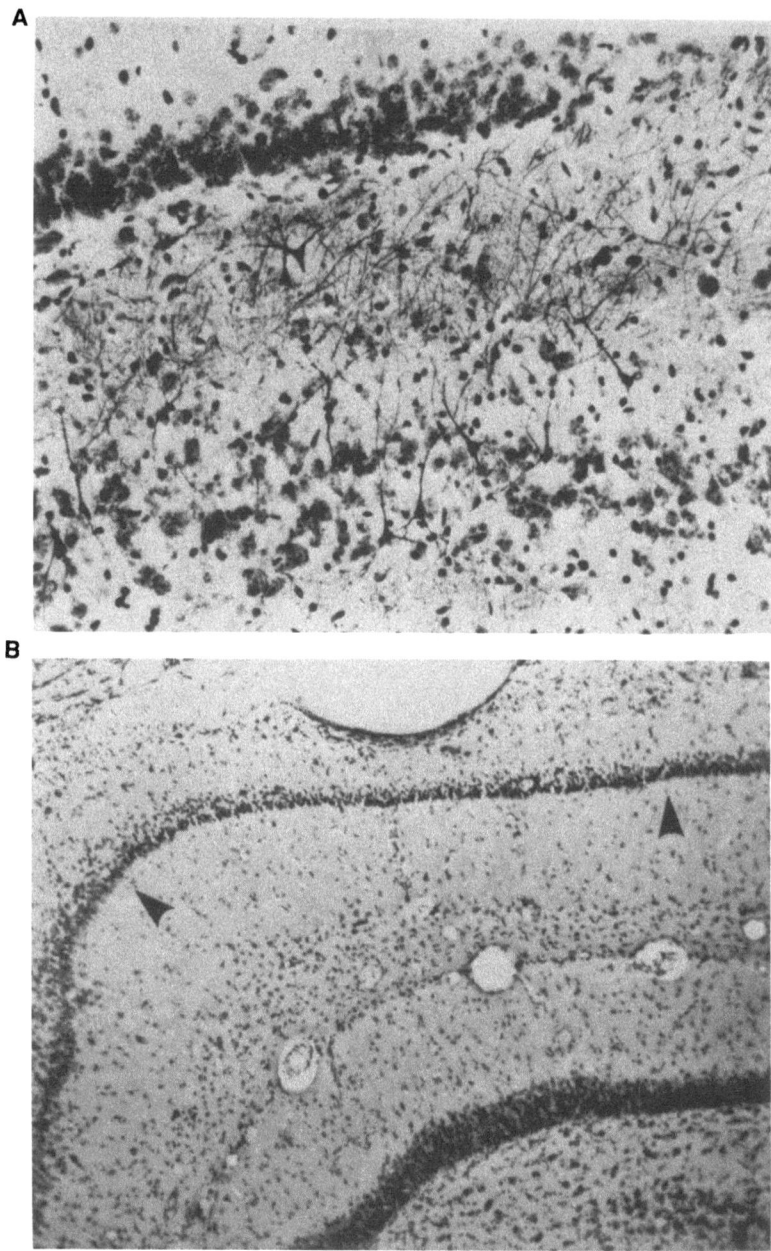

Fig. 2. Later occurring cell loss within the hippocampus.
A: Dorsal field CA3c, with Golgi-like staining of a few presumably degenerating neurons; Fink-Heimer stain, quenched 30 minutes after status onset.
B: Collapse of a segment (between arrowheads) within dorsal field CA1 after 45 minutes of status epilepticus (thionin stain).

dorsal neurons. Alternatively, since hippocampal mossy fibers are necessary for the expression of kainate toxicity there (Nadler and Cuthbertson, 1980), an earlier loss of dorsal hippocampal neurons may reflect the ratio of granule-to-pyramidal cells at 10:1 dorsally but only 1:1 ventrally (Gaarskjaer, 1978).

Finally, the death of brain cells from kainate-induced seizures is regionally organized along the time dimension in a hierarchy. Fifteen minutes of status epilepticus already had produced extensive amygdaloid damage, but at least 45 minutes of status were necessary before a comparatively modest number of hippocampal pyramidal cells were destroyed.

TIME RELATIONSHIP BETWEEN INCREASED CVP-p AND NEURONAL DEATH IN THE AMYGDALA AND HIPPOCAMPUS

Rats were injected with kainate as before and their behavior was rated as in the quenching experiments. Either 5 (n = 2) or 15 (n = 6) minutes after the onset of behavioral status epilepticus, the seizures were quenched with pentobarbital. As soon as anesthesia was achieved, the animals were prepared for the immunocytochemical demonstration of endogenous IgG as described above. Seven control cases were simultaneously prepared.

Only one of eight cases gave evidence of IgG-R within the amygdala; in that case, several faint spots were detected in the cortical nucleus. In contrast, IgG-R was well developed throughout the dorsoventral extent of the hippocampus in all eight cases. A distinct 'streak' of IgG-R was always present within stratum lucidum of the dorsal hippocampus (Fig. 3A). Additionally, stratum oriens of dorsal field CA1 contained IgG-R, and in half of these cases reaction product extended into stratum radiatum as well (Fig. 3B). Curiously, when it occurred, this latter reactivity ended abruptly at the border between stratum radiatum and stratum lacunosum-moleculare. In the ventral hippocampus, IgG-R took the form of spots in ventral field CA3 and the ventral subiculum. Such spots could occur more dorsally as well. As has been consistently observed, non-seizing animals exhibited IgG-R only in the circumventricular organs.

CONCLUSIONS

These experiments provide further information on the nature of increased CVP-p during kainate-induced motor seizures. Normally excluded from the brain, endogenous immunoglobulins are extravasated during kainate seizures in a regionally localized pattern. This pattern resembles that which occurs using HRP as the protein tracer (Ruth, 1984). The anatomic specificity of extravasation reaches a maximum in the dorsal hippocampus, where IgG-R is consistently restricted to particular strata within fields CA3 and CA1. A remarkable degree of local control over cerebrovascular permeability must exist to produce such confined extravasation patterns.

The results also indicate that within the hippocampus increased CVP-p precedes significant irreversible neuronal damage by at least 30 minutes. This time relationship is necessary if increased CVP-p has any role in the destruction of neurons there. In the amygdala, extravasation of IgG was not detected by these methods, at least prior to the onset of major damage there. At this time, a role for CVP-p in the obliteration of the amygdala by kainate seems doubtful. However, a moiety of immunoreactivity was probably destroyed by glutaraldehyde and, to a lesser extent, by paraformaldehyde. We found a total elimination of all IgG-R with glutaraldehyde concentrations in excess of 1%. Therefore, more IgG extravasation occurred

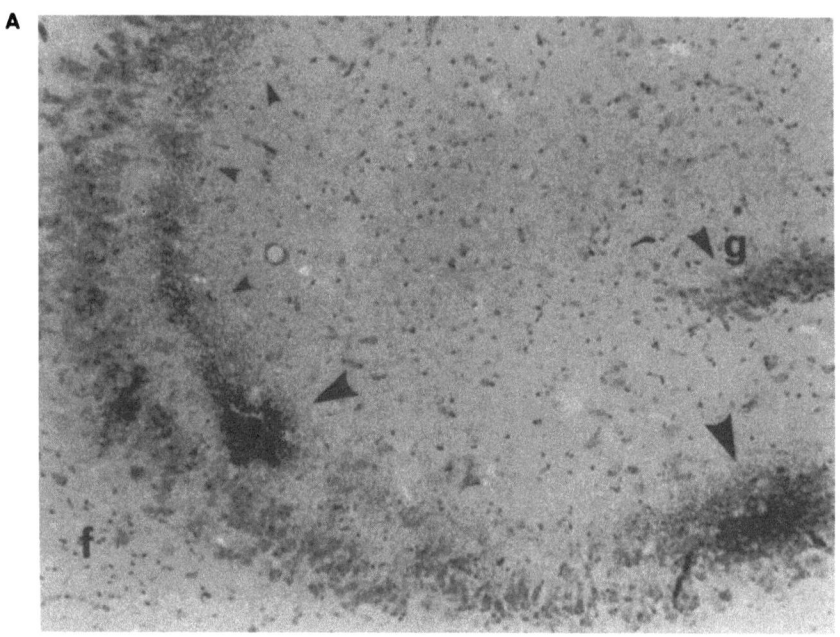

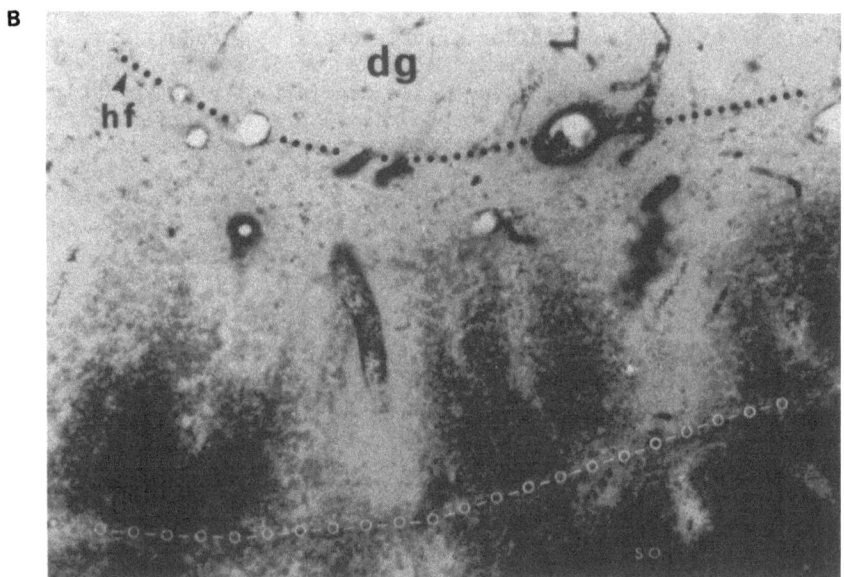

Fig. 3. Immunoglobulins in the hippocampus five minutes after status onset. A: Spots (large arrows) and streaks (small arrows) in stratum lucidum of dorsal field CA3. Top-right arrow: suprapyramidal tip of dentate granule cells (g). B: Extensive IgG-R in all strata of dorsal field CA1 except stratum lacunosum-moleculare. White circles/dashes: stratum pyramidale; dg (molecular layer of dentate gyrus); hf (hippocampal fissure); so (stratum oriens).

than the present methods detected; what is being visualized here may be
the areas of highest serum protein extravasation, rather than the only
areas where any extravasation occurred. Nevertheless, the outcomes suggest
that both common as well as unique molecular phenomena occur in a least
two kainate-susceptible regions, and that somewhat separate phenomena
may be involved in their eventual destruction.

The vasogenic edema associated with increased CVP-p provides one
potential mechanistic linkage between protein extravasation and neuronal
injury (Klatzo, 1967). Lassmann et al. (1984) provided evidence that
the destructive mechanism during kainate seizure was not vasogenic, but
rather cytotoxic, edema. The present results are consonant with this
view concerning the amygdaloid region. However, the presence of extravasated
serum protein well before hippocampal cell loss, and the lack of suitable
direct evidence on the contribution of this phenomenon to neuronal damage,
do not as yet permit a similar conclusion regarding the fate of the hippocampus. In addition, a consideration of the possible toxicity of the
various extravasated solutes themselves may be worthwhile (see also Remler
and Marcussen, 1984). With respect to one such solute, IgG, it is conceivable that the hippocampal membrane could contain antigenic determinants
for endogenous antibody; alternatively, the intense mechanical distortion
of hippocampal tissue during kainate seizures (Olney et al., 1979; Fuller
and Olney, 1981) could inadvertently expose previously hidden determinants.
Evidence continues to mount for the presence of cross-reacting autoantibodies
to brain tissue (cf. Schuller-Petrovic et al., 1983; Budka and Majdic,
1985; Stefansson et al., 1985) and could conceivably provide one teleologic
purpose for the blood-brain barrier to protein. If the extravasated antibody
was specifically bound to brain tissue (e.g., Alafuzoff et al., 1983),
and presuming that complement protein is also extravasated, immunologic
reactions become a possible contributor to hippocampal lysis. After such
speculation, it is important to return to the interpretive boundaries
of the present results; only two essential features of causality (i.e.,
increased CVP-p to endogenous protein occurs and occurs prior to hippocampal
cell loss) have been demonstrated here. Mechanistic linkages, if any,
connecting neuronal death to increased CVP-p and/or to vasogenic edema
from kainate-induced seizures still remain to be clarified.

REFERENCES

Alafuzoff, I., Adolfsson, R., Bucht, G., and Winblad, B., 1983, Albumin
    and immunoglobulin in plasma and cerebrospinal fluid, and blood-cerebro-
    spinal fluid barrier function in patients with dementia of Alzheimer
    type and multi-infarct dementia, J. Neurol. Sci., 60:465.
Baker, R.N., Cancilla, P., Pollack, P., and Frommes, B., 1971, The movement
    of exogenous protein in experimental cerebral edema, J. Neuropathol. Exp.
    Neurol., 30:668.
Bazan, N.G., 1970, Effects of ischemia and electroconvulsive shock on
    free fatty acid pool in the brain, Biochim. Biophys. Acta, 218:1.
Ben-Ari, Y., Tremblay, E., Riche, D., Ghilini, G., and Naquet, R., 1981,
    Electrographic, clinical and pathological alterations following systemic
    administration of kainic acid, bicuculline or pentetrazol; metabolic
    mapping using the deoxyglucose method with special reference to the
    pathology of epilepsy, Neuroscience, 6:1361.
Budka, H., and Majdic, O., 1985, Shared antigenic determinants between
    human hemopoietic cells and nervous tissue and tumors, Acta. Neuropathol.,
    67:58.
Chan, P.H., and Fishman, R., 1984, The role of arachidonic acid in vasogenic
    brain edema, Fed. Proc., 43:210.

Chan, P.H., Fishman, R., Caronna, J., Schmidley, J., Prioleau, G., and Lee, J., 1983, Induction of brain edema following intracerebral injection of arachidonic acid, Ann. Neurol., 13:625.

Collins, R.C., McLean, M., and Olney, J., 1980, Cerebral metabolic response to systemic kainic acid: 14-C-deoxyglucose studies, Life Sci., 27:855.

Fink, R.P., and Heimer, L., 1967, Two methods for selective silver impregnation of degenerating axons and their synaptic endings in the central nervous system, Brain Res., 4:369.

French, E.D., Aldinio, C., and Schwarcz, R., 1982, Intrahippocampal kainic acid seizures and local neuronal degeneration: relationships assessed in unanesthetized rats, Neuroscience, 7:2525.

Fuller, T.A., and Olney, J., 1981, Only certain anticonvulsants protect against kainate neurotoxicity, Neurobehav. Toxicol. Teratol., 3:355.

Gaarskjaer, F., 1978, Organization of the mossy fiber system of the rat studied in extended hippocampi: terminal area related to number of granule and pyramidal cells, J. Comp. Neurol., 178:49.

Haug, F.-M.S., 1976, Sulfide silver pattern and cytoarchitectonics of parahippocampal areas in the rat, Adv. Anat. Embryol. Cell Biol., 52:1.

Johansson, B., and Nilsson, B., 1977, The pathophysiology of the blood-brain barrier dysfunction induced by severe hypercapnia and by epileptic brain activity, Acta Neuropathol., 38:153.

Klatzo, I., 1967, Neuropathological aspects of brain edema, J. Neuropathol. Exp. Neurol., 24:1.

Klatzo, I., 1983, Disturbances in the blood-brain barrier in cerebrovascular disorders, Acta Neuropathol. (Suppl.), 7:81.

Lassmann, H., Petsche, U., Kitz, K., Baran, H., Sperk, G., Seitelberger, F., and Hornykiewicz, O., 1984, The role of brain edema in epileptic brain damage induced by systemic kainic acid injection, Neuroscience, 13:691.

Lothman, E., and Collins, R.C., 1981, Kainic acid induced limbic seizures: metabolic, behavioral, electroencephalographic and neuropathological correlates, Brain Res., 218:299.

Mihály, A., and Bozoky, B., 1984, Immunohistochemical localization of extravasated serum albumin in the hippocampus of human subjects with partial and generalized epilepsies and epileptiform convulsions, Acta Neuropathol., 65:25.

Mesulam, M.-M., 1976, The blue reaction product in horseradish peroxidase neurochemistry: incubation and visibility, J. Histochem. Cytochem., 24:1273.

Nadler, J.V., 1981, Kainic acid as a tool for the study of temporal lobe epilepsy, Life Sci., 29:2031.

Nadler, J.V., and Cuthbertson, G., 1980, Kainic acid neurotoxicity toward hippocampal formation: dependence on specific excitatory pathways, Brain Res., 195:47.

Nitsch, C., and Klatzo, I., 1983, Regional patterns of blood-brain barrier breakdown during epileptiform seizures induced by various convulsive agents, J. Neurol. Sci., 59:305.

Olney, J.W., Fuller, T., and deGubareff, T., 1979, Acute dendrotoxic changes in the hippocampus of kainate treated rats, Brain Res., 176:91.

Olney, J.W., Rhee, V., and Ho, O., 1974, Kainic acid: a powerful neurotoxic analog of glutamate, Brain Res., 77:507.

Rapoport, S.I., 1976, Blood-Brain Barrier in Physiology and Medicine, Raven Press, New York.

Remler, M.P., and Marcussen, W., 1984, The blood-brain barrier lesion and the systemic convulsant model of epilepsy, Epilepsia, 25:574.

Rodrigues de Turco, E.B., Morelli de Liberti, S., and Bazan, N., 1983, Stimulation of free fatty acid and diacylglycerol accumulation in cerebrum and cerebellum during bicuculline-induced status epilepticus. Effect of pretreatment with α-methyl-p-tyrosine and p-chlorophenylalanine, J. Neurochem., 40:252.

Ruth, R.E., 1984, Increased cerebrovascular permeability to protein during systemic kainic acid seizures, Epilepsia, 25:259.

Ruth, R.E., 1982, Kainic acid lesions of hippocampus produced iontophoretically: the problem of distant damage, Exp. Neurol., 76:508.

Schuller-Petrovic, S., Gebhart, W., Lassmann, H., Rumpold, A., and Kraft, D., 1983, A shared antigenic determinant between natural killer cells and nervous tissue, Nature, 306:179.

Schwob, J.E., Fuller, T., Price, J., and Olney, J., 1980, Widespread patterns of neuronal damage following systemic or intracerebral injections of kainic acid: a histological study, Neuroscience, 5:991.

Siesjö, B., Ingvar, M., Folbergrova, J., and Chapman, A., 1983, Local cerebral circulation and metabolism in bicuculline-induced status epilepticus: relevance for development of cell damage, in: Status Epilepticus: Mechanisms of Brain Damage and Treatment, A. Delgado-Escueda, C. Wasterlain, D. Trieman, and R. Porter, eds., Raven Press, New York, p. 217.

Sperk, G., Lassmann, H., Baran, H., Kish, S., Seitelberger, F., and Hornykiewicz, O., 1983, Kainic acid induced seizures: neurochemical and histopathological changes, Neuroscience, 10:1301.

Stefansson, K., Marton, L., Dieperink, M., Molnar, G., Schlaepfer, W., and Helgason, C., 1985, Circulating autoantibodies to the 200,000-dalton protein of neurofilaments in the serum of healthy individuals, Science, 228:1117.

Westergaard, E., 1980, Ultrastructural permeability properties of cerebral microvasculature under normal and experimental conditions after application of tracers, Adv. Neurol., 38:1.

Zucker, D., Wooten, G., and Lothman, E., 1983, Blood-brain barrier changes with kainic acid-induced limbic seizures, Exp. Neurol., 79:422.

ULTRASTRUCTURAL ANALYSIS OF RAT BRAIN TISSUE FOLLOWING SYSTEMIC
KAINATE ADMINISTRATION

H. Lassmann, H. Baran, U. Petsche, K. Kitz, G. Sperk,
O. Hornykiewicz and F. Seitelberger

Neurological Institute and Institute for Biochemical
Pharmacology, University of Vienna
Vienna, Austria

INTRODUCTION

Morphological studies in models of experimental epilepsy have mainly focused on pathohistological changes of nerve cells (Schwob et al., 1980; Ben-Ari et al., 1981). Although it is evident, that neurons are the primary target in the action of excitotoxins, recent evidence suggests that besides direct neurotoxicity other pathogenetic mechanisms play an important role in the development of postepileptic brain damage. It is thus interesting to note, that irreversible neuropathological changes following kainic acid (KA) induced seizures not only involve nerve cells, but also glia, myelin sheaths and blood vessels (Sperk et al., 1983). In our present study we describe the sequence of morphological events in the limbic system of rats, systemically injected with KA. Our findings indicate that cytotoxic brain edema and subsequent focal ischemia plays a central role in the pathogenesis of irreversible seizure induced brain lesions.

MATERIAL AND METHODS

Young adult Sprague Dawley rats were obtained from FIV (Himberg, Austria). KA (Sigma, 10 mg/kg) dissolved in saline and adjusted to pH 7.0 was injected subcutaneously. Control animals received an equal amount of saline. Behavioral changes were scored as described in detail earlier (Sperk et al., 1983; Lassmann et al., 1984). At various time points (1,2, 3,24,48 hours and 3 and 6 days) after KA injection animals, were anesthetized with thiopental (Pentothal Na, Sanabo, Austria) and perfused via the aorta with a mixture, containing 0.5% paraformaldehyde and 1.5% glutaraldehyde in 0.1 M phosphate buffer (pH 7.4). Brains were dissected and further immersed in 3% glutaraldehyde for an additional two hours. Coronal slices of brain tissue were then routinely embedded in paraffin and analyzed light microscopically. In addition, small tissue blocks of hippocampus and entorhinal/pyriform cortex were osmicated, embedded in epon and studied in the electron microscope.

Blood brain barrier damage was analyzed by studying the leakage of horse radish peroxidase (HRP, Boehringer Grade II, 265 mg/kg), injected intravenously 45 minutes prior to sampling of the animals. Peroxidase reaction product was visualized according to Reese and Karnovski (1967).

RESULTS

Earliest neuropathological changes in the hippocampus and the entorhinal/pyriform cortex of KA treated animals were noted one hour after injection of the drug (a few minutes after onset of generalized seizures). Scattered nerve cells appeared dark and condensed (Fig. 1). The intracytoplasmic organelles were normal with the exception of some swelling of endoplasmic reticulum and mitochondria (Fig. 1). In addition, increase in the volume of some cytoplasmic processes resulted in a microvacuolated appearance of the neuropil. These edematous changes were mainly found in the hippocampus (CA1, CA3, CA4), mainly involving the stratum pyramidale and stratum oriens, and in the layers 3-5 of the entorhinal/pyriform cortex. Cytoplasmic swelling was mainly confined to astrocytes, although some dendrites were dilated too.

During the following hours after KA injection, neuronal and dendritic changes were similar qualitatively and quantitatively as described above. However, swelling of astrocyte processes increased dramatically, reaching maximal intensity three hours after KA (Figs. 2, 3, and 4). In this stage, extreme vacuolation of the tissue was noted in the CA1 and CA3 sector of the hippocampus (mainly confined to the stratum pyramidale; Fig. 2) and in most parts of the entorhinal/pyriform cortex, sparing the molecular layer. Astrocyte swelling was most pronounced in perineuronal and perivascular areas (Figs. 2, 3, and 4). The blood vessels in affected regions showed narrow, slit-like lumina (Figs. 2, 3), frequently containing trapped

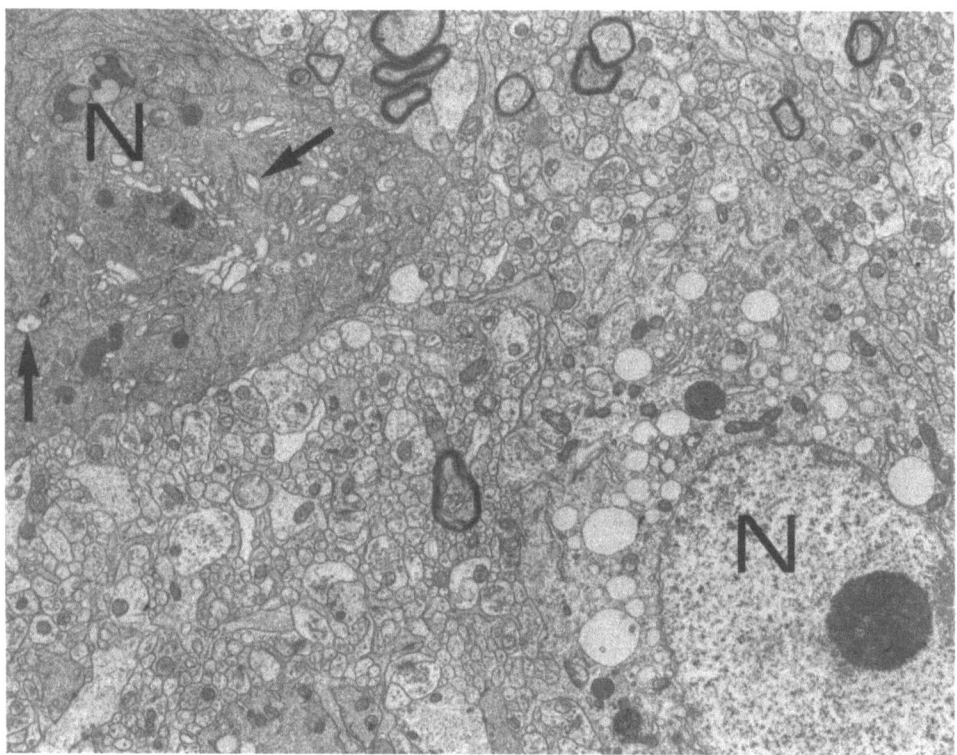

Fig. 1. CA1 region of the hippocampus of a rat one hour after systemic injection of KA. Two nerve cells (N), the upper one with dark, condensed cytoplasm. Some swelling of endoplasmic reticulum, golgi and mitochondria (arrows) EM x 7.300.

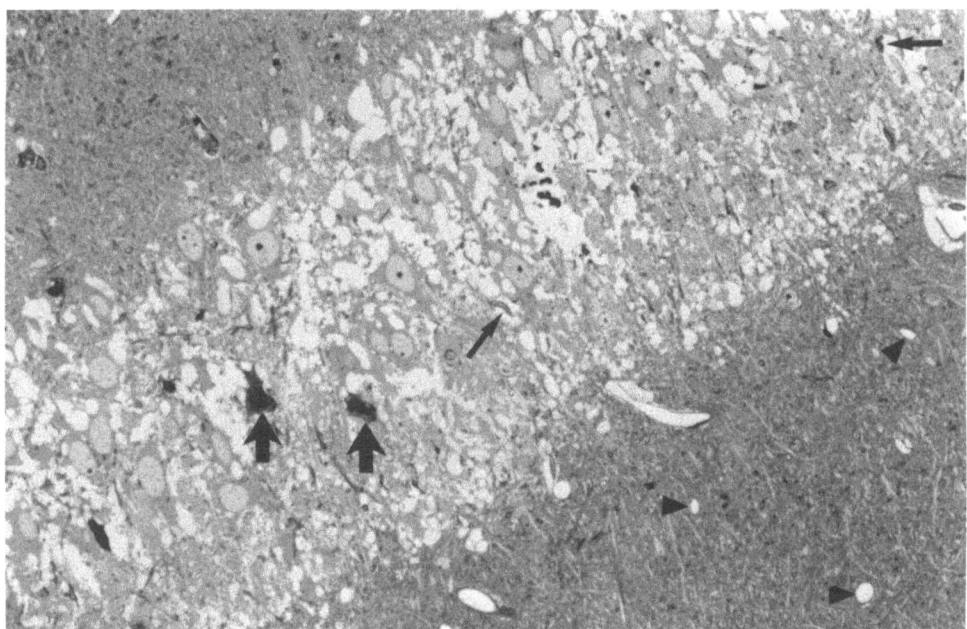

Fig. 2. CA1 region of the hippocampus of a rat three hours after systemic injection of KA. Extreme vacuolation of the tissue due to astrocyte swelling. Most of the nerve cells are well preserved, only two neurons appear dark and shrunken (thick arrow). Blood vessels in the edematous region are compressed with slit-like lumina (thin arrows). Vessels in the adjacent non edematous areas are normal and well perfused (triangles). Toluidine blue stained plastic section, 360x.

erythrocytes or serum components, in spite of optimal vascular perfusion of adjacent non edematous areas.

Although a moderate degree of blood brain barrier damage was observed during these early stages of the disease, it has to be emphasized, that HRP leakage in these animals was much less pronounced, compared to that found in classical models of vasogenic brain edema like encephalitis or osmotic opening of the blood brain barrier. Furthermore, peroxidase leakage was most pronounced in the thalamus with very little or absent extravasation of serum proteins in the areas of brain edema in the limbic system.

Late Changes

One day and later after KA injection, edematous changes were much less pronounced than during earlier stages. However, the regions which before showed pronounced edema now revealed evidence of incomplete tissue necrosis (Figs. 5, 6 and 7). Although nerve cells in these regions were not severely affected (Fig. 5), also other elements of the central nervous system like myelin or oligodendrocytes were involved (Figs. 5, 6). In addition, small perivascular hemorrhages, especially around drainage veins of the hippocampus and entorhinal/pyriform cortex were found (Fig. 7).

Not surprisingly, blood brain barrier damage in these late stages of the disease was accentuated in areas of tissue necrosis, with only little peroxidase leakage in other brain regions.

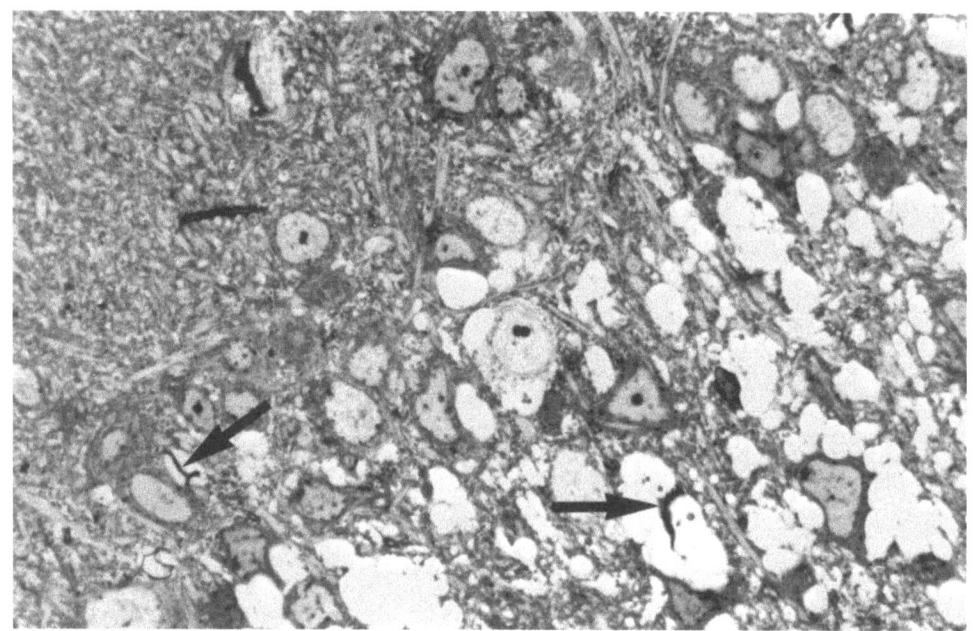

Fig. 3. CA1 region of the hippocampus three hours after KA injection. Extensive perineuronal and perivascular astrocyte swelling with compression of cerebral blood vessels (arrows). Toluidine blue stained plastic section, 900x.

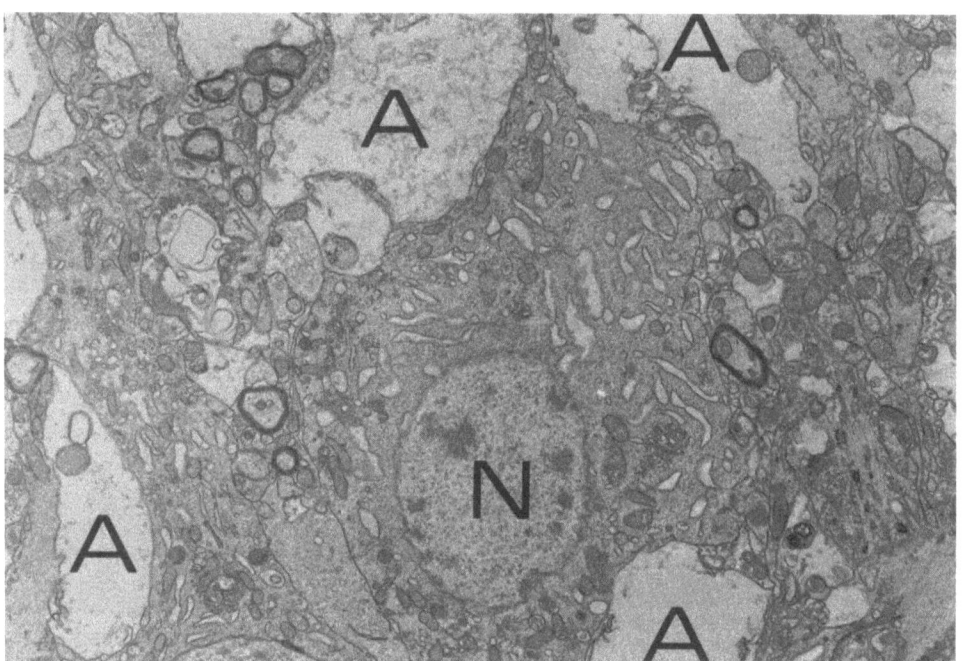

Fig. 4. CA1 region of the hippocampus, three hours after KA injection. Extensive swelling of astrocyte processes (A). Some condensation of nerve cell cytoplasm (N) with some dilation of endoplasmic reticulum and well preserved mitochondria. EM 5,400x.

Fig. 5. Entorhinal cortex, 48 hours after KA injection. Incomplete tissue necrosis with only few preserved nerve cells (arrows), degeneration of myelin sheaths, numerous macrophages and beginning gliosis. Toluidine blue stained plastic section, 360x.

DISCUSSION

Earliest morphological changes in the limbic system of rats one hour after KA injection consisted of shrinkage and condensation of nerve cells, with or without swelling of intracytoplasmic organelles, especially mitochondria (Sperk et al., 1983). These neuronal alterations are not unique for KA intoxication, but can be found, although in different topographical distribution, also in other models of experimental epilepsy (Blennow et al., 1978; Ben-Ari et al., 1981; Atillo et al., 1983) as well as in hypoglycemia (Auer et al., 1985) or ischemia (Spielmeyer, 1927). Recent evidence suggests that these changes are accompanied by an influx of calcium into the cytoplasm (Griffiths et al., 1983; Evans et al., 1984). In a detailed study of hypoglycemia, Auer et al. (1985) differentiated these changes from fully developed irreversible nerve cell damage, the latter accompanied by cellular acidophilia, by defects in nuclear and outer cell membranes and by kariolysis. It is interesting to note that these irreversible nerve cell changes were found in our experiments rarely during the early stages (1-3 hours after KA) but in considerable extent in later stages (1-6 days after KA). In fact, evidence accumulates that early nerve cell changes occurring during the first hours after KA injection are reversible since treament of animals in this stage with either diazepam (Evans et al., 1984) or mannitol (Lassmann et al., 1984) may prevent nerve cell degeneration.

The second hallmark of early KA lesions in the brain is the appearance of brain edema. This is reflected by massive swelling of astrocyte processes, which reaches maximal intensity three hours after KA injection. At this stage, focal brain edema, restricted to limited portions of the limbic system, is severe enough to focally impair local blood flow. Also, brain edema is not unique for KA intoxication but is found in a variable

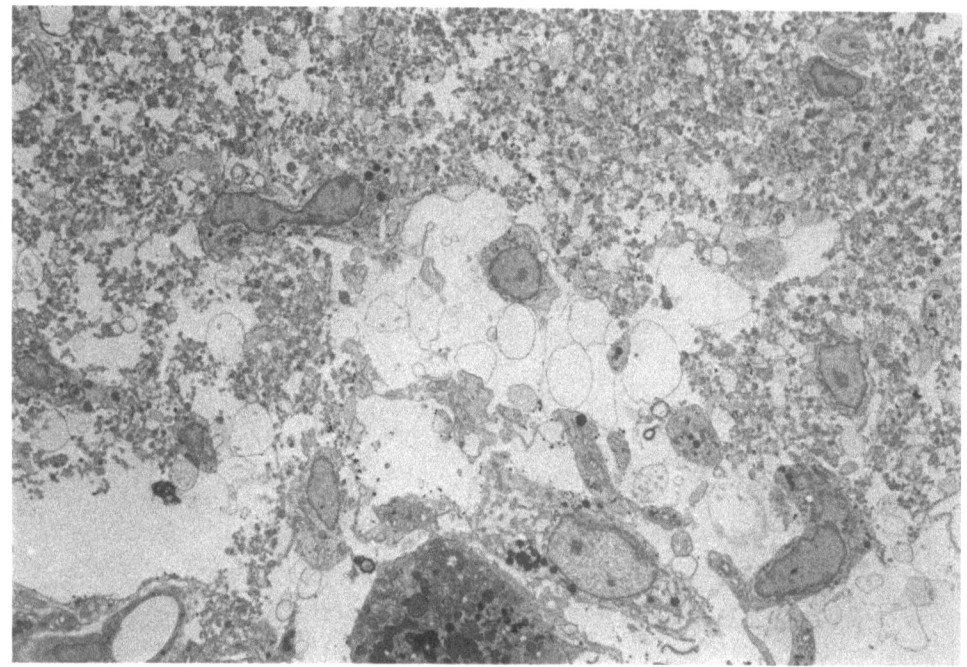

Fig. 6. Pyriform cortex, six days after KA injection. Incomplete, microcystic tissue necrosis with numerous macrophages and some astrocytes. In addition small cytoplasmic profiles, mainly axons and nerve terminals. EM 1,800x.

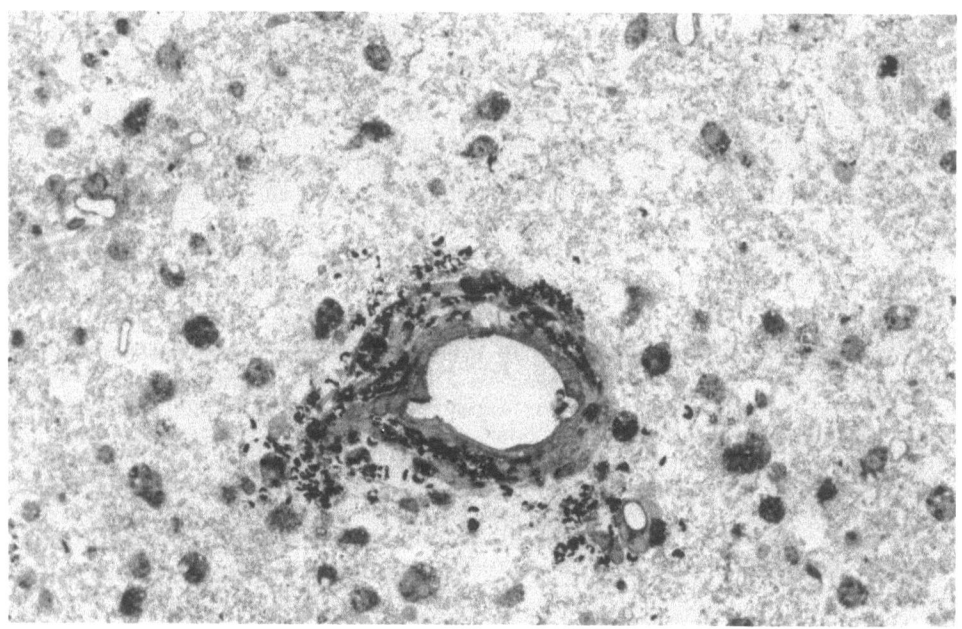

Fig. 7. CA1 region of hippocampus, 48 hours after KA with incomplete tissue necrosis and numerous macrophages. A small vein with perivascular hemorrhage. Toluidine blue stained plastic section, 500x.

degree and in different topographical distribution also in other models of experimental epilepsy (Ben-Ari et al., 1981; Atillo et al., 1983).

At later stages (1-6 days after KA) signs of irreversible brain damage, debris containing macrophages and reactive gliosis are found in the same areas, which earlier show extensive edematous changes.

The pathogenetic mechanisms leading to brain edema in this model are unresolved up to now. Blood brain barrier studies revealed that increased vascular permeability is mainly present in other areas than those showing the most severe edematous changes (Nitsch and Klatzo, 1983; Lassmann et al., 1984). Thus, leakage of serum proteins apparently contributes little to the development of edema in KA lesions. Overactivity of neurons may be involved by release of potassium into the extracellular space (Wade et al., 1975), resulting in ionic imbalance, or by liberating osmotically active or toxic metabolites (e.g., lactic acid). In addition, a direct action of KA or other excitotoxins on the permeability of astrocyte membranes has to be considered.

Nevertheless, cytotoxic brain edema appears to play an important role in the pathogenesis of irreversible brain damage in this model. Treatment of brain edema with mannitol, but not with dexamethasone, may completely prevent neuronal damage in the limbic system of KA animals (Lassmann et al., 1984; Baran et al., in preparation). Furthermore, mannitol treatment reveals an effect on seizure activity itself (Baran et al., in preparation). When KA animals were treated with a sufficient dose of mannitol, seizures were arrested within 30-60 minutes after onset of therapy. This indicates that generalized status epilepticus is only triggered by KA, its further propagation then depends on local tissue alterations, possibly induced by focal brain edema.

In conclusion, we propose the following pathogenetic mechanisms involved in the development of irreversible KA induced brain damage. Systemic application of KA leads to overexcitation of susceptible neurons. This results in liberation of potassium and toxic metabolites into the extracellular space and in an influx of calcium into the nerve cell cytoplasm. As seen from mannitol treatment experiments, these alterations appear to be reversible. Uptake of potassium and metabolites in astrocytes results in focal cytotoxic brain edema with impairment of local microcirculation. This further augments the disparity between metabolic needs of overactive nerve cells and the supply of oxygen and nutritional factors, finally resulting in irreversible cell damage. Nerve cell degeneration may lead to release of endogenous excitotoxins (e.g., glutamate) and by this may propagate epileptic activity as a self perpetuating circle.

ACKNOWLEDGEMENTS

We are greatly indebted to Ms. A. Cervenka, Mrs. H. Breitschopf and Ms. S. Katzensteiner for expert technical assistance and photographic work. This study was funded by Science Research Fund, Austria; Projects S 25/02 and S 25/06.

REFERENCES

Atillo, A. Söderfeldt, B., Kalimo, H., Olsson, Y., and Siesjö, B.K., 1983, Pathogenesis of brain lesions caused by experimental epilepsy: light and electron microscopic changes in the rat hippocampus following bicuculline induced status epilepticus, Acta Neuropathol., 59:11.

Auer, R.N., Kalimo, H., Olsson, Y., and Siesjö, B.K., 1985, The temporal evolution of hypoglycemic brain damage. I. Light and electron microscopic findings in the rat cerebral cortex, Acta Neuropathol., 67:13.

Ben-Ari, Y., Tremblay, E., Riche, D., Ghilini, G., and Naquet, R., 1981, Electrographic, clinical and pathological alterations following systemic administration of kainic acid, bicuculline or pentetrazole: metabolic mapping using deoxyglucose method with special reference to pathology in epilepsy, Neuroscience, 6:1361.

Blennow, G., Brierley, J.B., Meldrum B.S, and Siesjö, B.K., 1978, Epileptic brain damage. The role of systemic factors that modify cerebral energy metabolism, Brain, 101:687.

Evans, M.C., Griffiths, T., and Meldrum, B.S., 1984, Kainic acid seizures and the reversibility of calcium loading in vulnerable neurones in the hippocampus, Neuropath. Appl. Neurobiol., 10:285.

Griffiths, T., Evans, M.C., and Meldrum, B.S., 1983, Intracellular calcium accumulation in rat hippocampus during seizures induced by bicuculline or L-allylglycine, Neuroscience, 10:385.

Lassmann, H., Petsche, U., Kitz, K., Baran, H., Sperk, G., Seitelberger, F., and Hornykiewicz, O., 1984, The role of brain edema in epileptic brain damage induced by systemic kainic acid injection, Neuroscience, 13:691.

Nitsch, C., and Klatzo, I., 1983, Regional patterns of blood brain barrier breakdown during epileptiform seizures induced by various convulsive agents, J. Neurol. Sci., 59:305.

Reese, T.S., and Karnovski, M.J., 1967, Fine structural localization of a blood brain barrier to exogenous peroxidase, J. Cell. Biol., 34:207.

Schwob, J.E., Fuller, T., Price, J.L., and Olney, J.W., 1980, Widespread patterns of neuronal damage following systemic or intracerebral injections of kainic acid: a histological study, Neuroscience, 5:991.

Sperk, G., Lassmann, H., Baran, H., Kish, S.J., Seitelberger, F., and Hornykiewicz, O., 1983, Kainic acid induced seizures: neurochemical and histopathological changes, Neuroscience, 10:1301.

Spielmeyer, W., 1927, Die Pathogenese des epileptischen Krampfes. Histologischer Teil, Z. Ges. Neurol. Psychiat., 109:501.

Wade, J.G., Amtorp, O., and Sorensen, S.C., 1975, No flow state following cerebral ischemia. Role of increase in potassium concentration in brain interstitial fluid, Arch. Neurol., 32:381.

COMMENTARY - EXCITATORY AMINO ACIDS AND THE BLOOD-BRAIN BARRIER (BBB)

J.M. LeFauconnier and I. Klatzo

There are several aspects of the BBB which directly concern epileptic seizures. The first aspect applies to transport features and kinetics of passage of various convulsive agents across the barrier. The second aspect refers to mechanisms in seizure-induced opening of the BBB and its effect on development of post-ictal neuronal injury. Five presentations in this session dealt with some parts of these aspects and have contributed to enhancing our understanding of the role of the BBB in epileptic seizures.

Concerning the entry of excitatory amino acids into brain tissue, two presented papers (Lefauconnier et al.; Berger et al.) indicated a very low permeability of the blood-brain barrier to kainate whose influx constant was lower than that of glutamic acid or even sucrose. In the discussion it was suggested that this permeability could explain the delay between injection and the onset of the seizures, although it has been underscored that this delay was shorter when the electrical activity was recorded than when determined from the clinical onset. As a consequence of this low permeability, the concentration obtained in hippocampus after systemic injection of kainate was also low, both studies giving very similar results (250 nmol of kainate per kg wet weight after an i.v. injection of 4 mg/kg body weight). Thus, hypotheses concerning the mechanism responsible for limbic seizures and brain damage must operate in this range of concentration. In the discussion it was stated that kainate brain level at seizure onset was higher in 10-day old than in 30-day old rats. Furthermore, it was mentioned that according to recent studies young (3-week old) gerbils have considerably higher threshold to ischemic injury than adult ones (3-month old).

The measurement of blood-brain barrier permeability to glutamate gave a much higher influx constant than that of kainate. Uptake studies on isolated microvessels showed that the active uptake of glutamate by these cells had many characteristics of the uptake system of glial cells and that no receptor to acidic amino acids other than that of the transport system seemed to be present on the endothelial cell membranes. In the discussion, it was suggested that the direction of glutamate transport *in vivo* was probably mainly from brain to blood and that the measured influx could represent only blood-cytoplasm of the endothelial cell influx and not blood-brain influx.

Mechanisms and sequelae of the BBB dysfunction in epileptic seizures were considered in three presentations. With regard to the route of passage of macromolecules, the paper of Nitsch et al., using horseradish peroxidase (HRP) as a protein tracer, supported the contention that pinocytotic transport constitutes the main mechanism. As has been repeatedly demonstrated (Petito et al., 1977; Westergaard et al., 1978; Nitsch and Klatzo, 1983;

Suzuki et al., 1984), opening of the BBB in seizures requires elevation of systemic blood pressure (BP) and it is usually associated with considerable increase in regional cerebral blood flow (rCBF). On the other hand, there seems to be no direct correlation between rCBF level and barrier opening and with some convulsive agents even tri-fold increase of rCBF in particular brain regions may not be accompanied by BBB breakdown to protein tracers (Suzuki et al., 1984).

It appears then, that in an induction of the BBB in epileptic seizures two factors can be operative. The first one, 'hemodynamic', relates primarily to an increase of intraluminar pressure which affects directly the luminar surface of endothelial cells and may induce a vigorous pinocytotic transport. The second factor relates to changes in perivascular milieu in brain parenchyma, due to accumulation of various metabolites and release of neurotransmitters, which can further modulate the behavior of the BBB.

Two papers of this session (Lassmann et al.; Ruth) dealt with a role of BBB breakdown to proteins in the production of brain tissue damage, which follows kainic acid-induced seizures. The presentation by Lassmann et al. tended to reduce the role of BBB breakdown to proteins in the development of tissue damage, indicating that the HRP leakage in the limbic system was rather limited and became accentuated only at later stages when it was associated with tissue necrosis. Otherwise, the authors considered the swelling of astrocytes, leading to an intense cytotoxic edema and compression of microcirculation, as the main factor in the production of severe tissue damage.

On the other hand, Ruth considers the role of involvement of increased cerebrovascular permeability in kainate-induced brain tissue damage as an open question, since, using sensitive immunocytochemical methods, he was able to demonstrate a specific extravasation of immunoglobins in the dorsal hippocampus, which significantly preceded the onset of neuronal damage.

Although the effect of protein extravasation on the development of post-ictal neuronal injury remains obscure, it should be kept in mind that entry of proteins into extracellular spaces induces the vasogenic edema, which, otherwise, may resolve without significant tissue damage (Kuroiwa et al., 1985). On the other hand, entry of proteins into neurons, as it is frequently observed in epileptic seizures (Nitsch and Klatzo, 1983), must signify neuronal injury, which can be of reversible or irreversible nature.

REFERENCES

Kuroiwa, T., Cahn, R., Juhler, M., Goping, G., Campbell, G., and Klatzo, I., 1985, Role of extracellular proteins in the dynamics of vasogenic brain edema, Acta Neuropathol., 66:3.
Nitsch, C., and Klatzo, I., 1983, Regional patterns of blood-brain barrier breakdown during epileptiform seizures induced by various convulsive agents, J. Neurol. Sci., 59:305.
Petito, C.K., Schaefer, J.A., and Plum, F., 1977, Ultrastructural characteristics of the brain and the blood-brain barrier in experimental seizures, Brain Res., 127:251.
Suzuki, R., Nitsch, C., Fujiwara, K., and Klatzo, I., 1984, Regional changes in cerebral blood flow and blood-brain barrier permeability during epileptiform seizures and in acute hypertension in rabbits, J. Cereb. Blood Flow Metab., 4:96.

Westergaard, E., Hertz, M.M., and Bolwig, T.G., 1978, Increased permeability to horseradish peroxidase across cerebral vessels, evoked by electrically induced seizures in rat, *Acta Neuropathol.*, 41:73.

SESSION IV
EXCITATORY AMINO ACIDS: RECEPTOR INTERACTIONS

ANATOMICAL ORGANIZATION OF EXCITATORY AMINO ACID RECEPTORS

AND THEIR PROPERTIES

C.W. Cotman and D.T. Monaghan

Department of Psychobiology
University of California
Irvine, California 92717 U.S.A.

INTRODUCTION

The excitatory amino acids are the major class of excitatory neurotransmitter in the CNS and their actions are mediated by four or more physiologically-identified receptor classes. Electrophysiological experiments indicate that L-glutamate excitations are mediated by at least three agonist-defined receptors (N-methyl-D-aspartate (NMDA), kainate, (KA), and quisqualate (QA); Watkins and Evans, 1981; Cotman et al., 1981; McLennan, 1981). And a fourth class characterized by the antagonism of synaptic responses by L-2-amino-4-phosphonobutyric acid (L-AP4; Koerner and Cotman, 1981). In order to understand the role of these receptors in the operation of brain circuitry, several approaches are needed, all of which must ultimately be in accord. Electrophysiological studies are required to assess the functional properties and membrane conductance events, and biochemical studies are needed to characterize the detailed receptor properties and assess their regulation. These approaches have traditionally been used in studying the physiology and pharmacology of many receptors throughout the body. However, the CNS is extremely heterogeneous and multiple receptors exist in specific regions which may have different properties. Thus, in addition to traditional approaches, it is desirable to have a method which allows the direct visualization and study of biochemical properties of receptor classes in discrete brain areas. Autoradiography can be used for this purpose. We have found that when great attention is paid toward resolving the multiple receptor populations, the pharmacological properties in autoradiography closely parallel those obtained by neurophysiological analysis in the same receptor field. Until our recent studies, this was not clear and several contradictions existed. Many of these inconsistencies have now been resolved.

In this chapter we will describe our recent autoradiography results on the anatomical distribution of these excitatory amino acid receptors in the rodent brain and the detailed properties of receptor subtypes in specific areas. We have used L-[^{3}H]glutamate as a general ligand to label all sites with specific displacers to distinguish subtypes and have confirmed these results by the use of specific ligands for NMDA, KA and QA receptors, e.g., D-[^{3}H]AP5 (D-[^{3}H]2-amino-5-phosphonopentanoate), [^{3}H]KA and [^{3}H]AMPA ([^{3}H]α-amino-3-hydroxy-5-methyl-4-isoxazolepropionic acid). Our results indicate that, when carried out under appropriate ionic conditions, L-[³H] glutamate binds to each of the sites that are selectively labeled by D-[³H]AP5, [³H]KA, and [³H]AMPA. We will focus this discussion on the NMDA and KA receptors, discussing others only for comparison. NMDA receptors have attracted much interest recently because of their role in plastic events such as long term potentiation and pathological conditions such as epilepsy, ischemia and hypoglycemia. Kainate receptors are also of considerable interest because of their capacity for generating seizures and neuronal degeneration, their highly specific and often complementary to NMDA receptor anatomical organization, and their plasticity in neurodegenerative disease such as Alzheimer's disease (see below).

In addition to providing data on the distribution of specific receptors, autoradiography is also an extremely sensitive method for quantitatively examining binding sites. Thus, in our standard glutamate binding assay using purified synaptic plasma membranes, a workable signal requires approximately 10 fmole L-[³H]glutamate binding, and KA-sensitive and QA-sensitive glutamate binding sites fall below this level of detection. Although measurable, the NMDA sensitive population of glutamate binding sites exhibits a less than favorable signal to noise ratio (1:2). In contrast with our autoradiographic method (Monaghan et al., 1983), we can achieve a workable signal with approximately .001 fmole because of the increased sensitivity of autoradiography and the lower levels of nonspecific binding. Thus, we have been able to successfully analyze the three receptor classes using quantitative autoradiography. Furthermore, discrete brain regions can be discerned which contain almost pure populations of a single receptor class. For example, KA receptors, which are an undetectable fraction of total glutamate binding, have been shown to make up 80-90 percent of the binding in the stratum lucidum of the hippocampus. Thus, a site normally inaccessible to standard binding studies can be kinetically and pharmacologically characterized in a selected region of the brain by quantitative autoradiography.

NMDA RECEPTORS

Autoradiographic visualization of the NMDA receptor was first described using L-[³H]glutamate (Monaghan et al., 1983). NMDA-sensitive L-[³H]glutamate binding displays rapid and reversible binding. L-[³H]glutamate binds to NMDA sites with an association rate of 0.32 min^{-1} μM^{-1} and a dissociation rate of 0.32 min^{-1}. Equilibrium analysis indicates a K_d of 0.5 µM (Monaghan and Cotman, 1985).

Although L-[³H]glutamate labels several binding sites, we have found it possible to study the anatomical distribution of NMDA site binding selectively by using NMDA as a displacer. To study the pharmacological characteristics of the NMDA binding site, we have assessed the properties of L-[³H]glutamate binding sites in regions enriched in NMDA sites. NMDA agonists and antagonists interact with this binding with the appropriate potency, while acidic amino acids which are not potent at this receptor are relatively ineffective (Fig. 1A; Monaghan et al., 1983; Monaghan et al., 1985a). Thus, the NMDA antagonist D-α-aminoadipate and the NMDA agonists L-homocysteate, L-glutamate, L-aspartate, D-aspartate, D-glutamate, ibotenate, and NMDA are potent displacers of the NMDA-sensitive L-[³H]glutamate binding. Compounds which do not act at the NMDA receptor (Watkins and Evans, 1981; Davies and Watkins, 1983), most notably KA, QA, AMPA, and AP4, are poor displacers of NMDA-sensitive L[³H]glutamate binding (Fig. 1A).

NMDA has a powerful excitatory action on hippocampal and cortical pyramidal neurons. In iontophoretic studies it is one of the most powerful

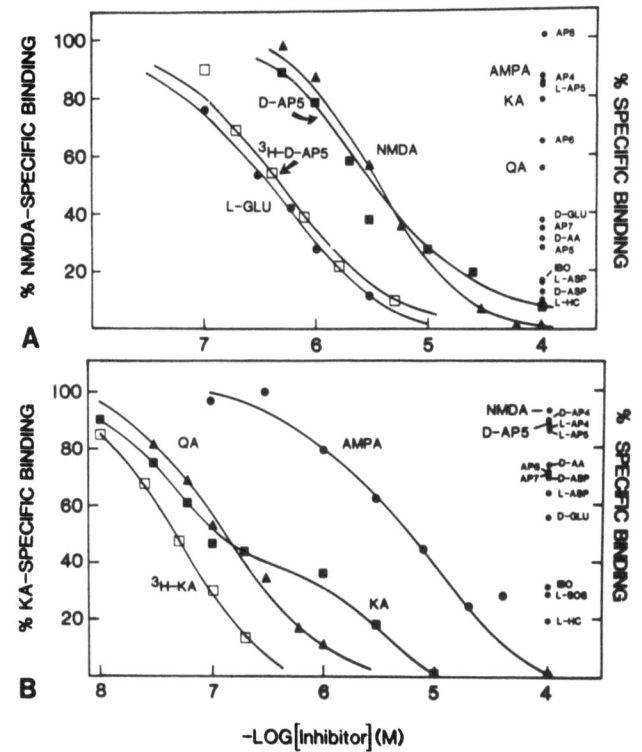

Fig. 1. Pharmacological specificity of NMDA-specific and KA-specific L-[^{3}H]glutamate binding sites. L-[^{3}H]glutamate binding in stratum radiatum of hippocampus (A) and stratum lucidum (B) exhibit acidic amino acid specificities corresponding to NMDA and KA receptors, respectively. Inhibition by compounds tested at multiple concentrations were expressed as % inhibition of maximal inhibition (approximately 80% of total specific binding in both cases). Compounds tested at 100 µM were expressed as % inhibition of total specific binding. Displacement curves are also shown for cold ligand displacement of D-[^{3}H]AP5 binding to stratum radiatum in (A). [^{3}H]KA binding (100 - % of B_{max}) to whole tissue sections is shown in (B). Abbreviations: D-GLU: D-glutamate; D-AA: D-α-aminoadipate; IBO: ibotenate L-ASP: L-aspartate; D-ASP: D-aspartate; L-SOS: L-serine-O-sulfate; L-HC: L-homocysteate. Data from Monaghan and Cotman, 1982; Monaghan et al., 1983, 1984b, 1985a.

stimulants of hippocampal pyramidal CA1 neurons. Several antagonists exist for this receptor. Of particular value in the identification of NMDA receptors is the phosphonic acid series with increasing carbon chain length from 2-amino-4-phosphonobutyrate (AP4) to 2-amino-8-phosphonooctanoate (AP8). The 5 and 7 carbon analogues (AP5 and 2-amino-7-phosphonoheptanoate, AP7) exhibit potent and selective NMDA receptor antagonism while the 4,

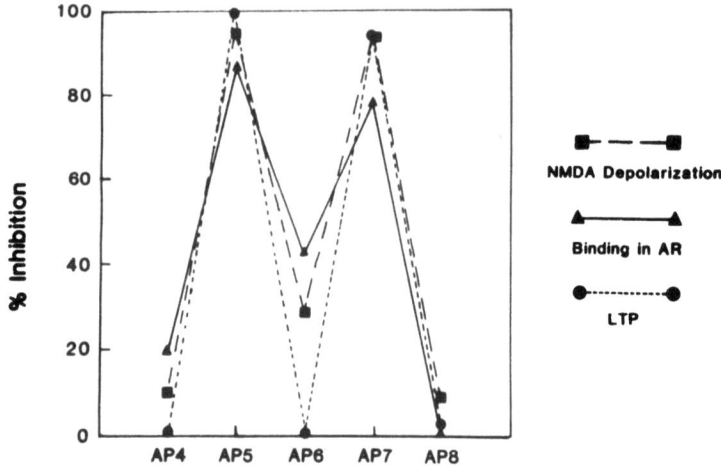

Fig. 2. Comparison between the inhibitory potencies of a series of ω-phosphonic acid glutamate analogs on NMDA-sensitive L-[^{3}H]glutamate binding, NMDA-induced focal depolarization, and long-term potentiation (LTP). All measurements were made in the stratum radiatum of CA1 hippocampus using 100 μM of the racemate. See text for abbreviations.

6 and 8 carbon analogs are poor NMDA antagonists (Evans et al., 1982; Watkins, 1984). Fig. 2 shows the excellent correspondence between NMDA induced depolarization and NMDA-senstive L-[^{3}H]glutamate binding in CA1. These data support the NMDA receptor identification of this binding site.

NMDA-sensitive L-[^{3}H]glutamate binding sites are found predominantly within the telencephalon (Monaghan et al., 1983, 1985a; Monaghan and Cotman, 1985; Fig. 3). Highest levels of binding are found in the stratum radiatum and stratum oriens of the hippocampal CA1 field (regio superior). It is interesting that the highest levels of binding are found in these fields because this area is among the most sensitive to ischemic injury and also where physiological plasticity is readily displayed. The corresponding layers in CA3 have moderately high levels of binding as does the inner layer of the dentate gyrus molecular layer. Within the hippocampus, relatively low levels of binding are found in the hilus and in the stratum lucidum. Cerebral cortex displays moderate to high levels of binding with higher densities found in layers I-III and Va. The higher concentration of NMDA sites in the outer layers are in agreement with results from electrophysiological studies by Pumain et al. (1984). Frontal, pyriform, anterior cingulate, and perirhinal regions contain higher levels of binding than do the parietal, posterior cingulate, and entorhinal cortices. Within the basal ganglia, the nucleus accumbens has the highest levels of NMDA sites, closely followed by the caudate nucleus. These regions have considerably higher levels of binding than does the globus pallidus. The anterior olfactory nuclei, olfactory accessory bulb, olfactory tubercles, and nucleus of the lateral olfactory tract all have relatively high levels of binding sites. Within the olfactory bulb, the highest concentration of these sites is found within the external plexiform layer.

In general, thalamic regions have moderate levels of binding sites whereas the hypothalamus has low levels. Within the thalamus, the anterior dorsal and certain midline nuclei (e.g., rhomboid and reuniens) have higher

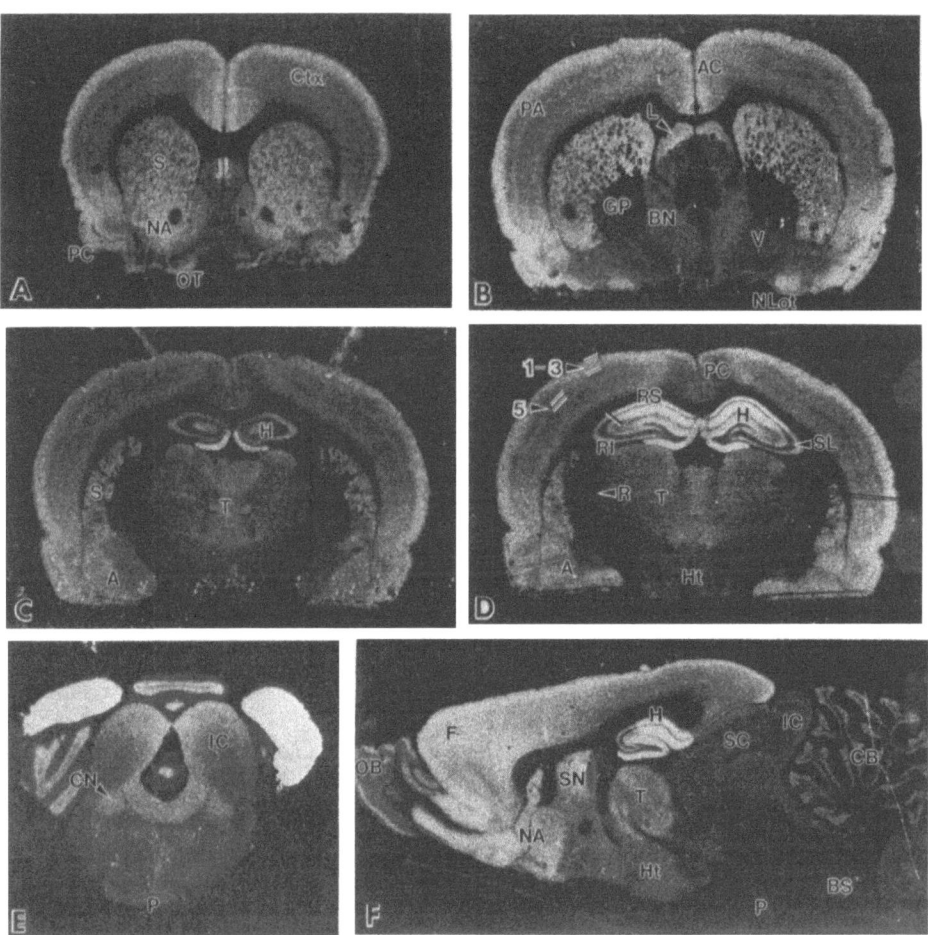

Fig. 3. Distribution of NMDA-sensitive L-[^{3}H]glutamate binding sites. Tritium sensitive film autoradiograms of L-[^{3}H]glutamate binding in rat brain using conditions which result in predominantly (>95%) NMDA-sensitive site labeling (Monaghan et al., 1985b; see also Monaghan and Cotman, 1985). Abbreviations: Ctx: cerebral cortex; S: striatum; NA: nucleus accumbens; PC: pyriform cortex; OT: olfactory tubercle; PA: parietal cortex; AC: anterior cingulate cortex; L: lateral septum; GP: globus pallidus; BN: bed nucleus of the stria terminalis; V: ventral pallidum; NLot: Nucleus of the lateral olfactory tract; H: hippocampus; T: thalamus; A: amygdala; 1-3: cortical layers I-III; 5: layer V; PC: posterior cingulate cortex; RS: regio superior (CA1); RI: regio inferior (CA3); SL: stratum lucidum; R: reticular nucleus of thalamus; Ht: hypothalamus; CN: cuneiform nucleus; IC: inferior colliculus; OB: olfactory bulb; F: frontal cortex; SN: septal nuclei; SC: superior colliculus; CB: cerebellum; P: pontine nuclei; BS: brain stem.

binding levels than do the reticular nucleus and the zona incerta. Midbrain and brain stem have an overall low density of NMDA-sensitive sites with certain regions having higher densities. These include the superficial gray and intermediate gray layers of the superior colliculus, dorsal medial inferior colliculus, dorsal raphe nucleus, central gray, cuniform nucleus, granule cell layer of the cochlear nucleus, medial vestibular nucleus, parabrachial nucleus, nucleus of the solitary tract, and the substantia gelatinosa of the spinal cord. The only ventral brain stem structure preferentially labeled was the inferior olive. Low levels are also found within the cerebellum, with the granule cell layer exhibiting more binding than does the molecular layer (for quantitative values see Monaghan and Cotman, 1985). Our data agree with Greenamyre et al. (1985), who reported a generally similar distribution of NMDA-sensitive sites in the areas studied, hippocampus, cerebellum, and cerebral cortex.

NMDA receptors may also be labeled by D-[^{3}H]AP5 (Olverman et al., 1984). Although D-[^{3}H]AP5 rapidly dissociates from its binding site (dissociation half-life of five seconds), autoradiograms prepared with this ligand display the appropriate pharmacological profile and a distinct anatomical specificity (Monaghan et al., 1984b). Consistent with the properties of the NMDA receptor, D-[^{3}H]AP5 binding sites exhibit higher affinity for L-glutamate, L-homocysteate, L-aspartate, D-AP5, and NMDA than for QA, L-AP4, KA, and AMPA. Since L-glutamate appears to have the same high affinity for the D-[^{3}H]AP5 binding site as it does for the NMDA-sensitive L-[^{3}H]glutamate binding site, and since the binding site density is also comparable, it seems likely that these two ligands are binding to the same site. This conclusion is reinforced by the observation that both ligands have their highest binding site density within the hippocampus and that within this structure they have identical distributions. In both autoradiographic and membrane fraction ligand binding experiments, AP5 more potently displaces D-[^{3}H]AP5 than NMDA-sensitive L-[^{3}H]glutamate binding (Fagg et al., 1983a; Monaghan et al., 1983, 1984b, 1985a, 1985b; Olverman, 1984).

Given that the highest binding is found in the hippocampal CA1 field, it would be expected that antagonists of NMDA receptors would block synaptic transmission in this area. We and others have evaluated this and much to our initial surprise NMDA antagonists have essentially no effect (Koerner and Cotman, 1982; Collingridge et al., 1983; Harris et al., 1984; Ganong et al., 1986). For example, the Schaffer collateral response recorded intracellularly is indistinguishable in the presence or absence of D-AP5 (Ganong et al., 1986). If, however, GABA mediated inhibition is blocked by picrotoxin and the membrane is depolarized, AP5 has a significant effect on the synaptic potential (see King and Dingledine, this volume). Thus, in healthy cells with no other drugs present, NMDA antagonists have little if any effect on the evoked synaptic potential.

The Schaffer collateral system in the hippocampus displays a synaptic analog of short term memory called long term potentiation (LTP). When this pathway is stimulated repetitively (100 Hz for 1 second), the EPSP shows a long lasting increase in its amplitude. _In vivo_ studies indicate that this process appears to be involved in the development of memory in the hippocampus (Anderson et al., 1985). Recently we have evaluated the possibility that NMDA receptors are involved in the formation of LTP at this synapse (Harris et al., 1984) following an initial report by Collingridge et al. (1983).

A series of ω-phosphono-α-carboxylic acids were tested as antagonists of excitatory amino acid depolarizations and long-term potentiation (LTP) in region CA1 of rat hippocampal slices (Harris et al., 1984). Fig. 2 shows that the inhibition of LTP displays the pharmacological characteristics of an NMDA receptor and that these properties correspond to those observed in autoradiograms. The 5- and 7-phosphono compounds (±AP5 and ±AP7) blocked NMDA depolarizations and prevented the induction of LTP of the synaptic field potential and population spike components of the Schaffer collateral response. ±AP5 and ±AP7 did not reduce kainate or quisqualate depolarizations and did not affect synaptic response amplitude. Furthermore, it was only D-isomers of AP5 which inhibited LTP and NMDA-induced depolarization with the appropriate potency. ±AP4, ±AP6 and ±AP8 did not block amino acid excitant responses or LTP. These results demonstrate that NMDA receptors present in hippocampal region CA1 are not necessary for normal synaptic transmission, but are involved in the initiation of long-term synaptic plasticity (Collingridge et al., 1983; Harris et al., 1984; Wigström and Gustafsson, 1984).

Thus, one of the roles of NMDA receptors is to serve as a neuromodulator for plasticity mechanisms in the hippocampus. In this way, these receptors serve to help adapt the organism to its environment. NMDA receptors also, however, appear to play a role in excitotoxic induced cell death associated with injury. NMDA injected directly into the brain destroys neurons and NMDA antagonists protect the brain against neuronal loss resulting from ischemia and hypoglycemia (Simon et al., 1984; Wieloch, 1985; see also Wieloch, this volume; Meldrum, this volume). It seems that these receptors are critically involved in determining the levels of excitation in the CNS. The distribution of NMDA receptors as determined by autoradiography predicts that plasticity and vulnerability are indeed properties of higher cortical systems and shows their relative organization.

KAINATE RECEPTORS

L-[^{3}H]Glutamate may also be used to label KA binding sites. L-Glutamate is known to have a high affinity for [^{3}H]KA binding sites (Simon et al., 1976; London and Coyle, 1979). However, given the relatively low maximal binding site density displayed by KA sites compared to the sum of the other glutamate binding sites, KA would not be expected to displace a significant amount of L-[^{3}H]glutamate binding in membrane preparations. Using quantitative autoradiography, we have demonstrated that L-[^{3}H]glutamate binding sites in the stratum lucidum of the rat hippocampus are predominantly KA binding sites (Monaghan et al., 1983; Monaghan et al., 1985a). As with [^{3}H]KA binding sites in membrane preparations, KA-sensitive L-[^{3}H]glutamate binding sites are more potently displaced by KA, QA, and L-glutamate than by D-glutamate and AMPA (Fig. 1B). Furthermore, these sites exhibit little displacement by 100 μM concentrations of NMDA, D-AP5, L-APB, L-aspartate, D-aspartate, and D-α-aminoadipate. These sites exhibit the same distribution as do [^{3}H]KA binding sites.

Highest concentrations of [^{3}H]KA binding are found in the stratum lucidum of the hippocampus, and relatively high levels are also found in the caudate/putamen, dentate gyrus inner molecular layer, and deep cerebral cortical layers (Layers V and VI; Monaghan and Cotman, 1982; Unnerstall and Wamsley, 1983). Relatively low levels are found in the septum, outer cortex and lower brain regions (Fig. 4). In many regions, KA and NMDA sites exhibit a complementary distribution. In the cerebral cortex, KA sites are concentrated in the deep layers in contrast to NMDA sites which are found to be highest in the outer layers. The region of highest KA sites density is the hippocampus. In the diencephalon, KA sites show higher levels in the remainder of thalamus; NMDA sites show the opposite distribution.

However, many regions do overlap, e.g., the basal ganglia, dentate gyrus inner molecular layer, anterior olfactory nuclei, nucleus of the lateral olfactory tract, and various brainstem nuclei. Interestingly, certain ventrally-located structures show a preferentially KA localization relative to NMDA sites. These include hypothalamus, mamillary bodies and pons. Thus, the signal enhancing NMDA receptor does not seem to be as important for ventrally located nuclei.

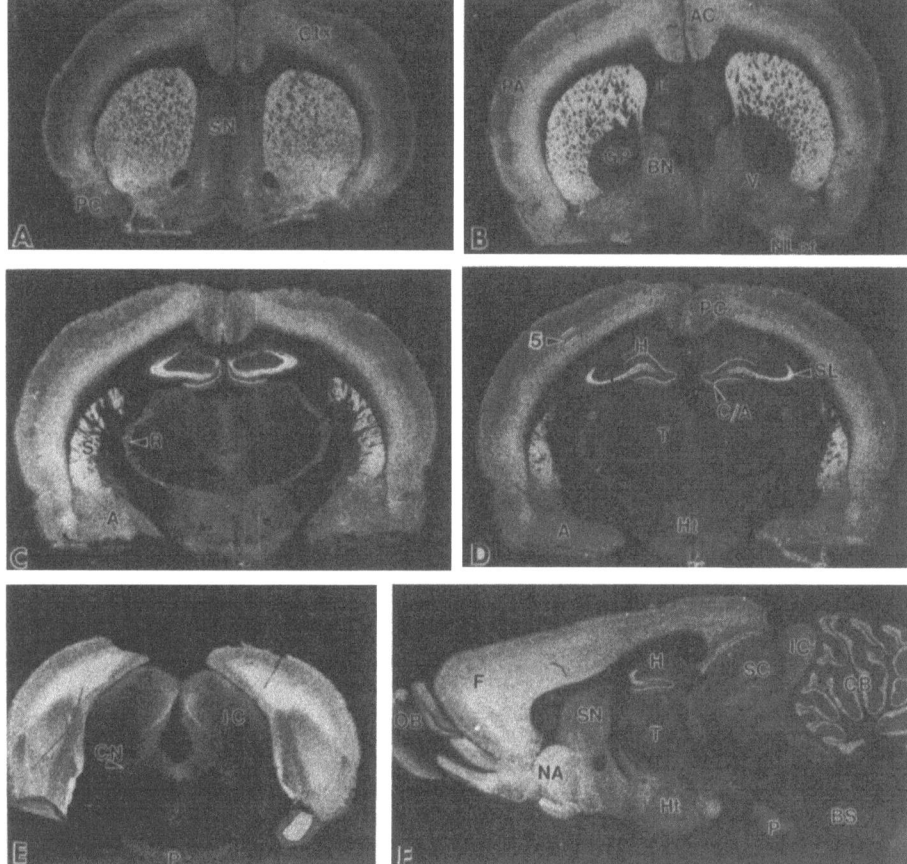

Fig. 4. Distribution of [^{3}H]KA binding sites in rat brain. A-E: Coronal plane; F: saggital plane. Autoradiographic procedure modified from Monaghan and Cotman (1982). C/A represents the commissural/associational termination zone of the dentate gyrus; for other abbreviations see the legend to Fig. 3.

In general, the distribution of [^{3}H]KA sites corresponds to those areas most sensitive to the excitotoxic actions of KA. For example, area CA3 is one of the most sensitive and also has the most dense distribution of binding sites (Nadler et al., 1978; Foster et al., 1981; Monaghan and Cotman, 1982; Unnerstall and Wamsley, 1982). However, the pyriform cortex does not correspond as well (Monaghan and Cotman, 1982).

Analysis of receptor function in CNS pathways such as mossy fiber input to CA3 should provide valuable insights to the normal function of this receptor. Taken together with the observation that these sites are enriched in purified synaptic junctions (Foster et al., 1981), KA receptors may not simply be a non-specific excitotoxin receptor but a distinct pathway-specific receptor. KA receptors appear to show a similar distribution in the primate and human brain, at least in the hippocampus, though the density of stratum lucidum is more similar to other fields in the hippocampus (Geddes et al., 1985; Tremblay et al., 1985; Monaghan et al., 1986).

OTHER RECEPTORS AND BINDING SITES

As with NMDA sites, the first report of AMPA site autoradiography was with the use of L-[^{3}H]glutamate (Monaghan et al., 1983). After displacing the NMDA-sensitive L-[^{3}H]glutamate binding, one can observe the AMPA-sensitive L-[^{3}H]glutamate binding sites over the pyramidal cell layer of the hippocampus. The pharmacological profile and anatomical distribution of these sites is similar to that found for the [^{3}H]AMPA binding sites (Monaghan et al., 1985a).

[^{3}H]AMPA is suitable for labeling of QA-type receptors in autoradiographic and isolated membrane preparations (Honoré et al., 1982; Monaghan et al., 1984a; Rainbow et al., 1984). In fact, AMPA is actually a preferable label since QA is not as selective for this so-called class of binding sites (see Monaghan and Cotman, 1986 for discussion). Highest densities of [^{3}H]AMPA binding sites are found over the hippocampal pyramidal cells and over the cell bodies of the indusium griseum, suggesting a neurochemical similarity between these developmentally related cells. High levels are found in the CA1 stratum radiatum and stratum oriens of the hippocampus. Outer cortical layers (I, II, and III) have higher levels of binding than layers V and VI, while layer IV has lower levels. In posterior sections, there is a dense band of binding found in layer V, probably corresponding to layer Vb. The caudate/putamen and the nucleus accumbens have more dense binding than does the globus pallidus, substantia innominata, and ventral pallidum. Basolateral, lateral, and posterior amygdaloid nuclei have higher levels of binding than the adjacent central, medial, and anterior cortical nuclei. This amygdaloid pattern is the same as that seen for the NMDA receptor and for other markers of excitatory amino acid neurotransmission (see Ottersen and Storm-Mathisen, this volume). Midbrain and brain stem have considerably lower levels of binding.

This distribution of [^{3}H]AMPA binding sites is largely similar to that of the NMDA sites with the exception of relatively higher concentrations of [^{3}H]AMPA binding in the hippocampal pyramidal cell layer, the indusium griseum, the lateral septum, the dentate gyrus hilus, and the layer V binding observed in posterior cortex. Also distinct from the NMDA-sensitive L-[^{3}H]glutamate binding are the relatively lower levels of [^{3}H]AMPA binding found in the thalamus, the Va layer of anterior cortex, the external plexiform layer of the olfactory bulb, and the granule cell layer of the cerebellum. Although there is a striking qualitative and quantitative agreement between the study of Monaghan et al. (1984a) and Rainbow et al. (1984), the latter study found relatively lower levels of binding over the hippocampal pyramidal cell layer. This could possibly be the result of the latter group's use of thicker sections (32 versus 6 μm) or the use of a longer rinse time (10 versus 0.5 minutes).

Thus, L-[^{3}H]glutamate appears to label binding sites in autoradiograms with the properties of NMDA, KA and QA receptors. Together these three receptors account for the majority of L-[^{3}H]glutamate binding when labeling is carried out in the absence of Ca^{2+} and Cl^-. Thus, as Fig. 5 illustrates, glutamate labels primarily NMDA sites in these conditions and these sites, plus KA and AMPA displaceable sites, account for over 90 percent of the total glutamate binding. L-AP4, the other clearly identified receptor, does not have a significant effect on glutamate binding under these conditions or, in fact, even in the presence of $CaCl_2$, which stimulates L-AP4 sensitive binding in membranes (Fagg et al., 1982; Mena et al., 1982). The L-AP4 displaceable glutamate binding sites in membranes are complex and the majority do not appear to correspond to the electrophysiologically characterized L-AP4 receptor site (see Monaghan and Cotman, 1986 for detailed discussion). It is now generally acknowledged that in the presence of $CaCl_2$, L-[^{3}H]glutamate binds primarily to a site with properties indicative

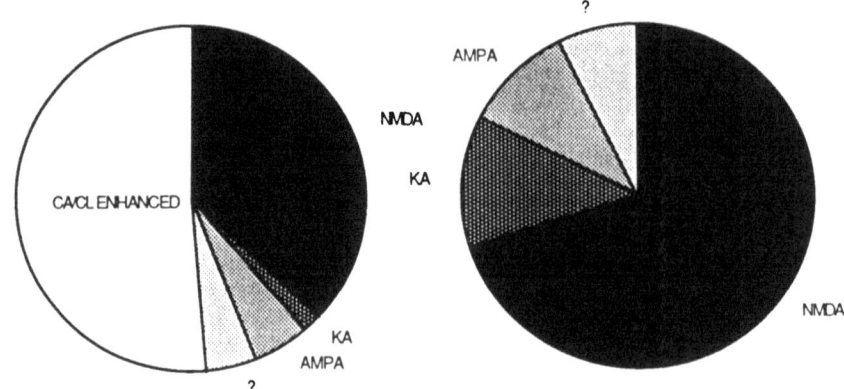

Fig. 5. Relative proportions of L-[³H]glutamate binding sites. L-[³H]glutamate binding to whole rat brain tissue sections, which are selectively displaced by NMDA (100 μM), KA (1 μM), and AMPA (5 μM) under differeing incubation conditions. Left: 30 minutes, 30°C, 2.5 mM Ca^{2+}, 20 mM Cl^-, 50 mM Tris-acetate, pH 7.2. Right: Optimal conditions for L-[³H]glutamate binding to NMDA-, KA-, and AMPA-sensitive binding sites in tissue sections, 10 minutes, 0°C, 50 mM Tris-acetate, pH 7.2.

of a transport site (Mena, 1984; Pin et al., 1984; Bridges et al., in press). This site constitutes the majority of binding in membranes, is Ca^{2+}/Cl^- dependent, and is freeze labile (Fagg et al., 1983). Thus, in the preparation of autoradiograms, since the brain tissue is frozen, most of the L-AP4 sites are removed from the total binding. Nevertheless, as illustrated in Fig. 5., Ca^{2+}/Cl^- addition results in significant additional binding sites in autoradiograms. In the hippocampus, these sites do not correspond to any receptor and may be due in part to labeling of binding sites in glial cells, unknown receptors, or the resealing of membranes and the accumulation of label in membrane compartments. In fact, these sites are relatively insensitive to almost all of the excitatory amino acid analogs tested (see Monaghan and Cotman, 1986 for more detailed discussion). The exact identity of the Ca^{2+}/Cl^--dependent sites is still under investigation. A further complication is that there exists still another Ca^{2+}/Cl^--dependent glutamate uptake/binding site in the cerebellum. Under identical conditions, the Ca^{2+}/Cl^--dependent glutamate sites in the cerebellum are readily displaced by low concentrations of quisqualate whereas the Ca^{2+}/Cl^- sites of the hippocampus exhibit little displacement (Monaghan and Cotman, 1986).

It should be pointed out that not all results with autoradiography appear to be consistent. This is because the relative labeling depends on the conditions employed. For example, the Cl^-/Ca^{2+}-dependent L-[³H] glutamate sites display a partial temperature dependency, with higher levels of binding observed at 30°C than at 0°C. In contrast, the levels of NMDA-sensitive binding are lower at 30°C than at 0°C (Monaghan and Cotman, 1985). Thus, total L-[³H]glutamate binding levels are not greatly affected by changes in temperature, but the proportions of the differing subtypes are changed. KA and AMPA populations represent a relatively small percentage of the L-[³H]glutamate binding, especially KA sites when Ca^{2+} is present (note: Ca^{2+} ions appear to selectively inhibit the higher affinity KA site; Monaghan et al., 1985b, 1986). Thus, in whole tissue sections, KA and AMPA compounds do not displace the majority of the binding. After consideration of the effects of temperature and Cl^- and Ca^{2+} ions upon L-[^{3}H]glutamate binding, the results obtained in the various autoradiographic studies are consistent (Greenamyre et al., 1983; Monaghan et al., 1983; Halpain et al., 1984). Of course, this emphasizes the importance of having direct labels (e.g., [^{3}H]KA, [^{3}H]AMPA, D-[^{3}H]AP5) for the receptor in addition to more non-specific ligand such as L-[^{3}H]glutamate, which require the use of selective displacers.

ACIDIC AMINO ACID RECEPTORS IN ALZHEIMER'S DISEASE

One of the goals of research on excitatory amino acids as neurotransmitters is to develop new approaches toward the treatment of various neurological disorders. Recently, we have examined brain tissues from patients with senile dementia of the Alzheimer's type collected at autopsy. It appears that these receptors are sufficiently stable to allow their study in these tissues (Geddes et al., 1985; Geddes et al., in preparation).

Recently, severe cell loss has been demonstrated in the entorhinal cortex and subiculum of Alzheimer's disease (AD) patients. This has been suggested to underly some of the memory deficits in this disease (Hyman et al., 1984). Axons, arising from the entorhinal cortex, provide the major cortical input to the hippocampus, terminating primarily on the granule cell dendrites in the outer 2/3 of the dentate gyrus molecular layer. In response to the loss of entorhinal input, the remaining inputs in the dentate molecular layer undergo a major reorganization. After clearing of the degenerating nerve terminals, the remaining afferent fibers, an extrinsic afferent from the septum and two intrinsic afferents, the commissural and associational systems, sprout to form new synapses with the denervated target cells. The commissural and associational (C/A) fibers, originating in the contralateral and ipsilateral dentate hilus, respectively, normally terminate in the inner 1/3 of the molecular layer adjacent to the zone of entorhinal innervation. Following an entorhinal lesion, this zone expands until it occupies the inner 1/2 of the molecular layer. Sprouting of the septal input is concentrated in the outer 1/2 of the molecular layer (see Cotman and Nieto-Sampedro, 1984, for details).

A high density of KA receptors is restricted to the inner 1/3 of the dentate molecular layer, corresponding to the terminal zone of the C/A system (Foster et al., 1981; Monaghan and Cotman, 1982). After a unilateral entorhinal lesion, the KA binding pattern in the dentate gyrus molecular layer on the control side (contralateral to the lesion) was similar to that previously observed in untreated rodents (Fig. 6c). A distinct band of high density KA binding sites, (70 ± 6 μm), occupied the inner 1/3 of the molecular layer corresponding to the C/A system terminal zone. On the deafferented side (ipsilateral to the lesion; Fig. 6d), the region of high density KA binding sites had expanded to 110 ± 6 μm (a 57% increase) such that it now occupied approximately 40% of the molecular layer (Geddes et al., 1985). This correlates with the magnitude of the C/A afferent fiber sprouting (Lynch et al., 1976). These data provide the first anatomical demonstration of receptor induction correlated to afferent sprouting.

The overall KA binding pattern observed in human hippocampal samples, obtained from non-demented patients, was very similar to that observed in the normal rat brain. A region with a high density of KA binding sites, measuring 150 ± 20 μm, occupied the inner 1/3 of the molecular layer (Fig. 6e). In AD patients, this region had expanded to 260 ± 30 μm (a 73% increase; Fig. 6f) so that it now occupied over 1/2 of the molecular layer. These results demonstrate that the human brain is capable of neuronal

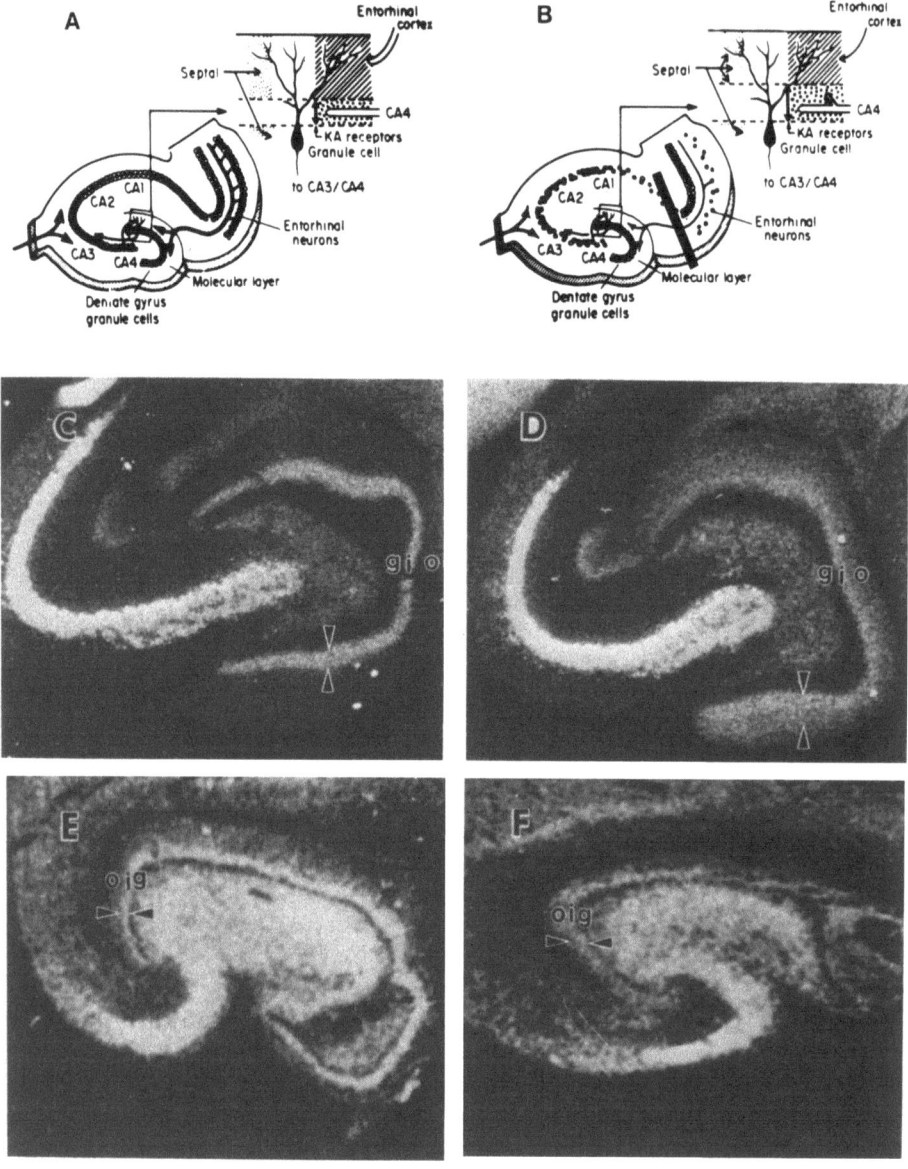

Fig. 6. (A) is a diagram of the normal hippocampus in rat. In (B), the loss of entorhinal cortical neurons (corresponding to some of the neuronal loss in Alzheimer's disease) elicits changes in the innervation pattern. Photographs are [^{3}H]kainate binding site autoradiograms of rat (C,D) and human hippocampus (E,F). Arrows indicate the region of high KA binding density corresponding to the termination zone of the CA4 commissural/associational fibers. This zone appears wider in rat hippocampus following an entorhinal lesion (D) in comparison to a control hippocampus (C). Likewise, the band of kainate binding in human hippocampus from an Alzheimer's disease patient (F) appears wider than in normal tissue (E). Abbreviations: g: granule cell layer of the dentate gyrus; i: inner molecular layer - commissural/associational terminal zone; o: outer molecular layer, terminal zone of entorhinal fibers.

plasticity in response to a pathologically-induced neuronal loss. Furthermore, in AD patients who have not had a severe loss of cholinergic input to the hippocampus, sprouting of cholinergic fibers was also observed (Geddes et al., 1985).

These results demonstrate that the neuronal loss in the entorhinal cortex of AD patients acts as a stimulus to a lesion in the rodent brain. It removes the perforant path input to the hippocampus and dentate gyrus, inducing a compensatory response from adjacent C/A system afferents and from the septal afferents, if present. The observed expansion of a receptor field and increase in activity of a transmitter-metabolizing enzyme are in marked contrast to the numerous reports of reductions in transmitter-related parameters in this disease. The rearranged inputs may facilitate transmission through the hippocampus (Cotman, Gibbs and Nieto-Sampedro, 1986). Manipulation of functional plasticity and excitatory amino acid transmission may be a useful therapeutic approach to the treatment of AD and other neurodegenerative disorders.

CONCLUSION

Autoradiographic analysis provides for the first time a detailed analysis on the distribution of NMDA, KA, and QA receptors in the brain. When the excitatory action of glutamate on CNS neurons was first described over a decade ago, it was thought that its action was too universal for it to play a major role as a neurotransmitter. It is now clear, however, that the various classes of glutamate receptor subtypes are highly localized, particularly in telencephalic regions and they are precisely organized. Autoradiography provides a powerful technique for study of detailed receptor properties in their respective pathways. It should also facilitate the identification of other endogenous acidic amino acids, which may be acting as neurotransmitters in these pathways (see Cuénod, this volume). Identification of an acidic amino acid receptor in a particular pathway indicates the structural properties of the transmitter. In turn, changes in these receptors in plasticity and disease may provide clues on the underlying events and perhaps suggest ways to correct malfunctions.

ACKNOWLEDGEMENTS

This work was supported by ARO grant DAMD 17-83-C-3189.

REFERENCES

Anderson, E., Baudry, M., and Morris, R.G.M., 1985, Stereospecific impairment of spatial learning and long-term potentiation in vivo by APV-5, an NMDA receptor antagonist, Neurosci. Lett. Suppl., 21:S52.
Bridges, R.J., Nieto-Sampedro, M., and Cotman, C.W., 1985, Stereospecific binding of L-glutamate to astrocyte membranes, Soc. Neurosci. Abstr., 11:110.
Bridges, R.J., Hearn, T.J., Monaghan, D.T., and Cotman, C.W., A comparison of AP4 receptors and ^{3}H-AP4 binding sites in the rat brain, Brain Res., in press.
Collingridge, G.L., Kehl, S.J., and McLennan, H., 1983, Excitatory amino acids in synaptic transmission in the Schaffer-commissural pathway of the rat hippocampus, J. Physiol., 334:33.
Cotman, C.W., Foster, A.C., and Lanthorn, T., 1981, An overview of glutamate as a neurotransmitter, Adv. Biochem. Psychopharmacol., 27:1.
Cotman, C.W., and Nieto-Sampedro, M., 1984, Cell biology of synaptic plasticity, Science, 225:1287.

Cotman, C.W., Gibbs, R.B., and Nieto-Sampedro, M., 1986, Synapse turnover in the adult central nervous system, in: The Dahlem Workshop Report: Neural and Molecular Mechanisms of Learning, Springer Verlag, New York, in press.

Davies, J., and Watkins, J.C., 1983, Role of excitatory amino acid receptors in mono- and polysynaptic excitation in the cat spinal cord, Exp. Brain Res., 49:280.

Evans, R.H., Francis, A.A., Jones, A.W., Smith, D.A.S., and Watkins, J.C., 1982, The effects of a series of ω-phosphonic-carboxylic amino acids on electrically evoked and excitant amino acid-induced responses in isolated spinal cord preparations, Br. J. Pharmacol., 75:65.

Fagg, G.E., Foster, A.C., Mena, E.E., and Cotman, C.W., 1982, Chloride and calcium ions reveal a pharmacologically distinct population of L-glutamate binding sites in synaptic membranes: correspondence between biochemical and electrophysiological data, J. Neurosci., 2:958.

Fagg, G.E., Foster, A.C., Mena, E.E., and Cotman, C.W., 1983a, Chloride and calcium ions separate L-glutamate receptors in synaptic membranes, Eur. J. Pharmacol., 88:105.

Fagg, G.E., Mena, E.E., Monaghan, D.T., and Cotman, C.W., 1983b, Freezing eliminates a specific population of L-glutamate receptors in synaptic membranes, Neurosci. Lett., 38:157.

Foster, A.C., Mena, E.E., Monaghan, D.T., and Cotman, C.W., 1981, Synaptic localization of kainic acid binding sites, Nature, 281:73.

Ganong, A.H., and Cotman, C.W., 1986, Kynurenic acid and quinolinic acid act at N-methyl-D-aspartate receptors in the rat hippocampus, J. Pharmacol. Exp. Ther., 236:293.

Geddes, J.W., Monaghan, D.T., Cotman, C.W., Lott, I.T., Kim, R., and Chiu, H., 1985, Plasticity of hippocampal circuitry in Alzeheimer's disease, Science, 230:1179.

Greenamyre, J.T., Young, A.B., and Penny, J.B., 1983, Quantitative autoradiography of L-[^{3}H]glutamate binding to rat brain, Neurosci. Lett., 37:155.

Greenamyre, J.T., Olson, J.M., Penny, J.B., and Young, A.B., 1985, Autoradiographic characterization of N-methyl-D-aspartate-, quisqualate-, and kainate-sensitive glutamate binding sites, J. Pharmacol. Exp. Therap., 233:254.

Halpain, S., Wieczorek, C.M., and Rainbow, T.C., 1984, Localization of L-glutamate receptors in rat brain by quantitative autoradiography, J. Neurosci., 4:2247.

Harris, E.W., Ganong, A.H., and Cotman, C.W., 1984, Long-term potentiation in the hippocampus involves activation of N-methyl-D-aspartate receptors, Brain Res., 323:132.

Honoré, T., Lauridsen, J., and Krogsgaard-Larsen, P., 1982, The binding of [^{3}H]AMPA, a structural analogue of glutamic acid, to rat brain membranes, J. Neurochem., 38:173.

Hyman, B.I., VanHoesen, G.W., Damasio, A.R., and Barnes, C.L., 1984, Alzheimer's disease: cell-specific pathway pathology isolates the hippocampal formation, Science, 225:1168.

Koerner, J., and Cotman, C.W., 1981, Micromolar L-2-amino-4-phosphonobutyric acid selectively inhibits perforant path synapses from lateral entorhinal cortex, Brain Res., 216:192.

Koerner, J.F., and Cotman, C.W., 1982, Response of Schaffer collateral-CA1 pyramidal cell synapses of the hippocampus to analogues of acidic amino acids, Brain Res., 251:105.

London, E.D., and Coyle, J.T., 1979, Specific binding of [^{3}H]kainic acid to receptor sites in rat brain, Molec. Pharmacol., 15:492.

Lynch, G., Gall, C., Rose, G., and Cotman, C.W., 1976, Changes in the distribution of the dentate gyrus associational system following unilateral or bilateral entorhinal lesion in the adult rat, Brain Res., 110:57.

McLennan, H., 1981, On the nature of the receptors for various excitatory amino acids in the mammalian central nervous system, Adv. Biochem. Psychopharmacol., 27:253.

Mena, E.E., Fagg, G.E., and Cotman, C.W., 1982, Chloride ions enhance L-glutamate binding to rat brain synaptic membranes, Brain Res., 243:378.

Mena, E.E., Whittemore, S.R., Monaghan, D.T., and Cotman, C.W., 1984, Ionic regulation of glutamate binding sites, Life Sci., 35:2427.

Monaghan, D.T., and Cotman, C.W., 1982, Distribution of [^{3}H]kainic acid binding sites in rat CNS as determined by autoradiography, Brain Res., 252:91

Monaghan, D.T., Holets, V.L., Toy, D.W., and Cotman, C.W., 1983, Anatomical distribution of four pharmacologically distinct L-[^{3}H]glutamate binding sites, Nature, 306:176.

Monaghan, D.T., Yao, D., and Cotman, C.W., 1984a, Distribution of [^{3}H]AMPA binding sites in rat brain as determined by quantitative autoradiography, Brain Res., 324:160.

Monaghan, D.T., Yao, D.T., Olverman, H.J., Watkins, J.C., and Cotman, C.W., 1984b, Autoradiography of D-[^{3}H]2-amino-5-phosphonopentanoate binding sites in rat brain, Neurosci. Lett., 52:253.

Monaghan, D.T., and Cotman, C.W., 1985, Distribution of NMDA-sensitive L-[^{3}H]glutamate binding sites in rat brain as determined by quantitative autoradiography, J. Neuroscience, 5:2909.

Monaghan, D.T., Yao, D., and Cotman, C.W., 1985a, L-[^{3}H]-glutamate binds to kainate-, NMDA-, and AMPA-sensitive binding sites: an autoradiographic analysis, Brain Res., 340:378.

Monaghan, D.T., Yao, D., Nguyen, L., and Cotman, C.W., 1985b, Excitatory amino acid binding sites: correspondence between autoradiographic and membrane fraction preparations, Soc. Neurosci. Abstr., 11:110.

Monaghan, D.T., and Cotman, C.W., 1986, Anatomical organization of NMDA, kainate, and quisqualate receptors, in: Excitatory Amino Acids, P.J. Roberts, J. Storm-Mathisen, and H.F. Bradford, eds., Macmillan, London, in press.

Monaghan, D.T., Nguyen, L., and Cotman, C.W., 1986, Distribution of [^{3}H]kainate binding sites in primate hippocampus is similar to the distribution of both Ca^{++}-sensitive and Ca^{++}-insensitive [^{3}H]kainate binding sites in rat hippocampus, Neurochem. Res., in press.

Nadler, J.V., Perry, B., and Cotman, C.W., 1978, Intraventricular kainic acid preferentially destroys hippocampal pyramidal cells, Nature, 271:676.

Olverman, H.J., Jones, A.W., and Watkins, J.C., 1984, L-glutamate has higher affinity than other amino acids for ^{3}H-D-AP5 binding sites in rat brain membranes, Nature, 307:460.

Pin, J.-P., Bockaert, J., and Recasens, M., 1984, The Ca^{++}/Cl^- dependent L-[^{3}H]glutamate binding: a new receptor or a particular transport process? FEBS Letters, 175:31.

Pumain, R., Kurceicz, I., Louvel, J., and Heinemann, U., 1984, Electrophysiological evidence for a differential localization of excitatory amino acid receptors in the rat neocortex, Neurosci. Lett. Suppl., 18:S433.

Rainbow, T.C., Wieczorek, C.M., and Halpain, S., 1984, Quantitative autoradiography of binding sites for [^{3}H]AMPA, a structural analogue of glutamic acid, Brain Res., 309:173.

Simon, J.R., Contrera, J.F., and Kuhar, M.J., 1976, Binding of [^{3}H]kainic acid, an analogue of L-glutamate, to brain membranes, J. Neurochem., 26:141.

Simon, R.P., Swan, J.H., Griffiths, T., and Meldrum, B.S., 1984, Blockade of N-methyl-D-aspartate receptors may protect against ischemic damage in the brain, Science, 226:850.

Tremblay, E., Repressa, A., and Ben-Ari, Y., 1985, Autoradiographic localization of kainic acid binding sites in the human hippocampus, Brain Res., 343:378.

Unnerstall, J.R., and Wamsley, J.K., 1983, Autoradiographic localization of high-affinity [^{3}H]kainic acid binding sites in the rat forebrain, Eur. J. Pharmacol., 86:361.

Watkins, J.,C., and Evans, R.H., 1981, Excitatory amino acid transmitters, Ann. Rev. Pharmacol. Toxicol., 21:165.

Watkins, J.C., 1984, Excitatory amino acids and central synaptic transmission, TIPS, 8:373.

Wieloch, T., 1985, Hypoglycemia-induced neuronal damage prevented by an N-methyl-D-aspartate antagonist, Science, 230:681.

Wigström, H., and Gustafsson, B., 1984, A possible correlate of the postsynaptic condition for long-lasting potentiation in the guinea pig hippocampus in vitro, Neurosci. Lett., 44:327.

HOMOCYSTEIC ACID, AN ENDOGENOUS AGONIST OF NMDA-RECEPTOR:

RELEASE, NEUROACTIVITY AND LOCALIZATION

M. Cuénod[1], K.Q. Do[1], P.L. Herrling[2], W.A. Turski[2], C. Matute[1] and P. Streit[1]

[1] Brain Research Institute, University of Zürich
August-Forelstr. 1, CH-8029 Zürich, Switzerland
[2] Wander Research Institute (a Sandoz Research Unit)
P.O. Box 2747, CH-3001 Bern, Switzerland

Many lines of evidence are pointing toward a role of the excitatory amino acid transmitters and their receptors in the mechanisms leading to epileptic seizure disorders. Various excitatory amino acids and structural analogs, such as kainic acid, quinolinic acid or ibotenic acid, initiate epilepticlike discharges when applied to nervous tissue (Ben-Ari et al., 1979; French et al., 1982; Schwarcz et al., 1984). Changes in excitatory amino acid content in cerebral tissue and cerebrospinal fluid have been reported to occur in various animal models and human epileptic conditions (Van Gelder et al., 1972; Morselli et al., 1981). Most convincing are the potent anticonvulsant effects of D-2-amino-7-phosphonoheptanoic acid (AP-7), a specific N-methyl-D-aspartic acid (NMDA) receptor antagonist, in several animal models of epilepsy (Croucher et al., 1982; Meldrum et al., 1983).

RELEASE OF ENDOGENOUS SULFUR CONTAINING AMINO ACIDS

Glutamate and aspartate are well established as excitatory amino acid transmitters in many pathways of the mammalian central nervous system. Some sensory afferents, various subcortical connections and most corticofugal neurons might use them (for review, see Fonnum et al., 1981; Cuénod et al., 1982; Cuénod and Streit, 1983; Streit, 1984; Cuénod et al., 1985). However, many important pathways, such as the thalamo-cortical one, have not been ascribed a known neurotransmitter. Furthermore, the presence of at least three pharmacologically distinct classes of receptors activated by excitatory substances suggests the possible existence of other endogenous agonists in addition to Glu and Asp.

Therefore, Do et al. (1986a) searched for endogenous neuroactive substances possibly involved in neurotransmission. In order to select them, advantage was taken of the fact that compounds involved in intercellular communication are released from neurons upon membrane depolarization in a Ca^{2+}-dependent manner. High pressure liquid chromatography (HPLC) analysis of the superfusates of nervous tissue slices for trace amount of amines, amino acids, oligopeptides and related molecules, allows to compare their efflux under resting and stimulated conditions. Compounds released upon depolarization can then be investigated in order to test if they are neuroactive.

Rat brain slices of neocortex, hippocampus, striatum, mesodiencephalon, cerebellum, pons-medulla and spinal cord were superfused with Earl's bicarbonate-buffered salt solution. The depolarization was induced by raising the [K^+] to 50 mM or by adding 33 µg/ml of veratrine. In parallel experiments, the [Ca^{2+}] was decreased from 2 mM to 0.1 mM and the [Mg^{2+}] increased from 1 mM to 12 mM. The superfusion solutions were reacted with 4-N,N-dimethylamino-azobenzene-4'-isothiocyanate (DABITC) according to a modification of the method described by Chang (1981). Besides the analysis of other common protein amino acids, the precolumn derivatization reverse phase HPLC developed allowed to resolve and to quantify simultaneously at the picomole level the sulfur containing amino acids, such as cysteine sulfinic acid (CSA), cysteic acid (CA), homocysteine sulfinic acid (HCSA) and homocysteic acid (HCA) in biological material. These very polar compounds coeluted with the void volume in the standard amino acid analytical methods. All four were shown to be present in TCA-extracts of rat neocortex.

The analysis of the superfusates of cortical slices, either under resting or K^+-depolarization conditions, gave the following results: the well-known release of GABA, Glu and Asp was confirmed. In addition, a discrete release of CSA (ratio 1.5), HCSA (1.9) as well as a rather important release of HCA (7.3) were observed, while the CA efflux was unaffected (Fig. 1). With low [Ca^{2+}] superfusion, these sulfur containing amino acids could not be detected anymore, either under resting or stimulation conditions.

The depolarization induced release of the sulfur containing amino acids was different in the various rat brain regions investigated (Fig. 1): CSA and HCSA were released moderately but significantly in neocortex, hippocampus and mesodiencephalon, and, for HCSA only, in striatum. A striking release of HCA was observed in all the regions, the neocortex and the hippocampus showing the highest values. In contrast, the CA efflux was practically unaffected by the depolarization except in the hippocampus and the mesodiencephalon where a very weak release was detected. When slices of all regions were superfused with 33 µg/ml of veratrine, the concentration of the four sulfur containing amino acids was similarly increased as described for K^+ depolarization.

These four sulfur containing amino acids have been reported by Watkins and his collaborators as well as other authors to exert an excitatory effect on CNS neurons (Curtis and Watkins, 1960; Wu and Dowling, 1978; Watkins and Evans, 1981; Mewett et al., 1983). Their release upon depolarization as endogenous compounds from slices of various brain regions suggests that they might be considered as transmitter candidates in the central nervous system (Do et al., 1986a). In the case of CSA, anatomical distribution (Baba et al., 1980; Ida and Kuriyama, 1983), binding to synaptic membranes (Recasens et al., 1982, 1983), and exogenous release (Iwata et al., 1982; Baba et al., 1983; Recasens et al., 1984) have been demonstrated, supporting the hypothesis of its transmitter role. Moreover, HCA which shows, as reported above, the most prominent release in many brain structures, should also be considered as a putative transmitter.

L-HOMOCYSTEIC ACID, AN NMDA-RECEPTOR PREFERRING AGONIST

To our knowledge, presence in nervous tissue and release of endogenous HCA have not been reported yet. Curtis and Watkins (1960) and Mewett et al. (1983) pointed out the excitatory, 2-amino-5-phosphono-valeric acid (APV) sensitive, action of HCA as well as its high affinity for the Glu binding site. Low and high affinity uptake of L-[^{35}S]-HCA was observed

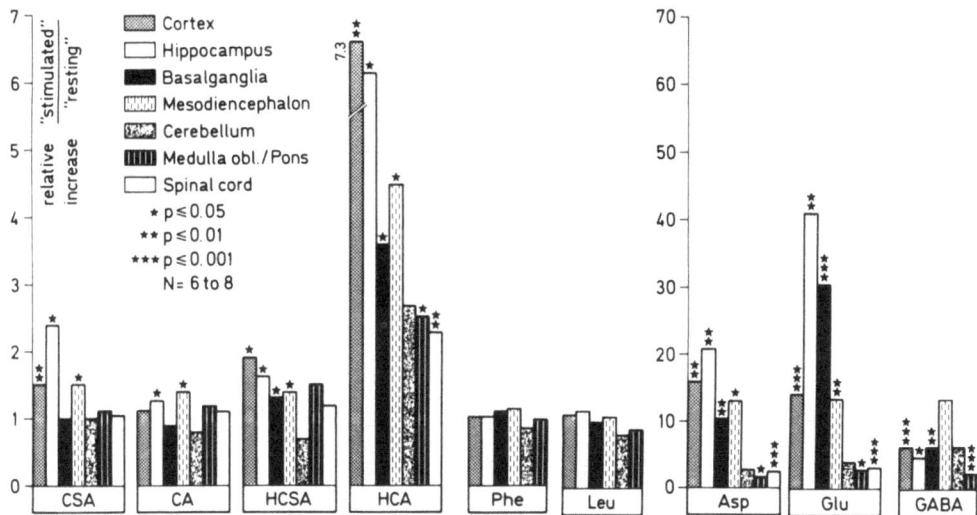

Fig. 1. Distribution of the relative increase of CSA, CA, HCSA, HCA, Phe, Leu, Asp, Glu and GABA, as released from slices of various rat brain regions. Each column represents the ratio of the efflux under stimulated (Earl's bicarbonate buffer solution containing 2 mM Ca^{2+}, 1 mM Mg^{2+} and 50 mM K^+) and under resting conditions (same solution but 3.75 mM K^+). Significance: unpaired Student's t-test performed with the absolute values. Note the clearcut K^+-induced increase of HCA, as well as of some other sulfur containing amino acids, and its confirmation for Asp, Glu and GABA. Phe and Leu are taken as control. (From Do et al., 1986a).

in rat cerebral cortex slices (Cox et al., 1977). Both Luini et al. (1984) and Baudry et al. (1983) observed that D,L-HCA was one of the most potent activators of the Na^{2+} efflux, and also suggested that it had a specificity for the NMDA receptor.

In view of the stimulation induced release of HCA in various rat brain regions, as reported above, the excitatory effect of L-HCA was tested in the cat caudate nucleus in vivo and the frog spinal cord in vitro (Herrling and Turski, 1986; Do et al., 1986b). Intracellular recording of caudate neurons (presumably of the medium spiny type) were performed in cats anesthetized with halothane, while the cortico-caudate pathway was stimulated through electrodes placed in the anterior sigmoid gyrus and agonists and/or antagonists were applied by microiontophoresis.

L-HCA applied to caudate neurons induced in most cells potential changes very similar to those elicited by microiontophoresis of NMDA: they consisted of an abrupt depolarization lasting for about 100 ms accompanied by bursts of numerous action potentials (Fig. 2A a,b). This firing pattern is very different from the one induced by microiontophoretic application of quisqualic acid (QUIS) (Fig. 2A c) or kainic acid (KA) which is characterized by short and fast depolarization with regularly spaced action potentials (Herrling et al., 1983; Herrling, 1985).

The long lasting depolarization and bursty firing induced by L-HCA was strongly depressed by the simultaneous application of D- or D,L-2-amino-7-phosphonoheptanoic acid (AP-7) (Fig. 2B a,b), a selective and potent NMDA antagonist (Evans et al., 1982; Perkins et al., 1982). In contrast,

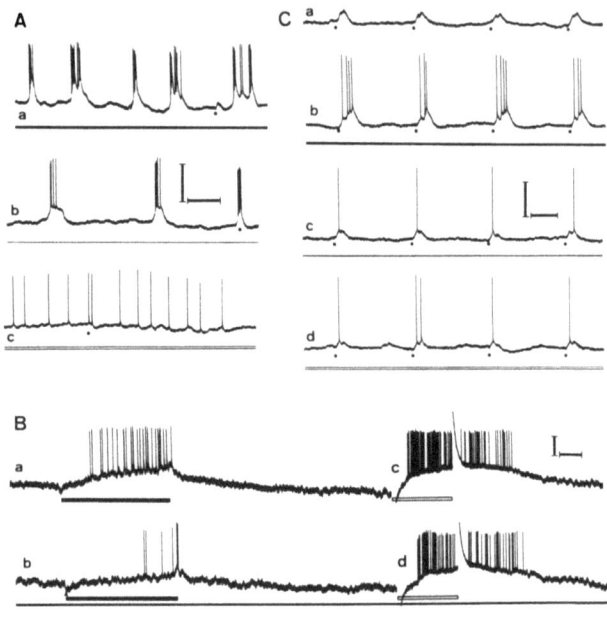

Fig. 2. Intracellular recordings in the cat caudate nucleus. A: Microiontophoretic application of (a) NMDA (-25 nA) or (b) L-HCA (-20 nA) induces a bursty firing pattern with abrupt de- and repolarizations of the membrane, while that of (c) quisqualate (-20 nA) elicits a regular firing pattern, without abrupt polarization changes. Calibration: 30 mV, positivity upwards, 300 ms.
B: The response induced by microiontophoretic application of (a) L-HCA (-50 nA) is prevented by (b) D-AP-7 (-20 nA), while that of (c) quisqualate (-150 nA) is not (d). Calibration: 20 mV, 2s.
C: (a) The EPSPs evoked by cortical electrical stimulation (dots) are (b) potentiated by iontophoretic application of L-HCA at -75 nA and many action potentials are induced, while the application of (c) L-CSA (-110 nA) or of (d) quisqualate (-50 nA), which induced the same membrane depolarization as L-HCA, are much less effective in potentiating the cortical response. Calibration: 30 mV, 150 ms. (From Do, Herrling, Streit, Turski and Cuénod, 1986b).

AP-7 had very little effect on the discharge elicited by QUIS application to the same cells (Fig. 2B c,d).

In order to obtain a more quantitative evaluation of the interactions of L-HCA with the excitatory amino acid receptors, the frog (Rana temporaria) spinal cord perfused in vitro with Ringer solution was used as model. Agonists and antagonists were applied in the perfusate and the direct current was recorded between the tip of the ventral roots and the hemisected body of the spinal cord. The depolarization induced by NMDA or L-HCA was strongly antagonized by D,L-AP-7, while that elicited by QUIS remained practically unaffected by AP-7. The displacement to the right of the dose-response curve was, however, not as pronounced for L-HCA as for NMDA, an observation suggesting that L-HCA might also interact, to a small extent, with non-NMDA receptors.

Lambert et al. (1981) and Mayer and Westbrook (1984, 1985) reported that NMDA receptor agonists, including L-HCA, cause a region of 'negative slope conductance' in the current-voltage relationship of spinal cord neurons, and suggested that, at low concentration, they might potentiate EPSPs. Following up this concept, the EPSP and the discharge induced by cortical stimulation in caudate neurons (Fig. 2C a) were compared in the absence or presence of either L-HCA or QUIS, applied by microiontophoresis. Under the influence of L-HCA (Fig. 2C b), the cortically evoked EPSPs were larger in amplitude and duration and the evoked action potentials were more numerous than under control conditions or during application of L-CSA or QUIS (Fig. 2C c,d).

Thus, L-HCA is a neuroactive endogenous substance which seems to affect predominantly the NMDA receptor, as both L-HCA and NMDA induce the same AP-7 sensitive depolarization and firing pattern in caudate neurons. Furthermore, it becomes a transmitter candidate for pathways possibly involved in the modulating potentiation of other inputs.

IMMUNOHISTOCHEMICAL LOCALIZATION OF HOMOCYSTEIC ACID

In order to test if HCA could be attributed to some defined CNS pathways, Matute and Streit (in preparation) attempted to localize presynaptic HCA by immunohistochemical methods.

Monoclonal antibodies (MAb) were developed against HCA coupled to bovine serum albumin (BSA) by means of glutaraldehyde (GA) according to the conjugation procedure devised by Storm-Mathisen et al. (1983). Hybridomas obtained by standard techniques were tested by an enzyme-linked immunosorbent assay (ELISA) for secretion of antibodies recognizing HCA-BSA but not Glu-BSA conjugates. Specificity of the MAb produced by a selected cell line was tested by serial dilution experiments in ELISA using BSA, GA-treated BSA, as well as various amino acid-BSA conjugates (HCA-, HCSA-, homocysteine (Hcys)-, Glu-, CA-, CSA-, Cys-, Asp-, GABA-, Met-, Tau-, Cyt-, GA-BSA, BSA). In the case of one particular antibody, the immunoreactivity for HCSA-, Hcys-, CA-, Glu-, Asp-BSA was about 16,000 time weaker than for HCA-BSA. The immunoreactivity with other amino acid-BSA-conjugates tested was even weaker (Fig. 3).

In semithin sections of GA-fixed rat brain, this specific MAb revealed dot-like structures, which could be interpreted as terminals in the most superficial layers of the cerebral cortex and in the caudoputamen (Fig. 4) among other areas. In contrast, the intermediate and deep cortical layers contained fewer dot-like structures, and the cerebellar cortex showed only background labeling.

These very preliminary observations, showing the selective and localized labeling of few structures, possibly nerve terminals, with an anti-HCA monoclonal antibody, support the idea that HCA is associated to some pathways of the central nervous system. The presumed HCA input to the striatum is unlikely to originate in the neocortex, as the response elicited in the cat by cortical stimulation is not antagonized by D-α-aminoadipate, an NMDA receptor antagonist (Herrling et al., 1983). Although other sources cannot be excluded, the central medial and parafascicular thalamic nuclei could be at the origin of the terminals observed. A projection terminating

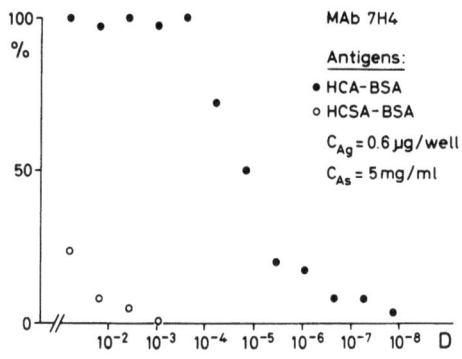 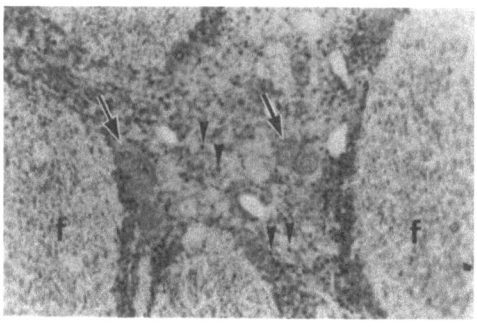

Fig. 3. Immunoreactivity of monoclonal anti-HCA antibody (MAb) 7H4 as determined by ELISA using glutaraldehyde coupled HCA- or closely related HCSA-BSA conjugates as antigens and serial dilutions (D) of the antibody. The values obtained for the HCSA-BSA conjugate are representative for the very low immunoreactivity found with BSA-conjugates of Hcys, Glu, CA, CSA, Cys, Asp, Met, Tau. C_{Ag}: concentration of conjugate added to wells of ELISA plates; C_{As}: protein concentration in ascitic fluid partially purified by ammonium sulfate precipitation.

Fig. 4. Distribution of HCA-like immunoreactivity in rat striatum as revealed by monoclonal anti-HCA antibody 7H4. Unlabeled antibody peroxidase-antiperoxidase (PAP) method; MAb 7H4: ascitic fluid diluted 1:200. Intensely stained dot-like structures (arrow-heads) suggest localization of HCA-like immunoreactivity in nerve terminals. f: unlabeled fiber bundles; arrows: unlabeled cell bodies. Total magnification x480.

in the superficial layers of the rodent neocortex could originate in one or more of the following thalamic nuclei: ventral anterolateral, lateral dorsal, ventromedial, reuniens, posterior, lateral posterior and magnocellular medial geniculate (Herkenham, 1980). Their termination on the distal ends of pyramidal apical dendrites places them in a favorable position to modulate the excitability of pyramidal cells, as induced by the stimulation of the nonspecific thalamic nuclei (Dempsey and Morison, 1942a,b; Morison and Dempsey, 1942; Hanbery and Jasper, 1953; Jasper, 1960, Purpura, 1959; Purpura and Shofer, 1964; Purpura et al., 1964). The presence of NMDA-sensitive glutamate binding sites in the striatum and in the superficial cortical layers of the rat (Monaghan and Cotman, 1985) correlates with the distribution of HCA-like immunoreactivity, an observation in agreement with the proposal that HCA acts on an NMDA receptor type. It is also worth mentioning that in the rat the projections from the nonspecific thalamic nuclei to the striatum (Streit, 1980) and to the sensorimotor cortex (Ottersen et al., 1983) are selectively labeled by the retrograde marker D-[^{3}H]Asp (Cuénod and Streit, 1983), which could be taken up by neurons using HCA in addition to those using Glu and Asp as possible mediators.

Taken together, these results on release, neuroactivity and localization of endogenous HCA are consistent with the hypothesis that it is a neurotransmitter in some CNS pathways, possibly the nonspecific thalamic projections, potentiating the excitability of neurons via their NMDA-receptors. This hypothesis could even be extended to include a role of such HCA-NMDA synapses in certain central plastic changes, as it has been shown that NMDA receptor antagonists can prevent the formation of long term potentation in hippocampal slices (Collingridge et al., 1983; Harris et al., 1984).

Furthermore, these neuronal systems are likely to be involved in the mechanisms of some seizure disorders, particularly in view of the fact that the selective NMDA receptor antagonist AP-7 has an anticonvulsant action (Croucher et al., 1982; Meldrum et al., 1983). In that context, it is worth mentioning that HCA has been reported to be present in excess in urine of homocystinuric patients, who indeed suffer from epileptic seizures (Mudd and Levy, 1983).

ACKNOWLEDGEMENTS

The authors wish to thank Dr. J.C. Watkins for many very helpful discussions and the gift of several reference amino acids. This work has been supported by grants 3.455.83 and 3.228.82 of the Swiss National Science Foundation and by the Dr. Eric Slack-Gyr-Foundation. The support of the Sandoz-Foundation, the 'Jubiläumsstiftung der Schweiz. Lebensversicherung- und Rentenanstalt', the 'Geigy-Jubiläums-Stiftung' and the 'Hartmann-Müller-Stiftung' is gratefully acknowledged.

REFERENCES

Baba, A., Yamagami, S., Mizuo, H., and Iwata, H., 1980, Microassay of cysteine sulfinic acid by an enzymatic cycling method, Anal. Biochem., 101:288.

Baba, A., Okumura, S., Mizuo, H., and Iwata, H., 1983, Inhibition by diazepam and γ-aminobutyric acid of depolarization-induced release of [^{14}C] cysteine sulfinate and [^{3}H]glutamate in rat hippocampal slices, J. Neurochem., 40:280.

Baudry, M., Kramer, K., Fagni, L., Recasens, M., and Lynch, G., 1983, Classification and properties of acidic amino acid receptors in hippocampus, Mol. Pharmacol., 24:222.

Ben-Ari, Y., Lagowska, J., Tremblay, E., and Le Gal La Salle G., 1979, A new model of focal status epilepticus: intraamygdaloid application of kainic acid elicits repetitive secondarily generalized convulsive seizures, Brain Res., 163:176.

Chang, J.-Y, 1981, Isolation and characterization of polypeptide at the picomole level, Biochem. J., 199:537.

Collingridge, G.L., Kehl, S.J., and McLennan, H., 1983, Excitatory amino acids in synaptic transmission in the Schaffer collateral-commissural pathway of the rat hippocampus, J. Physiol., 334:33.

Cox, D.W.G., Headley, M.H., and Watkins, J.C., 1977, Actions of L- and D-homocysteate in rat CNS: a correlation between low-affinity uptake and the time courses of excitation by microelectrophoretically applied L-glutamate analogues, J. Neurochem., 29:579.

Croucher, M.J., Collins, J.F., and Meldrum, B.S., 1982, Anticonvulsant action of excitatory amino acid antagonist, Science, 216:899.

Cuénod, M., Bagnoli, P., Beaudet, A., Rustioni, A., Wiklund, L., and Streit, P., 1982, Transmitter specific retrograde labeling of neurons, in: Cytochemical Methods in Neuroanatomy, V. Chan-Palay and S. L. Palay, eds., Alan R. Liss, Inc., New York, p. 17.

Cuénod, M., and Streit, P., 1983, Neuronal tracing using retrograde migration of labeled transmitter-related compounds, in: Methods in Chemical Neuroanatomy, A. Björklund and T. Hökfelt, eds., Elsevier, Amsterdam, p. 365.

Cuénod, M., Do, K.Q., Matute, C., and Streit, P., 1986, Identification of pathways for acidic amino acid transmitters and search for new candidates: sulphur containing amino acids, in: Excitatory Amino Acids, P.J. Roberts, J. Storm-Mathisen and H.F. Bradford, eds., Macmillan Press, London, in press.

Curtis, D.R., and Watkins, J.C., 1960, The excitation and depression of spinal neurones by structurally related amino acids, J. Neurochem., 6:117.

Dempsey, E.W., and Morison, R.S., 1942a, The production of rhythmically recurrent cortical potentials after localized thalamic stimulation, Amer. J. Physiol., 135:293.

Dempsey, E.W., and Morison, R.S., 1942b, The interaction of certain spontaneous and induced cortical potentials, Amer. J. Physiol., 135:301.

Do, K.Q., Mattenberger, M., Streit, P., and Cuénod, M., 1986a, In vitro release of endogenous excitatory sulphur containing amino acids from various rat brain regions, J. Neurochem., 46:779.

Do, K.Q., Herrling, P.L., Streit, P., Turski, W.A., and Cuénod, M., 1986b, In vitro release and electrophysiological effects in situ of homocysteic acid, an endogenous N-methyl-D-aspartic acid agonist, in the mammalian striatum, J. Neurosci., in press.

Evans, R.H., Francis, A.A., Jones, A.W., Smith, D.A.S., and Watkins, J.C., 1982, The effects of a series of omegaphosphonic alpha-carboxylic amino acid-induced responses in the isolated spinal cord preparation, Br. J. Pharmac., 75:65.

Fonnum, F., Søreide, A., Kvale, I., Walker, J., and Walaas, I., 1981, Glutamate in cortical fibers, in: Glutamate as a Neurotransmitter, G. DiChiara and G.L. Gessa, eds., Raven Press, New York, p. 29.

French, E.D, Aldinio, C., and Schwarcz, R., 1982, Intrahippocampal kainic acid, seizures and local neuronal degeneration: relationships assessed in unanesthetized rats, Neuroscience, 7:2525.

Hanberry, J., and Jasper, H.H., 1953, Independence of diffuse thalamic projection system shown by specific nuclear destructions, J. Neurophysiol., 16:252.

Harris, E.W., Ganong, A.H., and Cotman, C.W., 1984, Long-term potentiation in the hippocampus involves activation of N-methyl-D-aspartate receptors, Brain Res., 323:132.

Herkenham, M., 1980, Laminar organization of thalamic projections to the rat cortex, Science, 207:532.

Herrling, P.L., Morris, R., and Salt, T.E., 1983, Effects of excitatory amino acids and their antagonists on membrane and action potentials of cat caudate neurones, J. Physiol., 339:207.

Herrling, P.L., 1985, Pharmacology of the corticocaudate excitatory postsynaptic potential in the cat: evidence for its mediation by quisqualate- or kainate-receptors, Neuroscience, 14:417.

Herrling, P.L., and Turski, W.A., 1986, Interactions of sulphur-containing excitatory amino acids with membrane and synaptic potentials of cat caudate neurons, in: Excitatory Amino Acids, P.J. Roberts, J. Storm-Mathisen and H.F. Bradford, eds., Macmillan Press, London, in press.

Ida, S., and Kuriyama, K., 1983, Simultaneous determination of cysteine sulfinic acid and cysteic acid in rat brain by high-performance liquid chromatography, Anal. Biochem., 130:95.

Iwata, H., Yamagami, S., Mizuo, H., and Baba, H., 1982, Cysteine sulfinic acid in the central nervous system: uptake and release of cysteine sulfinic acid by a rat brain preparation, J. Neurochem., 38:1268.

Jasper, H.H., 1960, Unspecific thalamocortical relations, in: Handbook of Physiology: Neurophysiology, J. Field, H.W. Magoun and V.E. Hall, eds., American Physiological Society, Washington, p. 1307.

Lambert, J.D.C., Flatman, J.A., and Engberg, I., 1981, Actions of excitatory amino acids on membrane conductance and potential in motoneurones, in: Glutamate as a Neurotransmitter, G. DiChiara and G.L. Gessa, eds., Raven Press, New York, p. 205.

Luini, A., Goldberg, O., and Teichberg, V.I., 1984, An evaluation of selected brain constituants as putative excitatory neurotransmitters, Brain Res., 324:271.

Mayer, M.L., and Westbrook, G.L., 1984, Mixed-agonist action of excitatory amino acids on mouse spinal cord neurones under voltage clamp, J. Physiol., 354:29.

Mayer, M.L., and Westbrook, G.L., 1985, The action of N-methyl-D-aspartic acid on mouse spinal neurones in culture, J. Physiol., 361:65.

Meldrum, B.S., Croucher, M.J., Badman, G., and Collins, J.F., 1983, Antiepileptic action of excitatory amino acid antagonists in photosensitive baboon, Papio papio, Neurosci. Lett., 39:101.

Mewett, K.N., Oakes, D.J., Olverman, H.J., Smith, D.A.S., and Watkins, J.C., 1983, Pharmacology of the excitatory actions of sulphonic and sulphinic amino acids, in: CNS Receptors - From Molecular Pharmacology to Behavior, P. Mandel and F.V. DeFeudis, eds., Raven Press, New York, p. 163.

Monaghan, D.T., and Cotman, C.W., 1985, Distribution of N-methyl-D-aspartate-sensitive L-[^{3}H]Glutamate-binding sites in rat brain, J. Neurosci., in press.

Morison, R.S., and Dempsey, E.W., 1942, A study of thalamo-cortical relations, Amer. J. Physiol., 135:281.

Morselli, P.L., Lloyd, K.G., Löscher, W., Meldrum, B., and Reynolds, E.H., 1981, Neurotransmitters, Seizures and Epilepsy, Raven Press, New York.

Mudd, S.H., and Levy, H.L., 1983, Disorders of transsulfuration, in: The Metabolic Basis of Inherited Diseases, J.B. Stanbury, J.B. Wyngaarden, D.S. Fredrickson, J.L. Goldstein and M.S. Brown, eds., McGraw-Hill, New York, p. 522.

Ottersen, O.P., Fischer, B.O., and Storm-Mathisen, J., 1983, Retrograde transport of D-[^{3}H]aspartate in thalamo-cortical neurones, Neurosci. Lett., 42:19.

Perkins, M.N., Collins, J.F., and Stone, T.W., 1982, Isomeres of 2-amino-7-phosphonoheptanoic acid as antagonists of neural excitants, Neurosci. Lett., 32:65.

Purpura, D.P., 1959, Nature of electrocortical potentials and synaptic organizations in cerebral and cerebellar cortex, in: International Review of Neurobiology, C.C. Pfeiffer and J.R. Smythies, eds., Academic Press, New York, p. 47.

Purpura, D.P., and Shofer, R.J., 1964, Cortical intracellular potentials during augmenting and recruiting responses. I. Effects of injected hyperpolarizing currents on evoked membrane potential changes, J. Neurophysiol., 27:117.

Purpura, D.P., Shofer, R.J., and Musgrave, F.S., 1964, Cortical intracellular potentials during augmenting and recruiting responses. II. Patterns of synaptic activities in pyramidal and nonpyramidal tract neurons, J. Neurophysiol., 27:133.

Recasens, M., Varga, V., Nanopoulos, D., Saadoun, F., Vincendon, G., and Benavides, J., 1982, Evidence for cysteine sulfinate as a neurotransmitter, Brain Res., 239:153.

Recasens, M., Saadoun, F., Varga, V., DeFeudis, F.V., Mandel, P., Lynch, G., and Vincendon, G., 1983, Separate binding sites in rat brain synaptic membranes for L-cysteine sulfinate and L-glutamate, Neurochem. Int., 5:89.

Recasens, M., Fagni, L., Baudry, M., and Lynch, G., 1984, Potassium and veratrine-stimulated L-[^{3}H]cysteine sulfinate and L-[^{3}H]glutamate release from rat brain slices, Neurochem. Int., 6:325.

Schwarcz, R., Foster, A.C., French, E.D., Whetsell, W.O. Jr., and Köhler, C., 1984, Excitotoxic models for neurodegenerative disorders, Life Sci., 35:19.

Storm-Mathisen, J., Leknes, A.K., Bore, A.T., Vaaland, J.L., Edminson, P., Haug, F.-M.S., and Ottersen, O.P., 1983, First visualization of glutamate and GABA in neurones by immunocytochemistry, Nature, 301:517.

Streit, P., 1980, Selective retrograde labeling indicating the transmitter of neuronal pathways, J. Comp. Neurol., 191:429.

Streit, P., 1984, Glutamate and aspartate as transmitter candidates for systems of the cerebral cortex, in: Cerebral Cortex Vol. 2, E.G. Jones and A. Peters, eds., Plenum Publ. Corp., New York, p. 119.

Van Gelder, N.M., Sherwin, A.L., and Rasmussen, T., 1972, Amino acid content of epileptogenic human brain. Focal versus surrounding regions, Brain Res., 40:385.

Watkins, J.C., and Evans, R.H., 1981, Excitatory amino acid transmitters, Ann. Rev. Pharmacol. Toxicol., 21:165.

Wu, S.M., and Dowling, J.E., 1978, L-aspartate: evidence for a role in cone photoreceptor synaptic transmission in the carp retina, Proc. Natl. Acad. Sci. USA, 75:5205.

EXCITATORY AMINO ACID PATHWAYS IN THE BRAIN

O.P. Ottersen and J. Storm-Mathisen

Anatomical Institute, University of Oslo
Karl Johans gate 47
N-0162 Oslo 1, Norway

INTRODUCTION

The neurochemical techniques used to obtain the results reviewed here take advantage of the fact that chemical neurotransmission requires 1) the presence of the transmitter in the presynaptic element, 2) a synthesizing apparatus for transmitter replenishment, 3) a mechanism for transmitter release, and 4) a mechanism for terminating transmitter action, involving, in the case of excitatory amino acids (EAA), reuptake in the presynaptic element. These techniques alone do not give definitive proof of the transmitter identity, but positive results with any one of them in a neuronal system suggest that the possibility of EAA neurotransmission should be more closely investigated, including also electrophysiological and pharmacological methods.

The first part of this review addresses the usefulness and specificity of the various biochemical and histological techniques that are available for tracing EAA pathways. The second part gives a brief survey of the complex network of putative EAA connections that have emerged from studies with such methods. Most of the data are presented in tables. We have made efforts to cover all brain regions of the central nervous system (except the retina), but apologize if any important references should not be included.

METHODS

In the following, the methods have been classified according to the criterion (1-4 in Introduction) which they address.

Presence of Transmitter in the Presynaptic Element

Several lines of evidence suggest that the transmitter pools of glutamate (Glu) and aspartate (Asp) constitute a minor proportion, perhaps in the range of 20-45%, of the total amount of free Glu and Asp in the brain (Fonnum, 1984). The bulk of these amino acids serves roles in the intermediary metabolism or in protein synthesis or breakdown, and they would therefore be expected to occur in all neurons regardless of transmitter identity. This has been borne out in immunocytochemical experiments with antisera against conjugated Glu or Asp (Ottersen and Storm-Mathisen, 1984a, b, 1985; Storm-Mathisen and Ottersen, 1985). Nevertheless, our studies indicate that the levels of Glu and Asp vary considerably among neurons, Glu-like immunoreactivity (Glu-LI) being intermediate to high in, e.g., populations of putative Glu/Asp-ergic, cholinergic, and dopaminergic cells, and very low in many of the cell groups thought to use GABA for neural transmission. The scarcity of Glu-LI in the latter type of neurons agrees with biochemical data on Glu contents (Minchin and Fonnum, 1979; Korf and Venema, 1983) and may reflect consumption of Glu for GABA synthesis. In contrast, some GABA-ergic cells seem to be enriched in Asp (Ottersen and Storm-Mathisen, 1984c, 1985). The implication is that the presence of Glu or Asp cannot be regarded as a specific marker of Glu/Asp-ergic neurons. However, data on Glu and Asp contents may serve as a useful adjunct to other parameters and may be valuable in attempts to distinguish between Glu-ergic and Asp-ergic neurotransmission.

If Glu is generally enriched in neurons relative to glia (Patel and Hunt, 1985), any structure deprived of a major synaptic input would show a fall in Glu concentration consequent to the increased proportion of glia in the deafferented tissue. This effect would be enhanced by glial proliferation, and calls for caution in the interpretation of small lesion-induced decreases in Glu. The possibility also exists that changes in EAA levels after lesions may occur as a result of perturbation of metabolic activity.

By comparing the effects of manipulating Glu and Asp synthesis in normal tissue with the effects obtained in tissue deprived of its EAA input, it has been suggested that the transmitter pool has a much faster turnover than the non-transmitter pool (Engelsen and Fonnum, 1983; Fonnum et al., 1984; Wroblewski et al., 1985). This implies that measurement of turnover rates may facilitate the interpretation of data on endogenous contents and aid in identifying neurotransmitter function (Freeman et al., 1983). Different rates of synthesis may be one of the factors underlying our finding that in certain *in vitro* conditions the transmitter pool of Glu in the hippocampus seems to prevail whereas the metabolic pool is lost (Ottersen and Storm-Mathisen, 1985; Storm-Mathisen and Ottersen, 1985).

<u>Transmitter Synthesis</u>

In the search for specific enzyme markers for Glu/Asp-ergic neurons, the interest has been focused on aspartate aminotransferase (AAT) and phosphate-activated glutaminase (PAG), which are responsible for the interconversion of Glu and Asp and the formation of Glu from glutamine (Hertz et al., 1983). Whereas immunoreactivity for the cytoplasmic isoenzyme of AAT seems to be high in certain neurons thought to use Glu or Asp as transmitter, it is also enriched in some putative GABA-ergic cell populations (Wenthold and Altschuler, 1985), and biochemically recorded AAT has been reported to decrease after degeneration of cholinergic input (Sterri and Fonnum, 1980). It has been claimed that the level of this enzyme may be more closely correlated with metabolic activity than with transmitter identity (Bolz et al., 1985; Ross and Godfrey, 1985). The matter is further complicated by evidence suggesting that AAT and cysteine sulfinic acid transaminase may be the same protein (Recasens et al., 1980).

PAG appears to be more consistently enriched in putative EAA neurons than AAT (Wenthold and Altschuler, 1985). The high concentration of PAG in such neurons may reflect the high turnover rate of transmitter Glu. However, PAG-like immunoreactivity is found in most brain structures, in agreement with the ubiquity of metabolic Glu. Most published results have been obtained with antibodies against PAG purified from kidney (Altschuler et al., 1982). A different antiserum raised against brain

PAG shows a more uniform distribution of immunoreactivity, in particular no concentration in hippocampal mossy fibers (Svenneby and Storm-Mathisen, 1983, and unpublished observations).

In conclusion, current evidence suggests that none of the enzymes engaged in synthesis or breakdown of Glu or Asp are restricted to neurons using these amino acids for neurotransmission. However, in conjunction with data obtained with other methods, the enrichment of PAG in a neuron may serve as an indicator of transmitter identity.

Transmitter Release

The capacity to release preloaded [^{3}H]Glu or [^{3}H]Asp in a Ca^{2+}-dependent fashion is considered specific for Glu/Asp-ergic neurons and is a useful marker for tracing of EAA pathways. The widely held opinion that the release process, like the uptake process (see below), generally does not distinguish between Glu and Asp has stimulated the use of the metabolically inert D-[^{3}H]Asp to study both Asp and Glu release mechanisms. The tenability of this strategy should now be subject to renewed investigation since it has been reported that synaptic vesicles, isolated by precipitation with antibodies against synapsin I, show an ATP-dependent uptake of L-Glu but not of L- or D-Asp (Naito and Ueda, 1983, 1985). The possibilities remain, however, that release may occur independently of the vesicles (De Belleroche and Bradford, 1972), and that Asp-concentrating vesicles do exist but escaped detection by the procedure that was employed.

If, instead of D-[^{3}H]Asp, L-[^{3}H]Glu or L-[^{3}H]Asp are used as substrate for the release mechanism, one is faced with the problem that some of the radioactivity will end up in other molecules, necessitating an analysis of the released substances. However, the metabolic lability of the L-amino acids can be used to advantage. Thus, after preloading with [^{3}H]glutamine or other precursors, a proportion of the radioactivity can be recovered in the released Asp or Glu (e.g., Čanžek and Reubi, 1980). In this paradigm, Ca^{2+}-dependent release of labeled Asp or Glu will not only require a release mechanism but also the presence of a specific synthesizing apparatus.

Measuring the release of endogenous rather than preloaded amino acids should permit distinguishing Glu from Asp. However, this approach is technically difficult, particularly because it entails the sampling of very small quantities of amino acids. An additional problem, shared with release studies of labeled transmitters, is to unequivocally identify the nerve structures responsible for the release. For instance, it is likely that stimulation of the lateral olfactory tract not only leads to release from the stimulated fibers, but also from the target neurons in the olfactory cortex. To differentiate between these sources may be difficult (Collins et al., 1981). Interpretative problems of this nature may partly explain the controversy as to whether Glu or Asp is the likely transmitter of the lateral olfactory tract fibers (see Table 3). Another example is the release of Glu through the hippocampal alvear surface in response to stimulation of the perforant path (Crawford and Connor, 1973).

Transmitter Uptake

Methods based on Glu or Asp uptake are the most widely used methods for tracing EAA pathways. Sodium-dependent, high affinity neuronal uptake of Glu/Asp is thought to occur selectively in neurons using Glu or Asp as a transmitter, and can be recorded biochemically as well as autoradiographically after incubation with the radiolabeled transmitter in vitro. The proportion of uptake localized in glial elements may be as low as 13% (Storm-Mathisen and Iversen, 1979), but seems to vary with the exact conditions and with the region under study (Wilkin et al., 1982). At

Table 1. Efferent neocortical excitatory amino acid projections

Corticofugal projections to:

- caudatoputamen

    U_B: Divac et al. 1977; McGeer et al. 1977; Fonnum et al. 1981a[1]; Walaas 1981[1]; Young et al. 1981 (cat); Carter 1982; Kerkerian et al. 1983 (cat)

    C_{Glu}: Kim et al. 1977; Fonnum et al. 1981a; Druce et al. 1982; Hassler et al. 1982[1]; Sandberg et al. 1985

    C_{PAG}: Ward et al. 1982; Sandberg et al. 1985[2]

    R_{Glu}: Rowlands and Roberts 1980; Druce et al. 1982; Roberts et al. 1982

    R_{Asp}: Druce et al. 1982[3]

    R_L: Reubi and Cuénod 1979; Godukhin et al. 1980; Reubi et al. 1980[4]

    T: Streit 1980[1,5]

- nucleus accumbens

    U_B: Walaas 1981[6]; Christie et al. 1985b[7,8]

- olfactory tubercle

    U_B: Walker and Fonnum 1983b[8]

- amygdala

    U_B: Fischer et al. 1982a; Ottersen et al. 1986; Walker and Fonnum 1983a[9]

- thalamus[10]

    U_B, C_{Glu}: Fonnum et al. 1981a[11]; Walker and Fonnum 1983a

- dorsal subdivision of the lateral geniculate body
  (from visual cortex)

    U_B: Lund Karlsen and Fonnum 1978; Kvale and Fonnum 1983; Fosse et al. 1984[12] (cat)

    T: Baughman and Gilbert 1980 (cat), 1981[13] (cat)

- ventrolateral thalamic nucleus
  (from sensorimotor cortex)

    U_B, C_{Glu}: Bromberg et al. 1981 (cat); Young et al. 1981 (cat)

    U_B: Kerkerian et al. 1983 (cat)

- centromedian/parafascicular nuclei

    T: Wiklund and Cuénod 1984

- superior colliculus
  (from visual cortex)

    U_B: Lund Karlsen and Fonnum 1978; Fonnum et al. 1979; Kvale and Fonnum 1983; Fosse et al. 1984[14,15] (cat)

    T: Matute et al. 1984[15]

- inferior colliculus
  (from auditory cortex)

    C_{Glu}: Adams and Wenthold 1979[16] (cat)

- ventral tegmental area

    U_B, C_{Asp}: Christie et al. 1985a[8]

- nucleus ruber
  (from sensorimotor cortex)

    U_B: Bromberg et al. 1981 (cat); Young et al. 1981 (cat); Kerkerian et al. 1983 (cat)

Table 1 Part 2

- substantia nigra

    U_B: Carter 1982[17]; Fonnum et al. 1981b; Kerkerian et al. 1983[18] (cat); Abarca and Bustos 1985

    C_{Glu}:   Kornhuber et al. 1984

    R_L: Abarca and Bustos 1985

    T: Streit 1980[19]

- pontine nuclei

    U_A, U_B: Thangnipon et al. 1983[20]

    U_B: Young et al. 1981 (cat)

    R_L: Thangnipon and Storm-Mathisen 1981

- dorsal column nuclei
  (from sensorimotor cortex)

    T: Rustioni and Cuénod 1982[5]

- spinal cord (contralaterally)

    U_B: Young et al. 1981 (cat); Potashner and Tran 1985[21] (guinea pig)

    R_{Glu}:   Fagg et al. 1978 (cat)

    R_L: Potashner and Tran 1985[21] (guinea pig)

Associational projections

    R_{Glu}, R_{Asp}:   Hicks et al. 1985 (cat)

    T: Streit 1980; Baughman and Gilbert 1981 (cat); Fischer et al. 1982b; Ottersen et al. 1983

Commissural projections

    U_B: Fonnum et al. 1981a

    T: Streit 1980; Fischer et al. 1982b; Ottersen et al. 1983

The references in this and the following tables have been grouped according to methodological approach. U: Autoradiographic (U_A) or biochemical (U_B) recording of high affinity uptake of radiolabelled Asp or Glu. C: Biochemical analysis of endogenous contents of Glu (C_{Glu}) or Asp (C_{Asp}), or of activities of phosphate activated glutaminase (C_{PAG}) or aspartate aminotransferase (C_{AAT}). R: Measurement of released radiolabelled transmitter after preloading with transmitter or precursor (R_L), or of released endogenous Glu (R_{Glu}) or Asp (R_{Asp}). T: Autoradiographic mapping of retrograde axonal transport of D-[^{3}H]Asp after injections in vivo. I_{Glu}, I_{Asp}, I_{PAG}, I_{AAT}: Immunocytochemistry of Glu, Asp, glutaminase, or aspartate aminotransferase. Unless otherwise indicated the studies referred to under U, C, R, or I have based their conclusions on the effect of lesions, or, in the case of release studies, on the effects of lesion and/or stimulation of the relevant pathways. When conclusions have been drawn from normal material this is indicated by (N), e.g., $U_A(N)$. If another experimental animal than rat has been used the species is indicated.

1   Evidence was also obtained of a small excitatory amino acid projection to the contralateral striatum.
2   No decrease in this parameter was found by McGeer and McGeer (1979).
3   Axotomy induced a larger decrease in K^+-evoked release of Asp than of Glu (33% vs. 21%).
4   Labelled Asp formed from precursor asparagine was used as marker.
5   Retrogradely labelled cells after D-[^{3}H]Asp injections were mainly found in cortical layer V.
6   The excitatory amino acid input from the hippocampus seems to be quantitatively more important (Walaas and Fonnum, 1979).
7   The changes were restricted to the anterior part of the nucleus.
8   The lesions were made in the medial frontal cortex and probably included neocortical as well as allocortical areas.
9   The observation of a 31% decrease in amygdalar D-[^{3}H]Asp uptake after neocortical ablation is apparently in conflict with neuroanatomical data indicating a very sparse neocortical input to the amygdala in the rat (Ottersen, 1982). However, as judged from their drawing, Walker and Fonnum included in their neocortex several cortical areas that are usually regarded as allocortical and which project to the amygdala.
10 The samples for analysis consisted of the entire thalamus, or large parts of it.
11 A small decrease in endogenous Asp was also recorded.
12 A decreased uptake was also detected in the pulvinar.
13 Retrogradely labelled cells mainly occurred in layer VI.
14 The effects after area 17 ablation were restricted to the superficial (i.e., "visual") part of the superior colliculus.
15 Baughman and Gilbert (1981), in the cat, found no evidence of retrograde D-[^{3}H]Asp transport from the superior colliculus to the visual cortex.
16 The changes were small (10%) and restricted to 2 extracentral regions of the colliculus.
17 Changes were observed after lesions of the pregenual part, but not after lesions of the supracallosal part of the frontal cortex, in agreement with neuroanatomical tracing data (Beckstead, 1979).
18 The lesions were made in the sensorimotor cortex and the decrease mainly affected the lateral half of the substantia nigra.
19 Only a few cells were labelled.
20 The data were consistent with the existence of a small contingent of crossed corticopontine excitatory amino acid fibers.
21 These authors studied the effects of partial cordotomy, which leads to the interruption of several descending fiber systems.

Table 2. Excitatory amino acid projections in the hippocampal region[1]

From fascia dentata to CA3 and CA4
(mossy fiber system)

$U_A(N)$: Storm-Mathisen and Iversen 1979[2,3]; Heggli et al. 1981

C_{Glu}, C_{Asp}: Nadler et al. 1978[4]

R_{Glu}: Crawford and Connor 1973[5] (cat)

T: Storm-Mathisen and Wold 1981[6] (hamster); Fischer et al. 1982c, 1985[6]

$I_{Glu}(N)$: Storm-Mathisen et al. 1983; Ottersen and Storm-Mathisen 1985[7]

$I_{PAG}(N)$: Altschuler et al. 1985[2] (rat and guinea pig)

From CA4 to ipsi- and contralateral fascia dentata[8,9]

U_A: Storm-Mathisen 1981; Taxt and Storm-Mathisen 1984

R_{Asp}: Nadler et al. 1976, 1978

T: Fischer et al. 1982c, 1985

From CA3 to CA1[8,9]
(Schaffer collaterals)

U_A, U_B: Storm-Mathisen 1981; Taxt and Storm-Mathisen 1984[10]

U_B: Storm-Mathisen 1977[10]; Fonnum and Walaas 1978[10]

C_{Glu}, C_{Asp}: Storm-Mathisen 1978

R_{Glu}, R_{Asp}: Spencer et al. 1981

R_L: Malthe-Sørenssen et al. 1979 (guinea pig); Skrede and Malthe-Sørenssen 1981a[11] (guinea pig)

T: Fischer et al. 1982c, 1985

From CA3 to contralateral hippocampus[8]
(commissural projection)

U_A, U_B: Taxt and Storm-Mathisen 1984

U_B: Storm-Mathisen 1977

C_{Glu}: Nitsch et al. 1979a[12], 1979b (rabbit)

R_{Glu}, R_{Asp}: Nadler et al. 1976, 1978

R_L: Skrede and Malthe-Sørenssen 1981b (guinea pig)

T: Fischer et al. 1982c, 1985

From CA3 to lateral septum[8]

U_B, C_{Glu}: Fonnum and Walaas 1978; Fonnum et al. 1979; Zaczek et al. 1979; Walaas and Fonnum 1980

U_B: Storm-Mathisen 1978; Storm-Mathisen and Woxen Opsahl 1978

C_{Glu}: Nitsch et al. 1979a

R_L: Malthe-Sørenssen et al. 1980 (guinea pig)

T: Fischer et al. 1982c, 1985[13]

From hippocampus to entorhinal cortex

C_{Glu}: Nitsch et al. 1979a[14]

From CA1 to subiculum

U_A, U_B: Taxt and Storm-Mathisen 1984

From CA1 to amygdala

T: Fischer et al. 1982a; Ottersen et al. 1986

Table 2 Part 2

From the hippocampal formation[15] to mediobasal hypothalamus, nucleus of the diagonal band, and bed nucleus of stria terminalis

    U_B, C_{Glu}:   Walaas and Fonnum 1980[16]

From subiculum to nucleus accumbens

    U_B: Walaas and Fonnum 1979[17], 1980; Walaas 1981

    C_{Glu}:   Nitsch et al. 1979a; Walaas and Fonnum 1979, 1980

From subiculum to the mammillary body

    U_B: Storm-Mathisen 1978; Storm-Mathisen and Woxen Opsahl 1978; Walaas and Fonnum 1980

    C_{Glu}:   Nitsch et al. 1979a; Walaas and Fonnum 1980

    C_{Asp}:   Walaas and Fonnum 1980

From subiculum to amygdala

    T: Ottersen et al. 1986

From the entorhinal cortex to fascia dentata
(perforant path)

    U_A: Storm-Mathisen 1977, 1981; Taxt and Storm-Mathisen 1984

    U_B: Nadler et al. 1976, 1978; Storm-Mathisen 1977, 1981, 1982

    C_{Glu}:   Nadler and Smith 1981[18]

    C_{Asp}:   Di Lauro et al. 1981[18]

    R_{Glu}:   Nadler et al. 1976, 1978; Hamberger et al. 1978

    R_L: Dolphin et al. 1982[19]; Hamberger et al. 1979

    T: Fischer et al. 1985

From the entorhinal cortex to amygdala

    U_B, C_{Glu}:   Walker and Fonnum 1983a[20]

    T: Fischer et al. 1982a; Ottersen et al. 1986

---

1. Under this heading we include all efferent and intrinsic connections of the hippocampus proper, the fascia dentata, and the retrohippocampal allocortical areas.
2. Labelling of mossy fiber boutons was demonstrated both light- and electronmicroscopically.
3. The failure of Nadler et al. (1978) to detect significant changes in Glu uptake or release in the whole CA3 after granular cell degeneration is not inconsistent with a decrease in the mossy fiber zone, since this zone constitutes only a small proportion of CA3.
4. In the Purdue-Wistar rat, the decrease in endogenous amino acid levels in regio inferior, after X-irradiation-induced degeneration of the granule cells, was larger for Glu than for Asp. No change in Glu was found in the other rat strain studied (Sprague-Dawley).
5. The released Glu may have come from several of the fiber systems.
6. The labelling observed in the mossy fiber zone after D-[^{3}H]Asp infusions or injections involving the area dentata was probably due to anterograde transport subsequent to uptake in hilar collaterals.
7. The mossy fiber boutons can accumulate Glu-like immunoreactivity, but not Asp-like immunoreactivity, under certain conditions in vitro. However, both immunoreactivities are low in mossy fibers of animals fixed by perfusion.
8. The case for Glu or Asp as transmitter receives support from the finding in the in vitro hippocampal slice of normal rats that Glu/Asp high affinity uptake (Storm-Mathisen and Iversen, 1979) and Glu-like and Asp-like immunoreactivities (Ottersen and Storm-Mathisen, 1985) are concentrated in the termination zones of these pathways.
9. These fiber systems were also labelled after intraventricular infusions of D-[^{3}H]Asp (Storm-Mathisen and Wold, 1981).
10. A decreased uptake was also found corresponding to the termination of the "longitudinal associational bundle".
11. Repetitive electrical stimulation of the Schaffer collaterals in vitro was followed by an increase in the stimulus-evoked release of preloaded D-[^{3}H]Asp.
12. The changes were small and significant only when compared with sham-operated animals (values not significantly different from unoperated animals). Asp was not measured. In a similar experiment, Nadler et al. (1978) found no decrease in endogenous Asp or Glu contents. However, in contrast to Nitsch et al., these authors did not dissect the termination zones for separate analysis.
13. Terminal labelling in septum after D-[^{3}H]Asp injections in hippocampus is probably due to anterograde axonal transport subsequent to uptake by intrahippocampal axon collaterals.
14. The decrease in entorhinal Glu contents after bilateral hippocampal extirpation was small (15%).
15. As used here the term "hippocampal formation" includes the subiculum.
16. The nucleus of the diagonal band also showed a decrease in endogenous aspartate.
17. Injections of kainic acid into the nucleus accumbens led to a 45% decrease in [^{3}H]Glu uptake, suggesting the presence of excitatory amino acid interneurons.
18. The reason for the apparent discrepancy between the studies of Nadler and Smith, and Di Lauro et al., is not clear. Methodological differences may have played a role.
19. Long term potentiation of the perforant path led to an increased release of Glu synthesized from [^{3}H]glutamine.
20. The lesion also included the piriform cortex.

Table 3. Excitatory amino acid projections from allocortical areas[1] (except the hippocampal region)

From the olfactory bulb to the olfactory cortex[2]

C_{Glu}, C_{Asp}: Harvey et al. 1975 (guinea pig); Godfrey et al. 1980; Scholfield et al. 1983[3] (guinea pig); Sandberg et al. 1984 (guinea pig)

C_{Asp}: Collins 1979a,b[4]; Collins 1984[5]

C_{PAG}: Sandberg et al. 1984 (guinea pig)

R_{Asp}: Collins 1979b, 1980; Collins and Probett 1981, Collins et al. 1981 (rat and guinea pig)

R_{Glu}: Bradford and Richards 1976[6] (guinea pig)

R_L: Yamamoto and Matsui 1976 (guinea pig)

From the nucleus of the lateral olfactory tract to the olfactory bulb

T: Watanabe and Kawana 1984.

From layer III of the nucleus of the lateral olfactory tract to amygdala[7]

T: Fischer et al. 1982a; Ottersen et al. 1986

From the piriform cortex

to the olfactory tubercle: U_B: Fonnum et al. 1981b

to the olfactory bulb: T: Watanabe and Kawana 1984

to the thalamus: U_B, C_{Glu}, C_{Asp}: Walker and Fonnum 1983a

to the amygdala: U_B, C_{Glu}: Fonnum et al. 1981b; Walker and Fonnum 1983a[8]

T: Fischer et al. 1982a; Ottersen et al. 1986

1. Some of the studies referred to in Table 1 were based on lesions that included allocortical as well as neocortical areas.
2. Some pharmacological data are difficult to reconcile with a transmitter role of an excitatory amino acid in this pathway (review: Halász and Shepherd, 1983). It is also noteworthy that D-[³H]Asp uptake is low in the termination zone of the lateral olfactory tract fibers (Ottersen and Storm-Mathisen, 1984a).
3. There was also a decrease in the synthesis of Glu and Asp.
4. The decrease was most pronounced at the depth where the lateral olfactory tract fibers normally terminate.
5. The periamygdaloid cortex differed from the 4 other olfactory cortical regions tested by not showing a decrease in aspartate after bulbectomy. This finding points to a possible transmitter heterogeneity of lateral olfactory tract fibers.
6. The discrepant results of release experiments may be due to species differences or other factors (see Discussion in Collins 1985).
7. The projection was bilateral with an ipsilateral predominance.
8. The lesion also included the entorhinal cortex.

any rate, axotomy-induced losses of Glu/Asp uptake are difficult to explain in terms of effects on glial uptake. In most regions and experimental situations, the neuronal uptake pumps seem unable to distinguish between Glu and Asp. Exceptions have been reported; thus, rods in the goldfish retina (Marc and Lam, 1981), and rat cerebellar slices (Toggenburger et al., 1983) accumulate Glu in preference to Asp, and differential effects on Glu vs. Asp uptake have been demonstrated in response to interruption of the corticopontine pathway (Thangnipon et al., 1983) and after treatment with Li^+ ions (Peterson and Raghupathy, 1974). Differential retrograde transport of D,L-Glu and D-Asp was observed in the corticostriatal projection (Streit, 1980).

Decreases in uptake by as much as 80-90% have been reported in target areas of degenerating EAA pathways (Storm-Mathisen, 1977). The corresponding changes in Glu/Asp contents are usually far less pronounced (rarely exceeding 40%), substantiating the notion that uptake is the more sensitive marker (see above).

The nerve-terminal uptake pattern seen in autoradiograms of tissue incubated *in vitro* can be reproduced after infusion of the labeled transmitters *in vivo* at μM concentrations (Storm-Mathisen and Wold, 1981). If the animal is allowed to survive for a sufficient period of time after the injection, the tracer will be retrogradely transported in the axons from the terminals to the parent cell bodies, where it accumulates and can be detected by autoradiography (Streit, 1980; Cuénod et al., 1981, 1982). The notion is that the selectivity of the perikaryal labeling depends on the selectivity of the nerve terminal uptake process. This is the method of choice for tracing pathways which are not massive enough to be easily demonstrable with lesion techniques, and is uniquely suited for determining the precise origin of an EAA connection (e.g., the laminar origin of the various corticofugal EAA projections).

The results obtained with the retrograde axonal transport technique are not entirely congruent with other experimental data. Thus, injections of D-[^{3}H]Asp in the cerebellar cortex failed to label granular cells (except those likely to be damaged during injection; Cuénod et al., 1982; Wiklund et al., 1984). Further, whereas interruption of the corticofugal fibers to the cat superior colliculus led to a decrease in collicular Glu/Asp uptake (Fosse et al., 1984), these fibers failed to sustain retrograde transport of D-[^{3}H]Asp (Baughman and Gilbert, 1980, 1981). Possible explanations for these discrepancies include variable access of the injected label to the nerve terminal uptake sites, and different axonal transport capacities. The transport capacity may depend on intrinsic properties of the fibers as well as on their length and collateralization. The possibility of false positive results should be considered, particularly since the high concentrations of radiolabeled transmitter which are necessary to obtain retrograde axonal transport after *in vivo* injections might allow some uptake by low affinity mechanisms and also result in considerable spread. It is possible, in some fiber systems, to keep the concentration low. Labeling of inferior olivary neurons was thus obtained after prolonged superfusion of the cerebellar cortex with D-[^{3}H]Asp at concentrations favoring uptake by the high affinity system (Wiklund et al., 1982). Spread can be reduced by concentrating D-[^{3}H]Asp together with porous resin beads for subsequent implantation of the loaded particles in the brain, permitting sustained release of the tracer (Fischer et al., 1986).

Anterograde transport of D-[^{3}H]Asp has been reported in a number of fiber systems. Such transport may occur subsequent to uptake in collaterals terminating within the injection site (Storm-Mathisen, 1982; Wiklund et al., 1982, 1984). If the uptake mechanism is uniquely responsible for the anterograde transport, the selectivity should be as good as for the retrograde transport.

Table 4. Excitatory amino acid projections from subcortical structures

From the retina to the optic tectum

U_B: Henke et al. 1976[1,2] (pigeon); Bondy and Purdy 1977 (chick)

C_{Glu}, C_{Asp}: Fonnum and Henke 1982[1] (pigeon)

C_{Glu}: Yates and Roberts 1974 (frog)

R_{Asp}, R_{Glu}: Čanžek et al. 1981 (pigeon)

T: Beaudet et al. 1981[3,4] (pigeon)

From the anterior olfactory nucleus to the olfactory bulb

T: Watanabe and Kawana 1984[5]

From nucleus accumbens to substantia nigra

T: Streit 1980[6]

From caudatoputamen to substantia nigra

C_{Asp}: Taniyama et al. 1980[7] (cat); Korf and Venema 1983[8]

T: Streit 1980[9]

From "nonspecific" thalamic nuclei

to caudatoputamen: T: Streit 1980

to neocortex: T: Ottersen et al. 1983

to amygdala: T: Fischer et al. 1982a; Ottersen et al. 1986

From claustrum to neocortex

T: Fischer et al. 1982b

From the substantia nigra to the ventromedial thalamic nucleus

C_{Glu}: Fletcher et al. 1979

From mesencephalic and pontine periventricular grey, and parabrachial nuclei to the centromedian/parafascicular nuclei of the thalamus

T: Wiklund and Cuénod 1984

From the vestibular ganglion to the vestibular nuclei

U_B: Raymond et al. 1985 (cat)

T: Demêmes et al. 1984 (cat)

From the vestibular ganglion to the cerebellum

T: Cuénod et al 1982

From the spiral ganglion to the cochlear nucleus (auditory nerve fibers)

U_B: Potashner 1983 (guinea pig)

C_{Glu}, C_{Asp}: Wenthold and Gulley 1977 (guinea pig), 1978 (guinea pig); Wenthold 1978[10] (guinea pig)

C_{PAG}, C_{AAT}: Wenthold 1980 (guinea pig)

R_{Asp}, R_{Glu}: Wenthold 1979 (guinea pig)

R_L: Čanžek and Reubi 1980[11] (cat); Hansson et al. 1980[12] (guinea pig); Potashner 1983 (guinea pig)

T: Kane 1979[13] (cat); Oliver et al. 1983 (cat and guinea pig)

I_{AAT}: Altschuler et al. 1981 (guinea pig); Wenthold and Altschuler 1983 (guinea pig)

$I_{PAG}(N)$: Wenthold and Altschuler 1983 (guinea pig); Altschuler et al. 1984 (guinea pig)

Table 4 Part 2

<u>Granule cell/parallel fiber system in the cochlear nucleus</u>

    T: Oliver et al. 1983 (cat and guinea pig)

    $I_{PAG}(N)$, $I_{AAT}(N)$:   Wenthold and Altschuler 1983 (guinea pig)

<u>From the inferior olive to the cerebellum</u>
(climbing fiber system)

    C_{Asp}^{14}:   Nadi et al. 1977; McBride et al. 1978; Rea et al. 1980

    R_{Asp}^{14}:   Wiklund et al. 1982[15]; Toggenburger et al. 1983

    R_L: Toggenburger et al. 1983

    T: Wiklund et al. 1982, 1984; Künzle and Wiklund 1982 (turtle)

<u>Primary vagal afferent nerve fiber</u>

    U_B: Talman et al. 1980

    C_{Glu}, R_L: Perrone 1981; Reis et al. 1981

    R_L: Granata and Reis 1983

<u>Primary afferent dorsal root fibers</u>

    U_B: Roberts and Hill 1978[16]; Potashner and Tran 1985 (guinea pig)

    C_{Glu}, C_{Asp}: Johnston 1976[17]

    R_{Glu}, R_L: Roberts 1974[16]

    T: Cuénod et al. 1982; Hunt 1983

    $I_{PAG}(N)$:   Cangro et al. 1985[18]

<u>Spinal cord interneurons</u>

    C_{Asp}:   Davidoff et al. 1967[19] (cat); Homma et al. 1979[19] (cat)

    T: Rustioni and Cuénod 1982

---

1. Significant changes in these parameters were not obtained in a similar study in the same species but with shorter survival time (Beart, 1976).
2. Glu/Asp uptake in synaptosomes prepared from whole rat superior colliculi was not decreased after enucleation (Lund Karlsen and Fonnum 1978).
3. A subpopulation of retinotectal axons sustained retrograde as well as anterograde transport of D-[^{3}H]Asp after intratectal and intraocular injections, respectively. These results are consistent with those of Ehinger (1981), who reported that 5-10% of the cells of the retinal ganglion cell layer showed D-[^{3}H]Asp uptake, in pigeon, guinea-pig, and rabbit.
4. The anterogradely transported D-[^{3}H]Asp could be released Ca^{2+}-dependently in vitro by high K^+.
5. Bilateral projection.
6. D-[^{3}H]Asp injections into the substantia nigra also led to labelling of a few cells in other forebrain structures, viz., the bed nucleus of the stria terminalis, globus pallidus, and hypothalamus. Labelled cells were also found in the dorsal raphe nucleus; however, these were smaller than those labelled after [^{3}H]serotonin injections.
7. The decrease in Asp contents in the substantia nigra after cylindric lesions of the caudate nucleus could be due to concomitant interruption of corticonigral fibres.
8. This study was based on lesions with kainic acid.
9. Only scattered labelled cells were observed.
10. The decreases were most pronounced in the region where the primary auditory terminals are concentrated.
11. The tissue was preloaded with [^{14}C]Glu or [^{3}H]glutamine.
12. Acoustic stimulation in vivo increased [^{14}C]Glu or [^{3}H]Asp release from the cochlear nuclei. The released excitatory amino acids may have come from other fiber systems than the primary auditory terminals.
13. The metabolically labile L-[^{3}H]Asp and L-[^{3}H]Glu were used as tracers injected into the cochlea.
14. The climbing fibers were destroyed with 3-acetylpyridine.
15. A small decrease in Glu release was also found.
16. Changes were recorded in the dorsal column nuclei after lesion or stimulation of the dorsal column fibers.
17. The decrease was found in the dorsal column. Other groups have found no decrease in Asp or Glu after dorsal root transection in samples including the dorsal gray matter (Jones et al., 1974; Roberts and Keen, 1974).
18. About 30% of the small cells in the dorsal root ganglion were enriched in PAG.
19. Neuronal degeneration was induced by ischemia.

Since retrograde transport studies require relatively long survival times and do not allow analysis of the compound responsible for the radioactivity in the perikarya, it is essential to use the metabolically inert D-[^{3}H]Asp as a tracer rather than the more labile L-forms of Glu and Asp.

NERVE PATHWAYS

Tables 1-5 provide a survey of possible EAA pathways in the central nervous system and contain references grouped according to methodological approach. The fact that the different methods show different degrees of selectivity and sensitivity (see above) should be taken into account when critically judging the evidence favoring EAA neurotransmission in a particular pathway. The tables are self-explanatory and no attempts will be made here to review the extensive amount of data contained in them. Instead, we will briefly comment on some general principles that emerge from these data.

It can be safely concluded that EAA are likely to be important and perhaps the predominant transmitters in corticofugal projections (Table 1). Glu is more consistently and severely reduced than Asp after interruption of such projections irrespective of whether the main source of fibers is layer V (e.g., the corticostriatal pathway) or layer VI (e.g., some corticothalamic projections). So far, comparatively few EAA projections have been traced from subcortical structures. This may partly be due to the fact that such projections are generally less massive and less amenable to studies with lesion techniques than the cortico-fugal projections. It is to be expected that with extended use and refinements of the methods based on retrograde axonal transport of D-[^{3}H]Asp many more subcortical EAA pathways will be detected. Interestingly, 'non-specific' thalamic nuclei seem to project fibers with D-Asp transport capacity to all of its major targets, i.e., the cerebral cortex (Ottersen et al., 1983), caudatoputamen (Streit, 1980), and amygdala (Ottersen et al., 1986).

EAA have been suggested as transmitter candidates in a number of sensory pathways, including the lateral olfactory tract and the primary auditory and spinal afferent fibers. There is also evidence for EAA neurotransmission in the retinotectal axons of birds and of some other submammalian species (Table 4); in mammals, such fibers seem to constitute no more than a small proportion of the retinofugal projection. Of great clinical interest are the observations pointing to an EAA as the likely transmitter of the baroreceptor afferents.

The transmitter identity in central sensory pathways such as the medial lemniscus and the specific thalamocortical projections is still unresolved. This is remarkable when taking into account that these connections are among the most massive in the central nervous system. Other massive projections, including the mossy fiber system of the cerebellum[*], also lack obvious transmitter candidates. It is likely that further studies will show that Glu and Asp are only two of many small-molecular compounds that are released from nerve endings to act on EAA receptors (Luini et al., 1984; Bernstein et al., 1985; Cuénod et al., this volume; Coyle et al., this volume).

---

[*]Acetylcholine has been proposed as transmitter in a minor population of these fibers (Kan et al., 1978). Further, a small proportion of mossy fibers shows D-[^{3}H]Asp transport (Table 4).

Table 5: Efferent and intrinsic excitatory amino acid projections of the cerebellum

From the cerebellum to nucleus ruber

U_B: Nieoullon and Dusticier 1981a[1] (cat); Nieoullon et al. 1984[1] (cat)

From the cerebellum to the ventrolateral thalamic nucleus

U_B: Nieoullon et al. 1984 (cat)

Granule cell/parallel fiber system

U_B: Young et al. 1974 (hamster); Rohde et al. 1979

C_{Glu}: Young et al. 1974 (hamster); Valcana et al. 1972; Hudson et al. 1976 (mouse); McBride et al. 1976a (mouse), 1976b; Rea and McBride 1978; Roffler-Tarlov and Sidman 1978 (mouse); Rohde et al. 1979; Rea et al. 1981; Roffler-Tarlov and Turey 1982 (mouse)

R_{Glu}, R_L: Sandoval and Cotman 1978

[1] The decrease was restricted to the caudal part of the nucleus

ACKNOWLEDGEMENTS

Financial support is acknowledged from the Norwegian Research Council for Science and the Humanities, the Norwegian Council on Cardiovascular Disease, and the Norwegian Academy of Sciences.

REFERENCES

Abarca, J., and Bustos, G., 1985, Release of D-[^{3}H]aspartic acid from the rat substantia nigra: effect of veratridine-evoked depolarization and cortical ablation, Neurochem. Int., 7:229.

Adams, J.C., and Wenthold, R.J., 1979, Distribution of putative amino acid transmitters, choline acetyltransferase and glutamate decarboxylase in the inferior colliculus, Neuroscience, 4:1947.

Altschuler, R.A., Neises, G.R., Harmison, G.G., Wenthold, R.J., and Fex, J., 1981, Immunocytochemical localization of aspartate aminotransferase immunoreactivity in cochlear nucleus of the guinea pig, Proc. Natl. Acad. Sci. USA, 78:6553.

Altschuler, R.A., Mosinger, J.L., Harmison, G.G., Parakkal, M.H., and Wenthold, R.J., 1982, Aspartate aminotransferase-like immunoreactivity as a marker for aspartate/glutamate in guinea pig photoreceptors, Nature, 298:657.

Altschuler, R.A., Wenthold, R.J., Schwartz, A.M., Haser, W.G., Curthoys, N.P., Parakkal, M., and Fex, J., 1984, Immunocytochemical localization of glutaminase-like immunoreactivity in the auditory nerve, Brain Res., 291:173.

Altschuler, R.A., Monaghan, D.T., Haser, W.G., Wenthold, R.J., Curthoys, N.P., and Cotman, W., 1985, Immunocytochemical localization of glutaminase-like and aspartate aminotransferase-like immunoreactivities in the rat and guinea pig hippocampus, Brain Res., 330:225.

Baughman, R.W., and Gilbert, C.D., 1980, Aspartate and glutamate as possible neurotransmitters of cells in layer 6 of the visual cortex, Nature, 287:848.

Baughman, R.W., and Gilbert, C.D., 1981, Aspartate and glutamate as possible neurotransmitters in the visual cortex, J. Neurosci., 1:427.

Beart, P.M., 1976, An evaluation of L-glutamate as the transmitter released from optic nerve terminals of the pigeon, Brain Res., 110:99.

Beaudet, A., Burkhalter, A., Reubi, J.-C., and Cuénod, M., 1981, Selective bidirectional transport of [^{3}H]D-aspartate in the retino-tectal pathway, Neuroscience, 6:2021.

Beckstead, R.M., 1979, An autoradiographic examination of corticocortical and subcortical projections of the mediodorsal-projection (prefrontal) cortex in the rat, J. Comp. Neurol., 184:43.

Bernstein, J., Fisher, R.S., Zaczek, R., and Coyle, J., 1985, Dipeptides of glutamate and aspartate may be endogenous neuroexcitants in the rat hippocampal slice, J. Neurosci., 5:1429.

Bolz, J., Thier, P., and Brecha, N., 1985, Localization of aspartate aminotransferase and cytochrome oxidase in the cat retina, Neurosci. Lett., 53:315.

Bondy, S.C., and Purdy, J.L., 1977, Putative neurotransmitters of the avian visual pathways, Brain Res., 119:417.

Bradford, H.F., and Richards, C.D., 1976, Specific release of endogenous glutamate from piriform cortex stimulated in vitro, Brain Res., 105:168.

Bromberg, M.B., Penney, J.B., Jr., Stephenson, B.S., and Young, A.B., 1981, Evidence for glutamate as the neurotransmitter of cortico-thalamic and corticorubral pathways, Brain Res., 215:369.

Cangro, C.B., Sweetnam, P.M., Wrathall, J.R., Haser, W.B., Curthoys, N.P., and Neale, J.H., 1985, Localization of elevated glutaminase immunoreactivity in small DRG neurons, Brain Res., 336:158.

Čanžek, V., and Reubi, J.C., 1980, The effect of cochlear nerve lesion on the release of glutamate, aspartate and GABA from cat cochlear nucleus in vitro, Exp. Brain Res., 38:437.

Čanžek, V., Wolfensberger, M., Amsler, U., and Cuénod, M., 1981, In vivo release of glutamate and aspartate following optic nerve stimulation, Nature, 293:572.

Carter, C.J., 1982, Topographical distribution of possible glutamatergic pathways from the frontal cortex to the striatum and substantia nigra in rats, Neuropharmacology, 21:379.

Christie, M.J., Bridge, S., James, L.B., and Beart, P.M., 1985a, Excitotoxin lesions suggest an aspartatergic projection from rat medial prefrontal cortex to ventral tegmental area, Brain Res., 333:169.

Christie, M.J., James, L.B., and Beart, P.M., 1985b, An excitant amino acid projection from the medial prefrontal cortex to the anterior part of nucleus accumbens in the rat, J. Neurochem., 45:477.

Collins, G.G.S., 1979a, Effect of chronic bulbectomy on the depth distribution of amino acid transmitter candidates in rat olfactory cortex, Brain Res., 171:552.

Collins, G.G.S., 1979b, Evidence of a neurotransmitter role for aspartate and γ-aminobutyric acid in the rat olfactory cortex, J. Physiol., 291:51.

Collins, G.G.S., 1980, Release of endogenous amino acid neurotransmitter candidates from rat olfactory cortex slices: possible regulatory mechanisms and the effects of pentobarbitone, Brain Res., 190:517.

Collins, G.G.S., Anson, J., and Probett, G.A., 1981, Patterns of endogenous amino acid release from slices of rat and guinea-pig olfactory cortex, Brain Res., 204:103.

Collins, G.G.S., and Probett, G.A., 1981, Aspartate and not glutamate is the likely transmitter of the rat lateral olfactory tract fibres, Brain Res., 209:231.

Collins, G.G.S., 1984, Amino acid transmitter candidates in various regions of the primary olfactory cortex following bulbectomy, Brain Res., 296:145.

Collins, G.G.S., 1985, Excitatory amino acids as transmitters in the olfactory system, in: Excitatory Amino Acids, P.J. Roberts, J. Storm-Mathisen, and H.F. Bradford, Eds., Macmillan, London, in press.

Crawford, I.L., and Connor, J.D., 1973, Localisation and release of glutamic acid in relation to the hippocampal mossy fibre pathway, Nature, 244:442.

Cuénod, M., Beaudet, A., Čanžek, V., Streit, P., and Reubi, J.-C., 1981, Glutamatergic pathways in the pigeon and the rat brain, in: Glutamate as a Neurotransmitter, G. Di Chiara, and G.L. Gessa, eds., Raven Press, New York, p. 57.

Cuénod, M., Bagnoli, P., Beaudet, A., Rustioni, A., Wiklund, L., and Streit, P., 1982, Transmitter specific retrograde labeling of neurons, in:

Cytochemical Methods in Neuroanatomy, V. Chan-Palay and S.L. Palay, eds., Alan R. Liss, New York, p. 17.

Davidoff, R.A., Graham, L.T., Jr., Shank, R.P., Werman, R., and Aprison, M.H., 1967, Changes in amino acid concentrations associated with loss of spinal interneurons, J. Neurochem., 14:1025.

De Belleroche, J.S., and Bradford, H.F., 1977, On the site of origin of transmitter amino acids released by depolarization of nerve terminals in vitro, J. Neurochem., 29:335.

Demêmes, D., Raymond, J., and Sans, A., 1984, Selective retrograde labeling of neurons of cat vestibular ganglion with [^{3}H]D-aspartate, Brain Res., 304:188.

Di Lauro, A., Schmid, R.W., and Meek, J.L., 1981, Is aspartic acid the transmitter of the perforant pathway? Brain Res., 207:476.

Divac, I., Fonnum, F., and Storm-Mathisen, J., 1977, High affinity uptake of glutamate in terminals of corticostriatal axons, Nature, 266:377.

Dolphin, A.C., Errington, M.L., and Bliss, T.V.P., 1982, Long-term potentation of the perforant path in vivo is associated with increased glutamate release, Nature, 297:496.

Druce, D., Peterson, D., De Belleroche, J., and Bradford, H.F., 1982, Differential amino acid neurotransmitter release in rat neostriatum following lesioning of the cortico-striatal pathway, Brain Res., 247:303.

Ehinger, B., 1981, [^{3}H]-D-Aspartate accumulation in the retina of the pigeon, guinea-pig and rabbit, Exp. Eye Res., 33:381.

Engelsen, B., and Fonnum, F., 1983, Effects of hypoglycemia on the transmitter pool and the metabolic pool of glutamate in rat brain, Neurosci. Lett., 42:317.

Fagg, G.E., Jordan, C.C., and Webster, R.A., 1978, Descending fibre-mediated release of endogenous glutamate from the perfused cat spinal cord, in vivo, Brain Res., 158:159.

Fischer, B.O., Ottersen, O.P., and Storm-Mathisen, J., 1982a, Labelling of amygdalopetal and amygdalofugal projections after intra-amygdaloid injections of tritiated D-aspartate, Neuroscience, 7(Suppl.):S69.

Fischer, B.O., Ottersen, O.P., and Storm-Mathisen, J., 1982b, Axonal transport of D-[^{3}H]aspartate in the claustro-cortical projection, Neuroscience, 7(Suppl.):S69.

Fischer, B.O., Ottersen, O.P., and Storm-Mathisen, J., 1982c, Anterograde and retrograde axonal transport of D-[^{3}H]-aspartate (D-Asp) in hippocampal excitatory neurones, Neuroscience, 7(Suppl.):S68.

Fischer, B.O., Storm-Mathisen, J., and Ottersen, O.P., 1985, Hippocampal excitatory neurons. Anterograde and retrograde axonal transport of D-[^{3}H]aspartate, in: Excitatory Amino Acids, P.J. Roberts, J. Storm-Mathisen, and H.F. Bradford, eds., Macmillan, London, in press.

Fischer, B.O., Ottersen, O.P., and Storm-Mathisen, J., 1986, Implantation of D-[^{3}H]aspartate loaded gel particles permits restricted uptake sites for transmitter selective axonal transport, submitted.

Fletcher, A., James, T.A., Kilpatrick, I.C., MacLeod, N.K., and Starr, M.S., 1979, Neurochemical and electrophysiological evidence for GABAergic and glutamatergic nigro-thalamic neurones, Neurosci. Lett., Suppl.3:222.

Fonnum, F., and Walaas, I., 1978, The effect of intrahippocampal kainic acid injections and surgical lesions on neurotransmitters in hippocampus and septum, J. Neurochem., 31:1173.

Fonnum, F., Lund Karlsen, R., Malthe-Sørenssen, D., Skrede, K.K., and Walaas, I., 1979, Localization of neurotransmitters, particularly glutamate, in hippocampus, septum, nucleus accumbens and superior colliculus, Prog. Brain Res., 51:167.

Fonnum, F., Storm-Mathisen, J., and Divac, I., 1981a, Biochemical evidence for glutamate as neurotransmitter in the corticostriatal and corticothalamic fibres in rat brain, Neuroscience, 6:863.

Fonnum, F., Søreide, A. Kvale, I., Walker, J., and Walaas, I., 1981b, Glutamate in cortical fibers, in: Glutamate as a Neurotransmitter, G. Di Chiara, and G.L. Gessa, eds., Raven Press, New York, p. 29.

Fonnum, F., and Henke, H., 1982, The topographical distribution of alanine, aspartate, γ-aminobutyric acid, glutamate, glutamine, and glycine in the pigeon optic tectum and the effect of retinal ablation, J. Neurochem., 38:1130.

Fonnum, F., 1984, Glutamate: a neurotranmitter in mammalian brain, J. Neurochem., 42:1.

Fonnum, F., Fosse, V.M., and Allen, C.N., 1984, Identification of excitatory amino acid pathways in the mammalian nervous system, in: Excitotoxins, K. Fuxe, P. Roberts, and R. Schwarcz, eds., Plenum Press, New York, p. 3.

Fosse, V.M., Heggelund, P., Iversen, E., and Fonnum, F., 1984, Effects of area 17 ablation on neurotransmitter parameters in efferents to area 18, the lateral geniculate body, pulvinar and superior colliculus in the cat, Neurosci. Lett., 52:323.

Freeman, M.E., Lane, J.D., and Smith, J.E., 1983, Turnover rates of amino acid neurotransmitters in regions of rat cerebellum, J. Neurochem., 40:1441.

Godfrey, D.A., Ross, C.D., Carter, J.A., Lowry, O.H., and Matschinsky, F.M., 1980, Effect of intervening lesions on amino acid distributions in rat olfactory cortex and olfactory bulb, J. Histochem. Cytochem., 28:1157.

Godukhin, O.V., Zharikova, A.D., and Novoselov, V.I., 1980, The release of labeled L-glutamic acid from rat neostriatum in vivo following stimulation of frontal cortex, Neuroscience, 5:2151.

Granata, A.R., and Reis, D.J., 1983, Release of [^{3}H]L-glutamine acid (L-Glu) and [^{3}H]D-aspartic acid (D-Asp) in the area of nucleus tractus solitarius in vivo produced by stimulation of the vagus nerve, Brain Res., 259:77.

Halász, N., and Shepherd, G.M., 1983, Neurochemistry of the vertebrate olfactory bulb, Neuroscience, 10:579.

Hamberger, A., Chiang, G., Nylén, E.S., Scheff, S.W., and Cotman, C.W., 1978, Stimulus evoked increase in the biosynthesis of the putative neurotransmitter glutamate in the hippocampus, Brain Res., 143:549.

Hamberger, A.C., Chiang, G.H., Nylén, E.S., Scheff, S.W., and Cotman, C.W., 1979, Glutamate as a CNS transmitter. I. Evaluation of glucose and glutamine as precursors for the synthesis of preferentially released glutamate, Brain Res., 168:513.

Hansson, E., Jarlstedt, J., and Sellström, A., 1980, Sound-stimulated ^{14}C-glutamate release from the nucleus cochlearis, Experientia, 36:576.

Harvey, J.A., Scholfield, C.N., Graham, L.T., Jr., and Aprison, M.H., 1975, Putative transmitters in denervated olfactory cortex, J. Neurochem., 24:445.

Hassler, R., Haug, P., Nitsch, C., Kim, J.S., and Paik, K., 1982, Effect of motor and premotor cortex ablation on concentrations of amino acids, monoamines, and acetylcholine and on the ultrastructure in rat striatum. A confirmation of glutamate as the specific cortico-striatal transmitter, J. Neurochem., 38:1087.

Heggli, D.E., Aamodt, A., and Malthe-Sørenssen, D., 1981, Kainic acid neurotoxicity: effect of systemic injection on neurotransmitter markers in different brain regions, Brain Res., 230:253.

Henke, H., Schenker, T.M., and Cuénod, M., 1976, Effects of retinal ablation on uptake of glutamate, glycine, GABA, proline, and choline in pigeon tectum, J. Neurochem., 26:131.

Hertz, L., Kvamme, E., McGeer, E.G., and Schousboe, A., 1983, Glutamine, Glutamate and GABA in the Central Nervous System, Alan R. Liss, New York.

Hicks, T.P., Ruwe, W.D., Veale, W.L., and Veenhuizen, J., 1985, Aspartate and glutamate as synaptic transmitters of parallel visual cortical pathways, Exp. Brain Res., 58:421.

Homma, S., Suzuki, T., Murayama, S., and Otsuka, M., 1979, Amino acid and substance P contents in spinal cord of cats with experimental hindlimb rigidity produced by occlusion of spinal cord blood supply, J. Neurochem., 32:691.

Hudson, D.B., Valcana, T., Bean, G., and Timiras, P.S., 1976, Glutamic acid: a strong candidate as the neurotransmitter of the cerebellar granule cell, Neurochem. Res., 1:73.

Hunt, S.P., 1983, Cytochemistry of the spinal cord, in: Chemical Neuroanatomy, P.C. Emson, ed., Raven Press, New York, p. 53.

Johnston, G.A.R., 1976, Glutamate and aspartate as transmitters in the spinal cord, Adv. Biochem. Psychopharmacol., 15:175.

Jones, I.M., Jordan, C.C., Morton, I.K.M., Stagg, C.J., and Webster, R.A., 1974, The effect of chronic dorsal root section on the concentration of free amino acids in the rabbit spinal cord, J. Neurochem., 23:1239.

Kan, K.-S.K., Chao, L.-P., and Eng, L.F., 1978, Immunohistochemical localization of choline acetyltransferase in rabbit spinal cord and cerebellum, Brain Res., 146:221.

Kane, E.S., 1979, Central transport and distribution of labelled glutamic and aspartic acids to the cochlear nucleus in cats: an autoradiographic study, Neuroscience, 4:729.

Kerkerian, L., Nieoullon, A., and Dusticier, N., 1983, Topographic changes in high-affinity glutamate uptake in the cat red nucleus, substantia nigra, thalamus and caudate nucleus after lesions of sensorimotor cortical areas, Exp. Neurol., 81:598.

Kim, J.S., Hassler, R., Haug, P., and Paik, K., 1977, Effect of frontal cortex ablation on striatal glutamic acid level in rat, Brain Res., 132:370.

Korf, J., and Venema, K., 1983, Amino acids in the substantia nigra of rats with striatal lesions produced by kainic acid, J. Neurochem., 40:1171.

Kornhuber, J., Kim, J.S., Kornhuber, M.E., and Kornhuber, H.H., 1984, The cortico-nigral projection: reduced glutamate content in the substantia nigra following frontal cortex ablation in the rat, Brain Res., 322:124.

Künzle, H., and Wiklund, L., 1982, Identification and distribution of neurons presumed to give rise to cerebellar climbing fibers in turtle. A retrograde axonal flow study using radioactive D-aspartate as a marker, Brain Res., 252:146.

Kvale, I., and Fonnum, F., 1983, The effects of unilateral removal of visual cortex on transmitter parameters in the adult superior colliculus and lateral geniculate body, Develop. Brain Res., 11:261.

Luini, A., Tal, N., Goldberg, O., and Teichberg, V.I., 1984, An evaluation of selected brain constituents as putative excitatory neurotransmitters, Brain Res., 324:271.

Lund Karlsen, R., and Fonnum, F., 1978, Evidence for glutamate as a neurotransmitter in the corticofugal fibres to the dorsal lateral geniculate body and the superior colliculus in rats, Brain Res., 151:457.

Malthe-Sørenssen, D., Skrede, K.K., and Fonnum, F., 1979, Calcium-dependent release of D-[^{3}H]aspartate evoked by selective electrical stimulation of excitatory afferent fibers to hippocampal pyramidal cells in vitro, Neuroscience, 4:1255.

Malthe-Sørenssen, D., Skrede, K.K., and Fonnum, F., 1980, Release of D-[^{3}H]aspartate from the dorsolateral septum after electrical stimulation of the fimbria in vitro, Neuroscience, 5:127.

Marc, R.E., and Lam, D.M.K., 1981, Uptake of aspartic and glutamic acid by photoreceptors in goldfish retina, Proc. Natl. Acad. Sci. USA, 78:7185.

Matute, C., Waldvogel, H.J., Streit, P., and Cuénod, M., 1984, Selective retrograde labeling following D-[^{3}H]aspartate and [^{3}H]GABA injections in the albino rat superior colliculus, Neurosci. Lett., Suppl. 18:S190.

McGeer, E.G., and McGeer, P.L., 1979, Localization of glutaminase in the rat neostriatum, J. Neurochem., 32:1071.

McGeer, P.L., McGeer, E.G., Scherer, U., and Singh, K., 1977, A glutamatergic corticostriatal path?, Brain Res., 128:369.

Minchin, M.C.W., and Fonnum, F., 1979, The metabolism of GABA and other amino acids in rat substantia nigra slices following lesions of the striatonigral pathway, J. Neurochem., 32:203.

Nadi, N.S., Kanter, D., McBride, W.J., and Aprison, M.H., 1977, Effects of 3-acetylpyridine on several putative neurotransmitter amino acids in the cerebellum and medulla of the rat, J. Neurochem., 28:661.

Nadler, J.V., Vaca, K.W., White, W.F., Lynch, G.S., and Cotman, C.W., 1976, Aspartate and glutamate as possible transmitters of excitatory hippocampal afferents, Nature, 260:538.

Nadler, J.V., White, W.F., Vaca, K.W., Perry, B.W., and Cotman, C.W., 1978, Biochemical correlates of transmission mediated by glutamate and aspartate, J. Neurochem., 31:147.

Nadler, J.V., and Smith, E.M., 1981, Perforant path lesion depletes glutamate content of fascia dentata synaptosomes, Neurosci. Lett., 25:275.

Naito, S., and Ueda, T., 1983, Adenosine triphosphate-dependent uptake of glutamate into protein I-associated synaptic vesicles, J. Biol. Chem., 258:696.

Naito, S., and Ueda, T., 1985, Characterization of glutamate uptake into synaptic vesicles, J. Neurochem., 44:99.

Nieoullon, A., and Dusticier, N., 1981, Decrease in choline acetyltransferase and in high affinity glutamate uptake in the red nucleus of the cat after cerebellar lesions, Neurosci. Lett., 24:267.

Nieoullon, A., Kerkerian, L., and Dusticier, N., 1984, High affinity glutamate uptake in the red nucleus and ventrolateral thalamus after lesion of the cerebellum in the adult cat: biochemical evidence for functional changes in the deafferented structures, Exp. Brain Res., 55:409.

Nitsch, C., Kim, J.-K., Shimada, C., and Okada, Y., 1979a, Effect of hippocampus extirpation in the rat on glutamate levels in target structures of hippocampal efferents, Neurosci. Lett., 11:295.

Nitsch, C., Kim, J.-K., and Shimada, C., 1979b, The commissural fibers in rabbit hippocampus: synapses and their transmitter, Progr. Brain Res., 51:193.

Oliver, D.L., Potashner, S.J., Jones, D.R., and Morest, D.K., 1983, Selective labeling of spiral ganglion and granule cells with D-aspartate in the auditory system of cat and guinea pig, J. Neurosci., 3:455.

Ottersen, O.P., 1982, Connections of the amygdala of the rat. IV: Corticoamygdaloid and intraamygdaloid connections as studied with axonal transport of horseradish peroxidase, J. Comp. Neurol., 205:30.

Ottersen, O.P., Fischer, B.O., and Storm-Mathisen, J., 1983, Retrograde transport of D-[^{3}H]aspartate in thalamocortical neurones, Neurosci. Lett., 42:19.

Ottersen, O.P., and Storm-Mathisen, J., 1984a, Neurons containing or accumulating transmitter amino acids, in: Handbook of Chemical Neuroanatomy, A. Björklund, T. Hökfelt, and M.J. Kuhar, eds., Elsevier/North-Holland, Amsterdam, p. 141.

Ottersen, O.P., and Storm-Mathisen, J., 1984b, Glutamate- and GABA-containing neurons in the mouse and rat brain, as demonstrated with a new immunocytochemical technique, J. Comp. Neurol., 229:374.

Ottersen, O.P., and Storm-Mathisen, J., 1984c, Neurotransmitters in the hippocampal formation and related structures, Neurosci. Lett., Suppl. 18:S147

Ottersen, O.P., Fischer, B.O., Rinvik, E., and Storm-Mathisen, J., 1986, Putative amino acid transmitters in the amygdala, in: Excitatory Amino Acids and Epilepsy, R. Schwarcz and Y. Ben-Ari, eds., Plenum Press, London, in press.

Ottersen, O.P., and Storm-Mathisen, J., 1985, Different neuronal localization of aspartate-like and glutamate-like immunoreactivities in the hippocampus of cat, guinea pig, and Senegalese baboon (Papio papio), with a note on the distribution of GABA, Neuroscience, 16:589.

Patel, A.J., and Hunt, A., 1985, Concentration of free amino acids in primary cultures of neurones and astrocytes, J. Neurochem., 44:1816.

Perrone, M.H., 1981, Biochemical evidence that L-glutamate is a neurotransmitter of primary vagal afferent nerve fibers, Brain Res., 230:283.

Peterson, N.A., and Raghupathy, E., 1974, Selective effects of lithium on synaptosomal amino acid transport systems, Biochem. Pharmacol., 23:2491.

Potashner, S.J., 1983, Uptake and release of D-aspartate in the guinea pig cochlear nucleus, J. Neurochem., 41:1094.

Potashner, S.J., and Tran, P.L., 1985, Decreased uptake and release of D-aspartate in the guinea pig spinal cord after partial cordotomy, J. Neurochem., 44:1511.

Raymond, J., Nieoullon, A., Demêmes, D., and Sans, A., 1984, Evidence for glutamate as a neurotransmitter in the cat vestibular nerve: radioautographic and biochemical studies, Exp. Brain Res., 56:523.

Rea, M.A., and McBride, W.J., 1978, Effects of X-irradiation on the levels of glutamate, aspartate and GABA in different regions of the cerebellum of the rat, Life Sci., 23:2355.

Rea, M.A., McBride, W.J., and Rohde, B.H., 1980, Regional and synaptosomal levels of amino acid neurotransmitters in the 3-acetylpyridine deafferentated rat cerebellum, J. Neurochem., 34:1106.

Rea, M.A., McBride, W.J., and Rohde, B.H., 1981, Levels of glutamate, aspartate, GABA, and taurine in different regions of the cerebellum after X-irradiation-induced neuronal loss, Neurochem. Res., 6:33.

Recasens, M., Benzra, R., Basset, P., and Mandel, P., 1980, Cysteine sulfinate aminotransferase and aspartate aminotransferase isoenzymes of rat brain. Purification, characterization, and further evidence for identity, Biochemistry, 19:4583.

Reis, D.J., Granata, A.R., Perrone, M.H., and Talman, W.T., 1981, Evidence that glutamic acid is the neurotransmitter of baroreceptor afferents terminating in the nucleus tractus solitarius (NTS), J. Auton. Nerv. Syst., 3:321.

Reubi, J.C., and Cuénod, M., 1979, Glutamate release in vitro from corticostriatal terminals, Brain Res., 176:185.

Reubi, J.C., Toggenburger, C., and Cuénod, M., 1980, Asparagine as a precursor for transmitter aspartate in corticostriatal fibres, J. Neurochem., 35:1015.

Roberts, F., and Hill, R.G., 1978, The effect of dorsal column lesions on amino acid levels and glutamate uptake in rat dorsal column nuclei, J. Neurochem., 31:1549.

Roberts, P.J., 1974, The release of amino acids with proposed neurotransmitter function from the cuneate and gracile nuclei of the rat in vivo, Brain Res., 67:419.

Roberts, P.J., and Keen, P., 1974, Effect of dorsal root section on amino acids of rat spinal cord, Brain Res., 74:333.

Roberts, P.J., McBean, G.J., Sharif, N.A., and Thomas, E.M., 1982, Striatal glutamatergic function: modifications following specific lesions, Brain Res., 235:83.

Roffler-Tarlov, S., and Sidman, R.L., 1978, Concentrations of glutamic acid in cerebellar cortex and deep nuclei of normal mice and weaver, staggerer and nervous mutants, Brain Res., 142:269.

Roffler-Tarlov, S., and Turey, M., 1982, The content of amino acids in the developing cerebellar cortex and deep cerebellar nuclei of granule cell deficient mutant mice, Brain Res., 247:65.

Rohde, B.H., Rea, M.A., Simon, J.R., and McBride, W.J., 1979, Effects of X-irradiation induced loss of cerebellar granule cells on synaptosomal levels and the high affinity uptake of amino acids, J. Neurochem., 32:1431.

Ross, C.D., and Godfrey, D.A., 1985, Distributions of aspartate aminotransferase and malate dehydrogenase activities in rat retinal layers, J. Histochem. Cytochem., 33:624.

Rowlands, G.J., and Roberts, P.J., 1980, Specific calcium-dependent release of endogenous glutamate from rat striatum is reduced by destruction of the cortico-striatal tract, Exp. Brain Res., 39:239.

Rustioni, A., and Cuénod, M., 1982, Selective retrograde transport of D-aspartate in spinal interneurons and cortical neurons of rats, Brain Res., 236:143.

Sandberg, M., Bradford, H.F., and Richards, C.D., 1984, Effect of lesions of the olfactory bulb on the levels of amino acids and related enzymes in the olfactory cortex of the guinea pig, J. Neurochem., 43:276.

Sandberg, M., Ward, H.K., and Bradford, H.F., 1985, Effect of corticostriate pathway lesion on the activities of enzymes involved in synthesis and metabolism of amino acid neurotransmitters in the striatum, J. Neurochem., 44:42.

Sandoval, M.E., and Cotman, C.W., 1978, Evaluation of glutamate as a neurotransmitter of cerebellar parallel fibers, Neuroscience, 3:199.

Scholfield, C.N., Moroni, F., Corradetti, R., and Pepeu, G., 1983, Levels and synthesis of glutamate and aspartate in the olfactory cortex following bulbectomy, J. Neurochem., 41:135.

Skrede, K.K., and Malthe-Sørenssen, D., 1981a, Increased resting and evoked release of transmitter following repetitive electrical tetanization in hippocampus: a biochemical correlate to longlasting synaptic potentiation, Brain Res., 208:436.

Skrede, K.K., and Malthe-Sørenssen, D., 1981b, Differential release of D-[^{3}H]aspartate and [^{14}C]γ-aminobutyric acid following activation of commissural fibres in a longitudinal slice preparation of guinea pig hippocampus, Neurosci. Lett., 21:71.

Spencer, H.J., Tominez, G., and Halpern, B., 1981, Mass spectographic analysis of stimulated release of endogenous amino acids from rat hippocampal slices, Brain Res., 212:194.

Sterri, S.H., and Fonnum, F., 1980, Acetyl-CoA synthesizing enzymes in cholinergic nerve terminals, J. Neurochem., 35:249.

Storm-Mathisen, J., 1977, Glutamic acid and excitatory nerve endings: reduction of glutamic acid uptake after axotomy, Brain Res., 120:379.

Storm-Mathisen, J., 1978, Localization of putative transmitters in the hippocampal formation with a note on the connection to septum and hypothalamus, in: Functions of the Septohippocampal System, Ciba Foundation Symposium, 58 (New Series), Elsevier/Excerpta Medica/North Holland, Amsterdam, p. 49

Storm-Mathisen, J., and Woxen-Opsahl, M., 1978, Aspartate and/or glutamate may be transmitters in hippocampal efferents to septum and hypothalamus, Neurosci. Lett., 9:65.

Storm-Mathisen, J., and Iversen, L.L., 1979, Uptake of [^{3}H]glutamic acid in excitatory nerve endings: light and electronmicroscopic observations in the hippocampal formation of the rat, Neuroscience, 4:1237.

Storm-Mathisen, J., 1981, Autoradiographic and microchemical localization of high affinity glutamate uptake, in: Glutamate: Transmitter in the Central Nervous System, P.J. Roberts, J. Storm-Mathisen, and G.A.R. Johnston, eds., John Wiley and Sons, Chichester, p. 89.

Storm-Mathisen, J., and Wold, J.E., 1981, In vivo high-affinity uptake and axonal transport of D-[2,3-^{3}H]aspartate in excitatory neurons, Brain Res., 230:427.

Storm-Mathisen, J., 1982, Amino acid compartments in hippocampus: an autoradiographic approach, in: Neurotransmitter Interaction and Compartmentation, H.F. Bradford, ed., Plenum Press, New York, p. 395.

Storm-Mathisen, J., Leknes, A.K., Bore, A.T., Vaaland, J.L., Edminson, P., Haug, F.-M.Š., and Ottersen, O.P., 1983, First visualization of glutamate and GABA in neurones by immunocytochemistry, Nature, 301:517.

Storm-Mathisen, J., and Ottersen, O.P., 1985, Antibodies against amino acid transmitters, in: Neurohistochemistry Today, P. Panula, H. Päivärinta, and S. Soinila, eds., Alan R. Liss, New York, in press.

Streit, P., 1980, Selective retrograde labeling indicating the transmitter of neuronal pathways, J. Comp. Neurol., 191:429.

Svenneby, G., and Storm-Mathisen, J., 1983, Immunological studies on phosphate activated glutaminase, in: Glutamine, Glutamate, and GABA in the Central Nervous System, L. Hertz, E. Kvamme, E.G., McGeer, and A. Schousboe, eds., Alan R. Liss, New York, p. 69.

Talman, W.T., Perrone, M.H., and Reis, D.J., 1980, Evidence for L-glutamate as the neurotransmitter of baroreceptor afferent nerve fibers, Science, 209:813.

Taniyama, K., Nitsch, C., Wagner, A., and Hassler, R., 1980, Aspartate, glutamate and GABA levels in pallidum, substantia nigra, center median and dorsal raphe nuclei after cylindric lesion of caudate nucleus in cat, Neurosci. Lett., 16:155.

Taxt, T., and Storm-Mathisen, J., 1984, Uptake of D-aspartate and L-glutamate in excitatory axon terminals in hippocampus: autoradiographic and biochemical comparison with γ-aminobutyrate and other amino acids in normal rats and in rats with lesions, Neuroscience, 11:79.

Thangnipon, W., and Storm-Mathisen, J., 1981, K^+-evoked Ca^{2+}-dependent release of D-[^{3}H]aspartate from terminals of the cortico-pontine pathway, Neurosci. Lett., 23:181.

Thangnipon, W., Taxt, T., Brodal, P., and Storm-Mathisen, J., 1983, The cortico-pontine projection: axotomy-induced loss of high affinity L-glutamate and D-aspartate uptake, but not of GABA uptake, glutamate decarboxylase or choline acetyltransferase, in the pontine nuclei, Neuroscience, 8:449.

Toggenburger, G., Wiklund, L., Henke, H., and Cuénod, M., 1983, Release of endogenous and accumulated exogenous amino acids from slices of normal and climbing fibre-deprived rat cerebellar slices, J. Neurochem., 41:1606.

Valcana, T., Hudson, D., and Timiras, P.S., 1972, Effects of X-irradiation on the content of amino acids in the developing rat cerebellum, J. Neurochem., 19:2229.

Walaas, I., and Fonnum, F., 1979, The effect of surgical and chemical lesions on neurotransmitter candidates in the nucleus accumbens of the rat, Neuroscience, 4:209.

Walaas, I., and Fonnum, F., 1980, Biochemical evidence for glutamate as a transmitter in hippocampal efferents to the basal forebrain and hypothalamus in the rat brain, Neuroscience, 5:1691.

Walaas, I., 1981, Biochemical evidence for overlapping neocortical and allocortical glutamate projections to the nucleus accumbens and rostral caudatoputamen in the rat brain, Neuroscience, 6:399.

Walker, J.E., and Fonnum, F., 1983a, Regional cortical glutamergic and aspartergic projections to the amygdala and thalamus of the rat, Brain Res., 267:371.

Walker, J.E., and Fonnum, F., 1983b, Effect of regional cortical ablations on high-affinity D-aspartate uptake in striatum, olfactory tubercle and pyriform cortex of the rat, Brain Res., 278:283.

Ward, H.K., Thanki, C.M., Peterson, D.W., and Bradford, H.F., 1982, Brain glutaminase activity in relation to transmitter glutamate biosynthesis, Biochem. Soc. Trans., 10:369.

Watanabe, K., and Kawana, E., 1984, Selective retrograde transport of tritiated D-aspartate from the olfactory bulb to the anterior olfactory nucleus, pyriform cortex and nucleus of the lateral olfactory tract in the rat, Brain Res., 296:148.

Wenthold, R.J., and Gulley, R.L., 1977, Aspartic acid and glutamic acid levels in the cochlear nucleus after auditory nerve lesion, Brain Res., 138:111.

Wenthold, R.J., 1978, Glutamic acid and aspartic acid in subdivisions of the cochlear nucleus after auditory nerve lesion, Brain Res., 143:544.

Wenthold, R.J., and Gulley, R.L., 1978, Glutamic and aspartic acid in the cochlear nucleus of the waltzing guinea pig, Brain Res., 158:295.

Wenthold, R.J., 1979, Release of endogenous glutamic acid, aspartic acid and GABA from cochlear nucleus slices, Brain Res., 162:338.

Wenthold, R.J., 1980, Glutaminase and aspartate aminotransferase decrease in the cochlear nucleus after lesion of the auditory nerve, Brain Res., 190:293.

Wenthold, R.J., and Altschuler, R.A., 1983, Immunocytochemistry of aspartate aminotransferase and glutaminase, in: Glutamine, Glutamate, and GABA in the Central Nervous System, L. Hertz, E. Kvamme, E.G. McGeer and A. Schousboe, eds., Alan R. Liss, New York, p. 33.

Wenthold, R.J., and Altschuler, R.A., 1985, Immunocytochemical localization of enzymes involved in the metabolism of excitatory amino acids, in: Excitatory Amino Acids, P.J. Roberts, J. Storm-Mathisen, and H.F. Bradford, eds., Macmillan, London, in press.

Wiklund, L., Toggenburger, G., and Cuénod, M., 1982, Aspartate: possible neurotransmitter in cerebellar climbing fibers, Science, 216:78.

Wiklund, L., and Cuénod, M., 1984, Differential labelling of afferents to thalamic centromedian-parafascicular nuclei with [^{3}H]choline and D-[^{3}H]aspartate: further evidence for transmitter specific retrograde labelling, Neurosci. Lett., 46:275.

Wiklund, L., Toggenburger, G., and Cuénod, M., 1984, Selective retrograde labelling of the rat olivocerebellar climbing fiber system with D-[^{3}H]aspartate, Neuroscience, 13:441.

Wilkin, G.P., Garthwaite, J., and Balázs, R., 1982, Putative acidic amino acid transmitters in the cerebellum. II. Electron microscopic localization of transport sites, Brain Res., 244:69.

Wroblewski, J.T., Blaker, W.D., and Meek, J.L., 1985, Ornithine as a precursor of neurotransmitter glutamate: effect of canaline on ornithine aminotransferase activity and glutamate content in the septum of rat brain, Brain Res., 329:161.

Yamamoto, C., and Matsui, S., 1976, Effect of stimulation of excitatory nerve tract on release of glutamic acid from olfactory cortex slices in vitro, J. Neurochem., 26:487.

Yates, R.A., and Roberts, P.J., 1974, Effects of enucleation and intraocular colchicine on the amino acids of frog optic tectum, J. Neurochem., 23:891.

Young, A.B., Oster-Granite, M.L., Herndon, R.M., and Snyder, S.H., 1974, Glutamic acid: selective depletion by viral induced granule cell loss in hamster cerebellum, Brain Res., 73:1.

Young, A.B., Bromberg, M.B., and Penney, J.B., Jr., 1981, Decreased glutamte uptake in subcortical areas deafferented by sensorimotor cortical ablation in the cat, J. Neurosci., 1:241.

Zaczek, R., Hedreen, J.C., and Coyle, J.T., 1979, Evidence for a hippocampal-septal glutamatergic pathway in the rat, Exp. Neurol., 65:145.

SYNTHESIS AND RELEASE OF AMINO ACID TRANSMITTERS

F. Fonnum, R.H. Paulsen, V.M. Fosse, and B. Engelsen

Norwegian Defense Research Establishment, Division for
Environmental Toxicology, P.O. Box 25, N-2700-Kjeller

INTRODUCTION

GABA, glutamate and aspartate are considered as strong neurotransmitter candidates in mammalian brain. The metabolism of the three amino acids are coupled in a complex pattern (Fonnum, 1985). Glutamate is ubiquitously distributed in brain tissue. It occupies an important position as excitatory neurotransmitter, in brain metabolism and as precursor for the inhibitory neurotransmitter GABA. This implies that glutamate is compartmentalized in brain and present in a transmitter pool, in a neuronal metabolic pool, in a glial pool and in a GABAergic terminal pool. Aspartate is closely related metabolically and chemically to glutamate. Due to the low level of oxaloacetate in brain (van den Berg and Garfinkel, 1971), the equilibrium between glutamate, aspartate, oxaloacetate and 2-oxoglutarate may change with the metabolic status of the animal.

Aspartate and glutamate also play an important role as excitotoxic amino acids. Detailed morphological studies (Olney, 1969; Olney and Price, 1981) and biochemical studies have described the toxic effect of glutamate in the retina (Lund Karlsen and Fonnum, 1976) and hypothalamus (Walaas and Fonnum, 1978).

Recent investigations have shown considerably elevated release of the two endogenous excitotoxic amino acids glutamate and aspartate during experimental ischemia and hypoglycemia in rats (Benveniste et al., 1984; Wieloch et al., 1985) and even in human ventricular CSF during impaired brain circulation (Table 1; Engelsen et al., 1985). There are reasons to believe that the enhanced release of these excitotoxic amino acids may be responsible for the pathological changes in ischemia and hypoglycemia.

In the present investigation, we have therefore tried to define the sizes of different glutamate pools in brain by surgical and chemical manipulations. We have also tried pharmacologically to interfere with their metabolism to better understand their interrelationship.

The Experimental Model

The neurotransmitter contents of the different anatomical pathways in the basal ganglia (neostriatum, globus pallidus and substantia nigra) are well known (Fonnum and Walaas, 1979). Neostriatum receives a heavy

Table 1. Amino Acid Concentrations in Human Ventricular Cerebrospinal Fluid

Amino acids	Group A Mean ± SD	Group B BI	Group B RM
Glutamate	2.9 ± 2.2	38	22
Aspartate	0.2 ± 0.2	2.2	0.6
Glutamine	716 ± 243	643	599
GABA	0.4 ± 0.2	0.6	trace
Taurine	7.1 ± 3.0	14	6
Alanine	20 ± 10	27	27

Results are presented as: µM (mean ± SD).
Group A: 10 hydrocephalic control patients.
Group B: 2 patients (BI and RM) with signs of reduced cerebral perfusion (ischemia) due to increased CSF pressure.
From Engelsen et al. (1985).

glutamergic input from the cerebral cortex, particularly the frontal cortex (Fonnum et al., 1981). The corresponding inputs to globus pallidus and substantia nigra are very small if not absent (Fonnum et al., 1981). In neostriatum, GABA is present in local inhibitory neurons which project heavily to globus pallidus and substantia nigra (Fonnum et al., 1978).

The distribution of amino acid transmitters in the basal ganglia makes it possible to produce several interesting models to focus in more detail on the composition and turnover of the glutamergic and GABAergic terminals. Unilateral decortication allows us to compare neostriatum in the presence and absence of glutamergic terminals. Likewise, a unilateral lesion in neostriatum will destroy the GABAergic axons and terminals ipsilaterally in globus pallidus and substantia nigra. Kainic acid lesion of neostriatum destroys the many local neurons and leaves a region enriched in glial structures, a relatively increased proportion of glutamergic terminals and deprived of the many non-glutamergic neuronal structures. A hemitransection of the medial forebrain bundle will remove the monoaminergic projection to neostriatum and globus pallidus.

To these models we have used several different pharmacological tools which are expected to interfere with the amino acid transmitters. One of the most interesting conditions was obtained by administering rapidly acting insulin to study the effect of hypoglycemia on the amino acid content in the intact and decorticated neostriatum. In this way we have studied both the importance of glucose as a precursor for the transmitter amino acids and the turnover of glutamate in the transmitter and the metabolic pools in brain (Engelsen and Fonnum, 1983).

The hypoglycemic conditions were stimulated in neostriatal slices in vitro. The slices were preincubated with D-[^{3}H]aspartate and incubated (washed) for 30 minutes with Krebs-Ringer solution containing glucose. The slices were subsequently stimulated with potassium (50 mM) or veratridine (50 µM) in the presence (normoglycemic) or absence (hypoglycemic) of glucose. The release of endogenous amino acids and exogenous D-[^{3}H]aspartate were monitored. The involvement of calmodulin ion KCl-evoked release was asses-

Table 2. Amino Acid Concentrations in Rat Neostriatum

	Normoglycemic rats		'Severe' hypoglycemic rats	
	NO (n=17)	O (n=11)	NO (n=6)	O (n=6)
Glutamate	92.1 ± 10.7	70.5 ± 11.1***	48.0 ± 10.1	74.6 ± 11.3**
Aspartate	14.9 ± 2.9	12.7 ± 2.2*	54.4 ± 9.5	28.6 ± 5.2***
Glutamine	43.0 ± 7.1	55.7 ± 10.3***	6.0 ± 2.3	40.7 ± 15.9**
GABA	16.5 ± 2.7	15.7 ± 3.4	13.5 ± 2.7	18.1 ± 3.6
Taurine	68.7 ± 11.5	66.9 ± 12.9	88.6 ± 16.6	88.4 ± 14.9

The results are expressed as nmol/mg protein (mean ± SD).
n = number of samples
O = operated side
NO = non-operated side
Differences between non-operated and operated animals: * p < 0.05;
** p < 0.01; *** p < 0.001
From Engelsen and Fonnum, 1983.

sed as well. Thus, normo- and hypoglycemic slices were preincubated with the calmodulin antagonist trifluoroperazin (Cheung, 1980).

The effects of methionine sulphoximine, an inhibitor of glutamine synthetase (Cooper et al., 1983) on amino acid levels in the intact and decorticated neostriatum have been studied (Engelsen and Fonnum, 1985). This study was executed in order to obtain data on the importance of glutamine as a precursor for transmitter and metabolic pools of glutamate and for GABA synthesis. Fluorocitrate, which is taken up into glial cells, was injected into neostriatum to inhibit the Krebs cycle of the glial cells.

γ-vinyl GABA is believed to be a relatively specific inhibitor of GABA aminotransferase (Jung et al., 1977) and has been recommended for GABA turnover studies (Fonnum, 1985). The inhibition of GABA aminotransferase will block the catabolism of GABA and allowed us to estimate the proportion of the steady-state level of glutamine which is derived from GABA. It will also in principle inhibit the proposed GABA-glutamine-GABA cycle (van den Berg and Garfinkel, 1971). γ-vinyl GABA also allowed us to measure the GABA turnover rate under different experimental conditions such as hypoglycemia.

RESULTS AND DISCUSSION

Glutamate Pools

Selective degeneration of glutamergic terminals by surgical lesion was accompanied by a 20-30% decrease of the total glutamate content in neostriatum, lateral septum, nucleus accumbens and thalamus (Table 2; Lund Karlsen and Fonnum, 1978; Walaas and Fonnum, 1980; Fonnum et al., 1981). The transmitter pool of glutamate is therefore assumed to be 20-30% of the total glutamate pool. Electron microscopic studies of neostriatum after cortical ablation showed that degenerated terminals only occupied a few percent of the total volume in this region. The concentration of glutamate in glutamergic terminals is therefore very high.

The pool of glutamate which functions as a precursor for GABA can be estimated in a similar way after selective destruction of GABAergic terminals. In substantia nigra, the region with the highest content of GABAergic terminals, selective destruction of GABAergic terminals was accompanied by a small decrease in glutamate level (Minchin and Fonnum, 1979; Korf and Venema, 1983). The precursor pool of glutamate is therefore probably less than 10% of the total glutamate pool in any brain region. The concentration of glutamate in GABAergic terminals is therefore relatively low.

The proportion of glutamate in the glial cells can be estimated from the relative specific radioactivity of glutamine compared to glutamate after administering precursors which specifically label the glial cells. In several regions of cat brain, this ratio with acetate was 5, which should indicate that the glial pool is 20% of the total glutamate pool (Berl et al., 1961). This is probably an overestimation since acetate also enters the neuronal pool.

Intrastriatal injection of fluorocitrate reduced the glutamine content 80% and glutamate content about 30% after four hours. The reduction in glutamate could be an overestimation and comprise both the reduction in glial glutamate and also a reduction in terminal glutamate secondary to the loss of its precursor glutamine (Table 3).

The main pool of glutamate in the brain is probably linked to the general metabolism in neurons. In neostriatum, destruction of neuronal cell bodies with kainic acid was accompanied by 50% loss of glutamate (Nicklas et al., 1979). This is expected to account for neuronal structures except glutamergic and monoaminergic terminals.

<u>Synthesis and Release of Glutamate and GABA</u>

The three main precursors which have been suggested for glutamate are glucose, glutamine and 2-oxoglutarate. Particularly <u>in vitro</u>, glutamine is by far the most efficient precursor (Hamberger et al., 1979), but the case has not been clearly demonstrated <u>in vivo</u>.

Insulin-induced hypoglycemia was accompanied by an increase in aspartate and a decrease in glutamine and glutamate in neostriatum. During the development of hypoglycemia the level of glutamine was reduced prior to that of glutamate (Engelsen and Fonnum, 1983). The changes were more dramatic in the intact neostriatum than in the decorticated neostriatum. It is in this context interesting to recall that GABA was not reduced until a substantial reduction both in glutamate and glutamine had taken place (Table 2; Agardh et al., 1979).

Table 3. Effect of Intrastriatal Injection of Fluorocitrate (4 h, 0.4 µg) on Amino Acid Contents in Intact Neostriatum

	ASP	GLU	GLN	TAU	ALA	GABA
Control (8)	9 ± 3	78 ± 24	52 ± 19	94 ± 27	10 ± 4	16 ± 5
Injected (5)	6 ± 1	52 ± 6*	9 ± 6*	82 ± 8	18 ± 3*	14 ± 4

The results are expressed as nmol/mg protein (means ± SD).
Differences between control and injected group: *$p < 0.05$.

Table 4. Effects of Trifluoroperazin (TFP) on KCl-Evoked Transmitter Release from Striatum Slices

	Spontaneous		Evoked		
	0 μM TFP	50 μM TFP	0 μM TFP	50 μM TFP	%change
1. D-[^{3}H]Asp	2.6 ± 0.5	1.9 ± 0.6	12.3 ± 1.8	5.1 ± 1.2	-67
2. Aspartate	0.05 ± 0.03	0.06 ± 0.02	0.27 ± 0.05	0.09 ± 0.06	-86
Glutamate	0.09 ± 0.03	0.09 ± 0.03	1.05 ± 0.02	0.28 ± 0.04	-80
Taurine	0.33 ± 0.04	0.19 ± 0.02	0.53 ± 0.02	0.40 ± 0.02	+ 5
GABA	< 0.01	< 0.01	0.86 ± 0.13	0.50 ± 0.09	-42

Results are presented as: (1) % released of tissue radioactivity, and (2) nmol · mg protein^{-1} · 5 min^{-1} (mean ± SD, n = 6). The % change is computed from the respective differences between evoked and spontaneous release values.

The results can be interpreted as if the turnover of glutamate is higher in glutamergic terminals than in the other glutamate pools. Alternatively, the glutamergic input is the main driving force in neostriatal metabolism. Secondly, since glutamine was reduced prior to glutamate, this indicates a precursor-product relationship. It was noted, however, that glutamine was reduced quite substantially before there was any reduction in glutamate. This could mean that the role of glutamine as a precursor is not so important as the in vitro experiments here indicated. An alternative explanation is that the K_m for glutamate synthesis from glutamine is very low, so that the consequences of a glutamine reduction for glutamate synthesis is small. Thirdly, glutamine has been suggested as an important precursor for GABA synthesis in all kinetic schemes. This does not seem to be the case. GABA turnover was reduced during hypoglycemia from 29 to 20 μmol/g protein/h. This was much smaller than the proportional decrease in glutamine.

The rapid interchange which seemed to take place between aspartate and glutamate in the intact neostriatum during hypoglycemia, raises the question whether aspartate and glutamate may substitute for each other as neurotransmitters. This was further investigated by studying the release from neostriatal slices. When the slices were depolarized with KCl both under normoglycemic and hypoglycemic conditions, the release of glutamate, aspartate and GABA, but not that of taurine, glutamine and alanine, were Ca-dependent. The release evoked by veratridine for these amino acids was inhibited by TTX both in the normoglycemic and in the hypoglycemic conditions. Similarly, the KCl-evoked release of exogenous D-[^{3}H]aspartate and endogenous aspartate, glutamate and GABA was inhibited by the caldomulin-antagonist trifluoroperazin (Table 4). This was also the case under hypoglycemic conditions in vitro (Fosse and Fonnum, in preparation).

Under normoglycemic conditions, the release of glutamate was much higher than that of aspartate. Under hypoglycemic conditions, the release of aspartate increased on repetitive KCl-depolarization whereas the release of glutamate decreased (Table 5). Also, during veratridine depolarization, the release of aspartate compared to that of glutamate increased on repetitive stimulation under hypoglycemic conditions. The results indicate, therefore, that L-aspartate may substitute for L-glutamate during depolariza-

Table 5. Evoked Release of Endogenous Amino Acids from Neostriatal Slices

	50 mM potassium				Veratridine	
	Normoglycemic		Hypoglycemic		Hypoglycemic	
Amino Acids	SI	SIII	SI	SIII	SI	SIII
Aspartate	0.30	0.18	0.52	1.18	0.42	0.53
Glutamate	1.23	0.78	2.03	1.34	1.52	0.87
GABA	0.62	0.56	0.94	0.80	0.97	0.61

SI = 1st depolarization
SIII = 3rd depolarization
Results are expressed as nmol · mg protein^{-1} · 5 min^{-1}

tion of the presumable glutamergic terminals in neostriatal slices in vitro. This finding is in accordance with several other observations on the overlapping or similar activities displayed by aspartate and glutamate. They are taken up into glial cells and nerve terminals by the same HA uptake system and they show overlapping affinities towards different excitatory amino acid (i.e., quisqualate, N-methyl D-aspartate and kainic acid) receptors (Fonnum, 1984). Interestingly, stimulated release of GABA did not seem to be much affected by the hypoglycemic state.

The effect of methionine sulphoximine, administered i.p., on the amino acid levels in neostriatum was in accordance with those obtained from the hypoglycemic experiments. Two and four hours after methionine sulphoximine administration, glutamine was reduced in neostriatum. The reduction was most significant on the operated side. The reduction in glutamine was quite substantial before there was any reduction in glutamate. There was no significant change in the GABA level even though the glutamine level was decreased dramatically (Table 6).

The more pronounced decrease in glutamine on the decorticated side during methionine sulphoximine treatment could be due to a lower release

Table 6. The Effect of Methionine Sulphoximine on Amino Acid Contents in the Intact and Decorticated Neostriatum

		Control	MSO Treated		
			2h	4h	6h
			nmol/mg protein		
Glutamate	C	136.9	138.1	117.5*	80.2*
	L	105.6	101.0	101.8	95.6
Glutamine	C	76.1	84.3	30.1*	4.1*
	L	99.0	56.9*	9.4*	3.2*
GABA	C	28.0	27.5	29.2	22.1
	L	32.0	31.7	26.3	28.1

*$p < 0.05$ between control and experimental group.
Methionine sulphoximine was administered 250 mg/kg, i.p.
C: control group, L: decorticated group.
From: Engelsen and Fonnum (1985).

Table 7. Changes in Amino Acid Contents after Intrastriatal Injection of γ-vinyl GABA

Time	ALA	ASP	GLU	GLN	TAU	GABA
0 h (17)	9.2 ± 2.9	19.9 ± 4.2	109.4 ± 21.3	66.5 ± 9.7	128.6 ± 24.1	17.6 ± 3.2
2 h (11)	8.1 ± 2.4	13.4 ± 3.0*	93.5 ± 16.0*	48.6 ± 8.6*	126.5 ± 9.7	69.5 ± 6.9*
4 h (6)	7.6 ± 3.0	10.5 ± 1.3*	86.2 ± 12.8*	38.1 ± 4.6*	120.8 ± 17.9	109.6 ± 15.1*

Results are expressed as nmole/mg protein ± SD.
* $p < 0.05$

of transmitter glutamate, which is a precursor for glutamine, on the operated side. The reduction in glutamate which only occurred in the intact side, indicated a higher turnover on that side. Similar to the case with hypoglycemia, glutamine may be a precursor for glutamate, but apparently only a poor substrate for GABA synthesis.

When γ-vinyl GABA was injected into neostriatum, GABA aminotransferase was inhibited 80-90%. Under these conditions, there was an almost linear increase in GABA for four hours (Table 7). During this period, there was a substantial reduction in glutamine and a lesser decrease in glutamate and aspartate. The results can be explained by assuming that GABA after release is taken up into glial cells and converted to glutamate and later to glutamine. Inhibition of GABA and aminotransferase prevents this conversion. It is interesting in this respect that such a high proportion of glutamine should be derived from GABA.

Intrastriatal injection of fluorocitrate inhibits the metabolism of the glial cells by inhibiting the Krebs cycle (Clarke et al., 1970). Fluorocitrate seems to be much more specific for this purpose than the closely related fluoroacetate (Clarke and Berl, 1973). This may be due to the more specific uptake of citric acid than acetic acid into the glial cells. Already a short time after injection of fluorocitrate there was a substantial reduction in glutamine and an increase in alanine. Several hours later, there was a large decrease in glutamine, a smaller decrease in glutamate and aspartate, and no change in GABA (Table 3). The results should reflect the effect of inhibition of the glial cell metabolism on amino acids in neostriatum. It is well established from _in vitro_ experiments that fluorocitrate inhibits the formation of glutamine from [$1-^{14}C$] acetate almost completely (Cheng and Nakamura, 1972; Clarke and Berl, 1973). It is not known whether this is due to perturbed transport of glutamate into the glial cell or another mechanism. It could also be due to an inhibition of ATP formation since glutamine synthetase is an ATP-dependent enzyme.

CONCLUDING REMARKS

The use of the basal ganglia has proved to be valuable for the study of the synthesis and metabolism of amino acid transmitters. The main conclusions are:

1) The turnover of glutamate is higher in the intact than in the decorticated neostriatum. This shows the importance of the glutamate terminals for neostriatal activity.

2) Under experimental conditions like hypoglycemia, it seems that

at least in *in vitro* studies glutamate and aspartate may substitute for each other as transmitters.

3) The kinetic models for studying the metabolism of amino acids may have overestimated the role of glutamine as a precursor for GABA and possibly for glutamate.

4) 2-oxoglutarate is an alternative candidate for the synthesis of transmitter glutamate. It may well be that the synthesis of transmitter amino acids may differ under different experimental conditions and that both glucose, glutamine and 2-oxoglutarate are strong precursor candidates.

5) Trifluoroperazin inhibits the release of aspartate, glutamate and GABA from slices during normo- as well as hypoglycemic conditions indicating that their release may be calmodulin dependent.

REFERENCES

Agardh, C.D., Carlsson, A., Lundquist, M., and Siesjö, B.K., 1979, The effect of pronounced hypoglycemia on the monoamine metabolism in rat brain, Diabetes, 28:804.

Benveniste, H., Drejer, J., Schousboe, A., and Diemer, N.H., 1984, Elevation of the extracellular concentrations of glutamate and aspartate in rat hippocampus during transient cerebral ischemia monitored by intracerebral microdialysis, J. Neurochem., 43:1369.

Berl, S., Lajhta, A., and Waelsch, H., 1961, Amino acid and protein metabolism. VI. Cerebral compartments of glutamic acid metabolism, J. Neurochem., 7:186.

Cheng, S.-C., and Nakamura, R., 1970, A study on the tricarboxylic acid cycle and the synthesis of acetylcholine in the lobster nerve, Biochem. J., 118:451.

Cheung, W.Y., 1980, Calmodulin plays a pivotal role in cellular regulation, Science, 207:19.

Clarke, D.D., Nicklas, W.J., and Berl, S., 1970, Tricarboxylic acid cycle metabolism in brain: effect of fluoroacetate and fluorocitrate on the labeling of glutamate, aspartate, glutamine and γ-amino-butyrate, Biochem. J., 120:345.

Clarke, D.D., and Berl, S., 1973, Alteration in the expression of compartmentation : *in vitro* studies, in: Metabolic Compartmentation in the Brain, R. Balázs, and J.E. Cremer, eds., Macmillan, London. p. 97.

Cooper, A.J.L., Vergara, F., and Duffy, T.E., 1983, Cerebral glutamine synthetase, in: Glutamine, Glutamate, Aspartate and GABA in the Central Nervous System, L. Hertz, E. Kvamme, E.G. McGeer, and A. Schousboe, eds., Alan R. Liss, New York, p. 77.

Engelsen, B., and Fonnum, F., 1983, Effects of hypoglycemia on the transmitter pool and the metabolic pool of glutamate in rat brain, Neurosci. Lett., 42:317.

Engelsen, B., and Fonnum, F., 1985, The effect of methionine sulfoximine, an inhibitor of glutamine synthetase on the levels of amino acids in the intact and decorticated rat neostriatum, Brain Res., 338:165.

Engelsen, B.A., Fosse, V.M., Myrseth, E., and Fonnum, F., 1985, Elevated concentrations of glutamate and aspartate in human ventricular CSF (vCSF) during episodes of increased CSF pressure and clinical signs of impaired brain circulation, Neurosci. Lett., 42:97.

Fonnum, F., Gottesfeld, Z., and Grofova, I., 1978, Distribution of glutamate decarboxylase, choline acetyltransferase and aromatic animo acid decarboxylase in the basal ganglia of normal and operated rats. Evidence for striatopallidal, striatoentopeduncular and striatonigral GABAergic fibers, Brain Res., 143:125.

Fonnum, F., and Walaas, I., 1979, Localization of neurotransmitter candidates in neostriatum, in: The Neostriatum, I. Divac and R.G.E. Öberg, eds., Pergamon Press, New York, p. 53.

Fonnum, F., Søreide, A., Kvale, I., Walder, J., and Walaas, I., 1981, Glutamate in cortical fibers, in: Glutamate as a Neurotransmitter, G. DiChiara and G.L. Gessa, eds., Raven Press, New York, p. 29.

Fonnum, F., Storm-Mathisen, J., and Divac, I., 1981, Biochemical evidence for glutamate as neurotransmitter in the cortico-striatal and cortico-thalamic fibres in rat brain, Neuroscience, 6:863.

Fonnum, F., 1984, Glutamate: a neurotransmitter in mammalian brain, J. Neurochem., 42:1.

Fonnum, F., 1985, Determination of transmitter amino acid turnover, in: Neuromethods Vol. 3, Amino Acids, A.A. Boulton, G.B. Baker, and J.D. Wood, eds., Humana Press, Clifton, New Jersey, p. 201.

Hamberger, A., Chiang, G.H., Nylén, E.S., Scheff, S.W., and Cotman, C.W., 1979, Glutamate as a CNS transmitter. I. Evaluation of glucose and glutamine as precursors for the synthesis of preferentially released glutamate, Brain Res., 168:513.

Jung, M.J., Lippert, B., Metcalf, B.W., Böhlen, P., and Schechter, P., 1977, γ-Vinyl-GABA (4-aminohexionic acid), a new selective irreversible inhibitor of GABA-T: effects on brain GABA metabolism in mice, J. Neurochem., 29:797.

Korf, J., and Venema, K., 1983, Amino acids in the substantia nigra in rats with striatal lesions produced by kainic acid, J. Neurochem., 40:1171.

Lund-Karlsen, R., and Fonnum, F., 1976, The toxic effect of sodium glutamate on rat retina: changes in putative transmitters and their corresponding enzymes, J. Neurochem., 27:1437.

Lund-Karlsen, R., and Fonnum, F., 1978, Evidence for glutamate as a neurotransmitter in the corticofugal fibres to the dorsal lateral geniculate body and the superior colliculus in rats, Brain Res., 151:457.

Minchin, M.C.W., and Fonnum, F., 1979, The metabolism of GABA and other amino acids in rat substantia nigra slices following lesions of the striato-nigral pathway, J. Neurochem., 32:203.

Nicklas, W.J., Nunez, R., Berl, S., and Duvoisin, R., 1979, Neuronal glial contributions to transmitter amino acid metabolism: studies with kainic acid-induced lesions of rat striatum, J. Neurochem., 33:839.

Olney, J.W., 1969, Glutamate-induced retinal degeneration in neonatal mice: electron microscopy of the acutely evolving lesion, J. Neuropath. Exp. Neurol., 28:444.

Olney, J.W., and Price, M.T., 1981, Neuroendocrine effect of excitotoxic amino acids, in: Glutamate as a Neurotransmitter, G. DiChiara and G.L. Gessa, eds., Raven Press, New York, p. 423.

van den Berg, C.J., and Garfinkel, D., 1971, A simulation study of brain compartments, Biochem. J., 123:211.

Walaas, I., and Fonnum, F., 1978, The effect of parental glutamate treatment on the localization of neurotransmitters in the mediobasal thalamus, Brain Res., 153:549.

Walaas, I., and Fonnum, F., 1980, Biochemical evidence for glutamate as a transmitter in hippocampal efferents to the basal forebrain and hypothalamus in rat brain, Neuroscience, 5:1691.

Wieloch, T., Engelsen, B., Westerberg, E., and Auer, R., 1985, Lesions of cortico-striatal projections in the rat ameliorate hypoglycemic brain damage in the striatum, Neurosci. Lett., 58:25.

NA$^+$ FLUXES AS A TOOL TO IDENTIFY ANTICONVULSANT ANTAGONISTS OF

NEUROEXCITATION

V.I. Teichberg[1], M. Beaujean[1,2], P. David[1],
D. Eisenberg-Tamarin[1], U. Erez[1], H. Frenk[3], A. Luini[1],
G. Urca[4] and O. Goldberg[2]

The Departments of Neurobiology[1] and Organic Chemistry[2]
The Weizmann Institute of Science
The Departments of Psychology[3], Physiology and Pharmacology[4]
Tel Aviv University, Israel

INTRODUCTION

In spite of the availability of a relatively large number of antiepileptic drugs, the search for new, effective yet non-toxic antiepileptic drugs still goes on. Today as in the past, the prime strategy in this search remains the screening of potential drugs in two established seizure models, i.e., the maximal electroshock seizure test and the subcutaneous metrazol seizure threshold test. The compounds passing successfully the tests of long term efficacy, toxicity and pharmacokinetics undergo clinical trials and may eventually be recommended as drugs, often in spite of the partial or even lack of knowledge concerning the mechanism of their anticonvulsant action. The relative success of this screening approach can be judged by the fact that only three new antiepileptic drugs (carbamazepine, clonazepam and valproic acid) have been marketed in the last 25 years.

In contrast to the above, the alternative and more rational approach is to develop drugs acting as specific targets known to be involved in the epileptic phenomenon. Some of these targets have been identified: they include enzymes involved in the catabolism of GABA (Meldrum, 1979), the GABA-benzodiazepine-barbiturate macromolecular complex (Leeb-Lundberg et al., 1981), the presynaptic GABA uptake site (Croucher et al., 1983) and more recently the receptor of N-methyl-D-aspartate (NMDA-receptor) (Croucher et al., 1982; Meldrum et al., 1983a,b).

The NMDA-receptor is one of the four types of excitatory amino acid receptors (Luini et al., 1981; Watkins and Evans, 1981; Monaghan et al., 1983; Foster and Fagg, 1984) which are believed to mediate the bulk of the neuroexcitatory events in the CNS. In view of the fact that the prototype agonists of the above mentioned receptors, once injected in the brain, or intraperitoneally (Johnston, 1973) produce seizures, it is generally assumed that the various excitatory amino acid receptors are all involved in the epileptic phenomenon.

The roles played in epilepsy by the receptors of kainate (KA-receptor), quisqualate (Quis-receptor) and glutamate/aspartate (Glu/Asp-receptor) have not yet been clarified, mainly because of the lack of antagonists capable of blocking selectively the activation of these receptors in vivo.

The development of antagonists of excitatory amino acids and thereby possibly of novel anticonvulsant drugs can be rationally carried out provided that the pharmacological effects of newly synthesized compounds can be unambiguously related to their structure and to a specific site of action. This task is to some extent made difficult by the fact that most ligands of excitatory amino acid receptors appear to interact with more than one receptor. Nevertheless, one may overcome this difficulty in cases when the extent of overlap in the ligand structures recognized by the various receptors can be determined. A method allowing such a determination is the radioactive sodium (^{22}Na) efflux receptor assay (Luini et al., 1981; Teichberg et al., 1981).

The purposes of this chapter are: a) to present a typical example of the use of the radioactive Na efflux receptor assay in the detection and study of the structure-function relation of antagonists of excitatory amino acids; b) to analyze the anticonvulsant action of some of these antagonists, discuss the origin of their side effects and evaluate the therapeutic potential of the anticonvulsant excitatory amino acid antagonists.

^{22}NA$^+$ EFFLUX RECEPTOR ASSAY IN THE DETECTION AND STUDY OF STRUCTURE-FUNCTION RELATIONS OF ANTAGONISTS OF EXCITATORY AMINO ACIDS

The radioactive Na efflux receptor assay has been described in detail in several publications (Luini et al., 1981; Teichberg et al., 1980, 1981). In a typical experiment with radioactive Na-preloaded brain slices, the increase in radioactive Na efflux produced by each of the four typical excitatory amino acids (NMDA, Quis, KA and Glu) is measured in the absence and presense of a putative antagonist. The specificity and efficacy of the antagonist are derived from the extent of inhibition of the excitatory amino acid response. Fig. 1 shows, as an example, the results of experiments performed to test the possible antagonist properties of a series of newly synthesized γpeptides of kainic acid (Goldberg and Teichberg, 1985). As can be seen, some of the latter are rather selective antagonists of the NMDA-receptor (i.e., γKA-Glu and γKA-KA) and affect very little the responses to KA, Quis or Glu. The absence of any significant inhibition of the KA response is surprising since the tested compounds are all derived from kainic acid. These results indicate that the ligand specificity of the NMDA-receptor is rather loose, whereas that of the KA-receptor is very strict.

The differences observed in the extent of inhibition displayed by the various γ-kainyl-dipeptides are of interest since they reveal some of the structural features required of antagonists for a more favorable interaction with the NMDA-receptor. As seen from Fig. 1, the γ-dipeptides of KA with the β-amino acid β-Ala and with the γ-amino acid GABA, exhibit weak antagonism, whereas better inhibitory activities are observed for the γ-dipeptides of KA with the α-amino acids, namely, γKA-Gly, γKA-Tyr, γKA-Glu and γKA-KA. The latter two, which in addition to the α-amino acid skeleton possess an acidic side-chain, are the most efficient NMDA blockers in this series of compounds. We assume that it is the presence of an extra carboxyl group and not the length of the side chain which adds a significant contribution to the antagonist activity since the corresponding chain length is also present in the less active depeptide γKA-GABA. It is also of interest to compare the activity of γ-L-Glu-Gly to that of its rigid analog γKA-Gly. The fact that they inhibit to the same extent the NMDA response suggests that in both cases the ligand-NMDA-receptor interactions involve common structural features, i.e., the alpha-amino

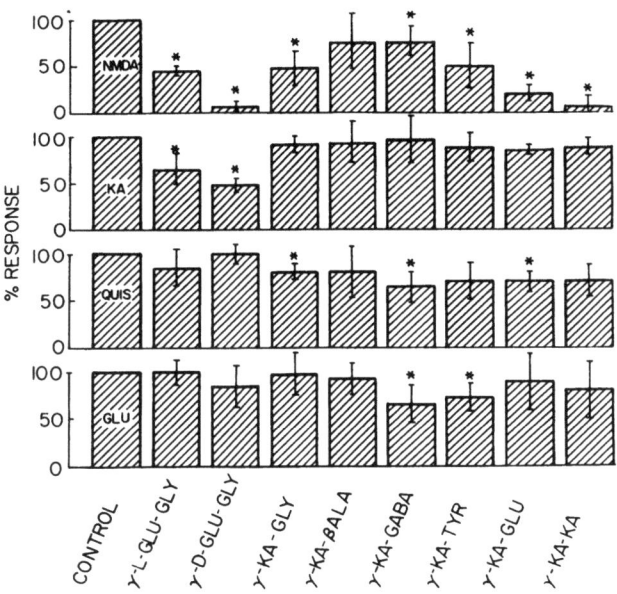

Fig. 1. Excitatory amino acid response in the presence of dipeptide derivatives of glutamate and kainate. The excitatory amino acids were applied to radioactive Na-preloaded striatal slices at the following concentrations: 30 μM NMDA, 100 μM KA, 100 μM Quis and 0.5 mM L-Glu. The dipeptide concentration was 1 mM. The bars represent mean values ± standard deviation. The significance of the difference from control mean was calculated by variance analysis: *p < 0.05.

acid skeleton and not the bulky pyrrolidine ring or the isopropenyl side chain. Furthermore, the NMDA-receptor appears to accomodate more readily amino acids in the D- rather than the L-configuration since γ-D-Glu-Gly has a stronger antagonist activity that γ-L-Glu-Gly.

ANTICONVULSANT ACTIVITY AND SIDE EFFECTS OF ANTAGONISTS OF EXCITATORY AMINO ACIDS

The anticonvulsant activity of antagonists of excitatory amino acids was measured using as animal models of epilepsy, the picrotoxin or pentylenetetrazole-induced seizures in CD1 mice and the picrotoxin-induced seizures in SPD rats. Fig. 2 shows that mice can be increasingly protected from pentylenetetrazole-induced seizures when increasing amounts of γ-kainylglutamate or γ-D-glutamylglycine (γDGG) are injected intracerebroventricularly.

As expected from the radioactive Na efflux receptor assay (see Fig. 1), the potency of γ-D-glutamylglycine both as antagonist of NMDA and KA and as anticonvulsant agent, is slightly stronger than that of γ-kainylglutamate. The median protective dose of γ-D-glutamylglycine calculated by the method of Litchfield and Wilcoxon (1949) is 0.1 μmol (0.06-0.15; 95% confidence limits) whereas that of γ-kainylglutamate is 0.23 μmol (0.11-0.44; 95% confidence limits).

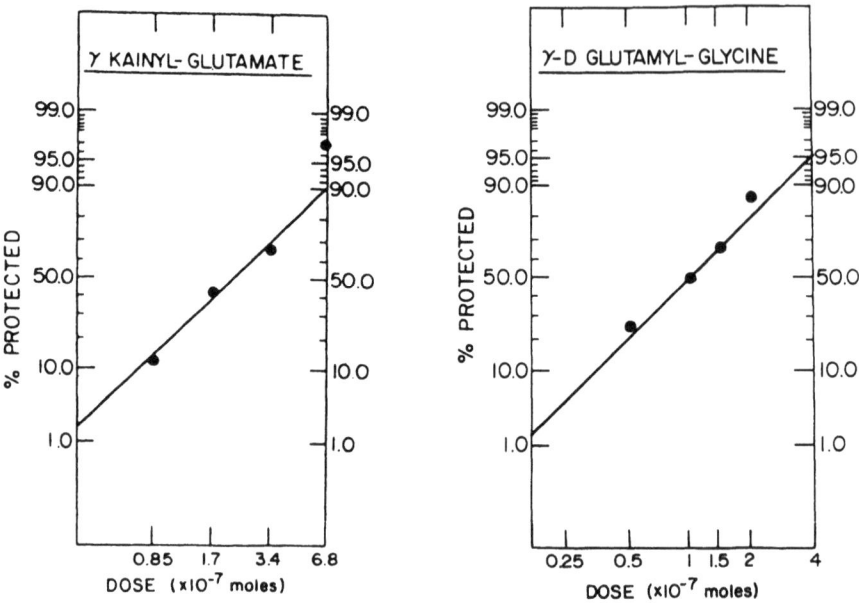

Fig. 2. Dose-dependent protection from pentylenetetrazole-induced convulsions in CD1 mice by intracerebroventricular injections of γ-kainylglutamate (left) and γ-D-glutamylglycine (right). Each point corresponds to ten mice. The results were calculated by the method of Litchfield and Wilcoxon (1949).

One ought to mention that the above antagonists produce beside their anticonvulsant activity several other effects. At a dose of 0.05 μmol, γ-glutamylglycine produced, upon i.c.v. administration, a reduction of muscle tone. With increasing doses, symptoms of tremor, loss of righting reflex and incoordinate movements appeared together with yawning and clonic contractions of one of the legs. With γ-kainylglutamate, the symptoms included also tachypnea and episodes of total immobility. Upon intrathecal injections, γ-D-glutamylglycine (0.025 μmol) produced a marked reduction in the tail tone and, in fifty percent of the cases, a paralysis of the hindlimbs. There were no effects on responses to touch and pinch but an increase in the tail withdrawal latency to hot water (50°C; Fig. 3). The latter effect can most probably be viewed as a consequence of the decrease in tail tone.

When γ-D-glutamylglycine (1 μmol) was injected i.c.v. in chronically implanted rats, the major effects observed were a strong decrease in motor activity and motor coordination, a decrease in muscle tone and in palpebral opening and an increase in yawning. At this dose, the animals were entirely protected from picrotoxin-induced seizures.

The reduction in muscle tone noticeable after each i.c.v. administration of γ-D-glutamylglycine was not systematically observed with other anticonvulsant excitatory amino acid antagonists. Injections of 2-amino-5-phosphonovalerate (APV, 0.01 μmol) or 2-amino-7-phosphonoheptanoate (APH, 0.01 μmol) into the ventricles of C57Bl/6 mice produced an increase in motor activity and hyperventilation and induced, in some cases, ataxia and explosive motor behavior. In DBA/2 mice, APH produced no toxic signs at doses protecting against audiogenic seizures and had a sedative effect at higher doses (Croucher et al., 1982). The i.c.v. administration of 3-hydroxy-2-quinoxa-

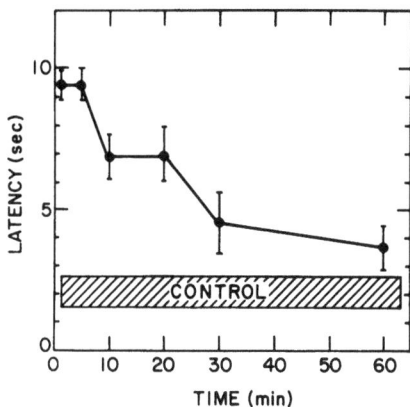

Fig. 3. Time dependent latency of tail withdrawal after intrathecal administration of 25 nmol γ-D-glutamylglycine. The peptide dissolved in saline (5 μl) was administered into the subarachnoidal space between L5 and L6 of female CD1 mice. The tail was immersed in a water bath at 50°C (Gamse and Teichberg, 1982).

linecarboxylic acid (HQC) at doses above 15 μmol produced ataxia and transient convulsive effects (Erez et al., 1985). Some animals displayed explosive motor behavior while others underwent transient sedation. These signs dissappeared five minutes after HQC administration while the anticonvulsant activity of HQC manifested itself for more than 25 minutes.

The results obtained with HQC can be explained if one assumes that the threshold for the anticonvulsant action of HQC is reached at HQC doses lower than those eliciting the side effects. Furthermore, the finding that high doses of HQC after i.c.v. administration produce immediate side effects indicates that the latter effects might be due to a blockade of some of the excitatory processes in the circumventricular areas whereas the anticonvulsant effects of HQC would predominate after its diffusion from the ventricles and its dilution in the extracellular space.

In an attempt to answer the question of the origin of the side effects produced by NMDA antagonists such as APV, APH and HQC upon i.c.v. administration, the pivaloyl derivative of the methyl ester of HQC was synthesized. Fig. 4 shows that this lipophilic compound, upon intraperitoneal administration, crosses readily the blood brain barrier and produces a delay in the time of occurrence of picrotoxin-induced tonic and clonic seizure up to a dose of 65 mg/kg. Larger doses of the HQC derivative were less effective, a phenomenon most probably due to the intrinsic toxicity of the compound.

It is of interest to mention that upon subdural injection onto the cortex of free moving rats, APV, γ-DGG and HQC were found to effectively suppress behavioral and electrographic seizure activity induced by strychnine, morphine and picrotoxin administered via the same route (Frenk et al., 1986). Moreover, the cortical application of the NMDA antagonists

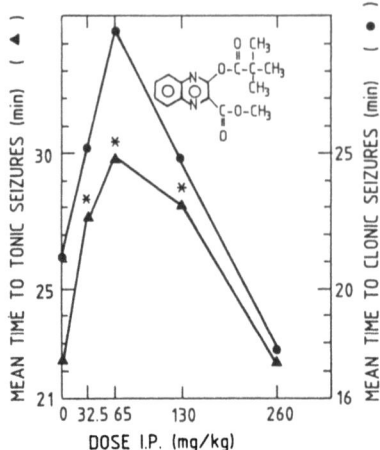

Fig. 4. Dose-dependent delay in the mean time to picrotoxin-induced tonic and clonic seizures by intraperitoneal injections of a lipophilic derivative of HQC. Each point represents the mean value obtained in a group of eight CD1 mice. The significance of the difference from control mean was calculated by variance analysis: *$p < 0.05$.

did not induce behavioral or electrographic changes, and the behavioral side efffects commonly observed following the intracerebroventricular administration of these compounds were absent.

From all the above, it appears thus that the anticonvulsant effects and side effects produced by antagonists of excitatory amino acids depend on intrinsic properties of the administered compounds, the routes of injection and possibly on the strains and species of animals under study. Several animal species are indeed genetically susceptible to seizures such as the photosensitive baboon and fowl or the mongolian gerbil (Löscher and Meldrum, 1984). Interestingly, seizures in gerbils are not sensitive to alterations in excitatory amino acid-mediated neurotransmission (Löscher, 1985). According to these facts, one expects that the final evaluation of the therapeutic potential of excitatory amino acid antagonists as anticonvulsant drugs in humans will depend on the availability of novel nontoxic compounds displaying anticonvulsant effects in several animal species including primates.

ACKNOWLEDGEMENTS

This work is supported by the Wills Foundation and the Esther A. and Joseph Klingenstein Fund. V.I.T. holds the Louis and Florence Katz-Cohen professorial chair in neuropharmacology.

REFERENCES

Croucher, M.J., Collins, J.F., and Meldrum, B.S., 1982, Anticonvulsant action of excitatory amino acid antagonists, Science, 216:988.

Croucher, M.J., Meldrum, B.S., and Krogsgaard-Larsen, P., 1983, Anticonvulsant activity of GABA uptake inhibitors and their prodrugs following central or systemic administration, Eur. J. Pharmacol., 89:217.

Erez, V., Frenk, H., Goldberg, O., Cohen, A., and Teichberg, V.I., 1985, Anticonvulsant properties of 3-hydroxy-2-quinoxalinecarboxylic acid, a newly found antagonist of excitatory amino acids, Europ. J. Pharmacol., 110:31.

Foster, A.C., and Fagg, G.E., 1984, Acidic amino acid binding sites in mammalian neuronal membranes: their characteristics and relationship to synaptic receptors, Brain Res. Rev., 7:103.

Frenk, H., Liban, A., Urca, G., and Teichberg, V.I., 1986, Absence of side effects in the anticonvulsant action of cortically applied antagonists of N-methyl-D-aspartate, Brain Res., in press.

Gamse, R., and Teichberg, V.I., 1982, unpublished results.

Goldberg, O., and Teichberg, V.I., 1985, Peptides derived from kainic acid as antagonists of N-methyl-D-aspartate-induced neuroexcitation in rat brain, Neurosci. Lett., 60:101.

Johnston, G.A.R., 1973, Convulsions induced in 10-day old rats by intraperitoneal injection of monosodium glutamate and related excitant amino acids, Biochem. Pharmacol., 22:137.

Leeb-Lundberg, F., Snowman, A., and Olsen, R.N., 1981, Interaction of anticonvulsants with the barbiturate-benzodiazepine-GABA receptor complex, Eur. J. Pharmacol., 72:125.

Litchfield, J.T., and Wilcoxon, F., 1949, A simplified method of evaluating dose-effects experiments, J. Pharmacol. Exp. Therap., 96:99.

Löscher, W., and Meldrum, B.S., 1984, Evaluation of anticonvulsant drugs in genetic animal models of epilepsy, Fed. Proc., 43:276.

Löscher, W., 1985, Influence of pharmacological manipulation of inhibitory and excitatory neurotransmitter systems on seizure behavior in the mongolian gerbil, J. Pharmacol. Exp. Therap., 233:204.

Luini, A., Goldberg, O., and Teichberg, V.I., 1981, Distinct pharmacological properties of excitatory amino acid receptors in the rat striatum: study by radioactive Na efflux assay, Proc. Natl. Acad. Sci. USA, 78:3250.

Meldrum, B.S., 1979, Convulsant drugs, anticonvulsants and GABA-mediated neuronal inhibition, in: GABA Neurotransmitters, P. Krogsgaard-Larsen, J. Scheel-Krüger and H. Kofod, eds., Munksgaard, Copenhagen, p. 390.

Meldrum, B.S., Croucher, M.J., Badman, G., and Collins, J.F., 1983, Antiepileptic action of excitatory amino acid antagonists in the photosensitive baboon, Papio papio, Neurosci. Lett., 39:101.

Meldrum, B.S., Croucher, M.J., Czuczwar, S.J., Collins, J.F., Curry, K., Joseph, M., and Stone, T.W., 1983, A comparison of the anticonvulsant potency of (+)2-amino-5-phosphonopentanoic acid and (+)2-amino-7-phosphonoheptanoic acid, Neuroscience, 9:925.

Monaghan, D.T., Holets, V.R., Toy, D.W., and Cotman, C.W., 1983, Anatomical distribution of four pharmacologically distinct (3)H-L-glutamate binding sites, Nature, 306:176.

Teichberg, V.I., Goldberg, O., Tal, N., and Luini, A., 1980, The interactions of excitatory amino acids with brain tissue: A study of amino acid-induced ion fluxes in rat striatal slices, in: Neurotransmitters and Their Receptors, U.Z. Littauer, Y. Dudai, I. Silman, V.I. Teichberg, and Z. Vogel, eds., John Wiley, New York, p. 349.

Teichberg, V.I., Goldberg, O., and Luini, A., 1981, The stimulation of ion fluxes in brain slices by glutamate and other excitatory amino acids, Molec. Cell. Biochem., 39:281.

Watkins, J.C., and Evans, R.H., 1981, Excitatory amino acid transmitters, Ann. Rev. Pharmacol. Toxicol., 21:165.

INVOLVEMENT OF EXCITATORY AMINO ACID RECEPTORS IN THE MECHANISMS

UNDERLYING EXCITOTOXIC PHENOMENA

A.C. Foster

Merck, Sharp and Dohme Ltd.
Neuroscience Research Centre, Terlings Park
Eastwick Road, Harlow, Essex, U.K.

ACIDIC AMINO ACIDS AS NEUROTRANSMITTERS

Acidic amino acids are thought to be the major class of excitatory neurotransmitters within the mammalian central nervous system (CNS; Watkins and Evans, 1981; Fagg and Foster, 1983; Fonnum, 1984). Due to their abundance, L-glutamate and L-aspartate are the endogenous excitatory amino acids which have received most attention, although a number of sulphur-containing analogs present in smaller quantities must also be considered as transmitter candidates (Watkins and Evans, 1981; Iwata et al., 1982; see chapter by Cuénod).

The receptors for excitatory amino acids have recently been divided into three subtypes (Table 1). The nomenclature which has evolved names each receptor subtype after the prototypical agonists N-methyl-D-aspartate (NMDA), quisqualate and kainate. However, it is apparent from radioligand binding experiments that these compounds do not show absolute specificity. Quisqualate has a high affinity for kainate binding sites (Slevin et al., 1983) and will displace NMDA-like sites in micromolar concentrations (Greenamyre et al., 1985; Fagg and Foster, unpublished observations), whereas kainate has a reasonably high affinity for quisqualate-like sites (Honoré et al., 1982). Therefore, the use of these agonist names when defining receptor subtypes is rather inappropriate and can be misleading. Consequently, in this chapter I will use a nomenclature which avoids agonist names (Foster and Fagg, 1984); thus , A1, A2 and A3 refer to the receptor subtypes whose pharmacology is defined in Table 1.

The receptor classification scheme originated from electrophysiological studies in which selective antagonist substances were used (Hicks et al., 1978; McLennan and Lodge, 1979; Watkins and Evans, 1981), and has subsequently received strong support from biochemical measurements of receptor function (Luini et al., 1981; Lehmann and Scatton, 1982). Quite recently, radioligand binding techniques have demonstrated the existence within mammalian CNS synaptic membranes of sites which have the pharmacological specificities of A1, A2 and A3 receptors (Fagg and Matus, 1984; Foster and Fagg, 1984). In addition, autoradiographical studies have revealed distinct anatomical locations for each receptor subtype within the mammalian CNS (Monaghan et al., 1983, 1985; Greenamyre et al., 1985; see chapter by Cotman).

Therefore, our current understanding is that excitatory amino acids

Table 1. Excitatory Amino Acid Receptor Subtypes

Receptor subtypes	A1 (NMDA)	A2 (Quisqualate)	A3 (Kainate)
Most selective agonists	NMDA Ibotenate trans-2,3-PDA	AMPA	Kainate Domoate
Most potent agonists	L-glutamate D-aspartate	Quisqualate L-glutamate	Kainate Domoate
Less selective agonists	L-glutamate	L-glutamate Kainate	L-glutamate Quisqualate
Most potent and selective antagonists	D-APV D-APH ASP-AMP		
Less selective antagonists	cis-2,3-PDA KYNA	cis-2,3-PDA KYNA GAMS	cis-2,3-PDA KYNA GAMS

Abbreviations of compound names as in text, except: AMPA, α-amino-3-hydroxy-5-methyl-isoxazole-4-propionic acid; D-APV, D-2-amino-5-phosphonovaleric acid; ASP-AMP, β-D-aspartyl aminomethyl phosphonate; GAMS, γ-D-glutamyl-aminomethylsulphonate.

fulfill their neurotransmitter role through at least three anatomically distinct receptors, which in turn suggests a multiplicity of function. What, then, would be the consequence of a malfunction of these systems?

EXCITATORY AMINO ACIDS AND HUMAN NEURODEGENERATIVE DISORDERS

It has been known for a number of years that excitatory amino acids can produce convulsions, seizures and neuronal degeneration (see Fuxe et al., 1983). The pattern of neurotoxicity caused by these compounds is termed 'axon-sparing'. Neurons which have their cell bodies and dendrites within the affected area are depleted, whereas axons which pass through or terminate there, and also non-neuronal elements such as glia, are spared. This type of lesion is rather similar to that which is observed in human neurodegenerative disorders such as Huntington's disease and temporal lobe epilepsy, and consequently, a pathological overactivity of excitatory amino acid systems has been proposed as an underlying cause of these disease states (Olney, 1980; Coyle et al., 1981; Schwarcz et al., 1984b).

Several lines of evidence point towards the involvement of excitatory amino acids in epileptic phenomena. Firstly, glutamate and its more potent analogs, e.g., kainic acid, produce seizures (see Ben-Ari, 1985) and can reproduce 'epileptic' brain damage in experimental animals (Nadler et al., 1978; Sloviter and Dempster, 1985). Secondly, electrophysiological studies indicate that the paroxysmal depolarizing shifts and burst-firing which can be evoked in the presence of bicuculline or low extracellular magnesium concentrations (and thought to be indicative of seizure-activity at the level of the neuron) are inhibited by A1 receptor antagonists (Hynes and Dingledine, 1984; Coan and Collingridge, 1985). Thirdly, the release of endogenous glutamate is enhanced at epileptic foci (Dodd et al., 1980).

The fourth, and perhaps most compelling, piece of evidence is that excitatory amino acid antagonists, particularly those selective for the A1 receptor, are anticonvulsants in a variety of animal models of seizure activity (Croucher et al., 1982; Meldrum et al., 1983; Croucher et al., 1984). It follows that thorough investigations into the mechanism of action of excitatory amino acids and the properties of their receptors are necessary to further evaluate the involvement of these systems in epilepsy and other neurodegenerative disorders and to point the way towards the design of new and improved therapeutic strategies. In the remainder of this chapter, I will discuss the neurodegenerative properties of excitatory amino acids in terms of their proposed mechanism of action.

EXCITOTOXICITY

The link between the neurotoxic properties of acidic amino acids and their excitatory effects was first recognized by Olney on the basis of a comparison of structure-activity relationships (Olney et al., 1971). He called these compounds 'excitotoxins' to indicate the connection between their excitatory and neurotoxic properties and put forward the 'excitotoxic hypothesis' to explain their mechanism of action. The hypothesis has three major facets:

(1) an excitatory mechanism underlies the neurotoxicity of acidic amino acids.

(2) specific receptors, located on the dendrosomal surface of susceptible neurons, mediate the excitatory effect.

(3) an excessive concentration of acidic amino acid at the receptor site evokes a state of continuous neuronal depolarization which depletes the cells energy stores in an attempt to maintain ionic gradients and/or results in a lethal alteration of the intracellular components.

In the time that has elapsed since the excitotoxic hypothesis was first proposed, considerable advances have been made in our understanding of excitatory amino acids and their receptors within the mammalian CNS. It is therefore appropriate to briefly review the hypothesis in the light of this new information.

Excitatory Mechanism

The correspondence between the potency of acidic amino acids as neuroexcitants and neurotoxic agents originally noted by Olney has been generally upheld as further agonists with varying potencies have been added to the list of excitotoxins (Schwarcz et al., 1978; Nadler et al., 1981a; Slevin et al., 1983). In addition, parallel variations of excitatory and neurotoxic potency have been demonstrated between brain regions for certain excitatory amino acids such as kainate and quinolinic acid (QUIN; deMontigny and Lund, 1980; Schwarcz and Köhler, 1983; Perkins and Stone 1983). A more direct proof has come from the observations that compounds which are selective antagonists of the excitatory effects of acidic amino acids can also block neurotoxicity (Olney et al., 1979; Schwarcz et al., 1982, 1983a). These comparisons of agonist and antagonist effects constitute the major evidence that an excitatory mechanism underlies excitotoxicity.

Specific Dendrosomal Receptors

As outlined above, structure-activity relationships and blockade by selective antagonists indicate that specific receptors mediate the excitotoxic effects of acidic amino acids. It is also clear that activation of each receptor subtype (A1, A2 or A3) can result in neurotoxicity. Olney proposed that the receptors responsible for excitotoxic effects are located on the dendrosomal surface since this is the site where the initial signs of neurodegeneration occur. The localization of excitatory amino acid binding sites as determined autoradiographically supports this idea since sites with the characteristics of A1, A2 and A3 receptors are found to be present within the dendritic regions of susceptible neurons (Foster et al., 1981; Monaghan et al., 1983, 1985; see chapter by Cotman). This is particularly clear in the hippocampal formation where A1 sites are abundant in strata radiatum and oriens of area CA1, the neurons of which are very sensitive to the neurotoxic effects of A1 agonists (Nadler et al., 1981a; Schwarcz et al., 1983b; Foster et al., 1983), whereas A3 sites have their highest density in stratum lucidum of CA3 and CA4 whose pyramidal cells are preferentially susceptible to the neurotoxic effects of kainic acid (Nadler et al., 1978). It should be noted, however, that such autoradiographical analyses cannot distinguish between a pre- or post-synaptic localization of binding sites. Little information is available from lesion studies concerning the sub-synaptic distribution of excitatory amino acid binding sites, and studies with A3 sites have proved equivocal (Monaghan and Cotman, 1982). However, membrane binding assays indicate that A1 and A2 sites are located within post-synaptic densities (Fagg and Matus, 1984). Therefore, the available evidence is consistent with the proposal that the receptors which mediate the excitotoxic effects of acidic amino acids are located within dendritic regions of susceptible neurons, in support of the excitotoxic hypothesis.

Neuronal Death

Detailed information concerning the precise events which lead to the excitotoxin-induced death of neurons is lacking at the present time. Certainly, electrophysiological studies indicate that excessive neuronal depolarization can be produced by excitatory amino acids from which a cell may never recover, supporting the idea that neurons can be 'excited-to-death'. In invertebrate muscle, excitoxicity caused by glutamate is only apparent when the normal desensitization process of the receptors is blocked (Duce et al., 1983). Desensitization is not a prominent feature of excitatory amino acid receptors on vertebrate neurons, a property which increases the likelihood of excitotoxic damage.

A lethal alteration of intracellular ion concentrations is suggested as a likely mechanism of excitotoxic neuronal death. Excitation of neurons by acidic amino acids is accompanied by large decreases of extracellular calcium apparently due to calcium entry into post-synaptic structures (Pumain and Heinemann, 1985), and electrophysiological studies indicate a calcium component to A1 receptor responses (Dingledine, 1983). Since intracellular accumulations of calcium are associated with several types of cell death (Farber, 1981), the implication is that excessive calcium entry could be the link between over-stimulation of excitatory amino acid receptors and neuronal degeneration. This possibility is supported by the observations that calcium binding protein (an intracellular calcium buffer) is lacking in neurons which are very sensitive to excitotoxins, e.g., CA3 pyramidal neurons (Baimbridge and Miller, 1982), and that calcium deposits are present in degenerating neurons following seizures (Griffiths et al., 1984). However, recent experiments using retinal slices and hippocampal neurons *in vitro* have demonstrated that removal of extracellular calcium does not affect excitotoxicity, whereas reduction of sodium or chloride is protective (Olney et al., 1984; Rothman, 1984). The suggestion is that excessive neuronal depolarization caused by influx of sodium is accompanied by a passive intracellular accumulation of chloride and that the resulting increase in intracellular osmolarity causes edema and cell lysis. The degree to which these *in vitro* experiments reflect the situation _in vivo_ remains to be evaluated, however it seems clear that changes in the internal ionic composition of neurons play a major role in excitotoxic phenomena.

Thus, our current understanding of the nature of excitatory amino acids and their receptors in the mammalian CNS supports the tenets of the excitotoxic hypothesis, and indeed this theory gives a satisfactory explanation of the basic aspects of excitotoxicity.

INDIRECT MECHANISMS OF EXCITOTOXICITY

There is good evidence, however, that factors other than direct neuronal depolarization contribute towards the neurotoxic effects of certain excitatory amino acids. This was first recognized using kainic acid, since the neurotoxicity of this compound in several areas of the mammalian brain is dependent upon intact afferent pathways (McGeer et al., 1978; Whetsell et al., 1979; Köhler et al., 1979; Coyle et al., 1981). For the most part, these neuronal pathways are thought to use glutamate or a related excitatory amino acid as transmitter, however lesions of cholinergic (Schwarcz and Köhler, 1980) and serotonergic (Berger et al., 1982) pathways are also reported to block kainate neurotoxicity. It seems certain that a component of kainate's neurotoxic effects is related to the seizures induced by this potent excitotoxin (Nadler et al., 1981b; Ben-Ari, 1985), and thus interruption of pathways critical for regenerative seizure activity will reduce the neurotoxicity.

Nevertheless, in the striatum, the evidence points strongly towards a specific interaction between kainate and the cortico-striatal terminals. In developmental studies, kainate neurotoxicity appears in parallel with glutamate uptake (Campochiaro and Coyle, 1978) and the synaptic contacts formed by invading cortical nerve terminals (Whetsell et al., 1979). Decortication does not reduce the sensitivity of striatal neurons to the excitatory effects of kainate (McLennan, 1980), but protects them from its neurotoxic action (McGeer et al., 1978; Biziere and Coyle, 1978), and this is reportedly restored by co-injection of L-glutamate (Biziere and Coyle, 1978). _In vitro_ experiments indicate that kainate can interact with excitatory amino acid-using nerve terminals in two ways, firstly by inhibiting uptake (Johnston et al., 1979) and secondly by enhancing the release of endogenous glutamate and aspartate (Ferkany et al., 1982). Coyle and his coworkers have put forward the proposal (Coyle, 1983) that the neurotoxicity of kainate is due to the combined effects of direct neuronal depolarization via A3 receptors coupled with an increased release of glutamate and aspartate which also act to depolarize the susceptible neurons.

The picture for A1 agonists was thought to be more straightforward, since the neurotoxic effects of ibotenic acid are not dependent on intact afferents (Köhler et al., 1979). However, recent studies with the endogenous A1 agonist quinolinic acid (QUIN) indicate that the situation is not so simple.

Quinolinic Acid

QUIN is a structural analogue of glutamate and aspartate (Fig. 1) which is present in animal and human brain tissue (Wolfensberger et al., 1983; Moroni et al., 1984). For many years, QUIN has been known to be present in mammalian peripheral organs as an intermediate in the biosynthesis of nicotinamide adenine dinucleotide from tryptophan (Nishizuka and Hayaishi, 1963). Peripheral QUIN is prevented from entering the CNS by the blood-brain barrier (Foster et al., 1984a), and the presence of

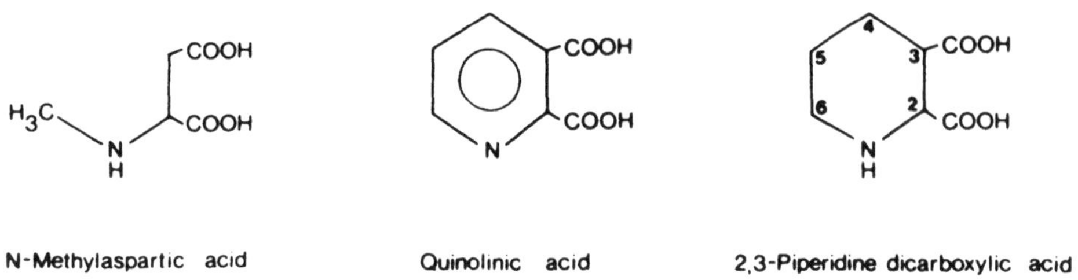

Fig. 1. Structures of N-methyl-aspartate, QUIN and 2,3-PDA.

both synthetic and degredative enzymes in CNS tissues (Foster and Schwarcz, 1984; Foster et al., 1985) indicate that it is synthesized intracerebrally. QUIN excites neurons through an action primarily at A1 receptors (Stone and Perkins, 1981; Herrling et al., 1983) and causes axon-sparing neuronal lesions when injected directly into certain CNS areas (Schwarcz et al., 1983b). Thus, QUIN is an excitotoxin which is produced within CNS tissue and as such is under investigation as a potential culprit in neurodegenerative disorders (Schwarcz et al., 1984b).

Characteristics of QUIN Excitotoxicity

In common with other compounds whose excitatory effects are mediated by A1 receptors, the neurodegeneration caused by QUIN can be potently blocked by the selective A1 antagonist D-2-amino-7-phophonoheptanoic acid (D-APH; Schwarcz et al., 1983a). This indicates that A1 receptors are intimately involved in QUIN's neurotoxic action, yet certain other features of QUIN's excitotoxicity are not characteristic of classical A1 agonists and indicate that indirect events are also involved in QUIN's actions.

QUIN shows a marked regional variation in terms of its excitotoxic effects, being very active in the hippocampus and striatum, but weak in the substantia nigra, cerebellum and hypothalamus (Perkins and Stone, 1983; Schwarcz and Köhler, 1983). This contrasts with the A1 agonist ibotenic acid which shows a more uniform spectrum of potency in these areas (Köhler and Schwarcz, 1983). In the rat hippocampus, QUIN acts preferentiallly to deplete the pyramidal neurons (particularly in area CA1 - Schwarcz et al., 1984a) whereas the granule cells of the dentate gyrus are only destroyed at high doses (Schwarcz et al., 1983b). This property is shared by the QUIN analogue cis-2,3-piperidine dicarboxylate (cis-2,3-PDA; Fig. 1) but not by other A1 agonists such as trans-2,3-PDA (Foster et al., 1983), ibotenic acid (Köhler and Schwarcz, 1983) or NMDA itself (Nadler et al., 1981a) which are equally toxic to both pyramidal and granule cells. Thus, a distinction can be made on the basis of hippocampal neuronal vulnerability between QUIN and cis-2,3-PDA on one hand and the 'classical' A1 agonists NMDA, ibotenate and trans-2,3-PDA on the other.

A similar distinction between the excitotoxic effects of these compounds is apparent when they are injected into the immature rat striatum. Both QUIN and cis-2,3-PDA, at doses which give rise to a profound loss of neurons in the adult, are without effect when injected into the 7 day old rat striatum (Foster et al., 1983), whereas trans-2,3-PDA (Foster et al., 1983), ibotenate (Steiner et al., 1984) and NMDA (Foster and Schwarcz, unpublished observations) show the same degree of neurotoxicity in both rat pup and adult striatum. In analogy with the situation for kainic acid (see above), the resistance of developing striatal neurons to QUIN's neurotoxic effects may be due to a lack of innervation by the cortico-striatal fibers which are thought to use an excitatory amino acid as transmitter. This idea is strongly supported by the observation that the removal of cortico-striatal nerve terminals following unilateral decortication in the adult rat completely abolishes the neurotoxic effects of QUIN (Schwarcz et al., 1984b). The neurotoxic effects of each of these compounds – QUIN, cis-2,3-PDA, trans-2,3-PDA, ibotenate and NMDA – can be blocked by the selective A1 antagonist D-APH (Schwarcz et al., 1982, 1983a; Foster et al., 1983), indicating that A1 receptors are fundamental to the expression of their excitotoxic effects. In that case, how is it possible to explain the apparent dependence of the neurotoxicity of QUIN and cis-2,3-PDA on afferent terminals, whereas trans-2,3-PDA, ibotenate and NMDA have no such dependence?

One simple explanation might be that in the absence of a cortical afferent, the post-synaptic A1 receptors in the striatum have an altered specificity such that they are no longer activated by QUIN and cis-2,3-PDA. This has been tested by examining binding sites which have the characteristics of A1 receptors visualized in sections of brain tissue by autoradiography. With L-[^{3}H]glutamate as the ligand, the majority of sites labeled using this technique have the characteristics of A1 receptors (Monaghan et al., 1983, 1985; see chapter by Cotman). When L-[^{3}H]glutamate binding sites are compared in adult and 7 day old rat brains (Fig. 2A and B), a similar pattern is observed with the highest density of sites occurring in the hippocampal CA1 area, cerebral cortex and striatum. Quantitative analyses (Table 2) indicate that in the 7 day old rat, the level of binding in the striatum and hippocampal area CA1 is approximately 50% of that in the adult. When a concentration of NMDA is included (100 μM) which maximally displaces all NMDA-sensitive sites, the majority of L-[^{3}H]glutamate binding is displaced in both the adult and 7 day old rat brain (Fig. 2C and D), the most notable exception being sites in the stratum lucidum and pyramidal cell layers of the hippocampus, which represent A3 and A2 sites, respectively (Monaghan et al., 1983, 1985). If, in the absence of corticostriatal afferents, the striatal A1 sites do not recognize QUIN, then a difference would be apparent between the ability of QUIN to displace L-[^{3}H]glutamate binding in the 7 day old versus the adult striatum. This is clearly not the case since QUIN (1 mM; Fig. 2E and F) is able to displace binding at both ages, a maximally effective concentration giving an identical pattern of residual binding to that observed with NMDA. By using a range of NMDA or QUIN concentrations, a measure of the affinity and maximum percentage of displaceable binding can be obtained, and these data indicate that in the striatum and hippocampus both the IC_{50} values for NMDA and QUIN and the percentage of sites displaced are equivalent at 7 days of age and in the adult (Table 2). Therefore, an alteration of the pharmacological specificity of A1 receptors is unlikely to explain the observed dependence of QUIN (and cis-2,3-PDA) neurotoxicity in the striatum upon the cortical input.

Kynurenic Acid

The actions of kynurenic acid (KYNA), a compound metabolically related to QUIN, provide a further distinction between the neurotoxicity caused by QUIN and other A1 agonists. KYNA, like QUIN, is a tryptophan metabolite (Coppini, 1975) but has been shown to antagonize excitatory amino acid responses and synaptic excitations in the CNS (Perkins and Stone, 1982; Ganong et al., 1983; Herrling, 1985). KYNA is an antagonist of A1, A2 and A3 receptors, although it shows some preference for the A1 sybtype (Ganong et al., 1983; Herrling, 1985), and is thus a 'broad spectrum' excitatory amino acid antagonist. In keeping with the excitotoxic hypothesis, KYNA is able to block the neurotoxicity caused by excitatory amino acids, but it shows a marked preference for QUIN (Foster et al., 1984b). In the rat striatum, the dose of KYNA necessary for half-maximal protection of the neurotoxicity of QUIN is 6-7 times lower than that required for NMDA and kainate, against which KYNA is equi-effective (Foster et al., 1984b).

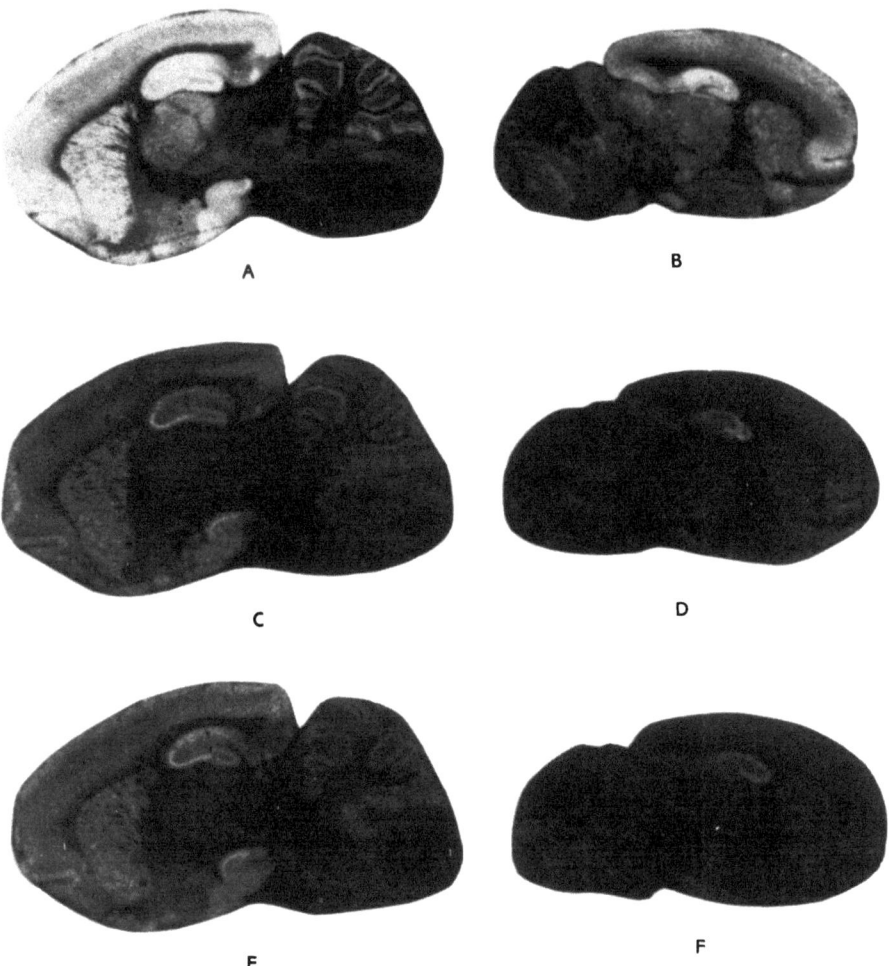

Fig. 2. Autoradiograms of L-[^{3}H]glutamate binding to adult rat and 7 day old rat pup brain: displacement by NMDA and QUIN. Autoradiograms were obtained by the method of Monaghan et al. (1983, 1985) using 100 nM L-[^{3}H]glutamate alone, or in the presence of 100 μM NMDA or 1 mM QUIN. A,C and E: adult; B,D and F: 7 day old pup. A,B: total binding; C,D: in presence of 100 μM NMDA; E,F: in presence of 1 mM QUIN. Images are the negatives of developed LKB ultrafilm, i.e., light areas represent distribution of radioactivity on section.

A Presynaptic Receptor for QUIN?

To attempt to explain the unusual excitotoxic effects of QUIN the following scenario was suggested (Foster et al., 1983; Schwarcz et al., 1984b). The apparent dependence of QUIN neurotoxicity on an excitatory amino acid-using input may indicate that the predominant action of QUIN is to increase the release of glutamate, aspartate or some related compound from the nerve terminals. The endogenous excitatory amino acid would then act on post-synaptic A1 receptors causing neuronal excitation. An excess of QUIN would cause a sustained release and consequent over-excita-

Table 2. Quantitative Data From L-[^{3}H]Glutamate Autoradiograms

Specific binding (nCi/g)

	ADULT	7 DAY OLD PUP
Striatum	3175.7 ± 239.8	1572.4 ± 73.8
Hippocampus	4156.3 ± 468.2	2872.7 ± 332.9

Inhibition Data

	ADULT		7 DAY OLD PUP	
	IC_{50} (μM)	%Imax	IC_{50} (μM)	%Imax
Striatum				
NMDA	2.2 ± 1.1	70.6 ± 6.8	1.7 ± 0.3	70.7 ± 1.7
QUIN	32.0 ± 8.5	65.9 ± 6.6	27.5 ± 9.7	71.5 ± 1.5
Hippocampus (CA1)				
NMDA	2.8 ± 0.5	78.2 ± 3.4	1.6 ± 0.6	81.5 ± 2.1
QUIN	35.8 ± 12.5	72.0 ± 4.4	21.7 ± 5.3	75.9 ± 1.6

Data were obtained from autoradiograms (see Fig. 2) and quantified using tritium standards exposed on the same film. IC_{50} and %Imax (maximum percent inhibition of specific binding) values were obtained by in iterative curve-fitting method. Values are means ± S.E.M. of three separate experiments.

tion of the post-synaptic neuron, resulting in cell death. A similar chain of events has been hypothesized to explain the neurotoxicity of kainic acid (Coyle, 1983). This scheme would explain why QUIN requires an intact excitatory amino acid input, and why QUIN excitotoxicity is blocked by A1 antagonists. The selective action of KYNA could be explained in this model by assuming that this compound can antagonize QUIN's actions at the presynaptic receptor in addition to the postsynaptic A1 receptor. Although this hypothesis accounts for all the observed neurotoxic effects of QUIN, as yet there is no evidence that QUIN has any presynaptic actions. QUIN does not affect the basal or K^+-evoked release of D-[^{3}H]aspartate from preloaded striatal slices (Foster and Schwarcz, unpublished observations). Similarly, no changes in extracellular glutamate concentration are apparent in the rat hippocampus following the injection of a neurotoxic dose of QUIN (Vezzani et al., 1985). Therefore, the possibility that a presynaptic receptor mediates the neurotoxic effects of QUIN remains to be substantiated.

A Postsynaptic Interaction With A2/A3 Receptors?

The interactions of QUIN with an excitatory amino acid input might also be expressed at the postsynaptic level (see Fig. 3). It seems clear from antagonist studies that the postsynaptic receptor which mediates the action of cortico-striatal nerve terminals is not of the A1 type, but is probably an A2 or A3 receptor (Herrling, 1985). The A1 receptors on striatal neurons may not play a role in direct excitatory synaptic events, but perhaps in more specialized forms of transmission, as has been suggested for A1 receptors in the hippocampus (Collingridge et al., 1983; Harris et al., 1984). The excitotoxic effects of potent A1 agonists such as NMDA, ibotenate and trans-2,3-PDA appear to be a result of a direct activation of A1 receptors, however the voltage-dependent block by magnesium of the channels operated by A1 receptors (Nowak et al., 1984) may present

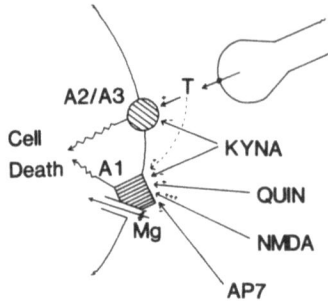

Fig. 3. Scheme to illustrate mechanism of QUIN neurotoxicity: hypothetical interaction between A1 receptor and excitatory amino acid input.

a barrier for expression of the excitotoxic effects of weak A1 agonists. Thus, QUIN and cis-2,3-PDA may receive help in overcoming the magnesium block from a tonic activation of the A2/A3 receptors mediating the depolarizing action of the cortico-striatal transmitter. This would explain the dependence of certain A1 agonists on an excitatory input and provides a reason for the preferential block of QUIN neurotoxicity by KYNA, since this antagonist is the most potent substance known to block the cortico-striatal input (Herrling, 1985). It would also explain why QUIN is equipotent with NMDA as a neurotoxin, but >10 times less potent at A1 receptors in binding studies (Fagg and Matus, 1984; Table 2). Such a postsynaptic interaction might be of considerable importance for excitotoxic mechanisms in neurodegenerative disorders. Thus an increase in the extracellular concentration of even a weak excitatory amino acid agonist (e.g., QUIN) may have profound neurotoxic effects if the activity of excitatory pathways is also increased, a situation which occurs in epileptic phenomena.

CONCLUDING REMARKS

The excitotoxic hypothesis appears to be a satisfactory explanation of the basic aspects of the neurodegenerative properties of acidic amino acids. The excitotoxicity of certain compounds such as kainate and QUIN, appears to be dependent upon a contribution from excitatory nerve terminals. This interaction may be expressed at the presynaptic level, but a synergistic relationship with ongoing excitatory transmission at the postsynaptic level is a distinct possibility. Given the evidence that A1 receptors are involved in neurodegenerative disorders such as epilepsy, ischemia and hypoglycemia (see chapters by Meldrum, Wieloch), further investigations into the excitotoxicity of endogenous A1 agonists such as QUIN may lead us towards an understanding of the etiology of these human pathologies.

ACKNOWLEDGEMENTS

I am grateful to Jacqueline Tredinnick, Alan Hudson and Roy Hammans for their assistance in the preparation of this chapter and to Dr. Robert Schwarcz for his continued interest and encouragement.

REFERENCES

Baimbridge, K.G., and Miller, J.J., 1982, Immunohistochemical localization of calcium-binding protein in the cerebellum, hippocampal formation and olfactory bulb of the rat, Brain Res., 245:223.

Ben-Ari, Y., 1985, Limbic seizure and brain damage produced by kainic acid: mechanisms and relevance to human temporal lobe epilepsy, Neuroscience, 14:375.

Berger, M., Sperk, G., and Hornykiewicz, O., 1982, Serotonergic denervation partially protects rat striatum from kainic acid toxicity, Nature, 299:254.

Biziere, K., and Coyle, J.T., 1978, Influence of cortico-striatal afferents on striatal kainic acid neurotoxicity, Neurosci. Lett., 8:303.

Campochiaro, P., and Coyle, J.T., 1978, Ontogenetic development of kainate neurotoxicity: corrrelates with glutamatergic innervation, Proc. Natl. Acad. Sci. USA, 75:2025.

Coan, E.J., and Collingridge, G.L., 1985, Magnesium ions block an N-methyl-D-aspartate receptor-mediated component of synaptic transmission in rat hippocampus, Neurosci. Lett., 53:21.

Collingridge, G.L., Kehl, S.J., and McLennan, H., 1983, The antagonism of amino acid-induced excitations of rat hippocampal CA1 neurons in vitro, J. Physiol., 334:19.

Coppini, D., 1975, Development of analytical methods for the detection and determination of kynurenic and xanthurenic acids, Acta Vitaminol. Enzymol., 29:35.

Coyle, J.T., Bird, S.J., Evans, R.H., Gulley, R.L., Nadler, J.V., Nicklas, W.J., and Olney, J.W., Excitatory amino acid neurotoxins: selectivity, specificity and mechanism of action, Neurosci. Res. Prog. Bull., 19:331.

Coyle, J.T., 1983, Neurotoxic action of kainic acid, J. Neurochem., 41:1.

Croucher, M.J., Collins, J.F., and Meldrum, B.S., 1982, Anticonvulsant action of excitatory amino acid antagonists, Science, 216:899.

Croucher, M.J., Meldrum, B.S., Jones, A.W., and Watkins, J.C., 1984, γ-D-glutamylaminomethylsulphonic acid (GAMS), a kainate and quisqualate antagonist, prevents sound-induced seizures in DBA/2 mice, Brain Res., 322:111.

deMontigny, C., and Lund, J.P., 1980, A microiontophorectic study of the action of kainic acid and putative neurotransmitters in the rat mesencephalic trigeminal nucleus, Neuroscience, 5:1621.

Dingledine, R., 1983, N-methyl aspartate activates voltage-dependent calcium conductance in rat hippocampal pyramidal cells, J. Physiol., 343:385.

Dodd, P.R., Bradford, H.F., Abdul-Ghani, A.S., Cox, D.W.G., and Continho-Netto, J., 1980, Release of amino acids from chronic epileptic and sub-epileptic foci in vivo, Brain Res., 193:505.

Duce, I.R., Donaldson, P.L., and Usherwood, P.N.R., 1983, Investigations into the mechanism of excitant amino acid cytotoxicity using a well-characterized glutamatergic system, Brain Res., 263:77.

Fagg, G.E., and Foster, A.C., 1983, Amino acid neurotransmitters and their pathways in the mammalian central nervous system, Neuroscience, 9:701.

Fagg, G.E., and Matus, A., 1984, Selective association of N-methyl-aspartate and quisqualate types of L-glutamate receptor with brain postsynaptic densities, Proc. Natl. Acad. Sci. USA, 81:6876.

Farber, J.L., 1981, The role of calcium in cell death, Life Sci., 29:1289.

Ferkany, J.W., Zaczek, R., and Coyle, J.T., 1982, Kainic acid stimulates amino acid release at presynaptic receptors, Nature, 298:757.

Fonnum, F., 1984, Glutamate: a neurotransmitter in mammalian brain, J. Neurochem., 42:1.

Foster, A.C., Mena, E.E., Monaghan, D.T., and Cotman, C.W., 1981, Synaptic localisation of kainic acid binding sites, Nature, 289:73.

Foster, A.C., Collins, J.F., and Schwarcz, R., 1983, On the excitotoxic properties of quinolinic acid, 2,3-piperidine dicarboxylic acids and structurally related compounds, Neuropharmacology, 22:1331.

Foster, A.C., and Fagg, G.E., 1984, Acidic amino acid binding sites in mammalian neuronal membranes: their characteristics and relationship to synaptic receptors, Brain Res. Rev., 7:103.

Foster, A.C., and Schwarcz, R., 1984, Synthesis of quinolinic acid by 3-hydroxyanthranilic acid oxygenase in rat brain tissue, Soc. Neurosci. Abstr., 10:11.4.

Foster, A.C., Miller, L.P., Oldendorf, W.H., and Schwarcz, R., 1984a, Studies on the disposition of quinolinic acid after intracerebral or systemic administration in the rat, Exp. Neurol., 84:428.

Foster, A.C., Vezzani, A., French, E.D., and Schwarcz, R., 1984b, Kynurenic acid blocks neurotoxicity and seizures induced in rats by the related brain metabolite quinolinic acid, Neurosci. Lett., 48:273.

Foster, A.C., Zinkand, W.C., and Schwarcz, R., 1985, Quinolinic acid phosphoribosyltransferase in rat brain, J. Neurochem., 44:446.

Fuxe, K., Roberts, P.J., and Schwarcz, R., 1983, Excitotoxins, Macmillan Press, London.

Ganong, A.H., Lanthorn, T.H., and Cotman, C.W., 1983, Kynurenic acid inhibits synaptic and acidic amino acid-induced responses in the rat hippocampus and spinal cord, Brain Res., 273:170.

Greenamyre, J.T., Olson, J.M.M., Penney, J.B., Jr., and Young, A.B., 1985, Autoradiographic characterization of N-methyl-D-aspartate-, quisqualate- and kainate-sensitive glutamate binding sites, J. Pharmacol. Exp. Therap., 233:254.

Griffiths, T., Evans, M.C., and Meldrum, B.S., 1984, Status epilepticus: the reversibility of calcium loading and acute neuronal pathological changes in the rat hippocampus, Neuroscience, 12:557.

Harris, E.W., Ganong, A.H., and Cotman, C.W., 1984, Long-term potentiation in the hippocampus involves activation of N-methyl-D-aspartate receptors, Brain Res., 323:132.

Herrling, P.L., Morris, R., and Salt, T.E., 1983, Effects of excitatory amino acids and their antagonists on membrane and action potentials of cat caudate neurons, J. Physiol., 339:207.

Herrling, P.L., 1985, Pharmacology of the cortico-caudate excitatory postsynaptic potential in the cat: evidence for its mediation by quisqualate- or kainate-receptors, Neuroscience, 14:417.

Hicks, T.P., Hall, J.G., and McLennan, H., 1978, Ranking of excitatory amino acids by the antagonists glutamate diethyl ester and D-α-aminoadipic acid, Canad. J. Physiol. Pharmacol., 56:901.

Honore, T., Lauridsen, J., and Krogsgaard-Larsen, P., 1982, The binding of [^{3}H]AMPA, a structural analogue of glutamic acid, to rat brain membrane, J. Neurochem., 36:173.

Hynes, M.A., and Dingledine, R., 1984, Attenuation of epileptiform burst firing in the rat hippocampal slice by antagonists of N-methyl-D-aspartate receptors, Soc. Neurosci. Abstr., 10:68.6.

Iwata, H., Yamagami, S., Mizio, H., and Baba, A., 1982, Cysteine sulphinic acid in the central nervous system: uptake and release of cysteine sulphinic acid by a rat brain preparation, J. Neurochem., 38:1268.

Johnston, G.A.R., Kennedy, S.M.E., and Twitchin, B., 1979, Action of the neurotoxin kainic acid on high affinity uptake of L-glutamic acid in rat brain slices, J. Neurochem., 32:121.

Köhler, C., Schwarcz, R., and Fuxe, K., 1979, Hippocampal lesions indicate differences between the excitotoxic properties of acidic amino acids, Brain Res., 175:366.

Köhler, C., and Schwarcz, R., 1983, Comparison of ibotenate and kainate neurotoxicity in rat brain: a histological study, Neuroscience, 8:819.

Lehmann, J., and Scatton, B., 1982, Characterization of the excitatory amino acid receptor-mediated release of [^{3}H]-acetylcholine from rat striatal slice, Brain Res., 252:77.

Luini, A., Goldberg, O., and Teichberg, V.I., 1981, Distinct pharmacological properties of excitatory amino acid receptors in the striatum: study by Na^+ efflux assay, Proc. Natl. Acad. Sci. USA, 78:3250.

McGeer, E.G., McGeer, P.L., and Singh, K., 1978, Kainate-induced degeneration of neostriatal neurons: dependence upon corticostriatal tract, Brain Res., 139:381.

McLennan, H., and Lodge, D., 1979, The antagonism of amino acid-induced excitation of spinal neurons in the cat, Brain Res., 169:83.

McLennan, H., 1980, The effect of decortication on excitatory amino acid sensitivity of striatal neurons, Neurosci. Lett., 18:313.

Meldrum, B.S., Croucher, M.J., Badman, G., and Collins, J.F., 1983, Antiepileptic action of excitatory amino acid antagonists in the photosensitive baboon, Papio papio, Neurosci. Lett., 39:101.

Monaghan, D.T., and Cotman, C.W., 1982, The distribution of [^{3}H]-kainic acid binding sites in the rat CNS as determined by autoradiography, Brain Res., 252:91.

Monaghan, D.T., Holets, V.R., Toy, D.W., and Cotman, C.W., 1983, Anatomical distributions of four pharmacologically distinct ^{3}H-L-glutamate binding sites, Nature, 306:176.

Monaghan, D.T., Yao, D., and Cotman, C.W., 1985, L-[^{3}H]-glutamate binds to kainate-, NMDA- and AMPA-sensitive binding sites: an autoradiographic analysis, Brain Res., 340:378.

Moroni, F., Lombardi, G., Moneti, G., and Aldinio, C., 1984, The excitotoxin quinolinic acid is present in the brain of several animal species and its cortical content increases during the aging process, Neurosci. Lett., 47:51.

Nadler, J.V., Perry, B.W., and Cotman, C.W., 1978, Intraventricular kainic acid preferentially destroys hippocampal pyramidal cells, Nature, 271:676.

Nadler, J.V., Evenson, D.A., and Cuthbertson, G.J., 1981a, Comparative toxicity of kainic acid and other acidic amino acids towards rat hippocampal neurons, Neuroscience, 6:2505.

Nadler, J.V., Evenson, D.A., and Smith, E.M., 1981b, Evidence from lesion studies for epileptogenic and non-epileptogenic neurotoxic interactions between kainic acid and excitatory innervation, Brain Res., 205:405.

Nishizuka, Y., and Hayaishi, O., 1963, Studies on the biosynthesis of nicotinamide adenine dinucleotide. I. Enzymatic synthesis of niacin ribonucleotides from 3-hydroxyanthranilic acid in mammalian tissues, J. Biol. Chem., 238:3369.

Nowak, L., Bregestovski, P., Ascher, P., Herbert, A., and Prochiantz, A., 1984, Magnesium gates glutamate-activated channels in mouse central neurons, Nature, 307:462.

Olney, J.W., Ho, O.L., and Rhee, V., 1971, Cytotoxic effects of acidic and sulphur containing amino acids on the infant mouse central nervous system, Exp. Brain Res., 14:61.

Olney, J.W., deGubareff, T., and LaBruyere, J., 1979, α-Aminoadipate blocks the neurotoxic action of N-methylaspartate, Life Sci., 25:537.

Olney, J.W., 1980, Excitotoxic mechanisms of neurotoxicity, in: Experimental and Clinical Neurotoxicology, P.S. Spencer and H.H. Schaumburg, eds., Williams and Wilkins, Baltimore, p. 272.

Olney, J.W., Price, M.T., Sanson, L., and LaBruyere, J., 1984, The ionic basis of excitotoxin-induced neuronal necrosis, Soc. Neurosci. Abstr., 10:11.8.

Perkins, M.N, and Stone, T.W., 1982, An iontophoretic investigation of the actions of convulsant kynurenines and their interaction with the endogenous excitant quinolinic acid, Brain Res., 247:184.

Perkins, M.N., and Stone, T.W., 1983, Quinolinic acid: regional variations in neuronal sensitivity, Brain Res., 259:172.

Pumain, R., and Heinemann, U., 1985, Stimulus- and amino acid-induced calcium and potassium changes in rat neocortex, J. Neurophysiol., 53:1.

Rothman, S.M., 1984, Excitatory amino acid neurotoxicity is produced by passive chloride influx, Soc. Neurosci. Abstr., 10:11.7.

Schwarcz, R., Scholz, D., and Coyle, J.T., 1978, Structure-activity relations for the neurotoxicity of kainic acid derivatives and glutamate analogues, Neuropharmacology, 17:145

Schwarcz, R., and Köhler, C., 1980, Evidence against an exclusive role of glutamate in kainic acid neurotoxicity, Neurosci. Lett., 19:243.

Schwarcz, R., Collins, J.F., and Parks, D.A., 1982, α-Amino-ω-phosphonocarboxylates block ibotenate but not kainate neurotoxicity in rat hippocampus, Neurosci. Lett., 33:85.

Schwarcz, R., and Köhler, C., 1983, Differential vulnerability of central neurons of the rat to quinolinic acid, Neurosci. Lett., 38:85.

Schwarcz, R., Whetsell, W.O. Jr., and Foster, A.C., 1983a, The neurodegenerative properties of intracerebral quinolinic acid and its structural analogue cis-2,3-piperidine dicarboxylic acid, in: Excitotoxins, K.Fuxe, P. Roberts and R. Schwarcz, eds., Macmillan Press, London, p. 122.

Schwarcz, R., Whetsell, W.O. Jr., and Mangano, R.M., 1983b, Quinolinic acid: an endogenous metabolite that produces axon-sparing lesions in rat brain, Science, 219:316.

Schwarcz, R., Brush, G.S., Foster, A.C., and French, E.D., 1984a, Seizure activity and lesions after intrahippocampal quinolinic acid injection, Exp. Neurol., 84:1.

Schwarcz, R., Foster, A.C., French, E.D., Whetsell, W.O. Jr., and Köhler, C., 1984b, Excitotoxic models for neurodegenerative disorders, Life Sci., 35:19.

Slevin, J.T., Collins, J.F., and Coyle, J.T., 1983, Analogue interactions with the brain receptor labeled by [^{3}H]kainic acid, Brain Res., 265:169.

Sloviter, R.S., and Dempster, D.W., 1985, 'Epileptic' brain damage is replicated qualitatively in the rat hippocampus by central injection of glutamate or aspartate but not by GABA or acetylcholine, Brain Res. Bull., 15:39.

Steiner, H.X., McBean, G.J., Köhler, C., Roberts, P.J., and Schwarcz, R., 1984, Ibotenate-induced neuronal degeneration in immature rat brain, Brain Res., 307:117.

Stone, T.W., and Perkins, M.N., 1981, Quinolinic acid: a potent endogenous excitant at amino acid receptors in CNS, Eur. J. Pharmacol., 72:411.

Vezzani, A., Ungerstedt, V., French, E.D., and Schwarcz, R., 1985, In vivo brain dialysis of amino acids and simultaneous EEG measurements following intrahippocampal quinolinic acid injection: evidence for a dissociation between neurochemical changes and seizures, J. Neurochem., 45:335.

Watkins, J.C., and Evans, R.H., 1981, Excitatory amino acid transmitters, Ann. Rev. Pharmacol. Toxicol., 21:165.

Whetsell, W.O. Jr., Ecob-Johnston, M.S., and Nicklas, W.J., 1979, Studies of kainate-induced caudate lesions in organotypic tissue culture, in: Advances in Neurology, Vol. 23, Huntington's Disease, Raven Press, New York, p. 645.

Wolfensberger, M., Amsler, U., Cuénod, M., Foster, A.C., Whetsell, W.O. Jr., and Schwarcz, R., 1983, Identification of quinolinic acid in rat and human brain tissue, Neurosci. Lett., 41:247.

COMMENTARY - EXCITATORY AMINO ACIDS AND RECEPTOR INTERACTIONS

C.W. Cotman and J. Storm-Mathisen

This session centered on some of the properties of excitatory amino acids as neurotransmitters in the CNS, e.g., their receptors, the pathways involved, and the possible nature of the endogenous NMDA (N-methyl-D-aspartate) ligand.

Autoradiographic data (Cotman) were presented illustrating the distribution of NMDA, kainate (KA) and quisqualate or AMPA receptor sites throughout the rodent brain. High levels of NMDA receptors are found in the hippocampus and cerebral cortex, indicating an association with highly specialized and integrative brain structures. The pharmacological properties of NMDA sensitive binding correspond quite closely to NMDA receptors studied neurophysiologically. ^{3}H-AMPA sites show a somewhat parallel distribution to NMDA sites suggesting both receptors may act in concert as, for example, has been proposed during long term potentiation. Importantly, the NMDA, QA, and KA receptor types appear to account for the majority of L-^{3}H-glutamate binding in autoradiograms when conditions of low calcium chloride are used. Excitatory amino acid-stimulated Na$^+$ influx into brain slices has also been used as an assay for study of excitatory amino acid receptors (Teichberg). This assay reveals, in addition to the three receptors, a fourth type and is useful in the screening of compounds which increase the membrane permeability to Na$^+$ ions.

An important question is what is the nature of the molecules released from putative excitatory amino acid terminals in addition to or instead of glutamate and/or aspartate? Cuénod described the use of an extremely sensitive HPLC method which allows resolution of many released endogenous excitatory amino acids. Superfusates from depolarized brain slices contain, in addition to glutamate or aspartate, cysteine sulfinic acid, homocysteine sulfinic acid, and homocysteic acid (HCA), the exact amounts depending on the brain area stimulated. Electrophysiological data suggests that L-HCA acts predominantly at the NMDA receptor. Preliminary results with monoclonal antibodies suggest that homocysteate may be localized in nerve terminals in the cerebral cortex and caudatoputamen, possibly of thalamic afferents. It was also noted that release and immunological data correlated quite well with the autoradiographically defined distribution of NMDA receptors. Thus the search for the endogenous NMDA receptor agonist is progressing.

The technique of immunocytochemical localization of amino acids is now well established (Storm-Mathisen). Thus metabolic compartmentation and functional changes in glutamate and aspartate can be studied. These substances seem concentrated in excitatory nerve endings in slices fixed in vitro, but not in brains of animals fixed by perfusion. Although the distribution obtained in the latter material would not be inconsistent with the estimated pool sizes of glutamate (see below), the apparent discrepancy is a major challenge for further studies. One possibility to bear in mind is that glutamate and/or aspartate derivatives or oligopeptides could be the real transmitters. Such a substance is N-acetyl aspartyl glutamate, which has recently been localized immunocytochemically in nerve endings (cf. Coyle, this volume).

The finding that the antibodies interact even with the free amino acids suggests that in the future such antibodies could be used as tools to modify synaptic events, to produce amino acid selective electrodes, or even that antidiotypic antibodies could be found which could be useful ligands for receptors and transport proteins.

For anatomical mapping of excitatory amino acid pathways, retrograde tracing by means of D-[^{3}H]aspartate seems to be the best approach. Many neurons have been thus identified as tentatively 'excitatory amino acid-ergic'. It is important to obtain further evidence by recording release and synaptic responses.

Based on denervation and various metabolic experiments in corpus striatum, rough estimates of the relative glutamate content in the different pools were put forward (Fonnum): excitatory nerve endings ~ 20%, GABA nerve endings < 10%, neuronal perikarya ~ 50%, glia ~ 20%. Although some of the assumptions involved evidently make the values uncertain, they may be useful for further work. These studies also emphasize the importance of glial function for supporting the formation of glutamine. Moreover, they suggest that transmitter glutamate is turning over faster than 'metabolic' glutamate, and that nerve endings may release glutamate and aspartate in varying proportions depending on the metabolic conditions.

Endogeneous excitatory amino acids are likely to play a role in producing excitotoxic lesions in brain (Foster). Quinolinic acid may have a special role in this context being a rigid aspartate analog which is formed and broken down in the brain (cf. Schwarcz, this volume). It also has other interesting features, including the existence of an endogenous antagonist, kynurenic acid. The findings suggest that other endogenous compounds with similar properties should be searched for and could be involved in the pathogenesis of disorders of the nervous system, and perhaps in normal defense against excitotoxic damage. It seems necessary to invoke interplay of more than one type of postsynaptic receptors, as well as of presynaptic receptors, to explain the differing properties of excitotoxic compounds.

SESSION V
EXCITATORY AMINO ACIDS AND SEIZURES:
NEUROCHEMICAL INTERRELATIONSHIPS

EXCITATORY AMINO ACID ANTAGONISTS AS NOVEL ANTICONVULSANTS

B. Meldrum

Neurology Department
Institute of Psychiatry
Denmark Hill, London SE5 8AF

INTRODUCTION

The convulsant effect of application of dicarboxylic amino acids to the cortex was first reported by Hayashi (1954). This observation suggests that antagonists of excitation induced by amino acid neurotransmitters might be anticonvulsant agents in some forms of epilepsy. Some rather weak and indeterminate anticonvulsant effects of glutamic acid diethyl ester were initially described (Freed and Michaelis, 1978; Freed, 1985). However, following the identification of potent and specific excitatory amino acid antagonists (Davies et al., 1982) it was shown that compounds that selectively antagonized excitation at the N-methyl-D-aspartate preferring receptor are anticonvulsant, with a potency matching that of the benzodiazepines when administered intracerebroventricularly (Croucher et al., 1982; Chapman et al., 1984). Testing in a wide range of animal models shows that NMDA antagonists provide a novel class of anticonvulsant agent with a broad spectrum of activity roughly equivalent to that of sodium valproate when administered systemically. The further observation that they can protect against ischemic brain damage has strengthened the concept that an excitotoxic mechanism is involved in such damage (Meldrum et al., 1982; Simon et al., 1984).

This chapter summarizes evidence relating to the anticonvulsant action of excitatory amino acid antagonists, emphasizing the specific effect of NMDA receptor blockade and discussing the possible cooperative contribution of blockade at other excitatory receptor subtypes. Critical focal sites of anticonvulsant action of excitatory amino acid antagonists are discussed. The therapeutic potential of NMDA antagonists is reviewed in 1) epilepsy, assessing known or possible neurological side effects and 2) in neurodegenerative disorders, particularly ischemic brain damage.

NMDA ANTAGONISTS AS ANTICONVULSANTS

The range of animal models of epilepsy in which NMDA antagonists have been shown to be anticonvulsant is illustrated in Table 1. It includes genetically determined syndromes of reflex epilepsy, chemically and electrically induced seizures.

In terms of minimal effective dose, the sound-induced seizures in

Table 1. Anticonvulsant Potency of Selective NMDA Antagonists
Administered Systemically in Animal Models of Epilepsy

Model	Effective Dose (mmol/kg) 2APH	2APP	Reference
Sound-induced seizures (DBA/2 mice) clonus:	0.05	0.3	Meldrum et al., 1983b
wild running:	0.19	1.1	
Photically-induced myoclonus (Papio papio)	0.3-1.0	1.0-3.3	Meldrum et al., 1983a
High pressure neurological syndrome (rats)	<1.0		Meldrum et al., 1983c
Electroshock seizures (rats) (current threshold increase)	0.24	0.48	Czuczwar et al., 1985
Kindled seizures (rat) amygdala	(0.5 µmol i.c.v.	0.5 µmol) i.c.v.	Peterson et al., 1983
Chemically-induced seizures 3-mercaptopropionic acid	0.165	0.33	Meldrum et al., 1983b
N-methyl-D-aspartate	0.165-0.33	0.33	Meldrum et al., 1983b
Pentylenetetrazol	1.2		Croucher et al., 1982

DBA/2 mice are the most sensitive to excitatory amino acid antagonists. This raises the possibility that the effect might be related to afferent transmission in the auditory pathway. However, this appears to be an unlikely explanation as the later seizure stages are more susceptible to protection than the initial stage (which is directly triggered by the sound). Even in the models in which rather high systemic doses are required to achieve a full anticonvulsant effect, such as the baboons with photically-induced myoclonus, there is still a significant margin between the anticonvulsant dose of 2-APH and the dose producing loss of muscle tone (the only overt neurological side effect).

Comparison of effective doses of 2-amino-7-phosphonoheptanoic acid (2-APH) and of 2-amino-5-phosphonopentanoic acid (2-APP) shows that 2-APH is consistently more potent (by a factor of at least 2). This difference is as great or greater when the intracerebroventricular route of administration is used (see Table 2) so that differential entry to the brain would not appear to be the explanation. However, such a difference in potency is not seen when studying antagonism of depolarization induced in the spinal cord by excitatory amino acids (Evans et al., 1982) or single unit responses to iontophoretic application of agonist and antagonist (Meldrum et al., 1983b) or even when an in vitro cortical slice preparation is used to assess 'anticonvulsant' potency (Lodge et al., 1985).

The relative potency of 2-APH compared with other anticonvulsants (such as benzodiazepines and sodium valproate) decreases by two orders of magnitude if administered systemically rather than intracerebroventricularly (Chapman et al., 1984). This is presumably due to limited entry into the brain from the blood. Nevertheless, studies with ^{3}H-2-APH have shown that the compound accumulates in the brain in a non-metabolized

Table 2. Comparative Potency of NMDA Antagonists Against Two Phases of the Sound-Induced Seizure Response in DBA/2 Mice Following Intracerebroventricular Injection

Compound	ED_{50} (µmol) Clonus	Wild running
β-D-aspartylaminomethyl phosphonate	0.0006	0.0028
2-amino-7-phosphonoheptanoate	0.0015	0.0037
γ-D-glutamylaminomethyl phosphonate	0.0018	0.0096
2-amino-5-phosphonovalerate	0.020	0.039
2-amino-6-phosphonohexanoate	0.14	0.17

References: Croucher et al., 1982; Jones et al., 1984

form with a time course that corresponds to the time course of its anticonvulsant action (Chapman et al., 1983). The *in vivo* concentration of 2-APH is close to the threshold concentration for NMDA antagonism in *in vitro* spinal cord studies (0.5 µM).

It is reasonable to assume that 2-APH acts on NMDA receptors in the neocortex, hippocampus or other regions with a low threshold for burst discharges to attenuate the initiation or spread of seizure activity. This interpretation is supported by *in vitro* studies in hippocampal slices reported at this meeting (Lodge et al., 1985; Hablitz, 1985; Swann and Brady, 1985; King and Dingledine, 1985). However, all these studies employed concentrations of 2-APH or 2-APP (5-100 µM) higher than are found *in vivo*. Studies employing focal microinjection of 2-APH in various seizure models in the rat have shown that certain midbrain sites are extremely sensitive to the anticonvulsant effect of 2-APH. In particular, limbic seizures induced by the intraperitoneal injection of pilocarpine (380 mg/kg) can be totally suppressed by the focal injection of 2-APH in the substantia nigra pars reticulata (Turski et al., 1986) or in the entopeduncular nucleus (Patel et al., 1986). These experiments show that the output of the basal ganglia system critically influences the spread of seizure activity within the limbic system. Sound-induced seizures in genetically epilepsy prone rats (University of Arizona, Wistar strain) can be totally suppressed by focal microinjection of 2-APH in the inferior colliculus, substantia nigra (pars reticulata) or the midbrain reticular formation (Millan et al., 1986). In this reflex seizure model, protective effects apparently occur on both the sensory input and motor output systems. In maximal electroshock seizures, intranigral injections of 2-APH protect against the tonic extensor component of the motor response (De Sarro et al., 1984).

NON-SELECTIVE (KAINATE/QUISQUALATE/NMDA) ANTAGONISTS AS ANTICONVULSANTS

Some compounds that are antagonists at all three receptor subtypes and show anticonvulsant activity are listed in Table 3. This indicates their relative potencies as antagonists at the three major receptor subtypes as estimated from spinal cord studies. It should be noted that there are no antagonists which are totally selective at the kainate-preferring receptor. However, γ-D-glutamylaminomethyl sulphonate (γ-D-GAMS), 1-(p-bromobenzoyl)-2,3-dicarboxylate and its chloro analog all preferentially block excitation due to quisqualate or NMDA. The anticonvulsant potency of 1-(p-bromobenzoyl)-2,3-dicarboxylate is apparently greater than could be accounted for by its antagonism of NMDA. However, there may be important cooperative effects between antagonism at different receptor

Table 3. Anticonvulsant Potency in DBA/2 Mice of Non-Selective Excitatory Amino Acid Antagonists

Compound	ED_{50} (μmol, i.c.v.) clonus	Receptor Subtype Activity[a]		
		NMDA	kainate	quisqualate
γ-D-glutamyl glycine[b]	0.046	+++	++	+(+)
cis-2,3-piperidine dicarboxylate[c]	0.017	++(+)	++(+)	+(+)
kynurenic acid[d]	0.090	+++	+	+
γ-D-glutamylaminomethyl sulphonate[e]	0.074	+	++(+)	++
1-(p-bromobenzoyl) piperazine -2,3-dicarboxylate[f]	0.005	++	+++	++
1-(p-chlorobenzoyl) piperazine -2,3-dicarboxylate[f]	0.015	++	++(+)	++

References: [a] Davies et al., 1982; Davies et al., 1984; Peet et al., 1983;
[b] Croucher et al., 1982
[c] Croucher et al., 1984b
[d] Mello and Meldrum, unpublished
[e] Croucher et al., 1984a
[f] Chapman et al., 1985

subtypes. The membrane conductance effects of kainate are enhanced in the presence of threshold concentrations of NMDA (Mayer and Westbrook, 1984). The depolarizing action of NMDA is enhanced (or overcome) by depolarization following activation at the glutamate/quisqualate receptor. Thus, combined effects on NMDA and non-NMDA receptors may be particularly effective. This would appear to be the case with kynurenic acid whose potency as an NMDA antagonist appears not to be sufficient to account for its anticonvulsant activity.

Glutamic Acid Diethylester

In contrast, GDEE which lacks NMDA antagonist activity, but shows some quisqualate antagonism in some, but not all, neurophysiological tests, protects against tonic-clonic seizures in diverse syndromes, including alcohol withdrawal convulsions, seizures induced by homocysteine and seizures induced by enhanced atmospheric pressure (Wardley-Smith and Meldrum, 1984; Freed, 1985). It is inactive in a wide range of models including audiogenic seizures in mice and electroshock. GDEE apparently exacerbates some kinds of seizure. It is certainly possible that post-synaptic blockade of glutamate/quisqualate receptors could lead to enhanced synaptic release of glutamate and greater action on NMDA receptors.

β-KAINIC ACID AND EFFECTS ON EXCITATORY AMINO ACID RELEASE

The lack of a specific antagonist acting selectively on the kainate receptor led to the synthesis and *in vivo* evaluation of a series of substituted analogues of α-kainate and β-kainate (Collins et al., 1984). Surprisingly, β-kainic acid proved to be a potent anticonvulsant, both by the intracerebroventricular route (ED_{50} against clonic phase of seizure response in DBA/2 mice = 0.09 μmol), and by the intraperitoneal route (ED_{50} against clonic phase = 0.66 mmol/kg). Coinjection of specific agonists and antagonists into the cerebral ventricles of mice, with estimation of the ED_{50}

for induction of clonus by the agonist, provides a method that clearly differentiates 2-APH from γ-D-GAMS (Turski et al., 1985). Thus, 2-APH (0.1 μmol) increases the ED_{50} of NMDA 153-fold and that of α-KA 6-fold, whereas γ-D-GAMS (1.0 μmol) increases the ED_{50} of α-KA 90-fold and that of NMDA 27-fold (see Fig. 1). β-KA (1.0 μmol) behaves like 2-APH, increasing the ED_{50} for NMDA 100-fold and that for α-KA only 16-fold.

In microiontophoretic studies, β-kainate appears to be a weak agonist at the α-kainate receptor, with very little if any antagonist activity at the NMDA, kainate or quisqualate receptors (Stone and Collins, 1985). However, the potassium-stimulated release of glutamate from rat hippocampal slices is halved in the presence of 1 mM β-kainate (Connick and Stone, 1985). Thus, it is necessary to consider whether the anticonvulsant actions of β-kainate could be explained by a presynaptic action modifying the release of excitatory amino acids. The data of Fig. 1 would appear to run contrary to such an explanation.

Presynaptic mechanisms controlling release of excitatory amino acids are not well understood. There is evidence for an adenosine receptor mechanism (Dolphin and Archer, 1983) a $GABA_B$ (baclofen) receptor and perhaps a GABA/benzodiazepine receptor (Meldrum and Chapman, 1983). However there is considerable uncertainty about the role of an autoreceptor. L-2-amino-4-phosphonobutyric acid apparently acts at a presynaptic receptor to depress synaptic transmission in the hippocampus (Fagg and Lanthorn, 1985). However this compound is not an effective anticonvulsant in test systems in which 2-APH and β-kainate are highly effective. Interestingly, it does inhibit the development of kindled seizures.

Decreased release of excitatory amino acids may play a role in the anticonvulsant action of several established anticonvulsants, such as benzodiazepines and barbiturates (Meldrum and Chapman, 1983), and some novel agents such as lamotrigine. We need to know more about the specificity of neurotransmitter release in relation to post-synaptic receptors before this approach can be pursued on a rational basis.

THERAPEUTIC POTENTIAL OF SPECIFIC EXCITATORY AMINO ACID ANTAGONISTS IN EPILEPSY

On the basis of the studies in animal models, excitatory amino acid antagonists with activity at the NMDA-preferring receptor provide a novel class of anticonvulsant agents. They appear likely to show a broad spectrum of anticonvulsant activity in human epilepsy. In terms of neurological toxicity, two effects require consideration. Firstly, muscle relaxation or weakness is apparent at dose levels 2-4 times those producing an anticonvulsant effect. This is also manifested as an antispastic action in animal models (Turski et al., 1985b). There is also the possibility of impairment of some cognitive processes, particularly spatial learning (Anderson et al., 1985), an effect possibly related to the action of NMDA antagonists on long-term potentiation. Both these side effects appear comparable to the side effects of anticonvulsant benzodiazepines.

However, none of the compounds described here are optimal from the point of view of pharmacokinetics. The most potent compound, 2-APH, is not active by the oral route. Also, it is likely that a more lipophilic analog or derivative would penetrate the blood-brain barrier better and be active at lower systemic doses.

Finally, we do not know whether the optimal anticonvulsant will be an antagonist only at the NMDA receptor, or at a subtype of NMDA receptor (Perkins and Stone, 1983) or might possess a particular combination of

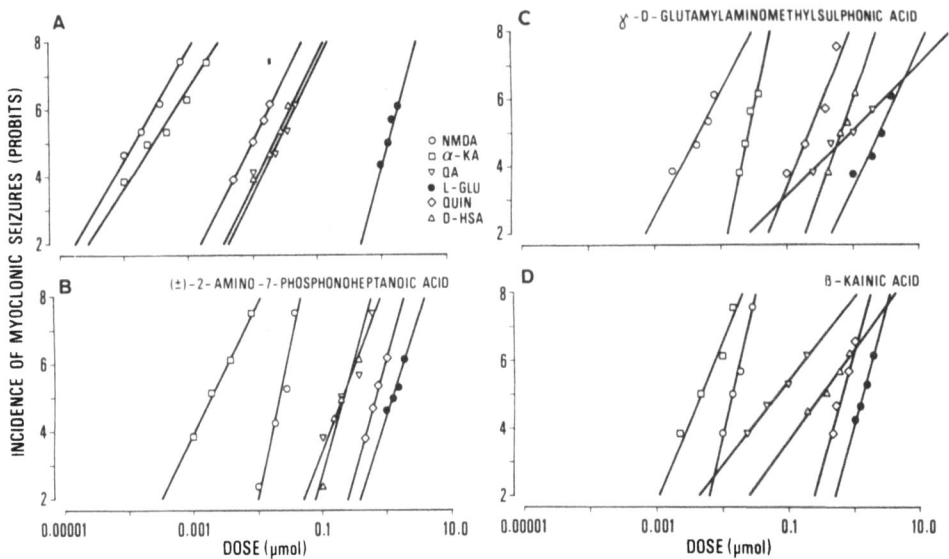

Fig. 1. Probit-log dosage regression curves for myclonic seizures induced by amino acid agonists in mice (NMDA = N-methyl-D-aspartate; α-KA = α-kainate; QA = quisqualate; L-Glu = L-glutamate; QUIN = quinolinic acid; D-HSA = D-homocysteine sulphinate. The incidence of myclonic seizures (probit-transformed percentages) is plotted against the log doses of excitatory amino acid agonists and the data fitted by linear regression analysis. All drugs were administered into the lateral cerebral ventricle of Swiss S mice in a volume of 0.5 µl, and the animals observed for 30 minutes. A: Agonists alone; B: Agonist + 2-APH, 0.1 µmol; C: Agonist + γ-D-GAMS, 1.0 µmol; D: Agonist + β-kainic acid, 1.0 µmol. (Data from Turski et al., 1985a).

actions at the NMDA receptor and at other sites. Indeed, different syndromes of epilepsy might respond to different blends of antagonist potency.

THERAPEUTIC POTENTIAL OF EXCITATORY AMINO ACID ANTAGONISTS IN NEURODEGENERATIVE DISORDERS

The neurotoxic actions of N-methyl-D-aspartate in the hypothalamus can be blocked by the systemic administration of D-2APP (Olney et al., 1981) and focal hippocampal lesions induced by ibotenate or quinolinate can be blocked by co-injection of 2-APH (Schwarcz and Meldrum, 1985).

The first pharmacological evidence implicating the action of endogenous excitatory transmitters acting on the NMDA receptor in the pathogenesis of neurodegenerative disorders was provided by Simon et al. (1984a). They showed that ischemic neuronal degeneration in the rat hippocampus induced by a transient episode of forebrain ischemia could be prevented by the local injection of 2-APH. The excitotoxic mechanism appears to be related to the induction of burst firing in selectively vulnerable neurons during the phase of reperfusion. This leads to a massive intracellular accumulation of calcium (Simon et al., 1984b) in the vulnerable neurons, and secondary cell death. Excessive activity at NMDA receptors may underly patterns of selective cell loss seen in a number of neurodegenerative disorders. This suggests a possible therapeutic role for excitatory amino acid antagonists in a wide range of neurological disorders that at present lack pharmacotherapy (Schwarcz and Meldrum, 1985).

REFERENCES

Anderson, E., Baudry, M., Lynch, G., and Morris, R.G.M., 1985, Selective impairment of learning and blockade of long-term potentiation by an N-methyl-D-aspartate receptor antagonist, APV-5, J. Physiol., 367:31P.

Chapman, A.G., Collins, J.F., Meldrum, B.S., and Westerberg, E., 1983, Uptake of a novel anticonvulsant compound, 2-amino-7-phosphonoheptanoic [4,5-^{3}H] acid into mouse brain, Neurosci. Lett., 37:75.

Chapman, A.G., Croucher, M.J., and Meldrum, B.S., 1984, Evaluation of anticonvulsant drugs in DBA/2 mice with sound-induced seizures, Arzneim. Forschung, 34:1261.

Chapman, A.G., Hart, G.P., Meldrum B.S., Turski, L., and Watkins, J.C., 1985, Anticonvulsant activity of two novel piperazine derivatives with potent kainate antagonist activity, Neurosci. Lett., 55:325.

Collins, J.F., Dixon, A.J., Badman, G., DeSarro, G., Chapman, A.G., Hart, G.P., and Meldrum, B.S., 1984, Kainic acid derivatives with anticonvulsant activity, Neurosci. Lett., 51:371.

Connick, J.H., and Stone, T.W., 1985, β-kainic acid inhibits glutamate release from rat brain slices, IRCS Med. Sci., 13:824.

Croucher, M.J., Collins, J.F., and Meldrum, B.S., 1982, Anticonvulsant action of excitatory amino acid antagonists, Science, 216:899.

Croucher, M.J., Meldrum, B.S., Jones, A.W., and Watkins, J.C., 1984a, γ-D-Glutamylaminomethyl-sulphonic acid (GAMS), a kainate and quisqualate antagonist, prevents sound-induced seizures in DBA/2 mice, Brain Res., 322:111.

Croucher, M.J., Meldrum, B.S., and Collins, J.F., 1984b, Anticonvulsant and proconvulsant properties of a series of structural isomers of piperidine dicarboxylic acid, Neuropharmacology, 23:467.

Czuczwar, S.J., Cavalheiro, E.A., Turski, L., Turski, W.A., and Kleinrok, Z., 1985, Phosphonic analogues of excitatory amino acids raise the threshold for maximal electroconvulsions in mice, Neurosci. Res., 3:86.

Davies, J., Evans, R.H., Jones, A.W., Smith, D.A.S., and Watkins, J.C., 1982, Differential activation and blockade of excitatory amino acid receptors in the mammalian and amphibian central nervous system, Comp. Biochem. Physiol., 72C:211.

Davies, J., Jones, A.W., Sheardown, M.J., Smith, D.A.S., and Watkins, J.C., 1984, Phosphono dipeptides and piperazine derivatives as antagonists of amino acid induced and synaptic excitation in mammalian and amphibian spinal cord, Neurosci. Lett., 52:79.

De Sarro, G., Meldrum, B.S., and Reavill, C., 1984, Anticonvulsant action of 2-amino-7-phosphonoheptanoic acid in the substantia nigra, Eur. J. Pharmacol., 106:175.

Dolphin, A.C., and Archer, E.R., 1983, An adenosine agonist inhibits and a cyclic AMP analogue enhances the release of glutamate but not GABA from slices of rat dentate gyrus, Neurosci. Lett., 43:49.

Evans, R.H., Francis, A.A., Jones, A.W., Smith, D.A.S., and Watkins, J.C., 1982, The effects of a series of ω-phosphonic α-carboxylic amino acids on electrically-evoked and excitant amino acid-induced responses in isolated spinal cord preparations, Br. J. Pharmac., 75:65.

Fagg, G.E., and Lanthorn, T.H., 1985, Cl^-/Ca^{2+}-dependent L-glutamate binding sites do not correspond to 2-amino-4-phosphonobutanoate-sensitive excitatory amino acid receptors, Br. J. Pharmacol., 86:743.

Freed, W.J., and Michaelis, E.K., 1978, Glutamic acid and ethanol dependence, Pharmacol. Biochem. Behav., 8:509.

Freed, W.J., 1985, Selective inhibition of homocysteine-induced seizures by glutamic acid diethyl ester and other glutamate esters, Epilepsia, 26:10.

Hablitz, J.J., 1985, Suppression of synaptically evoked epileptiform discharges in the hippocampus by NMDA antagonists, Epilepsia, 26:512.

Hayashi, T., 1954, Effects of sodium glutamate on the nervous system, Keio J. Med., 3:183.

Jones, A.W., Croucher, M.J., Meldrum, B.S., and Watkins, J.C., 1984, Suppression of audiogenic seizures in DBA/2 mice by two new dipeptide NMDA receptor antagonists, Neurosci. Lett., 45:157.

King, G.L., and Dingledine, R., 1985, Role for N-methyl-D-aspartate receptors in epileptiform bursting, Epilepsia, 26:509.

Lodge, D., Martin, D., and Aram, J.A., 1985, Comparison of DL-2-amino-5-phosphono-valerate (AP5) and DL-2-amino-7-phosphonoheptanoate (AP7) as anticonvulsants and N-methyl-D-aspartate antagonists in vitro, Epilepsia, 26:507.

Mayer, M.L., and Westbrook, G.L., 1984, Mixed-agonist action of excitatory amino acids on mouse spinal cord neurones under voltage clamp, J. Physiol. 354:29.

Meldrum, B.S., Griffiths, T., and Evans, M., 1982, Hypoxia and neuronal hyperexcitability - a clue to mechanisms of brain protection, in: Protection of Tissue Against Hypoxia, A. Wauquier, M. Borgers and W.K. Amery, eds., Elsevier/North Holland, p. 275.

Meldrum, B.S., and Chapman, A.G., 1983, Excitatory amino acids and anticonvulsant drug action, in: Glutamine, Glutamate, and GABA in the Central Nervous System, L. Hertz, E. Kvamme, E. McGeer and A. Schousboe, eds., Alan R. Liss Inc., New York, p. 625.

Meldrum, B.S., Croucher, M.J., Badman, G., and Collins, J.F., 1983a, Antiepileptic action of excitatory amino acid antagonists in the photosensitive baboon, Papio papio, Neurosci. Lett., 39:101.

Meldrum, B.S., Croucher, M.J., Czuczwar, S.J., Collins, J.F., Curry, K., Joseph, M., and Stone, T.W., 1983b, A comparison of the anticonvulsant potency of ($\pm$)2-amino-5-phosphonopentanoic acid and ($\pm$)2-amino-7-phosphonoheptanoic acid, Neuroscience, 9:925.

Meldrum, B.S., Wardley-Smith, B., Halsey, M., and Rostain, J. C., 1983c, 2-Amino-7-phosphonoheptanoic acid protects against the high pressure neurological syndrome, Eur. J. Pharmacol., 87:501.

Millan, M.H., Faingold, C.L., and Meldrum, B.S. 1986, Intranigral injection of 2-amino-7-phosphonoheptanoic acid protects against audiogenic seizures in genetically epilepsy prone rats, in: Advances in Epileptology, P. Wolf, ed., Raven Press, New York.

Olney, J.W., Labruyere, J., Collins, J.F., and Curry, K., 1981, D-aminophosphonovalerate is 100 fold more powerful than D-alpha-aminoadipate in blocking N-methylaspartate neurotoxicity, Brain Res., 221:207.

Patel, S., Millan, M.H., Mello, L.M., and Meldrum, B.S., 1986, 2-amino-7-phosphonoheptanoic acid (2-APH) infusion into entopeduncular nucleus protects against limbic seizures in rats, Neurosci. Lett., 64:226.

Peet, M.J., Leah, J.D., and Curtis, D.R., 1983, Antagonists of synaptic and amino acid excitation of neurons in the cat spinal cord, Brain Res., 266:83.

Perkins, M.N., and Stone, T.W., 1983, Pharmacology and regional variations of quinolinic acid-evoked excitations in the rat central nervous system, J. Pharmacol. Exp. Therap., 226:551.

Peterson, D.W., Collins, J.F., and Bradford, H.F., 1983, The kindled amygdala model of epilepsy: anticonvulsant action of amino acid antagonists, Brain Res., 275:169.

Schwarcz, R., and Meldrum, B.S., 1985, Excitatory amino acid antagonists provide a therapeutic approach to neurological disorders, Lancet, ii:140.

Simon, R.P., Swan, J.H., Griffiths, T., and Meldrum, B.S., 1984a, Blockade of N-methyl-D-aspartate receptors may protect against ischemic damage in the brain, Science, 226:850.

Simon, R.P., Griffiths, T., Evans, M.S., Swan, J.H., and Meldrum, B.S., 1984b, Calcium overload in selectively vulnerable neurons of the hippocampus during and after ischemia: an EM study in the rat, J. Cerebr. Blood Flow Metab., 4:350.

Stone, T.W., and Collins, J.F., 1985, β-kainic acid is not an amino acid antagonist, J. Pharm. Pharmacol., 37:668.

Swann, J.W., and Brady, R.J., 1985, NMDA antagonists block penicillin-induced after discharges in immature CA3 hippocampal neurons, Epilepsia, 26:512.

Turski, L., Collins, J.F., and Meldrum, B.S., 1985a, Is β-kainic acid an N-methyl-D-aspartate antagonist?, Brain Res., 336:162.

Turski, L., Schwarz, M., Turski, W.A., Klockgether, T., Sontag, K.H., and Collins, J.F., 1985b, Muscle relaxant action of excitatory amino acid antagonists, Neurosci. Lett., 53:321.

Turski, L., Cavalheiro, E.A., Turski, W.A., and Meldrum, B.S., 1986, Excitatory neurotransmission within substantia nigra pars reticulata regulates threshold for seizures produced by pilocarpine in rats: effects of intranigral 2-amino-7-phosphonoheptanoate and N-methyl-D-aspartate, Neuroscience, 18:61.

Wardley-Smith, B., and Meldrum, B.S., 1984, Effect of excitatory amino acid antagonists on the high pressure neurological syndrome in rats, Eur. J. Pharmacol., 105:351.

THE HYPEREXCITED BRAIN: GLUTAMIC ACID RELEASE AND FAILURE OF INHIBITION

N.M. van Gelder

CRSN/Dép. de physiologie, Faculté de médecine
Université de Montréal, C.P. 6128, succursale A
Montréal, Québec, H3C 3J7

There no longer exists much doubt that epilepsy is associated with a rather specific set of biochemical alterations (Delgado-Escueta and Greenberg, 1984). Nevertheless, considerable controversy still surrounds the significance of these alterations, touching on several areas of importance. Perhaps the most crucial issue to be resolved is the question whether the biochemical changes found are in consequence of the cerebral dysfunction or whether, indeed, such alterations contribute directly to the cause of the condition. Traditionally, a diagnosis of epilepsy and its classification to type is most commonly confirmed by electroencephalographic evidence. However, the initial reason for arriving at such a diagnosis is usually occasioned only after an individual has complained of periodic and unpredictable episodes of uncontrollable movements, sensations, vegetative or emotional 'storms', or other inappropriate outward expressions of autonomous cerebral activity. Hence, an abnormal or epileptiform EEG activity per se, whether focal or diffuse, in most instances is no cause for a positive and clear diagnosis of epilepsy (Hockaday and Whitty, 1969).

An EEG activity of this type, without being accompanied by clinical manifestations in the form of seizures, is therefore more prevalent among the general population than is the incidence of epilepsy (Cloninger et al., 1982). For this reason, it would appear imperative to isolate from the biochemical parameters accompanying epilepsy those changes which are associated with the predisposition or an interictal period, as opposed to the neurochemical events clearly associated with a seizure in progress. One may suspect that, in part, the biochemical changes in the brain of an individual who never exhibits seizures despite the presence of epileptiform EEG activity will overlap and, possibly, are exaggerated during an epileptic episode. On the other hand, it appears equally plausible that the metabolic condition giving rise to such a predisposition has to be joined at a specific moment to one or more additional biochemical alterations, in order to produce a seizure.

To separate from among the wealth of data obtained over the past 20 years with regard to these two sets of biochemical correlates, it is necessary to adopt different investigative and experimental strategies. In one case, the strategy is intended to examine the neurochemical and biochemical events which are associated with an abnormal EEG suggesting increased cerebral excitation, but which is uncomplicated by accompanying seizure activity. The second strategy should be directed to uncovering those neurochemical changes common to any seizure state, irrespective of the origin or the manner in which the seizure state was elicited.

THE BIOCHEMISTRY OF AN EPILEPTIC PREDISPOSITION

Electrophysiologically, an epileptic tendency in neural tissue is defined by an exaggerated discharge of certain neurons on normal afferent input and, most importantly, an enhanced probability that such a discharge recruits immediately neighboring neurons, to produce an abnormally hypersynchronous discharge pattern (Traub and Wong, 1982; Johnston and Brown, 1984). That a particular type of cerebral cytoarchitecture aids in promoting or resisting the biochemical influences responsible for this hypersynchrony, is evident from the fact that a number of brain regions are far more receptive to the phenomenon than are others. A notable example of opposite extremes is represented by two neural structures which on first appearance are equivalent in terms of cytoarchitectural orderliness and which both receive important GABA mediated inhibitory input. Yet abnormal hypersynchronous activity in the cerebellum is very rare whereas the hippocampus, in contrast, demonstrates a very high susceptibility towards epileptiform discharges (Prince and Connors, 1984).

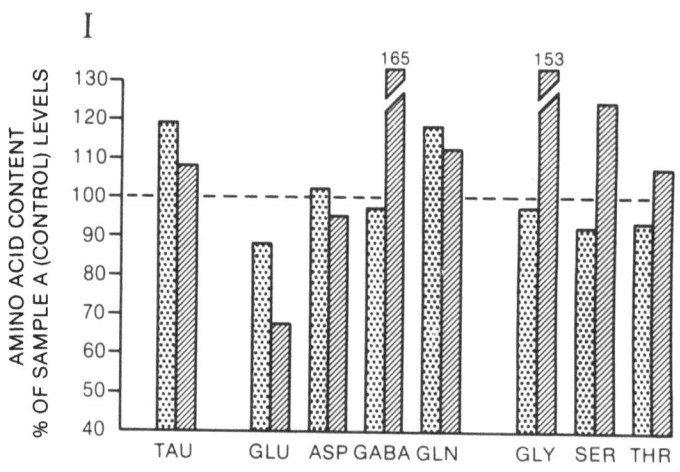

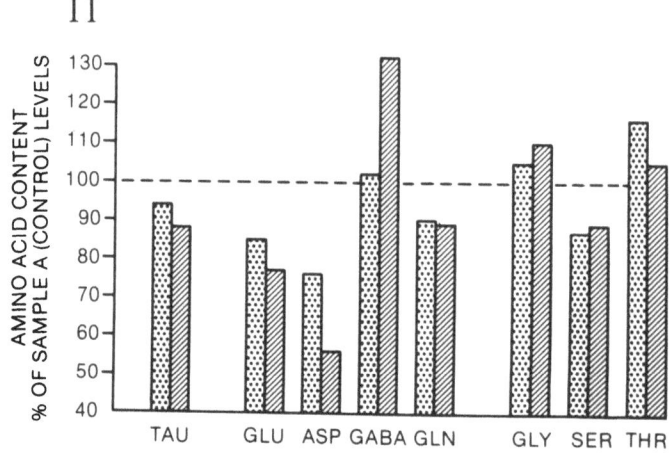

Several seizure-free conditions are associated with a higher than average occurrence of abnormal or epileptiform EEG activity. Prominently among these are trauma in susceptible brain regions, Spreading Depression of Leao (1944), and 3/second spike-and-wave epilepsy. The first is common to all individuals, the second exhibits a certain familial incidence in relation to migraine (Lance, 1981), whereas the third seems to demonstrate that deviating EEG phenomena are more heritable than epilepsy itself (Andermann, 1982; Cloninger et al., 1982). These three conditions thus should demonstrate, despite a varying origin, a common biochemical denominator as a cause for an EEG activity reflecting a tendency for abnormal neuronal recruitment and a lowered discharge threshold. The experimental counterpart of these conditions would be the period preceding a genuine epileptic condition in a non-traumatized brain. Since the latter condition, being experimental, is the most easily amenable to investigative manipulation and data acquisition, this model can serve as a guide to examine the possibility whether the other three conditions occurring in human subjects share certain neurochemical parameters.

Penicillin Spike-and Wave Epilepsy

Intramuscular injections of penicillin in the cat produce a gradual increase in the amplitude of the evoked visual potential which eventually culminates into spike-and-wave epilepsy. Onset of true epileptic electrographic activity is preceded by the return to normal amplitude of the evoked visual potential (van Gelder et al., 1983). In this model, the electrocorticogram can therefore serve to distinguish between a pre-epileptic period, marked by an increased tendency for hypersynchronous discharge on normal synaptic input, and an epileptic period marked by autonomous, high frequency synchronous discharges. Analysis of cortical tissue sampled during the pre-epileptic period and during the period of epileptic discharges (Fig. 1) reveals a gradual change in the metabolism of glutamic acid and its amino acid derivatives. The pattern of changes observed appear quite consistent, with the exception of those seen in taurine, which is neither metabolized nor synthesized in cat, man and most other mammalian species. This amino acid thus may serve as an indicator of the ability by neural

Fig. 1. Pre-epileptic (dotted) and spike-and-wave (hatched) amino acid changes in cat cortex following intramuscular injection of penicillin. Percent change from control (= damaged cortex). I: The pre-epileptic period is marked by a near normal taurine content, a fall in glutamic acid with a reciprocal increase of glutamine; GABA and aspartic acid levels are unchanged. This condition of increased cortical excitability demonstrates one aspect of neuronal release of taurine and glutamic acid, which are then captured by surrounding glial elements where glutamic acid is metabolized to glutamine (see Fig. 2). During the epileptic phase (hatched bars), more glutamic acid is released and, following transformation to glutamine in glia and its transfer to GABA synthetic sites, the released glutamic acid is used to elevate GABA levels. II: The pre-epileptic period (dotted bars) is more advanced, in that taurine and glutamate release is no longer matched by glial uptake. This is evident from taurine and glutamate loss from the tissue without a reciprocal increase in glial glutamine. GABA synthesis now becomes more dependent on local aspartic acid stores being transaminated to glutamic acid (see Fig. 3). When spike-and-wave activity is well established, these features of a failing glial capture rate in the face of increased taurine and glutamic acid release are accentuated (for details, see van Gelder et al., 1983).

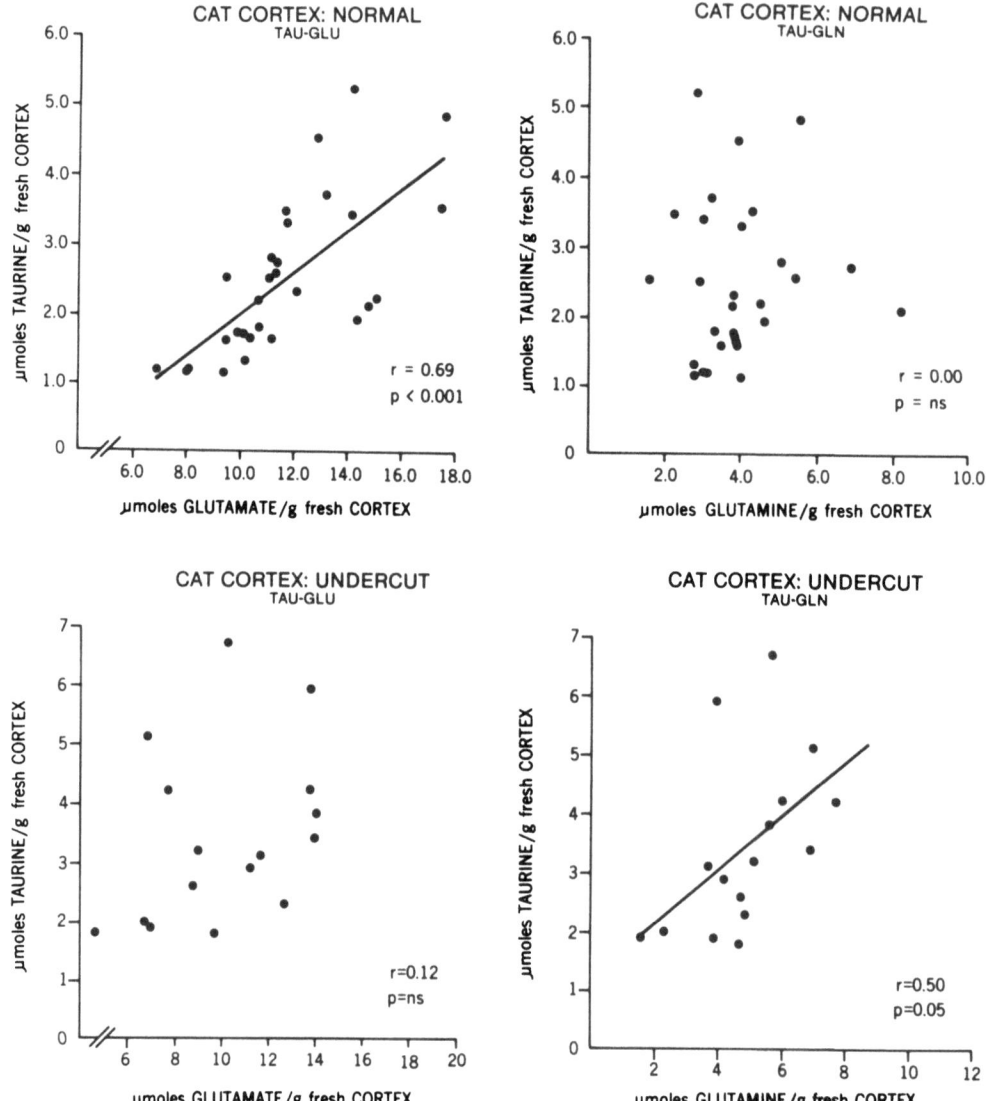

Fig. 2. A close association exists under natural physiological conditions between glutamic acid and taurine in the cat cortex. As excitation increases in a chronically undercut cortex, the two amino acids are released from neuronal structures and captured by surrounding glial elements, where glutamic acid is amidated to glutamine. Hence, a stronger association between taurine and glutamine becomes apparent. As hyperexcitation progresses into a chronic seizure state, when much of the released taurine and glutamate can no longer be all captured by the glial compartment, the close correlation between glutamic acid and taurine becomes reestablished but at a lower tissue content of both amino acids (see van Gelder, 1982; van Gelder et al., 1983). Data for undercut cortex from Koyama and Jasper, 1977.

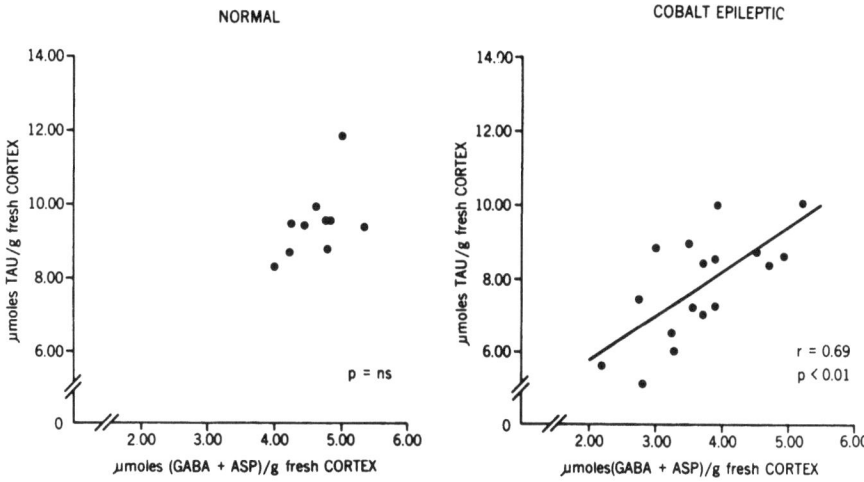
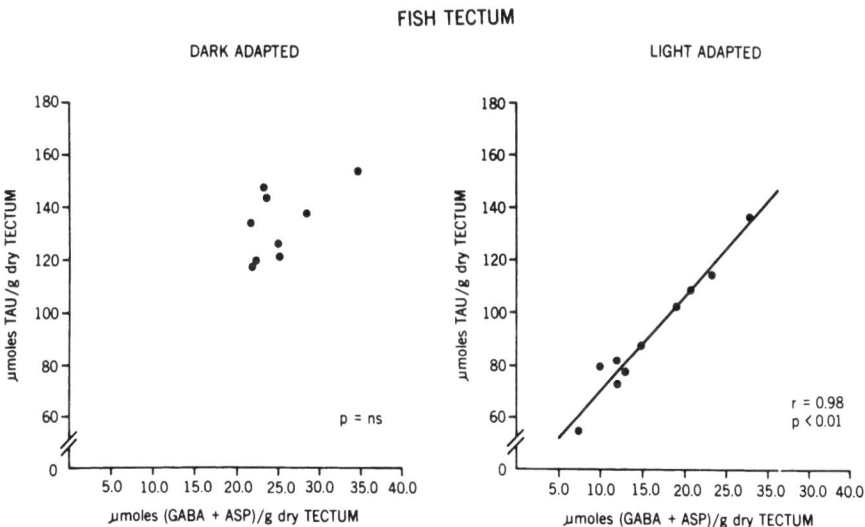

Fig. 3. In species with a high neural tissue content of taurine, a large fraction of the amino acid is stored in non-neuronal structures from which it seems to be preferentially released when a chronically hyperexcitable condition becomes established. This loss unmasks a taurine compartment closely associated with structures containing GABA, where aspartic acid serves as an emergency substrate (see Fig. 1). These and other data strongly suggest that the non-neuronal structures represent certain astrocytes which normally are able to store large amounts of excess taurine. This ability is diminished in hyperexcitable states, either because of metabolic alteration or destruction of these cells (see van Gelder and Drujan, 1980; van Gelder, 1982).

tissue to capture and store certain amino acids, released on stimulation. Evidence that taurine is mostly sequestered in neurons, and upon release is captured by glia, has been summarized elsewhere (van Gelder, 1978; 1981).

Typically, two types of amino acid profile changes can be obtained during a pre-epileptic period (dotted bars, Fig. 1). In the first type (I), taurine is somewhat elevated or normal, a glutamic acid decrease is accompanied by an increase of glutamine, while aspartate and GABA remain relatively unaltered. The second type of change (II) shows a decrease of taurine, a greater decrease of glutamate now accompanied by diminished glutamine, whereas GABA remains at normal levels as aspartic acid content falls.

The interpretation of these findings are straightforward once the metabolic redistribution of taurine and glutamic acid between elements of the synaptic complex have been taken into consideration (van Gelder, 1983a and Fig. 2). The type I changes reflect a brain in which taruine and glutamic acid are released (van Gelder, 1982) in increased amounts but are taken up in glia where glutamic acid is amidated to glutamine by glutamine synthetase, an exclusive enzyme in glia; the process is accompanied by glial edema (Pappius and Elliott, 1956; Benjamin and Quastel, 1974). GABA synthesis, derived from glutamine transport to the nerve terminal and deamidation to glutamate (Nicklas et al., 1975), remains more or less intact. In the type II change, a taurine decrease indicates a dissociation between neuronal loss of taurine and glutamate, and the capture ability of the surrounding glial envelope. As a result, glutamine levels decrease since these are mostly maintained by transfer of neuronal glutamate to glia. In order to maintain GABA levels in the face of a glutamine deficit, aspartate in the terminals becomes increasingly transaminated to glutamate to provide a substrate for GABA (see Fig. 3). Since a glial uptake of exogenous glutamate is accompanied by edema (Pappius and Elliott, 1956), type II changes probably are accompanied by a better interstitial flow through, thus accounting for the loss of taurine and glutamate from the region sampled (see van Gelder, 1983a). Note, however, that this implies increased extracellular levels of excitatory glutamic acid concentrations in the affected areas.

As the pre-epileptic phase becomes transformed into an electrographic seizure state, this pattern of change continues, with one important difference (Fig. 1, hatched bars). During this phase, GABA levels increase significantly at the expense of aspartate, indicating either a blocked release mechanism, or feedback mechanisms which signal the need for increased GABA synthesis because loss of inhibitory effectiveness or increased excitations mimics a GABA deficit. Inhibition of GABA-T without diminished GABA action presents an alternative explanation.

Among these changes it is evident that an increased release of taurine and glutamic acid (Orrego et al., 1976; Dodd et al., 1980; Lehmann et al., 1985) are early neurochemical correlates of a beginning hyperexcitable state. If so, non-epileptic individuals judged to demonstrate increased cerebral excitability because of abnormal EEG activity, should have these metabolic changes in common, if one wishes to attribute an exaggerated taurine and glutamate release as a cause for increased neuronal excitation and synchronous discharges. Loss of cellular taurine increases Ca^{2+} mobility (see van Gelder, 1983b) whereas extracellular glutamic acid represents one of the most effective endogenous excitatory agents which also mediates its effect in part by enhancing Ca^{2+} mobility (Puil, 1981). Both phenomena are therefore certainly able to account for increasing synchronous neural excitation in the region where this process occurs. These changes may not be sufficient, however, to cause by themselves a seizure state.

Brain Injury

Limited brain trauma, hypoxia, temporary anoxia and ischemia are all conditions known to increase the probability of periodic, spontaneous seizure incidences in an individual. In the affected regions one consistently encounters temporary edema, a permanent increase in the number of a certain type of astroglia and dendritic deformations (Reulen et al., 1976; Scheibel et al., 1983; Mihály and Bozoky, 1984). In many instances of brain trauma, such regions demonstrate a lowered glutamic acid decarboxylase activity, suggesting a defective GABA mediated synaptic inhibition (Ribak et al., 1982). There is no doubt that soon after the function of GABA in the CNS became apparent, many experimental studies indicated that a GABA deficiency in brain will lead to seizures. Less often quoted, but nevertheless an equally consistent finding, are the observations that an increase of GABA will also enhance the risk of seizures (in section below). In both situations the seizures, when provoked, are not self limited and, if not corrected, very frequently lead to 'status epilepticus' and death. In contrast, the hallmark of epilepsy is the sporadic and limited nature of the seizure incident. Evidently then, a permanent failure of the GABA system following injury probably cannot totally account for epilepsy (Engel et al., 1981; Prince and Connors, 1984).

All forms of brain injury, including those mentioned above, are associated with a decreased ability of neural tissue to sequester glutamic acid in neurons (Benveniste et al., 1984; Hirsch and Gibson, 1984). This is demonstrated by a decrease of tissue glutamic acid and an increased release of the amino acid on stimulation (van Gelder, 1978; Dodd et al., 1980). With respect to taurine, the most common finding is either an unchanged tissue content or a decrease (see Durelli and Mutani, 1983). Release, as it is for all endogenously stored amino acids with electrophysiological activity, is increased under these circumstances (Oja et al., 1985; Lehmann et al., 1985). A fall in taurine is especially apparent in rats and mice where endogenous cerebral taurine levels are unusually high for a mammalian species (5 and 10 micromoles/g respectively as opposed to 2-3 micromoles/g in cat or primate; see however Durelli et al., 1977). The amino acid also decreases in damaged neural tissue of the fish which contains very high endogenous taurine levels (van Gelder and Drujan, 1980; van Gelder, 1982). The chronic biochemical changes found in an injured brain, and those observed during the penicillin induced pre-epileptic period, while not identical and quantitatively different, are therefore nevertheless similar (van Gelder et al., 1983).

Spreading Depression

The vascular changes preceding certain migraine episodes produce a transient vasoconstriction involving usually the basilar arterial circulation (Lauritzen, 1986). Spreading depression may occur, which is accompanied by cerebral edema as evidenced by the increased electrical resistance which invariably occurs in the region where spreading depression is observed (van Harreveld and Fifkova, 1971; Nicholson, 1983).

No endogenous substance other than glutamic acid and its exogenous applications in millimolar concentrations to the cortex is known to mimic spreading depression. High extracellular K^+ facilitates eliciting a spreading depression by glutamic acid (van Harreveld and Fifkova, 1971; Nicholson, 1983). Increased CO_2 inspiration, experimentally will transform spreading depression into seizure activity (van Harreveld and Ochs, 1957). The reverse of this situation is encountered in certain epileptic individuals having seizures originating in the visual cortex. In these patients, a seizure is consistently followed by a migraine attack, suggesting that in the case of this special type of epilepsy a metabolic continuum exists between the two cerebral dysfunctions (Gastaut and Zifkin, 1986).

Vasoconstriction and/or temporary ischemia are capable of massive release of glutamic acid, especially when a large area is affected (Bosley et al., 1983; Lehmann et al., 1985). The glial uptake of such large quantities of glutamic acid in an attempt to mitigate the depolarizing effect of the amino acid leads to neural edema (Pappius and Elliott, 1956). Moreover, a combination of reduced 'flush out' capacity due to vasoconstriction in a cortical region and edematous conditions in a tightly packed cellular cytoarchitecture, as is found in the visual cortex (or retina), would seem to make this area especially susceptible to spreading depression as a consequence of extracellular glutamic acid accumulation. When vasoconstriction is transformed into vasodilation, by either increased CO_2 levels or lactic acid accumulation, the resulting improvement of the circulation may sufficiently lower glutamic acid levels in dense cellular structures to change a depolarizing block in neurons to partial depolariziation, thereby causing epileptic discharges to appear in such susceptible regions (Goldensohn, 1969).

Hypoxic conditions also promote increased GABA levels (Elliott, 1965), and at least one particular GABA agonist, beta-phenyl GABA (Buu and van Gelder, 1974), tends to produce rhythmic 'epileptiform' ECoG activity. These data raise a possibility that increased inhibition in a normal cortex may contribute to the reported incidences of disrhythmic EEG activity among a number of migraine patients (Selby and Lance, 1960). On the other hand, a chronic tendency in a migrainous cortex to leak glutamic acid or a glial defect of glutamic acid uptake as a familial trait, would promote persisting enhanced Ca^{2+} mobility, slightly edematous conditions, and an enhanced susceptibility to regional synchronized depolarization. This cerebral environment also accounts better for the apparent hypersensitive vascular response in migrainous individuals, the EEG changes observed, and explains why environmental or hormonal factors which facilitate Ca^{2+} mobility serve as particular triggers for migraine attacks in susceptible families.

Primary Epilepsy: 3/Second Spike-and-Wave

The genetic contribution to this form of epilepsy (Andermann, 1982) leads to suspect that the EEG abnormalities are facilitated by a permanent alteration in the cerebral environment. The fact that this type of EEG is also not uncommon in relatives, without clinical signs, indicates that the cerebral modifications are rather subtle. Being genetically directed, a grossly distorted synaptic configuration, specific transmitter abnormalities, or a deficit of an isolated enzyme, would be expressed throughout the CNS. Since no one single brain region exhibits true exclusivity in terms of such parameters, a genetic defect of this type might be expected to affect brain function far more drastically than is seen in individuals with a 3/sec. spike-and-wave EEG. Indeed, in individuals where such global type of genetic, or even environmentally induced, changes are observed, the effects are usually quite devastating with respect to brain function. On the other hand, the genetic contribution as a cause for epilepsy suggests that any cell within or outside the CNS should incorporate the metabolic marker of such a contribution. Hence, one may anticipate that the metabolic alterations promoting epileptiform EEG activity in neural tissue will be reflected in some form outside the CNS, e.g., in urine, blood, liver function, etc.

Up until now, at least four amino acid studies involving serum from epileptic patients and/or their relatives, as well as one in epileptic beagle dogs, have borne out that the metabolic aberrations caused by a genetic contribution are not confined to the CNS (Monaco et a., 1975;

van Gelder et al., 1975; van Gelder et al., 1980; van Gelder, 1981; Huxtable et al., 1983). These studies, carried out in different laboratories with different patient and control populations, were to a lesser or greater extent flawed and open to criticism. All the more impressive, therefore, that despite such reservations each one of these studies have indicated a metabolic abnormality involving serum glutamic acid, or less consistently, taurine (see reviews: van Gelder, 1981; Durelli and Mutani, 1983; Huxtable et al., 1983). The dominant finding was a disturbed serum glutamic acid to taurine ratio suggesting, at a minimum, a greater than normal release of glutamic acid in such individuals, epileptic patients as well as their relatives. In part, the divergence in results can be explained further by taking into consideration that no precise kidney threshold for taurine seems to exist in individuals, as there is normally for glutamic acid; taurine excretion is mostly governed by tissue needs or retention capacity (Chesney et al., 1985) which, as was indicated by above cited data, appears defective during pre-epileptic conditions. In some epileptic patients, glutamic acid urine excretion appears abnormally high (van Gelder et al., 1975).

From investigations of this type, which are extremely laborious and difficult to carry out, nevertheless the suggestion is clear that a carefully controlled, long range follow-up study of epileptic patients will eventually reveal that the genetic factors mediating a predisposition to epilepsy are metabolically expressed within the CNS by an enhanced tendency in neural tissue for the extracellular accumulation of glutamic acid and a decreased retention of taurine in cells which normally sequester this amino acid. In this respect, these findings thus match those situations, environmentally created or experimentally induced, which also enhance a lowering of the seizure threshold. Furthermore, close relatives of epileptic patients would also carry the biochemical trait and one can anticipate a higher incidence of epileptiform EEG activity among this group than is the norm in the overall population.

A lowering of the convulsive threshold can also be induced by a decreased monoamine content (Chauvel et al., 1982; Hiramatsu, 1983; Sherwin and van Gelder, 1986) although it should be pointed out that not all conditions associated with a deficient catecholamine content appear to cause an enhanced susceptibility to seizures. Notably among the exceptions are Parkinson's Disease and the administration of 6-hydroxydopamine. On the other hand, the well known interaction of the catecholamines with glucose-glutamic acid metabolism (Lejhon et al., 1969; Nicklas et al., 1975), the decrease in the epileptic focus of alpha-1, but not beta, receptors combined with an increase of tyrosine hydroxylase (Brière et al., 1986), do indicate that a lowered monoamine effectiveness in the cortex may cause a decrease in the seizure threshold (Hopkins and Johnston, 1984). One likely consequence of a cerebral catecholamine deficiency would be a general effect on the energy metabolism, resulting in a decreased tissue amino acid retention, since such retention has been clearly demonstrated to be energy dependent (Benveniste et al., 1984; Lehmann et al., 1985).

GABA METABOLISM AND SEIZURES

The moment GABA was proven to exhibit potent inhibitory properties when applied to neurons (Bazemore et al., 1957; Kuffler and Edwards, 1958), and was demonstrated to represent the principle inhibitory substance in inhibitory nerves of crustaceans (Kravitz et al., 1962), it was only a short step to accept the proposition that seizures arise as a consequence of a GABA deficiency in the CNS and, in particular, in the cerebral cortex (Iwama and Jasper, 1957; Elliott and Jasper, 1959). This proposition was rapidly substantiated by the general observation that in all situations in which GABA levels were found to be low in the cortex, the subject had exhibited seizures before sample removal. Especially convincing were seizure states associated with a pyridoxal phosphate (B6) deficiency, since mere reversal of the deficiency could not only abolish the convulsions but this abolition appeared invariably to coincide with the restoration of normal cerebral GABA levels (Hunt et al., 1954; Killam and Bain, 1957; Matsuda et al., 1979).

From these and many other arguments, it now has become axiomatic that seizure activity and a lowered GABA activity go together (Krnjević, 1983). Nevertheless, in the course of accepting this concept as a well proven principle, somehow much of the work carried out during the 1960's seemed to become ignored. Summarized by Elliott (1965), those studies indicated that the majority of agents inducing increases of GABA by as much as five-fold were poor anticonvulsants. Moreover, most of such agents proved to be toxic to the liver, in addition to causing severe seizures when administered in what may be considered moderate doses. These seizures appear extremely resistant to conventional anticonvulsant therapy, and occur in the presence of very high cerebral GABA levels (personal observations).

Possibly the most widely investigated among the many substances tested has been aminooxyacetic acid (Baxter and Roberts, 1961), which increases GABA by inhibition of GABA aminotransferase (GABA-T). At most, this pyridoxal phosphate complexer exerts a small anticonvulsant activity within the first two hours after administration when GABA levels are still only fractionally higher than normal (Kuriyama et al., 1966). At the time, the lack of effect of substances like aminooxyacetic acid were attributed to their rather non-specific action of removing pyridoxal phosphate from the apoenzyme. Since many other transaminases as well as decarboxylases, including glutamic acid decarboxylase (GAD), also require B6 as a cofactor, the minimal success with AOAA was thought due to its rather nonselectivity for GABA-T. However, with the advent of GABA-T inhibitors acting specifically on the apoenzyme, and thus being by necessity also weak GABA agonists (van Gelder, 1969; Buu and van Gelder, 1974), it has become quite apparent that merely indiscriminately raising cerebral GABA levels without targeting specific brain structures (see Iadarola et al., 1979; Mirski et al., 1984), does not seem to constitute a successful anticonvulsant therapy strategy (Fromm et al., 1984); more potent agonists of GABA also do not appear as promising as was hoped.

<u>Elevated GABA Levels and Metabolic Alterations</u>

High cerebral GABA levels are almost always associated with descriptive reports of 'sedation'. However, as early as 1958 when in Elliott's laboratory at least a dozen cats received intraventricular injections of GABA (van Gelder, 1959; Ph.D. Thesis), it was apparent that a general rise of GABA in the brain causes severe disturbances of motor coordination without loss of consciousness or simple reflexes. The inability of animals to maintain normal motor activity leads to suspect that the term 'sedation' in this instance might be more aptly substituted by the word 'dazed'. Certainly there is no question that the animals retain a keen awareness of their surroundings, give every sign of being fearful, and remain reactive to even mild noxious stimuli. The onset of the incoordination and its persistence are closely correlated with the duration of the GABA elevation in the CNS.

In connection with the GABA-T inhibitors, the associated seizures occurring during the presence of very high GABA levels seem to make their appearance at the moment when glutamic acid levels are beginning to decrease (van Gelder, 1969). Although one has always assumed that such a decrease is evidence that GABA metabolism provides an important source for glutamate resynthesis via regeneration of 2-oxoglutaric acid from GABA transamination (Elliott and Jasper, 1959), on hindsight this glutamic acid decrease may well serve as a trigger for the onset of seizures, since the decrease may reflect excessive release of glutamic acid from GABA terminals as it is no longer decarboxylated to GABA. The work of Martin and collaborators (Porter and Martin, 1984) reveals a rather complex inhibitory feedback mechanism in GAD, involving not only high GABA levels as such but also a change in affinity of pyridoxal phosphate to the substrate-apoenzyme complex, in combination with ATP. In this context, it is of interest that congenital high cerebral B6 levels, rather than low, may under certain circumstances contribute to an enhanced seizure predisposition (Norris et al., 1985).

Denner and Wu (1985) have now described two species of GAD in the CNS with differing affinities for their co-factor. Quite likely, the cellular localizations of these two GAD enzymes are distinct, raising the possibility that the presence of GABA is not always exclusively associated with its release from the nerve terminal (see Fig. 3), and that a certain function of GABA serves a purpose other than that of an inhibitory transmitter.

The general observation that GABA increases Cl^- permeability (Kuffler and Edwards, 1958), suggests among others that rising GABA levels in the synaptic environment might strongly influence the $Cl^- - HCO_3^-$ exchange, which is an important mechanism to control glial edema and tissue pH (van Gelder, 1983a; Fischel and Medzihradsky, 1985). At least one type of seizure, that caused by kainic acid, appears mediated by an effect on carbonic anhydrase and involves glial edema (Krespan et al., 1982; Takano et al, 1984). High Cl^- levels are also reported to alter the affinity of GABA to its receptors (Madtes, 1984) and GABA release may eventually change the Cl^- equilibrium potential of post- or pre-synaptic structures by acting on the Cl^- permeability. Since synaptic GABA release is in part apparently governed by the synaptic membrane potential and not only by Ca^{2+} influx (Sihra et al., 1984), the release from terminals and uptake of GABA into glia may, by means of changing Cl^- redistribution, gradually provoke 'desensitization' of the receptor, inhibition of synaptic GABA release, and glial engorgement. An inhibition of GABA-T would only accentuate these phenomena as would GABA agonists, by promoting increased GABA levels in structures of the synaptic complex.

Storage of GABA and Physiological Availability

By 1960, it became apparent that structurally intact cortex can capture large amounts of exogenous GABA (Elliott and van Gelder, 1960). Both this extra GABA taken up and endogenous GABA appear to exist as two distinct pools within neural tissue, since homogenization of the tissue in isotonic saline media or, in general, destruction of the cytoarchitecture releases approximately only 30-40% of the total GABA content (Elliott, 1965). Lack of glucose, adverse energy conditions such as hypoxia or homogenization in non-ionic media will release practically all endogenous GABA, indicating that the retention of GABA in the tissue requires a physiologically intact environment. While the use of radiolabeled GABA demonstrates uptake of the isotope, in isolated organelles or synaptosomes, calculations of the actual amounts which are taken up in such fractionated tissue reveal that the exogenous GABA captured can only represent an extremely small percentage of the endogenous pool of GABA present at the time. Aside from occasionally neglecting isotope exchange, such experiments are also carried out with GABA of very high specific activity and when counts are translated into the actual amounts of GABA taken up by tissue fractions, the quantities occluded in this manner are very small indeed.

Intact neural tissue does not under natural physiological conditions release much GABA, although Jasper et al. (1965) have reported some GABA release in sleeping cats, but not during seizures (Koyama and Jasper, 1977). Thus, GABA appears localized in neural tissue as two separate intracellular fractions - one firmly sequestered, the other loosely occluded; little GABA at any time appears to exist extracellularly (see Dodd et al., 1980). When GABA-T inhibitors increase endogenous GABA levels, this increase seems to occur primarily in the loosely occluded fraction (Elliott and van Gelder, 1960). That fraction appears less responsive to normal electrophysiological conditions of release, since it is for the most part insensitive to Ca^{2+} but does remain responsive to K^+ mediated depolarization (Bedwani et al., 1984). In contrast, those agents or conditions which enhance seizure resistance, predominantly influence the strongly sequestered GABA fraction.

Findings of this type indicate that changes of total cerebral GABA levels, even when increased, do not necessarily imply that existing mechanisms of GABA release or synaptic action are intact. It does point, however, to a possible explanation why high endogenous GABA levels may at times contribute to seizure manifestations, if not being directly implicated in originating such seizure activity. When they occur, the changes in GABA levels during convulsions are large, suggesting that well in advance of the seizure appearance a considerable alteration of GABA levels must have taken place. This hints at either a threshold level, beyond which GABA mediated inhibition fails, or that the changes reflect a metabolic effect of biochemical phenomena more directly responsible for promoting seizure conditions. The latter possibility seems more likely since, certainly, in some human epileptic foci neither GABA (van Gelder et al., 1975), its synthesizing enzyme GAD (Sherwin et al., 1984), or one type of GABA receptor (Sherwin et al., 1985), appear very different from those found in non-spiking tissue from the same patient.

CONCLUSION

Many epileptic patients do not exhibit continuous seizure activity but, on the contrary, only occassionally demonstrate the clinical manifestations of a chronic hyperexcitable condition. If a failure of GABA mediated inhibition is responsible for the seizures, it would signify that the system is at most only temporarily affected by altered metabolic conditions. In addition, the enhancement of excitatory synaptic potentials combined with greater cellular interconnectivity, reported to occur during the onset of hippocampal seizures (Traub and Wong, 1982; Prince and Connors, 1984), are easier to explain by an increased glutamic acid efflux than by a chronic deficit in GABA mediated inhibition, since neither substrate, enzyme nor storage mechanisms of GABA demonstrate sufficiently large changes to explain these findings.

An enhanced glutamic acid release, matched only partially by a greater rate of glial capture, will increase local neural hyperexcitability, glial edema and promote Ca^{2+} mobility, mediated by external glutamic acid. Continuing GABA release with the accompanying increased Cl^- permeability may accentuate this condition even further by affecting the function of carbonic anhydrase localized in glia. The consequence is a local environment exhibiting partial depolarization, a decrease in the extracellular space, greater intercellular connectivity and a shift to increased glycolytic metabolism in conjuction with a lowering of the pH. This environment is not unlike that seen in injured neural tissue or in a migrainous brain and under these conditions such areas may demonstrate abnormal EEG activity even when no seizure or migraine incident ensues.

As glial capture rates for glutamate decrease, the glutamine supply for the synthesis of GABA becomes diminished. This activates the local reserves of aspartic acid but the GABA synthesized from glutamate thus formed may occur at a site which differs from that normally occupied by GAD, responsible for the synthesis of synaptically released GABA. Failure of GABA inhibition releases the spontaneous and synaptically driven bursting activity of certain cells triggered by their enhanced endogenous excitation caused by glutamate; these cells then form the nucleus for autonomous focal activity and seizures.

Despite indications that raising cerebral GABA levels overall may not constitute an effective antiseizure therapy, this should not be interpreted to signify that high GABA levels during seizure states are a sign of intact synaptic inhibition in the CNS. Compounds designed to selectively raise GABA levels in specific brain regions can thus prove useful as antiseizure agents. On the other hand, by altering glial metabolism to offset extracellular accumulation of glutamic acid, it may be possible in certain cases to eliminate the zones of hyperexcitability created by this phenomenon.

REFERENCES

Andermann, E., 1982, Multifactorial inheritance of generalized and focal epilepsy, in: Genetic Basis of the Epilepsies, V. E. Anderson, W.A. Hauser, J.K. Penry, and C.F. Sing, eds., Raven Press, New York, p. 351.
Bazemore, A.W., Elliott, K.A.C., and Florey, E., 1957, Isolation of Factor I, J. Neurochem., 1:334.
Baxter, C.F., and Roberts, E., 1961, Elevation of gamma-aminobutyric acid in brain: selective inhibition of gamma-aminobutyric-alpha-ketoglutaric acid transaminase, J. Biol. Chem., 236:3287.
Bedwani, J.R., Songra, A.K., and Trueman, C.J., 1984, Influence of aminooxyacetic acid on potassium-evoked release of [^{3}H]gamma-aminobutyric acid from slices of rat cerebral cortex, Neurochem. Res., 9:1101.
Benjamin, A.M., and Quastel, J.H., 1974, Fate of L-glutamate in the brain, J. Neurochem., 23:457.
Benveniste, H., Drejer, J., Schousboe, A., and Diemer, N.H., 1984, Elevation of the extracellular concentrations of glutamate and aspartate in rat hippocampus during transient cerebral ischemia monitored by intracerebral microdialysis, J. Neurochem., 43:1369.
Bosley, T.M., Woodhams, P.L., Gordon, R.D., and Balázs, R., 1983, Effects of anoxia on the stimulated release of amino acid neurotransmitters in the cerebellum in vitro, J. Neurochem., 40:189.
Brière, R., Sherwin, A.L., Robitaille, Y., Olivier, A., Quesney, L.F., and Reader, T.A., 1986, Alpha-1 adrenoceptors are decreased in human epileptic foci, Ann Neurol., 19:26.
Buu, N.T., and van Gelder, N.M., 1974, Biological actions in vivo and in vitro of two gamma-aminobutyric acid (GABA) analogues: beta-chloro GABA and beta-phenyl GABA, Br. J. Pharmacol., 52:401.
Chauvel, P., Trottier, S., Nassif, S., and Dedek, J., 1982, Une altération des afférences noradrénergiques est-elle en cause dans les épilepsies focales?, Rev. E.E.G. Neurophysiol., 12:1.
Chesney, R.W., Gusowski, N., Dabbagh, S., and Padilla, M., 1985, Renal cortex taurine concentrations regulate renal adaptive response to altered dietary intake of sulfur amino acids, in: Taurine: Biological Aspects and Clinical Perspectives, S.S. Oja, L. Ahtee, P. Kontro and M.K. Paasonen, eds., Alan R. Liss, Inc., New York, p. 33.
Cloninger, C.R., Rice, J., Reich, T., and McGurfin, P., 1982, Genetic analysis of seizure disorders as multidimensional threshold characters, in: Genetic Basis of the Epilepsies, V.E. Anderson, W.A. Hauser, J.K. Penry, and C.R. Sing, eds., Raven Press, New York, p. 291.

Delgado-Escueta, A.V., and Greenberg, D., 1984, The search for epilepsies ideal for clinical and molecular genetic studies, Ann. Neurol., 16 (Suppl.):S1.
Denner, L.A., and Wu, J.-Y., 1985, Two forms of rat brain glutamic acid decarboxylase differ in their dependence on free pyridoxal phosphate, J. Neurochem., 44:957.
Dodd, P.R., Bradford, H.F., Abdul-Ghani, A.S., Cox, D.W.G., and Coutinho-Netto, J., 1980, Release of amino acids from chronic epileptic and sub-epileptic foci in vivo, Brain Res., 193:505.
Durelli, L., Mutani, R., Quattrocolo, G., Delsedime, M., Buffa, C., Fassio, F., Valentino, C., and Fumero, S., 1977, Relationships between electro-encephalographic pattern and biochemical picture of the cobalt epileptogenic lesion after cortical superfusion with taurine, Exp. Neurol., 54:489.
Durelli, L, and Mutani, R., 1983, The current status of taurine in epilepsy, Clin. Neuropharmacol., 6:37.
Elliott, K.A.C., and Japser, H.H., 1959, Gamma-aminobutyric acid, Physiol. Rev., 39:383.
Elliott, K.A.C., and van Gelder, N.M., 1960, The state of Factor I in rat brain: the effects of metabolic conditions and drugs, J. Physiol. 153:423.
Elliott, K.A.C., 1965, Gamma-aminobutyric acid and other inhibitory substances, Brit. Med. Bull., 21:70.
Engel, J., Ackermann, R., Caldecott-Hazard, S., and Kuhl, D., 1981, Epileptic activation of antagonistic systems may explain parodoxical features of experimental and human epilepsy: a review and hypothesis, in: Kindling 2, J. Wada, ed., Raven Press, New York, p. 193.
Fischel, S.V., and Medzihradsky, F., 1985, Assessment of membrane permeability in primary cultures of neurons and glia in response to osmotic perturbation, J. Neurosci. Res., 13:369.
Fromm, G.H., Terrence, C.F., and Chattha, A.S., 1985, Differential effect of antiepileptic and non-epileptic drugs on the reticular formation, Life Sci., 35:2665.
Gastaut, H., and Zifkin, B.G., 1986, Benign epilepsy of childhood with occipital spike and wave complexes: correlations with other primary epilepsies and with migraine, in: Migraine and Epilepsy, F. Andermann, and E. Lugaresi, eds., Butterworth, Boston, in press.
Goldensohn, E.S., 1969, Experimental seizure mechanisms, in: Basic Mechanisms of the Epilepsies, H.H. Jasper, A.A. Ward, and A. Pope, eds., Little, Brown and Co., Boston, p. 289.
Hiramatsu, M., 1983, Brain 5-hydroxytryptamine level, metabolism, and binding in E1 mice, Neurochem. Res., 8:1163.
Hirsch, J.A., and Gibson, G.E., 1984, Selective alteration of neurotransmitter release by low oxygen in vitro, Neurochem. Res., 9:1039.
Hockaday, J.M., and Whitty, C.W.M., 1969, Factors determining the electro-encephalogram in migraine: a study of 560 patients, according to clinical type of migraine, Brain, 92:769.
Hopkins, W.F., and Johnston, D., 1984, Frequency-dependent noradrenergic modulation of long-term potentiation in the hippocampus, Science, 226:350.
Hunt, A.D., Stokes, J., McGrory, W.W., and Stroud, H.H., 1954, Pyridoxine dependency, Pediatrics, 13:140.
Huxtable, R.J., Laird, H., Lippincott, S.E., and Walson, P., 1983, Epilepsy and the concentrations of plasma amino acids in humans, Neurochem. Int., 5:125.
Iadarola, I., Raines, A., and Gale, K., 1979, Differential effects of n-dipropylacetate and amino-oxyacetic acid on gamma-aminobutyric acid levels in discrete areas of the rat brain, J. Neurochem., 33:1119.
Iwama, K., and Jasper, H.H., 1957, The action of gamma-aminobutyric acid upon cortical electrical activity in the cat, J. Physiol., 138:365.

Jasper, H.H., Khan, R.T., and Elliott, K.A.C., 1965, Amino acids released from the cerebral cortex in relation to its state of activation, Science, 147:1448.

Johnston, D., and Brown, T.H., 1984, The synaptic nature of the paroxysmal depolarizing shift in hippocampal neurons, Ann. Neurol., 16(Suppl.):S65.

Killam, K.F., and Bain, J.A., 1957, Convulsant hydrazides 1: in vitro and in vivo inhibition of vitamin B6 enzymes by convulsant hydrazides, J. Pharmacol. Exp. Therap., 119:255.

Koyama, I., and Jasper, H., 1977, Amino acid content of chronic undercut cortex of the cat in relation to electrical afterdischarge: comparison with cobalt epileptogenic lesions, Can. J. Physiol. Pharmacol., 55:523.

Kravitz, E.A., Potter, D.D., and van Gelder, N.M., 1962, Gamma-aminobutyric acid distribution in the lobster nervous system: CNS, peripheral nerves and isolated motor and inhibitory axons, Biochem. Biophys. Res. Commun., 7:231.

Krespan, B., Berl, S., and Nicklas, W.J., 1982, Alterations in neuronal-glial metabolism of glutamate by the neurotoxin kainic acid, J. Neurochem., 38:509.

Krnjević, K., 1983, GABA-mediated inhibitory mechanisms in relation to epileptic discharges, in: Basic Mechanisms of Neuronal Hyperexcitability, H.H. Jasper, and N.M. van Gelder, eds., Alan R. Liss, Inc., New York, p. 249.

Kuffler, S.W., and Edwards, C., 1958, Mechanism of gamma aminobutyric acid (GABA) action and its relation to synaptic inhibition, J. Neurophysiol., 21:586.

Kuriyama, K., Roberts, E., and Rubinstein, M.K., 1966, Elevation of gamma-aminobutyric acid in brain with amino-oxyacetic acid and susceptibility to convulsive seizures in mice: a quantitative reevaluation, Biochem. Pharmacol., 15:221.

Lance, J.W., 1981, Pathophysiology of the migraine syndrome, in: Current Concepts in Migraine, Ayerst Lab. Publ., p. 5.

Lauritzen, M., Trojaborg, W., and Olesen, J., 1981, EEG during attacks of common and classical migraine, Cephalogia, 1:63.

Lauritzen, M., 1986, Cerebral blood flow in migraine and spreading depression, in: Migraine and Epilepsy, F. Andermann, and E. Lugaresi, eds., Butterworth, in press.

Leao, A.A.P., 1944, Pial circulation and spreading depression of activity in the cerebral cortex, J. Neurophysiol., 7:391.

Lehmann, A., Hagberg, H., Nyström, B., Sandberg, M., and Hamberger, A., 1985, In vivo regulation of extracellular taurine and other neuroactive amino acids in the rabbit hippocampus, in: Taurine: Biological Actions and Clinical Perspectives, S.S. Oja, L. Ahtee, P. Kontro, and M.K. Paasonen, eds., Alan R. Liss, Inc., New York, 289-311.

Lejhon, H.B., and Jackson, S.G., 1969, Regulation of mitochondrial glutamic dehydrogenase by divalent metals, nucleotides, and alpha-ketoglutarate, J. Biol. Chem., 244:5346.

Madtes, P., 1984, Chloride ions preferentially mask high-affinity GABA binding sites, J. Neurochem., 43:1434.

Matsuda, M., Abe, M., Hoshino, M., and Sakurai, T., 1979, Gamma-aminobutyric acid in subcellular fractions of mouse brain and its relation to convulsions, Biochem. Pharmacol., 28:2785.

Mihály, A., and Bozoky, B., 1984, Immunohistochemical localization of extravasated serum albumin in the hippocampus of human subjects with partial and generalized epilepsies and epileptiform convulsions, Acta Neuropathol., 65:25.

Mirski, M.A., and Ferrendelli, J.A., 1984, Interruption of the mammillothalamic tract prevents seizures in guinea pigs, Science, 226:72.

Monaco, F., Mutani, R., Durelli, L., and Delsedime, M., 1975, Free amino acids in serum of patients with epilepsy: significant increase in taurine, Epilepsia, 16:245.

Nicholson, C., 1983, Regulation of the ion microenvironment and neuronal excitability, in: *Basic Mechanisms of Neuronal Hyperexcitability*, H.H. Jasper, and N.M. van Gelder, eds., Alan R. Liss, Inc., New York, p. 185.

Nicklas, W.J., Berl, S., and Clarke, D.D., 1975, Relationship between amino acid and catecholamine metabolism in brain, in: *Metabolic Compartmentation and Neurotransmission: Relation to Brain Structure and Function*, S. Berl, D.D. Clarke, and D. Schneider, eds., Plenum Press, New York, p. 497.

Norris, D.K., Murphy, R.A., and Chung, S.H., 1985, Alterations of amino acid metabolism in epileptogenic mice by elevation of brain pyridoxal phosphate, *J. Neurochem.*, 44:1403.

Oja, S.S., Korpi, E.R., Halopainen, I., and Kontro, P., 1985, Mechanisms of stimulated taurine release from nervous tissue, in: *Taurine: Biological Actions and Clinical Perspectives*, S.S. Oja, L. Ahteen, P. Kontro, and M.K. Paasonen, eds., Alan R. Liss, Inc., New York, p. 237.

Orrego, F., Miran, R., and Soldate, C., 1976, Electrically induced release of labelled taurine, alpha- and beta-alanine, glycine, glutamate and other amino acids from rat neocortical slices in vitro, *Neuroscience*, 1:325.

Pappius, H.M., and Elliott, K.A.C., 1956, Water distribution in incubated slices of brain and other tissue, *Can. J. Physiol. Pharmacol.*, 34:1007.

Porter, T.G., and Martin, D.L., 1984, Evidence for feedback regulation of glutamate decarboxylase by gamma-aminobutyric acid, *J. Neurochem.*, 43:1464.

Prince, D.A., and Connors, B.W., 1984, Mechanisms of epileptogenesis in cortical structures, *Ann. Neurol.*, 16(Suppl.):S59.

Puil, E., 1981, S-glutamate: its interactions with spinal neurons, *Brain Res. Rev.*, 3:229.

Reulen, H.J., Graham, R., Fenske, A., Tsuyumu, M., and Klatzo, I., 1976, The role of tissue pressure and bulk flow in the formation and resolution of cold-induced edema, in: *Dynamics of Brain Edema*, H.M. Pappius, and W. Feindel, eds., Springer-Verlag, Berlin, p. 103.

Ribak, C.E., Bradburne, R.M., and Harris, A.B., 1982, A preferential loss of GABAergic, symmetric synapses in epileptic foci: a quantitative ultrastructural analysis of monkey neocortex, *J. Neurol Sci.*, 2:1725.

Scheibel, A.B., Paul, L., and Fried, I., 1983, Some structural substrates of the epileptic states, in: *Basic Mechanisms of Neuronal Hyperexcitability*, H.H. Jasper and N.M. van Gelder, eds., Alan R. Liss, Inc., New York, p. 109.

Selby, G., and Lance, J.W., 1960, Observations on 500 cases of migraine and allied vascular headache, *J. Neurol. Neurosurg. Psychiat.*, 23:23.

Sherwin, A., Quesney, F., Gautheir, S., Olivier, A., Robitaille, Y., McQuaid, P., Harvey, C., and van Gelder, N.M., 1984, Enzyme changes in actively spiking areas of human epileptic cerebral cortex, *Neurology*, 34:927.

Sherwin, A.L., and van Gelder, N.M., 1986, Amino acid and catecholamine markers of metabolic abnormalities in human focal epilepsy, in: *Basic Mechanisms of the Epilepsies*, A.V. Delgado-Escueta, A.A. Ward, D.M. Woodbury, and A.J. Porter, eds., Raven Press, New York, in press.

Sihra, T.S., Scott, I.G., and Nichols, D.G., 1984, Ionophore A23187, verapamil, protonophores, and veratridine influence the release of gamma-aminobutyric acid from synaptosomes by modulation of the plasma membrane potential rather than the cytosolic calcium, *J. Neurochem.*, 43:1624.

Takano, T., Kaneko, Y., Kumashiro, H., Sugai, N., and Oosaki, T., 1984, Kainate seizure and carbonic anhydrase (CAH) reaction in the hippocampal structures, *Neurosciences* (Kobe), 10:309.

Traub, R.D., and Wong, R.K.S., 1982, Cellular mechanisms of neuronal synchronization in epilepsy, *Science*, 216:745.

van Gelder, N.M., Sherwin, A.L., Sacks, C., and Andermann, F., 1975, Biochemical observations following administration of taurine to patients with epilepsy, Brain Res., 94:297.

van Gelder, N.M., 1978, Taurine, the compartmentalized metabolism of glutamic acid, and the epilepsies, Can. J. Physiol. Pharmacol., 56:362.

van Gelder, N.M., and Drujan, B.D., 1980, Alterations in the compartmentalized metabolism of glutamic acid with changed cerebral conditions, Brain Res., 200:443.

van Gelder, N.M., Janjua, N.A., Metrakos, K., MacGibbon, B., and Metrakos, J.D., 1980, Plasma amino acids in 3/sec spike-and-wave epilepsy, Neurochem. Res., 5:659.

van Gelder, N.M., 1981, The role of taurine and glutamic acid in the epileptic process: a genetic predisposition, Rev. Pure Appl. Pharmacol. Sci., 2:293.

van Gelder, N.M., 1982, Changed taurine-glutamic acid content and altered nervous tissue cytoarchitecture, Adv. Expt. Med. Biol., 139:239.

van Gelder, N.M., 1983a, Metabolic interactions between neurons and astroglia: glutamine synthetase, carbonic anhydrase and water balance, in: Basic Mechanisms of Neuronal Excitability, H.H. Jasper, and N.M. van Gelder, eds., Alan R. Liss, Inc., New York, p. 5.

van Gelder, N.M., 1983b, A central mechanism of action for taurine: osmoregulation, bivalent cations and excitation threshold, Neurochem. Res., 8:687.

van Gelder, N.M., Siatitsas, I., Ménini, C., and Gloor, P., 1983, Feline generalized penicillin epilepsy: changes of glutamic acid and taurine parallel the progressive increase in excitability of the cortex, Epilepsia, 24:200.

Van Harreveld, A., and Ochs, S., 1957, Electrical and vascular concomitants of spreading depression, Am. J. Physiol., 189:159.

Van Harreveld, A., and Fifkova, E., 1971, Effects of glutamate and other amino acids on the retina, J. Neurochem., 18:2145.

ANTI-EXCITOTOXIC ACTIONS OF TAURINE IN THE RAT HIPPOCAMPUS STUDIED
IN VIVO AND IN VITRO

E.D. French, A. Vezzani, W.O. Whetsell, Jr.[1], and R. Schwarcz

Maryland Psychiatric Research Center, University of Maryland
Baltimore, Maryland
[1]Department of Pathology, Vanderbilt University Medical Center
Nashville, Tennessee

INTRODUCTION

The role of neuroactive amino acids in seizure phenomena has been a subject of intensive study in the past (Morselli et al., 1981; Perry and Hansen, 1981; Huxtable et al., 1983). An additional impetus to this area of research followed the observation that exogenous excitatory amino acids, such as kainic acid, can produce in animals an electroencephalographic and neuropathological profile reminiscent of that found in human patients with temporal lobe epilepsy (Nadler et al., 1978; Pisa et al., 1980; Lothman and Collins, 1981; Sloviter and Damiano, 1981; French et al., 1982; Ben-Ari, 1985). Thus, it was reasonable to conclude that endogenous excitatory amino acids bearing structural similarities to kainate might play pivotal roles in the etiology of seizure disorders. In this regard, glutamate, aspartate, and, more recently, quinolinate (QUIN), endogenous excitants of central nervous tissue, have been suggested as factors involved in initiating events leading to seizures (Lapin, 1978; Coutinho-Netto et al., 1981; Nitsch et al., 1983; Smialowski, 1983; Schwarcz et al., 1984). This conjecture has been indirectly validated by the recent observations that excitatory amino acid antagonists possess anticonvulsant activity in a number of animal models of epilepsy (Croucher et al., 1982; Meldrum et al., 1983; Schwarcz et al., 1984).

Neuroinhibitory amino acids, such as GABA, glycine and taurine, also have been identified as prominent contributing components in processes associated with seizure phenomena (Van Gelder, 1972; Ribak et al., 1979; Haug and Nitsch, 1982; Huxtable et al., 1983). The fact that physiological studies continue to provide evidence that these amino acids depress the firing rate of central neurons has supported the argument that these substances may function to maintain a homeostatic level of neuronal excitability and under conditions of hyperexcitability serve as endogenous anticonvulsants (Barbeau et al., 1975; Phillis, 1978; Huxtable, 1981; Toth et al., 1983; Lehmann et al., 1984). Indeed, a number of investigations have shown that seizure activity can be diminished and in some cases suppressed by increases in the cerebral tissue levels of these neuroinhibitory substances (Enna et al., 1981; Lerma et al., 1984).

Recently, we began to examine the central dynamics of several amino acids during QUIN-induced ictal phenomena. By modifying the recently devised method of in vivo brain dialysis (Ungerstedt et al., 1982), we were able to directly measure the extracellular content of several amino acids during various phases of seizure activity induced by an intrahippocampal injection of the endogenous excitotoxin QUIN. Extracellular levels of taurine were found to be markedly increased following the direct application of QUIN into the hippocampus (Vezzani et al., 1985). Interestingly, the 2.5-fold increase in extracellular taurine occurred only in the injected hippocampus. The dialysate content of taurine from the non-injected side remained unchanged even though electrographic seizure activity was identical in both hippocampi. Moreover, extracellular taurine levels showed significant increases prior to the observance of EEG seizures. Thus, it was concluded that taurine released into the extracellular space may be an immediate response by the brain to control or modify the excitotoxic actions of endogenous excitatory substances such as QUIN. We now report on studies designed to further elucidate a possible role of taurine in the control of QUIN-induced excitatory and neurodegenerative phenomena.

IN VIVO BRAIN DIALYSIS

In a recent analysis of changes induced by intrahippocampally administered QUIN, we found a strong correlation between dose and the severity of seizure activity (Schwarcz et al., 1984). Using the technique of in vivo brain dialysis, we now sought to determine if QUIN-induced changes in extracellular taurine levels also followed a similar dose-response relationship. For this purpose, hippocampal extracellular taurine levels were measured in rats given unilateral injections of either 5 or 20 μg QUIN. These doses were previously shown to be markedly different with respect to the severity of seizure activity and extent of neuronal degeneration (Schwarcz et al., 1984).

The preparation, implantation and perfusion of the in vivo brain dialysis assembly has been described in complete detail elsewhere (Vezzani et al., 1985). Briefly, a flow-through hollow dialysis fiber was bent into a loop and attached to bipolar recording electrodes and a cannula guide. This assembly was then inserted into the dorsal hippocampus of an anesthetized (chloral hydrate) stereotaxically mounted rat (175-

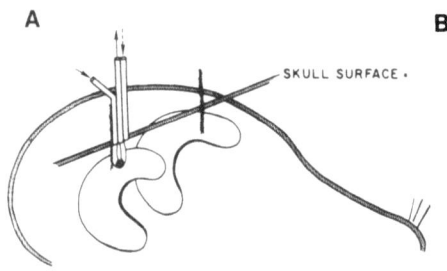 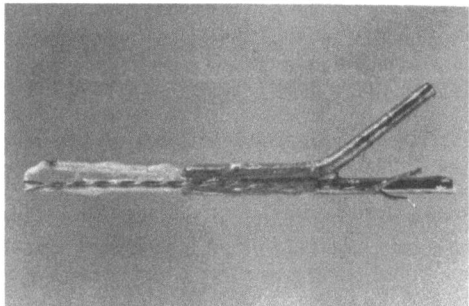

Fig. 1. A) Diagrammatic representation of the dialysis fiber/electrode assembly implanted into the dorsal hippocampus of the rat. The assembly is also fitted with a cannula guide tube used for the intrahippocampal injection of drugs. The inlet and outlet portions of the fiber are indicated by the solid arrows. The 3 mm section open for dialysis is shown by the black portion at the tip of the loop. The dorsal hippocampus contralateral to the fiber perfusion assembly is implanted with a bipolar recording electrode only. B) Photomicrograph enlargement (3.5x) of the actual fiber/electrode assembly. Dialysis occurs in the region between the arrows at the tip of the fiber.

250 g). A second bipolar electrode was implanted into the contralateral dorsal hippocampus (Fig. 1). Electrode connections were made to a multi-pin socket and secured to the skull by dental acrylic. The following day, the unanesthetized freely behaving rat was placed in a plexiglass recording chamber and attached through a lead socket to an EEG machine. The inlet portion of the fiber was then attached to a perfusion pump and Krebs-Ringer bicarbonate (pH 7.4) pumped through the fiber at a rate of 2 µl/minute. Ten-minute perfusate fractions were transported from the fiber outlet via a short tubing to collection tubes. EEG recordings and sample collections were taken for 60 minutes prior to and for 180 minutes following the intrahippocampal injection of 5 and 20 µg QUIN (1 µl). When kynurenic acid (KYNA) was tested for its ability to block QUIN-induced biochemical changes, the above procedure was the same except that KYNA (68 µg) and QUIN (20 µg) were co-injected into the dorsal hippocampus.

Extracellular taurine levels were measured in dialysate fractions collected at the site of the QUIN injection. The profile of extracellular taurine concentrations measured in the perfusates collected from 0-30 and 30-120 minutes following intrahippocampal QUIN is shown in Fig. 2. Both the 5 and 20 µg doses elicited increases in the extracellular content of taurine in the injected hippocampus. As previously reported, the number of animals demonstrating seizures and the number of seizures per animal

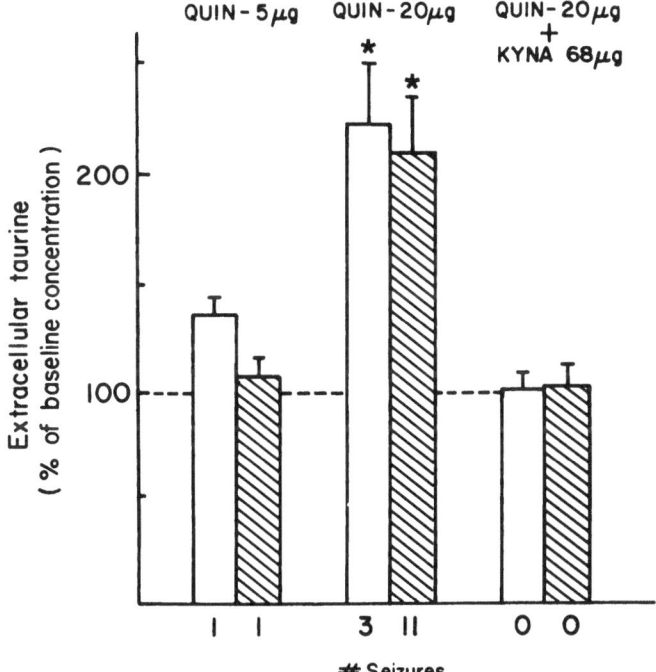

Fig. 2. Effects of QUIN on extracellular taurine content and seizure activity and blockade of QUIN effects by kynurenic acid (KYNA). The taurine change occurring during the first 30 minutes (open bars) following drug injection and subsequent 90 minutes (cross-hatched bars) is expressed as a percentage (mean ± S.E.M.) of the average taurine content of six samples immediately preceding the injection. *$p < 0.05$.

increased from 5 to 20 µg QUIN (Schwarcz et al., 1984).

The ability of KYNA to block the actions of NMDA-agonists (including QUIN) has been shown in a number of experimental paradigms (Ganong et al., 1983; Foster et al., 1984). In order to ascertan if the taurine response to the QUIN injection was receptor mediated, KYNA was therefore co-injected with QUIN and extracellular taurine levels and seizure activity was measured. Indeed, the presence of KYNA completely prevented the QUIN-induced increase in extracellular taurine (Fig. 2). Moreover, as previously reported, KYNA also antagonized the seizures and neuronal degeneration caused by QUIN (Foster et al., 1984). Thus, it appears that taurine's release into the extracellular space is receptor mediated and triggered by the presence of QUIN. Hypothetically, taurine may then act to modulate hippocampal neuronal excitability as well as serve as a neuroprotective agent by attenuating QUIN-induced neurotoxicity. In the next set of experiments, microiontophoretic methods were therefore used to determine the ability of taurine to antagonize QUIN-induced excitation of hippocampal neurons.

IONTOPHORESIS OF TAURINE

The presence of high concentrations of taurine in the brain has led to the speculation that this amino acid plays an important role in the modulation of neuronal activity (Oja and Kontro, 1978; Phillis, 1978; Huxtable, 1981). Indeed, a number of studies employing iontophoretic methods have found that taurine consistently depresses the firing rate of central neurons (Curtis and Johnston, 1974; Krnjević and Puil, 1976; McBride and Frederickson, 1978). In addition, taurine is also able to block the neuronal actions of the endogenous excitatory amino acids, glutamate and aspartate (Curtis and Johnston, 1974; Kurachi et al., 1983). Recent evidence indicates that QUIN, a tryptophan metabolite present in brain, is a particularly potent excitant of hippocampal neurons (Perkins and Stone, 1983). Since this brain area also contains high concentrations of taurine (Phillis, 1978; Nitsch et al., 1983), it was important to determine what effect taurine may have against the excitatory actions of QUIN in the hippocampus.

All recordings were made in male Sprague-Dawley rats (250-350 g) anesthetized with chloral hydrate, mounted in a stereotaxic frame, and maintained at a body temperature of 37.5°C. Using bregma, the coordinates for recording from the dorsal hippocampus were 2.0-3.0 mm posterior and lateral and 2.2-3.8 mm below the dura. Single-unit extracellular recordings and iontophoretic applications of drugs were made through a 5-barrel micropipette with an overall tip diameter of 4-7.5 µm and a resistance of 1-3 MΩ. Three of the barrels contained either taurine (0.2 M, pH 8.5), QUIN (0.1 M, pH 6.0), L-glutamate (0.1 M, pH 8.0) or acetylcholine (0.5 M, pH 4.0). Of the remaining two barrels, one contained 3 M NaCl for current neutralization while the recording barrel contained 2% pontamine sky blue in 0.5 M sodium acetate. L-glutamate and QUIN were ejected with negative currents while taurine and acetylcholine were ejected as cations. Application of drugs was accomplished by an automated iontophoresis circuit with automatic current balancing. Unitary action potentials were recorded as previously described (French and Siggins, 1980). QUIN-induced increases of firing rates were considered antagonized if taurine decreased the excitations by a least 50%, the blockade could be repeated at least twice, and there was less than 5 nA current imbalance during the challenge.

In three separate experiments, microiontophoretically applied taurine was tested on eight dorsal hippocampal pyramidal cells directly excited by QUIN. In every instance taurine effectively blocked the QUIN-induced

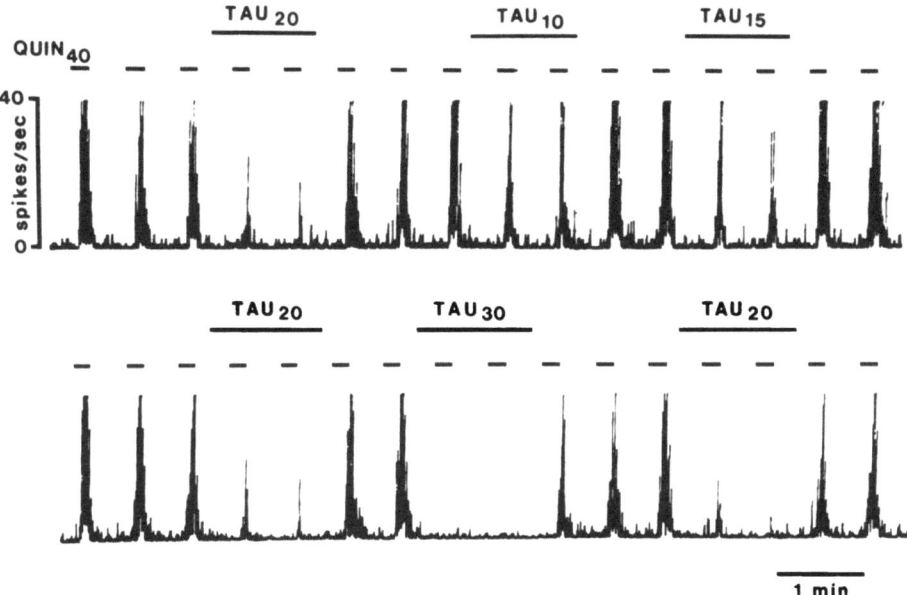

Fig. 3. Ratemeter record of QUIN-induced excitation of a hippocampal pyramidal neuron and its blockade by taurine (TAU). Numbers over bars indicate nanoamperes of ejection current and length of bars indicate duration of drug ejection. Note the dose response relationship of the TAU blockade.

excitations. Taurine's antagonistic actions were maximal shortly after onset of ejection as was the return of cellular activity to pre-taurine levels following cessation of taurine release from the micropipette (Fig. 3). In those instances where L-glutamate or acetylcholine were applied to the same neuron as QUIN, taurine likewise blocked their stimulatory actions. Taurine also inhibited the firing rate of spontaneously active hippocampal neurons (records not shown).

These data show that taurine can effectively control neuronal excitability in a region of the brain that is highly susceptible to QUIN's excitatory actions.

ACUTE AND CHRONIC TAURINE ADMINISTRATION

Based upon the results from the iontophoretic experiments, it seemed plausible that taurine, by blocking the excitatory actions of QUIN, could prevent both the seizure and neuropathological phenomena caused by intrahippocampal QUIN injections. To test this hypothesis, the effects of QUIN were assessed in animals pretreated acutely or chronically with intrahippocampal infusions of taurine.

Four groups of rats were prepared for bilateral EEG recordings and an injection cannula implanted for the unilateral intrahippocampal administration of drugs. The 'acute treatment' groups received an infusion (0.2 µl/minute) of a solution containing 640 mM taurine or equimolar saline into the dorsal hippocampus 30 minutes prior to the injection of QUIN (20 µg). Following QUIN, perfusion with taurine or saline was recommenced until the end of the recording period (60 minutes). The 'chronic treatment' groups had an intrahippocampal infusion cannula implanted, which was connected to a subdural Alzet^R minipump containing 640 mM taurine or 640 mM

Table 1. Quantitative Assessment of Quinolinic Acid-Taurine Interactions in the Rat Hippocampus

Treatment	Latency to Seizure Onset (min)[a]	Animals Seizing (%)	Hippocampal Nerve Cell Loss[a]
I. Acute Saline + QUIN (4)	46.2 ± 13.2	100	3.2 ± 0.3
II. Acute Taurine + QUIN (6)	10.2[b]	33	2.8 ± 0.8
III. Chronic Saline + QUIN (9)	22.8 ± 4.0	100	3.0 ± 0.6
IV. Chronic Taurine + QUIN (6)	16.2 ± 5.3	100	3.3 ± 0.2

[a] mean ± S.E.M.
[b] only two animals had seizures
Number of animals per group is given in parentheses.

saline. After five days of continuous infusion (1 µl/hour), the rats were challenged with an intrahippocampal injection of 20 µg QUIN. Perfusion with taurine or saline was maintained for the remaining delivery time of the seven-day minipump. EEG recordings were visually inspected for electrographic seizure activity and morphological analysis of hippocampal tissue was performed on thionin stained 30 µm cryostat sections (French et al., 1982). The extent of hippocampal nerve cell loss was rated according to the following scale: 0 = no specific damage; 1 = specific damage restricted to the injection site; 2 = pyramidal cell loss, but no granule cell loss; 3 = subtotal loss of pyramidal and granule cells; 4 = total loss of both pyramidal and granule cells.

Acute administration of taurine substantially diminished QUIN-induced seizure activity (Table 1). In particular, the number of animals presenting with EEG seizures was only 33% compared to 100% of the animals with the acute saline pretreatment. Although taurine pretreatment definitely attenuated the severity of seizure activity, it appeared to provide no protection against QUIN-induced hippocampal nerve cell loss. In contrast to the acute effects of taurine, chronic pretreatment with this amino acid failed to alter either the seizures or neuropathological consequences of QUIN administration (Table 1).

The inability of taurine to fully antagonize QUIN's effects in an in vivo situation may be related to the fact that cerebral mechanisms exist for the removal of this amino acid from the synapse (Schmid et al., 1975; Hruska et al., 1978; Lombardini, 1978). In an effort to test this supposition we proceeded to examine interactions between QUIN and taurine in an experimental system in which metabolic and dispositional factors are minimized.

TAURINE-QUINOLINIC ACID INTERACTIONS IN ORGANOTYPIC HIPPOCAMPAL CULTURES

Organotypic hippocampal cultures maintained in vitro for 21-25 days have been shown to be markedly disrupted by the presence of QUIN when examined both at the light and electron microscopic level (Whetsell and Schwarcz, 1983). Moreover, the neuropathological changes induced by QUIN in these cultures can be completely prevented by the co-administration of a specific excitatory amino acid antagonist. Thus, this in vitro test system can be successfully implemented for investigating mechanisms related to the prevention of QUIN-induced excitotoxic nerve cell degeneration.

The methodological details for the growth, maintenance and development of hippocampal cultures have been described previously (Whetsell and Schwarcz, 1983). Briefly, hippocampal cultures maintained $\underline{in\ vitro}$ from 21-25 days were exposed to culture medium containing either QUIN (10^{-3}M), QUIN (10^{-3}M) plus taurine (10^{-3} to 10^{-1}M), or taurine alone (10^{-3} to 10^{-1}M) for up to 22 hours. At the end of the drug exposure period, the cultures were fixed and embedded in Spur's embedding medium. From these tissue blocks, 0.5 μm sections were cut and stained with aqueous toluidine blue for light microscopic examination, or 500 Å-unit sections were cut and stained with uranyl acetate-lead citrate for electron microscopic study.

As previously shown, hippocampal cultures exposed to QUIN underwent severe disruption and swelling of the neuropil (Fig. 4). Electron microscopic examination of these cultures revealed swelling of postsynaptic elements as the earliest morphological manifestation of QUIN-toxicity; at later times, severe generalized disruption of neuropil and destruction of neurons was evident such that by 22 hours of QUIN-exposure few synaptic profiles could be identified. In contrast, sibling cultures simultaneously exposed to QUIN plus taurine showed only scattered disruption of neuropil or destruction of neurons at 22 hours of exposure (Fig. 4). Cultures exposed to taurine alone showed no morphological alteration and appeared identical to sibling cultures which were not exposed to either QUIN or taurine.

These data demonstrate that in this $\underline{in\ vitro}$ model, under conditions where sufficient extracellular levels of the amino acid can be maintained, taurine is indeed capable of providing substantial protection against the neurodegenerative effects of QUIN.

DISCUSSION

Overabundant endogenous excitants are currently receiving increasing attention as pathogens of seizure disorders (this volume). The fact that excitatory amino acid antagonists have proved to be effective anticonvulsants in several animal models of epilepsy constitutes a key argument in support of such a notion (Croucher et al., 1982). Recently, an endogenous excitatory amino acid has been identified which, in animals, can cause and sustain a degree of neuronal excitation sufficient to elicit a syndrome similar to that observed in temporal lobe epilepsy. This compound, QUIN, is a tryptophan metabolite which is present in brain at a concentration of approximately 1 μM (Wolfensberger et al., 1983; Moroni et al., 1984). QUIN, when injected into the rat hippocampus, causes pronounced limbic seizures and selective, 'axon-sparing' neuronal degeneration (Schwarcz et al., 1983, 1984). Since convulsants and epileptic phenomena have been shown to alter the content of neuroinhibitory amino acids (e.g., GABA, glycine, taurine) which may enable the brain to exert control of excitability (see introduction), it became imperative to determine what effect intracerebral injections of QUIN may have on those compounds. By adapting the recently developed method of $\underline{in\ vivo}$ brain dialysis, we were able to measure local changes in extracellular content of several amino acids following a convulsive dose of QUIN. Of the amino acids measured, we found that the levels of taurine, in dialysates from the $\underline{injected}$ hippocampus, showed by far the most prominent change in response to QUIN (Vezzani et al., 1985). Using a similar experimental paradigm, others have reported that exogenous excitotoxins, such as kainate and NMDA, also produce a pronounced elevation in hippocampal extracellular taurine concentrations (Lehmann et al., 1983, 1985). Importantly, the QUIN-induced effect is mediated via specific membrane receptors which are sensitive to blockade by KYNA, a known excitatory amino acid receptor antagonist. Notably, KYNA, which is also metaboli-

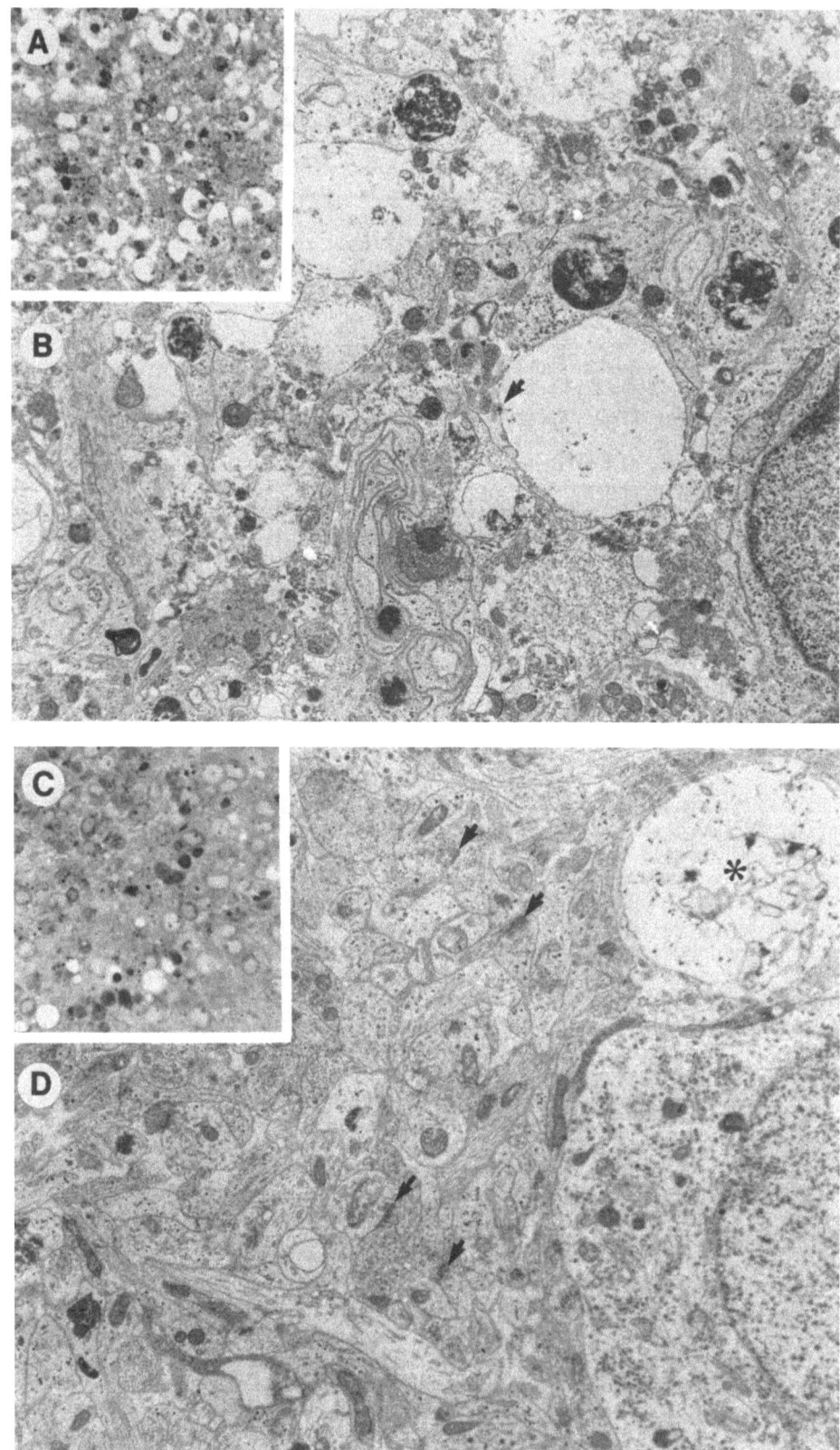

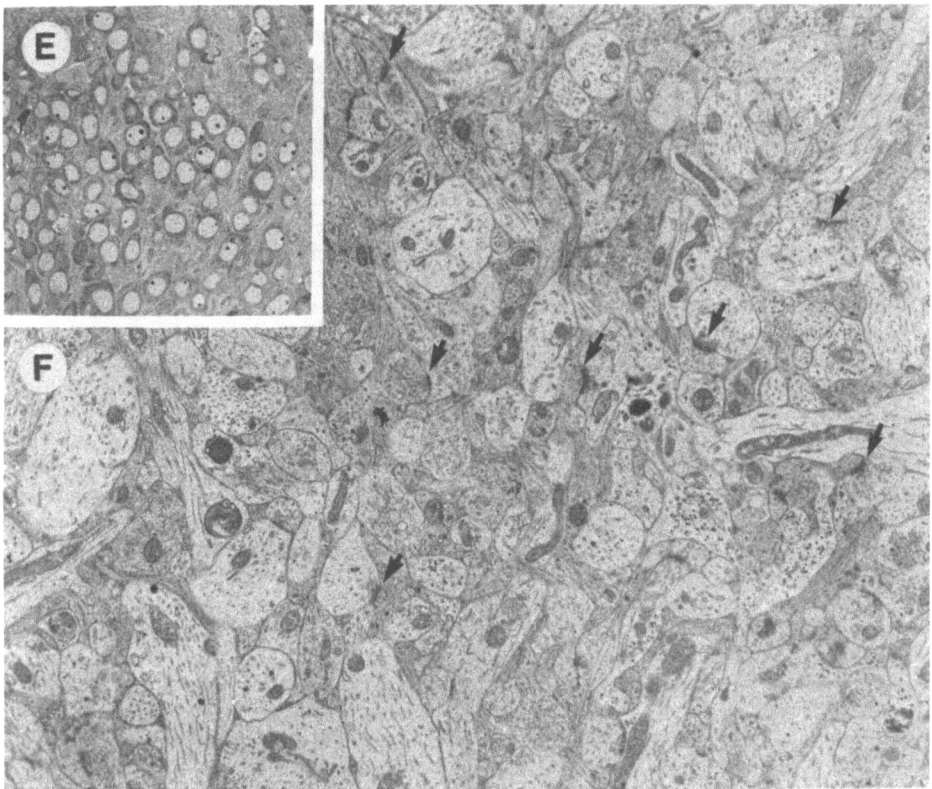

Fig. 4. A. Hippocampal culture at 21 days <u>in vitro</u> after exposure to QUIN 10^{-3}M for 22 hours (light microscopy, fixed, stained with toluidine blue, original magnification = 200x).

B. Ultrastructural appearance of the same culture as in A. Note marked swelling of numerous components of the neuropil. No definitive synapses can be identified, but one structure suggestive of a degenerating postsynaptic element is indicated (arrow). Original magnification = 14,500x.

C. Hippocampal culture at 21 days <u>in vitro</u> after exposure to QUIN 10^{-3}M plus taurine 10^{-2}M for 22 hours (light microscopy, fixed, stained with toluidine blue, magnification = 200x).

D. Ultrastructural appearance of the same culture as in C. Note generally well-preserved neuropil and scattered synaptic profiles (arrows). Occasional swollen structures are seen in the neuropil (*) which correspond to the vacuoles seen by light microscopy; occasional degenerating nerve cells are also seen either by light microscopy or electron microscopy. Original magnification = 16,000x.

E. Hippocampal culture at 21 days <u>in vitro</u> after exposure to taurine 10^{-2}M for 24 hours (light microscopy, fixed, stained with toluidine blue, original magnification = 200x).

F. Ultrastructural appearance of the same culture as in E. There are no detectable morphological alterations in these cultures compared to normal controls (untreated). Note scattered synaptic complexes (arrows). Magnification = 16,000x.

cally derived from tryptophan, can also block QUIN-induced seizures and neurotoxicity. A functional imbalance between QUIN and KYNA has thus been proposed as a causative factor in epileptic disorders (Foster et al., 1984).

The high concentration of taurine in the mammalian central nervous system has led to a substantial amount of speculation concerning its role in cerebral mechanisms (Huxtable, 1981; Van Gelder, 1983). Early investigations, carried out at the single cell level using iontophoretic techniques, found that taurine, like GABA, potently inhibits the activity of neurons in a variety of brain areas (Krnjević and Puil, 1976; Curtis and Johnston, 1974, Phillis, 1978). Ironically, little is known about the effects of taurine in the hippocampus, a region containing high concentrations of this amino acid (Phillis, 1978), and a structure exquisitely sensitive to the convulsant properties of QUIN. As first reported here, iontophoretic application of taurine completely antagonized the excitatory actions of QUIN on presumptive hippocampal pyramidal neurons. Since taurine also blocked spontaneous activity and the excitatory actions of glutamate and acetylcholine, however, its effects are obviously related to a more general inhibition of neuronal activity. Nevertheless, it is likely that increases in extracellular taurine levels, induced by endogenous convulsants like QUIN, would serve to control local multicellular hyperexcitability and thereby prevent the spread of seizure activity. In fact, our data show that _acute_ administration of a very high concentration of taurine did reduce the incidence to QUIN-induced seizures. In contrast, chronic application of taurine proved to be ineffective. Since much lower concentrations were employed in the experiments in which taurine was infused chronically, it seems reasonable to conclude that a critical high extracellular level is required in order to block the seizures elicited by 20 µg QUIN. Mechanisms for the cerebral uptake of the amino acid have been described and may be responsible for the inactivation of slowly infused taurine (Schmid et al., 1975; Hruska et al., 1978; Lombardini, 1978).

There exists an extensive literature linking taurine and seizure disorders (Van Gelder, 1972; Van Gelder and Courtois, 1972; Wheler et al., 1977; Haug and Nitsch, 1982; Huxtable et al., 1983; cf. also Van Gelder, this volume). All available evidence, including trials in human epileptics (Mutani et al., 1974), indicates that taurine has mild anticonvulsant activity. This contention is also supported by the fact that experimental taurine depletion lowers seizure thresholds (Bonhaus et al., 1985). In addition, animals with a genetically determined dysfunction of taurine transport processes show a markedly increased susceptibility to seizures (Bonhaus and Huxtable, 1984).

Since the excitotoxic hypothesis is based on the premise that blockade of neuroexcitatory phenomena should also prevent neurodegeneration (Olney, 1983), it seemed plausible that taurine's ability to block QUIN-induced excitation would be accompanied by an ability to block QUIN's neurotoxic effects. However, neither acute nor chronic intracerebral administration of taurine provided protection against QUIN-induced neurotoxicity. These negative _in vivo_ findings might best be explained by the fact that the extracellular concentration of taurine could not be maintained for a prolonged period of time at a level sufficiently high to prevent the relatively slowly progressing QUIN-induced degenerative process. In order to circumvent the limitations imposed by the metabolic and dispositional factors controlling extracellular taurine levels _in vivo_, organotypic hippocampal cultures were used to assess taurine's ability to block QUIN-induced neurotoxicity. As demonstrated at both the light microscopic and ultrastructural levels, the neuropathological consequences resulting from exposure to QUIN could indeed be substantially attenuated in the _in vitro_ situation.

In conclusion, the data presented here suggest that the release of taurine following the intrahippocampal application of QUIN may reflect a selective tissue reaction to suddenly increased amounts of the endogenous excitotoxin. While taurine is clearly a mild endogenous anticonvulsant (see above), seizure phenomena per se are unable to trigger increases in extracellular taurine levels (Vezzani et al., 1985). More importantly, rapid elevations in extracellular taurine concentrations could therefore prove beneficial because of taurine's anti-neurodegenerative effects, i.e., taurine may constitute an endogenous protective factor in the early stages of neuronal degeneration (triggered by QUIN or another endogenous excitotoxin). In accordance with this interpretation, we have recently observed that extracellular taurine levels in the cerebellum, a structure not damaged by QUIN (Schwarcz and Köhler, 1983), are not elevated following a local injection of the excitotoxin (manuscript in preparation). By inference, a dysfunctional central taurinergic system or an imbalance between taurine and QUIN in the brain may play a role in the mechanisms underlying neurodegenerative pheonomena.

ACKNOWLEDGEMENTS

This work was supported by USPHS Grants NS 16102 and NS 20509. We gratefully acknowledge the excellent technical assistance of Deborah Parks and Karen Anderson during the course of this project. We also thank Peggy Johnson for help in manuscript preparation.

REFERENCES

Barbeau, A., Inoue, N., Tsukada, Y., and Butterworth, R.F., 1975, The neuropharmacology of taurine, Life Sci., 17:669.

Ben-Ari, Y., 1985, Limbic seizure and brain damage produced by kainic acid; mechanisms and relevance to human temporal lobe epilepsy, Neuroscience, 14:375.

Bonhaus, D.W., and Huxtable, R.J., 1984, Seizure-susceptibility and decreased taurine transport in the genetically epileptic rat, Neurochem. Int., 6:365.

Bonhaus, D.W., Pasantes-Morales, H., and Huxtable, R.J., 1985, Actions of guanidinoethane sulfonate on taurine concentration, retinal morphology and seizure threshold in the neonatal rat, Neurochem. Int., 7:263.

Coutinho-Netto, J., Abdul-Ghani, A.S., Collins, J.F., and Bradford, H.F., 1981, Is glutamate a trigger factor in epileptic hyperactivity?, Epilepsia, 21:289.

Croucher, M.J., Collins, J.F., and Meldrum, B.S., 1982, Anticonvulsant action of excitatory amino acid antagonists, Science, 216:899.

Curtis, D.R., and Johnston, G.A.R., 1974, Amino acid transmitters in the mammalian central nervous system, Ergeb. Physiol., 69:97.

Enna, S.J., Kondell, D.A., and Browner, M., 1981, Differential effects of γ-vinyl GABA on chemically induced seizures, in: Neurotransmitters, Seizures and Epilepsy, P.L. Morselli, K.G., Lloyd, W. Löscher, B. Meldrum and E.H. Reynolds, eds., Raven Press, New York, p. 107.

Foster, A.C., Vezzani, A., French, E.D., and Schwarcz, R., 1984, Kynurenic acid blocks neurotoxicity and seizures induced in rats by the related brain metabolite quinolinic acid, Neurosci. Lett., 48:273.

French, E.D., and Siggins, G.R., 1980, An iontophoretic survey of opioid peptide actions in the rat limbic system: in search of opiate epileptogenic mechanisms, Regulatory Peptides, 1:127.

French, E.D., Aldinio, C., and Schwarcz, R., 1982, Intrahippocampal kainic acid, seizures and local neuronal degeneration: relationships assessed in unanesthetized rats, Neuroscience, 7:2525.

Ganong, A.H., Lanthorn, T.H., and Cotman, C.W., 1983, Kynurenic acid inhibits synaptic and acidic amino acid-induced responses in the rat hippocampus and spinal cord, Brain Res., 273:170.

Haug, P., and Nitsch, C., 1982, Increase in taurine content before onset of seizures induced by a glutamate decarboxylase inhibitor, Exp. Brain Res., 48:463.

Hruska, R.E., Huxtable, R.J., and Yamamura, H.I., 1978, High-affinity, temperature sensitive, and sodium dependent transport of taurine in rat brain, in: Taurine and Neurological Disorders, A. Barbeau and R.J. Huxtable, eds., Raven Press, New York, p. 109.

Huxtable, R.J., 1981, Insights on function: metabolism and pharmacology of taurine in the brain, in: The Role of Peptides and Amino Acids as Neurotransmitters, J.B. Lombardini and A. Kenny, eds., Alan R. Liss, Inc., New York, p. 53.

Huxtable, R.J., Laird, H., Lippincott, S.E., and Walson, P., 1983, Epilepsy and the concentrations of plasma amino acids in humans, Neurochem. Int., 5:125.

Krnjević, K., and Puil, E., 1976, Electrophysiological studies on actions of taurine, in: Taurine, R. Huxtable and A. Barbeau, eds., Raven Press, New York, p. 179.

Kurachi, M., Yoshihara, K., and Aihara, H., 1983, Effect of taurine on depolarizations induced by L-glutamate and other excitatory amino acids in the isolated spinal cord of the frog, Jap. J. Pharmacol., 33:1247.

Lapin, I.P., 1978, Stimulant and convulsive effects of kynurenines injected into brain ventricles in mice, J. Neural Trans., 42:37.

Lehmann, A., Hagberg, H., and Hamberger, A., 1984, A role for taurine in the maintenance of homeostasis in the central nervous system during hyperexcitation?, Neurosci. Lett., 52:341.

Lehmann, A., Lazarewicz, J.W., and Zeise, M., 1985, N-methylaspartate-evoked liberation of taurine and phosphoethanolamine in vivo: site of release, J. Neurochem., 45:1172.

Lerma, J., Herreras, O., Herranz, A.S., Munoz, D., and del Rio, R.M., 1984, In vivo effects of nipecotic acid on levels of extracellular GABA and taurine, and hippocampal excitability, Neuropharmacology, 23:595.

Lombardini, J.B., 1978, High-affinity transport of taurine in the mammalian central nervous system, in: Taurine and Neurological Disorders, A. Barbeau and R. Huxtable, eds., Raven Press, New York, p. 119.

Lothman, E.W., and Collins, R.C., 1981, Kainic acid-induced limbic seizures: metabolic, behavioral, electroencephalographic and neuropathological correlates, Brain Res., 218:299.

McBride, W.J., and Frederickson, R.C.A., 1978, Neurochemical and neurophysiological evidence for a role of taurine as an inhibitory neurotransmitter in the cerebellum of the rat, in: Taurine and Neurological Disorders, A. Barbeau and R.J. Huxtable, eds., Raven Press, New York, p. 415.

Meldrum, B.S., Croucher, M.J., Badman, G., and Collins, J.F., 1983, Antiepileptic action of excitatory amino acid antagonists in the photosensitive baboon, Papio papio, Neurosci. Lett., 39:101.

Moroni, F., Lombardini, G., Carla, V., and Moneti, G., 1984, The excitotoxin quinolinic acid is present and unevenly distributed in the rat brain, Brain Res., 295:352.

Morselli, P.L., Lloyd, K.G., Löscher, W., Meldrum, B., and Reynolds, E.H., 1981, Neurotransmitters, Seizures and Epilepsy, Raven Press, New York.

Mutani, R., Bergamini, L., Delsedine, M., and Durelli, L., 1974, Effects of taurine in chronic experimental epilepsy, Brain Res., 79:330.

Nadler, J., Perry, B.W., and Cotman, C.W., 1978, Intraventricular kainic acid preferentially destroys hippocampal pyramidal cells, Nature, 271:676.

Nitsch, C., Schmude, B., and Haug, P., 1983, Alterations in the content of amino acid neurotransmitters before the onset and during the course of methoxypyridoxine-induced seizures in individual rabbit brain regions, J. Neurochem., 40:1571.

Oja, S.S., and Kontro, P., 1978, Neurotransmitter actions of taurine in the central nervous system, in: Taurine and Neurological Disorders, A. Barbeau and R.J. Huxtable, eds., Raven Press, New York, p. 181.

Olney, J.W., 1983, Excitotoxins: an overview, in: Excitotoxins, K. Fuxe, P. Roberts and R. Schwarcz, eds., Macmillan, London, p. 82.

Perkins, M.N., and Stone, T.W., 1983, The pharmacology and regional variations of quinolinic acid evoked excitations in the rat CNS, J. Pharmacol. Exp. Ther., 226:551.

Perry, T.L., and Hansen, S., 1981, Amino acid abnormalities in epileptogenic foci, Neurology, 31:872.

Phillis, J.W., 1978, Overview of neurochemical and neurophysiological actions of taurine, in: Taurine and Neurological Disorders, A. Barbeau and R.J. Huxtable, eds., Raven Press, New York, p. 289.

Pisa, M., Sanberg, P.R., Corcoran, M.E., and Fibiger, H.C., 1980, Spontaneously recurrent seizures after intracerebral injections of kainic acid in rat: a possible model of human temporal lobe epilepsy, Brain Res., 200:481.

Ribak, C.E., Harris, A.B., Vaughan, J.E., and Roberts, E., 1979, Inhibitory GABAergic terminals decrease at sites of focal epilepsy, Science, 205:211.

Schmid, R., Sieghart, W., and Karobath, M., 1975, Taurine uptake in synaptosomal fractions of rat cerebral cortex, J. Neurochem., 25:5.

Schwarcz, R., and Köhler, C., 1983, Differential vulnerability of central neurons of the rat to quinolinic acid, Neurosci. Lett., 38:85.

Schwarcz, R., Whetsell, W.O. Jr., and Mangano, R.M., 1983, Quinolinic acid: an endogenous metabolite that produces axon-sparing lesions in rat brain, Science, 219:316.

Schwarcz, R., Brush, G.S., Foster, A.C., and French, E.D., 1984, Seizure activity and lesions after intrahippocampal quinolinic acid injection, Exp. Neurol., 84:1.

Sloviter, R.S., and Damiano, B.P., 1981, On the relationship between kainic acid-induced epileptiform activity and hippocampal neuronal damage, Neuropharmacology, 20:1003.

Smialowski, A., 1983, Excitatory effect of intrahippocampal injection of glutamic acid on rabbit EEG, J. Neural Trans., 58:205.

Toth, E., Lajtha, A., Sarhan, S., and Seiler, N., 1983, Anticonvulsant effects of some inhibitory neurotransmitter amino acids, Neurochem. Res., 8:291.

Ungerstedt, U., Herrera-Marschitz, M., Jungnelius, U., Stahle, L., Tossman, U., and Zetterström, T., 1982, Dopamine synaptic mechanisms reflected in studies combining behavioral recordings and brain dialysis, Adv. Dopamine Res., 37:219.

Van Gelder, N.M., 1972, Antagonism by taurine of cobalt induced epilepsy in cat and mouse, Brain Res., 47:157.

Van Gelder, N.M., and Courtois, A., 1972, Close correlation between changing content of specific amino acids in epileptogenic cortex of cats and severity of epilepsy, Brain Res., 43:477.

Van Gelder, N.M., 1983, A central mechanism of action for taurine, Neurochem. Res., 8:687.

Vezzani, A., Ungerstedt, U., French, E.D., and Schwarcz, R., 1985, In vivo brain dialysis of amino acids and simultaneous EEG measurements following intrahippocampal quinolinic acid injection: evidence for a dissociation between neurochemical changes and seizures, J. Neurochem., 45:335.

Wheler, G.H.T., Osborne, R.H., Bradford, H.F., and Davison, A.N., 1977, Uptake studies of taurine in vivo and its effects on the course of experimental focal epilepsy in rats, Brain Res., 136:535.

Whetsell, W.O. Jr., and Schwarcz, R., 1983, Mechanisms of excitotoxins examined in organotypic cultures of rat central nervous system, in: Excitotoxins, K. Fuxe, P. Roberts, and R. Schwarcz, eds., Macmillan, London, p. 207.

Wolfensberger, M., Amsler, U., Cuénod, M., Foster, A.C., Whetsell, W.O. Jr., and Schwarcz, R., 1983, Identification of quinolinic acid in rat and human brain tissue, Neurosci. Lett., 41:247.

ALTERATIONS IN EXTRACELLULAR AMINO ACIDS AND Ca^{2+} FOLLOWING EXCITOTOXIN ADMINISTRATION AND DURING STATUS EPILEPTICUS

A. Lehmann, H. Hagberg, J.W. Lazarewicz[1], I. Jacobson and A. Hamberger

Institute of Neurobiology, University of Göteborg
Göteborg, Sweden
[1]Medical Research Centre, Polish Academy of Sciences
Warsaw, Poland

Neuroexcitatory amino acids are considered to act as transmitters in several pathways of the mammalian brain. Very recently, disturbances in glutamatergic and/or aspartatergic systems have been associated with various pathological states. Thus, there is evidence to suggest that an uncontrolled release of excitatory amino acids and a subsequent postsynaptic overexcitation is involved in the induction of neuronal damage which occurs after ischemia (Benveniste et al., 1984; Simon et al., 1984; Hagberg et al., 1985; Wieloch et al., 1985) and hypoglycemia (Sandberg et al., 1985; Wieloch, 1985). Similar mechanisms are proposed to operate during status epilepticus-induced neuronal necrosis (Griffiths et al., 1982). No direct measurements of extracellular amino acids during status epilepticus have been reported and we have therefore addressed this issue using the brain dialysis method.

The mechanism of action of excitatory amino acids presently attracts considerable interest. In particular, the relationships between excitotoxins and Ca^{2+} homeostasis has been studied intensely since this may be involved in the necrotizing and/or epileptogenic actions of these compounds. In order to shed some light on the site(s) of action of excitotoxins, we describe herein effects of kainate (KA), N-methyl-DL-aspartate (NMA), dihydrokainate (DKA) and glutamate on Ca^{2+} fluxes in vivo and in vitro. The relation between NMA receptor activation and taurine release is also discussed.

EXPERIMENTAL PROCEDURES

A semipermeable thin dialysis tubing, glued between plastic tubings, was implanted into the dorsal hippocampus of adult rabbits. Its length was 10 mm, the outer diameter 0.3 mm and the MW cut 3000 Dalton. On the day after surgery, the tubing was perfused with Krebs-Ringer bicarbonate buffer (KRB) at 2.5 µl/min. The perfusate was sampled in ten minute fractions and analyzed with respect to primary amines using an HPLC system (Lindroth and Mopper, 1979). When stable concentrations of amino acids had been achieved, NMA (5.0 mM) was included and perfusate samples were collected for 160 minutes. In some cases, Ca^{2+} was excluded and NMA was included in Ca^{2+}-free medium. The effect of Co^{2+} (6 mM) on NMA-induced amino acid release was tested in Ca^{2+}-free media. D-2-amino-5-phosphonovaleric acid (5 mM) was perfused in normal KRB and in 5 mM NMA-containing KRB.

In other experiments, status epilepticus was induced in paralyzed, ventilated rabbits by means of intravenous injections of KA (10 mg/kg) or bicuculline (BC; 0.6 mg/kg). Blood pressure, blood gases and pH were monitored and adjusted when necessary.

When extracellular calcium (Ca^{2+}) in the hippocampus was to be assessed, the dialysis tubing was perfused with Ca^{2+}-free KRB to facilitate the analysis of Ca^{2+} changes. The effect of NMA was tested at 5 mM, with or without taurine (1 or 10 mM). Ca^{2+} in the perfusate was determined with Ca^{2+} selective minielectrodes (Hagberg et al., 1984).

The effect of KA on Ca^{2+} was tested after a preloading period with $^{45}Ca^{2+}$: the tubing was subsequently perfused with Ca^{2+}-free KRB for 150 minutes and KA (1 mM) was added in the presence or absence of 2 μM tetrodotoxin (TTX). Samples were collected for radioactivity measurements.

For release studies, rabbit hippocampal synaptosomes preloaded with [^{3}H] taurine were superfused with KRB (with 10 mM glucose). The effects of additions of 2 mM NMA as well as of 65 mM KCl to the superfusion medium was monitored. The released radioactivity was counted in a Beckman LS 9000 counter.

Synaptosomes were prepared from rabbit cerebral cortex and hippocampus (Hajos, 1975). Synaptosomal uptake of $^{45}Ca^{2+}$ was determined as described by Lazarewicz et al. (1985). The effect of NMA on synaptosomal Ca^{2+} accumulation was studied with 0.5, 2.0 or 5.0 mM NMA in KRB. The effects of KA, DKA and glutamate were studied over a wide concentration (0.1 - 5.0 mM). The effects of these excitants were also tested in 65 mM K^+-containing KRB.

EFFECTS OF NMA ON AMINO ACID EFFLUX IN VIVO AND IN VITRO

Perfusion of the rabbit hippocampus with NMA via the dialysis tubing elicited a marked increment of taurine and phosphoethanolamine (PEA) in the extracellular space (Table 1). The effect on other amino acids was comparatively small. The specificity of this effect was clearly concentration dependent with respect to NMA as well as to perfusion time. NMA-induced release of taurine and PEA was suppressed when the NMA antagonist APV was co-perfused with NMA at equimolar concentrations (Table 1).

Administration of NMA in nominally Ca^{2+}-free media suppressed the taurine-PEA response by approximately 50%. A further corroboration of the Ca^{2+}-sensitivity of this effect was the finding that Co^{2+} addition to the medium suppressed the NMA-elicited efflux of taurine and PEA (Table 1).

In vitro, the efflux of [^{3}H] taurine from hippocampal synaptosomes was unaffected by NMA. The addition of 65 mM K^+ media, however, markedly enhanced taurine efflux (Fig. 1).

EFFECTS OF KA AND NMA ON Ca^{2+} FLUXES IN VIVO AND IN VITRO

Perfusion of the dialysis tubing with KA resulted in a drop in $^{45}Ca^{2+}$ (Fig. 2). This decrease was unaffected by TTX. However, in isolated synaptosomes, KA had a stimulatory effect on $^{45}Ca^{2+}$ uptake which was only

Table 1. Effects of Ca^{2+} Omission, Co^{2+} and APV Addition on NMA Evoked Release of Taurine and PEA

(min)	NMA Tau	NMA PEA	$NMA-Ca^{2+}$(a) Tau	$NMA-Ca^{2+}$(a) PEA	$NMA-Ca^{2+}+Co^{2+}$(b) Tau	$NMA-Ca^{2+}+Co^{2+}$(b) PEA	NMA+APV (a) Tau	NMA+APV (a) PEA
0	100	100	100	100	159_c	130_f	100_d	100_d
40	1,028	1,597	502_e	681_e	305_f	312_f	119_d	143_d
80	1,130	2,552	738_e	$1,498_e$	361_f	392_f	158_d	212_d
120	1,257	2,572	708_e	$1,502_e$	337_f	330_f	157_d	230_d
160	1,056	2,278	605_f	$1,138_f$	332_f	281_f	160_d	222_d

Ca^{2+} omission or APV perfusion had no effect on extracellular taurine and PEA. Therefore 100% represents the steady-state level after perfusion with Ca^{2+}-free or APV-containing KRB. Co^{2+} enhanced extracellular taurine and PEA and stable concentrations were attained 80 min after commencement of Co^{2+} perfusion (6mM). In this case the time zero values represent those obtained after 80 min of perfusion with Ca^{2+}-free, Co^{2+}-containing medium. The results are expressed relative to the steady-state levels during perfusion with Ca^{2+}-free KRB. The concentrations of NMA and APV were 5mM.
(a) Significance tested against NMA
(b) Significance tested against $NMA-Ca^{2+}$
(c) $p<0.032$; $^d p<0.018$; $^e p<0.016$; $^f p<0.008$ (Mann-Whitney U test)

25% of that obtained with 65 mM K^+. The degree of depolarization of the synaptosomes did not affect KA-stimulated uptake of $^{45}Ca^{2+}$. Although both glutamate and DKA enhanced $^{45}Ca^{2+}$ uptake into synaptosomes, their potency was considerably lower than that of KA (Fig. 3).

NMA induced a decrement in Ca^{2+} *in vivo* in a similar manner to KA (Fig. 4). This decrease was sensitive to exogenously applied taurine. Incubation of synaptosomes in NMA-containing KRB did not affect the accumulation of $^{45}Ca^{2+}$ (Fig. 5).

STATUS EPILEPTICUS

Intravenous injection of KA induced epileptiform EEG activity in the rabbit hippocampus within two minutes. In all cases, the seizure activity lasted for the entire experimental period (40 minutes). BC epilepsy was initiated 5-10 seconds after the administration of the drug. BC seizures were generally of shorter duration (20 minutes) than KA seizures, and a supportive dose had to be given.

The physiological status of the animals was maintained within acceptable limits. KA was better tolerated than BC since some of the BC-treated rabbits developed hypotension and became slightly hypoxic. KA induced epilepsy was accompanied by an elevation in extracellular taurine (Fig. 6), PEA and alanine (Fig. 7). In contrast, BC epilepsy was associated with increments in PEA and alanine (Fig. 7) while taurine was unaffected (Fig. 6). Alterations in neurotransmitters such as glutamate (Fig. 6), aspartate (data not shown) or GABA (data not shown) could not be detected in either case.

DISCUSSION

Investigations of the effect of excitotoxins on amino acid release have so far been performed mainly with *in vitro* preparations. Such studies

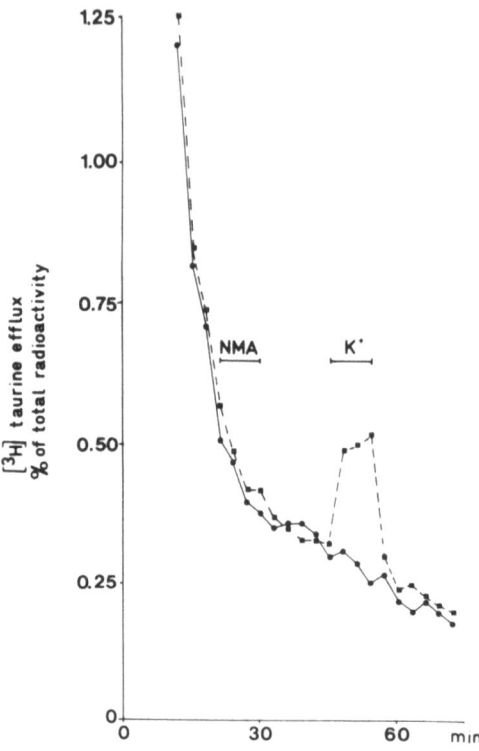

Fig. 1. Effects of NMA (2 mM) and K^+ on the release of [^{3}H]taurine from hippocampal synaptosomes. Synaptosomes loaded with [^{3}H] taurine (2.56 μM, 5 μCi) per ml sample were superfused at 0.5 ml/min with KRB medium containing 2 mM NMA. 65 mM K^+ was later applied in six-minute pulses as indicated by bars (dashed line). The controls (solid line) were superfused with KRB throughout the experiment.

have disclosed that KA induces a selective enhancement of glutamate and aspartate liberation, a property not shared by other excitatory amino acids (Ferkany and Coyle, 1983). The _in vivo_ effect of excitatory cytocides differs markedly from the _in vitro_ response. Thus, _in vivo_ administration of KA to the hippocampus evokes a small increase in glutamate release whereas taurine and PEA increase considerably in the extracellular space (Lehmann et al., 1983). The present study demonstrates that NMA also induces a massive release of taurine and PEA. With regard to the origins of the releasable taurine pool, nerve terminals are probably not the site of release as NMA, in contrast to K^+, was ineffective in releasing [^{3}H] taurine from synaptosomes.

In vivo administration of NMA suppressed the concentration of Ca^{2+}. Since the NMA-provoked release of taurine and PEA was partially Ca^{2+}-dependent and since NMA triggers Ca^{2+} spikes in hippocampal dendrites (Dingledine, 1983), we suggest that taurine and PEA are released from

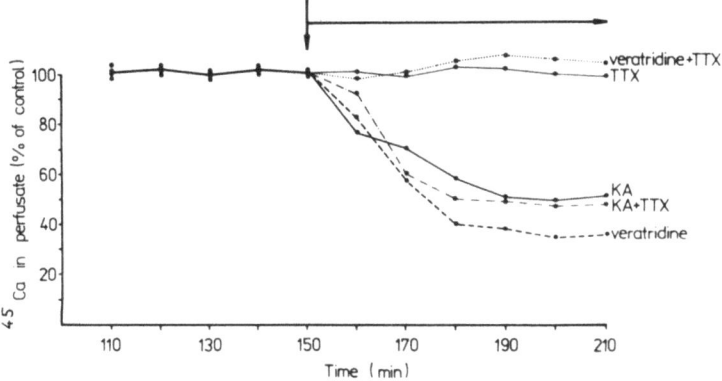

Fig. 2. Effects of 2 µM tetrodotoxin on 1 mM kainate and on 0.1 mM veratridine-evoked drop in perfusate $^{45}Ca^{2+}$ concentration. After equilibration perfusion, the experimental media were introduced as indicated by arrows. Samples were collected every 10 minutes.

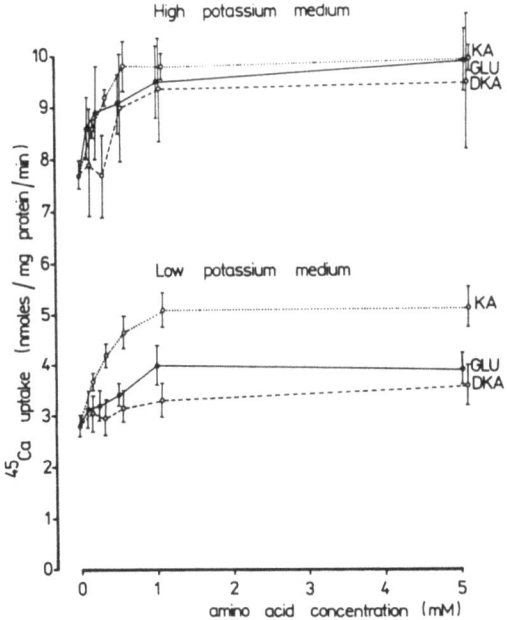

Fig. 3. Effect of glutamate, kainate and dihydrokainate on $^{45}Ca^{2+}$ uptake by synaptosomes. Ca^{2+} uptake by rabbit brain synaptosomes was tested either in normal Krebs-Ringer bicarbonate or in 65 mM K^+-containing medium. Amino acids, present at the final concentrations indicated, were added together with $^{45}Ca^{2+}$ and uptake was followed for 1 minute at 30°C.

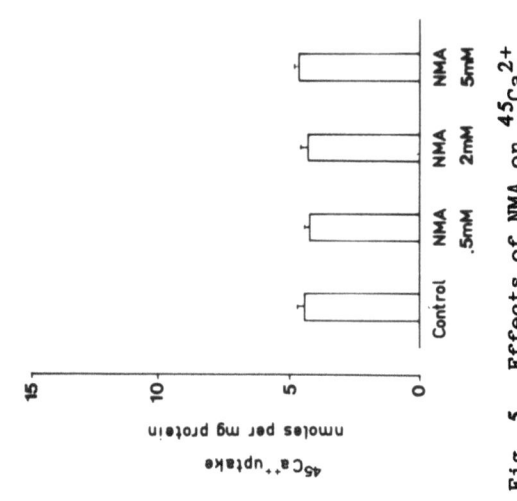

Fig. 5. Effects of NMA on $^{45}Ca^{2+}$ uptake by rabbit hippocampal synaptosomes. Synaptosomes were incubated for 1 minute at 30°C in KRB medium containing 1.2 mM $^{45}CaCl_2$ and 0.5, 2.0 or 5.0 mM NMA.

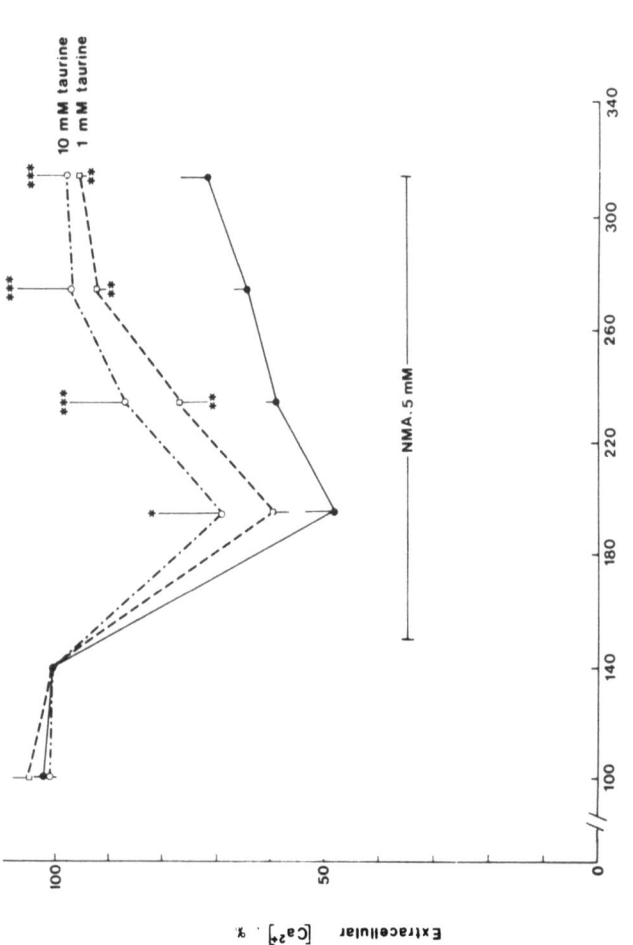

Fig. 4. The effect of NMA on extracellular Ca^{2+} concentration in the rabbit hippocampus. The dialysis fiber was perfused at 2.5 µl/min with nominally Ca^{2+}-free medium until a stable baseline had been achieved. The perfusion fluid was then exchanged for KRB containing 5 mM NMA or, in some experiments, 5 mM NMA with 1 or 10 mM taurine.

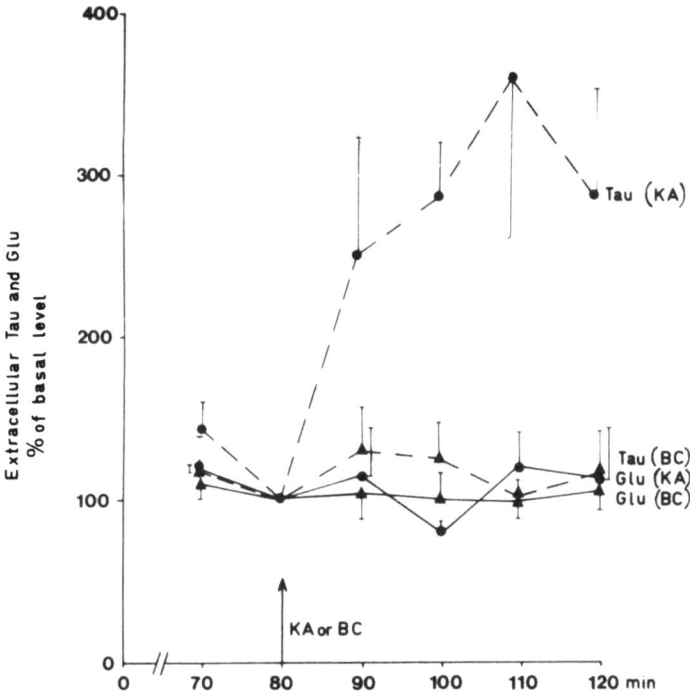

Fig. 6. Effect of kainic acid (KA) and bicuculline (BC) initiated seizures on extracellular taurine (Tau) and glutamate (Glu). KA (10 mg/kg) or BC (0.6 mg/kg) was injected i.v. as indicated and four ten-minute fractions were sampled.

dendrites. However, at this point, glial cells cannot be excluded as another possible source. Application of KA in vivo stimulated the cellular uptake of $^{45}Ca^{2+}$ in a TTX-insensitive fashion. While NMA had no effect on synaptosomal $^{45}Ca^{2+}$ accumulation, KA displayed a stimulatory effect on synaptosomal $^{45}Ca^{2+}$ uptake. Hence, a portion of the decrement in extracellular Ca^{2+} observed during KA administration in vivo might be attributed to nerve terminal Ca^{2+} sequestration. However, preliminary studies have shown that there is no difference between the KA-evoked drop in extracellular Ca^{2+} in the intact striatum and in the cortico-striatal deafferentated striatum (Butcher, Lazarewicz and Hamberger, unpublished results). The minimal effect on synaptosomal Ca^{2+} uptake by glutamate and DKA is compatible with their low efficacy at the KA receptor.

What is the possible biological significance of excitotoxin-induced postsynaptic Ca^{2+} fluxes? Based on circumstantial evidence, a number of authors have suggested that cytosolic Ca^{2+} accumulation may be responsible for the neuronal necrosis observed after excitotoxin application (Griffiths et al., 1982; Berdichevsky et al., 1983; Jancso et al., 1984). Very recently, however, direct evidence has emerged indicating that Ca^{2+} does not play a fundamental role in excitotoxicity (Price et al., 1985; Rothman, 1985). Also, inhibitory amino acids such as GABA and taurine are able to block excitotoxin-induced Ca^{2+} uptake (Zanotto and Heinemann, 1983; the present study) but apparently not the toxicity (Olney and Price, 1980; unpublished own observations). It should, however, be mentioned that taurine can protect against excitotoxicity in certain models (see below).

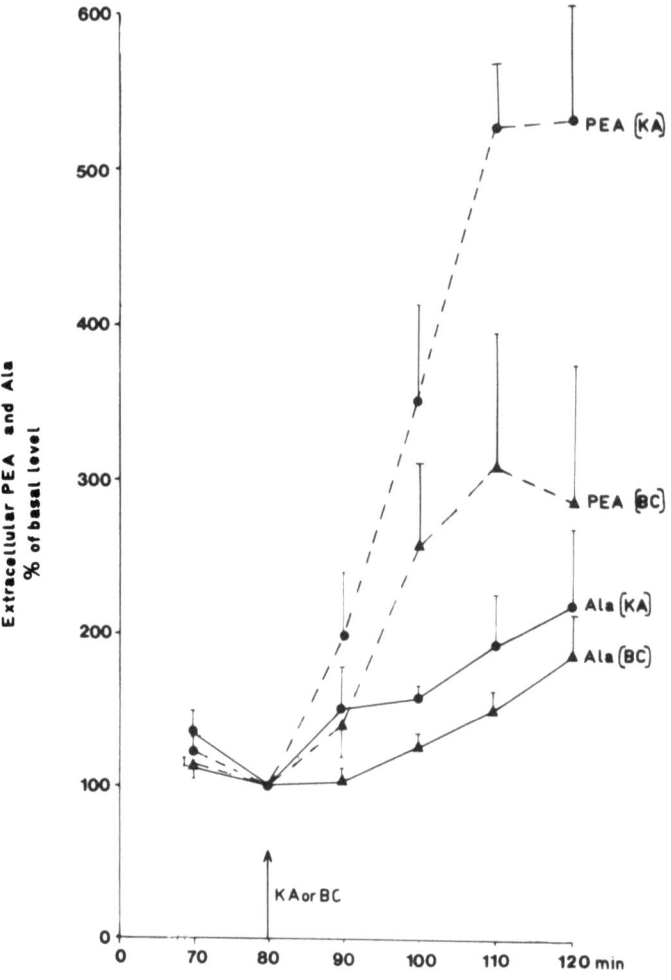

Fig. 7. Effects of kainic acid and bicuculline evoked status epilepticus on extracellular phosphoethanolamine (PEA) and alanine (Ala). The experimental conditions are described in Fig. 6.

Postsynaptic Ca^{2+} influx may be related to the initiation of burst firing (Lux and Heinemann, 1978; Pumain et al., 1983). The effects of excitatory amino acids on Ca^{2+} fluxes might therefore have a bearing on their epileptogenic actions. In this context, it is interesting to note that taurine partially inhibits the NMA-induced decrease in extracellular Ca^{2+}. Since NMA liberates taurine, this amino acid may serve as an endogenous anticonvulsant. Indeed, centrally infused taurine exerts a potent antiepileptic action in a variety of models (Durelli and Mutani, 1983). However, Vezzani et al. (1985) have generated evidence for a divergence between taurine release and seizures provoked by quinolinic acid. The relation between taurine release and excitotoxins continues to intrigue and may or may not be of physiological importance.

The effect of status epilepticus on CNS tissue has been described with respect to nerve cell injury, alterations in energy metabolism and changes in intracellular amino acids. No attempt has been made to monitor extracellular amino acids in this condition. Our present data imply that the alteration in the extracellular amino acid pattern is dependent on the convulsogenic drug employed. Both KA- and BC-induced status epilepticus enhances the release of PEA and alanine whereas KA stimulates, in addition, the liberation of taurine. The notion that endogenous taurine might serve an anticonvulsant and/or cytoprotective function is supported by the finding that taurine prevents both the convulsive and necrotizing effect of KA in the hippocampus (Fariello et al., 1982) and in the striatum (Sanberg et al., 1979). Elevated levels of intracellular alanine, as caused by KA and BC, is a prevalent metabolic response in epileptic tissue (Chapman et al., 1977, 1985). Our results imply that there is an overflow of alanine from intra- to extracellular compartments during status epilepticus. The prominent PEA liberation observed during KA and BC induced seizures is possibly a reflection of the absence of accumulative reuptake processes for this primary amine. The significance of the PEA release is not yet understood. We have recently found that high doses of PEA, injected intracerebroventricularly, have an epileptogenic effect in rats (Lehmann and Huxtable, unpublished results). Whether this effect is purely phenomenological has still to be determined. However, the possibility that PEA released during status epilepticus could exacerbate the seizures might be speculated. The unaltered levels of glutamic acid during KA and BC epilepsy was not expected as elevated glutamate release from epileptic tissue has been reported frequently (Koyama, 1972; Dodd and Bradford, 1976; Dodd et al., 1980; Peterson et al, 1983). Moreover, an enhanced level of extracellular glutamate during status epilepticus may be anticipated since the major excitatory circuits of the hippocampus and dentate gyrus are thought to be glutamatergic. These negative results, obtained with the dialysis tubing system, are probably caused by rapid clearing of the excess of extracellular glutamate by the uptake systems. In this context it may be pertinent to note that the effects of sustained seizures on extracellular amino acids are very modest as compared with those observed during glycolysis inhibition (Sandberg et al., 1985) or ischemia (Hagberg et al., 1985). These conditions are characterized by dramatic overflow of neuroactive amino acids such as glutamate, aspartate, taurine and GABA. Inclusion of uptake blockers in the perfusion fluid employed in the in vivo experiments may conceivably be necessary to unmask an exaggerated release of certain neuroactive amino acids during status epilepticus.

REFERENCES

Benveniste, H., Drejer, J., Schousboe, A., and Diemer, N.H., 1984, Elevations of the extracellular concentrations of glutamate and aspartate in the rat hippocampus during transient cerebral ischemia monitored by intracerebral microdialysis, J. Neurochem., 43:1369.

Berdichevsky, E., Riveros, N., Sanchez-Armass, S., and Orrego, F., 1983, Kainate, N-methylaspartate and other excitatory amino acids increase calcium influx into rat brain cortex cells in vitro, Neurosci. Lett., 36:75.

Chapman, A.G., Meldrum, B.S., and Siesjö, B.K., 1977, Cerebral metabolic changes during prolonged epileptic seizures in rats, J. Neurochem., 28:1025.

Chapman, A.G., Cheetham, S.E, Hart, G.P., Meldrum, B.S., and Westerberg, E., 1985, Effects of two convulsant β-carboline derivatives, DMCM and β-CCM$_3$ on regional neurotransmitter amino acid levels and on in vitro D-[^{3}H]-aspartate release in rodents, J. Neurochem., 45:370.

Dingledine, R., 1983, N-methylaspartate activates voltage-dependent calcium conductance in rat hippocampal pyramidal cells, J. Physiol., 343:385.

Dodd, P.R. and Bradford, H.F., 1976, Release of amino acids from the maturing cobalt-induced epileptic focus, Brain Res., 111:377.

Dodd, P.R., Bradford, H.F., Abdul-Ghani, A.S., Cox, D.W.G., and Couthino-Netto, J., 1980, Release of amino acids from chronic epileptic and subepileptic foci in vivo, Brain Res., 193:505.

Durelli, L., and Mutani, R., 1983, The current status of taurine in epilepsy, Clin. Neuropharmacol., 6:37.

Fariello, R.G., Golden, G.T., and Pisa, M., 1982, Homotaurine (3-aminopropane-sulfonic acid; 3-APS) protects from the convulsant and cytotoxic effect of systemically administered kainic acid, Neurology, 32:241.

Ferkany, J.W., and Coyle, J.T., 1983, Kainic acid selectively stimulates the release of endogenous excitatory acidic amino acids, J. Pharmacol. Exp. Ther., 225:399.

Griffiths, T., Evans, M.C., and Meldrum, B.J., 1982, Intracellular sites of early calcium accumulation in the rat hippocampus during status epilepticus, Neurosci Lett., 30:329.

Hagberg, H., Lehmann, A., and Hamberger, A., 1984, Inhibition by verapamil of ischemic Ca^{2+} uptake in the rabbit hippocampus, J. Cereb. Blood Flow Metab., 4:297.

Hagberg, H., Lehmann, A., Sandberg, M., Nyström, B., Jacobson, I., and Hamberger, A., 1985, Ischemia-induced shift of inhibitory and excitatory amino acids from intra- to extracellular compartments, J. Cereb. Blood Flow Metab., 5:000.

Hajos, F., 1975, An improved method for the preparation of synaptosomal fractions of high purity, Brain Res., 93:485.

Jancso, G., Karcsu, S., Kiraly, E., Szebeni, A., Toth, L., Bacsy, E., Joô, F., and Parducz, A., 1984, Neurotoxin induced nerve cell degeneration: possible involvement of calcium, Brain Res., 295:211.

Koyama, I., 1972, Amino acids in the cobalt-induced epileptogenic and nonepileptogenic cat's cortex, Can. J. Physiol. Pharmacol., 50:740.

Lazarewicz, J.W., Noremberg, K., Lehmann, A., and Hamberger, A., 1985, Effects of taurine on calcium binding and accumulation in rabbit hippocampal and cortical synaptosomes, Neurochem. Int., 7:421.

Lehmann, A., Isacsson, H., and Hamberger, A., 1983, Effects of in vivo administration of kainic acid on the extracellular amino acid pool in the rabbit hippocampus, J. Neurochem., 40:1314.

Lindroth, P., and Mopper, K., 1979, High-performance liquid chromatographic determination of subpicomole amounts of amino acids by precolumn fluorescence derivatization with o-phthaldialdehyde, Anal. Chem., 51:1667.

Lux, H.D., and Heinemann, U., 1978, Ionic changes during experimentally induced seizure activity, in: Contemporary Clinical Neurophysiology, W.A., Cobb and H. Van Duijn, eds., Elsevier, Amsterdam, p. 289.

Olney, J.W., and Price, M.T., 1980, Neuroendocrine interactions of excitatory and inhibitory amino acids, Brain Res. Bull., 5(Suppl.2):361.

Peterson, D.W., Collins, J.F., and Bradford, H.F., 1983, The kindled amygdala model of epilepsy: anticonvulsant action of amino acid antagonists, Brain Res., 275:169.

Price, M.T., Olney, J.W., Samson, L., and Labruyere, J., 1985, Calcium influx accompanies but does not cause excitoxin-induced neuronal necrosis in retina, Brain Res. Bull., 14:369.

Pumain, R., Kurcewicz, I., and Louvel, J., 1983, Fast extracellular calcium transients: involvement in epileptic processes, Science, 222:177.

Rothman, S.M., 1985, The neurotoxicity of excitatory amino acids is produced by passive chloride influx, J. Neurosci., 6:1483.

Sanberg, P.R., Staines, W., and McGeer, E.G., 1979, Chronic taurine effects on various neurochemical indices in control and kainic acid-lesioned neostriatum, Brain Res., 161:367.

Sandberg, M., Nyström, B., and Hamberger, A., 1985, Metabolically derived aspartate-elevated extracellular levels in vivo in iodacetate poisoning, J. Neurosci. Res., 13:489.

Simon, R.P., Swan, J.H., Griffiths, T., and Meldrum, B.S., 1984, Blockade of N-methyl-D-aspartate receptors may protect against ischemic damage in the brain, Science, 226:850.

Vezzani, A., Ungerstedt, U., French, E.D., and Schwarcz, R., 1985, In vivo brain dialysis of amino acids and simultaneous EEG measurements following intrahippocampal quinolinic acid injection: evidence for a dissociation between neurochemical changes and seizures, *J. Neurochem.*, 45:335.

Wieloch, T., 1985, Excitatory amino acids and hypoglycemic and ischemic neuronal damage, *Acta Neurol. Scand.*, 72:252.

Wieloch, T., Lindvall, O., Blomqvist, P., and Gage, F.H., 1985, Evidence for amelioration of ischemic neuronal damage in the hippocampal formation by lesions of the perforant path, *Neurol. Res.*, 7:24.

ACIDIC PEPTIDES IN BRAIN: DO THEY ACT AT PUTATIVE GLUTAMATERGIC SYNAPSES?

J.T. Coyle, R. Blakely, R. Zaczek, K.J. Koller, M. Abreu,
L. Ory-Layollée, R. Fisher, J.M.H. ffrench-Mullen, and
D.O. Carpenter[1]

Departments of Neuroscience, Pharmacology, Psychiatry and
Neurology
The Johns Hopkins University School of Medicine
Baltimore, Maryland 21205
[1]Division of Laboratories and Research
New York State Department of Health
Albany, New York 12201

Glutamic acid (GLU) and/or aspartic acid (ASP) are considered the most likely candidates as the predominant excitatory neurotransmitters in the mammalian brain (Curtis and Johnston, 1974; Cotman et al., 1981; Watkins and Evans, 1981; Fonnum, 1984). Nevertheless, certain reservations about this inference remain because these amino acids are involved in several metabolic pathways, including protein synthesis, and because they exhibit rather uniform excitatory effects on brain neurons. Furthermore, reports of inconsistencies between the pharmacology of ionophoretically applied GLU/ASP and that of the endogenous excitatory neurotransmitter released at putative GLU/ASP synapses have appeared. For example, Hori et al. (1981) demonstrated that α-amino-phosphono-butyric acid (APB) antagonized the effects of the excitatory neurotransmitter released by the lateral olfactory tract (LOT) but not the excitatory effects of ionophoretically applied GLU and ASP, which have been proposed as neurotransmitters for the LOT based upon their selective uptake and evoked release (Bradford and Richards, 1976; Collins, 1978). Shiells et al. (1981) have observed differences in the neurophysiologic effects of GLU on retinal bipolar cells from that of the endogenous neurotransmitter released by the photoreceptors, which is reputed to be GLU.

Nearly 30 years ago, Tallon et al. (1956) first reported the existence in brain of high concentrations of N-acetyl-aspartic acid (NAA) and several larger acidic peptides containing NAA. Subsequently, nearly a dozen N-blocked acidic peptides containing GLU and/or ASP have been demonstrated to be endogenous to the mammalian brain (Reichelt and Kvamme, 1973; Sinichkin et al., 1977). Nevertheless, the physiologic role of these acidic brain peptides has remained obscure. In screening brain extracts for acidic peptides that interact with brain membrane receptors for GLU, we found that N-acetyl-aspartyl-glutamate (NAAG) exhibited high affinity for GLU receptors and had potent epileptogenic effects when injected into the hippocampus (Zaczek et al., 1983). Accordingly, we proposed that NAAG might be an endogenous excitatory neurotransmitter. This chapter will review recent findings relevant to the hypothesis that NAAG and a growing

Table 1. Regional Levels of NAAG in Rat Brain

	nmol/mg protein
Spinal Cord	23 ± 3
Medulla	16 ± 1
Hypothalamus	9 ± 1
Dorsal Root Ganglia	9 ± 2
Thalamus	8 ± 1
Hippocampus	4 ± 0.2
Striatum	2 ± 0.1
Cortex	3 ± 0.2
Pituitary	0.7 ± 0.1

number of related endogenous acidic peptides might serve as the neurotransmitters at subsets of putative GLU/ASP neurons.

NEURONAL LOCALIZATION AND REGIONAL DISTRIBUTION OF NAAG IN BRAIN

The levels of NAA and NAAG were measured in 14 brain regions in peripheral tissues of the rat with an isocratic anion exchange HPLC method (Lenda, 1981; Koller et al., 1984). The levels of NAA were extremely low in peripheral tissues, such as the heart (0.3 ± 0.1 nmol/mg protein) and liver (1.7 ± 0.6 nmol/mg protein), whereas the brain levels were at least 20-fold greater. In the brain, the lowest concentration of NAA was observed in the pons and the highest level in the frontal cortex, consistent with a modest caudal to rostral gradient. With this sensitive assay, NAAG was undetectable in heart and liver and exhibited a much more robust regional variation than NAA, with a rostral to caudal gradient of distribution (Table 1). Thus, the concentration of NAAG in spinal cord was nearly 10-fold greater than in the striatum. The 10-fold higher concentration of NAA than NAAG and the poor correlation in their regional distributions do not seem to be consistent with the hypothesis that NAA serves as a precursor for NAAG or is merely a metabolite of it.

The levels of NAA and NAAG were measured in several regions in whole brains of rats from 15 days of gestation to adulthood. The levels of NAAG in whole brain increased 3-fold between 15 days of gestation and birth, which was 90% of the adult level (Koller and Coyle, 1984a). Whereas the level of NAAG in the spinal cord increased progressively from two days to four weeks after birth, in forebrain regions, the levels increased 2- to 3-fold between two and eight days after birth and then declined substantially to the adult level. In contrast, in whole brain and in several regions examined after birth, the levels of NAA rose steadily from 15 days of gestation to adulthood with a 30-fold increase in whole brain concentration during this time frame. The results of this study demonstrate the lack of correlation in the development of NAA and NAAG levels in brain and suggest that NAAG is localized in a cell system that matures early.

To assess the possible neuronal localization of NAAG, the effects of selective brain lesions were examined (Koller et al., 1984). Unilateral excitotoxin lesions of the corpus striatum with kainic acid reduced NAAG levels by 30% bilaterally. Unilateral decortication to induce degeneration of the excitatory cortico-striatal pathway resulted in a significant 28% decrease in NAAG ipsilateral to the lesion. Since the concentration of NAAG is highest in the spinal cord, it was of interest to determine whether

Table 2. Effect of Spinal Transection on NAAG Levels

	nmol/mg protein
Control	23 ± 1
Caudal 1	12 ± 1*
Caudal 2	14 ± 1*
Caudal 3	15 ± 1*
Rostral 1	14 ± 1*
Rostral 2	16 ± 1*
Rostral 3	17 ± 1*

Spinal cords were transected at mid-thorax, and NAAG was measured in 60 mg sections seven days after lesion. Results are the mean ± S.E.M. of five preparations.
*$p < 0.01$ versus control.

NAAG might be associated with ascending or descending pathways. After a mid-thoracic section, the levels of NAAG were consistently and significantly reduced by 37% to 51% in three segments caudal to the lesion, and by 25% to 40% in three segments rostral to the lesion (Table 2). In contrast, significant reductions in the level of NAA were restricted to those segments closest to the transection and were unaffected in the segments most distant from the lesion. The results of these studies are consistent with the interpretation that NAAG is associated with neuronal elements affected by these specific lesions.

RECEPTOR INTERACTIONS OF NAAG

The initial observation that prompted our interest in NAAG was its partial displacement with high affinity of the chloride dependent specific binding of [^{3}H]-L-GLU to brain membranes (Zaczek et al., 1983). A more extensive study indicated that this partial displacement at GLU receptors exhibited an uneven regional distribution in rat brain with the percentage of NAAG displaceable sites ranging from 61% in the thalamus to 40% in the cerebral cortex and hippocampus (Koller and Coyle, 1984b). The absolute density of these NAAG sites varied over 3-fold, with the greatest amount being in cortex and the least in spinal cord. Nevertheless, the affinity of NAAG for this subpopulation of GLU receptors remained relatively constant among regions at approximately 250 nM. Notably, related peptides including NAA, N-acetyl-glutamyl-aspartate, N-acetyl-glutamate and α-L-glutamyl-L-aspartate exhibited negligible inhibition of [^{3}H]-GLU binding at 0.1 mM.

With the recent availability of [^{3}H]-NAAG (46.6 Ci/mmol; New England Nuclear Corp.), it has been possible to study directly its receptor recognition sites through ligand binding techniques (Koller and Coyle, 1984c). [^{3}H]-NAAG bound in a specific, reversible and saturable fashion to brain membranes with optimal binding occurring in the presence of chloride ions and at 37°C. Saturation isotherms with [^{3}H]-NAAG revealed an apparent K_D of 380 nM and a B_{max} of 31 pmoles/mg protein in forebrain membranes, values which corresponded closely with those obtained through the displacement of [^{3}H]-L-GLU with unlabeled NAAG (Table 3). Similarly, the K_I values for serine-O-sulfate, quisqualate, ibotenate and GLU at the site labeled with [^{3}H]-NAAG were similar to those obtained at the site labeled by [^{3}H]-L-GLU (Koller and Coyle, 1985). Amino-phosphono analogs of glutamate, which act as inhibitors of excitatory neurotransmission and/or the excitatory effects of ionophoretically applied GLU analogs (Watkins and Evans, 1981), exhibited μM affinities for the recognition site labeled by [^{3}H]-NAAG.

Table 3. Comparison of the Pharmacology of Cl$^-$-dependent [^{3}H]-Glutamate Sites and [^{3}H]-NAAG Sites

	[^{3}H]-GLU	[^{3}H]-NAAG
	K_I (µM)	
L-Glutamate	1.0	0.5
L-Aspartate	4.4	3.0
NAAG	0.31*	0.38
NAA	>100	>100
AG	1.1	0.4
Quisqualate	0.6	0.4
Kainate	>100	>100

*NAAG displaces only 60% of the [^{3}H]-GLU under these conditions.

Notably, APB, which exerts selective antagonistic effects to the excitatory action of NAAG applied to LOT receptive pyramidal cells in the pyriform cortex, had the highest affinity, whereas α-amino-phosphono-valerate and α-amino-phosphono-heptanoate, antagonists at NMDA receptors, had lower affinities. Whereas no specific binding of the ligand was observed in membranes obtained from peripheral tissues, within the CNS, the thalamus exhibited the greatest amount of binding, and the cerebral cortex had the least. The density of the binding sites in rat whole brain had a developmental profile consistent with that of several other neurotransmitter receptor sites that have been examined. While the apparent K_D of the ligand for the receptor remained relatively constant from birth to adulthood, the density of sites rose 20-fold during the same time span with a 3-fold increase between day two and day eight after birth and with a 5-fold increase over the subsequent two weeks of post-natal development.

The distribution of specific binding sites for [^{3}H]-NAAG has also been examined using the _in vitro_ autoradiographic technique as modified from the method of Greenamyre et al. (1984). Scatchard analysis indicated that [^{3}H]-NAAG bound to a single site in brain sections with an apparent K_D of approximately 300 nM consistent with the findings obtained with brain homogenates (Abreu et al., 1985). The binding of [^{3}H]-NAAG was heterogeneously distributed in a pattern similar to that seen with the binding of [^{3}H]-GLU. The highest density of binding was found in the hippocampus, in which a distinct laminar pattern was apparent with the greatest binding observed in the stratum moleculare of the dentate gyrus and in the CA1 region of the stratum oriens, the granule cell layer, the dentate gyrus and the subiculum. Most regions of the cerebral cortex were heavily labeled, with a dense band of binding in the external cortical layers. Other forebrain structures, including the caudate-putamen, accumbens, anterior olfactory nuclei and olfactory tubercle demonstrated moderate to high levels of binding. Intermediate level of binding was observed in the thalamus and superior colliculus, while lower levels were apparent in the cerebellum, substantia nigra and substantia gelatinosa of the spinal cord.

NEUROPHYSIOLOGIC EFFECTS OF NAAG

A critical issue for ascribing a potential neurotransmitter role for NAAG was a characterization of its neurophysiologic effects. Intrahippocampal injection of NAAG revealed its convulsant properties which were equivalent in potency to quisqualic acid and could not be accounted for on the basis of proteolysis to free glutamate (Zaczek et al., 1983). Studies were carried out with Carpenter and his colleagues exploiting the perfused pyriform cortical slice preparation with extracellular recordings made on the pyramidal cells that receive monosynaptic excitatory input from the LOT (ffrench-Mullen et al., 1985). Although the LOT afferents have been proposed to utilize GLU/ASP as their neurotransmitter, Hori et al. (1981) demonstrated that bath applied APB blocked the excitatory effects of the neurotransmitter release by stimulation of the LOT but not by ionophoretically applied GLU or ASP. NAAG purified to homogeneity from rat brain was excitatory to all 50 pyramidal cells in the pyriform cortex onto which it was applied. Application was associated with rapid onset and rapid offset of excitation; and ejection currents up to 100 nA often resulted in spontaneous discharge and loss of spike amplitude, suggestive of an excitotoxic action.

Pharmacologic studies indicated that bath-applied APB antagonized the excitatory effects of LOT stimulation or ionophoretically applied NAAG. APB did not block the effects of ionophoretically applied GLU or ASP but did reverse the effects of NMDA in less than a quarter of the recorded units. Notably, the time course of recovery to the excitatory effects of NAAG with wash-out of the APB coincided with the return of the response to LOT stimulation. In contrast, amino-phosphono-valeric acid was ineffective against the effects of LOT stimulation or ionophoresis of NAAG, but blocked the excitatory effects of ASP and NMDA, and a fourth of the units responding to ionophoretically applied GLU.

In the perfused hippocampal slice preparation, Bernstein et al. (1985) compared the responses of CA1 pyramidal cells to pressure application of GLU, aspartyl-glutamate (AG) and NAAG, using both intracellular and extracellular recording techniques. Whereas GLU and AG were found to excite all cells tested, NAAG depolarized only 75% of the cells surveyed. When applied in sufficient amounts, the response to NAAG was quite rapid, approaching that of an EPSP and faster than the GLU response elicited by corresponding doses at the same site. Nevertheless, the NAAG-responsive sites were found much less frequently than those for GLU, suggesting a higher degree of selectivity with NAAG. The topographic sensitivity to NAAG and AG on the pyramidal cell differed considerably from that to GLU. Although GLU was equipotent in eliciting neuronal cell firing in CA1 pyramidal cells when applied at the neuronal perikarya or to the proximal, middle or distal dendrites, the latency to firing with AG was 3-4 times greater in the cell layer than in the dendrites. Similarly, with intracellular recording techniques, pressure ejected NAAG evoked no response or hyperpolarization when applied to the CA1 somata, although it caused depolarizing responses 75% of the time in the dendritic layers. Taken together, these results further support the excitatory actions of NAAG and the de-acetylated form, AG, but suggest a much higher degree of specificity than observed with GLU, with the distribution of responsive sites localized to the neuronal dendrites, consistent with the organization of excitatory inputs.

METABOLISM OF NAAG

A general feature for neurotransmitters is the presence in brain of processes to rapidly terminate their action at the synapse after release. Most simple neurotransmitters, such as the catecholamines and GABA, have presynaptic high-affinity uptake processes that serve this function, whereas there is now growing evidence for the existence of specific peptidases that cleave neuropeptides in brain (Dua et al., 1985). To address the metabolic disposition of NAAG, we have exploited two radiolabeled forms

Table 4. Comparison of the Pharmacology of the Transport of [^{3}H]-NAAG Radiolabel and [^{3}H]-GLU in Synaptosomal Preparations

	[^{3}H]-GLU	[^{3}H]-NAAG
	K_I (µM)	
L-Cysteic Acid	5	6
D-Aspartic Acid	12	14
L-Glutamic Acid	14	12
L-Aspartic Acid	16	16
Quisqualic Acid	>100	2
NAAG	>100	6
AG	>100	12

of the peptide: [^{3}H]-NAAG with the radiolabel on the terminal GLU and [^{14}C]-NAAG with the radiolabel on the acetyl group. Studies with intracerebral injection of these two radiolabeled forms of NAAG indicated a time dependent increase in free [^{3}H]-GLU and [^{14}C]-NAA. We then examined the disposition of radiolabeled NAAG in synaptosomal preparations (Blakely et al., 1985). Synaptosomes prepared from forebrain and incubated with [^{3}H]-NAAG in Krebs buffer accumulated tritium in a time dependent fashion with linearity maintained up to 10 minutes and an equilibrum achieved by 15 to 20 minutes. Incubation carried out at 4°C or 37°C with choline chloride substituted for NaCl in the buffer or after osmotic lysis of the synaptosomes resulted in negligible accumulation of tritium. Furthermore, veratrine (1 mM) and ouabain (1 mM), both of which disrupt the transmembrane Na$^+$-gradient, markedly inhibited the accumulation of tritium.

HPLC analysis of the tritium accumulated in the synaptosomes revealed that it co-chromatographed with GLU whereas no time-dependent accumulation of [^{3}H]-NAAG or [^{3}H]-AG was observed. In support of the conclusion that NAAG itself was not taken up intact was the observation that [^{14}C]-NAAG did not accumulate in a time dependent fashion in the synaptosomal preparation. Kinetic analysis revealed an apparent K_T for the transport of [^{3}H]-GLU derived from [^{3}H]-NAAG of 5.3 µM with a V_{max} of 2.9 nmoles/mg/hr. In regional comparison of the kinetics of transport of [^{3}H]-GLU and [^{3}H]-NAAG, it was found that the velocity of the latter process represented a variable percentage of the former process, ranging from 3.9% in the cerebellum to 10.5% in the hypothalamus. In pharmacologic studies, we found that substances that inhibited the synaptosomal uptake of [^{3}H]-GLU (Balcar and Johnston, 1972) exhibited comparable potencies in inhibiting the accumulation of tritium from [^{3}H]-NAAG (Table 4). However, several substances, including quisqualic acid, NAAG itself, α-glutamyl-glutamate and α-aspartyl-glutamate were potent inhibitors (K_Is = 2.0, 5.9, 11 and 12 µM, respectively) of tritium accumulation from [^{3}H]-NAAG, but had negligible effects (K_Is > 0.1 mM) for the transport of [^{3}H]-GLU.

To determine whether the cleavage of [^{3}H]-GLU from [^{3}H]-NAAG might be coupled with the transport process, a concentration of [^{3}H]-GLU equivalent to that calculated to be liberated from [^{3}H]-NAAG during the incubation procedure was used in subsequent assays. When this amount of [^{3}H]-GLU was added to the incubation medium, less than 35% of the [^{3}H]-GLU was accumulated in synaptosomes as compared to that when [^{3}H]-GLU was the substrate. This finding indicates that [^{3}H]-GLU generated by proteolysis of [^{3}H]-NAAG does not equilibrate in the incubation medium and suggests a tight linkage between peptide cleavage and subsequent transport of the [^{3}H]-GLU derived from [^{3}H]-NAAG (Fig. 1).

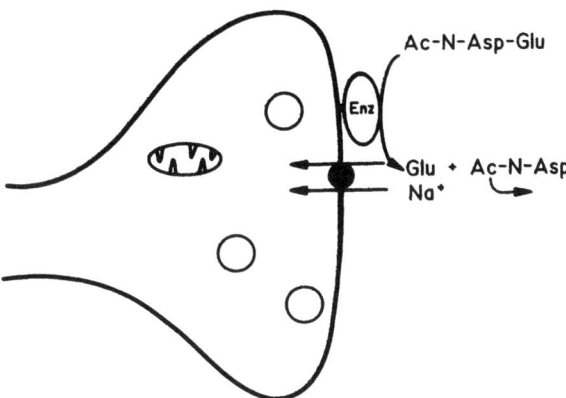

Fig. 1. Schematic representation of the metabolism of NAAG. A peptidase in close proximity to a GLU transporter cleaves GLU from NAAG to yield NAA. The GLU is then transported by a Na^+-dependent process into the nerve terminal.

CONCLUSION

Evidence accumulated over the last few years has provided increasing support for the hypothesis that NAAG may serve as an endogenous excitatory neurotransmitter. The peptide is concentrated in brain and is not detectable in peripheral non-neuronal tissue (Lenda, 1981; Koller et al., 1984), although significant amounts are present in the dorsal root ganglia (Ory-Lavollée, Blakely and Coyle, in preparation). Within brain, NAAG has a markedly uneven regional distribution, and the results from lesion studies are consistent with a neuronal localization. Preliminary findings with polyclonal antiserum with high specificity for NAAG (NAA, AG and GLU having affinities 1000-fold lower than NAAG) using the avidinbiotin peroxidase immunocytochemical technique have revealed selective staining of certain neuronal perikarya and their dendritic and axonal extensions in the brain stem (Blakely et al., in preparation). Independently, Cangro et al. (1985) have described specific immunocytochemical staining for NAAG in a subpopulation of neuronal perikarya within the dorsal root ganglia.

The potential neurotransmitter role of NAAG derives from findings from ligand binding and neurophysiologic experiments. Ligand binding studies indicate that NAAG interacts with a subpopulation of brain receptor recognition sites labeled with $[^3H]$-GLU (Zaczek et al., 1983; Koller and Coyle, 1984b). Only a portion of the $[^3H]$-GLU bound under incubation conditions containing chloride was displaced by NAAG but with a higher affinity than GLU itself; and NAAG did not displace specifically bound $[^3H]$-GLU under chloride-free conditions. The fraction of chloride dependent GLU receptors sensitive to NAAG varied among brain regions. Receptors directly labeled with $[^3H]$-NAAG exhibited kinetics identical to those defined by the displacement of $[^3H]$-GLU by unlabeled NAAG. The pharmacolgic characteristics of the recognition sites labeled with $[^3H]$-NAAG are similar to those reported in the GLU receptor subtype designated 'A-4' according to the nomenclature of Foster and Fagg (1984).

The metabolic studies with NAAG support the existence of a GLU transport process, which is closely linked to a specific peptidase that cleaves GLU from NAAG. This linkage between GLU cleavage and sodium-dependent uptake is reminiscent of the association between acetylcholinesterase hydrolysis of acetylcholine to choline and its sodium-dependent uptake process. As our pharmacologic studies cannot distinguish the sodium-dependent uptake process for [^{3}H]-GLU itself from that linked to [^{3}H]-GLU cleaved from NAAG, which may represent only 3% to 10% of the synaptosomal glutamate transport, it is likely that this NAAG-related subpopulation of transport sites would accumulate [^{3}H]-GLU in a fashion indistinguishable from the bulk of the sodium-dependent GLU uptake sites. Thus, prelabeling techniques could not distinguish the NAAG-related GLU transport sites from those unrelated to NAAG metabolism. With regard to studies of the depolarization-evoked release of endogenous GLU from neurons, the specific peptidase associated with the subpopulatin of NAAG-related terminals might well cleave GLU from released NAAG, leading to the misinterpretation that GLU itself is released. Furthermore, techniques commonly used to measure the evoked release of endogenous amino acids depend upon derivitization of the free amino group with O-phthalaldehyde or dansylation for fluorescence detection. Since NAAG is N-blocked, its presence in the effluent in evoked release studies cannot be detected by these methods.

Again in an analogy to the cholinergic system, choline itself has direct agonist effects at both muscarinic and nicotinic receptors, although acetylcholine is obviously the endogenous agonist (Krnjević and Reinhardt, 1979; Holz and Senter, 1981). Similarly, GLU has broad and uniform excitatory effects on brain neurons and is capable of activating pharmacologically distinct subsets of excitatory receptors (Watkins and Evans, 1981). A possible explanation for this broad action of GLU is that GLU and/or ASP are common components of a group of substances like NAAG which interact with discrete excitatory receptors. Accordingly, GLU/ASP in sufficient concentration may exert agonist effects by binding to that domain of the excitatory receptors, which recognize the GLU/ASP present within the endogenous ligands. Such an interpretation is not inconsistent with the observation that molecules substantially larger than GLU/ASP, such as γ-D-glutamyl-glycine or amino-phosphono-heptanoic acid, are potent antagonists at subsets of excitatory receptor subtypes.

Clearly, the proposal that NAAG is an endogenous neurotransmitter must remain highly tentative at the present. An essential requirement is the demonstration of calcium-dependent evoked release of NAAG from identified excitatory nerve terminals containing NAAG. Such studies are now feasible with the availability of specific antisera for radioimmunoassay. Furthermore, mechanisms involved in NAAG synthesis - be they exclusively enzymatic or mRNA-dependent - remain to be determined. It must be emphasized that if NAAG is an excitatory neurotransmitter it likely represents only a small portion of the putative GLU/ASP terminals based upon its skewed regional brain distribution and the kinetics of the [^{3}H]-NAAG uptake-cleavage process. In this regard, it is important to note that several acidic peptides containing GLU and/or ASP are endogenous to brain (Reichelt and Kvamme, 1973) and their physiologic roles remain to be determined. We propose that these peptides as well as other acidic peptide containing endocoids such as leukotrienes merit study for their potential role as endogenous neurotransmitters released at putative GLU/ASP synapses.

ACKNOWLEDGEMENTS

This research was supported by NIH research grants including NS13584 and RSDA Type II, MH00125 to J.T. Coyle. We thank Ms. Deborah A. Culp for her assistance in manuscript preparation.

REFERENCES

Abreu, M.E., Blakely, R.D., and Coyle, J.T., 1985, Distribution of [^{3}H]-N-acetyl-aspartyl glutamate binding sites in rat brain: a quantitative in vitro autoradiographic study, Soc. Neurosci. Abstr., 11:108.

Balcar, V.J., and Johnston, G.A.R., 1972, The structural specificity of high affinity uptake of L-glutamate and L-aspartate by rat brain slices, J. Neurochem., 19:2657.

Bernstein, J., Fisher, R.S., Zaczek, R., and Coyle, J.T., 1985, Dipeptides of glutamate and aspartate may be endogenous neuroexcitants in the rat hippocampal slice, J. Neurosci., 5:1429.

Blakely, R., Ory-Lavollée, L., Thompson, R., and Coyle, J.T., 1985, A high affinity synaptosomal uptake system involving N-acetyl-aspartyl-glutamate, Soc. Neurosci. Abstr., 11:108.

Bradford, H.F., and Richards, C.D., 1976, Specific release of endogenous glutamate from pyriform cortex stimulated in vitro, Brain Res., 105:168.

Cangro, C.B., Garrison, D.E., Luongo, P.A., Trackmiller, M.E., Namboodiri, M.A.A., and Neale, J.H., 1985, First immunohistochemical demonstration of N-acetyl-aspartyl-glutamate in specific neurons, Soc. Neurosci. Abstr., 11:108.

Collins, C.G.S., 1978, Evidence of neurotransmitter role for aspartate and gamma-aminobutyric acid in the rat olfactory cortex, J. Physiol., 291:51.

Cotman, C.W., Foster, A.C., and Lanthorn, T.H., 1981, An overview of glutamate as a neurotransmitter, Adv. Biochem. Psychopharmacol., 27:1.

Curtis, D.R., and Johnston, G.A.R., 1974, Amino acid transmitters in the mammalian CNS, Ergeb. Physiol., 69:97.

Dua, A.K., Pinsky, C., and LaBella, F.S., 1985, Peptidases that terminate the action of enkephalins. Consideration of physiological importance for amino-, carboxy-, endo-, and pseudoenkephalinase, Life Sci., 37:985.

ffrench-Mullen, J.M.H., Koller, K., Zaczek, R., Coyle, J.T., Hori, N., and Carpenter, D.O., 1985, N-Acetyl-aspartyl glutamate: possible role as the neurotransmitter of the lateral olfactory tract, Proc. Natl. Acad. Sci. USA, 82:3897.

Fonnum, F., 1984, Glutamate: a transmitter in mammalian brain, J. Neurochem., 42:1.

Foster, A.C., and Fagg, G.E., 1984, Acidic amino acid binding sites in mammalian neuronal membranes: their characteristics and relationship to synaptic receptors, Brain Res. Rev., 7:103.

Greenamyre, J.T., Olson, J.M., Penney, J.B., and Young, D.B., 1985, Autoradiographic characterization of N-methyl-D-aspartate, quisqualate and kainate-sensitive binding sites, J. Pharmacol. Exp. Ther., 238:254.

Holz, R., and Senter, R.A., 1981, Choline stimulates nicotinic receptors on adrenal medullary chromaffin cells to induce catecholamine secretion, Science, 214:466.

Hori, N., Auter, C.R., Braitman, D.J., and Carpenter, D.O., 1981, Lateral olfactory tract transmitter: glutamate, aspartate or neither? Cell Mol. Neurobiol., 1:115.

Koller, K.J., Zaczek, R., and Coyle, J.T., 1984, N-Acetyl-aspartyl-glutamate; regional levels in rat brain and the effects of brain lesions as determined by a new HPLC method, J. Neurochem., 43:1136.

Koller, K.J., and Coyle, J.T., 1984a, Ontogenesis of N-acetyl-aspartate and N-acetyl-aspartyl-glutamate in rat brain, Devel. Brain Res., 15:137.

Koller, K.J., and Coyle, J.T., 1984b, Characterization of the interactions of N-acetyl-aspartyl-glutamate with [^{3}H]-L-glutamate receptors, Eur. J. Pharmacol., 98:193.

Koller, K.J., and Coyle, J.T., 1984c, Specific labelling of brain receptors with [^{3}H]-N-acetyl-aspartyl-glutamate, Eur. J. Pharmacol., 104:193.

Koller, K.J., and Coyle, J.T., 1985, The characterization of the specific binding of [^{3}H]-N-acetyl-aspartyl-glutamate to rat brain membranes, J. Neurosci., 5:2882.

Krnjević, K., and Reinhardt, W., 1979, Choline excites cortical neurons, Science, 296:1321.

Lenda, K., 1981, Ion exchange liquid chromatography of N-acetyl-aspartic acid and some N-acetyl-aspartyl peptides, J. Liqu. Chromatogr., 4:863.

Reichelt, K.L., and Kvamme, E., 1973, Histamine-dependent formation of N-acetyl-aspartyl peptides in mouse brain, J. Neurochem., 21:849.

Shiells, R.A., Falk, G., and Naghshineh, S., 1981, Action of glutamate and aspartate analogues on rod horizontal and bipolar cells, Nature, 294:592.

Sinichkin, A., Sterri, S., Edminson, P.D., Reichelt, K.L., and Kvamme, E., 1977, In vivo labelling of acetyl-aspartyl peptides in mouse brain from intracranially and intraperitonealy administered acetyl-L-[α^{14}C]-aspartate, J. Neurochem., 29:425.

Tallon, H.H., Moore, S., and Stein, W.H., 1956, N-Acetyl-L-aspartic acid in brain, J. Biol. Chem., 224:257.

Watkins, J.C., and Evans, R.H., 1981, Excitatory amino acid transmitters, Ann. Rev. Pharmacol. Toxicol., 21:165.

Zaczek, R., Koller, K.J., Cotter, R., Heller, D., and Coyle, J.T., 1983, N-Acetyl-aspartyl-glutamate: an endogenous peptide with high affinity for a brain 'glutamate' receptor, Proc. Natl. Acad. Sci. USA, 80:1116.

COMMENTARY - ACIDIC EXCITATORY AMINO ACIDS AND SEIZURES:

NEUROCHEMICAL INTERRELATIONSHIPS

J.T. Coyle and N. van Gelder

Although virtually all anti-epileptic agents have been developed through behavioral screening techniques that do not address the molecular sites of action of the drugs, current evidence indicates that the majority of the clinically effective anti-epileptics, for which mechanisms have been defined, in some way enhance inhibitory, e.g., GABAergic, neurotransmission in brain (Enna and Beutler, 1985). These include the barbiturates, benzodiazepines, and sodium valproate. The fundamental defect involved in seizures at an operational level undoubtedly reflects an imbalance between inhibitory and excitatory processes. That most anticonvulsants address only the inhibitory component of this equation underlines the potential pharmacologic significance of agents that may have specific antagonistic effects at subsets of the excitatory receptors. Furthermore, a better appreciation of the excitatory contribution may allow for better understanding of the structural and metabolic factors that are responsible for seizure foci and seizure propagation. These issues were the primary focus of this session.

A particularly potent and specific antagonist of NMDA receptors, amino-phosphonoheptanoic acid (APH), has recently been identified (McLennan, 1982). An intriguing aspect of this compound is that it occurs in a family of omega-phosphono analogs of glutamate in which the butyrate (APB) derivative blocks a different subset of excitatory receptors than the valerate and heptanoate derivatives, which are NMDA antagonists, whereas the hexanoate derivative has negligible effects (Foster and Fagg, 1984). This structural specificity bespeaks the importance of chain length and further reinforces the concept of receptor heterogeneity involved in excitatory mechanisms. Although specific binding sites in brain membranes have been defined for [^{3}H]-APB and [^{3}H]-APH, the relationship of these to neurophysiologically defined receptors remains unclear (Ferkany and Coyle, 1983; Robinson et al., 1985).

The studies by Meldrum and his colleagues have demonstrated that APH exhibits potent anticonvulsant activity when injected intracerebrally (Croucher et al., 1982). The much lower efficacy of the agent with systemic administration obviously reflects its relative exclusion by the blood brain barrier due to its charge characteristics. Nevertheless, APH antagonizes sound induced seizures in DBA/2 mice, electro-shock induced tonic seizures, limbic seizures and photically induced myoclonus in the baboon. Dose-response studies indicate that with intracerebral injection, APH and beta-D-aspartyl-amino-ethyl phosphonate are among the most potent anticonvulsants known. These observations clearly indicate the potential therapeutic utility of agents with NMDA antagonistic effects in the treatment of seizure disorders and further implicate NMDA receptors in the pathophysiology of certain forms of epilepsy.

The demonstration that agents that antagonize acidic excitatory amino acid neurotransmission in brain have potent anticonvulsant effects provides further evidence of a critical role played by acidic excitatory amino acids in the pathophysiology of epilepsy and raises questions about the interaction of these systems with their contravailing inhibitory systems. Accordingly, measurement of the synaptic neurochemical indices contributing to the balance between excitatory and inhibitory neurotransmission represents a productive strategy for characterizing the neurometabolic events associated with seizures and seizure susceptibility. Van Gelder has examined the evolution in neurometabolic alterations from the pre-convulsant to the seizure state in brain. These studies have revealed counterintuitive findings based on steady state analysis that indicate an increase of the inhibitory neurotransmitters, taurine and GABA, in the face of a depletion of glutamate. However, in the context of dynamics of neurotransmitter turnover, the reduction of glutamate and increase in GABA is consistent with an accelerated turnover of the latter. In association with reduction in the levels of taurine, which may be associated with increased influx of calcium, these metabolic alterations lead to a dominance of excitatory events consistent with epileptogenesis.

One of the limitations of measuring tissue levels of putative amino acid neurotransmitters is that they are located in functionally and physiologically heterogeneous compartments. Measurement of the amino acids in the extracellular space with the _in vivo_ dialysis technique offers opportunities for examining the dynamics of the release of these substances under various physiologic conditions. Consistent with results of van Gelder, studies reported by Lehmann and his colleagues and by French and his colleagues indicate that taurine efflux in brain is regulated by excitatory processes. For example, Lehmann et al. demonstrated a striking release of taurine in the rabbit hippocampus after treatment with kainic acid but not after treatment with the GABA receptor antagonist bicuculline, another epileptogenic agent. Similarly, French et al. observed an increased efflux of taurine from rat hippocampus after local injection of the convulsant quinolinic acid. Taurine release was not elevated in the contralateral hippocampus although it exhibited concurrent seizures. Co-administration of the antagonist, kynurenic acid, prevented the hippocampal seizures, neurotoxicity and taurine release associated with quinolinate injection. Together, these studies suggest a role for taurine in response not so much to persistant excitation or seizures but rather in response to agonal events associated with excitotoxin induced neurotoxicity. These results suggest that elevated CSF concentrations of taurine may be an indicator of excitotoxic processes such as may occur in Huntington's disease, epilepsy and anoxia.

While glutamate is generally accepted as a major excitatory neurotransmitter in brain, Coyle and his colleagues have expressed concern about an overinclusive view of its role. This concern prompted the search for additional substances in brain that might interact with subsets of excitatory amino acid receptors as defined by ligand binding techniques. A family of small acidic peptides has been well described in brain, of which N-acetyl-aspartyl-glutamate (NAAG) has been most extensively characterized. This peptide is restricted to brain and localized to neurons, exhibits an uneven regional distribution and interacts with glutamate receptors. Ligand binding and neurophysiologic studies suggest that it acts as a subset of excitatory receptors which are specifically antagonized by APB. Injection of this peptide into the hippocampus produces a seizure pattern similar to that resulting from injection of the conformationally restricted analog of glutamate, quisqualic acid. Notably, quisqualic acid has high affinity for the receptor subtype labeled with NAAG.

Taken together, these studies indicate that endogenous substances, which have inhibitory or excitatory neurophysiologic effects, are altered in terms of their levels and their release under conditions associated with seizures and excitotoxin-induced neurodegeneration. While GABA and glutamate may be the dominant neurotransmitters in the yin-yang relationship between excitation and inhibition, it is becoming evident that other substances must be factored into this equation. Although taurine levels may be involved in modulating seizure susceptibility, the efflux of this inhibitory amino acid is dramatically increased under excitotoxin conditions. A new class of acidic neuropeptides with excitatory effects, of which NAAG is the best characterized, can produce seizures with intracerebral injection; notably, NAAG levels are altered in response to seizure kindling.

REFERENCES

Croucher, M.J., Collins, J.F., and Meldrum, B.S., 1982, Anticonvulsant action of excitatory amino acid antagonists, Science, 216:899.

Enna, S.J., and Beutler, J.A., 1985, The GABA receptor as a site for antiepileptic drug action, in: Epilepsy and GABA Receptor Agonists, G. Bartholini, L. Bossi, K.G. Lloyd and P.L. Morselli, eds., Raven Press, New York, p. 195.

Ferkany, J.W., and Coyle, J.T., 1983, Specific binding of [^{3}H]-2-amino-7-phosphono heptanoic acid to rat brain membranes in vitro, Life Sci., 33:1295.

Foster, A.C., and Fagg, G.E., 1984, Acidic amino acid binding sites in mammalian neuronal membranes. Their characteristics and relationship to synaptic receptors, Brain Res. Rev., 7:103.

McLennan, H., 1982, 2-amino-7-phosphono heptanoic acid as an amino acid antagonist, Can. J. Physiol., 60:91.

Robinson, M.B., Crooks, S.L., Johnson, R.L., and Koerner, J.F., 1985, Displacement of DL-[^{3}H]-2-amino-phosphono butanoic acid ([^{3}H]APB) binding with methyl substituted APB analogues and glutamate agonists, Biochemistry, 24:2401.

SESSION VI
MECHANISMS OF EPILEPTOGENESIS

SYNAPTIC EVENTS UNDERLYING SPONTANEOUS AND EVOKED PAROXYSMAL DISCHARGES

IN HIPPOCAMPAL NEURONS

D. Johnston, P.A. Rutecki, and F.J. Lebeda

Neuroscience Program, Section of Neurophysiology
Dept. of Neurology, Baylor College of Medicine
1 Baylor Plaza
Houston, Texas 77030

INTRODUCTION

   The cellular mechanisms responsible for the paroxysmal depolarizing shift (PDS) in cortical neurons have been studied extensively for over 20 years (Kandel and Spencer, 1961; Matsumoto and Ajmone Marsan, 1964a,b; Prince, 1966, 1968; Dichter and Spencer, 1969a,b; Matsumoto et al., 1969; Ayala et al., 1970; Lebovitz et al., 1971; Ayala et al., 1973; Wong et al., 1974; Prince, 1978; Schwartzkroin and Prince, 1978; Wong and Prince, 1978, 1979; Schwartzkroin and Wyler, 1980; Johnston and Brown, 1981; Wong and Prince, 1981; Traub and Wong, 1982; Wong and Schwartzkroin, 1982; Traub and Wong, 1983; Johnston and Brown, 1984b,c; Rutecki et al., 1985). The PDS, which represents the intracellular event that occurs during the interictal discharge - the signature of epileptogenic cortex - is a sudden depolarization of the neuron that triggers a series of action potentials (Matsumoto and Ajmone Marsan, 1964a,b). The membrane events that produce the sudden depolarization are still somewhat controversial (Prince, 1978; Johnston and Brown, 1981, 1984b, 1986; Alger, 1984). In previous work, both voltage-dependent and synaptic-conductance changes have been proposed to explain the origin of the PDS (Matsumoto and Ajmone Marsan, 1964b; Prince, 1966, 1968; Dichter and Spencer, 1969a,b; Ayala et al., 1973; Wong et al., 1974; Wong and Prince, 1978, 1979; Johnston and Brown, 1981; Wong and Prince, 1981; Traub and Wong, 1982; Wong and Schwartzkroin, 1982; Traub and Wong, 1983). With the recent application of voltage-clamp techniques to cortical neurons (Johnston et al., 1980), it has been shown that a large synaptic conductance change occurs in neurons during the interictal discharge (Johnston and Brown, 1981). This finding, as well as certain other results (Johnston and Brown, 1981, 1984b, 1986), provides the strongest evidence to date that large excitatory postsynaptic potentials (EPSPs) are involved in the generation of epileptiform activity in the cortex.

   The first voltage-clamp studies of the PDS were performed on CA3 neurons in the hippocampus (Johnston and Brown, 1981). We recently extended those findings by applying the same techniques to the study of the PDS in CA1 neurons and by comparing spontaneous and orthodromically evoked PDSs in both cell types. This report will summarize our recent findings. We will also describe a simple computer model of a neuronal network that, under certain conditions, generates rhythmic organized discharges. Using this model, we are able to ask certain general questions regarding the development of epileptiform activity in a neuronal network.

METHODS

Hippocampal slices maintained in vitro were used for these investigations. The methods are described in detail in previous publications (Johnston et al., 1980; Johnston and Brown, 1981; Brown and Johnston, 1983). Hippocampi were removed from either adult rats or guinea pigs and 400–600 μm thick slices were cut perpendicular to the septotemporal axis using a McIlwain tissue chopper. The slices were transferred to a recording chamber similar in design to that of Haas et al. (1979).

Intracellular recording microelectrodes were filled with either 4 M KAc, 2 M Cs_2SO_4, or 2 M CsCl and had tip resistances in the range of 20–50 MΩ. A single electrode clamp system (SEC) was used for all intracellular recordings. Technical details of the SEC and its use in these types of experiments are described in previous publications (Brown and Johnston, 1983; Johnston and Brown, 1983, 1984a).

RESULTS

Recent data from the CA3 subfield have shown that two types of intracellularly recorded burst discharges can occur in the same set of neurons (Johnston and Brown, 1984c). The two bursts are illustrated in Fig. 1. The first type (Fig. 1A) is an endogenous burst that results from voltage-dependent conductance changes. This burst is sustained by membrane currents endogenous to the recorded neuron and is independent of the activity of other neurons (Hablitz and Johnston, 1981; Wong and Schwartzkroin, 1982).

The second type of burst (Fig. 1B) occurs after the application of a convulsant drug, after elevating extracellular potassium above about 7 mM (Rutecki et al., 1985), or after intense stimulation of orthodromic pathways. This burst is considered epileptiform because it represents the synchronous discharge of a group of neurons, and a prominent extracellular field response can be recorded during the burst. By analogy to in vivo studies (Matsumoto and Ajmone Marsan, 1964a), we have called the membrane depolarization that triggers the burst the PDS. It is also a network-driven burst; that is, in contrast to the endogenous burst, the PDS is dependent on the interactions among a network of neurons and is only defined in a context in which there is synchronization among a group of cells (Johnston and Brown, 1984b).

Because of the superficial similarity of the two types of bursts, it was thought by several workers (Prince, 1978; Wong and Prince, 1979; Schwartzkroin and Wyler, 1980; Alger, 1984) that they shared common mechanisms, that each was an endogenous burst, and that an earlier hypothesis, which stated that the PDS was a large compound excitatory synaptic potential (Ayala et al., 1973), was not viable. Johnston and Brown (1981) derived four key predictions to differentiate the two hypotheses (endogenous burst hypothesis and giant synaptic potential hypothesis). Because of the development of single electrode voltage clamp techniques and improved recording methods (Johnston et al., 1980), the direct testing of these predictions became possible. The results from several investigations using CA3 neurons in the hippocampal slice supported strongly the giant synaptic potential hypothesis (Johnston and Brown, 1981, 1984b, Lebeda et al., 1982).

In the CA1 subfield, endogenous bursts are not a prominent feature of the pyramidal neurons (Wong et al., 1974; Wong and Prince, 1978, 1979). However, PDS-like events in CA1 neurons are common under similar sets

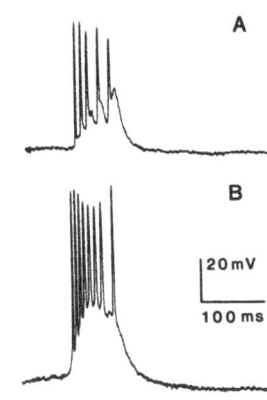

Fig. 1. Similarity in appearance of endogenous and network-driven bursts in hippocampal neurons. (A) Endogenous burst recorded intracellularly from a CA3 pyramidal neuron. The slice was bathed in normal saline solution, and an extracellular field microelectrode recorded no synchronous activity associated with the intracellular bursts. (B) Network-driven burst (PDS) recorded intracellularly from a CA3 pyramidal neuron. The slice was bathed in 3.4 mM of penicillin. An extracellular field microelectrode recorded a discharge associated with each intracellular burst. The discharge represented the synchronous firing of a large population of CA3 neurons (not shown). A and B are from different cells. (From Johnston and Brown, 1984c).

of experimental conditions as in the CA3 region. Recently, we were able to test the predictions of the two hypotheses in CA1 neurons as well as extend earlier findings from the CA3 region.

The first prediction is that if the PDS is a network-driven, large EPSP, then the frequency of occurrence of spontaneous PDSs and the probability of orthodromically activating a PDS should be independent of the membrane potential of the recorded neuron. In contrast, if the PDS is an endogenous burst triggered by a normal strength EPSP, then the frequency of occurrence or probability of activating a PDS should be highly dependent on the membrane potential (Johnston and Brown, 1981). For this first

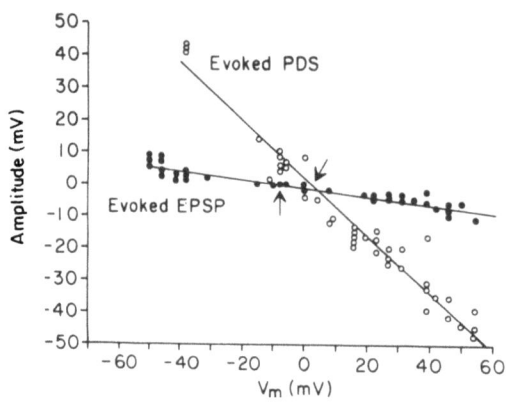

Fig. 2. Comparison of amplitudes of Schaffer collateral evoked EPSPs and PDSs. EPSPs and PDSs were recorded intracellularly from a CA1 pyramidal neuron in a slice bathed in normal saline plus 50 μM picrotoxin. The Schaffer collaterals were stimulated at 0.2 Hz. PDSs were elicited by about 50% of the stimuli. DC current was passed through the intracellular microelectrode, and the amplitudes of the EPSPs and PDSs were measured as a function of the membrane potential. The arrows indicate the measured reversal potentials (-8.1 mV and 2.3 mV for the EPSP and PDS, respectively). Each symbol represents a single measurement. Straight lines and reversal potential values were obtained from a least-squares-fit calculation.

prediction, we found that the frequency of occurrence of spontaneous PDSs and the probability of orthodromically activating PDSs in both CA1 and CA3 neurons were independent of the membrane potential - results that support the giant synaptic potential hypothesis.

The second prediction is that if the PDS is a large EPSP, then the amplitude of the PDS should be a monotonic function of the membrane potential. In contrast, if the PDS is a synaptically triggered endogenous burst, the amplitude of the PDS should be a nonmonotonic function of membrane potential (Johnston and Brown, 1981). The third prediction follows directly from the second: if the PDS is a large EPSP, then its amplitude should reverse in polarity at depolarized potentials. In contrast, if the PDS were an endogenous burst, its amplitude should reduce to zero, but never reverse, with depolarization of the membrane potential (Johnston and Brown, 1981).

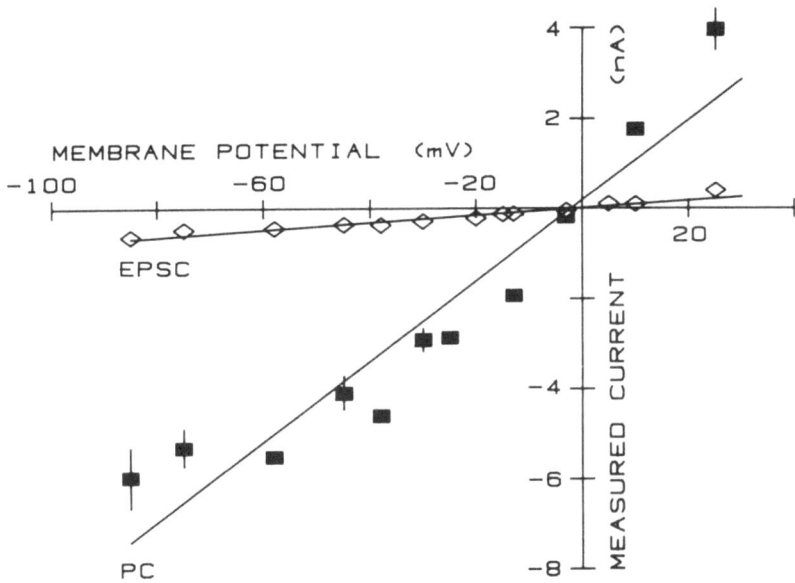

Fig. 3. Relative conductance changes associated with Schaffer collateral evoked EPSPs and PDSs. Intracellular recordings were made from a CA1 neuron in a slice bathed in normal saline plus 10 µM picrotoxin. Schaffer collaterals were stimulated at 0.5 Hz, and the intensity was adjusted so that a PDS was elicited by about every third stimulus. The neuron was voltage clamped using an SEC (see methods). The amplitudes of the currents were plotted as a function of the holding potential. Each point represents the mean of at least five trials. Error bars were not included unless they were bigger than the symbols. Straight lines were obtained from a least-squares calculation. The reversal potentials for both currents were about −2 mV. The conductance changes were 8.1 and 89 nS for the excitatory postsynaptic current (EPSC) and the paroxysmal current (PC), respectively.

For the second and third predictions, we stimulated the Schaffer collaterals and evoked a normal EPSP, as well as PDSs, and measured the amplitude of each as a function of the membrane potential. The results from one such experiment are illustrated in Fig. 2. The amplitude of the PDS varied as a monotonic function of the membrane potential and reversed in polarity at approximately zero mV. These results also support the giant synaptic potential hypothesis.

The fourth prediction requires the use of voltage clamp techniques (Johnston and Brown, 1981, 1984b,c). If the PDS is comprised of a large EPSP, then the conductance change associated with the PDS must be larger than the conductance change associated with normal EPSPs. Although the actual synaptic conductances in CA1 neurons cannot be measured accurately because the electrotonic locations of the synapse are only approximately known, we were able to compare the relative conductances of normal and PDS-associated synaptic events. The results from one such experiment are illustrated in Fig. 3. The measured conductance change associated with the Schaffer-collateral EPSP was 8.1 nS compared to 89 nS for the PDS. Both reversed in polarity at about −2 mV. The measured conductance change of the PDS was therefore about 10 times larger than that associated with a normal amplitude EPSP. These results provide convincing evidence in favor of the giant synaptic potential hypothesis.

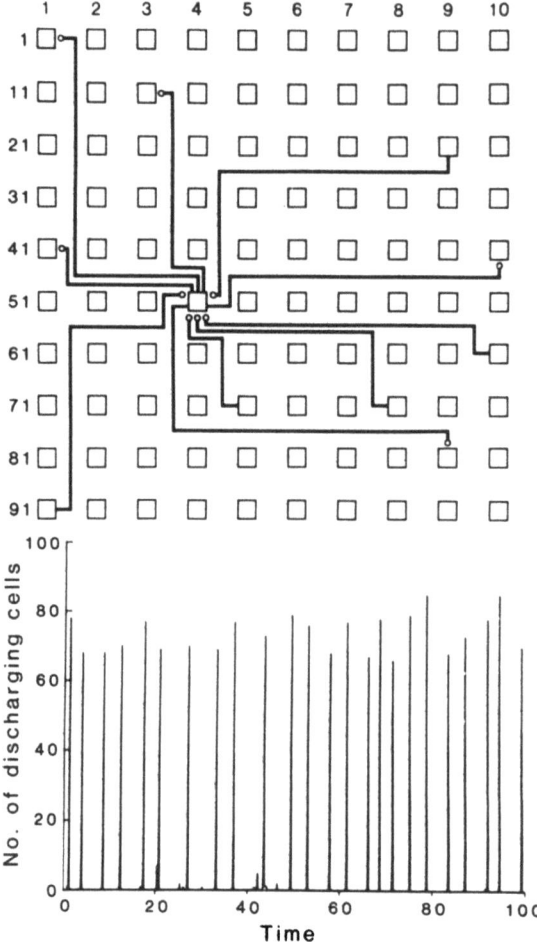

Fig. 4. Diagram of model neuronal network and periodic output. Top panel: an array of model neurons with the synaptic inputs and outputs shown only for a typical cell. None of the cells in this model network had pacemaker activity, but acted as threshold detectors. Initially, only spontaneously occurring EPSPs were present in the stimulation. Summation of these events above a preset threshold triggered an action potential that evoked EPSPs in follower cells. Bottom panel: simulation output showing the number of cells firing an action potential as a function of time.

DISCUSSION

The results presented here and elsewhere (Johnston and Brown, 1981; Lebeda et al., 1982; Johnston and Brown 1984b,c; Rutecki et al., 1985) have been consistent regardless of the hippocampal cell type studied or

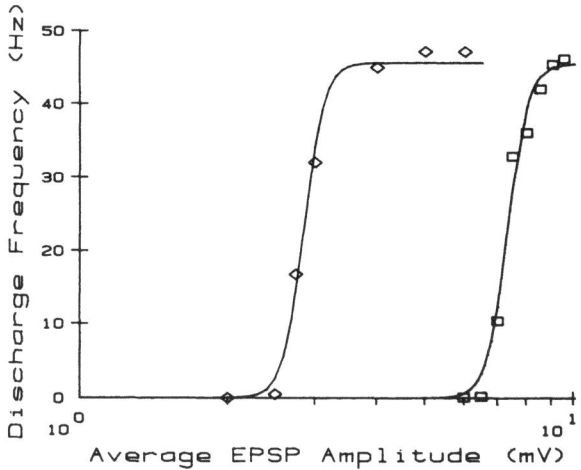

Fig. 5. Plot of the discharge frequency versus the average EPSP amplitude. A parallel, rightward shift in the input-output curve was produced by raising each cell's steady-state threshold by 5 mV. Solid curves were obtained from a nonlinear least-squares-fit program in which the slopes were identical.

whether the PDS occurred spontaneously or was evoked orthodromically. The four key predictions of the giant synaptic potential hypothesis have been confirmed in these cases. Of particular importance is that larger-than-normal strength excitatory synaptic events have now been demonstrated to exist in both CA1 and CA3 neurons and are associated with epileptiform activity.

NEURONAL NETWORK MODEL

To put epileptiform activity in the hippocampus into a more theoretical framework, we have constructed a simple computer model of a network of neurons. The primary goal was to determine whether synchronous, periodic activity could be generated by a population of non-bursting neurons, i.e., a network of interconnected cells that was not driven by a periodic intrinsic or extrinsic source. Although the latter characteristic has been used in other modeling studies (Anninos and Cyrulnik, 1977), it was deemed important to know whether or not a self-organizing process (Nicolis and Prigogine, 1977; Fogelman-Soulie, 1985) could emerge from this system of normally quiescent cells.

The network consisted of an array of threshold detectors (top of Fig. 4). The only ongoing neuronal activity in the system of 100 randomly connected cells was the presence of spontaneously occurring EPSPs. The EPSP amplitudes were obtained from a Gaussian distribution, while the frequency of their occurrence was generated from a Poisson distribution. Linear addition of EPSP amplitudes was used as a first approximation of cellular integration. Once a preset threshold was reached, an action potential was mimicked by setting the membrane potential to 0 mV for 1 msec. To simulate a relative refractory period, the threshold was then raised and subsequently allowed to decay as an exponential function of time to its steady state level (Segundo et al., 1968).

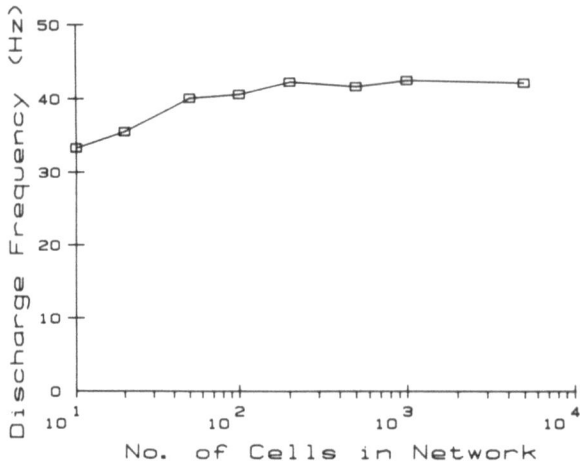

Fig. 6. Discharge frequency versus size of network. The number of cells in the network was varied from 10 to 5000. The frequency of discharge remained relatively constant over this range of network size. The percentage of cells in the network active during each discharge also remained relatively constant (data not shown). The important variables for determining discharge frequency therefore appear to be EPSP amplitude and threshold level.

The simulation output shown in the lower part of Fig. 4 illustrates the network response using a mean of 5 interconnections per cell. Sixty-five to 85% of the cells in the network fired synchronously in a relatively periodic manner. A wide range of output frequencies could be obtained with only small changes in the EPSP amplitude (Fig. 5). Raising the steady-state threshold by 5 mV produced a parallel, rightward shift in the network's nonlinear input-output relation. Although the frequency of discharges was not quite sensitive to EPSP amplitude and threshold level, it was relatively insensitive to the number of cells in the network (Fig. 6).

The results of these simulations are interesting from two standpoints. First, they support the position (Johnston and Brown, 1986) that there is no theoretical requirement for a periodic intrinsic or extrinsic source to produce periodic activity in a network of neurons. Second, they provide several experimentally testable predictions (frequency versus network size; frequency versus EPSP amplitude and threshold level). The details of the model can also be made more realistic as information is obtained regarding the basic membrane and synaptic properties of hippocampal neurons.

ACKNOWLEDGEMENTS

This work was supported by NIH grants NS11535, NS18295, and NS15772, by the Grass Foundation, and by USAMRDC DAMD 17-82-C-2254. We thank Judy Walker for technical help and Diane Jensen for secretarial assistance.

REFERENCES

Alger, B.E., 1984, Hippocampus: electrophysiological studies of epileptiform activity in vivo, in: *Brain Slices*, R. Dingledine, ed., Plenum Press, New York, p. 155.

Anninos, P.A., and Cyrulnik, R., 1977, A neural net model for epilepsy, *J. Theor. Biol.*, 66:695.

Ayala, G.F., Matsumoto, H., and Gumnit, R.J., 1970, Excitability changes and inhibitory mechanisms in neocortical neurons during seizures, *J. Neurophysiol.*, 33:73.

Ayala, G.F., Dichter, M., Gumnit, R.J., Matsumoto, H., and Spencer, W.A., 1973, Genesis of epileptic interictal spikes. New knowledge of cortical feedback systems suggests a neurophysiological explanation of brief paroxysms, *Brain Res.*, 52:1.

Brown, T.H., and Johnston, D., 1983, Voltage clamp analysis of the mossy fiber synaptic input to hippocampal pyramidal neurons, *J. Neurophysiol.*, 50:487.

Dichter, M., and Spencer, W.A., 1969a, Penicillin-induced interictal discharge from the cat hippocampus. I. Characteristics and topographical features, *J. Neurophysiol.*, 32:649.

Dichter, M., and Spencer, W.A., 1969b, Penicillin-induced interictal discharges from the cat hippocampus. II. Mechanisms underlying origin and restriction, *J. Neurophysiol.*, 32:663.

Fogelman-Soulie, F., 1985, Stable core in discrete iterations, in: *Dynamical Systems and Cellular Automata*, J. Demongeot, E. Goles, and M. Tchuente, eds., Academic Press, New York, p. 27.

Haas, H.L., Schaerer, B., and Vosmansky, M., 1979, A simple perfusion chamber for the study of nervous tissue slices in vitro, *J. Neurosci. Meth.*, 1:323.

Hablitz, J.J., and Johnston, D., 1981, The endogenous nature of spontaneous bursts in hippocampal pyramidal neurons, *Cell. Molec. Neurobiol.*, 1:325.

Johnston, D., Hablitz, J.J., and Wilson, W.A., 1980, Voltage clamp discloses slow inward current in hippocampal burst-firing neurones, *Nature*, 286:391.

Johnston, D., and Brown, T.H., 1981, Giant synaptic potential hypothesis for epileptiform activity, *Science*, 211:294.

Johnston, D., and Brown, T.H., 1983, Interpretation of voltage clamp measurements in hippocampal neurons, *J. Neurophysiol.*, 50:464.

Johnston, D., and Brown, T.H., 1984a, Biophysics and microphysiology of synaptic transmission in hippocampus, in: *Brain Slices*, R. Dingledine, ed., Plenum Press, New York, p. 51.

Johnston, D., and Brown, T.H., 1984b, Mechanisms of neuronal burst generation, in: *Electrophysiology of Epilepsy*, P.A. Schwartzkroin and W. Wheal, eds., Academic Press, New York, p. 277.

Johnston, D., and Brown, T.H., 1984c, The synaptic nature of the paroxysmal depolarizing shift in hippocampal neurons, *Ann. Neurol.*, 16:S65.

Johnston, D., and Brown, T.H., 1986, Control theory applied to neural networks illuminates synaptic basis of interictal epileptiform activity, in: *The Basic Mechanisms of the Epilepsies*, A.V. Delgado-Escueta, A.A. Ward, and D.M. Woodbury, eds., Raven Press, New York, in press.

Kandel, E.R., and Spencer, W.A., 1961, Excitation and inhibition of single pyramidal cells during hippocampal seizure, *Exp. Neurol.*, 4:162.

Lebeda, F.J., Hablitz, J.J., and Johnston, D., 1982, Antagonism of GABA-mediated responses by d-tubocurarine in hippocampal neurons, *J. Neurophysiol.*, 48:622.

Lebovitz, R.M., Dichter, M., and Spencer, W.A., 1971, Recurrent excitation in the CA3 region of cat hippocampus, *Intern. J. Neurosci.*, 2:99.

Matsumoto, H., and Ajmone Marsan, C., 1964a, Cortical cellular phenomena in experimental epilepsy: interictal manifestations, *Exp. Neurol.*, 9:286.

Matsumoto, H., and Ajmone Marsan, C., 1964b, Cortical cellular phenomena in experimental epilepsy: ictal manifestations, *Exp. Neurol.*, 9:305.

Matsumoto, H., Ayala, G.F., and Gumnit, R.J., 1969, Neuronal behavior and triggering mechanisms in cortical epileptic focus, J. Neurophysiol., 32:688.
Nicolis, G., and Prigogine, I., 1977, Self-Organization in Nonequilibrium Systems, Wiley InterScience, New York.
Prince, D.A., 1966, Modification of focal cortical epileptogenic discharge by afferent influences, Epilepsia, 7:181.
Prince, D.A., 1968, The depolarization shift in 'epileptic' neurons, Exp. Neurol., 21:467.
Prince, D.A., 1978, Neurophysiology of epilepsy, Ann. Rev. Neurol., 1:395.
Rutecki, P.A., Lebeda, F.J., and Johnston, D., 1985, Epileptiform activity during changes in extracellular potassium in the hippocampus, J. Neurophysiol., 54:1363.
Schwartzkroin, P.A., and Prince, D.A., 1978, Cellular and field potential properties of epileptic hippocampal slices, Brain Res., 147:117.
Schwartzkroin, P.A., and Wyler, A.R., 1980, Mechanisms underlying epileptiform burst discharge, Ann. Neurol., 7:95.
Segundo, J.P., Perkel, D.H., Wyman, H., Hegstad, H., and Moore, G.P., 1968, Input-output relations in computer-simulated nerve cells, Kybernetic, 4:157.
Traub, R.D., and Wong, R.K.S., 1982, Cellular mechanism of neuronal synchronization in epilepsy, Science, 216:745.
Traub, R.D., and Wong, R.K.S., 1983, Synchronized burst discharge in disinhibited hippocampal slice. II. Model of cellular mechanism, J. Neurophysiol., 49:442.
Wong, R.K.S., Prince, D.A., and Basbaum, A.I., 1974, Intradendritic recordings from hippocampal neurons, Proc. Natl. Acad. Sci. USA, 76:986.
Wong, R.K.S., and Prince, D.A., 1978, Participation of calcium spikes during intrinsic burst firing in hippocampal neurons, Brain Res., 159:385.
Wong, R.K.S., and Prince, D.A., 1979, Dendritic mechanisms underlying penicillin-induced epileptiform activity, Science, 204:1228.
Wong, R.K.S., and Prince, D.A., 1981, Afterpotential generation in hippocampal pyramidal cells, J. Neurophysiol., 45:86.
Wong, R.K.S., and Schwartzkroin, P.A., 1982, Pacemaker neurons in the mammalian brain: mechanisms and function, in: Cellular Pacemakers, Vol. 1, D.O. Carpenter, ed., John Wiley, New York, p. 237.

INWARD CURRENTS IN CAT NEOCORTICAL NEURONS STUDIED IN VITRO

W.E. Crill, P.C. Schwindt, J.A. Flatman, C.E. Stafstrom, and W. Spain

Department of Physiology and Biophysics
University of Washington
Seattle, Washington 98195

INTRODUCTION

Two distinct types of clinical recurrent seizures or epilepsy have been identified: those that begin focally in cortex (partial seizures) and those that appear to begin synchronously in both hemispheres (generalized seizures). Because most experimentalists use focal physical or chemical techniques to initiate seizures, our concepts about epileptic mechanisms are, therefore, more applicable to clinical partial epilepsy.

Some insight into the mechanisms of focal epilepsy can be gained by considering several of the well recognized characteristics of clinical and experimental epilepsy. Between seizures, the EEG or electrocorticogram often records sharp transient responses that reflect the synchronized discharge of a population of neurons in the focus. This is the interictal EEG spike. Intermittently and largely for unknown reasons the focal spike is associated with an afterdischarge or prolonged burst that signals the onset of a focal seizure or ictal response. The seizure activity can spread from an active experimental or naturally occurring focus to involve normal cortex causing a secondarily generalized convulsion. Many different pharmacological and physical agents initiate focal seizures in the laboratory (Kopeloff et al., 1947; Kandel and Spencer, 1961; Goldensohn and Purpura, 1963; Matsumoto and Ajmone Marsan, 1964a,b; Pollen and Ajmone Marsan, 1965; Schwindt et al., 1984). They necessarily do not act by the same mechanism, but they all do cause intense activation of neurons in the focus. The responses of both single cells and populations of neurons are remarkably similar among different models (Kandel and Spencer, 1961; Goldensohn and Purpura, 1963; Matsumoto and Ajmone, 1964a,b; Pollen and Ajmone Marsan, 1965; Prince and Futamachi, 1970). For example, at the cellular level all models of epilepsy are associated with a prolonged depolarization (DS) usually with superimposed high frequency repetitive firing. Some substances such as penicillin and strychnine cause the rapid development of epileptic behavior (Kopeloff et al., 1947; Kandel and Spencer, 1961; Fuortes and Nelson, 1963; Matsumoto and Ajmone Marsan, 1964a; Pollen and Ajmone Marsan, 1965; Dichter and Spencer, 1969a; Matsumoto et al., 1969; Schwindt and Crill, 1981); other modes of experimental seizure production take much longer to have effect (Kopeloff et al., 1947). The magnitude and duration of the depolarization shift varies among cell types and with the epileptogenic agent. DSs are more prominent in penicillin induced seizures (Prince, 1968; Dichter and Spencer, 1969a,b; Matsumoto et al., 1969; Prince, 1969). At the population level, synchronization of neuronal discharges is always part of experimental epileptiform activity. During the ictal phase, a more persistent depolarization of neurons occurs, often blocking the spike generating mechanism of individual cells. It is likely that the bursting in the focus, regardless of the cause, induces epileptic activity in the surrounding cortex and this change can be self-sustained. These observations are best explained by assuming that epileptogenic activity is induced by intense synchronized neuronal discharge and that the mechanisms necessary to support and sustain an epileptic seizure normally must be present in the central nervous system of mammals. Presumably the high frequency repetitive firing of the cells in the focus or the synchronized firing evoked by electrical stimulation 'releases' the mechanism leading to seizure activity.

During the first decade of investigation into cellular epileptic behavior, most experimental attention was devoted to the mechanisms underlying the DS (Prince, 1968; Dichter and Spencer, 1969a,b; Matsumoto et al., 1969; Prince, 1969; Ayala et al., 1970, 1973). Those early studies were largely descriptive, but nevertheless have given us guidelines that must be explained by any proposed mechanism for epileptogenesis. The normal excitable and synaptic properties of central neurons were not understood in detail. Our concepts were based on a generalization of axon properties (Hodgkin and Huxley, 1952) to central cells and the mechanisms of postsynaptic inhibition and excitation studied primarily in spinal motoneurons (Eccles, 1957). It is not surprising that controversies, such as whether or not the DS was a synaptic potential or an active cell response, arose (Prince, 1968; Dichter and Spencer, 1969a; Matsumoto et al., 1969; Prince, 1969; Ayala et al., 1970, 1973). Because the techniques available for studying active responses and synaptic mechanisms were limited, the arguments could not be settled by experimentation. Recently, with the development of *in vitro* techniques such as tissue culture (Fischbach and Nelson, 1977) and tissue slice preparations (Yamamoto, 1972; Schwartzkroin, 1975, 1977) and the application of voltage clamp (Barrett et al., 1980; Barrett and Crill, 1980; Brown and Griffith, 1983a,b; Constanti and Galvan, 1983) and patch clamp (Neher and Sakmann, 1976; Hamill et al., 1981) to central cells, our understanding of normal mechanisms has leaped forward. At present we still do not understand any central neuronal cell type in detail. We do, however, know that the active and synaptic properties of neurons are much more complex than the sodium-potassium mechanism in axons (Adams, 1982; Crill and Schwindt, 1983). Numerous classes of ionic channels have been described in the neuron types studied so far. Now, there is no shortage of identified membrane and synaptic mechanisms to explain paroxysmal DSs and synchronized epileptic behavior. As different mechansims causing persistent depolarization are identified, the questions asked by the experimental epileptologist have expanded. We must not only ask which mechanisms are responsible for the DS and bursting behavior in the focus but also which intrinsic mechanisms allow the spread of seizure activity?

Here, we briefly review the membrane properties of cat neocortical cells. We will emphasize the inward currents present in normal neurons. These currents must play a role in experimentally induced prolonged depolarizations amd normal high frequency firing. They are likely, therefore, to be important in epileptic behavior.

METHODS (Stafstrom et al., 1984b)

Cats were anesthetized with sodium pentobarbital (40 mg/kg) and a prism of neocortex and subcortical white matter including the pre- and postcruciate cortex was removed. A mechanical tissue chopper was used to make slices about 500μ thick. Slices were either placed in an oxygenated holding chamber or in a Haas-type recording chamber (Haas et al., 1979) for the experimental study. Slices were perfused with artificial CSF at pH 7.4 (in mM): 130 NaCl, 3 KCl, 2 $CaCl_2$, 2 $MgCl_2$, 1.25 NaH_2PO_4, 26 $NaHCO_3$ and 10 dextrose. When $CoCl_2$, $MnCl_2$ or $CdCl_2$ replaced $CaCl_2$, the NaH_2PO_4 was omitted. Most recordings were with microelectrodes filled with 2.7 M KCl (10-30 MΩ). Either a Dagan model 8100 or an Axoclamp-2 was used for intracellular recording and the injection of current into the neuron in either the bridge or switch current clamp mode (Wilson and Goldner, 1975). Focal electrical stimulation was applied directly to white matter under the recording area in the cortex to excite neurons both antidromically and synaptically. Microelectrodes were used to impale cells 1.2 to 1.5 mm below the pial surface in layer V of the cortex. The input resistance of 102 neurons was 11.7 $\pm$ 7.1 MΩ. Neurons stained with Lucifer yellow were large pyramidal shaped neurons (Stafstrom et al., 1984b). Impaled neurons were voltage clamped using the single electrode voltage clamp (SEVC) with the current injection and voltage-sampling modes set to alternate at a frequency of 3 kHz with a 50% duty cycle.

RESULTS

Transient Inward Sodium Current

The SEVC does not respond fast enough to measure the currents responsible for action potential depolarization. The spike in neocortical cells, like all other mammalian central neurons examined so far, is blocked by TTX (Connors et al., 1982; Stafstrom et al., 1984b, 1985; Fig. 1). We, therefore, assume that neocortical cells have a transient voltage dependent sodium conductance mechanism responsible for the all-or-nothing upstroke of the action potential. As expected, persistent depolarization will inactivate the spike generating mechanism (Stafstrom et al., 1985).

Calcium Current

Cat neocortical neurons have a high threshold calcium conductance mechanism (Connors et al., 1982; Stafstrom et al., 1985). It is not activated until the soma is depolarized by about 40 mV above resting potential. The identified calcium conductance system is probably located both in the soma and dendritic membranes. In the presence of TTX, alone no all-or-nothing calcium response is evoked by large intracellularly applied currents (Fig. 1B). If, however, the voltage-dependent potassium currents also are blocked by TEA (or other potassium channel blockers) large depolarizing currents evoke high threshold calcium spikes. These spikes are blocked by either cobalt or manganese ions. Barium ions easily traverse calcium channels and block some potassium currents (Hagiwara, 1975). When Ba^{2+} is substituted for extracellular Ca^{2+} intracellular stimulation of neocortical neurons evokes prolonged barium spikes in the presence of TTX. The barium spikes are also blocked by Cd^{2+} (Stafstrom et al., 1985). Presumably the channels for the high threshold calcium current are distributed both in the soma and beyond the region of voltage control, in the dendrites. Since the normal action potential of neocortical cells has no persistent phase of depolarization and even with the transient sodium mechanism blocked 'calcium spikes' cannot be evoked by depolarizing stimuli without blocking potassium currents, we assume the potassium currents in these cells prevent the high threshold calcium conductance from generating an all-or-nothing component to the active responses in normal neocortical cells. Nevertheless, even the short duration sodium spike will still open some calcium channels but the large outward currents prevent the calcium current from beccoming net inward current and evoking an all-or-nothing calcium spike. Calcium

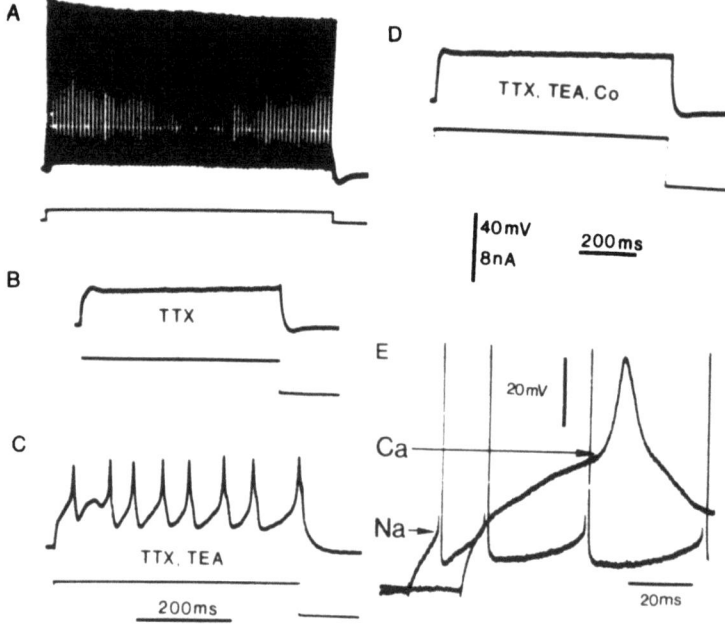

Fig. 1. Sodium and calcium spikes in neocortical neurons. (A-D) Responses of a single neocortical neuron (top trace) to injected intracellular current (bottom trace). (A) Normal repetitive firing; (B) Response in presence of TTX; (C) Calcium spikes evoked in presence of TTX and TEA; (D) Blocked calcium spikes with the addition of cobalt. (E) Superimposed traces of response to injected current in normal bathing solution (Na) and in solution containing TTX and TEA (Ca). Arrows mark the spike threshold. From Stafstrom et al., 1985.

entry still occurs and can signal the complex series of intracellular and membrane responses known to require calcium as a second messenger. The charge carrying properties of calcium current in neocortical cells, nevertheless, could still be an important putative epileptic mechanism. Any process that caused a relative increase in the current through calcium channels either directly or secondarily by a reduction in outward currents can lead to prolonged neuronal depolarization. For example, blocking some of the potassium channels with TEA or cesium induces bursting behavior in these neurons.

Persistent Sodium Current

Persistent subthreshold depolarizing responses to injected current in guinea pig cerebellar Purkinje cells (Llinas and Sugimori, 1980), neocortical neurons (Connors et al., 1982) and hippocampal pyramidal cells (Hotson et al., 1979) are partially blocked by TTX. These observations suggested the presence of noninactivating sodium conductance that is activated in the subthreshold range of membrane potentials. Susequently, voltage clamp experiments in these neuron types revealed a noninactivating TTX sensitive subthreshold current, INa_p (Stafstrom et al., 1982; Sugimori and Llinas, 1983; Stafstrom et al., 1985; French and Gage, 1986). Fig. 2 shows the subthreshold inward rectification in cat neurons located in layer V of

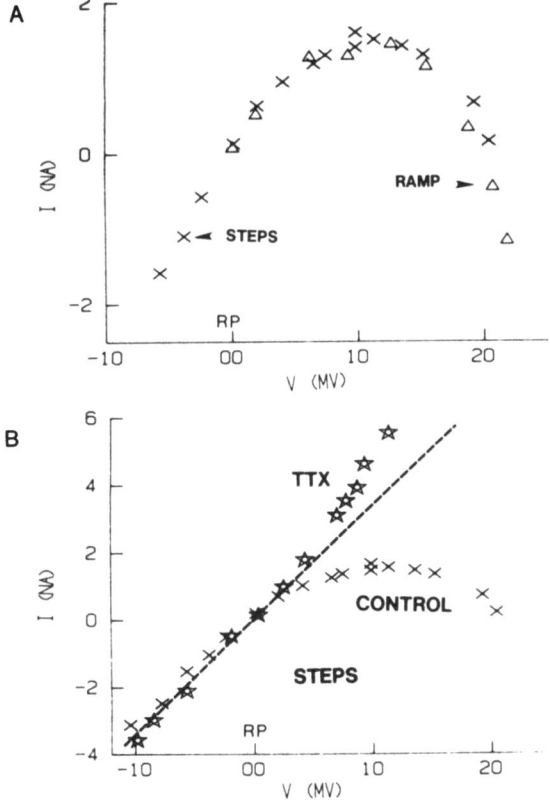

Fig. 2. I-V curves from neocortical neurons in the presence and absence of TTX. (A) I-V curves from the same neocortical neuron measured from voltage clamp ramp and step commands. (B) I-V curves before (x) and after the application of 10^{-6} M TTX (stars). From Stafstrom et al., 1985.

the sensorimotor cortex. In neocortical neurons, INa_p is activated in less than 3 msec (the maximum resolution of the SEVC). Although direct examination of INa_p with the SEVC is difficult at large depolarization because the transient sodium current is activated and cannot be controlled, slow or persistent depolarization of the neuron inactivates the transient sodium current revealing INa_p at membrane potentials 40 mV positive to rest. Persistent all-or-nothing depolarizations of the neuron 50 mV above rest occur in the presence of Co^{2+} to block calcium currents and TEA to block potassium currents. This persistent depolarization is TTX sensitive and strongly suggests the presence of a INa_p at these large depolarizations. INa_p is mixed with outward potassium currents at all potentials where it is activated. Several mechanisms could explain the apparent absence of inactivation in INa_p. Incomplete inactivation of classical sodium channels at large depolarizations would cause the sodium current to persist. Because INa_p is present when the spike generating mechanism is inactivated by depolarization, it is unlikely that INa_p is only a window current caused by overlap of the m_∞ and h_∞ curves. This mechanism could be evaluated by directly measuring the steady state activation and inactivation of

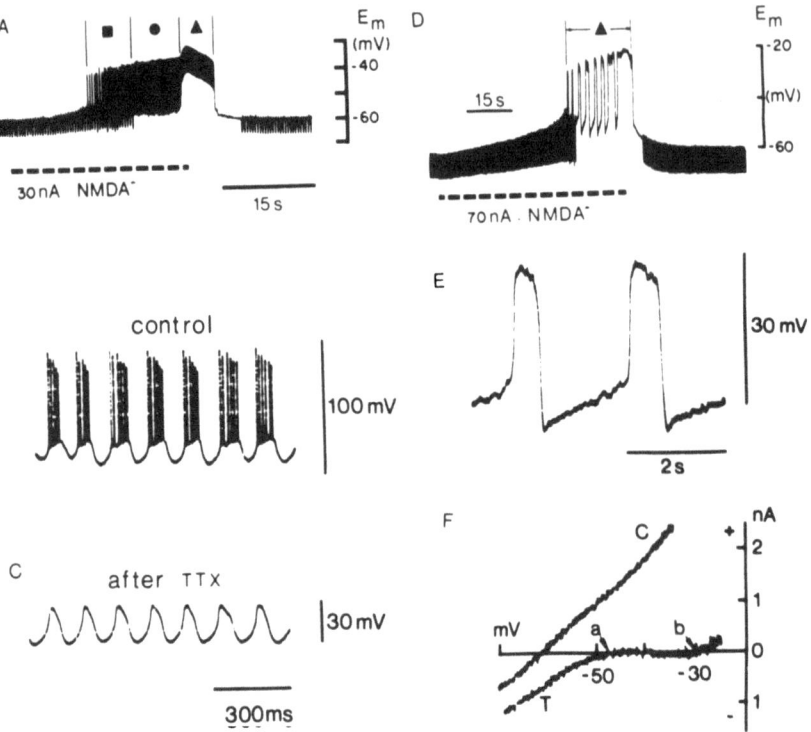

Fig. 3. Effects of iontophoretically applied NMDA to neocortical neurons. (A) Chart recorded traces of response of neocortical neuron to NMDA. Downward deflections are response to brief 0.4 nA current pulses to measure input conductance. Square shows period of slow depolarization; circle shows rhythmic depolarization shifts and triangle shows period of bistable membrane potential behavior. (B-F) Records from a different neuron. (B) NMDA induced rhythmic bursting and (C) the response to TTX. (D) Traces of response of the same cell showing bistable membrane potential behavior in presence of TTX, TEA, cobalt and calcium free bathing solution. (E) Fast oscilloscope traces of the episodic depolarizations shown in (D). (F) I-V relationship from the same neuron before (C) and after the iontophoresis of NMDA. a and b mark stable membrane potentials. From Flatman et al., 1983.

the transient sodium current. Most investigators favor the presence of two populations of sodium channels with different inactivation properties. INa_p has a higher sensitivity to intracellular QX-314 than INa. QX-314 is a lidocaine derivative that blocks sodium currents. Similar slowly inactivating sodium channels have been described in squid axon (Gilly and Armstrong, 1984).

NMDA Induced Inward Current

One type of receptor for excitatory amino acids is activated preferentially by N-methyl-D-aspartate (NMDA; Curtis et al., 1960; Davies et al., 1980). Following the bath or iontophoretic application of NMDA, rhythmic bursts are recorded from neurons in neocortical slices (Fig. 3). Each burst is characterized by regular large persistent depolarizations with superimposed spikes. The burst of action potentials is blocked by TTX, leaving the underlying rhythmic slow depolarizations. SEVC studies reveal the induction of a highly voltage dependent inward current by NMDA. The NMDA inward current reverses at about +15 mV and is probably carried by several cations. The slow NMDA induced depolarizations and the bistable behavior signifying a negative slope conductance are not blocked in the presence of TTX, TEA, low $[Ca^{2+}]_o$, and cobalt (Fig. 3D,E). The NMDA current is decreased in the presence of low $[Na^+]_o$. Patch clamp studies by others have revealed that the voltage dependent inward current is caused by a voltage dependent block of the NMDA current by magnesium ions (Nowak et al., 1984).

DISCUSSION

The large neurons in layer V of cat neocortex have diversity of conductance mechanisms. We have only discussed the inward currents. In addition to the classic sodium and potassium currents responsible for action potential generation, these neurons normally have two identified persistent inward conductance mechanisms. One is a high threshold calcium conductance and the other is a sodium conductance activated in the sub-threshold region. Although the precise functional role of either current is not known, indirect evidence suggests that INa_p is necessary for the transduction of synaptic depolarization into high frequency repetitive firing (Schwindt and Crill, 1980; Stafstrom et al., 1985). The calcium current does not appear to generate active responses in normal Betz cells. Either current can cause prolonged depolarization if the persistent inward currents are allowed to dominate the cells behavior. Normally, the upstroke of the action potential is associated with activation of INa, INa_p and some ICa. INa is rapidly inactivated by depolarization, but INa_p persists as does ICa. Depolarization also activates several types of potassium channels. The potassium currents plus the outward leakage current are much larger than the sum of the persistent inward currents and cause rapid spike repolarization. Following a single action potential a long afterhyperpolarization occurs because the time constant for removal of potassium conductance activation near rest is much longer than for removal of activation of the persistent sodium conductance and calcium conductance.

Continued depolarization of the cell by either tonic synaptic input or directly applied depolarizing current evokes repetitive firing and the interspike membrane potential trajectory remains positive to the resting potential. The mean potential of the interspike trajectory enters the range of membrane potential where INa_p is activated (Fig. 4D). Consequently, the amount of activated INa_p increases with progressively larger depolarizing synaptic or injected currents. The persistent activation of INa_p during the interspike trajectory will add with depolarizing synaptic current or applied current causing rapid depolarization of the neuron and high frequency repetitive firing. As noted above, once membrane potential reaches values where the I-V curve has a negative slope conductance, an all-or-nothing reponse is initiated. The inward current for this response is first carried by INa_p and then INa.

It should be evident from the foregoing discussion that output of the neuron is a function of the balance of outward and inward currents during the interspike interval. Any mechanism that causes an increase in inward currents relative to outward currents will increase the frequency of firing in response to a given synaptic input (Schwindt and Crill, 1982; Stafstrom et al., 1984a). The gain of pathways mediated by excitatory connections will be increased and also inhibitory circuits will be less effective. This can occur if the conductance for INa_p or ICa is increased

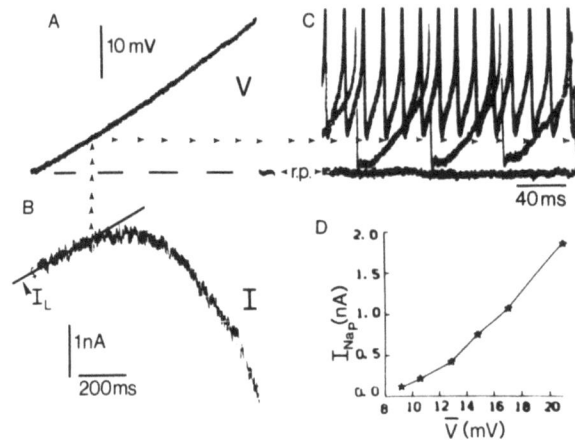

Fig. 4. Relationship of interspike membrane trajectory and activation of INa_p. All traces are from the same neocortical neuron. (A) Slowly rising voltage clamp ramp command. Dashed line is resting potential. (B) Membrane current measured during the slow ramp command. I_L is the associated membrane current. (C) Superimposed traces of resting potential and repetitive firing at two rates in response to injected current. Top of the spikes are clipped. The arrows in A-C identify the membrane potential where INa_p is first identified. (D) Plot of the magnitude of INa_p at various measured mean membrane interspike potentials. From Schwindt and Crill, 1982.

by either adding to the number of channels in the membrane or increasing the individual channel conductance. An increase in $[Na]_o$ or $[Ca]_o$ or a decrease in $[Na]_i$ or $[Ca]_i$ will increase inward current by increasing the driving potential for sodium or calcium. Alternatively, a relative increase in inward currents will occur if the outward repolarizing currents are decreased either by a decrease in conductance or a change in driving potential.

We do not know the role of these putative mechanisms in epilepsy as it spreads throughout normal neocortex. However, we do know that the underlying mechanisms are in place to add to the generation of self-sustained bursting in neocortex. For example, a number of investigators have shown an increase in $[K]_o$ during epileptiform activity (Somjen and Lothman, 1974; Fisher et al., 1976; Somjen, 1984). It does not appear to precede the onset of synchronized bursting so it is difficult to argue that a decrease in potassium driving potential (and associated decrease in IK) is the whole explanation for epileptic activity. Nevertheless, this mechanism can lead to increased firing in response to excitatory synaptic inputs and in the extreme case could be a factor in the occurrence of self-sustained prolonged depolarization of neurons (Schwindt and Crill, 1980). It could also contribute to a synchronization of activity in cortical neurons. For completeness, we should also mention that increased extracellular potassium is associated with a large increase in spontaneous synaptic activity in neocortical slices. We do not know any details about the membrane properties of synaptic terminals but it does appear that any mechanism that causes prolonged depolarization of the terminal will in turn cause increased release of neurotransmitters.

One of the most interesting recent discoveries about central neuron membrane and synaptic mechanisms is the identification of changes in voltage-dependent currents associated with the formation of transmitter-receptor complexes. Muscarinic agonists block one of the potassium channels in cortical neurons (Brown and Adams, 1980; Adams, 1982; Adams et al., 1982). For reasons described above, this will cause an increase in the action potential frequency to an excitatory input. Activation of the NMDA receptor induces a voltage dependent inward current that will have a similar effect upon a neuron's behavior (Flatman et al., 1983, 1986). For the first time we are at a stage in the scientific investigation of mammalian neuronal and synaptic mechanisms where testable precise hypotheses can be formulated concerning epileptic mechanisms.

ACKNOWLEDGEMENTS

We thank Gregg Hinz and Rebecca Gerlach for their technical assistance and Patrick Roberts and Robin Ferguson for preparation of this manuscript. This study was supported by NIH grant NS 16792.

REFERENCES

Adams, P.R., 1982, Voltage-dependent conductances of vertebrate neurones, Trends Neurosci., 5:116.
Adams, P.R., Brown, D.A., and Constanti, A., 1982, M-currents and other potassium currents in bullfrog sympathetic neurons, J. Physiol., 330:537.
Ayala, G.F., Matsumoto, H., and Gumnit, R.J., 1970, Excitability changes and inhibitory mechanisms in neocortical neurons during seizures, J. Neurophysiol., 33:73.
Ayala, G.F., Dichter, M., Gumnit, R.J., Matsumoto, H., and Spencer, W.A., 1973, Genesis of epileptic interictal spikes. New knowledge of cortical feedback systems suggests a neurophysiological explanation of brief paroxysms, Brain Res., 52:1.
Barrett, E.F., Barrett, J.N., and Crill, W.E., 1980, Voltage-sensitive outward currents in cat motoneurones, J. Physiol., 304:251.
Barrett, J.N. and Crill, W.E., 1980, Voltage clamp of cat motoneurone somata: properties of a fast inward current, J. Physiol., 304:231.
Brown, D.A., and Adams, P.R., 1980, Muscarinic suppression of a novel voltage-sensitive K^+ current in vertebrate neurone, Nature, 283:673.
Brown, D.A., and Griffith, W.H., 1983a, Calcium-activated outward current in voltage-clamped hippocampal neurones of the guinea-pig, J. Physiol., 337:287.
Brown, D.A., and Griffith, W.H., 1983b, Persistent slow inward calcium current in voltage-clamped hippocampal neurones of the guinea pig, J. Physiol., 337:303.
Connors, B.W., Gutnick, M.J., and Prince, D.A., 1982, Electrophysiologic properties of neocortical neurons in vitro, J. Neurophysiol., 48:1302.
Constanti, A., and Galvan, M., Fast inward-rectifying current accounts for anomalous rectification in olfactory cortex neurons, J. Physiol., 335:153.
Crill, W.E., and Schwindt, P.C., 1983, Active currents in mammalian neurons, Trends Neurosci., 6:236.
Curtis, D.R., Phillis, J.W., and Watkins, J.C., 1960, The chemical excitation of spinal neurones by certain acidic amino acids, J. Physiol., 150:656.

Davies, J., Evans, R.H., Francis, A.A., Jones, A.W., and Watkins, J.C., 1980, Excitatory amino acid receptors in vertebrate central nervous system, in: Neurotransmitters and Their Receptors, U.Z. Littauer, Y. Dudai, V.I. Silman, V.I. Silman, V.I. Teichberg, and Z. Vogel, eds., Wiley, London, p. 333.

Dichter, M., and Spencer, W.A., 1969a, Penicillin-induced interictal discharges from cat hippocampus. I. Characteristics and topographical features, J. Neurophysiol., 32:649.

Dichter, M., and Spencer, W.A., 1969b, Penicillin-induced interictal discharges from cat hippocampus. II. Mechanisms underlying origin and restriction, J. Neurophysiol., 32:663.

Eccles, J.C., 1957, The Physiology of Nerve Cells, Johns Hopkins Press, Baltimore.

Fischbach, G.D., and Nelson, P.G., 1977, Cell culture, in: Neurobiology, Handbook of Physiology, Section 1, Volume I, E. Kandel, ed., American Physiological Society, Bethesda, p. 719.

Fisher, R.S., Pedley, T.A., Moody, Jr., W.J., and Prince, D.A., The role of extracellular potassium in hippocampal epilepsy, Arch. Neurol., 33:76.

Flatman, J.A., Schwindt, P.C., and Crill, W.E., 1986, The induction and modification of voltage-sensitive responses in cat neocortical neurons by N-methyl-D-aspartate, Brain Res., 363:62.

Flatman, J.A., Schwindt, P.C., Crill, W.E., and Stafstrom, C.E., 1983, Multiple actions of N-methyl-D-aspartate on cat neocortical neurons in vitro, Brain Res., 266:169.

French, C.R., and Gage, P.W., 1986, A threshold sodium current in pyramidal cells in rat hippocampus, Brain Res., in press.

Fuortes, M.G.F., and Nelson, P.G., 1963, Strychnine: its action on spinal motoneuron spikes during rhythmic firing, Science, 140:806.

Gilly, W.F., and Armstrong, C.M., 1984, Threshold channels - a novel type of sodium channel in squid giant axon, Nature, 309:448.

Goldensohn, E.S., and Purpura, D.P., 1963, Intracellular potentials of cortical neurons during focal epileptogenic discharges, Science, 139:840.

Haas, H.L., Schaerer, B., and Vosmansky, M., 1979, A simple perfusion chamber for the study of nervous tissue slices in vitro, J. Neurosci. Meth., 1:323.

Hagiwara, S., 1975, Ca^{2+}-dependent action potentials, in: Membranes: A Series of Advances, G. Eisenman, ed., Dekker, New York, p. 359.

Hamill, O.P., Marty, A., Neher, E., Sakmann, B., and Sigworth, F.J., 1981, Improved patch-clamp techniques for high-resolution current recording from cells and cell-free membrane patches, Pflügers Arch., 391:85.

Hodgkin, A.L., and Huxley, A.F., 1952, A quantitative description of membrane current and its application to conduction and excitation in nerve, J. Physiol., 117:500.

Hotson, J.R., Prince, D.A., and Schwartzkroin, P.A., 1979, Anomalous inward rectification in hippocampal neurons, J. Neurophysiol., 42:889.

Kandel, E.R., and Spencer, W.A., 1961, Electrophysiology of hippocampal neurons. II. After potentials and repetitive firing, J. Neurophysiol., 24:243.

Kopeloff, L.M., Barrera, S.E., and Kopeloff, N., Recurrent convulsive seizures in animals produced by immunologic and chemical means, Am. J. Psych., 98:881.

Llinas, R., and Sugimori, M., 1980, Electrophysiological properties of in vitro Purkinje cell somata in mammalian cerebellar slices, J. Physiol., 305:171.

Matsumoto, H., and Ajmone Marsan, C., 1964a, Cortical cellular phenomena in experimental epilepsy: interictal manifestations, Exp. Neurol., 9:286.

Matsumoto, H., and Ajmone Marsan, C., 1964b, Cortical cellular phenomena in experimental epilepsy: ictal manifestations, Exp. Neurol., 9:305.

Matsumoto, H., Ayala, G.F., and Gumnit, R.J., 1969, Neuronal behavior and triggering mechanism in cortical epileptic focus, J. Neurophysiol., 32:668.
Neher, E., and Sakmann, B., 1976, Single-channel currents recorded from membrane of denervated frog muscle fibers, Nature, 260:779.
Nowak, L., Bregestovski, P., Ascher, P., Herbet, A., and Prochiantz, A., 1984, Magnesium gates glutamate-activated channels in mouse central neurons, Nature, 307:462.
Pollen, D.A., and Ajmone Marsan, C., 1965, Cortical inhibitory post-synaptic potentials and strychninization, J. Neurophysiol., 28:342.
Prince, D.A., 1968, The depolarization shift in 'epileptic' neurons, Exp. Neurol., 21:467.
Prince, D.A., 1969, Electrophysiology of 'epileptic' neurons: spike generation, Electroenceph. Clin. Neurophysiol., 26:476.
Prince, D.A., and Futamachi, K.J., 1970, Intracellular recordings from chronic epileptogenic foci in the monkey, Electroenceph. Clin. Neurophysiol., 29:496.
Schwartzkroin, P.A., 1975, Characterisitcs of CA1 neurons recorded intracellularly in the hippocampal in vitro slice preparation, Brain Res., 85:423.
Schwartzkroin, P.A., 1977, Further characteristics of hippocampal CA1 cells in vitro, Brain Res., 128:53.
Schwindt, P.C., and Crill, W.E., 1980, Role of a persistent inward current in motoneuron bursting spinal seizures, J. Neurophysiol., 43:1296.
Schwindt, P.C., and Crill, W.E., 1981, Voltage-clamp study of cat spinal motoneurons during strychnine-induced seizures, Brain Res., 204:226.
Schwindt, P.C., and Crill, W.E., 1982, Factors influencing motoneuron rhythmic firing: results from a voltage-clamp study, J. Neurophysiol., 48:875.
Schwindt, P.C., Spain, W., and Crill, W.E., 1984, Epileptogenic action of tungstic acid gel on cat lumber motoneurons, Brain Res., 291:140.
Somjen, G.G., and Lothman, E.W., 1974, Potassium, sustained focal potential shifts, and dorsal root potentials of the mammalian spinal cord, Brain Res., 69:153.
Somjen, G.G., 1984, Interstitial ion concentration and the role of neuroglia in seizures, in: Electrophysiology of Epilepsy, P. Schwartzkroin and H. Wheal, eds., Academic Press, London, p. 303.
Stafstrom, C.E., Schwindt, P.C., and Crill, W.E., 1982, Negative slope conductance due to a persistent subthreshold sodium current in cat neocortical neurons in vitro, Brain Res., 236:221.
Stafstrom, C.E., Schwindt, P.C., and Crill, W.E., 1984a, Repetitive firing in layer V neurons from cat neocortex in vitro, J. Neurophysiol., 52:264.
Stafstrom, C.E., Schwindt, P.C., Flatman, J.A., and Crill, W.E., 1984b, Properties of subthreshold response and action potential recorded in layer V neurons from cat sensorimotor cortex in vitro, J. Neurophysiol., 52:244.
Stafstrom, C.E., Schwindt, P.C., Chubb, M.C., and Crill, W.E., 1985, Properties of persistent sodium conductance and calcium conductance of layer V neurons from cat sensorimotor cortex in vitro, J. Neurophysiol., 53:153.
Sugimori, M., and Llinas, R., 1983, Voltage clamping of Purkinje cells in vitro: a study of guinea pig cerebellar slices, Soc. Neurosci. Abstr., 9:681.
Wilson, W.A., and Goldner, M.M., 1975, Voltage clamping with a single microelectrode, J. Neurobiol., 6:411.
Yamamoto, C., 1972, Activation of hippocampal neurons by mossy fiber stimulation in thin brain sections in vitro, Exp. Brain Res., 14:423.

SYNCHRONIZATION OF PYRAMIDAL CELL FIRING BY EPHAPTIC CURRENTS IN HIPPOCAMPUS IN SITU

K. Krnjević, T. Dalkara and C. Yim

Departments of Anaesthesia Research and Physiology
McGill University, 3655 Drummond St.
Montreal (Quebec), Canada H3G 1Y6

EPHAPTIC INTERACTIONS

Once it became clear that excitable cells generate significant electrical discharges, the idea that cell-to-cell communication is normally mediated by electrical currents came very naturally. Originally, such a mechanism was not conceived as requiring any specialized low resistance junctions, and electrical transmission was thus thought to be the predominant method of junctional transmission (Eccles, 1936).

That such transmission indeed takes place, however, was not easy to prove. One approach was to look for transmission across artificial synapses, made by placing axons or muscle fibers very close together. Such preparations were named 'ephapses' by Arvanitaki (1942), who made a close study of their properties. Because of its convenience, the term 'ephaptic' has been widely used to describe electrical interactions that involve no specialized low resistance junctions (such as gap junctions), whether occurring physiologically or at artificial junctions. More recently, some authors have preferred the more cumbersome expression electrical field effect interactions for physiological events, reserving 'ephaptic' for artificial ones (Korn and Faber, 1980). In the present paper, ephaptic is used in its wider sense.

Earlier History

Significant changes in excitability caused by activity in adjacent fibers have been studied quite extensively. Katz and Schmitt (1940) obtained very convincing evidence that the passage of an electrical impulse generated a sequence of anodal, cathodal and again anodal actions on neighboring fibers. These tended to slow down the velocity of faster impulses and accelerated slower ones: the overal effect was to increase the synchronization of axonal firing.

Under normal conditions, ephaptic currents in nerve trunks are probably insufficient to excite inactive fibers - but they may do so when excitability is abnormally high, because of local injury, ischemia, or the application of drugs (Arvanitaki, 1942), or because the insulation between axons is diminished by demyelination (Rasminsky, 1980).

In the CNS, early demonstrations of ephaptic actions depended on the use of convulsant drugs like caffeine (Gerard and Libet, 1940) or strychnine (Bremer, 1941). After Furukawa and Furshpan's (1963) discovery of an inhibitory ephaptic action on Mauthner cells (in fish), such electrical actions were taken more seriously, though it was assumed that they would be of relatively little significance in the mammalian CNS. This seemed to be supported by the demonstration of only weak, though unequivocal, interactions between spinal motoneurons in cats (Nelson, 1966).

In the hippocampus, as first pointed out by Green (1964), the situation should be exceptionally favorable for ephaptic actions, because of the unusually high density and regular alignment of nerve cell bodies and dendrites, as well as the scarcity of glia in the cellular layers. For two decades, however, this possibility remained unexplored, if not indeed ignored. It became highly topical only very recently, and quite fortuitously, as an outcome of experiments on hippocampal slices maintained in Ca^{2+}-free solutions.

In Hippocampal Slices

One important advantage of _in vitro_ studies of hippocampal slices is that various phenomena can be readily analyzed in the absence of synaptic transmission, by the use of low-Ca^{2+}, high-Mg^{2+} solutions. A curious result of prolonged incubation in such solutions is that slices become hyperexcitable: instead of single responses, _bursts_ of population spikes tend to be evoked by single electrical shocks or may even arise 'spontaneously'. Because population spikes reflect the simultaneous firing of many neurons (Andersen et al., 1971), which obviously could not be explained by _synaptic_ interactions, Taylor and Dudek (1982) and Jefferies and Haas (1982) proposed - and indeed provided evidence - that a synchronization of firing came about through the ephaptic action of large extracellular fields. Taylor and Dudek (1984a,b) further demonstrated that these extracellular population spikes generated substantial _transmembrane_ potentials that had all the properties expected of a passive, ephaptic event: they were largely independent of membrane potential, smoothly graded with the intensity of stimulation, and not accompanied by a conductance change. Moreover, their brief time course was very different from that of potentials transmitted by electrotonic junctions, already extensively studied by the same authors (Dudek et al., 1983).

Experiments on the Hippocampus in situ

Though of great interest, observations on slices exposed to low-Ca^{2+}, high-Mg^{2+} solutions might be of little relevance for physiological activity of the hippocampus _in situ_. Experiments on anesthetized rats were therefore performed in our laboratory, first in the CA3 region, where intracellular recordings are obtained with greater ease than in CA1. The results (Taylor et al., 1984) were very much in accordance with those on slices. Subthreshold antidromic volleys consistently generated depolarizing transmembrane potentials (equal to 39% of the extracellular field, on the average), which could be shown to cause a significant enhancement of excitability.

In recent experiments, we have focused on pyramidal cells in area CA1 (Yim et al., 1986). This region is especially relevant, because in slices exposed to low Ca^{2+} and high Mg^{2+}, the highly synchronized 'bursting' activity is predominantly observed in CA1 and very much less in CA3.

PRESENT OBSERVATIONS

In Sprague-Dawley rats, maintained under full anesthesia with urethane (1.5 g/kg i.p.), bipolar stimulating electrodes were inserted stereotaxically into the alveus and the fimbria (Fig. 1), as described in greater detail elsewhere (Ben-Ari et al., 1981; Yim et al., 1986). Intracellular records were obtained with glass micropipettes filled with 3 M KCl or K acetate, having tip resistances of 20-50 MΩ. Small currents (of constant intensity) could be injected intracellularly through the microelectrode to control cell excitability.

<u>Antidromic Fields and Related Transmembrane Potentials</u>

Stimulation of the alveus provided a convenient means of evoking antidromic responses from CA1 pyramidal cells (Fig. 1). Extracellular recordings in the pyramidal cell layer are characterized by a very brief (1 ms) negative population spike at an extremely short latency (0.5-1.2 ms), followed by a much slower positive wave (but no orthodromic spikes in contrast to records in CA3). The largest antidromic fields are recorded over a limited range of depth (Fig. 2), corresponding to that of the pyramidal cell layer. At greater depths, the potential reverses to a positive one, indicating failure of propagation along apical dendrites.

Penetration of the microelectrode into a pyramidal cell was signaled by the appearance of a substantial resting potential (up to -70 mV), and a prominent hyperpolarizing IPSP in response to subthreshold alvear stimulation. Because of Cl$^-$ leakage from the microelectrode, IPSPs tended to diminish, and after some minutes they were typically seen as large positive potentials (Fig. 4B). Of course, if the intensity of alvear stimulation exceeded the threshold of the axon of the penetrated cell, a large, all-or-none antidromic spike was recorded.

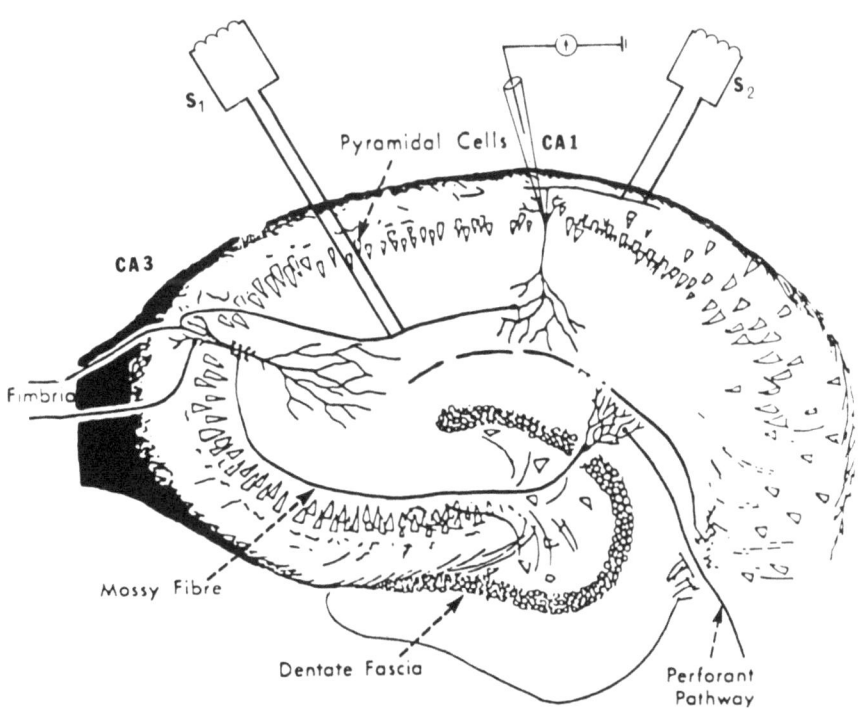

Fig. 1. Diagram of hippocampus, illustrating placement of stimulating electrodes in fimbria (S_1) for orthodromic, and in alveus (S_2) for antidromic activation of CA1 neurons.

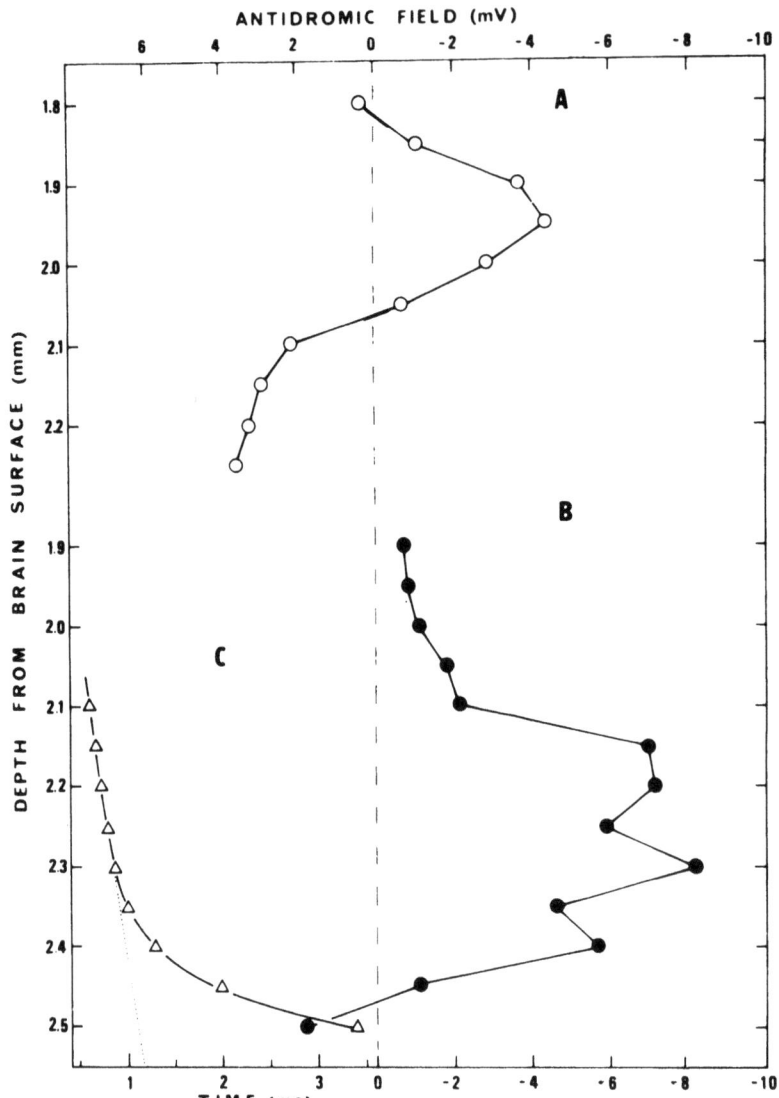

Fig. 2. Extracellular antidromic fields, at various depths in CA1, are shown by open and closed circles (A, B, for two separate experiments); the intensity of alvear stimulation was 10 V throughout. Triangles indicate <u>latency</u> of negative population spike recorded at various depths in experiment B: note very pronounced deceleration below 2.4 mm.

Some characteristics of the antidromic spike are important for the interpretation of the ephaptic potentials. As illustrated in Fig. 3, antidromic spikes appeared after a very brief and constant latency; and they could follow alvear stimulation at relatively high frequencies (up to 50 Hz). Further evidence of a high safety factor of antidromic invasion is the low susceptibility to block by hyperpolarizing current (Fig. 3D, G). These observations are very much in keeping with previous demonstrations that the CA1 antidromic population spike readily withstands high frequencies of stimulation (Leung, 1979) or the application of depressant agents (Sperti et al., 1967).

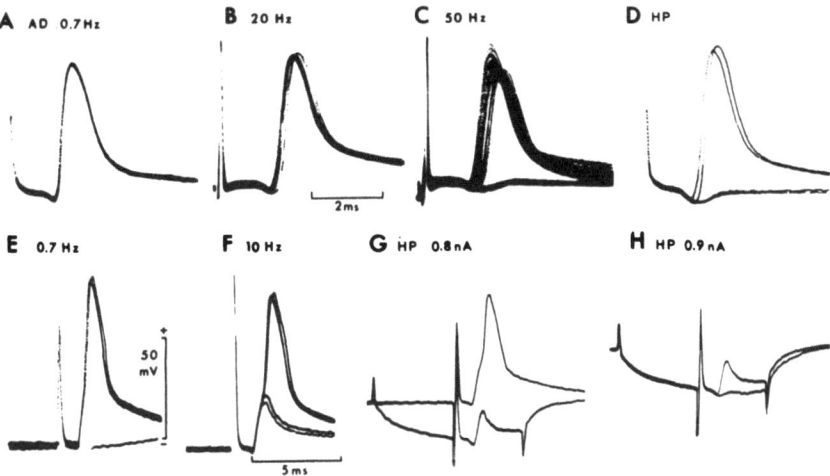

Fig. 3. Antidromic invasion of CA1 neurons has a high safety factor: superimposed traces are from two different cells, A-D and E-H, respectively. A and E are short latency spikes, at low frequency of stimulation (straddling threshold in E); note characteristic depolarizing after-potential and minimal 'jitter' of latency. Spikes of upper cell show no IS/SD discontinuity, even during stimulation at 20 Hz and 50 Hz (B, C) or at 0.7 Hz combined with hyperpolarization (D) (which caused some all-or-none failures of invasion). Spikes in F-H show partial block of invasion, leaving only smaller (IS) component during stimulation at 10 Hz (F) or during hyperpolarization (G). Stronger hyperpolarization (H) could eliminate even the smaller (IS) component.

When alvear stimulation was subthreshold for a given neuron, the negative population spike recorded inside the cell was consistently smaller than the field recorded just outside. This implies the existence of a transmembrane potential of opposite polarity, which indeed could be revealed by subtraction of the extracellular field from the intracellular record. Two examples of such fields recorded inside and outside neurons, and the corresponding transmembrane potential are illustrated in Fig. 4. The transmembrane potentials were graded in amplitude, according to the size of the extracellular field. On the average, they amounted to 41% (S.E. 3.9) of the extracellular potential. The very brief latency of the transmembrane potential made it unlikely that it could be a synaptic potential: this conclusion was reinforced by its lack of sensitivity to very large hyper- or depolarizations (by currents of as much as 4 nA). Another point of interest is that intraglial recordings (identified by their characteristic high and stable resting potential, inexcitability, as well as absence of synaptic potentials) showed no consistent ephaptic transmembrane potentials (Fig. 4C).

Functional Significance of Ephaptic Action

To prove that the observed transmembrance potentials can enhance excitability, they were evoked in conjunction with intracellular injections of subthreshold depolarizing pulses. A subthreshold excitatory action was then revealed by the appearance of full, all-or-none spikes, coinciding with the antidromic population spike. This test was done most easily

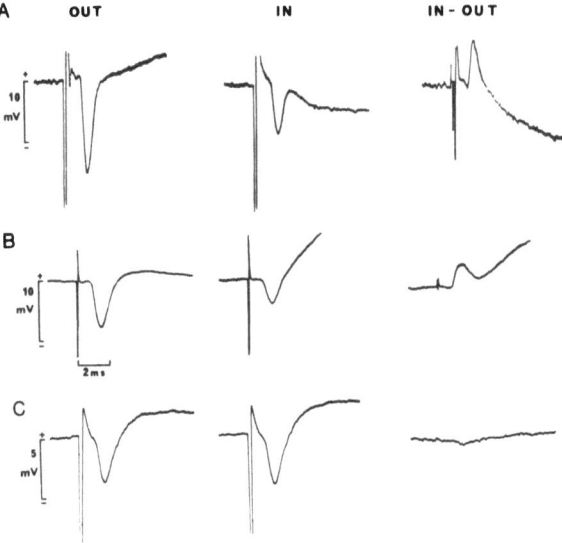

Fig. 4. Examples of antidromic fields in CA1, recorded just extracellularly (Out) and intracellularly (In), and the corresponding transmembrane potentials obtained by subtracting trace 'Out' from trace 'In'. A and B were obtained from two different neurons, and C from a non-responsive cell, presumed to be a glia.

when alvear stimulation evoked large antidromic fields and correspondingly intense ephaptic currents. Good examples of such interactions can be seen in Fig. 5. The top traces illustrate (on lower gain) the relatively large fields evoked in two different neurons by subthreshold alvear stimulation. The next set of traces (below) illustrates the subthreshold depolarizing effects of weak depolarizing current pulses (note that each record was obtained by the superimposition of several traces). However, when the same subthreshold antidromic volley and current pulses were given together (lowest traces), spikes were quite regularly elicited at the same latency as the antidromic field.

Even minute antidromic fields and transmembrane potentials can significantly raise the firing probability. This can be shown by bringing the neuron very close to its firing threshold. For example, in Fig. 6A, the maximal subthreshold antidromic volley generated only a very small field (at arrow). Although 0.1 nA depolarizing pulses were able to excite the neuron, the spikes occurred only after a latency of 8-10 ms. Clear ephaptic firing was revealed by the sharp reduction in spike latency (to 5-6 ms) when identical antidromic volleys were given concurrently with the current pulses. Similar observations made on a number of different cells indicated a highly significant, logarithmic relation between the transmembrane potential and increase in firing probability (Fig. 7).

DISCUSSION

These data provide strong evidence that significant ephaptic interactions can occur in the CA1 stratum pyramidale, *in situ*. As illustrated

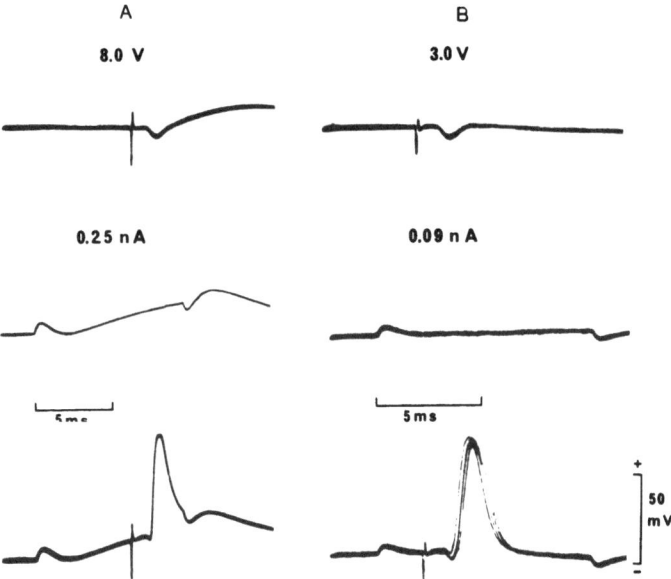

Fig. 5. Strong facilitation of firing of two CA1 neurons (A, B) by substantial antidromic <u>fields</u>. In upper traces are antidromic fields evoked by alvear stimulation of subthreshold intensity (indicated); middle traces show 10 ms injections of subthreshold depolarizing current (also indicated, transients at start and end of currents are capacitive artifacts); in lower traces, concurrent applications of same subthreshold antidromic stimuli and depolarizing pulses regularly evoke firing. Note that several traces are superimposed in each record. Antidromic field in neuron A is shown at higher gain in Fig. 4B.

in Fig. 3, antidromic invasion of CA1 neurons consistently had a high safety factor, in agreement with previous studies (Sperti et al., 1967; Leung, 1979). The facilitation of firing by subthreshold depolarizing currents therefore cannot be ascribed to a relief of antidromic conduction failure, but rather to the excitatory effect of ephaptic currents, responsible for the observed brief depolarizations.

These ephaptic transmembrane potentials – expressed as a fraction of the extracellular antidromic field – were almost identical to the corresponding potentials observed in CA3 (Taylor et al., 1984). Therefore, the much greater tendency for the CA1 region to generate bursts of firing <u>in vitro</u> must be accounted for by some other difference, perhaps in firing threshold.

<u>Factors Mainly Responsible for a Large Ephaptic Potential</u>

Several attempts have been made to model ephaptic interactions in homogeneous nerve trunk (Katz and Schmitt, 1940; Clark and Plonsey, 1971;

Markin, 1973). Although the hippocampus is hardly such a homogeneous structure, the very regular, parallel arrangement of fibers, cells and dendrites in the pyramidal cell layer provides some justification for analyzing the data by the Clark and Plonsey model, at least in a first approximation. One can then calculate what fraction of the extracellular field would appear as a transmembrane potential for different values of fiber (or cell) diameter, membrane resistance and capacity, and action potential conduction velocity. According to this model, ephaptic potentials are enhanced when the diameter increases or when the conduction velocity slows down. Both are prominent features of the stratum pyramidale: in the first place, there is a marked increase in diameter in the axon-soma region; secondly, the velocity of the antidromic spike becomes very much

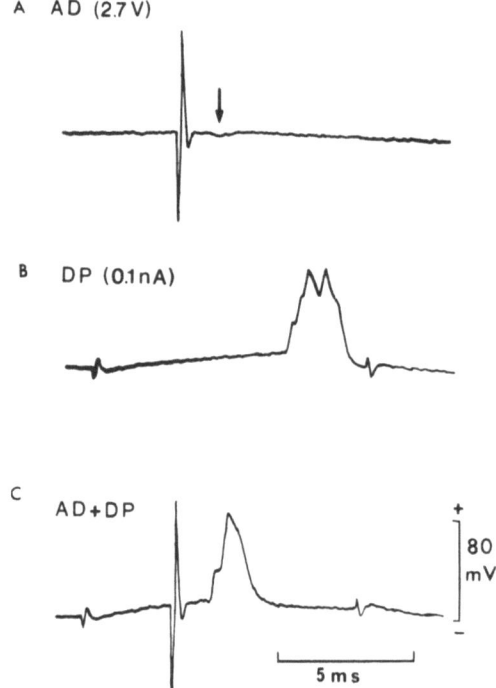

Fig. 6. A higher background of excitability is needed to demonstrate ephaptic excitation by small antidromic fields. In this figure, each record is an <u>average</u> of 6 original oscilloscope traces. Antidromic threshold was 2.8 V for this cell. Just-subthreshold stimulation at 2.7 V evoked only a small field (at arrow) and a corresponding transmembrane potential of 0.6 mV. Depolarizing pulse (0.1 nA, for 10 ms) given alone caused excitation after 8-10 msec, but when subthreshold alvear stimulus was also applied, (C), there was a high incidence of spikes at a consistently earlier latency, synchronous with the antidromic field.

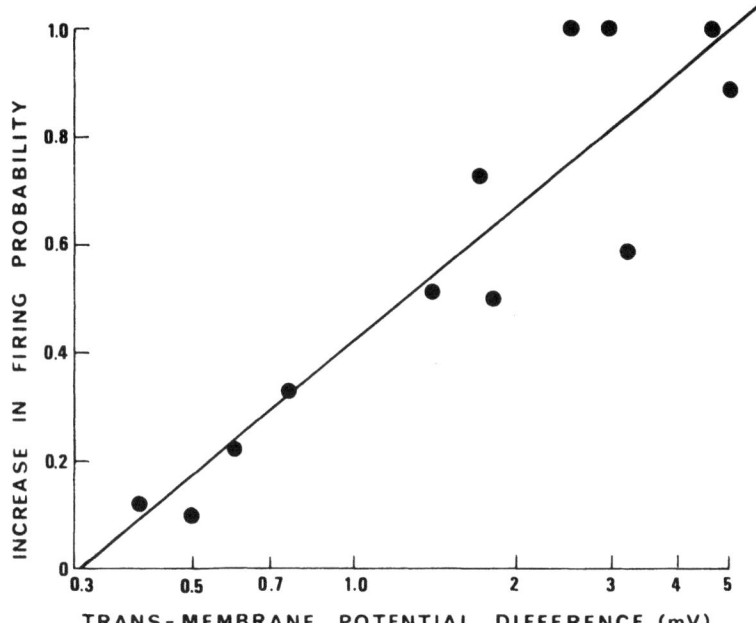

Fig. 7. Graph indicating highly significant correlation between increased incidence of firing (evoked by just-subthreshold antidromic fields) and the logarithm of the corresponding ephaptic transmembrane potential (r = 0.980).

slower at the pyramidal layer (Leung, 1979), tending towards zero in the apical dendrites (Fig. 2) where further conduction may fail. Together with an unusually restricted free extracellular pathway (Morris and Krnjević, 1980; Jefferys, 1984; Heinemann, this volume) - at least partly owing to the close packing of the cells - these characteristics of the hippocampus provide optimal conditions for strong ephaptic interactions.

Wider Significance of Ephaptic Currents

Under what conditions can one expect ephaptic actions to become a significant factor in the spread of excitation? In order to evoke ephaptic firing of hippocampal cells, their excitability had to be raised artificially by injecting small depolarizing currents. The requirements of substantial population spikes and a high excitability may seem very unphysiological. It must be remembered, however, that our experiments were performed under full anesthesia, and therefore under conditions of abnormally low excitability. In a freely behaving animal, hippocampal cells show plenty of 'spontaneous' activity (Ranck, 1973; O'Keefe and Nadel, 1978) and are presumably highly responsive to appropriate stimuli. We have shown elsewhere (Dalkara et al., 1986) that antidromic population spikes in CA3 can be sharply potentiated by local applications of acetylcholine (in the hippocampus) or by medial septal stimulations that activate a cholinergic input (Krnjević and Ropert, 1982). Various characteristics indicate that this effect is probably caused by a facilitation of ephaptic firing, in every way comparable to that obtained with subthreshold intracellular depolarizing currents (but of course acting on a group of cells rather than a single one).

In view of the small size of fields that are capable of causing a

10% increase in firing probability (just over 1 mV according to Fig. 7), one can speculate that under more natural conditions, 'spontaneous' fields may at times be sufficient to facilitate local cell firing when excitability is much enhanced, e.g., by cholinergic modulation from the septum. The ephaptic action might then serve to amplify synchronized discharges. Of course, its role would become far more predominant in pathological states, in which reduced inhibitory control permits the generation of paroxysmal depolarizations: with the appearance of large fields, powerful ephaptic currents could rapidly recruit large numbers of cells in brief but massive synchronized discharges, reinforcing more classical synaptic actions (Traub et al., 1985). Ephaptic currents cannot account for the <u>initiation</u> of excessive activity. But once this has occurred, they would greatly contribute to its growth and spread.

ACKNOWLEDGEMENTS

This research was financially supported by the Canadian Medical Research Council.

REFERENCES

Andersen, P., Bliss, T.V.P., and Skrede, K.K., 1971, Unit analysis of hippocampal population spikes, Exp. Brain Res., 13:208.
Arvanitaki, A., 1942, Effects evoked in an axon by the activity of a contiguous one, J. Neurophysiol., 5:89.
Ben-Ari, Y., Krnjević, K., Reiffenstein, R.J., and Reinhardt, W., 1981, Inhibitory conductance changes and action of GABA in rat hippocampus, Neuroscience, 6:2445.
Bremer, F., 1941, Le tetanos strychnique et le mecanisme de la synchronisation neuronique, Arch. Int. Physiol., 51:211.
Clark, J.W., and Plonsey, R., 1971, Fiber interaction in a nerve trunk, Biophys. J., 11:281.
Dalkara, T., Krnjević, K., Ropert, N., and Yim, C.Y., 1986, Chemical modulation of ephaptic activation of CA3 hippocampal pyramids, Neuroscience, 17:361.
Dudek, F.E., Andrew, R.D., MacVicar, B.A., Snow, R.W., and Taylor, C.P., 1983, Recent evidence for and possible significance of gap junctions and electrotonic synapses in the mammalian brain, in: Basic Mechanisms of Neuronal Hyperexcitability, H.H. Jasper and N.M. van Gelder, eds., Alan R. Liss, Inc., New York, p. 31.
Eccles, J.C., 1936, Synaptic and neuro-muscular transmission, Ergeb. Physiol., 38:339.
Furukawa, T., and Furshpan, E.J., 1963, Two inhibitory mechanisms in the Mauthner neurons of goldfish, J. Neurophysiol., 26:140.
Gerard, R.W., and Libet, B., 1940, The control of normal and 'convulsive' brain potentials, Am. J. Psychiatry, 96:1125.
Green, J.D., 1964, The hippocampus, Physiol. Rev., 44:561.
Jefferys, J.G.R., and Haas, H.L., 1982, Synchronized bursting of CA1 hippocampal pyramidal cells in the absence of synaptic transmission, Nature, 300:448.
Jefferys, J.G.R., 1984, Current flow through hippocampal slices, Soc. Neurosci. Abstr., 10:1074.
Katz, B., and Schmitt, O.H., 1940, Electric interaction between two adjacent nerve fibers, J. Physiol., 97:471.
Korn, H., and Faber, D.S., 1980, Electrical field effect interactions in the vertebrate brain, Trends Neurosci., 3:6.
Krnjević, K., and Ropert, N., 1982, Electrophysiological and pharmacological characteristics of facilitation of hippocampal population spikes by stimulation of the medial septum, Neuroscience, 7:2165.

Leung, L.-W.S., 1979, Potentials evoked by alvear tract in hippocampal CA1 region of rats. I. Topographical projection, component analysis, and correlation with unit activities, J. Neurophysiol., 42:1557.

Markin, V.S., 1973, Electrical interaction of parallel non-myelinated nerve fibers. III. Interaction in bundles, Biophysics, 18:324.

Morris, M.E., and Krnjević, K., 1980, Slow diffusion of Ca^{2+} in the rat's hippocampus, Can. J. Physiol. Pharmacol., 59:1022.

Nelson, P.G., 1966, Interaction between spinal motoneurons of the cat, J. Neurophysiol., 29:275.

O'Keefe, J., and Nadel, L., 1978, The Hippocampus as a Cognitive Map, Clarendon Press, Oxford.

Ranck, J.B., 1973, Studies on single neurons in dorsal hippocampal formation and septum in unrestrained rats. Part I. Behavioral correlates and firing repertoires, Exp. Neurol., 41:461.

Rasminsky, M., 1980, Ephaptic transmission between single nerve fibers in the spinal nerve roots of dystrophic mice, J. Physiol., 305:151.

Sperti, L., Gessi, T., and Volta, F., 1967, Extracellular potential field of antidromically activated CA1 pyramidal neurons, Brain Res., 3:343.

Taylor, C.P., and Dudek, F.E., 1982, Synchronous neural afterdischarges in rat hippocampal slices without active chemical synapses, Science, 218:810.

Taylor, C.P., and Dudek, F.E., 1984a, Excitation of hippocampal pyramidal cells by an electrical field effect, J. Neurophysiol., 52:126.

Taylor, C.P., and Dudek, F.E., 1984b, Synchronization without active chemical synapses during hippocampal afterdischarges, J. Neurophysiol., 52:143.

Taylor, C.P., Krnjević, K., and Ropert, N., 1984, Facilitation of hippocampal CA3 pyramidal cell firing by electrical fields generated antidromically, Neuroscience, 11:101.

Traub, R.D., Dudek, F.E., Snow, R.W., and Knowles, W.G., 1985, Computer simulations indicate that electrical field effects contribute to the shape of the epileptiform field potential, Neuroscience, 15:947.

Yim, C.Y., Krnjević, K., and Dalkara, T., 1986, Ephaptically-generated potentials in CA1 neurons of the rat's hippocampus in situ, J. Neurophysiol., in press.

EXCITATORY AMINO ACIDS AND REGENERATIVE ACTIVITY IN CULTURED NEURONS

J.F. MacDonald, J.H. Schneiderman, and Z. Miljkovic

Playfair Neuroscience Unit, University of Toronto
Toronto Western and Wellesley Hospitals
Toronto, Ontario Canada

INTRODUCTION

Hayashi (1954) originally demonstrated that the application of glutamic acid to the cortical surface of dogs evoked convulsions. Later, excitatory amino acids (EAA), such as L-glutamic and L-aspartic acid, were proposed as possible transmitter candidates suggesting that dysfunction of these transmitters might contribute to convulsions.

Excitation by these compounds, and by structurally related analogs, is mediated by two distinct conductance mechanisms. One of these behaves as if it were activating voltage gated channels (although this is not so, see below) and the other is relatively independent of membrane potential (MacDonald and Porietis, 1982; MacDonald and Wojtowicz, 1982; Flatman et al., 1983). EAA demonstrate a spectrum of abilities to activate one or the other of these mechanisms (Mayer and Westbrook, 1984) with N-methyl-D-aspartic (NMDA) and L-aspartic (L-ASP) at the voltage dependent and kainic (KAI), quisqualic and L-3-oxyalylamino-2-aminopropionic (β-ODAP; MacDonald and Morris, 1984) acids at the voltage independent end. Recently, the voltage dependency has been ascribed to a voltage dependent block of one of the underlying channels by physiological concentrations of Mg^{2+} (Mayer et al., 1984; Nowak et al., 1984) and, therefore, the channels are not themselves voltage gated.

The link between EAA and convulsant activity has been strengthened by the demonstration of anticonvulsant properties of EAA blockers. Some of these blockers [i.e., Mg^{2+}, previously used in preeclampsia; 2-amino-5-phosphonovaleric (2-APV), 2-amino-7-phosphonoheptanoic acids (Meldrum et al., 1983); ketamine (McCarthy et al., 1965)] also act specifically against EAA either with regard to receptor type (i.e., antagonizing NMDA and not KAI receptors) or conductance mechanism (reducing the conductance susceptible to voltage dependent block by Mg^{2+} ions and not the voltage independent mechanism). Because the NMDA receptor and/or the voltage dependent conductance mechanism appears to be the target of these antagonists, there might be some reason to suspect that the voltage dependent behavior of the EAA response contributes more directly to regenerative activity or bursting of neurons. One possibility is that EAA either accentuate the voltage dependent currents responsible for bursting (Flatman et al., 1983; MacDonald and Schneiderman, 1985) and/or potentiate Ca^{2+} currents (Dingledine, 1983), which could contribute to the generation of bursting in central neurons.

We report here our attempts to examine the relationship between intrinsic bursting (voltage dependent inward currents) and voltage dependent EAA. In addition, we have examined the actions of the NMDA blockers ketamine and phencyclidine (PCP; Anis et al., 1983; Harrison and Simmonds, 1985) on EAA currents.

METHODS AND RESULTS

Intracellular recordings (1 M KCl) from cultured (dissociated) spinal cord neurons, in the presence of TTX (1.3 mM Ca^{2+}), have demonstrated that few neurons (about 1%) are capable of generating burst-like activity or Ca^{2+} action potentials. Applications of voltage dependent EAA (but not voltage independent) can induce the appearance of regenerative bursts and seem to increase the probability of observing Ca^{2+} action potentials (see Fig. 1). These observations suggest that voltage dependent EAA responses can interact with the endogenous and voltage dependent conductances of the membrane.

Two Electrode Voltage Clamp

In cultured spinal cord neurons, under conventional two-electrode voltage clamp (1 M KCl or 1 M CsCl; and with 5 mM Ca^{2+} and 25 mM tetraethylammonium (TEA) added to the bathing solution), depolarizing voltage steps are associated with inward currents and slow inward current tails. These inward currents are depressed by lowering extracellular Ca^{2+} concentration and blocked by divalent cations such as Cd^{2+} which suggests a dependence upon Ca^{2+} fluxes (MacDonald and Schneiderman, 1985; although the ion conductances involved were not identified). The slow current tails are not necessarily Ca^{2+} current but are thought to be a Ca^{2+} activated chloride current (Owen et al., 1984). These current tails and the sustained inward currents during such steps are potentiated by applications of voltage dependent EAA (MacDonald and Schneiderman, 1985; see also Mayer and Westbrook, 1985) in a dose dependent fashion and suppressed by coapplications of Cd^{2+} (Fig. 2). Sufficiently high doses of L-ASP were also capable of suppressing the current tails (not shown).

The decay of these tail currents consisted of multiple components. L-ASP often accentuated the most slowly decaying components (i.e., time constants of seconds) as did moving the holding potential in the depolarizing direction. Furthermore, these slower tail components gradually decayed with repetition of the voltage steps, an action which mirrored the frequency dependent decay of Ca^{2+} spikes in these neurons. Frequency dependent decay is primarily expressed as a reduction in the duration of the plateau phase of the Ca^{2+} action potential and likely represents the activity of currents beyond the spatial control of the voltage clamp (MacDonald and Schneiderman, unpublished).

Whole Cell Voltage Clamp

In order to reduce the trauma to the membrane resulting from insertion of two electrodes, and thus improve our chances of observing Ca^{2+} and related currents, we have also employed whole cell recording (140 mM CsCl, 1 mM $CaCl_2$, 11 mM EGTA, 10 mM HEPES, pH 7.2; extracellular 5 mM Ca^{2+}, 25 mM TEA) in conjunction with a discontinuous (20 to 25 KHz) switching voltage clamp (@Axo-clamp 2). Cell patches were voltage clamped (-70 to -90 mV) prior to breaking into the cell. We observed both a fast inward current and a sustained inward current during depolarizing voltage steps (Fig. 3) and these currents demonstrated an anticipated washout during

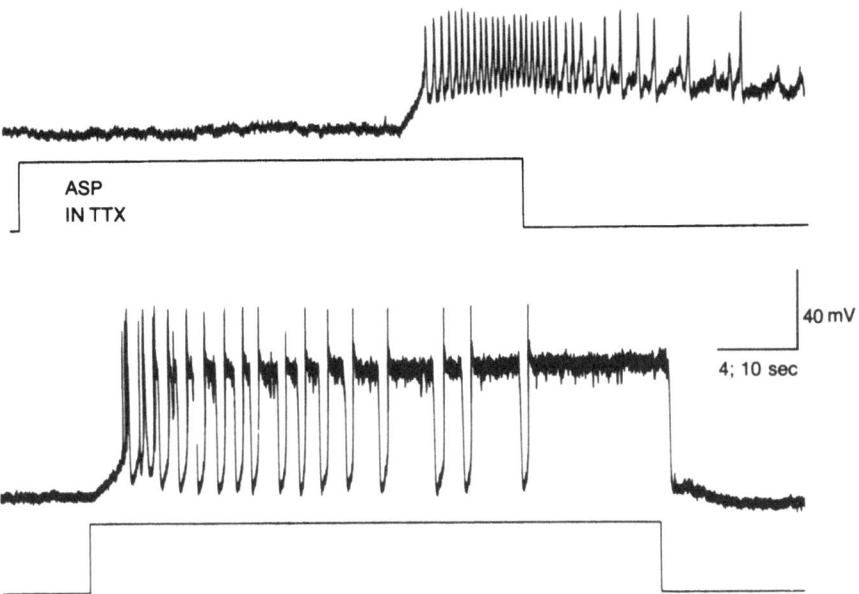

Fig. 1. Conventional current clamp recordings from cultured spinal cord neurons (1.3 mM Ca^{2+}, 0.9 mM Mg^{2+} in extracellular solution). Upper: This recording was made in the presence of TTX. L-ASP (50 µM) was applied by gradually increasing the pressure until a response was observed (the indicator of when the pressure was on is shown below the trace of voltage). L-ASP was able to evoke a series of small spikes (perhaps Ca^{2+} spikes), whereas applications of quisqualic acid to the same neuron were unable to evoke any spikes even though comparable depolarizations were produced by both drugs (not shown). Lower: Recording from another neuron (in the absence of TTX). L-ASP (100 µM) evokes oscillating bursts (note time scale in 10 seconds for this trace). Action potentials ride on the rising phase of the burst and are inactivated during the plateau of the burst. Neither passive depolarization of the neuron nor applications of KAI were capable of inducing such bursting in this neuron (not shown).

the recording period. However, in the majority of neurons tail currents were complicated by the presence of current plateaus (Fig. 3). These were identified by a clearly non-experimental decay which followed repolarization of the membrane. Furthermore, they underwent a decay in amplitude and duration which was a reflection of the frequency dependent decay of plateau of the spike under current clamp. Applications of L-ASP had complex actions upon such currents, causing, in small iontophoretic doses, prolongation of the current plateaus but a small depressant effect upon the fast inward current (Fig. 4). However, in higher doses L-ASP caused suppression of all voltage activated inward currents. Applications of β-ODAP or KAI depressed inward currents and failed to potentiate the plateau currents. In some cases, application of EAA (particularly when applied to the processes of the neuron) was able to evoke a current-spike followed by strong suppression of voltage activated inward currents. These results demonstrated that poor spatial clamp of the neurons was attained under these recording conditions and abrogated any quantitative analysis of the currents.

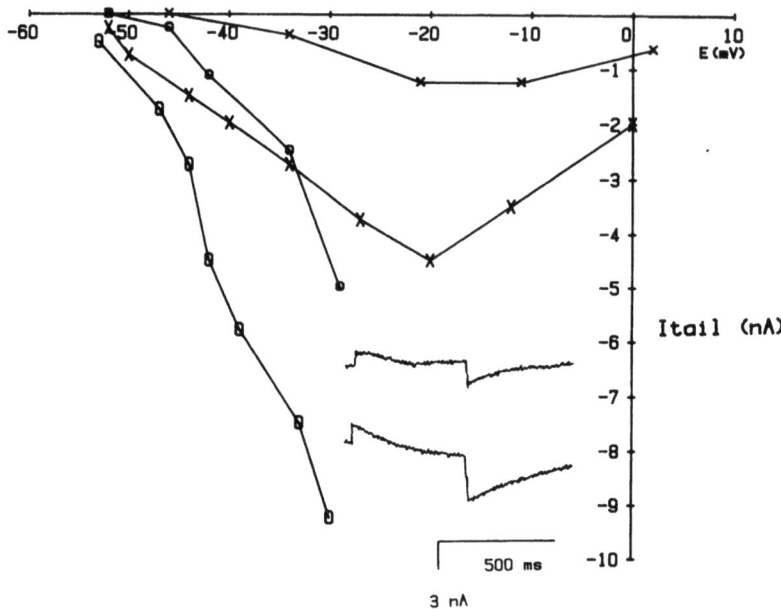

Fig. 2. Conventional two-electrode voltage clamp of a cultured spinal cord neuron (TTX, 5 mM Ca^{2+}, 25 mM TEA added to the extracellular solution). Voltage steps (500 ms) from a holding potential of -58 mV were made before (small circles) and during (large circles) the iontophoretic application of L-ASP. The amplitude of inward tail current (see insert) was plotted against the command voltage. L-ASP potentiated these inward tails. Following the application of 250 μM Cd^{2+} to the bathing solution, the steps were repeated before (small crosses) and during (large crosses) the same iontophoretic application of L-ASP. Although Cd^{2+} strongly depressed the tail currents, L-ASP was still able to cause potentiation (a greater potentiation on a percentage basis). These tail currents are likely carried by chloride (see text).

Actions of Ketamine and PCP on Amino Acid Responses

EAA were applied by pressure to cultured hippocampal neurons (dissociated from fetal mice, two weeks in culture) but all other test drugs and ions were introduced by perfusing the entire bath with the appropriate solutions (0 Mg^{2+} added, 1.3 mM Ca^{2+}). No attempt was made to washout ketamine or PCP because of the lipophilic properties of these drugs.

Cells (10 μM or less in diameter) were voltage clamped (whole cell) in two ways either by breaking the patch under current clamp and then switching to voltage clamp (holding -30 to -70 mV) or by voltage clamping the patch to a hyperpolarized holding potential (-70 to -90 mV) prior to breaking through. Brief pressure applications of EAA were directed at the soma of neurons and holding potentials varied over a range of -120 mV to +100 mV.

Returning the holding potential to hyperpolarized values following depolarizations was sometimes associated with a depression in responses to EAA and may be a consequence of Ca^{2+} related inactivation of these

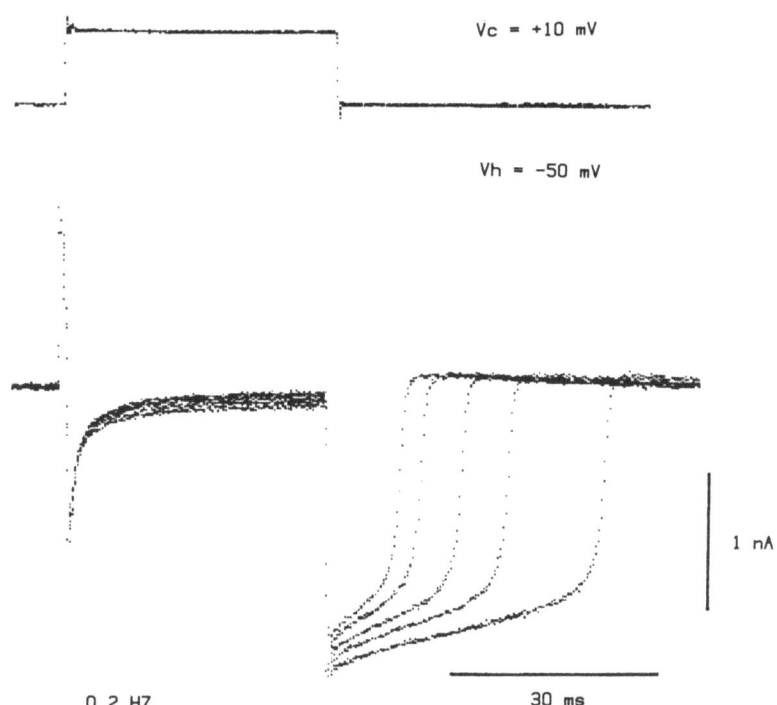

Fig. 3. A whole cell discontinuous voltage clamp of a cultured spinal cord neuron demonstrating frequency dependent decay of current plateaus. The voltage-step consisted of moving from −50 mV (Vh) to +10 mV (Vc), was repeated five times at 0.2 Hz. Superimposed current and voltage traces are shown. During the step, a fast inward current was initially observed which decayed to a relatively sustained value. The return to holding potential was followed by current plateaus which decreased in duration and amplitude with each repetition of the step.

responses described by Mayer and Westbrook (1985). This complicated the analysis of drug actions and made it difficult at times to assess blocking actions. We could usually convince ourselves of a drug effect or lack thereof by: 1) repeating the control current-holding potential relationship several times until reproducible; 2) by holding the potential at a hyperpolarized value and perfusing the test drug while repeating the EAA application at a constant interval; 3) making a comparison with an EAA which should not have been susceptible to this voltage dependent block (i.e., KAI or β-ODAP).

As previously reported (Mayer and Westbrook, 1985), the evoked currents could be relatively independent of holding potential over a range from −80 to +80 mV provided Mg^{2+} was not added to the extracellular or intracellular recording solutions. A voltage dependent blockade was evident after solutions containing divalent cations such as Mg^{2+} were bath perfused (not shown). In contrast, the blockade by 2-APV was similar at all holding potentials (Fig. 5).

In the absence of added Mg^{2+}, ketamine (10, 50 or 100 μM) and PCP (1, 10, 50 μM) depressed responses to L-ASP and NMDA in a voltage (i.e., see Fig. 5) and concentration dependent fashion. In this respect, ketamine

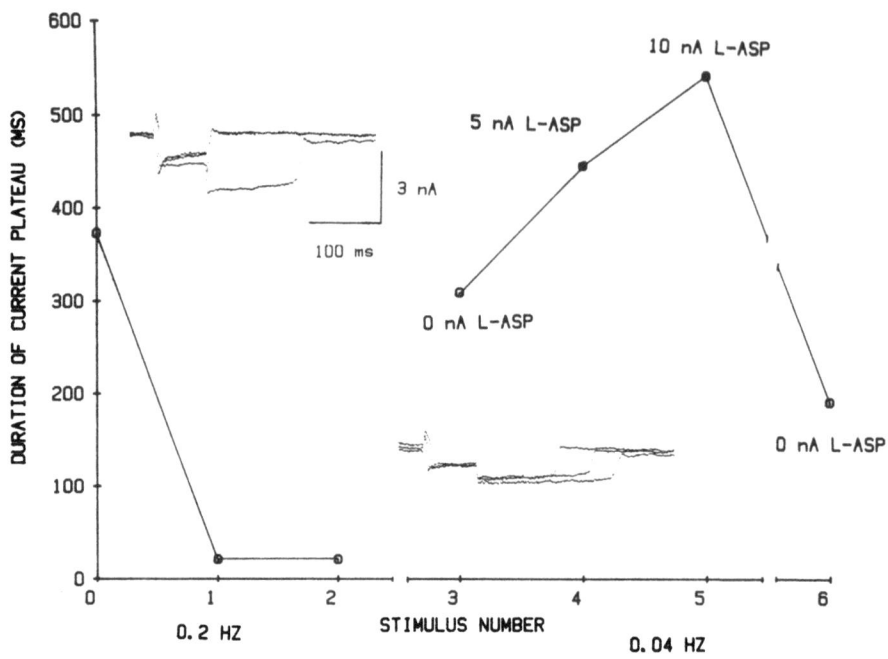

Fig. 4. A whole cell discontinuous voltage clamp of a cultured spinal cord neuron (TTX, 5 mM Ca^{2+}, 25 mM TEA added to extracellular solution). This neuron possessed a prominent current plateau (see inserts) as well as a fast inward current which decayed to a sustained level during the voltage step. Three superimposed traces are given on the left to illustrate the responses to the first three voltage steps (holding -48 mV, command -8 mV) repeated at a frequency of 0.2 Hz. The durations of the current plateaus (end of the step to end of the plateau) are plotted against stimulus number. After waiting several minutes, the same step was made at a lower frequency (0.4 Hz) to avoid decay of the plateau. L-ASP was applied by microiontophoresis to the region of the soma for the second and third stimuli of this series. Insets of the currents (right) show superimposed the control and the two responses during applications of L-ASP. L-ASP increased the duration of the current plateaus but did not potentiate voltage dependent inward currents evoked during the step. β-ODAP was unable to prolong currents in the neuron. Note the wash-out of currents (left to right inserts) that occurred over this period of recording.

and PCP were similar to divalent cations. L-ASP and NMDA currents were strongly depressed at potentials more hyperpolarized than 0 mV and unchanged at more depolarized potentials (Fig. 5). PCP was the more potent causing a similar degree of block at 10 μM as observed with 50 μM ketamine and possessing a threshold dose near 1 μM. In contrast, responses to KAI and β-ODAP were no more dependent upon holding potential in the presence than in the absence of these drugs.

The degree of the block of ketamine and PCP was strongly influenced by the number of applications of EAA. For example, if a neuron was bathed in ketamine, but had not been exposed to NMDA or L-ASP (holding at a negative

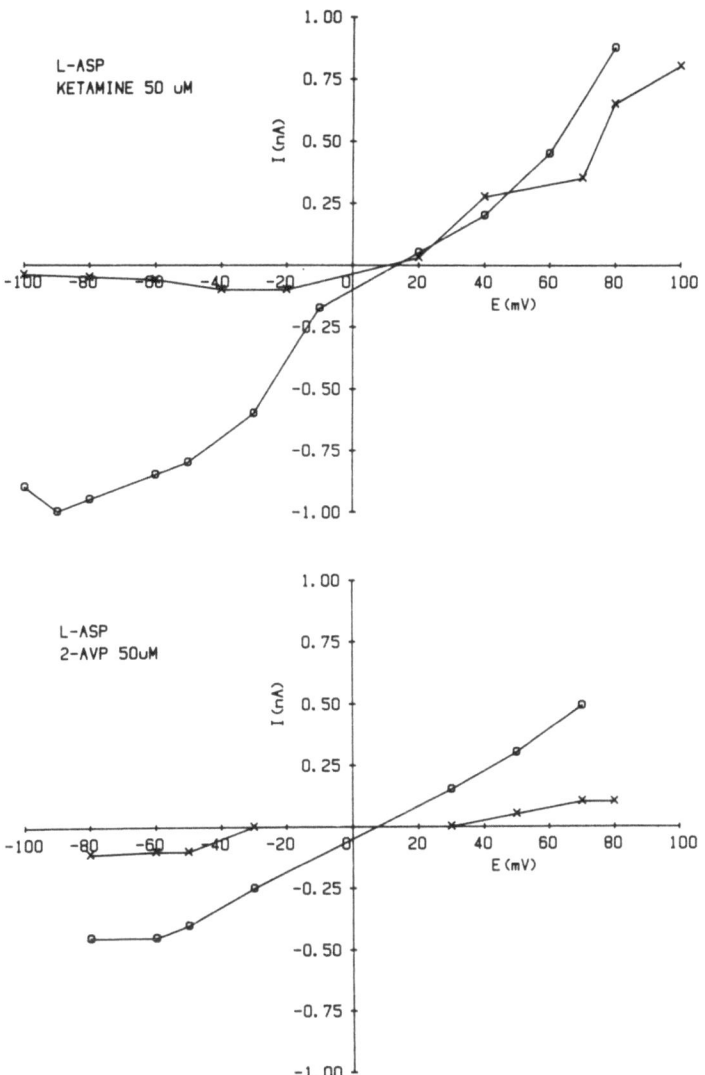

Fig. 5. Current responses of two hippocampal neurons to single applications of L-ASP made at different holding potentials before (circles) and following perfusion with ketamine (upper, crosses) or 2-APV (lower, crosses). Upper: The control curve is relatively linear with a reversal potential of about +15 mV. Following the introduction of ketamine, a series of five applications (0.2 Hz) was made at a holding potential of −70 mV, which permitted further development of the blockade (see Fig. 6 legend) and then single applications were made at hyperpolarized potentials before moving to depolarized values. The cell was held at +60 mV and five applications made before moving to other depolarized values of potential. A marked rectification at hyperpolarized potentials was observed in the presence of ketamine. Lower: The perfusion of 2-APV depressed responses to L-ASP at all holding potentials and no use dependency was observed (see Fig. 6 legend).

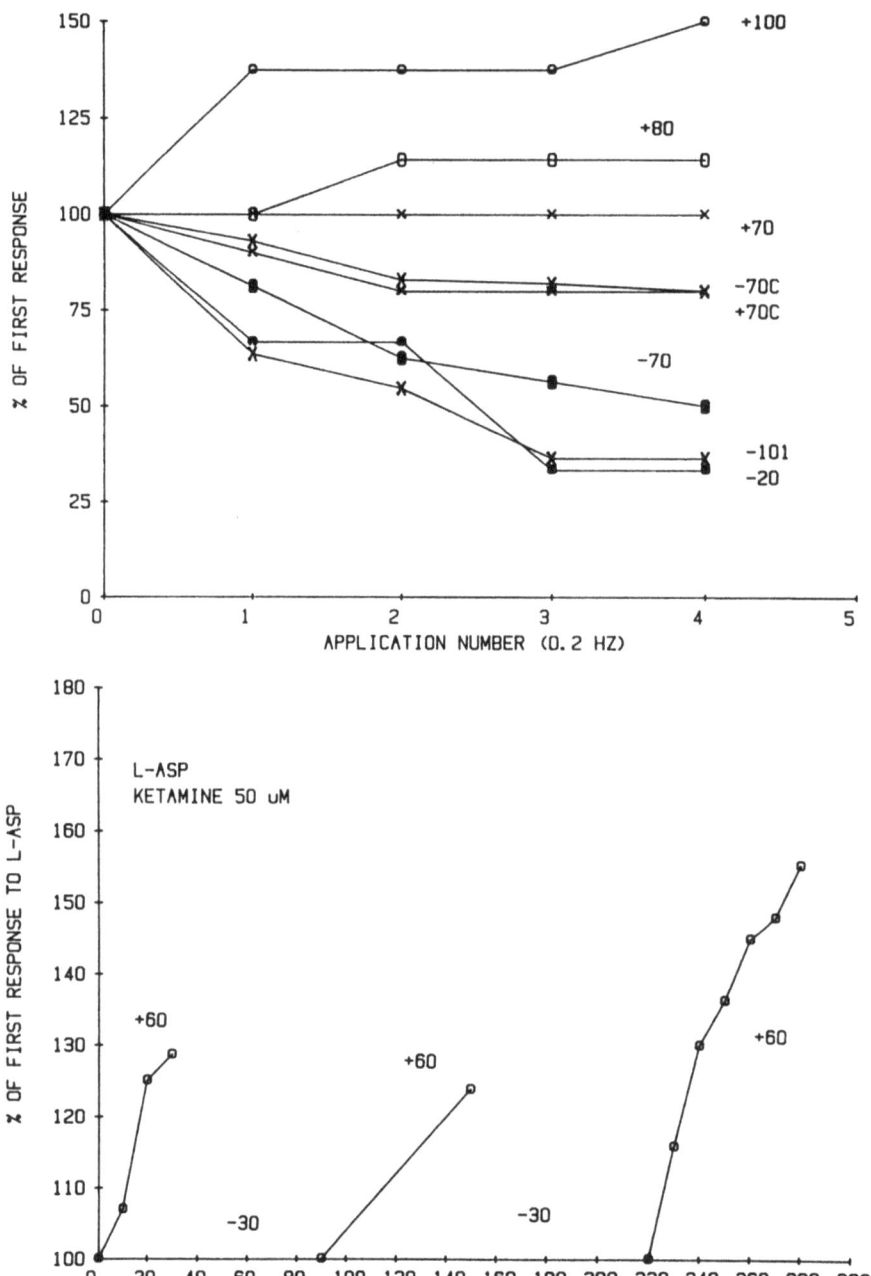

value), repetitive applications of these amino acids were associated with a marked and progressive decline in the peak amplitude. This depression did not recover provided holding potential was kept at hyperpolarized values (i.e., -50 to -100 mV) and therefore this decay was distinct from 'desensitization' (Fig. 6) which was observed prior to perfusion of ketamine or PCP and which recovered relatively quickly (i.e., in about 10 to 15 seconds). The use dependent decay induced by ketamine and PCP was related to the holding potential. If holding potential was moved from a depolarized value to a hyperpolarized value, the use dependent decay was again observed (Fig. 6). Upon returning to a depolarized potential, the decay reversed giving an increment in peak current (Fig. 6). Therefore, the magnitude of any current was dependent upon the previous history of the holding potential. Responses to β-ODAP and KAI did not undergo a use dependent block in the presence of ketamine and PCP nor did they demonstrate 'desensitization'.

DISCUSSION

The voltage dependent EAA mechanism has more potential for interaction with intrinsic bursting than its voltage independent counterpart. For example, although an application of KAI may depolarize, and thus bring membrane potential towards a threshold where regenerative potentials are possible, the steady state current evoked by KAI could attenuate bursting rather than contribute to it (Fig. 7). On the other hand, the voltage dependent (in the presence of Mg^{2+}) response to NMDA could interact by:
1) providing a primary source of inward current for bursting itself;

Fig. 6. Current responses (whole cell voltage clamp) of two cultured hippocampal neurons to pressure applications of 50 μM L-ASP. Upper: Peak responses to L-ASP were expressed as percentages of first response to five sequential applications (0.2 Hz) and plotted against application number for a series of holding potentials before (-70C(mV), +70C(mV); controls) and following perfusion with ketamine (50 μM). The membrane was alternated between hyperpolarized and depolarized values. For example, in the control the holding potential was kept at -70 mV and five applications made. Holding was then switched to +70 mV and about 30 seconds permitted for relaxation of a voltage dependent outward current and the series repeated. A similar alternation was used following perfusion with ketamine. This paradigm was used to illustrate the development of voltage dependent block of the response by ketamine (and its unblock depolarized potentials). If the membrane was not returned to depolarized potentials, the responses to L-ASP simply decayed during the first series of applications and then demonstrated little or no spontaneous recovery. This was not the case for controls where recovery occurred within 10 to 15 seconds. Lower: Responses to L-ASP (in the presence of 50 μM ketamine), expressed as a percentage of the first response of each series plotted against actual time. The first series of four applications was done with holding at +60 mV (previous holding was +40 mV) and then holding was moved to -30 mV for a period of 60 seconds. Holding was brought back to +60 mV and two applications, separated by 60 seconds, were made. Following another 60 second interval at -30 mV, the holding was again set at +60 mV and a series of seven applications given such that the final response occurred 60 seconds following the first. Note that the number of applications of L-ASP made at +60 mV determined the degree of increment, suggesting use dependent as well as voltage dependent unblock of the channels.

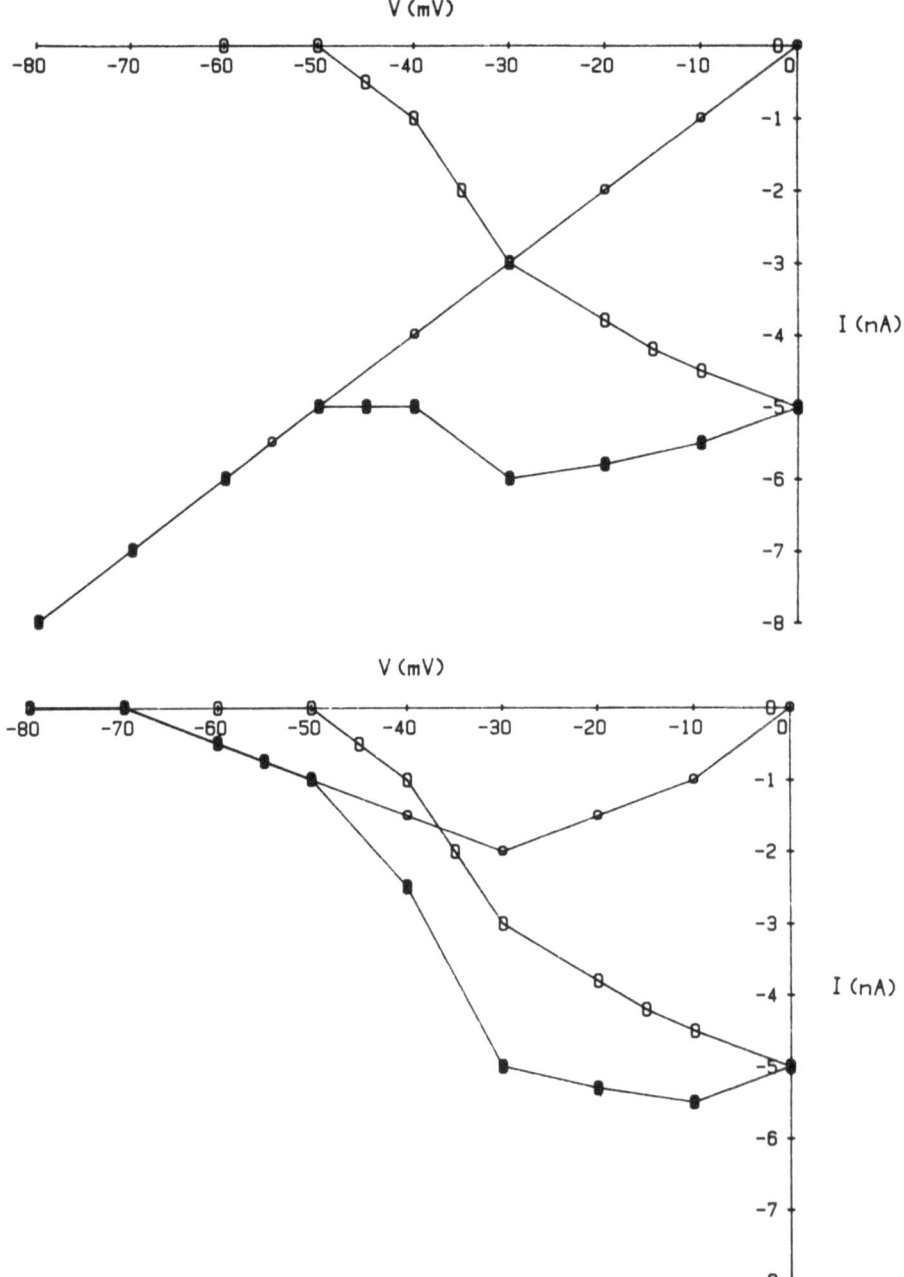

2) sum with already present voltage dependent inward currents important for bursting (i.e., Ca^{2+} currents) and indirectly accentuate their actions; and 3) potentiate Ca^{2+} currents directly (i.e., an action on the gating of Ca^{2+} channels).

Our results suggest poor spatial clamp was attained by us with regard to Ca^{2+} and Ca^{2+} activated currents, regardless of which technique we employed (two electrode voltage clamp or whole cell with discontinuous voltage clamp). This may be related to the relatively large neurons studied (in order to achieve two electrode voltage clamp or to avoid too rapid a washout of Ca^{2+} currents during whole cell recordings). The potentiation of voltage dependent inward currents (presumably Ca^{2+} and Ca^{2+} activated chloride) by L-ASP was associated with the appearance and/or prolongation of a component which appeared to be the plateau portion of a poorly clamped Ca^{2+} spike (MacDonald and Schneiderman, unpublished). Using whole cell discontinuous voltage clamp, the contamination of inward current tails by current plateaus was more obvious than in the case of two electrode voltage clamp. However, regardless of which technique was employed, L-ASP accentuated only these exceptionally slow components. Therefore, we interpret potentiation by amino acids such as L-ASP to be a consequence of facilitated transmission of poorly voltage clamped Ca^{2+} spike into the clamped region of the soma and cannot provide evidence for a direct action on Ca^{2+} currents or Ca^{2+} dependent currents. The effectiveness of EAA currents in contributing to bursting and Ca^{2+} currents would seem to be a consequence of their summation with endogenous voltage dependent currents.

The depression of inward currents and tails by higher doses of L-ASP by β-ODAP and KAI can likely be attributed to the inactivation of Ca^{2+} currents and accelerated wash-out following the elicitation of Ca^{2+} spikes in poorly voltage clamped regions of the neuron (i.e., consequential elevation of intracellular Ca^{2+}).

EAA channels appear to be relatively non-selective cation channels. Nevertheless, it is possible that some differences in selectivity to Ca^{2+} may exist between the two types of channels. For example, small differences in reversal potentials have been reported in the presence of high extracellular Ca^{2+} concentrations (MacDonald and Morris, 1984), and a considerable separation has been reported in sodium poor bathing solutions when intracellular Ca^{2+} may be elevated (Mayer and Westbrook, 1985). If a Ca^{2+}

Fig. 7. The plots show hypothetical currents measured under voltage clamp (steady state) versus command potentials. A voltage dependent inward current (large open circles, upper and lower) has a threshold near -50 mV and a negative slope conductance in a range from -50 to 0 mV (only one quadrant is shown). Such a current might support all-or-none activity in the neuron under current clamp. Other membrane currents are ignored for the purpose of simplicity. Lower: The voltage independent current evoked by an EAA (small open circles) is shown as a linear curve with a reversal potential of 0 mV. The addition curve of these two currents (large closed circles) illustrates one possible interaction between the EAA and bursting. The threshold is now shifted about 10 mV in the depolarized direction and the negative slope region is almost eliminated, making it impossible for the voltage dependent inward current to support regenerative activity. Upper: The current evoked by the EAA is now subject to a voltage dependent block (small open circles) and the two curves summed (large filled circles). Under this condition, the threshold is shifted by 20 mV in the hyperpolarizing direction and the negative slope value is increased. Therefore, bursting would be accentuated.

flux were to occur through the NMDA channel, it is possible that some modulation of Ca^{2+} currents (or dependent currents) would occur secondarily as a consequence of changes in intracellular concentration near the membrane.

Action of Ketamine and PCP

Ketamine and PCP caused a voltage dependent block of NMDA and L-ASP currents but had little effect upon the those to KAI and β-ODAP. The block reversed at depolarized potentials and was, at least to some degree, use dependent. Such evidence suggests that the channel activated by the NMDA receptor is blocked by ketamine and PCP as well as by Mg^{2+}. The use dependency of this block at hyperpolarized holding potentials (and its unblock at depolarized potentials) suggests that one component of the block is directed at the open channel. However, responses to the first application of L-ASP or NMDA (at hyperpolarized potentials) were also depressed, which suggests the block may have been established during (or perhaps before) the initial response.

The highly lipophilic properties of these drugs, and thus their probable dissolution in the membrane, could point towards a modification of either the NMDA receptor or channel from an intra-membranous site (Harrison and Simmonds, 1985). In contrast, our results seem to parallel those found for the nicotinic channel of the end plate where ketamine and PCP also cause a voltage dependent (and likely use dependent) block of the channels (Albuquerque et al., 1980). We conclude that ketamine and PCP exert a selective depression of EAA responses by blocking the NMDA channel. In contrast, the block by 2-APV was not voltage dependent and this result is consistent with an antagonism of the receptor.

Under conditions where the presence of voltage dependent EAA might contribute to bursting, an appropriate antagonist would be expected to attenuate this activity in several ways. If the NMDA receptor is antagonized (i.e., by 2-APV), the EAA response is simply diminished at all voltages. On the other hand, the channel is subject to a further voltage dependent blockade (i.e, by ketamine, PCP, or increased extracellular concentrations of Mg^{2+}) so that the effects could be more complex. Bursting might actually be more predominant in the presence of the blocker (Fig. 7) because the total proportion of voltage dependent inward current would be larger (even though the total inward current contributed by the EAA is reduced). Once the block was sufficient (at hyperpolarized potentials), the EAA might fail to make a significant contribution to the bursting.

ACKNOWLEDGEMENTS

This work was supported by the Medical Research Council of Canada. Z. Miljkovic wishes to thank the Savoy Foundation for assistance. We wish to thank Dr. B.E. Haskell for the gift of β-ODAP.

REFERENCES

Albuquerque, E.X., Tsai, M.C., Aronstam, R.S., Witkop, B., Eldefrawai, A.T., and Eldefrawai, M.E., 1980, Phencyclidine interactions with the ionic channel of the acetylcholine receptor and electrogenic membrane, Proc. Natl. Acad. Sci. USA, 77:1224.

Anis, N.A., Berry, S.C., Burton, N.R., and Lodge, D., 1983, The dissociative anaesthetics, ketamine and phencyclidine, selectiviely reduce excitation of central mammalian neurones by N-methyl-aspartate, Br. J. Pharmacol., 79:565.

Dingledine, R., 1983, N-methyl aspartate activates voltage-dependent calcium conductance in rat hippocampal pyramidal cells, *J. Physiol.*, 343:385.

Flatman, J.A., Schwindt, P.C., Crill, W.E., and Stafström, C.E., 1983, Multiple actions of N-methyl-D-aspartate on cat neocortical neurons in vitro, *Brain Res.*, 266:169.

Harrison, N.L., and Simmonds, M.A., 1985, Quantitative studies on some antagonists of N-methyl-D-aspartate in slices of rat cerebral cortex, *Br. J. Pharmacol.*, 85:381.

Hayashi, T., 1954, Effects of sodium glutamate on the nervous system, *Keio J. Med.*, 3:183.

MacDonald, J.F., and Porietis, A.V., 1982, DL-Quisqualic and L-aspartic acids activate separate excitatory conductances in cultured spinal cord neurons, *Brain Res.*, 245:175.

MacDonald, J.F., and Wojtowicz, J.M., 1982, The effects of L-glutamate and its analogues upon the membrane conductance of central murine neurons in culture, *Can. J. Physiol. Pharmacol.*, 60:282.

MacDonald, J.F., and Morris, M.E., 1984, Lathyrus excitotoxin: mechanism of neuronal excitation by L-2-oxyalylamino-3-amino- and L-3 oxyalylamino-2-amino propionic acid, *Exp. Brain Res.*, 57:158.

MacDonald, J.F., and Schneiderman, J.H., 1985, L-Aspartic acid potentiates 'slow' inward current in cultured spinal cord neurons, *Brain Res.*, 296:350.

Mayer, M.L., and Westbrook, G.L., 1984, Mixed-agonist action of excitatory amino acids on mouse spinal cord neurones under voltage clamp, *J. Physiol.*, 354:29.

Mayer, M.L., Westbrook, G.L., and Guthrie, P.B., 1984, Voltage-dependent block by Mg^{2+} of NMDA responses in spinal cord neurones, *Nature*, 309:261.

Mayer, M.L., and Westbrook, G.L., 1985, The action of N-methyl-D-aspartic acid on mouse spinal neurones in culture, *J. Physiol.*, 361:65.

McCarthy, D.A., Chen, G.M., Kaump, D.H., and Ensor, C., 1965, General anaesthetic and other pharmacological properties of 2-(o-chlorophenyl)-2-methylaminocyclohexanone HCl (CI-581), *J. New Drugs*, 5:21.

Meldrum, B.S., Croucher, M.J., Badman, G., and Collins, J.F., 1983, Antiepileptic action of excitatory amino acid antagonists in the photosensitive baboon, *Papio Papio*, *Neurosci. Lett.*, 39:101.

Nowak, L., Bregestovski, P., Ascher, P., Herbet, A., and Prochiantz, A., 1984, Magnesium gates glutamate-activated channels in mouse central neurones, *Nature*, 307:462.

Owen, D.G., Segal, M., and Barker, J.L., 1984, A Ca-mediated Cl-conductance is present in cultured spinal neurones, *Nature*, 311:567.

LONG-TERM ALTERATIONS IN AMINO ACID-INDUCED IONIC CONDUCTANCES IN CHRONIC

EPILEPSY

R. Pumain, J. Louvel and I. Kurcewicz

Unité de Recherches sur l'épilepsie, Inserm U97
2 ter rue d'Alésia, F-75014 Paris, France

SUMMARY

Extracellular free sodium $(Na^+)_o$ and calcium $(Ca^{2+})_o$ concentration changes were measured in the rat motor cortex, using ion-selective microelectrodes. During ionophoretic applications of excitatory amino acids, decreases in $(Na^{2+})_o$ and in $(Ca^{2+})_o$ were observed. Ca^{2+} signals were not or very little modified by applications of tetrodotoxin while Na^+ signals were slightly depressed, up to 20%. Laminar profile analysis revealed that, while the magnitude of Na^+ signals was rather constant throughout the cortex, Ca^{2+} signals were largest in upper cortical layers. Lesioning and pharmacological experiments indicated that the corresponding permeabilities were most probably located on apical dendrites of pyramidal tract neurons. The relative amplitude of Na^+ and Ca^{2+} signals induced by the release of the glutamate agonists N-methyl-D-aspartate, quisqualate and kainate and the shape of the laminar profile of such responses indicated that different ionic permeabilities located on different neurons underlie such responses. Similar experiments performed on chronic epileptogenic motor foci in rats indicated that the amino acid-induced ionic responses were altered. The significance of such alterations for epileptogenesis is discussed.

INTRODUCTION

Since the classical experiments by Matsumoto and Ajmone-Marsan (1964) and later by Prince (1968) and by Ayala et al. (1972), the cellular hallmark of epileptogenesis was considered to be the huge intracellular depolarizations of long duration (the paroxysmal depolarization shift or PDS) which occurred simultaneously in most neurons, in synchrony with the corresponding paroxysmal field potentials. However, subsequent investigations, this time addressed to chronic models of epilepsy (Prince and Futamachi, 1968, 1970; Pumain, 1981, 1982), revealed that such a feature was not ubiquitous and that other types of abnormal neuronal behavior could be observed. Large depolarizations, with a variable morphology were observed in a small proportion of neurons only. Such depolarizations were not triggered through intracellular or antidromic stimulations but often appeared following orthodromic activations (Pumain, 1981, 1982). Thus, it became apparent that chronic alterations of the excitability of a restricted subpopulation of neurons took place within the chronic epileptic tissue, probably exaggerated at dendritic sites. Such neurons could synaptically trigger epileptic activity in the whole neuronal population. Indeed, it has been shown that under definite conditions, a small number of abnormally-behaving neurons may induce paroxysmal behavior in a large population of neurons (Miles and Wong, 1983).

Neuronal excitability is governed, among other factors, by various membrane conductances which involves principally sodium, calcium and potassium ions. The density and distribution of the corresponding ionic channels may vary from neuron to neuron and also possibly over time. Thus, nerve cells in culture possess a larger density of calcium channels than differentiated ones (Fukuda and Kameyama, 1979). Moreover, lesions in the central nervous system may alter nerve cell excitability. For instance, axotomized motoneurons acquire progressively the capability to produce dendritic regenerative activity (Kuno and Llinas, 1970). Similar alterations may take place in chronic models of epilepsy, in which lesions of the nervous tissue often develop with time. Therefore, we decided to determine whether the distribution of ionic responses, essentially for calcium and sodium, were altered in chronic epileptic tissue. Since, as aforementioned, only a small proportion of neurons behave abnormally in such a preparation, a screening method appeared to be necessary: due to the restricted extracellular space, activity-related neuronal ionic currents give rise to changes in the extracellular ionic concentrations, even though a small number of neurons are involved. Such changes can be detected using ion selective microelectrodes. Ionic currents were induced through local ionophoretic release of the excitatory amino acids glutamate (Glu) and aspartate (Asp), which are thought to play a major transmitter role in the central nervous system. To further characterize the receptors involved, some of their analogs were also applied (Watkins, 1981).

The ionic responses in chronic foci were compared to responses obtained in normal cortex and in cortices in which the pyramidal tract neurons had degenerated, due to a previous chronic lesion of the pyramidal tract, performed at the bulbar level (Kurcewicz and Pumain, 1983; Pumain and Heinemann, 1985). The results show that marked alterations of the distribution of the amino acid-evoked calcium responses occur in chronic epileptic tissue. Such alterations may play an important role in the generation of epileptic activity.

METHODS

Chronic focal epilepsy was produced, as described elsewhere (Pumain, 1981), through topical application of minute quantities of cobalt powder upon the sensorimotor cortex of rats. Eight to ten days later, the animals show EEG as well as motor signs of epileptic activity. Under volatile anesthesia, the focus was then made available for investigation, and the microelectrode recordings were performed within the cortical surface surrounding the cobalt-induced lesion by 0.3 to 1.3 mm, where the largest paroxysmal field potentials were recorded. Details concerning the surgery of the pyramidal tract lesions are described elsewhere (Pumain and Heinemann, 1985). Conventional double-barrelled ion selective electrodes (Lux and Neher, 1973) were glued to five-barrelled ionophoresis electrodes, so that the inter-tip distance was usually about 10 µm. For sodium measurements and since the available sodium resin is not very selective against calcium (Simon et al., 1984), the calcium and sodium electrodes were attached together to the ionophoretic electrode, to record simultaneously the separate signals. The true sodium concentration changes were subsequently computed using the Nikolski formalism. The selectivity coefficients were determined for each electrode assembly from the calibration curves, using the fixed interference method. A retaining current of 15 nA, maintained throughout the applications, was uniformly applied to all the ionophoretic channels.

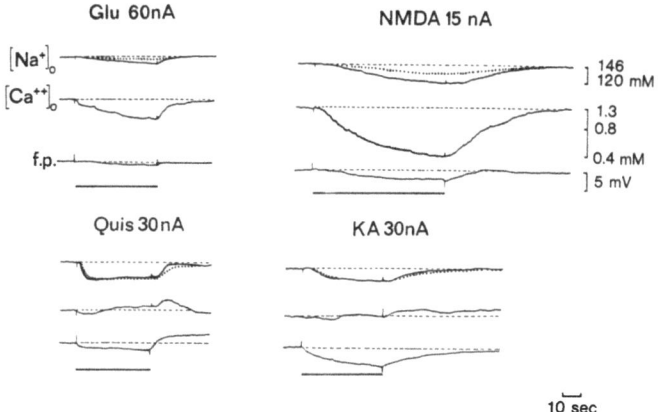

Fig. 1. Effects of ionophoresis of excitatory amino acids on $(Ca^{2+})_o$ and $(Na^+)_o$ in rat motor cortex. On the sodium traces, the dotted lines represent the true sodium signals, computed from measured sodium and calcium traces (see Methods). Recording depth: 200 μm.

RESULTS

Amino Acid-Evoked Ionic Changes in Normal Cortex

Excitatory amino acid receptors are usually classified into three categories (Watkins, 1981), although the list may not be exhaustive (Monaghan et al., 1983): the quisqualate (Quis), the kainate (Ka) and the N-methyl-D-aspartate (NMDA) sensitive ones. During application of NMDA, DL-Homocysteate (DLH), Glu and Asp, the extracellular calcium concentration decreased (Fig. 1) proportionally to the dose. For large ionophoretic doses, more than 90% of the extracellular calcium could disappear, from a baseline of about 1.25 mM, to a level of 0.1 mM or less. During prolonged application, the calcium signals reached a plateau, which remained stable throughout the application. The rank of potency of the various amino acids was: NMDA > DLH > Glu > Asp. Such calcium signals could be blocked by inorganic calcium antagonists (manganese or cobalt), but were little affected by applications of tetrodotoxin (Zieglgänsberger and Puil, 1972; Heinemann and Pumain, 1981), which prevents the generation of the sodium-dependent action potentials. Since calcium uptake into glial cells is probably negligible (Hösli et al., 1981; Bowman and Kimelberg, 1984), these results suggest that the corresponding ionic channels are located mainly on postsynaptic membranes. The calcium signals were not evenly distributed in the cortex: the laminar profiles of the responses showed particularly large decreases in the upper cortical layers (layers II-III), and a second maximum deeper, in layer V (Pumain and Heinemann, 1985). This pattern is quite reproducible and nearly identical for all the four amino acids mentioned above as shown in Fig. 2. The picture was quite different for Quis and for Ka. Quis-induced calcium signals were very often biphasic: the initial decrease faded away and was often followed by an increase in $(Ca^{2+})_o$ even if the application was maintained. An additional increase could often be observed after cessation of application (Fig. 1). At the very cortical surface, only increases in $(Ca^{2+})_o$ were usually observed. Both the decreases and the increases were maximum in the upper layers. The calcium decreases evoked by Quis were not very large, far smaller than those evoked by NMDA. The Ka responses were usually monophasic, showing increases in the upper layers and decreases in deeper layers (Fig. 2). One characteristic of

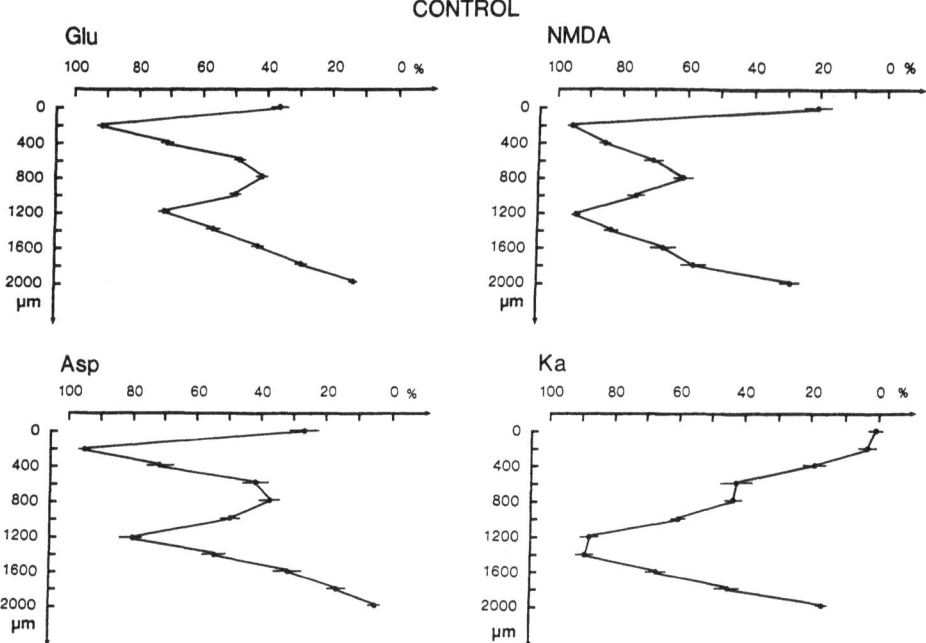

Fig. 2. Laminar profile of $(Ca^{2+})_o$ changes induced through ionophoresis of glutamate (Glu), aspartate (Asp), N-methyl-D-aspartate (NMDA) and kainate (Ka) in sensorimotor cortex of normal rats. Each point represents mean of at least ten measurements. Horizontal bars indicate $\pm$ SE. Laminar profiles are expressed at each depth as means of percentages of the maximal $(Ca^{2+})_o$ changes for each electrode track. For Ka, only the decreases in $(Ca^{2+})_o$ are represented on the graph.

the Ka-evoked responses was the very long duration of the recovery phase.

All the amino acids induced decreases in $(Na^+)_o$: the signals were usually monotonous, and their magnitude depended on the dose. The largest signals observed were in the order of 50 mM (from a baseline value of about 146 mM to approximately 100 mM) and no increases in $(Na^+)_o$ were detected. It is not likely that a sodium-dependent uptake mechanism takes a large part in such $(Na^+)_o$ decreases since ionophoresis of other amino acids (for instance GABA) which are also taken into cells through sodium-dependent systems never produced any significant reduction in $(Na^+)_o$.

The application of tetrodotoxin onto the cortical surface reduced the sodium signals by at most 20% (Zieglgänsberger and Puil, 1972).

Both the NMDA-evoked calcium and sodium responses were nearly completely abolished by low doses of the specific NMDA antagonist 2-amino-5-phosphono-valerate (2-APV; Davies and Watkins, 1982), while the Quis and Ka responses were not affected. High doses of 2-APV were necessary to antagonize the Glu or Asp responses.

A very characteristic feature was that the ratio of calcium to sodium signals was quite different for the various amino acids: this ratio was much larger for NMDA than for Quis or Ka, Glu and Asp being in between. As shown on Fig. 1, a large $(Na^+)_o$ decrease (30 mM) is accompanied by an increase in $(Ca^{2+})_o$ for Quis while at the same place, NMDA evoked a smaller $(Na^+)_o$ decrease (15 mM) with a large $(Ca^{2+})_o$ decrease.

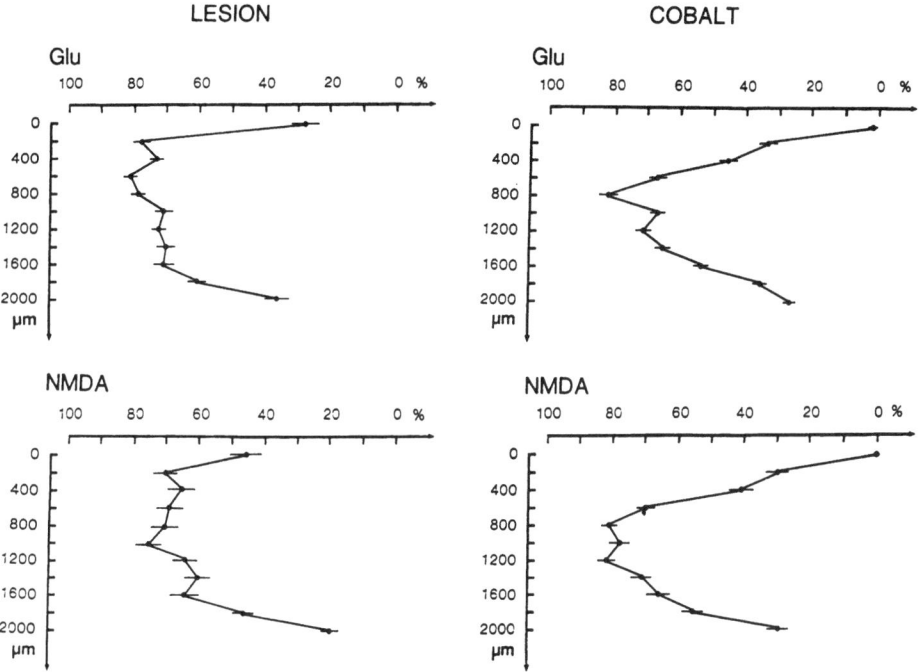

Fig. 3. Laminar profiles of $(Ca^{2+})_o$ decreases induced through ionophoresis of Glu and NMDA in sensorimotor cortex of rats bearing chronic unilateral lesion of the pyramidal tract (Lesion) or chronic cobalt foci (Cobalt). Otherwise, as in Fig. 2.

Effects of Pyramidal Tract Lesions on the Amino Acid-Evoked Ca^{2+} Changes

In order to further characterize the neuronal elements on which the conductances responsible for such extracellular ionic changes were located, we induced a selective degeneration of the pyramidal tract neurons by transecting on one side the pyramidal tract at the bulbar level (Kalil and Schneider, 1975; Kurcewicz and Pumain, 1983; Pumain and Heinemann, 1985). One to several months after such a procedure, similar amino acid-induced laminar profiles were obtained on the corresponding motor cortex. The results are shown on Fig. 3: for NMDA, Glu, Asp, and DLH, the Ca^{2+} signals were practically constant throughout the cortex. Furthermore, for these amino acids, the maximum Ca^{2+} changes were reduced by one-half to two-thirds. In constrast, the laminar profiles for the Ca^{2+} responses induced by Quis or Ka, were, if any, very little modified and the maximum values of the Ca^{2+} changes were not significantly altered.

$(Ca^{2+})_o$ and $(Na^+)_o$ Changes in the Chronic Cobalt Focus

Amino acid-induced changes in $(Ca^{2+})_o$ were compared in the epileptic focus and in the contralateral cortex, using the same electrodes in one particular experiment. The records obtained in the contralateral cortex were compared to records taken from control animals and no difference could be observed. The contralateral cortex was therefore considered as normal. Within the lesion, hardly detectable signals were obtained throughout the cortex. In the area surrounding the lesion where the maximal EEG paroxysms occurred, large calcium signals were recorded. In this area, the laminar profiles for NMDA, DLH, Glu and Asp, were profoundly altered in a similar manner: the superficial responses (in the first

200 to 300 μm of the electrode track) were small, if any, while large signals were observed from layer III to layer V, with usually a maximum at depths around 800 μm (Fig. 3), where the calcium signals were minimum in the control situation. The maximal reductions in $(Ca^{2+})_o$ were often larger in the cobalt foci when compared to the normal cortex. For Quis and Ka, the superficial responses were also nearly abolished. Concerning Quis, below the superficial layers, the calcium decreases were small, as in the control situation. For Ka, the laminar profile for the calcium decreases was not obviously altered, nor the maximal responses very different.

DISCUSSION AND CONCLUSIONS

The prominent difference concerning the ratio of the calcium to the sodium signals evoked through NMDA on one hand and through Ka and Quis on the other, indicates that the NMDA-operated channels possess a larger calcium permeability than the Quis- or Ka-operated ones and are probably not very selective among cations (Mayer and Westbrook, 1984; Nowak et al., 1984). Alternatively, NMDA may directly activate calcium channels (Dingledine, 1983). However, it is unlikely that entry of calcium through voltage-sensitive channels can entirely account for such a difference: very large Quis- or Ka-induced $(Na^{2+})_o$ decreases were often not accompanied by any calcium signal, while at the same recording site, a small NMDA-induced sodium signal was accompanied by a large reduction in $(Ca^{2+})_o$. Furthermore, in preliminary experiments, perfusion of the cortex with a high potassium solution did not produce any modification of the amino acid-induced calcium signals. Although such calcium entry through the NMDA-coupled ionophores may not be the most important factor in terms of neuronal depolarizations, it may modify the membrane excitability (Krnjević and Lisiewicz, 1972) or the cellular metabolism.

The increases in $(Ca^{2+})_o$ observed during application of Quis or Ka represent an intriguing feature. Such increases may be due to a shrinkage of the extracellular space induced by net influx of ions and subsequently of water into the cells during the amino acid-induced depolarizations. However, metabolic processes may also be implicated. This view is supported by the observation that, very superficially, relatively large $(Ca^{2+})_o$ increases could be measured while only very small changes were apparent on the concomitant sodium record.

The results of the pyramidal tract lesion experiments indicate that the conductances responsible for the large Ca^{2+} signals are located on the apical dendrites of the pyramidal tract neurons (Stafstrom et al., 1985) and on the cell body or on the basal dendrites of these neurons. They further indicate that the receptors responsible for the calcium responses evoked through Glu, Asp or DLH applications are likely of the NMDA type (Flatman et al., 1983; Olverman et al., 1984) although some pharmacological differences may appear. The receptors for Quis and Ka which give rise to calcium responses do not appear to be preferentially located on the pyramidal tract cells but on separate neuronal elements.

The results obtained from the cobalt focus show that a long-term modification of the amino acid-activated calcium conductances occur in this chronic model of epilepsy (Pumain and Heinemann, 1982). Such calcium conductances appear to be activated essentially through NMDA-type receptors, and therefore these data indicate that a long-term reorganization of these receptors also develops during the maturation of the focus. Since in the normal cortex, the NMDA calcium responses stem mostly from the pyramidal tract cells, one can speculate that such a reorganization concerns principally the same neurons. Thus, since the largest calcium responses are recorded in the focus at depths where, in the normal cortex, they are minimal, we can speculate that the corresponding conductances develop at soma-near sites on pyramidal tract cells. Such a feature could account for the large depolarizations observed, in the cobalt focus, to occur in pyramidal tract neurons during the epileptic EEG spikes (Pumain, 1981).

Large extracellular calcium decreases have been described in acute (Heinemann and Louvel, 1983), as well as in chronic (Heinemann et al., 1981) models of epilepsy. Thus, it appears that calcium-dependent mechanisms may be important in a variety of epileptic phenomena. The involvement of NMDA receptors in epilepsy described in the cobalt focus, may be a more general feature of the chronic epilepsies. For instance, abnormally large $(Ca^{2+})_o$ calcium decreases have been observed in the course of generalized seizures in the photosensitive baboon Papio papio (Pumain et al., 1985), and specific antagonists at the NMDA receptors have proven to be anticonvulsants in this model (Croucher et al., 1982). Furthermore, in slices of human epileptic cortices, we have observed that applications of NMDA induced very large extracellular calcium decreases (Louvel and Pumain, unpublished observations). In addition, similar features may occur in other pathological conditions, such as in ischemia (Simon et al., 1984).

ACKNOWLEDGEMENTS

The skillful asistance of Ms. L. Olive in the preparation of the manuscript is gratefully acknowledged.

REFERENCES

Ayala, G.F., Dichter, M., Gumnit, R.J., Matsumoto, H., and Spencer, W.A., 1972, Genesis of epileptic interictal spikes. New knowledge of cortical feedback systems suggests a neurophysiological explanation of brief paroxysms, Brain Res., 52:1.
Bowman, C.L., and Kimelberg, H.K., 1984, Excitatory amino acids directly depolarize rat brain astrocytes in primary culture, Nature, 311:656.
Croucher, M.J., Collins, J.F., and Meldrum, B.S., 1982, Anticonvulsant action of excitatory amino acid antagonists, Science, 216:889.
Davies, J., and Watkins, J,C., 1982, Actions of D and L-forms of 2-amino-5-phosphonovalerate and 2-amino-4-phosphonobutyrate in the cat spinal cord, Brain Res., 235:378.
Dingledine, R., 1983, N-methyl aspartate activates voltage-dependent calcium conductance in rat hippocampal pyramidal cells, J. Physiol., 343:385.
Flatman, J.A., Schwindt, P.C., Crill, W.E., and Stafstrom, C.E., 1983, Multiple actions of N-methyl-D-aspartate on cat neocortical neurons in vitro, Brain Res., 266:169.
Fukuda, J., and Kameyama, M., 1979, Enhancement of Ca spikes in nerve cells of adult mammals during neurite growth in tissue culture, Nature, 279:546.
Heinemann, U., Konnerth, A., and Lux, H.D., 1981, Stimulation induced changes in extracellular free calcium in normal cortex and chronic alumina cream foci of cats, Brain Res., 213:246.
Heinemann, U., and Louvel, J., 1983, Changes in $(Ca^{2+})_o$ and $(K^+)_o$ during repetitive electrical stimulation and during pentetrazol induced seizure activity in the sentorimotor cortex of cats, Pflügers Arch., 398:310.
Heinemann, U., and Pumain, R., 1981, Effects of tetrodotoxin on changes in extracellular free calcium induced by repetitive electrical stimulation and iontophoretic application of excitatory amino acids in the sensorimotor cortex of cats, Neurosci. Lett., 21:87.

Hösli, L., Hösli, E., Andres, P.F., and Landolt, H., 1981, Evidence that the depolarization of glial cells by inhibitory amino acids is caused by an efflux of K^+ from neurones, Exp. Brain Res., 42:43.

Kalil, K., and Schneider, G.E., 1975, Retrograde cortical and axonal changes following lesions of the pyramidal tract, Brain Res., 89:15.

Krnjević, K., and Lisiewicz, A., 1972, Injections of calcium ions into spinal motoneurones, J. Physiol., 225:363.

Kuno, M., and Llinas, R., 1970, Enhancement of synaptic transmission by dendritic potentials in chromatolysed motoneurones of the cat, J. Physiol., 210:807.

Kurcewicz, J., and Pumain, R., 1983, Excitatory amino acid activated calcium conductances are located on distal parts of apical dendrites of corticospinal neurones in rat motor cortex, J. Physiol., 345:73P.

Lux, H.D., and Neher, E., 1973, The equilibration time course of $(K^+)_o$ in cat cortex, Exp. Brain Res., 17:190.

Matsumoto, H., and Ajmone-Marsan, C., 1964, Cortical cellular phenomena in experimental epilepsy: ictal manifestations, Exp. Neurol., 9:305.

Mayer, M.L., and Westbrook, G.L., 1984, Mixed-agonist action of excitatory amino acids on mouse spinal cord neurones under voltage clamp, J. Physiol. 354:29.

Miles, R., and Wong, R.K.S., 1983, Single neurones can initiate synchronized population discharge in the hippocampus, Nature, 304:371.

Monaghan, D.T., Holets, V.R., Toy, D.W., and Cotman, C.W., 1983, Anatomical distributions of four pharmacologically distinct 3H-L-glutamate binding sites, Nature, 306:176.

Nowak, L., Bregestovski, P., Ascher, P., Herbet, A., and Prochiantz, A., 1984, Magnesium gates glutamate-activated channels in mouse central neurones, Nature, 307:462.

Olverman, H.J., Jones, A.W., and Watkins, J.C., 1984, L-glutamate has higher affinity than other amino acids for (3H)-D-AP5 binding sites in rat brain membranes, Nature, 307:460.

Prince, D.A., 1968, The depolarization shift in 'epileptic' neurons, Exp. Neurol., 21:467.

Prince, D.A., and Futamachi, K.J., 1968, Intracellular recordings in chronic focal epilepsy, Brain Res., 11:681.

Prince, D.A., and Futamachi, K.J., 1970, Intracellular recordings from chronic epileptogenic foci in the monkey, Electroenceph. Clin. Neurophysiol., 29:496.

Pumain, R., 1981, Electrophysiological abnormalities in chronic epileptogenic foci: an intracellular study, Brain Res., 219:445.

Pumain, R., 1982, Intracellular potentials of cortical neurons in a chronic epileptogenic focus, in: Physiology and Pharmacology of Epileptogenic Phenomena, M.R. Klee, H.D. Lux and E.J. Speckmann, eds., Raven Press, New York, p. 65.

Pumain, R., and Heinemann, U., 1982, Intracellular potential and extracellular calcium changes in chronic epilepsy, in: Advances in Epileptology: XIIIth Epilepsy International Symposium, H. Akimoto, H. Kazamatsuri, M. Seino and A. Ward, eds., Raven Press, New York, p. 497.

Pumain, R., and Heinemann, U., 1985, Stimulus- and amino acid-induced calcium and potassium changes in rat neocortex, J. Neurophysiol., 53:1.

Pumain, R., Menini, C., Heinemann, U., Louvel, J., and Silva-Barrat, C., 1985, Chemical transmission is not necessary for epileptic seizures to persist in the baboon Papio papio, Exp. Neurol., 89:250.

Simon, R.P., Swan, J.H., Griffiths, T., and Meldrum, B.S., 1984, Blockade of N-methyl-D-aspartate receptors may protect against ischemic damage in the brain, Science, 226:850.

Stafstrom, C.E., Schwindt, P.C., Chubb, M.C., and Crill, W.E., 1985, Properties of persistent sodium conductance and calcium conductance of layer V neurons from cat sensorimotor cortex in vitro, J. Neurophysiol., 53:153.

Steiner, R.A., Oehme, M., Ammann, D., and Simon, W., 1979, Neutral carrier sodium ion-selective microelectrode for intracellular studies, *Anal. Chem.*, 51:351.

Watkins, J.C., 1981, Pharmacology of excitatory amino acid transmitters, in: *Amino Acid Neurotransmitters*, F.V. De Feudis and P. Mandel, eds., Raven Press, New York. p. 205.

Zieglgänsberger, W., and Puil, E.A., 1972, Tetrodotoxin interference of CNS excitation by glutamic acid, *Nature New Biol.*, 239:204.

EXCITATORY AMINO ACIDS AND EPILEPSY-INDUCED CHANGES IN EXTRACELLULAR

SPACE SIZE

U. Heinemann

Max Planck Institute of Psychiatry,
Department of Neurophysiology
Am Klopferspitz 18a, D-8033 Planegg, F.R.G.

SUMMARY

Convulsant and stimulus induced seizures are associated with Ca, Na, K and Cl concentration changes in the extracellular space (ES), which are a resultant of transmembrane ionic fluxes and of changes in the ES size. The ES decreases on average by 30% during a single seizure. An analysis of the causes of ES size changes reveal a large contribution from the spatial glia K buffer mechanism which may account for up to 60% of the ES decreases. NaCl and KCl uptake into cells as well as increases in intracellular osmolarity due to anaerobic glycolysis contribute less to the local cytotoxic edema but account for a net gain of osmotic active particle at the site of the focus. Excitatory amino acids such as glutamate, aspartate, N-methyl-D-aspartate (NMDA), kainate and quisqualate also lead to Na, Cl and eventually Ca uptake into cells and to release of K with dose dependent decreases in $[Na]_o$, $[Ca]_o$ and $[Cl]_o$, increases in $[K]_o$ and transient decreases in ES size by up to 80% which are possibly associated with a net reduction of osmotically active particles. The predominant cause for this cytotoxic edema is NaCl uptake into cells but spatial K buffering through glial cells also contributes to this type of edema. The possible consequences of the various ion movements and the changes in osmolarity as well as ES size for tissue vulnerability are discussed.

INTRODUCTION

Excessive seizure activity as in status epilepticus can lead to selective cell loss in hippocampus (Meldrum and Brierley, 1973), neocortex (Meldrum and Brierley, 1973), thalamus (Collins and Olney, 1982) and other parts of the central nervous system (Sperk et al., 1983; Lassman et al., 1984). Electronmicroscopic investigations have shown that glial cells, presynaptic endings and dendrites in areas of excitatory transmission are particularly swollen, whereas somata of nerve cells can either shrink or swell (Meldrum and Brierley, 1973; Collins and Olney, 1982). Size changes of cells are caused by water shifts across the plasma cell membranes. They reflect in changes of the extracellular space (ES) size. These can be determined by measuring extracellular resistivity. For this current is applied to the ES and the resultant voltage changes are measured under the assumption that the applied current remains predominantly in the ES.

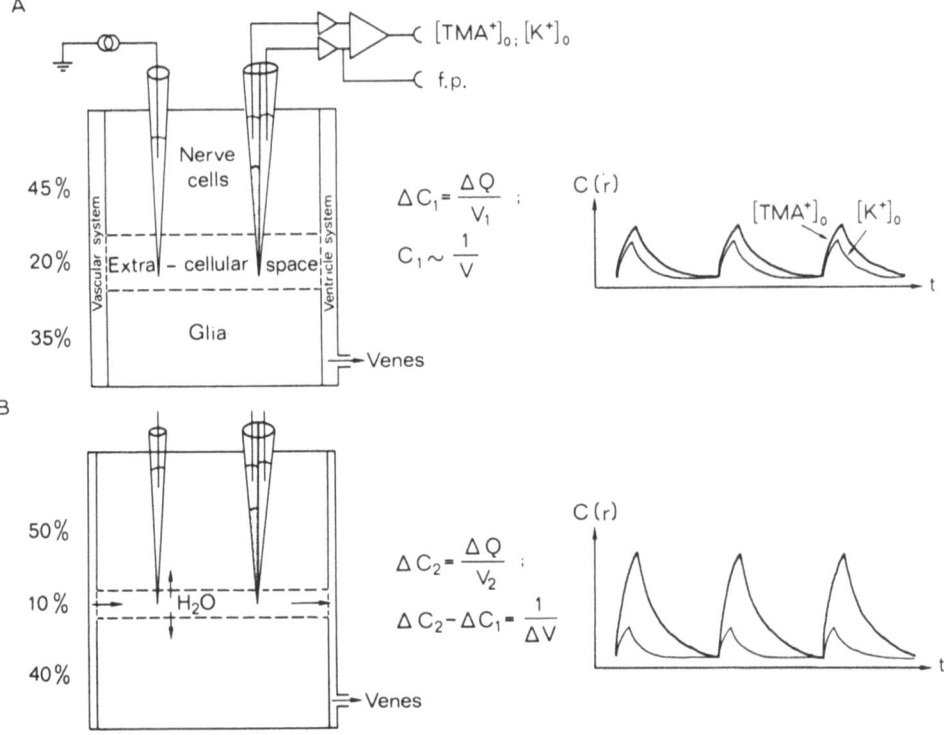

Fig. 1. Effects of extracellular space size on iontophoretically induced ionic signals. A: Normal extracellular space (ES), B: Reduced ES. Volume fractions of nerve cells, ES and glia indicated in %. The concentration of the probe ion tetramethylammonium (TMA) upon iontophoretic application with constant current pulses was measured at an interval of about 50 μm with a K sensitive double barreled ion selective/reference microelectrode. K signals are somewhat smaller than TMA signals because of K uptake into glia and nerve cells. Differences in size of TMA signals indicate changes in ES size. Differences in size of iontophoretic signals are the result of ES size changes and K uptake processes.

METHODS

Water shifts across the nerve cells also lead to concentration changes of ions which remain in the ES. Based on this consideration, techniques have been developed in which a membrane impermeable ion is iontophoretically applied with constant current pulses of constant time at constant intervals. The resulting ion concentration changes were measured with ion selective microelectrodes (Fig. 1). The observed concentration changes follow instantaneous point source diffusion kinetics in aqueous media (Lux and Neher, 1973) with the addition that the volume fraction of the ES and its tortuosity, i.e., the prolongation of the diffusion path by impermeable elements, are taken into account (Nicholson and Phillips, 1981). When cells swell, the volume fraction of the ES decreases (Fig. 1B) and the tortuosity increases. This later effect is, however, negligable. The decrease of the ES leads to an enhancement of iontophoretically induced ionic signals from which the change in the ES size can be calculated (Dietzel et al., 1980; see also Connors et al., 1982). We have employed tetramethylammonium (TMA). Brain cells are normally impermeable for TMA in the time domain of a few seconds (Dietzel et al., 1980; Nicholson and Phillips, 1981).

The TMA concentration change was detected with a nearby (tip separations less than 50 μM) ion selective microelectrode prepared and tested as described by Lux and Neher (1973). The Corning 477113 K sensitive resin was used. The electrodes have a detection limit for TMA of 5×10^{-6} M (Dietzel et al., 1980). Thus TMA concentration changes could be kept at levels which have no significant pharmacological action.

The additional use of other ion selective microelectrodes permits the determination of extracellular ionic changes during seizure activity and iontophoretic application of excitatory amino acids. These transmembrane ionic fluxes can be related to the changes in ES size (Dietzel et al., 1982; Lux et al., 1986).

RESULTS AND DISCUSSION

Ionic Changes During Epileptic Seizures

Fig. 2 illustrates measured K, Na, Cl and Ca concentration changes in the somatosensory cortex of cats during stimulus induced convulsions as well as changes in tissue reactivity and ES space size (Dietzel et al., 1980, 1982; Somjen, 1980; Dietzel et al., 1982; Pumain and Heinemann, 1985; for review see Heinemann et al., 1986; Lux et al., 1986). It becomes immediately apparent that during seizure activity the ES can locally shrink by more than 30%. Such a shrinkage is caused by water fluxes into nerve and glial cells following an increase in intracellular osmolarity. As far as it is caused by metabolic cleavage of large particles into small ones (for example, glucose in lactate, ATP in ADP and phosphate, bicarbonate accumulation), it should reflect in a net increase in tissue osmolarity. This is indeed observed. The sum of changes in $[Na]_o$, $[K]_o$ and $[Cl]_o$ as well as of bicarbonate suggest an increase in tissue osmolarity by up to 20 mosm. This would account for an about 8% reduction of the ES (Dietzel et al., 1982).

The fact that glial cells swell, which possess predominantly K permeable channels, suggests a role of extracellular K accumulation in the generation of the edema. Indeed, local application of K from an independent iontophoresis electrode or superfusion of the cortical surface with K enriched solutions induce a shrinkage of the ES. This was found both in normal cortical tissue as well as in gliotic brain tissue which develops, for example, after topical application of alumina cream (Dietzel et al., 1980, 1982; Heinemann and Dietzel, 1984). Two mechanisms are apparently involved (Fig. 3A); homogeneous elevation of K around tissue cultured astrocytes and in normal cortex leads to K accumulation inside glia and swelling which is possibly due to a Cl coflux (Bourke and Nelson, 1972; Kettenmann et al., 1983; Walz and Hertz, 1983; Kimelberg and Ransom, 1986). A second mechanism operates only when $[K]_o$ is inhomogeneously elevated. It is related to spatial K buffering through the glial cell syncytium. Therefore we compared the laminar distribution of ES size changes in relation to changes in $[K]_o$ during repetitive electrical stimulation of the cortical surface of the thalamus (Dietzel et al., 1980). The results show that the maximal increase in $[K]_o$ occurs in about layer III/IV with larger increases in the supra- than in the infragranular layers (Fig. 3B, 3C). The decrease in ES size is largest in middle cortical layers and smaller in supragranular layers whereas a widening of the ES can be observed in infragranular layers. The spatial K buffer mechanism can account for this laminar profile (Fig. 3A).

At the site of maximal K accumulation, glial cells will be depolarized most (Orkand et al., 1966; Kettenmann et al.; 1983; Orkand et al., 1986). This depolarization spreads along the surface of spatially extended

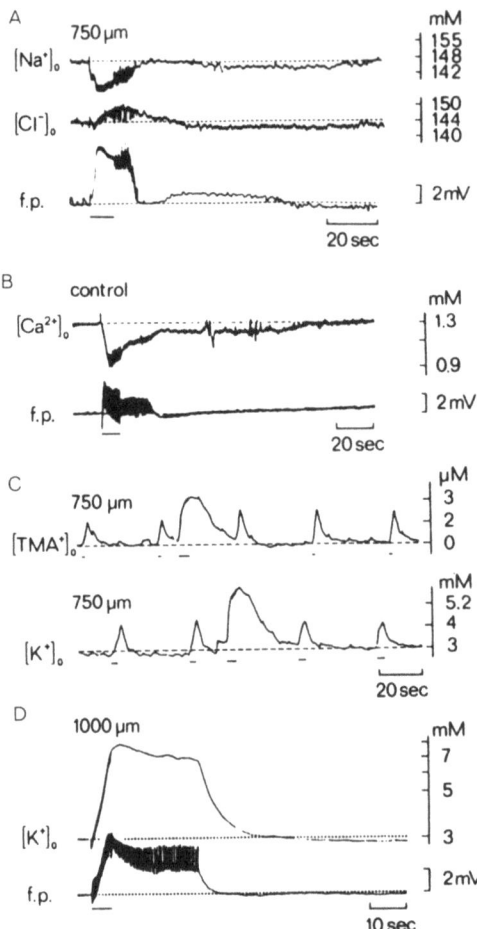

Fig. 2. Ionic and ES changes during stimulus induced seizures (20 Hz, 10 seconds, 0.1 ms). Horizontal bars indicate the duration of stimulation. Note that TMA signals are enhanced after a convulsion whereas iontophoretically induced K signals are reduced in amplitude. fp indicates field potentials.

(Somjen, 1983) and electrically coupled glial cells (Gutnick et., 1981). This leads to a disturbance of the electrochemical equilibrium, in that the glial cell depolarization is less than expected from the transmembrane K gradient and more at remote areas (Orkand et al., 1966). To reestablish equilibrium, K enters glial cells at sites of maximal K accumulation and it is released at remote areas. In that way K is spatially redistributed. Of course this charge transfer leads to the generation of extracellular potentials which are negative at the site of K uptake and less negative or even positive at remote areas (Somjen, 1983). The K inward current into glia at sites of maximal K accumulation and the remote K outward current from glia requires a compensatory current in the ES. This current is carried by Na and Cl ions, which are the prominent charge carriers

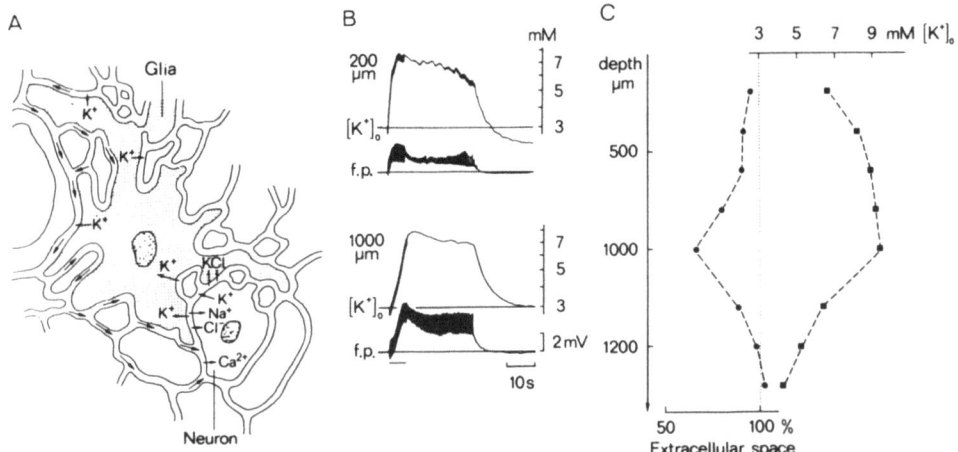

Fig. 3. Role of spatial K buffering in changes of the ES. A: Schematic drawing of ionic fluxes during a seizure. B: Sample recordings from a laminar profile of $[K]_o$ and field potential recordings. C: Average laminar profiles of changes in $[K]_o$ and ES size.

in the ES (Dietzel et al., 1980). Because of the different charges, this leads to a depletion of Cl and an accumulation of Na at the site of K uptake. For 100 K ions leaving the ES, 40 Na are replaced, but 60 Cl ions are lost. This effect leads to a reduction in osmotically active particles and hence to a water flux into cells. This current transport effect applies of course also to the intracellular space with anion accumulation at the site of K entry. Thereby it leads at the site of maximal K uptake to an increase in intracellular osmolarity. This mechanism explains the general glial cell swelling and it may account to some extent for the swelling of neuronal elements as well. It even explains the shrinkage of nerve cells deep in the cortex because the opposite effects occur where K is released, and where the ES becomes wider (Fig. 3C).

Based on this consideration, it is possible to estimate the K transport through glial cells by determining the transmembrane currents. This can be achieved by applying the current source density analysis to recorded extracellular field potentials. I. Dietzel (in preparation) has done that and found that during an electrically induced seizure 0.3 to 0.5 mM/l/s of K ions are cleared away from the site of maximal K accumulation (Lux et al., 1986). This mechanism would account for about **40%** of the observed ES reduction. This leaves some 30 to 40% which are accounted for by NaCl uptake into nerve cells as a consequence of increased transmitter release and by KCl uptake into glial cells. Transmitter and metabolite uptake, and the $Na/H/HCO_3/Cl$ countertransport mechanism (Kimelberg and Ransom, 1986) will also contribute to the generation of epileptic edema. However they may be less important since their transport capacities can cause in erythrocytes only a **10%** change within minutes. During seizures the time course of edema development is much faster.

Unfortunately, the evidence for this mechanism is rather circumstantial. We have therefore attempted to verify it more directly. One preparation of interest is the border zone between normal and gliotic tissue produced by topical application of alumina cream (Heinemann and Dietzel, 1984). Within this border zone, spontaneous epileptiform activity is seen accompanied by rises in $[K]_o$. Large rises in $[K]_o$ can be evoked by direct cortical surface stimulation. These become smaller and smaller as the electrode is moved into the gliotic tissue. Within the gliotic tissue, the spatial buffer capacity appears not to be disturbed as indicated by three lines of evidence. Microapplication of K by pressure or iontophoresis produce similar negative potential shifts as in normal tissue. Artificial local K application evokes a similar decrease in ES size as in normal cortical tissue. Finally, constant current through the cortical tissue leads to a larger than expected spatial K transport because some of the current is passing through glia and thereby redistributes K (Gardner-Medwin, 1983). This redistribution is similar in size in normal and gliotic tissue. Within the gliotic tissue, excitatory amino acid application does not lead to changes in $[K]_o$ and $[Na]_o$. Hence there are only few, if any, nerve cells. Consequently, the rises in $[K]_o$ within the gliotic tissue during repetitive electrical stimulation or seizure activity represent spatial K redistribution through glial cells. The spatial buffer hypothesis then predicts that the ES should shrink at the site of maximal K release and become wider within the gliotic tissue. This is indeed the case.

A second line of evidence comes from investigations on *in vitro* hippocampal slices (Heinemann et al., 1983). Synaptic transmission in this preparation can be blocked by lowering of extracellular Ca to levels near 0.2 mM (Konnerth and Heinemann, 1983). Generation of spontaneous spreading epileptiform activity (Haas and Jefferys, 1984; Yaari et al., 1986) can be delayed by elevating Mg to 7 to 9 mM. Hence hippocampal pyramidal cells can only be activated by antidromic stimulation. This induces ringing epileptiform field potentials in area CA1 which rapidly accomodate during repetitive antidromic stimulation (Fig. 4B). Repetitive stimulation leads, as in normal hippocampus, to largest rises in $[K]_o$ in the pyramidal cell layer (Krnjević et al., 1982). Because of the short duration of the action potentials, these presumably lead in the dendrites only to capacitive currents and to little K release from dendrites. Indeed, under this condition $[Na]_o$ and $[Ca]_o$ changes are usually observed with antidromic stimulation only within 50 to 100 μm from the pyramidal cell layer. Hence a considerable part of $[K]_o$ elevation must result from spatial redistribution of K through glial cells. Estimates of this K redistribution by three dimensional current source density analysis from the negative slow field potentials in the PC layer and their respective positive or less negative potentials in other layers (Fig. 4C) have revealed that about 19 mM/l/s can be spatially redistributed. This is associated with an about 10% decrease of the ES in SP and a considerable widening of ES size in SR (Fig. 4D). The relative small decrease of ES size in area CA1 may be related to the fact that the ES size in SP comprises only about 5% in contrast to the about 15% in SR and more in neocortex and cerebellum. Also the small distances and the steep osmotic gradients permit for a sufficient water diffusion in the ES to counteract larger ES size changes.

Obviously this mechanism has a number of beneficial effects. The decrease in ES size counteracts the loss of Cl and Na at sites of maximal neuronal activity and Na is brought in addition to the site of use. The same applies to Ca. Model calculations based on glial membrane potentials, tranfer of K as measured with the CSD and assumptions on intracellular osmolarity increases have been able to predict ionic changes in the ES to a very accurate degree (Lux et al., 1986). The elegant three compartmental analysis of Orkand et al. (1986), where [Na], [Cl] and [K] changes were measured in the neuronal, glial and ES compartment of the honey bee retina, also have demonstrated that spatial K buffering leads to a significant change in ES size and confirmed that the balance of ionic changes only operates when spatial K buffering is taken into account.

Thus, glial swelling appears to depend largely on KCl uptake and spatial K buffering. The decrease in cell somata size and the dendritic

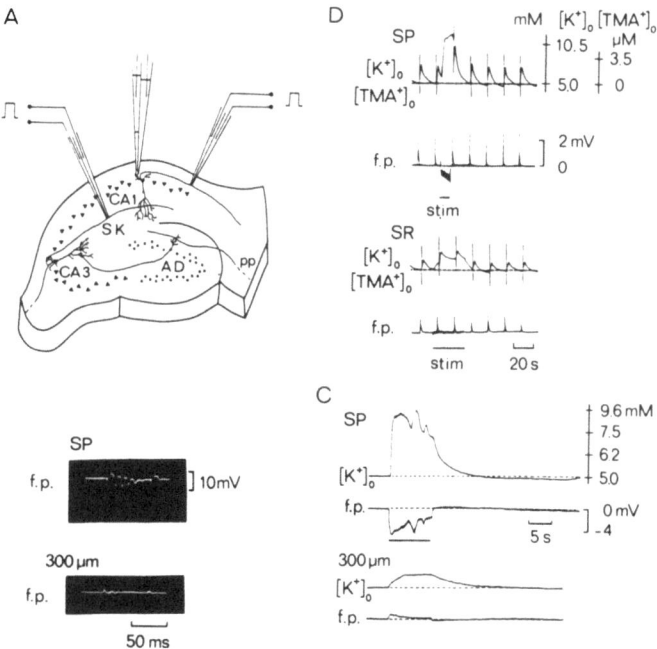

Fig. 4. Field potential and [K] recordings during antidromic stimulation in SP and SR of <u>in vitro</u> hippocampal slices under conditions of blocked chemical synaptic transmission (Perfusion with low Ca and 7 mM Mg Ringer). A: Methods, B: Antidromically elicited series of population spikes in stratum pyramidale (SP) and stratum radiatum (SR) at a distance of 300 μm from SP. C: Rises in $[K]_o$ during repetitive antidromic stimulation. D: Rises in $[K]_o$ and in [TMA] elicited by antidromic stimulation and by regular iontophoretic application of TMA. Note that the ES shrinks in SP and becomes wider in SR.

swelling may be related to the same mechanism. This does, however, not explain why dendritic spines are particularly prone to swelling. These are normally the targets for excitatory amino acid neurotransmission presumably mediated by glutamate and/or aspartate. We therefore performed a study (Heinemann and Pumain, in preparation) in which the Na, Cl, Ca and K concentration changes during excitatory amino acid application were measured together with changes in the ES size (see also Heinemann and Pumain, 1980; Engberg et al., 1983; Heinemann et al., 1986; Lambert and Heinemann, 1986). Fig. 5 demonstrates typical findings during quisqualate and NMDA application. Both excitatory amino acids lead to considerable drops in $[Cl]_o$, $[Na]_o$ and $[Ca]_o$ as well as to increases in $[K]_o$. Interesting differences are found when the relative size of $[Ca]_o$ decreases is compared for similar sized Na signals. Then it becomes apparent that NMDA facilitates much more Ca entry into cells than quisqualate and kainate (which act on diffferent receptors, Watkins, 1984). The same conclusion is reached when similar sized membrane depolarizations were compared to changes in $[Ca]_o$ (Lambert and Heinemann, 1986). Such experiments in <u>in vitro</u> hippocampal slices revealed that NMDA provoked a much larger Ca entry into cells

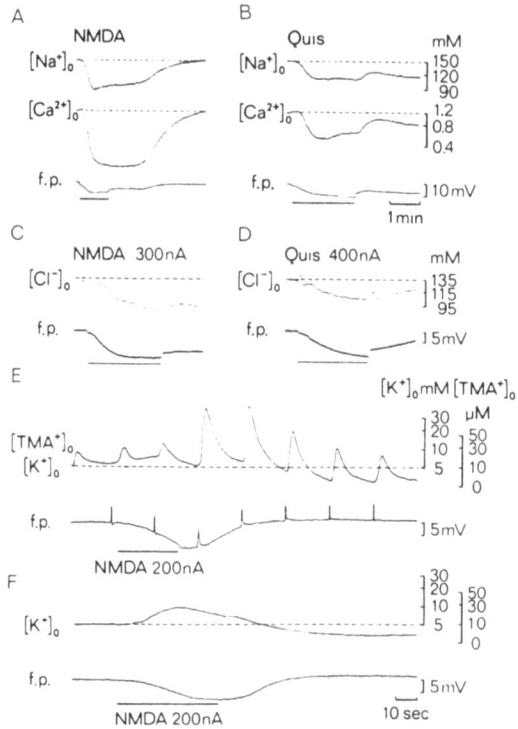

Fig. 5. Changes in the cellular microenvironment and ES size induced by iontophoretic application of N-methyl-D-aspartate and quisqualate. Note the discrepancy between changes in TMA signals and extracellular field potentials induced by iontophoretic application of TMA.

than quisqualate and kainate. Studies in that preparation also showed that quisqualate and kainate provoked Ca uptake into cells only when the neurons were strongly depolarized, while NMDA was effective in causing Ca uptake already near rest. Substitution experiments, where increasing amounts of Na were replaced by Tris revealed that NMDA activated ionophores possess a larger Ca permeability than quisqualate activated ones. A further difference is the sensitivity to organic and inorganic Ca antagonists which block quisqualate induced Ca signals more easily than NMDA induced ones. All these findings suggest that NMDA ionophores have a larger Ca permeability than non NMDA operated ionophores. This finding is in line with single channel conductance measurements (Nowak et al., 1984).

The question arose whether NMDA directly activates a voltage sensitive channel (Dingledine, 1983) and whether the response is principally carried by Ca ions. To test the latter question we investigated also intracellular Na concentration changes during iontophoretic application of NMDA and quisqualate in spinal motoneurons and found that both excitatory amino acids produced large increases in intracellular Na concentration (Heinemann, Engberg, Flatman and Lambert, manuscript in preparation).

The rises in $[K]_o$ and decreases in $[Na]_o$ and $[Cl]_o$ were both in hippocampus (e.g., Zanotto and Heinemann, 1983 and in preparation) and neocortex much larger than the $[Ca]_o$ changes, suggesting that the main charge carrier is Na but that K and Cl fluxes are evoked. If the ionic changes are added up to find the osmotic changes, then a decrease of extracellular osmolarity is expected. If electroneutrality requirements are taken into account, a small increase in bicarbonate concentration can be expected. However, these are not sufficient to cause isomolarity. Thus net shifts of osmotic particles into cells must have occurred. It was of interest to measure the ES size changes under these conditions. Typical changes of TMA signal during iontophoretic application of NMDA in cat cortex are illustrated in Fig. 5. They reveal that the ES transiently decreases by up to 80%. During both application of NMDA or kainate in vitro, this large decrease in the size of the ES is also described but it is transient in nature. The ES decrease recovers partially and reaches a steady state near 50%. It is of interest to note that the ES size changes were much smaller in the hippocampal PC layer where the volume fraction is small. Similar ES size changes were also noted by Engberg et al. (1983) in the cat spinal cord and by Heinemann and Pumain (1980) in cat neocortex. The question arose to what extent these ES size changes are caused by a net uptake of NaCl into nerve cells. The maximum decreases in $[Na]_o$ and $[Cl]_o$ were in the order of 30 mosm. When the steady state maximum decrease in ES size by 50% is taken into account, nerve cells must have taken up 300 mosm of Na and Cl. That certainly explains the steady state ES decrease. The transient component, however, cannot be explained on this base and it is likely that, particularly in case of iontophoretic application from a point source, spatial K buffering plays an additional role. With respect to the cytotoxic action of excitatory amino acids, two hypotheses can be formulated. It has been suggested that nerve cell degeneration depends on an excessive load with Ca during excitation (Farber, 1981; Siesjö, 1981; Harris and Symon, 1984). Alternatively, nerve cell degeneration may depend on the decrease in ES size, the load with Na and Cl and the increased demand for oxygen, which possibly cannot be met under these conditions.

At present, this question is difficult to answer. As a first approach, however, we have studied the effects of aspartate and kainate perfusion on hippocampal slices maintained in vitro. Both elicit Ca decreases, when iontophoretically or bath applied (Evans et al., 1984; Griffiths et al., 1982; Marciani et al., 1982; Pumain and Heinemann, 1985). When Ca was present in the perfusate during 20 minutes of 0.5 mM kainate or 20 mM aspartate application, antidromically and orthodromically evoked cellular responses disappeared while Ca transiently decreased. Field potentials did not recover for five hours with the exception of afferent volleys. However, when kainate or aspartate were applied in low Ca media with blocked synaptic transmission and ongoing epileptiform activity (Wong and Traub, 1983; Haas and Jefferys, 1984), there was partial recovery of synaptically evoked responses to about 50% of the original size. This suggests that Ca is a cofactor in cell death, induced by excitatory amino acids and probably also during status epilepticus, where the situation is aggravated by local edema and its effects on local oxygen supply (Marciani et al., 1982; Lassmann et al., this volume). Since spatial K buffering natures late during development (Connors et al., 1982), and since K accumulation facilitates Ca load of nerve cells, it is expected that perinatal brain vulnerability due to oxygen lack is dependent on similar mechanisms.

ACKNOWLEDGEMENTS

Supported by DFG grants He 1128/2-3, 3-2 and SFB 220. The skillful technical assistance of Ms. G. Schuster and B. Meinhadt in the experiments and preparation of the MS is gratefully acknowledged.

REFERENCES

Bourke, R.S., and Nelson, L.R., 1972, Further studies on the K^+-dependent swelling of primate cerebral cortex in vivo: the enzymatic basis of the K^+-dependent transport of chloride, J. Neurochem., 19:663.

Collins, R.C., and Olney, J.W., 1982, Focal cortical seizures cause distant thalamic lesions, Science, 218:177.

Connors, B.W., Ransom, B.R., Kunis, D.M., and Gutnick, M.J., 1982, Activity-dependent K^+ accumulation in the developing rat optic nerve, Science, 216:1341.

Dietzel, I., Heinemann, U., Hofmeier, G., and Lux, H.D., 1980, Transient changes in the size of extracellular space in the sensorimotor cortex of cats in relation to stimulus-induced changes in potassium concentration, Exp. Brain Res., 40:432.

Dietzel, I., Heinemann, U., Hofmeier, G., and Lux, H.D., 1982, Stimulus-induced changes in extracellular Na^+ and Cl^- concentration in relation to changes in the size of the extracellular space, Exp. Brain Res., 46:73.

Dingledine, R., 1983, N-methyl aspartate activates voltage-dependent calcium conductance in rat hippocampal pyramidal cells, J. Physiol., 343:385.

Engberg, I., Flatman, J.A., Lambert, J.D.C., and Lindsay, A., 1983, An analysis of bioelectrical phenomena evoked by microiontophoretically applied excitotoxic amino acids in the feline spinal cord, in: Excitotoxins, K. Fuxe, P.J. Roberts and R. Schwarcz, eds., Macmillan Press, London, p. 170.

Evans, M.C., Griffiths, T., and Meldrum, B.S., 1984, Kainic acid seizures and the reversibility of calcium loading in vulnerable neurones in the hippocampus, Neuropathol. Appl. Neurobiol., 10:285.

Farber, J.L., 1981, The role of calcium in cell death, Life Sci., 29:1289.

Gardner-Medwin, A.R., 1983a, A study of the mechanisms by which potassium moves through brain-tissue in the rat, J. Physiol., 355:353.

Gardner-Medwin, A.R., 1983b, Analysis of potassium dynamics in mammalian brain tissue, J. Physiol., 335:393.

Griffiths, T., Evans, M.C., and Meldrum, B.S., 1982, Intracellular sites of early Ca^{2+} accumulation in the rat hippocampus during status epilepticus, Neurosci. Lett., 30:323.

Gutnick, M.J., Connors, B.W., and Ransom, B.R., 1981, Dye-coupling between glial cells in the guinea pig neocortical slice, Brain Res., 213:486.

Haas, H.L., and Jefferys, G.R., 1984, Low-calcium field burst discharges of CA1 pyramidal neurones in rat hippocampal slices, J. Physiol., 354:185.

Harris, R.J., and Symon, I., 1984, Extracellular pH, potassium and calcium activities in progressive ischaemia of rat cortex, J. Cereb. Blood Flow Metab., 4:178.

Heinemann, U., and Pumain, R., 1980, Extracellular calcium activity changes in cat sensorimotor cortex induced by iontophoretic application of amino acids, Exp. Brain Res., 40:247.

Heinemann, U., Neuhaus, S., and Dietzel, I., 1983, Aspects of K regulation in normal and gliotic brain tissue, in: Cerebral Blood Flow, Metabolism and Epilepsy, M. Moulinier and B.S. Meldrum, eds., John Libbez and Co., London, p. 271.

Heinemann, U., and Dietzel, I., 1984, Extracellular potassium concentration in chronic alumina cream foci of cats, J. Neurophysiol., 52:421.

Heinemann, U., Konnerth, A., Pumain, R., and Wadman, W., 1986, Alterations of extracellular calcium sinks in chronic epileptic brain tissue, in: Basic Mechanisms of the Epilepsies: Cellular and Molecular Aspects, A.V. Delgado-Escueta, D.M. Woodbury, and K. Penry, eds., Raven Press, New York, in press.

Kimelberg, H.K., and Ransom, B.R., 1986, Physiological aspects of astrocytic swelling, in: Astrocytes, S. Federoff and A. Vernadakis, eds., Academic Press, New York, in press.

Konnerth, A., and Heinemann, U., 1983, Effects of GABA on presumed presynaptic Ca^{2+} entry in hippocampal slices, Brain Res., 270:185.

Kettenmann, H., Sonnhof, U., and Schachner, M., 1983, Exclusive K^+ dependence of the membrane potential in cultured oligodendrocytes, J. Neurosci., 3:506.

Krnjević, K., Morris, M.E., Reiffenstein, R.J., and Ropert, N., 1982, Depth distribution and mechanism of changes in extracellular K^+ and Ca^{2+} concentrations in the hippocampus, Can J. Physiol. Pharmacol., 6:1958.

Lambert, J.D.C., and Heinemann, U., 1986, The involvement of Ca^{2+} and Mg^{2+} in responses of hippocampal CA1 neurones to excitatory amino acids, in: Ca^{2+} Electrogenesis and Neuronal Functioning, U. Heinemann, M. Klee, E. Neher, and W. Singer, eds., Springer-Verlag, Heidelberg, Tokyo, New York, in press.

Lassmann, H., Petsche, U., Kitz, K., Baran, H., Sperk, G., Seitelberger, F., and Hornykiewicz, O., 1984, The role of brain edema in epileptic brain damage induced by systemic kainic acid injection, Neuroscience, 13:691.

Louvel, J., 1986, Effects of a calcium channel blocker in 'normal' and 'epileptic' hippocampal slices, in: Ca^{2+} Electrogenesis and Neuronal Functioning, U. Heinemann, M. Klee, E. Neher, and W. Singer, eds., Springer Verlag, Heidelberg, Tokyo, New York, in press.

Lux, H.D., and Neher, E., 1973, The equilibrium time course of $[K]_o$ in cat cortex, Exp. Brain Res., 17:190.

Lux, H.D., Heinemann, U., and Dietzel, I., 1986, Ionic changes and changes in the size of the extracellular space in brain cortex during epileptic activity, in: Basic Mechanisms of the Epilepsies: Cellular and Molecular Aspects, A.V. Delgado-Escueta, D.M. Woodbury and K. Penry, eds., Raven Press, New York, in press.

Marciani, M.G., Louvel, J., and Heinemann, U., 1982, Aspartate induced changes in extracellular free calcium in vitro hippocampal slices of rats, Exp. Brain Res., 238:272.

Meldrum, B.S., and Brierley, J.B., 1973, Prolonged epileptic seizures in primates: ischaemic cell change and its relation to ictal physiological events, Arch. Neurol., 28:10.

Nicholson, C., and Phillips, J.M., 1981, Ion diffusion modified by tortuosity and volume fraction in the extracellular microenvironment of the rat cerebellum, J. Physiol., 321:225.

Nowak, L., Bregestovski, P., Ascher, P., Herbet, A., and Prochiantz, A., 1984, Magnesium gates glutamate-activated channels in mouse central neurones, Nature, 307:462.

Orkand, R.K., Nicholls, J.G., and Kuffler, S.W., 1966, Effect of nerve impulses on the membrane potential of glial cells in the central nervous system of amphibia, J. Neurophysiol., 29:788.

Orkand, R.K., Cole, J.A., and Tsacopolous, M., 1986, The role of the glial cells in ion homeostasis in the retina of the honeybee drone, in: Calcium Electrogenesis and Neuronal Function, U. Heinemann, M. Klee, E. Neher and W. Singer, eds., Springer-Verlag, Heidelberg, Tokyo, New York, in press.

Pumain, R., and Heinemann, U., 1985, Stimulus-evoked and amino acid induced ionic changes in rat neocortex, J. Neurophysiol., 53:1.

Siesjö, B.K., 1981, Cell damage in the brain: a speculative synthesis, J. Cereb. Blood Flow Metab., 1:155.

Somjen, G.G., 1980, Stimulus-evoked and seizure-related responses of extracellular calcium activity in spinal cord compared to those in cerebral cortex, J. Neurophysiol., 44:617.

Somjen, G.G., 1983, Electrogenesis of sustained potentials, Prog. Neurobiol., 1:201.

Sperk, G., Lassmann, H., Baran, H. Kish, S.J., Seitelberger, F., and Hornykiewicz, O., 1983, Kainic acid induced seizures: neurochemical and histopathological changes, Neuroscience, 10:1301.

Walz, W., and Hertz, L., 1983, Functional interactions between neurons and astrocytes. II. Potassium homeostasis at the cellular level, Neurobiology, 20:133.

Watkins, J.C., 1984, Excitatory amino acids and central synaptic transmission, TIPS, 5:373.

Wong, R.K.S., and Traub, R.D., 1983, Synchronized burst discharge in disinhibited hippocampal slice. I. Initiation in CA2-CA3 region, J. Neurophysiol., 49:442.

Yaari, Y., Konnerth, A., and Heinemann, U., 1986, Nonsynaptic epileptogenesis at the mammalian hippocampus in vitro. II. Role of extracellular potassium, J. Neurophysiol., in press.

Zanotto, L., and Heinemann, U., 1983, Aspartate and glutamate induced reductions in extracellular free calcium and sodium concentration in area CA1 of in vitro hippocampal slices of rats, Neurosci. Lett., 35:79.

COMMENTARY - MECHANISMS OF EPILEPTOGENESIS

K. Krnjević and U. Heinemann

DISCUSSION POINTS

Questions were raised concerning the linear I/V relations and lack of voltage-dependent conductance changes observed by Johnston during both paroxysmal depolarizing shifts (PDS) and EPSPs, which suggest that NMDA receptors are not involved. Dingledine in particular had observed a substantial depression of PDS amplitude by amino phosphonovaleric acid, a well-known NMDA antagonist. The conclusion that the PDS is a 'network' property was queried by Ascher, who wondered whether a single neuron, with many terminals on a given cell, might not be able to generate a giant EPSP.

The characteristics of the persistent, low-threshold non-inactivating Na current (INa_p), described by Crill, prompted the question whether it might be produced by the persistence of extracellular glutamate, at a low concentration. One argument against this possibility is the TTX sensitivity of INa_p.

Questions that were asked about the apparent ephaptic phenomena observed by Krnjević in the hippocampus, *in situ*, included whether they could be accounted for either by gap junctions or by failure of antidromic invasion. Various lines of evidence against these possibilities were reviewed. It was emphasized that there was no obvious difference between CA1 and CA3 with regard to ephaptic potentials. Significant ephaptic interactions may well occur in other parts of the brain, though quite possibly of much smaller magnitude than in the hippocampus, where extracellular fields are much larger than, for example, in the neocortex.

The interesting block of NMDA-activated ionic channels by ketamine reported by MacDonald led to some discussion concerning the possible use of ketamine for the treatment of convulsive disorders.

Preliminary data obtained by Pumain with Mg^{2+}-electrodes indicate a resting, free Mg^{2+}-level in the extracellular space of about 2 mM. Because of the very high Ca^{2+}-sensitivity of these electrodes, quantitative estimates of changes in $[Mg^{2+}]_o$, during activity or during excitatory amino acid action, would not be readily obtained because of the large concurrent falls in $[Ca^{2+}]_o$; though significant falls in $[Mg^{2+}]_o$ were in fact reported by Heinemann, which he thought could reinforce appreciably the voltage dependence of NMDA actions reported by Ascher and colleagues.

SESSION VII
EXCITATORY AMINO ACIDS: PHYSIOLOGICAL STUDIES

EVIDENCE FOR THE ACTIVATION OF THE N-METHYL-D-ASPARTATE RECEPTOR DURING EPILEPTIFORM DISCHARGE

G.L. King and R. Dingledine

Department of Pharmacology
University of North Carolina at Chapel Hill
Chapel Hill, N.C. 27514

INTRODUCTION

N-methyl-D-aspartate (NMDA) receptors appear to be the most easily characterized class of excitatory amino acid receptors in the central nervous system. Receptor activation by NMDA is blocked by certain phosphonate compounds and is sensitive to the extracellular Mg^{2+} concentration, characteristics lacking in 'non-NMDA' receptors such as for quisqualate and kainate (for review, see Watkins, 1984). Although NMDA receptors are abundant in the hippocampus (Monaghan et al., 1983; Greenamyre et. al., 1985), putative NMDA antagonists have little or no effect on excitatory synaptic transmission in this structure under normal conditions (see Coan and Collingridge, 1985). Several investigators have therefore suggested that NMDA receptors are activated under abnormal or unusual conditions (Harris et al., 1984; Coan and Collingridge, 1985; Crunelli et al., 1985). For example, NMDA antagonists prevent long-term potentiation (Collingridge et al., 1983b), protect against CNS ischemia (Simon et al., 1984) and suppress some seizures (for review, see Meldrum, 1984). In addition, iontophoretic application of NMDA onto either mammalian neocortical or caudate neurons produces a pattern of burst discharge similar to that observed in an epileptic focus (Flatman et al., 1983; Herrling et al., 1983).

Our laboratory has recently shown that activation of an NMDA receptor in CA1 pyramidal neurons of the hippocampus produces a voltage-dependent depolarization that can reach threshold for Ca^{2+} spikes (Dingledine, 1983). Since the CA1 dendritic region shows both a high density of NMDA receptors (Monaghan et al., 1983; Greenamyre et al., 1985) and slow Ca^{2+} spikes that may contribute to the epileptiform discharge (Wong and Prince, 1979), we reasoned that NMDA receptor activation might occur in hippocampal slices treated with either bicuculline or picrotoxin to block inhibitory postsynaptic potentials (IPSPs). To test this hypothesis, we challenged the epileptiform burst in this 'disinhibited' model with the D- and L-isomers of 2-amino-5-phosphonovaleric acid (APV), an NMDA antagonist, and several other excitatory amino acid antagonists.

All intra- and extracellular recordings reported here were from CA1 pyramidal neurons of rat hippocampal slices, in the presence of either bicuculline (10-100 μM) or picrotoxin (100 μM), except where noted. Experiments were performed in either an interface or submersion chamber and compounds were applied in known concentration by either perfusion or pressure ejection. The specific methods are detailed elsewhere (Valentino and Dingledine, 1982; King et al., 1985) and some of these results have appeared in abstract form (Hynes and Dingledine, 1984; King and Dingledine, 1985).

EXCITATORY SYNAPTIC PHYSIOLOGY IN THE ABSENCE OF INHIBITORY SYNAPTIC TRANSMISSION

Fig. 1 illustrates the effect of APV on an epileptiform burst evoked by Schaffer-commissural system stimulation in the presence of picrotoxin. The major effect of APV is an attenuation of the burst duration, primarily the late portion encompassing the paroxysmal depolarizing shift (PDS). A similar effect was observed on spontaneously occurring bursts, and in the presence of bicuculline. Since the later, rather than the early phase of the PDS was attenuated, we tested for the possibility that APV acted

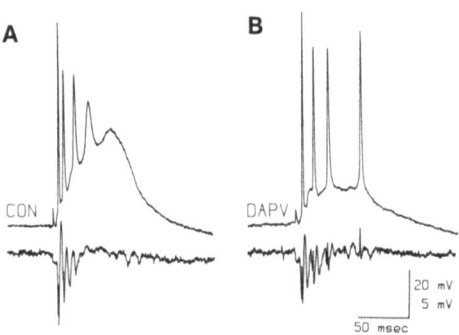

Fig. 1. APV reduces the PDS in disinhibited hippocampal slices. In the presence of 100 μM picrotoxin, epileptiform bursts were recorded in CA1 in response to 0.1 Hz stimulation of the Schaffer-commissural system. Upper: Intracellular recording from neuron with -61 mV resting membrane potential and 40 MΩ input resistance; electrode filled with 2M KMeSO$_4$. Lower: Nearby extracellular recording; high frequency components within this trace represent coupling artifacts between the recording electrodes. All traces are single sweeps. A. Control. B. After pressure ejection of 100 μM D-APV onto the slice surface.

Table 1. Effects of Micromolar APV on Properties of CA1 Pyramidal Neurons

Measure	Response to APV	n
Extracellular		
Input-Output Curve		
Prevolley to EPSP[a]	no change	4
EPSP to population spike[a]	no change	4
1st Population spike amplitude	no change	10
Coastline bursting index	decreased	12
Intracellular		
Input resistance	no change	24
Membrane potential	no change	31
Ca^{2+} spikes [b,c]	no change	6
Cs-TEA burst[c]	no change	10
Spike frequency adaptation[c]	no change	6
NMA depolarizing response[b]	decreased	9
Synaptic PDS	decreased	13
EPSP amplitude	decreased[d]	12

For these experiments we used D- or DL-APV in the presence of bicuculline or picrotoxin except where noted.
[a] normal perfusate
[b] tetrodotoxin
[c] in response to a depolarizing current pulse
[d] greater effect at more depolarized potentials
Cs-TEA: cesium-chloride and tetraethylammonium

on some intrinsic membrane mechanism(s) suggested to contribute to this portion of the burst. Our results are summarized in Table 1. We found that 100 μM APV did not alter a train of action potentials produced by intracellular depolarizing current, thus APV does not impair the neuron's ability to fire a short train of action potentials. APV also did not affect an intrinsic Ca^{2+} spike produced by intracellular depolarizing current in the presence of TTX, but did abolish the Ca^{2+} spike response to iontophoresed NMA. In addition, APV partially reduced the long-duration burst in response to orthodromic stimulation, but not depolarizing current pulses, when the microelectrodes were filled with TEA-CsCl. Thus, the action of APV appears specific for an event that occurs during synaptic transmission. We also found no change in either membrane potential or input resistance in response to APV.

To determine whether APV attenuated the excitatory postsynaptic potential (EPSP) or otherwise adversely affected synaptic transmission, we recorded extracellular input-output curves of both the prevolley-EPSP and EPSP-population spike relationship in normal physiological conditions. These results, like those of Koerner and Cotman (1982) and Collingridge et al. (1983b), showed that APV has no effect on either monosynaptic excitatory synaptic transmission or the threshold for excitation in these cells. However, in the presence of picrotoxin, 10 μM APV reduced the amplitude of intracellularly-recorded EPSPs produced by weak Schaffer-commissural system stimulation (subthreshold for a population spike), a result also reported by Hablitz and Langmoen (1984). In all eight neurons tested, this reduction in EPSP amplitude by APV was greater at more depolarized membrane potentials (Fig. 2). Since NMA responses are voltage-dependent

(Dingledine, 1983; Nowak et al., 1984; Mayer and Westbrook, 1985), this further suggests that APV acts on an NMDA receptor in these conditions. These data also suggest that the postsynaptic receptor responsible for the initiation of the epileptiform burst is distinct from that receptor responsible for the conventional EPSP.

EXCITATORY AMINO ACID PHARMACOLOGY IN THE ABSENCE OF INHIBITORY SYNAPTIC TRANSMISSION

The effect of APV on the extracellular concomitant of the intracellular burst was similar to that seen in Fig. 1. That is, micromolar concentrations of APV reduced the duration of the extracellular burst, with the predominant observation a reduced amplitude of the last few spikes in the burst (Fig. 3A, 5). In order to quantitate this drug action, we used an on-line computer program to measure the length of the waveform in the epileptiform burst (coastline bursting index) following the second population spike. This method of analysis is similar to a coastline measurement and is illustrated in Fig. 3. We also simultaneously measured the amplitude of the first

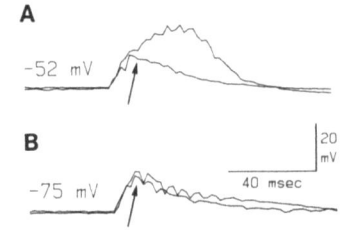

Fig. 2. APV reduces EPSPs in a voltage-dependent manner. Intracellular recording from CA1 neuron with -58 mV resting membrane potential and 45 MΩ input resistance. EPSPs were evoked by a weak constant-current orthodromic stimulus at 0.1 Hz in the presence of 100 μM picrotoxin. Membrane potential was varied by passing constant-current through the microelectrode; electrode filled with a combination of 2 M $KMeSO_4$ and 50 mM QX-314, a quaternary lidocaine derivative that blocks sodium spikes. All traces are single sweeps. Arrows in both A and B indicate the EPSP after perfusion with 10 μM D-APV.

population spike. All experiments were done in a submersion chamber to facilitate rapid drug delivery to the slice. Drug effects were observed within 30 seconds of administration and plateau effect was stable within two minutes. For all experiments, the Schaffer-commissural system was stimulated at 0.1 Hz at an intensity sufficient to elicit at least five population spikes.

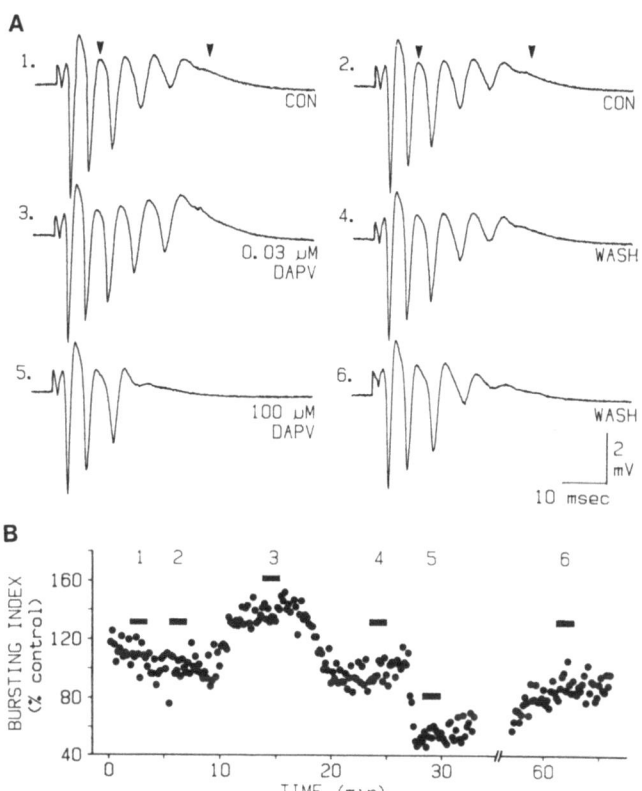

Fig. 3. Method for determining excitatory amino acid antagonist effect on coastline bursting index in disinhibited slices. A. Sample records of CA1 population spikes evoked by 0.1 Hz Schaffer-commissural system stimulation and exposed to different concentrations of D-APV. The coastline bursting index was measured as the length of the burst waveform line between the two arrowheads. Each numbered trace represents an average of ten sweeps and corresponds to the number in B. B. Sample record of coastline bursting index (filled circles), plotted as per cent control baseline (in this instance, the mean value before and after perfusion with 0.03 μM D-APV). Each black bar denotes those coastline bursting-index points averaged to produce the corresponding traces in A. D-APV (0.03 μM) was perfused between 9 and 17 minutes, and 100 μM between 27 and 32 minutes.

Fig. 4 is a cumulative dose-response curve for the effects of the D- and L-isomers of APV on the coastline bursting index. That the D-isomer was more effective than the L-form in reducing the coastline bursting index is reminiscent of APV stereospecificity noted by McLennan (1982), and further supports the notion of NMA receptor activation during an epileptiform burst. That nanomolar concentrations of both D- and L-APV increased the intensity of the coastline bursting index (see also Fig. 3B, D) was surprising. It is possible that APV acts to activate and then block the NMDA receptor, in a manner similar to that observed for nicotine at the ACh receptor. Alternatively, APV could be acting at another excitatory amino acid receptor. Collingridge et al. (1983a) reported that iontophoretically applied D-APV, more so than DL-APV, could potentiate iontophoretic responses to kainate and quisqualate.

In order to further characterize the specificity of NMDA involvement in the epileptiform burst in disinhibited slices, we tested several other excitatory amino acid antagonists at a fixed 100 µM concentration on the coastline bursting index. The results are summarized in Table 2. As seen here, D-APV was the most effective antagonist in reducing the coastline

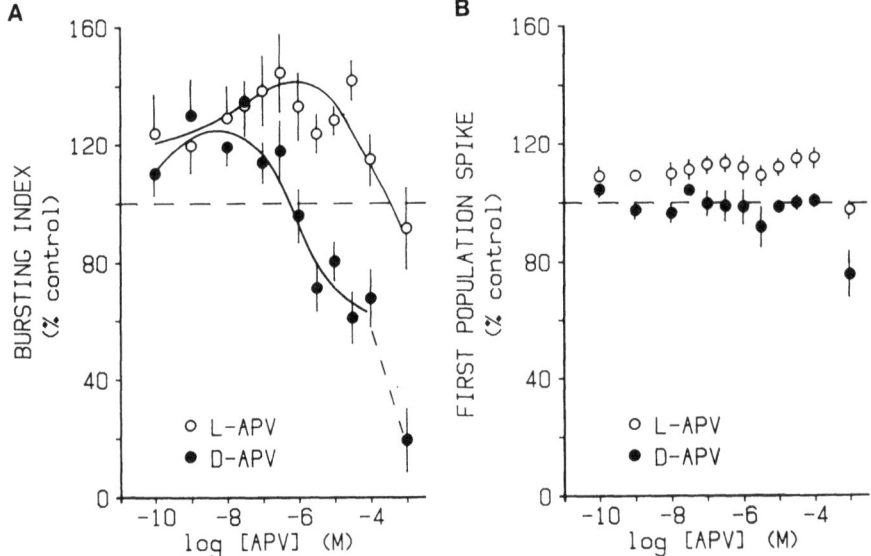

Fig. 4. Dose-response curves for D- and L-APV effect on coastline bursting index. Both coastline bursting index (A) and the amplitude of the first population spike (B) are expressed as percent control in response to sequentially increasing concentrations of APV. All values were averaged for a period of 2-5 minutes in response to 0.1 Hz stimulation of the Schaffer-commissural system. Plotted values represent the mean $\pm$ 1 S.E.M. from 6-11 slices. CA2/3 was removed from slices to prevent activation of this region by Schaffer-commissural stimulation, which in turn could reexcite CA1 cells (Hablitz, 1984). Methods are described in Fig. 3. A. The coastline bursting index is increased by nanomolar and decreased by micromolar concentrations of D-APV; L-APV is less than 100X as potent as D-APV in reducing the coastline bursting index. B. The amplitude of the first population spike is unchanged from control by D-APV except at concentrations greater than 100 µM, whereas that amplitude is greater than control for almost all concentrations of L-APV.

Table 2. Attenuation of Evoked Epileptiform Burst by Excitatory Amino Acid Antagonists

Antagonist	Bursting Index (% Control)	1st Pop. Spike (% Control)	N
D-APV	67.8 ± 9.8	100.9 ± 2.0	9
DL-APV	78.8 ± 5.4	99.9 ± 1.6	7
L-APV	111.1 ± 8.2	112.1 ± 4.7	7
γ-DGG	92.0 ± 6.3	93.4 ± 3.2	7
DAP	100.9 ± 9.4	94.7 ± 3.3	7
GDEE	100.1 ± 3.6	96.1 ± 3.1	7
GAMS	100.5 ± 4.4	98.8 ± 2.4	7

Bursts of five or more population spikes were evoked in the CA1 region in the presence of either bicuculline (10-100 μM) or picrotoxin (100 μM). The coastline bursting index (see text and Fig. 3) was measured in control conditions and then after perfusion with an excitatory amino acid antagonist at a concentration of 100 μM. The number of slices tested is shown as N and the values represent the mean ± 1 S.E.M.

D-APV: D-2-amino-5-phosphonovaleric acid
DL-APV: DL-2-amino-5-phosphonovaleric acid
L-APV: L-2-amino-5-phosphonovaleric acid
γ-DGG: γ-D-glutamylglycine
DAP: α,ε-diaminopimelic acid
GDEE: L-glutamic acid diethyl ester
GAMS: γ-D-glutamylaminomethyl sulphonic acid

bursting index. In addition, the rank order of potency for these agents was similar to that proposed by Watkins and Evans (1981) for action at an NMDA receptor. An unexpected finding was the increased amplitude of the first population spike in the presence of L-APV, although this was not dose related (see Fig. 4B).

A ROLE FOR THE NMDA RECEPTOR DURING REDUCED INHIBITORY SYNAPTIC TRANSMISSION

The principal finding of these studies is that, in disinhibited hippocampal slices, the late portion of an epileptiform burst recorded from CA1 pyramidal neurons is attenuated stereospecifically by micromolar concentrations of the NMDA antagonist, D-APV. APV appears to be selective for only synaptic events (Table 1) and is more effective than other excitatory amino acid antagonists in its action (Table 2). These results strongly suggest that NMDA receptor activation plays an important role in the manifestation of an epileptiform burst in the disinhibited hippocampal slice. The question then arises as to what suppresses the expression of NMDA receptor activation under normal (i.e., non-epileptic) conditions?

Based on our results, we propose the following model for events during an epileptiform burst (Fig. 5). Under normal physiological conditions, activation of the Schaffer-collateral axons produces the normal EPSP-IPSP sequence in CA1 cells that results from release of excitatory and inhibitory neurotransmitters (Fig. 5A,B,C). The feed-forward dendritic inhibitory current (Alger and Nicoll, 1982) shunts the dendritic excitatory current and prevents dendritic membrane depolarization to a potential that would activate the NMDA receptor. In disinhibited slices, however, the postsynaptic inhibitory current is reduced, and the dendritic membrane depolarizes to a potential that would activate the NMDA receptor. Application of an NMDA antagonist would partially suppress this response and facilitate termination of the burst.

This model provides an explanation for the observation that NMDA receptors do not contribute to normal synaptic transmission (Koerner and Cotman, 1981; Collingridge et al., 1983b) but do contribute to its more plastic properties. The NMDA receptor activation that has been implicated in LTP (Collingridge et al., 1983b) might be explained as a consequence of reduced GABA-ergic inhibition in response to tetanic stimuli. McCarren and Alger (1985) have observed that at stimulus frequencies used to induce LTP, the IPSP conductance is reduced. This reduction in inhibition could, in turn, result in NMDA receptor activation. The degree to which NMDA receptor activation is involved in other models of epileptogenesis remains to be studied. Our results further support the idea that a PDS results from a giant EPSP (Johnston and Brown, 1981; Rutecki et al., 1984).

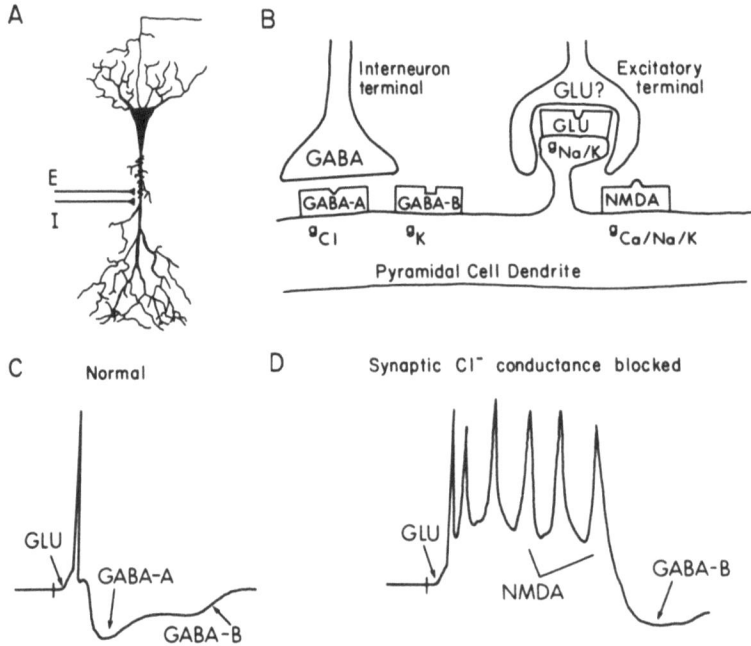

Fig. 5. Model illustrating the contribution of NMDA receptor activation to the epileptiform burst in disinhibited slices. A. Drawing of CA1 pyramidal neuron with dendritic excitatory (E) and inhibitory (I) synapses. B. Magnification of pyramidal cell dendritic region illustrated in A. The putative neurotransmitter receptors associated with each nerve terminal are in bold face type (GABA: γ-amino-butyric acid; GLU: glutamate; NMDA: N-methyl-D-aspartate). The ionic conductances associated with activation of each receptor are shown beneath each. C., D. Representation of electrophysiological consequences of activating specific receptor subtypes under normal (C) and disinhibited (D) conditions. The normal GABA-A response (an IPSP) is blocked with bicuculline or picrotoxin, which permits the expression of NMDA receptor activation. Further details in text.

In summary, we have shown that micromolar concentrations of D-APV have a specific and selective effect in reducing part of the epileptiform burst in disinhibited pyramidal cells. In addition, the potency of D-APV action is greater than that of any other excitatory amino acid antagonist tested. That nanomolar concentrations of APV, however, increase the duration of burst-firing of pyramidal neurons, suggests the involvement of a different receptor. We believe these data suggest that NMDA receptor activation can contribute to the PDS in epileptiform discharge and propose a model for how this occurs. We conclude that synaptic excitation, as well as intrinsic membrane properties, can contribute to the full expression of epileptiform bursting in disinhibited hippocampal pyramidal cells. In addition, the postsynaptic receptor responsible for the epileptiform burst is distinct from that which mediates the conventional EPSP.

ACKNOWLEDGEMENTS

We gratefully acknowledge the support of the National Institutes of Health (NS 17771, DA 12360 and NS 06953).

REFERENCES

Alger, B.E., and Nicoll, R.A., 1982, Feed-forward dendritic inhibition in rat hippocampal pyramidal cells studied in vitro, J. Physiol., 328:105.

Coan, E.J., and Collingridge, G.L., 1985, Magnesium ions block an N-methyl-D-aspartate receptor-mediated component of synaptic transmission in rat hippocampus, Neurosci. Lett., 53:21.

Collingridge, G.L., Kehl, S.J., and McLennan, H., 1983a, The antagonism of amino acid-induced excitations of rat hippocampal CA1 neurones in vitro, J. Physiol., 334:19.

Collingridge, G.L., Kehl, S.J., and McLennan, H., 1983b, Excitatory amino acids in synaptic transmission in the Schaffer collateral-commissural pathway of the rat hippocampus, J. Physiol., 334:33.

Crunelli, V., Forda, S., and Kelly, J.S., 1985, Excitatory amino acids in the hippocampus: synaptic physiology and pharmacology, TINS, 8:26.

Dingledine, R., 1983, N-methyl aspartate activates voltage-dependent calcium conductance in rat hippocampal pyramidal cells, J. Physiol., 343:385.

Flatman, J.A., Schwindt, P.C., Crill, W.E., and Stafström, C.E., 1983, Multiple actions of N-methyl-D-aspartate on cat neocortical neurons in vitro, Brain Res., 266:169.

Greenamyre, J.T., Olson, J.M.M., Penny, Jr., J.B., and Young, A. B., 1985, Autoradiographic characterization of N-methyl-D-aspartate, quisqualate- and kainate-sensitive glutamate binding sites, J. Pharmacol. Exp. Therap., 233:254.

Hablitz, J.J., 1984, Picrotoxin-induced epileptiform activity in hippocampus: role of endogenous versus synaptic factors, J. Neurophysiol., 51:1011.

Hablitz, J.J., and Langmoen, I.A., 1984, Possible NMDA receptor mediation of synaptic transmission in the hippocampal CA1 region, Soc. Neurosci. Abstr., 10:415.

Harris, E.W., Ganong, A.H., and Cotman, C.W., 1984, Long-term potentiation in the hippocampus involves activation of N-methyl-D-aspartate receptors, Brain Res., 323:132.

Herrling, P.L., Morris, R., and Salt, T.E., 1983, Effects of excitatory amino acids and their antagonists on membrane and action potentials of cat caudate neurons, J. Physiol., 339:207.

Hynes, M.A., and Dingledine, R., 1984, Attenuation of epileptiform burst firing in the rat hippocampal slice by antagonists of N-methyl-D-aspartate receptors, Soc. Neurosci. Abstr., 10:229.

Johnston, D., and Brown, T.H., 1981, Giant synaptic potential hypothesis for epileptiform activity, Science, 211:294.

King, G.L., and Dingledine, R., 1985, Micromolar and nanomolar concentrations of D-2-amino-5-phosphonovalerate have opposite effects on epileptiform burst-firing in the rat hippocampal slice, Soc. Neurosci. Abstr., 11:103.

King, G.L., Knox, J.J., and Dingledine, R., 1985, Reduction of inhibition by benzodiazepine antagonist, Ro15-1788, in the rat hippocampal slice, Neuroscience, 15:371.

Koerner, J.F., and Cotman, C.W., 1982, Response of Schaffer collateral-CA1 pyramidal cell synapses of the hippocampus to analogues of acidic amino acids, Brain Res., 251:105.

Mayer, M.L., and Westbrook, G.L., 1985, The action of N-methyl-D-aspartic acid on mouse spinal neurones in culture, J. Physiol., 361:65.

Meldrum, B., 1984, Amino acid neurotransmitters and new approaches to anticonvulsant action, Epilepsia (Suppl. 12), 25:S140.

McCarren, M., and Alger, B.E., 1985, Use-dependent depression of IPSPs in rat hippocampal pyramidal cells in vitro, J. Neurophysiol., 53:557.

McLennan, H., 1982, The isomers of 2-amino-5-phosphonovalerate as excitatory amino acid antagonists - a reappraisal, Eur. J. Pharmacol., 79:135.

Monaghan, D.T., Holets, V.R., Toy, D.W., and Cotman, C.W., 1983, Anatomical distributions of four pharmacologically distinct ^{3}H-L-glutamate binding sites, Nature, 306:176.

Nowak, L., Bregestowski, P., Ascher, P., Herbet, A., and Prochiantz, A., 1984, Magnesium gates glutamate-activated channels in mouse central neurones, Nature, 307:462.

Rutecki, P.A., Lebeda, F.J., and Johnston, D., 1984, Elevated extracellular potassium- and 4-aminopyridine-induced epileptiform activity in CA3 hippocampal neurons, Soc. Neurosci. Abstr., 10:1.

Simon, R.P., Swan, J.H., Griffiths, T., and Meldrum, B.S., 1984, Blockade of N-methyl-D-aspartate receptors may protect against ischemic damage in the brain, Science, 226:850.

Valentino, R.J., and Dingledine, R., 1982, Pharmacological characterization of opioid effects in the rat hippocampal slice, J. Pharmacol. Exp. Therap., 223:502.

Watkins, J.C., 1984, Excitatory amino acids and central synaptic transmission, TIPS, 5:373.

Watkins, J.C., and Evans, R.H., 1981, Excitatory amino acid transmitters, Ann. Rev. Pharmacol. Toxicol., 21:165.

Wong, R.K.S., and Prince, D.A., 1979, Dendritic mechanisms underlying penicillin-induced epileptiform activity, Science, 294:1228.

EFFECTS OF KAINATE ON CA1 HIPPOCAMPAL NEURONS RECORDED IN VITRO

E. Cherubini, C. Rovira, M. Gho and Y. Ben-Ari

LPN, CNRS, Gif-sur-Yvette, F-91190 France
and INSERM-U29, 123 Bd de Port-Royal
Hôpital de Port-Royal, 75014 Paris, France

The endogenous excitatory amino acids aspartate and glutamate are considered the most likely candidates as putative transmitters in the mammalian CNS. They act on one or more receptors. On the basis of electrophysiological studies, using mainly spinal cord preparations, three classes of excitatory amino acid receptors have been proposed: NMDA, quisqualate and kainate (Watkins and Evans, 1981).

Kainate has received considerable attention because of its convulsant properties. Systemic or intracerebral injection of KA produces a seizure and brain damage syndrome in which the hippocampus and other limbic structures play a central role and which constitutes a good model for human temporal lobe epilepsy (Ben-Ari, 1985).

In the hippocampal slice preparation, kainate induces convulsant activity in CA1 and CA3 fields (Robinson and Deadwyler, 1981), however the mechanisms of action are controversial. Thus, Collingridge et al. (1983) found that superfusion or iontophoretic application of kainate produces a large and long lasting potentiation of the population spikes evoked in CA1 by Schaffer collateral-commissural stimulation and a reduction of the field EPSP recorded in stratum radiatum. These authors suggest that a potentiation of synaptic activity may underlie the ability of kainate to induce hippocampal seizures and brain damage (also see Aitken, 1985). On the other hand a presynaptic failure of the GABA mediated inhibition has been proposed as a crucial factor in the onset of kainate induced epileptiform activity (Fisher and Alger, 1984).

In these studies, cellular excitability is enhanced consequently to changes which occur in surrounding neurons and is mediated by synaptic interactions. But neuronal excitability also depends on intrinsic membrane properties of individual neurons in the absence of synaptic or electrical inputs (Hablitz and Johnston, 1981). The purpose of this study was to investigate membrane properties of CA1 hippocampal neurons, in the absence of synaptic activity.

We have used hippocampal slices with the technique described by Pepper and Henderson (1980). Stable long lasting intracellular recordings were made from 62 CA1 hippocampal neurons for periods up to four hours. The cells had resting membrane potentials greater than −55 mV (range −55 to −72 mV), action potentials from 68 mV to 105 mV, input resistance from

20 to 75 MΩ and time constants from 9.8 to 12 ms. The cells often discharged spontaneously at 1-2 Hz. Spontaneous synaptic activity was usually present (EPSPs or reversed IPSPs, with KCl electrodes); no attempt was made to investigate these properties in detail. The effects of kainate were tested on the passive and active membrane properties of these neurons. Superfusion of kainate (50-200 nM) in 30% of the cells induced a small depolarization of the membrane potential (5-10 mV) associated with a decrease in input resistance (<15%).

KAINATE REDUCES ANOMALOUS RECTIFICATION

The voltage-current (V-I) relationship obtained by measuring the transmembrane voltage deflections produced by hyperpolarizing current pulses of long duration (>200 ms) is not linear for values of membrane potential 15-20 mV more negative than rest. A drop in the hyperpolarizing electrotonic potential appears. This phenomenon is reminiscent of the anomalous rectification described by Hoston et al. (1979) and depends on the activation of a specific current called IQ (Halliwell and Adams, 1982). Kainate (50-200 nM) depressed, in all the cells tested (N = 6) the 'sag' in the electrotonic potential at values of membrane potential more negative than -70 mV making the V/I curve more linear (Fig. 1).

In two cases, this effect was associated with a small decrease in input resistance (<15%). This effect of kainate was still present when TTX (1 μM) was added to the bathing solution. It was also seen in the presence of TEA (5 mM) or when Ca^{2+} was replaced by Ba^{2+} (2 mM).

KAINATE REDUCES CALCIUM-MEDIATED EVENTS

An intracellular depolarizing current pulse of long duration (usually 400 ms) induces action potentials which are not maintained throughout the pulse but decline or accomodate. The end of the pulse is followed by an afterhyperpolarization (AHP). This accomodation depends mainly on the activation of some specific K^+ conductances such as the M-current and the Ca^{2+}-activated potassium conductance (Madison and Nicoll, 1984).

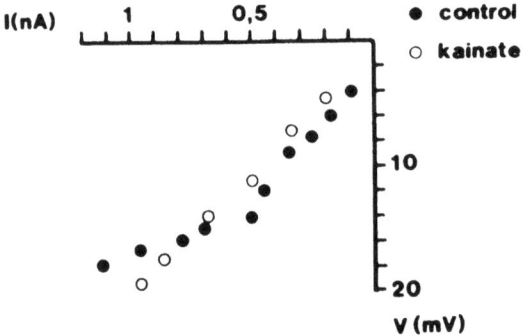

Fig. 1. Reduction of the anomalous rectification by kainate. V/I curve in control conditions (black circles) and during superfusion of kainate (150 nM, open circles). Resting potential: -71 mV.

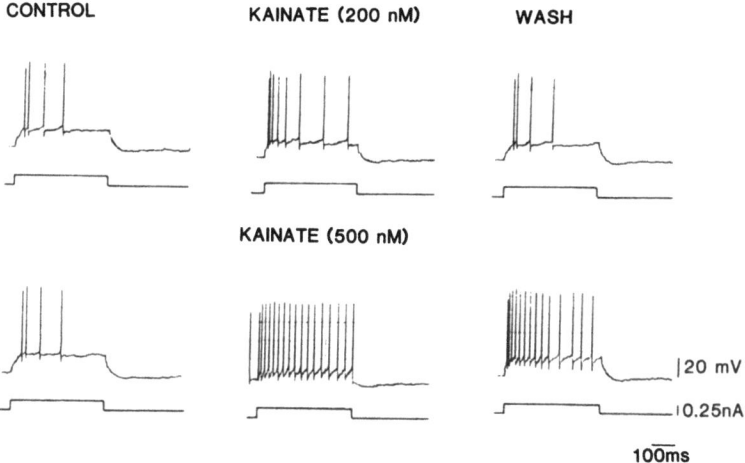

Fig. 2. Dose/dependent blockade of the accomodation by kainate. Response to a depolarizing current pulse (lower traces). No recovery is obtained after superfusion of 500 nM kainate.

Superfusion of kainate (100-500 nM) reduced or completely abolished spike accomodation in a dose-dependent way (Fig. 2). Kainate also reduced the duration of the AHP. In two cases, the amplitude and duration of the AHP were unchanged despite the increased number of action potentials. The action of kainate began within 30-40 seconds after the drug had reached the tissue and achieved its maximum effect within 2-3 minutes. Recovery was obtained within 20-30 minutes. With larger doses of kainate (300-500 µM), a complete washout was not found.

Since both accomodation and AHP are calcium mediated events, we have examined the effects of the toxin in the presence of selective calcium channel blockers. Bath application of Mn^{2+} (1-2 mM) strongly reduced the AHP and occasionally reduced the accomodation. Mn^{2+} blocked calcium entry into the cell since it abolished the calcium action potentials elicited in the presence of TTX (1 µm). In the presence of Mn^{2+}, the effects of kainate on the membrane potential, input resistance, spike accomodation and AHP were reduced or completely blocked. Similar observations were made in two neurons in which kainate was tested (and shown to reduce the AHP) before the application of Mn^{2+}.

Ca^{2+} spikes are readily generated in the presence of TTX (1 µM) and tetraethylammonium (TEA; 5 mM). In these conditions, depolarizing current pulses evoke one or more slow action potentials that are followed by an AHP. The AHP is quite variable in amplitude and duration and in many cases is characterized by two components, a fast peak (reached within a few milliseconds) and a slow one (decay in a few seconds; e.g., Gustafsson and Wigström, 1981). This second component is readily blocked by Mn^{2+} (Gustafsson and Wigström, 1981) or Cd^{2+} (Madison and Nicoll, 1984); divalent cations also block the calcium action potentials (Madison and Nicoll, 1984).

Bath application of kainate (50-200 nM) reduced the amplitude and duration of the AHP, without initially altering the number or shape of the calcium action potentials. The slow component of the AHP sensitive to calcium channel blockers was usually more affected by kainate. The reduction of the AHP (Fig. 3) was maximal within 2-3 minutes after the

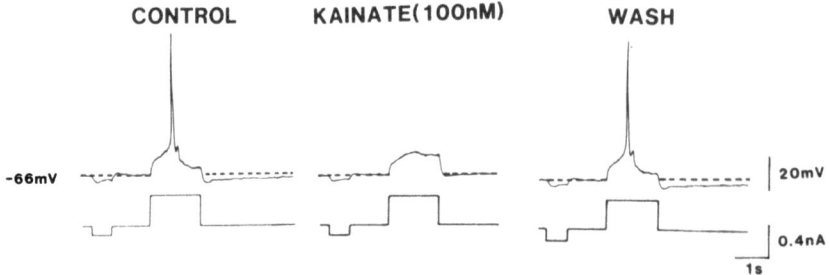

Fig. 3. Kainate blocks the calcium action potential and the subsequent AHP elicited by depolarizing current pulses (lower traces) in the presence of TTX (1 µM) and TEA (5 mM). The effect of kainate is associated with a small change in input resistance.

toxin had reached the bath. In 30% of the cases we observed initially (1-2 minutes) a temporary increase in the duration of the calcium action potentials. This was, however, always followed by a reduction of the number and size of the Ca^{2+} spikes. Eventually a complete blockade of these was achieved (Fig. 3).

At this stage, it was not possible to evoke an action potential even by increasing the strength of stimulation. The AHP that followed the action potential was still severely depressed. These effects were observed whether or not kainate caused any change in resting membrane potential or resistance. But at the doses used, a small depolarization (<10 mV) was observed only in 30% of the cases. In these circumstances, the action potentials were elicited at the same resting potential, after restoring it by passing a steady hyperpolarizing current through the recording electrodes. The effects of this drug washed out very slowly; sometimes 30-50 minutes were necessary to have a complete recovery. Kainate also reduced the calcium action potentials recorded with CsCl filled electrodes in the presence of TTX (1 µM). Kainate was applied after 20-30 minutes of stable recordings once the spikes had reached a maximal amplitude and duration. In these conditions (N = 4), there was no initial increase in the spike duration suggesting that this increase is probably due to a blockade of a potassium conductance.

NMDA ENHANCES CALCIUM MEDIATED EVENTS

To obtain some indications on the specificity of the effects of KA, we have tested the action of NMDA and compared these effects with those obtained with kainate. Superfusion of NMDA (1-2 µM) induced an increase in the number and amplitude of the Ca^{2+} evoked action potentials as well as subsequent AHP (Fig. 4 and Table 1).

The effect of NMDA was maximal within 2-3 minutes and washed out quite quickly in 10-20 minutes. NMDA caused always a membrane depolarization of 10-20 mV, and the action potentials were evoked always after restoring the membrane potential at the same level. In the presence of the specific NMDA antagonist D-APV (10-30 µM), the effects of NMDA were completely blocked, while D-APV had no effects by itself. D-APV did not antagonize the response to kainate.

Ba^{2+} can be substituted for Ca^{2+} in carrying inward currents and suppressing outward K^+ currents (Gormann and Hermann, 1979). In eight neurons, we substituted Ba^{2+} (2 mM) for Ca^{2+}. In these conditions, and

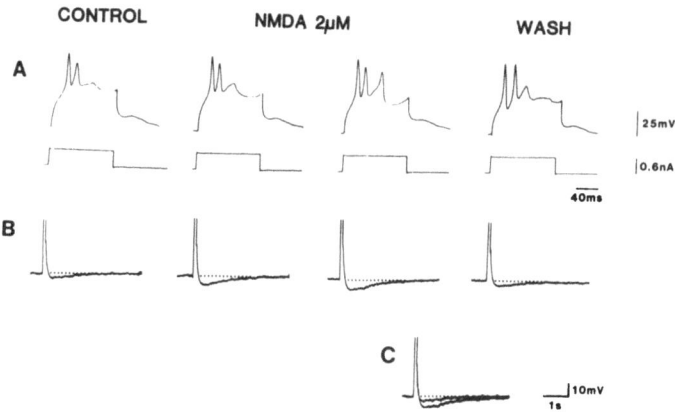

Fig. 4. NMDA enhances the number of calcium spikes and the amplitude and duration of the AHP.
A: Calcium spikes elicited in the presence of TTX (1 μM) and TEA (2 mM) by depolarizing current pulses (lower traces).
B: Same as A at a different time scale.
C: Comparison of the AHP in control and during superfusion of NMDA (2 μM).

in the presence of extracellular TTX (1 μM), spontaneous bursts occurred after a few minutes. These consisted of several regenerative events, triggered by a slow membrane depolarization or depolarizing current pulses. After 10 to 20 minutes in Ba^{2+} solution, injection of a steady depolarizing current triggered a long lasting action potential. This consisted of fast oscillations that terminated in a plateau lasting several seconds. The plateau potential was terminated by injecting an hyperpolarizing current. The voltage values for the threshold of spike activation and for the plateau were quite constant in different neurons, respectively, between -60 and -48 mV and +0 and +10 mV. These spikes were completely and reversibly abolished by superfusion of Co^{2+} (2 mM). When Ba^{2+} (2 mM) was applied for longer periods of time, a decrease in the number of the fast oscillations preceding the plateau occurred. These became progressively smaller. Then a steady depolarizing current or a depolarizing pulse evoked only one regenerative event followed by a long-lasting plateau (Fig. 5). Presumably at this stage the blockade of the outward K^+ currents by Ba^{2+} was more efficient.

Table 1. Effects of Kainate and NMDA on the AHP Following a Calcium Action Potential

	Control	Kainate (200 nM)	N	Control	NMDA (2 μM)	N
Peak amplitude (mV)	5.15 ± 0.8	3 ± 0.75	4	4.5 ± 0.5	7.7 ± 1.25	5
Time to 80% decay	2.1 ± 0.5	1.3 ± 0.4	4	1.7 ± 0.1	2.9 ± 0.45	5

The numbers refer to the mean and standard error of the mean. Peak amplitude refers to the second component of the AHP.

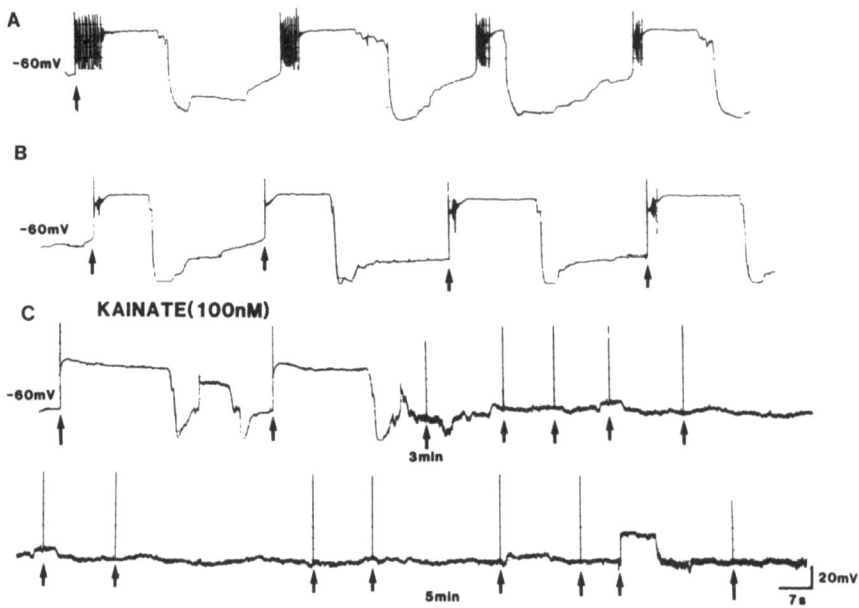

Fig. 5. Irreversible blockade of Ba^{2+} spikes by kainate. A: Plateau depolarizations induced by depolarizing current pulses (arrows) or slow depolarizing current in a pyramidal neuron bathed in (2 mM) extracellular Ba^{2+}, 0 Ca^{2+} and 1 µM of TTX. The plateau is preceded by fast oscillations. In B, several minutes after application of Ba^{2+} (>15), the oscillations are reduced and a depolarizing current pulse evokes only few fast events followed by a plateau. The plateau is terminated by a hyperpolarizing current. C: Superfusion of kainate in three minute completely blocks the Ba^{2+} spikes. In this neuron, kainate induces a 5 mV depolarization of the membrane potential to its initial value.

Superfusion of kainate (50-100 nM) produced a reduction of the oscillations preceding the plateau, and a reduction or a blockade of the plateau. The threshold for the Ba^{2+} action potentials was not changed. Usually 30-40 minutes of washing were required in order to obtain again spontaneous bursts or plateau. In contrast to KA, NMDA (2 µM) did not alter the Ba^{2+} mediated events (Fig. 6).

DISCUSSION

Two main actions of kainate on CA1 hippocampal neurons are described in the present study. The first is a reduction of a calcium sensitive potassium conductance and the second is a reduction or a complete block of the Ca^{2+} action potentials. While the first action will enhance excitability and probably contribute to kainate induced epileptogenesis, the second unexpected effect may relate to the neurotoxic effects of KA but its mechanisms are yet to be elucidated.

Effect of Kainate on Calcium-Dependent Potassium Conductance

The attenuation by KA of spike accomodation and of the amplitude and duration of the AHP occur with little modifications in the passive membrane properties of these neurons. The following observations suggest that the effects of KA are related to a partial block of calcium activated

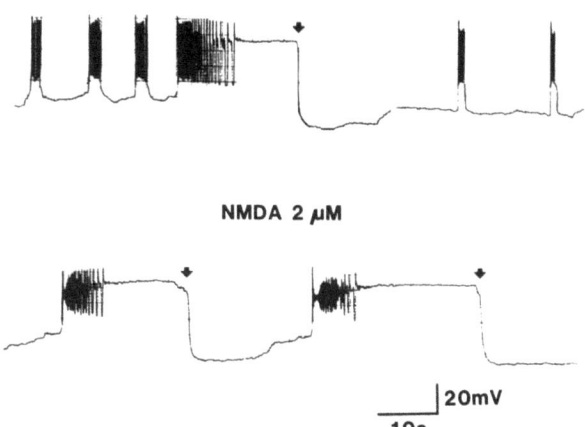

Fig. 6. NMDA does not block the Ba^{2+} spikes. Spontaneous bursting and plateau spikes in a CA1 neuron bathed in 2 mM Ba^{2+}, 1 µM TTX and 0 calcium. Spike repolarization is induced by injecting a steady hyperpolarizing current (arrows).

potassium conductance. First, kainate markedly reduces only the late component of the AHP without affecting the early component. Second, this effect is similar to that induced by the Ca^{2+} blocker Mn^{2+} and is prevented in the presence of Mn^{2+}. This suggests that both kainate and Mn^{2+} block the K^+ current activated by the intracellular calcium and the firing rate increases following the blockade of the AHP. A reduction of the AHP following a train of action potentials by kainate has been observed earlier in the hippocampal CA3 region (Robinson and Deadwyler, 1981). The reduction of AHP by kainate precedes the reduction in the number or the shape of the calcium action potentials. As a matter of fact, there is occasionally a temporary increase of the calcium spikes at a time when AHP is reduced. This raises the possibility that kainate enhances calcium sequestration in an intracellular pool, after calcium entry into the cell. In this respect, kainate could act in the same way as noradrenaline (Madison and Nicoll, 1982) corticotrophin releasing factor (Aldenhoff et al., 1983), histamine (Haas et al., 1983), in the hippocampus or muscarine and substance P in the myenteric plexus and sympathetic ganglia (North and Tokimasa, 1983; North et al., 1983; Tokimasa, 1984). All these substances enhance cell excitability by reducing calcium sensitive K^+ conductance. We do not know which K^+ current (IM or IC) is preferentially involved in the action of kainate, but IM seems unlikely, since this current is insensitive to calcium channel blockers (Hallivell and Adams, 1982) while the effect of kainate is blocked by Mn^{2+}.

Kainate also reduced another K^+ current IQ which is activated at values of membrane potential more negative than resting potential (Halliwell and Adams, 1982). Kainate reduces this current in the same way as intracellular Cs^+, making the V/I curve almost linear below -70 mV. In this respect, kainate acts similarly to L-glutamate in horizontal cells dissociated from goldfish retina (Kaneko and Tachibana, 1984) in which glutamate seems to act through kainate or quisqualate type of receptors (Ishida and Neyton, 1985). In motoneurons (Barrett and Crill, 1980) and in the heart (Di Francesco and Ojeda, 1980), this K^+ current is physiologically activated during the afterhyperpolarization that follows the action potentials. The resulting slow activating current will slow down the rate at which the neurons can be depolarized at the threshold. The partial block of this current by kainate will enhance cell excitability by increasing the firing rate and this effect would be additive to the partial block of the calcium sensitive potassium conductance.

Without appropriate antagonists it is not possible to assess the specificity of KA actions. NMDA induces opposite effects: namely an increase in the number and the duration of the calcium spikes and an enhancement of the amplitude and duration of the following AHP. Also, the specific NMDA receptor antagonist D-APV blocks the effects of NMDA but not those of kainate.

The mechanisms of action of NMDA will not be discussed in detail in the present report (e.g., Dingledine, this volume). According to Dingledine (1983), NMDA receptor is coupled with a voltage dependent Ca^{2+} conductance the activation of which produce Ca^{2+} entry into the cell. However, Nowak et al. (1984) have shown that the activation of the NMDA receptor is coupled to a voltage independent channel, the voltage dependency being only related to the blockade by Mg^{2+}. In these conditions, the enhancement of the number and duration of calcium spikes by NMDA could be explained mainly by the presence of a negative slope conductance in the current/voltage relationship of the NMDA current. In keeping with this, the effects of NMDA on Ca^{2+} mediated events were absent in slices kept for several hours in Mg^{2+}-free media (Cherubini et al., unpublished observations).

Effect of Kainate on Calcium Spikes

We have consistently observed that kainate reduces or completely blocks calcium or barium action potentials. Several possibilities can account for these effects. We cannot exclude the possibility that KA acts at distant (dendritic) sites to produce a reduction in the resistance not seen at the somatic level; this shunt would effectively reduce the generation of action potentials. It is possible that like glutamate in the retina (Tachibana, 1985) KA increases the conductance to different cations, including Ca^{2+}, in a nonselective way. This could explain why kainate induced a depolarization associated sometimes with an increase in membrane conductance. We do not know which ions carry the kainate current in the hippocampus, but Ca^{2+} seems to be relevant, since in the presence of Mn^{2+} the effects of kainate - notably the depolarization - are not observed. Calcium entry is also enhanced as a consequence of a decreased K^+ conductance; this will delay spike repolarization and increase their duration. The enhancement of free intracellular calcium may induce an inactivation of the Ca^{2+} channels. This mechanism of Ca^{2+} inactivation has been demonstrated in Aplysia neurons (Tillotson, 1979; Kramer and Zucker, 1985).

Alternatively, kainate may act as a blocking agent of open ion channels. First it will open the channels, thus increasing permeability to different cations. Then it will plug the channel itself, preventing the passage of ions, in the same way as already proposed for the nicotinic receptors at the neuromuscular junction (Adams, 1976; for a review see also Colquhoun, 1978).

Whatever the exact mechanism of action could be, the present study shows that KA shares with several modulators and transmitters the capacity to reduce or block calcium-dependent potassium conductance but in addition the toxin has a rather unique capacity of reducing calcium action potentials. A better evaluation of this effect could not only underlie KA neurotoxicity but be of general interest with regard to epileptogenesis.

ACKNOWLEDGEMENT

Supported by an INSERM Grant to E.C.

REFERENCES

Adams, P.R. 1976, Drug blockade of open end-plate channels, J. Physiol., 260:531.
Aitken, P.G., 1985, Kainic acid and penicillin: differential effects on excitatory and inhibitory interactions in the CA1 region of the hippocampal slice, Brain Res., 325:261.
Aldenhoff, J.B., Gruol, D.L., Rivier, J., Vale, W., and Siggins, G.R., 1983, Corticotropin releasing factor decreases postburst hyperpolarizations and excites hippocampal neurons, Science, 221:875.
Barrett, E.F., Barret, J.N., and Crill, W.E., 1980, Voltage sensitive outward currents in cat motoneurones, J. Physiol., 304:251.
Ben-Ari, Y., 1985, Limbic seizure and brain damage produced by kainic acid: mechanisms and relevance to human temporal lobe epilepsy, Neuroscience, 14:375.
Collingridge, G.L., Kehl, S.J., Loo, R., and McLennan, H., 1983, Effects of kainic and other amino acids on synaptic excitation in rat hippocampal slices: 1. Extracellular analysis, Exp. Brain Res., 52:170.
Colquhoun, D., 1981, The kinetics of conductance changes at nicotinic receptors of the muscle end-plate and of ganglia, in: Drug Receptors and Their Effectors, N.J.M. Birdsall, ed., MacMillan, London, p. 107.
Di Francesco, D., and Ojeda, C., 1980, Properties of the current in the sino-atrial node of the rabbit compared with those of the current iK2 in Purkinje fibres, J. Physiol., 308:353.
Dingledine, R., 1983, N-methyl aspartate activates voltage-dependent calcium conductance in rat hippocampal pyramidal cells, J. Physiol., 343:385.
Fisher, R.A., and Alger, B.E., 1984, Electrophysiological mechanisms of kainic acid induced epileptiform activity in the rat hippocampal slice, J. Neurosci., 4:1323.
Gormann, A.L.F., and Hermann, A., 1979, Internal effects of divalent cations on potassium permeability in molluscan neurones, J. Physiol., 296:393.
Gustafsson, B., and Wigström, H., Evidence for two types of afterhyperpolarization in CA1 pyramidal cells in the hippocampus, Brain Res., 206:462.
Haas, H.L., and Konnerth, A., 1983, Histamine and noradrenaline decrease calcium activated potassium conductance in hippocampal pyramidal cells, Nature, 302:432.
Hablitz, J.J., and Johnston, D., 1981, Endogenous nature of spontaneous bursting in hippocampal pyramidal neurons, Cell. Mol. Neurobiol., 1:325.
Halliwell, J.V., and Adams, P.R., 1982, Voltage-clamp analysis of muscarinic excitation in hippocampal neurons, Brain Res., 250:71.
Hoston, J.R., Prince, D.A., and Schwartzkroin, P.A., 1979, Anomalous inward rectification in hippocampal neurones, J. Neurophysiol., 42:889.
Ishida, A.T., and Neyton, J., Quisqualate and L-glutamate inhibit retinal horizontal-cell responses to kainate, Proc. Natl. Acad. Sci. USA, 82:1837.
Kaneko, A., and Tachibana, M., 1985, Effects of L-glutamate on the anomalous rectifier potassium current in horizontal cells of carassius auratus retina, J. Physiol., 358:169.
Kramer, R.H., and Zucker, R.S., 1985, Calcium-induced inactivation of calcium current causes the interburst hyperpolarization of Aplysia bursting neurones, J. Physiol., 362:131.
Madison, D.V., and Nicoll, R.A., 1982, Noradrenaline blocks accomodation of pyramidal cell discharge in the hippocampus, Nature, 299:636.
Madison, D.V., and Nicoll, R.A., 1984, Control of the repetitive discharge of rat CA1 pyramidal neurones in vitro, J. Physiol., 354:319.

North, R.A., Morita, K., and Tokimasa, T., 1983, Peptide actions on autonomic nerves, in: Systemic Role of Regulatory Peptides, S.R. Bloom, J.M. Polak, and E. Lindenlaub, eds., Springer-Verlag, Stuttgart-New York, p. 77.

North, R.A., and Tokimasa, T., 1983, Depression of calcium-dependent potassium conductance by muscarinic agonists, J. Physiol., 342:253.

Nowak, L., Bregestovski, P., Asher, P., Herbet, A., and Prochiantz, A., 1984, Magnesium gates glutamate-activated channels in mouse central neurones, Nature, 307:462.

Pepper, C., and Henderson, G., 1980, Opiates and opioid peptides hyperpolarize locus coeruleus neurons in vitro, Science, 209:394.

Robinson, J.H., and Deadwyler, S.A., 1981, Kainic acid produces depolarization of CA3 pyramidal cells in the in vitro hippocampal slice, Brain Res., 221:117.

Tachibana, M., 1985, Permeability changes induced by L-glutamate in solitary retinal horizontal cells isolated from Carassius Auratus, J. Physiol., 358:153.

Tillotson, D., 1979, Inactivation of Ca conductance dependent on entry of Ca ions in molluscan neurons, Proc. Natl. Acad. Sci. USA, 76:1497.

Tokimasa, T., 1984, Muscarinic agonists depress calcium-dependent gk in bullfrog sympathetic neurons, J. Autonom. Nerv. Syst., 10:107.

Watkins, J.C., and Evans, R.H., 1981, Excitatory amino acid transmitters, Ann. Rev. Pharmacol. Toxicol., 21:165.

BLOCKADE BY D-AMINOPHOSPHONOVALERATE OR Mg^{2+} OF EXCITATORY AMINO ACID-INDUCED RESPONSES ON SPINAL MOTONEURONS IN VITRO

A. Nistri and A.E. King

Department of Pharmacology, St. Bartholomew's Hospital
Medical College, University of London, Charterhouse Square
London EC1M 6BQ, Great Britain

INTRODUCTION

L-glutamate and its structurally-related analogs are known to evoke strong excitation of vertebrate central neurons (Nistri and Constanti, 1979; Puil, 1981). Pharmacological studies have shown that there are probably distinct receptor populations mediating responses to different excitatory amino acids (Watkins and Evans, 1981; McLennan, 1983): one receptor class is preferentially activated by N-methyl-D-aspartate (NMDA) while a separate receptor type selectively binds quisqualate. While it is not fully clear if kainate operates via a third receptor system, it seems that the endogenously occurring amino acids L-glutamate and L-aspartate have mixed agonist properties being able to bind NMDA as well as quisqualate receptors.

Current understanding of the pharmacology of excitatory amino acid receptors is, to a large extent, derived from the use of selective receptor antagonists. A powerful blocker of NMDA receptors is D-aminophosphono-valerate (D-APV; Evans et al., 1982); divalent cations, particularly Mg^{2+}, are also strong and selective depressants of NMDA depolarizations (Ault et al., 1980). It is therefore feasible using pharmacological tools to probe the location and function of the NMDA receptor sites in different areas of the central nervous system. In our laboratory, we were particularly interested in the receptor profile of spinal motoneurons since a large body of experimental evidence supporting the excitatory amino acid receptor classification was gathered from work on the spinal cord. Earlier studies (Watkins and Evans, 1981; Evans et al., 1982) employed extracellular recording techniques which provided a quantitative analysis of drug receptor interaction, but little insight into the membrane mechanisms responsible for these effects. Hence with intracellular recordings from amphibian spinal motoneurons in vitro we wished to explore the sensitivity of synaptic transmission and excitant-induced responses to D-APV and Mg^{2+}. All our experiments were conducted at low temperature ($7^{\circ}C$) to reduce considerably the cellular uptake systems for amino acids (Davidoff and Adair, 1975).

METHODS

Experiments were carried out on frogs (Rana temporaria) previously acclimatized at $7^{\circ}C$. After decerebration and removal of the spinal cord, a longitudinal parasagittal slice was prepared and placed in a superfusion chamber at 7°C. Pairs of lumbar ventral and dorsal roots were drawn into sidechambers and stimulated via miniature suction electrodes. The preparation was continuously superfused with precooled oxygenated Ringer solution of the following composition (mM): NaCl 111; KCl 2.5; $NaHCO_3$ 17; NaH_2PO_4 0.1; $CaCl_2$ 2; glucose 4. This solution was assumed to be 'Mg^{2+} free' as it would only contain Mg^{2+} in trace amounts as a contaminant; since the slice was superfused for many hours or even days with this solution the extracellular levels of unbound Mg^{2+} were also presumed to be negligible.

Motoneurons, identified by their all-or-none antidromic spikes, were recorded with 3 M KCl microelectrodes also used to pass current pulses for monitoring cell input conductance. Responses were amplified, displayed on an oscilloscope and stored on FM tape. All compounds were admitted to the recording chamber by superfusion through separate flowlines. Further details of the methods are given in Nistri and Arenson (1983) and Nistri et al. (1985).

RESULTS

Motoneurons had resting membrane potential of 65 ± 1 mV (mean ± S.E.M.) and input conductance of 92 ± 6 nS. The antidromic spike overshoot was 15-20 mV and cells often displayed synaptic noise resulting from interneuronal activity. In control Ringer solution, glutamate (2 mM), aspartate

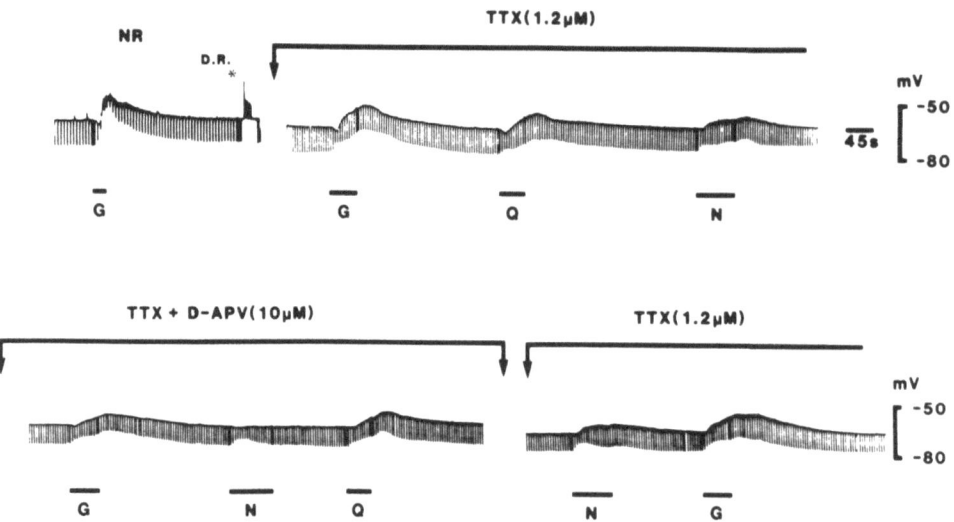

Fig. 1. Membrane potential records obtained from a single motoneuron showing the effect of D-APV on responses to glutamate (G; 2 mM), quisqualate (Q; 30 μM) and NMDA (N; 30 μM); the durations of application are represented by horizontal bars. Upper traces: NR indicates a glutamate response in control Ringer and is followed by amino acid responses after 35 minutes exposure to 1.2 μM tetrodotoxin (TTX); note dorsal root-evoked excitatory potentials (*DR) in NR. Lower traces: responses after superfusions with a TTX and 10 μM D-APV containing solution and subsequent recovery after D-APV removal. The downward deflections are hyperpolarizing electrotonic potentials evoked by intracellular current injection (-0.87 nA and -0.75 nA for top and bottom tracings, respectively). (From Corradetti et al., 1985).

(2 mM), NMDA (30 μM) and quisqualate (30 μM) were applied to produce matched depolarizations of about 10 mV. These were within the apparently linear part of the cell current/voltage relation (Fig. 4) and avoided activation of delayed rectification. In quiescent cells, changes in input conductance at the peak of the amino acid depolarizations were within 20% of control values. In cells which were spontaneously very active or produced intense discharges in response to excitatory amino acids, changes in input conductance were very large (see Fig. 1 top) or even unmeasurable as the size of the electrotonic potentials became too small to be determined with accuracy. In Mg^{2+}-free Ringer, repetitive or burst firing was usually not typical of excitatory amino acid responses on motoneurons (cf. also Nistri and Arenson, 1983). However, when ventral horn interneurons were recorded and their amino acid responses matched in amplitude to those of motoneurons, either steady state or burst firing was elicited by quisqualate or NMDA, respectively (glutamate induced firing patterns which were mixtures of those evoked by quisqualate and NMDA).

In an attempt to block indirect effects of amino acids, spinal slices were exposed to tetrodotoxin (TTX; 0.6-6 μM) which abolished regenerative spike activity. Fig. 1 shows that excitatory amino acids retained their depolarizing effects with relatively small input conductance changes. Interestingly, at the peak of these depolarizations some synaptic voltage noise (thickening of the trace) could often be detected, suggesting that excitatory amino acids may presynaptically induce release of endogenous transmitters via a mechanism independent from Na^+-mediated regenerative responses.

Actions of D-APV on Motoneurons

Bath application of D-APV (0.5-10 μM) often induced slight membrane hyperpolarization (2-3 mV) with no change in input conductance as judged by the amplitude of hyperpolarizing electrotonic potential or by the slope of current/voltage plots. The responses to NMDA, glutamate, aspartate or quisqualate were antagonized to various degrees as shown in Fig. 2. Full antagonism of NMDA depolarizations without reduction in the quisqualate response was achieved with 10 μM D-APV (Figs. 1 and 2). Glutamate or aspartate-induced depolarizations were depressed by about 40%; in other words, antagonism saturation became apparent if selected agonists were employed. When input conductance effects were analyzed, results from experiments with or without TTX were essentially similar: the relatively modest conductance increases induced by NMDA, quisqualate or glutamate were not significantly altered by D-APV.

Concentrations of D-APV which reduced NMDA and glutamate responses as summarized in Fig. 2 also depressed the amplitude of monosynaptic excitatory postsynaptic potentials (EPSPs) elicited by low strength dorsal root stimuli (<1.5 times threshold). Maximal depression of these EPSPs was 60% in the presence of 10 μM D-APV; polysynaptic EPSP amplitudes were not significantly diminished although the decay time of these EPSPs was shortened by 38%. Similar results were previously obtained with the less potent NMDA antagonist D-aminoadipate (Arenson et al., 1984).

Effects of Mg^{2+} on Motoneurons

Recent studies on mammalian cultured neurons have shown that low concentrations of Mg^{2+} block the NMDA receptor-activated ion channels (Nowak et al., 1984; Mayer and Westbrook, 1985). We therefore decided to investigate how Mg^{2+} would affect motoneuronal responses in an adult preparation with a well preserved architectural neuronal network. On frog motoneurons, high extracellular concentrations of Mg^{2+} (5-11 mM) are required to depress synaptic transmission, hyperpolarize the neuronal

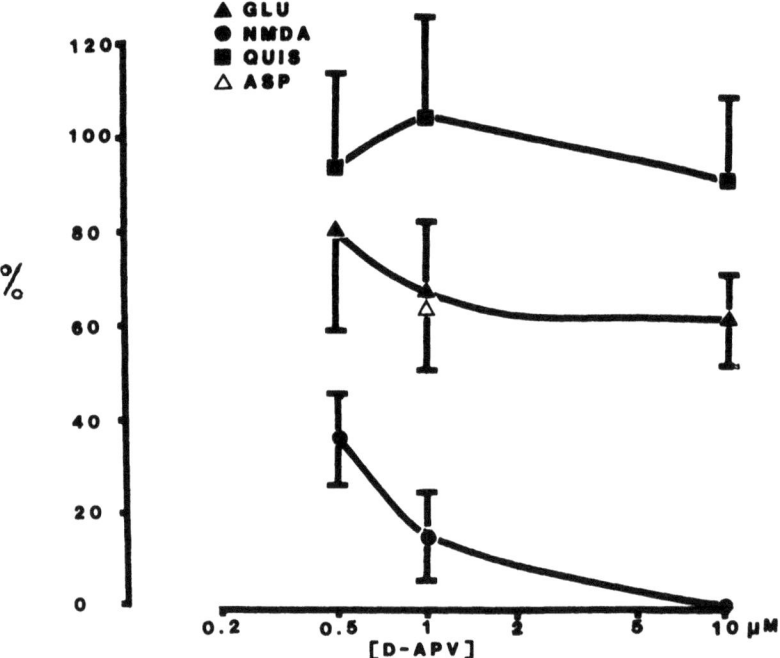

Fig. 2. Plot of increasing concentrations of D-APV (abscissa; log scale) versus amplitude of depolarizations (ordinate; linear scale, expressed as % of controls) produced by glutamate (GLU), NMDA, quisqualate (QUIS) and aspartate (ASP). Each point is the mean ± S.E.M. of 4-16 responses. Single and double asterisks near symbols indicate a statistically significant difference from control ($p < 0.01$ and $p \leq 0.05$, respectively). (From Corradetti et al., 1985).

membrane and reduce excitability (Erulkar et al., 1974): these phenomena stem from a combination of Ca^{2+} antagonism and increased surface charge density. We were therefore surprised to find a large decrease in EPSP amplitude with Mg^{2+} concentrations as low as 0.25 mM. In Fig. 3, polysynaptic EPSPs (indicated by arrows) were induced by dorsal root stimuli at threshold for firing (left) or below threshold (right). In the presence of 0.25 mM Mg^{2+} (which produced a 3 mV hyperpolarization), the EPSP on the left hand side was smaller and shorter, and elicited a single action potential while the EPSP produced by weaker stimuli (on the right) was strongly depressed. Both mono- and polysynaptic EPSPs were reversibly reduced in amplitude and duration by Mg^{2+} whose threshold dose was below 0.1 mM. Even the antidromic spike late afterhyperpolarization, considered to be due to the activation of a Ca^{2+} dependent K^+ conductance (Barrett and Barrett, 1976), was somewhat depressed by 0.25 mM Mg^{2+} and blocked by 1 mM Mg^{2+}. Current/voltage relations were not significantly altered by this divalent cation (Fig. 4) since the shape and slope of these plots did not vary appreciably. Fig. 3 also shows the depression of the glutamate depolarization by 0.25 mM Mg^{2+}; a second longer application of glutamate restored the response amplitude. A plot of increasing Mg^{2+} concentrations vs. the amino acid-induced depolarizations reveals theat 0.25 mM Mg^{2+} fully blocked NMDA actions, reduced the glutamate effect but did not substantially affect quisqualate responses (Fig. 5).

Amino acid actions on input conductance before and after application

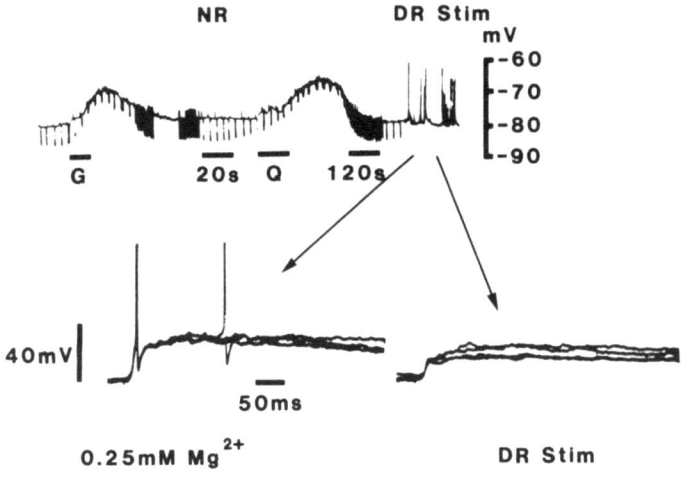
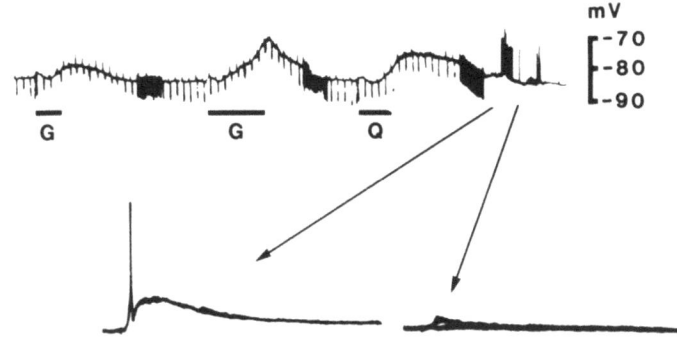

Fig. 3. Effect of Mg^{2+} (0.25 mM) on glutamate (G; 2 mM) and quisqualate (Q; 30 μM) induced depolarizations and synaptic excitation of a frog motoneuron. In each pair of tracings, top is a chart record of membrane potential and bottom shows oscilloscope recordings of dorsal root stimulation (DR stim) evoked EPSPs (indicated by arrows). Note depression of EPSPs and glutamate response in Mg^{2+} solution. A second larger application of glutamate reverses the Mg^{2+} block. NR = control Ringer.

of Mg^{2+} were variable, undoubtedly a consequence of indirect components contributing to the measured event. In fact, quisqualate-induced conductance changes as expected were not altered by Mg^{2+} Ringer. However, in the case of NMDA only 4/7 cells displayed the anticipated smaller conductance change in the presence of Mg^{2+}. With glutamate, 6/9 cells showed larger conductance increases while the remaining cells showed the opposite.

In order to eliminate as far as possible indirect contributions from interneuronal activity, motoneurons, after functional identification, were bathed in TTX media, tested for their responsiveness to amino acids and then exposed to Mg^{2+} solutions. The three excitants displayed a variable degree of sensitivity to Mg^{2+}: NMDA was the most susceptible to antagonism while quisqualate was hardly affected (Fig. 6). One interesting observation was that any voltage noise present at the peak of amino acid depolarizations

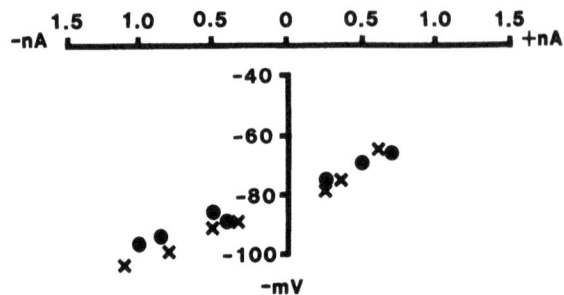

Fig. 4. Current/voltage relation for a single frog motoneuron in control or Mg^{2+} (0.25 mM) Ringer.

in TTX media was abolished by addition of Mg^{2+}. The effects of Mg^{2+} on the input conductance changes were complex and not always proportional to the Mg^{2+} concentration (0.1-1 mM) or the cell membrane potential within the range investigated. In some cells (see for instance Fig. 7), all three excitants clearly increased input conductance in TTX Ringer and (with the exception of quisqualate) less so in 0.1 mM Mg^{2+} solution. However, the mean conductance increases evoked in TTX media were small (Fig. 6B; see also Corradetti et al., 1985; Nistri et al., 1985) and only slightly depressed by Mg^{2+} with apparently little preference towards any of the three excitants (Fig. 6B).

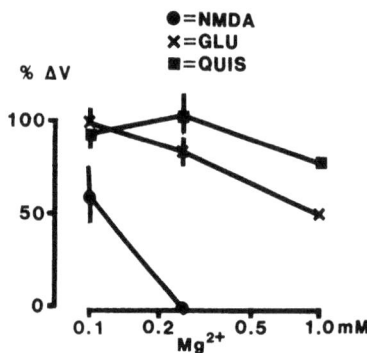

Fig. 5. Plot of increasing concentrations of Mg^{2+} vs. amplitude of excitant-induced depolarizations (ΔV expressed as % of controls in normal Ringer solution). GLU: glutamate; QUIS: quisqualate.

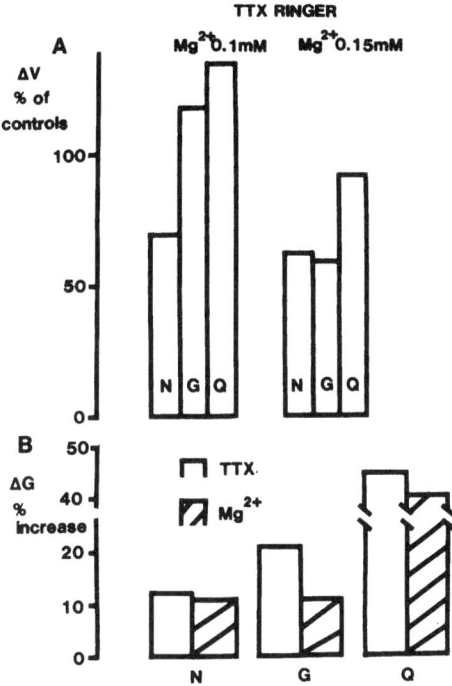

Fig. 6. A: Effects of 0.1 and 0.15 mM Mg^{2+} on NMDA (N; 30 µM), glutamate (G; 2 mM) and quisqualate (Q; 30 µM) evoked depolarizations in TTX containing Ringer. B: Comparison of input conductance increases during excitant responses in TTX (open columns) or Mg^{2+} (0.1-1 mM; hatched columns) Ringer. Abbreviations as in A.

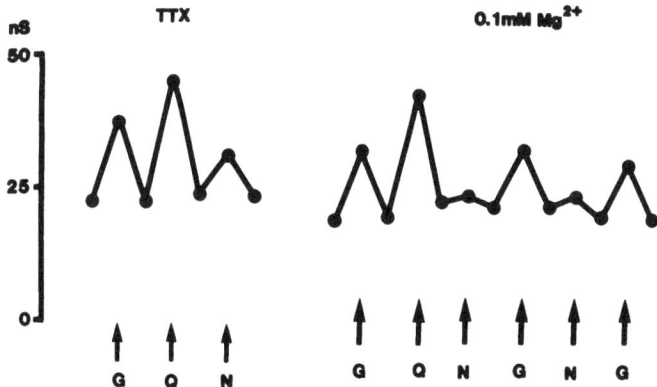

Fig. 7. Graph of input conductance increases elicited by excitants (abbreviations as in Fig. 6A) on a single frog motoneuron in TTX or Mg^{2+} solution.

DISCUSSION

The main scope of the present study was a characterization of the pharmacological activity of D-APV and Mg^{2+} as excitatory amino acid blockers on spinal motoneurons in vitro. Previous extracellular studies suggested that while D-APV and Mg^{2+} are both preferential antagonists of NMDA-mediated responses, their cellular site of action is distinct (Watkins and Evans, 1981). Our present intracellular investigation has been aimed at clarifying these mechanisms. On frog motoneurons, D-APV was indeed a selective NMDA receptor antagonist and this property, coupled to the drug's intrinsic potency and water solubility, makes it an ideal substance to probe the location and function of NMDA receptor sites in the brain and spinal cord. This notion has been put into practice by observing a significant antagonism of monosynaptic EPSPs on spinal motoneurons, a finding which implicates NMDA sensitive receptors in these fast synaptic transmission processes. Since D-APV reduced almost in parallel the glutamate (and aspartate) depolarization as well as the monosynaptic EPSP, it is likely that one (or both) of these amino acids is the natural excitatory transmitter of such synaptic responses.

Whereas it seems clear that quisqualate acted via a receptor class distinct from that for NMDA (cf. D-APV resistant quisqualate responses), glutamate and aspartate were probably mixed agonists with affinity for both the NMDA and the quisqualate receptor (Watkins and Evans, 1981). In contrast to the results of Mayer and Westbrook (1984), the conductance change produced by glutamate was not significantly altered by D-APV. However, this is not so surprising since our experiments were performed in the absence of Mg^{2+}, a divalent cation which strongly influences the conductance change produced by amino acids (Nowak et al., 1984; Mayer and Westbrook, 1985).

Although the NMDA depressant action of Mg^{2+} (and other divalent cations) had been known for some time (Ault et al., 1980), it was only recently interpreted as due to blockade of activated ion channels over a large range of membrane potential in mammalian cultured neurons (Nowak et al., 1984; Mayer and Westbrook, 1985). This 'plugging' action of Mg^{2+} may account for the apparent lack of input conductance increase during the NMDA depolarization and may be the reason for the strong voltage-sensitivity of NMDA-opened channels. Glutamate, as an agonist at the NMDA receptor, would share part of the NMDA sensitivity to Mg^{2+}, while quisqualate would be relatively resistant to it. A corollary of this notion is that, in the absence of Mg^{2+}, NMDA, glutamate and quisqualate should all elicit consistent input conductance increases, in spite of their different receptor sites.

In the present work on frog motoneurons, the role of Mg^{2+} appears to be complex and not fully accommodated within the framework of this theory. For instance, we, like others who work on brain slice neurons (Crunelli and Mayer, 1984), often found little input conductance increase induced by these excitants even in Mg^{2+} free TTX media. Contamination of bathing solutions by trace amounts of Mg^{2+} cannot be entirely excluded and might explain the case for NMDA but it cannot fully account for the results with quisqualate. Another possibility is that some intrinsic voltage-sensitive membrane channels might mask large conductance increases due to the excitants; nevertheless the relatively linear current/voltage relations (see Fig. 4) and the lack of clear conductance changes following manipulations of extracellular Na^+ (Nistri et al., 1985) or use of K^+ channel blockers (Arenson and Nistri, 1985) do not lend support to this hypothesis. Perhaps as far as frog motoneurons in vitro are concerned the comparatively modest conductance increase in Mg^{2+} free Ringer may be accounted for by the rather small changes in Na^+/K^+ permeability ratio which has been demonstrated with ion-sensitive microelectrodes during glutamate applications (Bührle and Sonnhof, 1983)

A blocking action of Mg^{2+} on the NMDA receptor complex (also activated by glutamate) may account for the observed antagonism of NMDA and glutamate depolarizations but the mean input conductance increases due to these agonists were only slightly smaller than their respective controls even in TTX. Several factors should be borne in mind when considering this anomalous result. Firstly, the degree of Mg^{2+} antagonism to the NMDA and glutamate-induced conductance responses will depend on the severity of the NMDA channel block produced by the various concentrations of Mg^{2+} (Mayer and Westbrook, 1985). The situation for glutamate is even more complicated since as the concentration of Mg^{2+} increases the interaction of glutamate with the quisqualate receptor will be revealed. Second, the low temperature of the present experiments may have considerably altered channel kinetics. Furthermore, even low concentrations of Mg^{2+} displayed a variety of effects including the ability to suppress TTX-resistant voltage noise induced by the excitants. A plausible reason for this phenomenon is that even in TTX media excitants may stimulate the presynaptic release of unidentified endogenous transmitters whose effects would be additive to the post-synaptically generated response. This view is supported by neurochemical demonstrations of amino acids influencing directly Ca^{2+} dependent transmitter release from brain slices (Roberts and Anderson, 1979; Ferkany and Coyle, 1983; Johnston and Lodge, 1984) via a Mg^{2+}-sensitive receptor class which may belong to a category of presynaptic receptors regulating transmitter release in the brain (see Chesselet, 1984, for review). Perhaps also important is the rather high sensitivity (possibly higher than hitherto suspected) of neuronal Ca^{2+} mechanisms to Mg^{2+} as suggested by the depression of the spike Ca^{2+} dependent late AHP during superfusion with low concentrations of Mg^{2+}. Interestingly, the quisqualate conductance response was also slightly reduced by Mg^{2+}, implying that a small portion of the quisqualate action was via receptors/channels analogous to those for NMDA. In keeping with this possibility is the loss of selectivity with higher concentrations of antagonists.

Considerations about the presynaptic site of actions of Mg^{2+} should be seen as complementary to the action of this cation on the postsynaptic membrane. Indeed, the generalized reduction in mono- and polysynaptic EPSPs may suggest that the postsynaptic blocking action of this divalent cation was superimposed on a presynaptic effect. Furthermore, full block of NMDA receptors by D-APV did not alter motoneuronal spike configuration or produce widespread depression of EPSPs.

In conclusion, the present results fully support a powerful influence of Mg^{2+} on central neurons and suggest that even small changes in the extracellular levels of this cation may profoundly alter motoneuronal activity. For example, if the amount of Mg^{2+} in the synaptic cleft was rapidly elevated, a sustained yet reversible depression of mono- and polysynaptic reflex arches may easily occur.

ACKNOWLEDGEMENTS

This work is supported by a grant from the National Fund for Research into Crippling Diseases (Action Research). We thank Miss C.A. Brown for secretarial assistance, and Mr. R. Croxton and Mr. G. Davis for photography and technical assistance. A.E.K. is a Postdoctoral Fellow of the M.R.C.

REFERENCES

Arenson, M.S., Berti, C., King, A.E., and Nistri, A., 1984, The effect of D-α-aminoadipate on excitatory amino acid responses recorded intracellularly from motoneurones of the frog spinal cord, Neurosci. Lett., 49:99.

Arenson, M.S., and Nistri, A., 1985, The effects of potassium channel blocking agents on the responses of in vitro frog motoneurones to glutamate and other excitatory amino acids: an intracellular study, Neuroscience, 14:317.

Ault, B., Evans, R.H., Francis, A.A., Oakes, D.J., and Watkins, J.C., 1980, Selective depression of excitatory amino acid-induced depolarizations by magnesium ions in isolated spinal cord preparations, J. Physiol., 307:413.

Barrett, E.F., and Barret, J.N., 1976, Separation ot two voltage sensitive potassium currents and the demonstration of a tetrodotoxin-resistant calcium current in frog motoneurones, J. Physiol., 255:737.

Bührle, C.P., and Sonnhof, U., 1983, The ionic mechanisms of excitatory action of glutamate upon the membranes of motoneurones of the frog, Pflügers Arch., 396:154.

Chesselet, M.F., 1984, Presynaptic regulation of neurotransmitter release in the brain, Neuroscience, 12:347.

Corradetti, R., King, A.E., Nistri, A., Rovira, C., and Sivilotti, L., 1985, Pharmacological characterization of D-aminophosphonovaleric acid antagonism of amino acid and synaptically evoked excitations on frog motoneurones in vitro: an intracellular study, Br. J. Pharmacol., 86:19.

Crunelli, V., and Mayer, M.L., 1984, Mg^{2+} dependence of membrane resistance increases evoked by NMDA in hippocampal neurones, Brain Res., 311:392.

Davidoff, R.A., and Adair, R., 1975, High affinity amino acid transport by frog spinal cord slices, J. Neurochem., 24:545.

Erulkar, S.D., Dambach, G.E., and Mender, D., 1974, The effect of magnesium at motoneurones of the isolated spinal cord of the frog, Brain Res., 66:413.

Evans, R.H., Francis, D.A., Jones, A.W., Smith, D.A.S., and Watkins, J.C., 1982, The effects of a series of ω-phosphonic α-carboxylic amino acids on electrically evoked and excitant amino acid-induced responses in isolated spinal cord preparations, Br. J. Pharmacol., 75:65.

Ferkany, J.W., and Coyle, J.T., 1983, Kainic acid selectively stimulates the release of endogenous excitatory amino acids, J. Pharmacol. Exp. Ther., 225:399.

Johnston, G.A.R., and Lodge, D., 1984, Ketamine and magnesium selectively block the N-methylaspartate-evoked release of acetylcholine from rat cortex slices in vitro, J. Physiol., 349:15P.

Mayer, M.L., and Westbrook, G.L., 1984, Mixed-agonist action of excitatory amino acids on mouse spinal cord neurones under voltage clamp, J. Physiol. 354:29.

Mayer, M.L., and Westbrook, G.L., 1985, The action of N-methyl-D-aspartic acid on mouse spinal neurones in culture, J. Physiol., 361:65.

McLennan, H., 1983, Receptors for the excitatory amino acids in the mammalian central nervous system, Progr. Neurobiol., 20:251.

Nistri, A., and Constanti, A., 1979, Pharmacological characterization of different types of GABA and glutamate receptors in vertebrates and invertebrates, Progr. Neurobiol., 13:117.

Nistri, A., and Arenson, M.S., 1983, Differential sensitivity of spinal neurones to amino acids: an intracellular study on the frog spinal cord, Neuroscience, 8:115.

Nistri, A., Arenson, M.S., and King, A., 1985, Excitatory amino acid-induced responses of frog motoneurones bathed in low Na^+ media: an intracellular study, Neuroscience, 14:921.

Nowak, L., Bregestovski, P., Ascher, P., Herbet, A., and Prochiantz, A., 1984, Magnesium gates glutamate-activated channels in mouse central neurones, Nature, 307:462.

Puil, E., 1981, S-Glutamate: its interactions with spinal neurons, Brain Res. Rev., 3:229.

Roberts, P.J., and Anderson, S.D., 1979, Stimulatory effect of L-glutamate and related amino acids on [^{3}H] dopamine release from rat striatum: an in vitro model for glutamate actions, J. Neurochem., 32:1539.

Watkins, J.C., and Evans, R.H., 1981, Excitatory amino acid transmitters, Ann. Rev. Pharmacol. Toxicol., 21:165.

THE MEMBRANE ACTION OF EXCITATORY AMINO ACIDS ON CULTURED MOUSE SPINAL

CORD NEURONS

G.L. Westbrook and M.L. Mayer

Laboratory of Developmental Neurobiology
National Institute of Child Health and Human Development
National Institutes of Health
Bethesda, Maryland 20892 U.S.A.

INTRODUCTION

Although the role of excitatory amino acids as neurotransmitters and as excitotoxins has been a topic of intense investigation, progress has been slowed by a lack of understanding of the membrane action of these compounds. Both endogenous agonists such as L-glutamate and L-aspartate and exogenous compounds such as kainate and quisqualate excite essentially all neurons within the vertebrate central nervous system. This seeming 'non-specific' excitation combined with the lack of agreement concerning the conductance changes accompanying L-glutamate depolarizations (as measured with current clamp recording) have complicated the analysis of the conductance mechanisms activated by excitatory amino acids.

However, two recent developments have led to a rapid increase in knowledge of these conductance mechanisms. These are the development of selective antagonists by Watkins and colleagues (see Watkins, 1981 for review), thus demonstrating the existence of at least three receptor subtypes in the vertebrate spinal cord, and the recognition of differences in the voltage dependence of the conductances associated with activation of the various receptor types (Engberg et al., 1979; MacDonald and Wojtowicz 1982; MacDonald and Porietis, 1982).

We have used voltage clamp methods in cultured mammalian spinal cord neurons to investigate the conductance mechanisms underlying the action of L-glutamate, L-aspartate and a number of their analogs. Our results demonstrate that the conductances linked to receptors preferentially activated by N-methyl-D-aspartate (NMDA) and L-aspartate are voltage sensitive as a consequence of voltage-dependent block by physiological concentrations of magnesium ions. In contrast, the conductances linked to kainate and quisqualate receptors are relatively voltage-insensitive. The ionic permeability also differs between the voltage-sensitive and -insensitive conductances in that shifts in the reversal potential of responses evoked by NMDA (i.e., the voltage-dependent conductance) with changes in extracellular $[Ca^{2+}]$ suggest that the NMDA channel is permeable to Ca^{2+} as well as monovalent cations.

MATERIALS AND METHODS

Primary dissociated cultures of spinal cord neurons taken from 13 day mouse embryos (C57BL/6) were prepared as previously described (Mayer and Westbrook, 1984). After two to four weeks in culture, spinal cord neurons were impaled with two independent microelectrodes for voltage clamp recording. In later experiments, the patch electrode technique was used for whole-cell voltage clamping using a discontinuous single-electrode voltage clamp (Axoclamp, Axon Instru.). The normal recording medium contained (mM): 143, NaCl; 4.8, KCl; 0.1-5, $CaCl_2$; 10, glucose; 10, HEPES; and 0-5, $MgCl_2$. However, in the absence of added $MgCl_2$ ('Mg-free'), there was a residual Mg^{2+} concentration of 3-8 µM as measured by atomic absorption spectrometry. This Mg^{2+} contamination may contribute to the residual voltage-sensitivity of NMDA responses in Mg^{2+}-free medium. Low sodium medium was prepared by isoosmotic substitution of choline chloride for NaCl. Changes in divalent cation concentrations were made by isoosmotic addition to the recording medium. Microelectrodes were usually filled with 1 M CsCl; patch electrodes contained (mM): 140, CsCl; 2, $MgCl_2$; 1.1, EGTA; 10, HEPES. Tetrodotoxin (1 µM) was added to the recording medium to block action potential generation and reduce synaptic activity.

Excitatory amino acids were dissolved in recording medium and applied by pressure ejection from extracellular micropipettes positioned near the soma of individual neurons. Brief pressure pulses (10-500 msec) evoked currents with rapid risetime and brief duration. For sustained responses (i.e., 30-60 seconds during a series of voltage jumps), repetitive pressure pulses were used and the amplitude of the current was maintained by adjusting the duration of the pressure pulse. All experiments were performed at room temperature (25-28°C).

RESULTS

A Voltage-Sensitive Conductance

NMDA, a selective agonist for one acidic amino acid receptor (Watkins, 1981), activates a voltage-sensitive conductance in spinal cord neurons both in vivo (Engberg et al., 1979) and in vitro (MacDonald et al., 1981; Mayer and Westbrook, 1984). Voltage recording in spinal cord neurons bathed in Mg^{2+}-containing medium demonstrates that the depolarization evoked by NMDA is accompanied by an apparent resistance increase (Fig. 1A). This results from the voltage-dependence of the response to NMDA since the hyperpolarization associated with the current pulses used to measure the membrane resistance acts to shut off the conductance activated by NMDA. This voltage-dependent behavior can be better characterized under voltage clamp. At holding potentials near -60 mV, NMDA evokes an inward current in cultured spinal cord neurons. In the presence of a steady inward current evoked by NMDA, the membrane current-voltage (I-V) relationship (generated from a series of 30 msec voltage jumps) is non-linear (triangles, Fig. 1C) and merges with the membrane I-V relationship in the absence of NMDA (open squares, Fig. 1C) at membrane potentials more negative than -80 mV. In other words, the agonist-activated conductance decreases with hyperpolarization.

I-V relationships of currents activated by NMDA, L-aspartate and L-homocysteate show such a negative slope conductance at membrane potentials negative to -30 mV in medium containing normal concentrations of Mg^{2+} (near 1 mM). This negative slope conductance underlies the apparent membrane resistance increases that accompany depolarizations evoked by these agonists. In contrast, kainate and quisqualate, selective agonists at other acidic amino acid receptors, behave as more classical excitatory agonists and

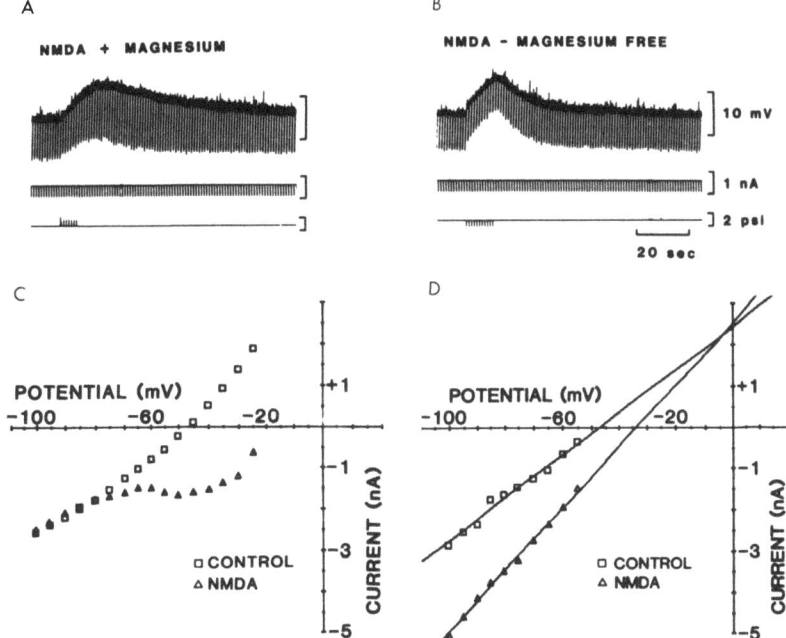

Fig. 1. A,B: The apparent membrane resistance change associated with NMDA-evoked depolarizations varies with the extracellular Mg^{2+} concentration. Upper trace: membrane potential. Hyperpolarizing current pulses (middle trace) were used to monitor membrane resistance. Bottom trace shows record of pressure pulses used to apply NMDA. C,D: I-V relationship of two neurons bathed in medium containing 1 mM Mg^{2+} and 0 mM Mg^{2+}, respectively. A series of voltage jumps before (open squares) and during (triangles) a steady inward NMDA current were used to generate the I-V plot. Note that the non-linearity of the response in C disappears when Mg^{2+} is removed from the extracellular solution (D). Responses in C and D were obtained under two-electrode voltage clamp; holding potential was -60 mV.

have a relatively linear current-voltage relationship (MacDonald and Porietis, 1982; Mayer and Westbrook, 1984).

Since the negative slope characteristic of the currents evoked by NMDA resembles regenerative sodium and calcium currents, it has been suggested that NMDA acts by modulating either a voltage-dependent sodium or calcium conductance (MacDonald et al., 1982; Dingledine, 1983; Flatman et al., 1983) rather than by activating a unique channel linked to an NMDA receptor. However, based on earlier observations that Mg^{2+} antagonizes the depolarizing action of NMDA on frog motoneurons (Evans et al., 1977; Ault et al., 1980), we have examined the behavior of NMDA responses in Mg^{2+}-free medium.

Under current clamp, depolarizations evoked by NMDA in Mg^{2+}-free medium now result in a membrane resistance decrease in marked contrast to responses in the presence of extracellular Mg^{2+} (compare Figs. 1A and 1B). As shown under voltage clamp in Fig. 1D, removal of Mg^{2+} results in loss of the negative slope behavior of the NMDA-activated conductance.

The membrane I-V relationship in the presence of NMDA (Fig. 1D, triangles) is nearly linear with the extrapolated regression line intersecting the control membrane I-V relationship (Fig. 1D, open squares) at -5 mV, the reversal potential for the NMDA response. Thus, the voltage-sensitivity of the NMDA response is due to a voltage-dependent action of physiological concentrations of Mg^{2+}.

Dose-ratio experiments have suggested that antagonism of NMDA responses by Mg^{2+} occurs at a different site than the site of action of competitive receptor antagonists (Evans and Watkins, 1978). In addition, block by Mg^{2+} of currents evoked by NMDA shows uncompetitive kinetics (Mayer and Westbrook, 1985). Voltage-dependence and uncompetitive kinetics are properties consistent with the effects of a charged molecule acting as a channel blocker, and thus suggest that Mg^{2+} enters and blocks NMDA channels under the influence of the membrane electric field. Preliminary calculations (based on steady-state chord conductance measurements) suggest that, in the presence of 1 mM Mg^{2+}, the blocking site is approximately halfway through the membrane (Mayer and Westbrook, 1985). Such a channel-blocking action by Mg^{2+} of current flow through single channels linked to NMDA receptors and activated by L-glutamate has recently been demonstrated (Nowak et al., 1984). Several other divalent cations including Ni^{2+} and Mn^{2+} also show voltage-dependent block of currents evoked by NMDA; however, Cd^{2+}, despite its extremely potent blocking action on voltage-sensitive Ca^{2+} currents, is only a weak blocker of NMDA responses (Mayer et al., 1984; Mayer and Westbrook, 1985).

Even in the absence of added Mg^{2+}, I-V plots of currents evoked by NMDA often show some voltage-sensitivity at membrane voltages negative to -70 mV (e.g., see Fig. 1, Mayer and Westbrook, 1985). This could be due either to a weak blocking action by Ca^{2+} in the recording medium (usually 2-5 mM) or to contamination by micromolar concentrations of Mg^{2+} in the 'Mg^{2+}-free medium' (see methods). To investigate this question further, we have recently examined the current-voltage relationship of responses evoked by NMDA over an extended range of membrane voltages (-220 mV to +30 mV) in medium containing the minimum possible divalent cations, 100 µM Ca^{2+} and no added Mg^{2+}. $[Ca^{2+}]_o$ lower than 100 µM resulted in irreversible deterioration of the neurons within 10 minutes, and has the additional complication of allowing Na^+ flux through Ca^{2+} channels (e.g., see Almers et al., 1984). Under these conditions, I-V plots of currents evoked by NMDA show a region of negative slope (or occasionally zero slope) at membrane voltages between -80 mV and -150 mV, but with further hyperpolarization the slope conductance increases again to a positive value (Fig. 2). This may be due to permeation of NMDA channels by the 'blocking' divalent cations, either Ca^{2+} or residual Mg^{2+} when the membrane electric field is sufficiently large, i.e., greater than -150 mV.

L-Glutamate

Voltage recording of responses to L-glutamate in vertebrate central neurons can be associated with either increases, decreases or no apparent change in membrane conductance (e.g., see Bernardi et al., 1972; Altmann et al., 1976; Segal, 1981; Hablitz and Langmoen, 1982). Various hypotheses to explain these observations have included an action of agonist at dendritic sites remote from the somatic recording electrode, a simultaneous increase in a sodium and decrease in a potassium conductance, or activation of a voltage-sensitive conductance. Under voltage clamp, I-V plots of currents evoked by L-glutamate on cultured spinal cord neurons consistently show a voltage-sensitivity intermediate between the highly voltage-dependent behavior of NMDA and L-aspartate and the rather voltage-insensitive behavior of kainate and quisqualate (Mayer and Westbrook, 1984). This is manifested as a region of zero slope conductance at membrane voltages between -30

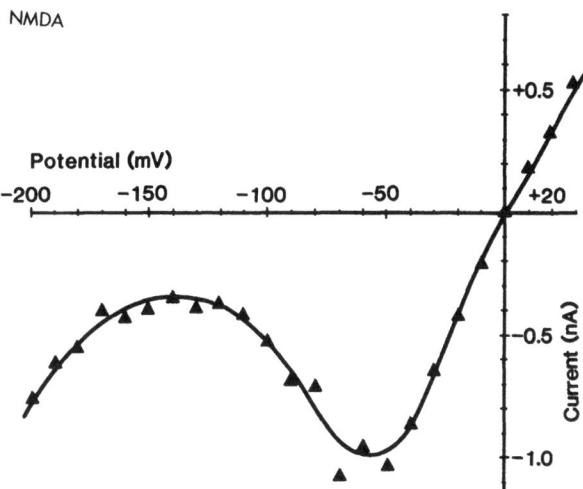

Fig. 2. Current-voltage relationship of NMDA-activated conductance of a spinal cord neuron in recording medium containing 100 μM Ca^{2+} and no added Mg^{2+}. I-V plot was constructed from a series of 30 msec voltage jumps during a steady inward current evoked by NMDA (50 μM); I-V relationships in the presence and absence of NMDA were digitally subtracted to give the plot of the agonist-activated conductance. Although the NMDA-activated conductance is relatively linear between −70 and 0 mV in this 'Mg^{2+}-free' medium, further hyperpolarization shows a region of negative slope which is relieved negative to −150 mV. Whole-cell patch recording with a discontinuous one-electrode voltage clamp.

mV and −80 mV (Fig. 3A, solid line). As expected for a conductance with zero slope, there is little or no membrane conductance change associated with the L-glutamate evoked current as measured by a 15 mV voltage jump under voltage clamp (Fig. 3B1). Since L-glutamate-evoked depolarizations result in activation of other voltage-dependent conductances (e.g., Ca^{2+} and K^+), a region of zero slope for the agonist-activated conductance permits the slope conductance of the membrane to determine the nature of the measured conductance change. This then could account for the variable changes in membrane resistance associated with L-glutamate-induced depolarizations.

Based on the use of selective antagonists, Watkins (1981) has suggested that L-glutamate may activate more than one type of acidic amino acid receptor. In the presence of the selective NMDA antagonist, (±)2-amino-5-phosphonovalerate (APV), responses to L-glutamate show a nearly linear I-V relationship (Fig. 3A, dashed line) and are associated with a clear increase in membrane conductance (Fig. 3B2). The difference current, i.e., the portion of the L-glutamate-evoked response that is blocked by APV, is highly voltage dependent (Fig. 3A, dotted line) and is indistinguishable from I-V plots of NMDA or L-aspartate in the same recording medium. That L-glutamate can activate NMDA receptors is supported by the findings

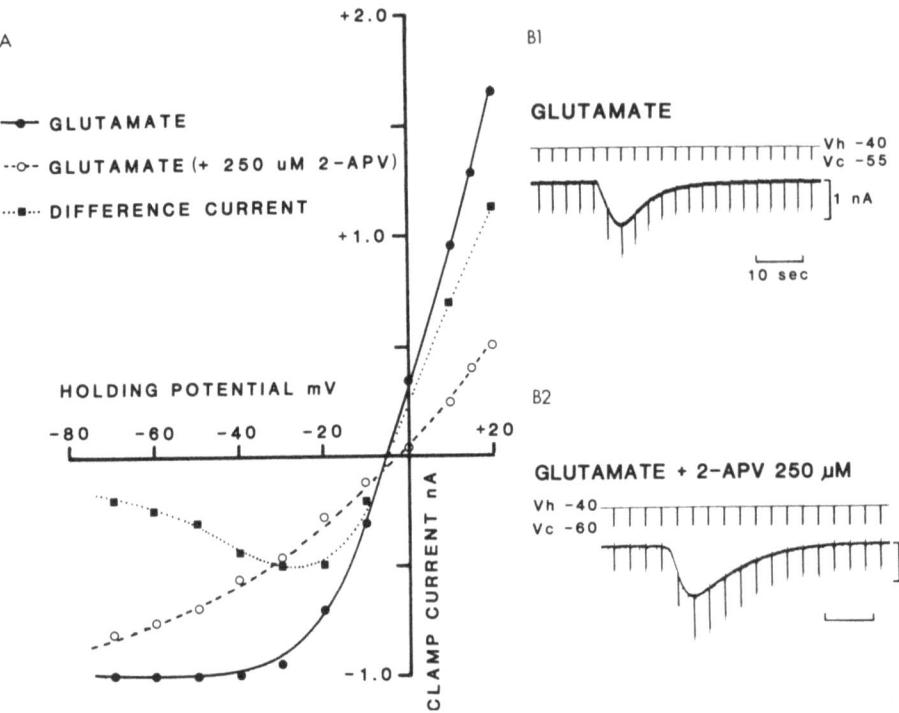

Fig. 3. A: Current-voltage plot of L-glutamate responses evoked in the same neuron under voltage clamp. During diffusion of (±)-2-APV from a large-tipped pipette containing 250 μM antagonist, the L-glutamate response became less voltage-sensitive. Subtraction of the control (solid line) from the response in the presence of APV (dashed line) was plotted as the difference current (dotted line). B1, B2: Conductance changes were measured in two different neurons with 200 msec voltage steps from a holding potential of -40 mV. In B1, the membrane conductance was 43 nS before and 46 nS at the peak of a 1 nA current evoked by L-glutamate. In B2, in the presence of 2-APV, the conductance increased from 28 nS to 55 nS during a similar amplitude L-glutamate current.

of Nowak et al. (1984) who reported the behavior of single channels activated by NMDA or L-glutamate. L-glutamate is also a potent displacer of the binding of ^{3}H-D(-)-2-amino-5-phosphonopentanoic acid, consistent with an action at NMDA receptors (Olverman et al., 1984). Thus, the intermediate voltage sensitivity of responses evoked by L-glutamate can be explained by simultaneous activation of a voltage-sensitive conductance, selectively activated by NMDA, and a voltage-insensitive conductance of the kainate/quisqualate type. At higher doses of L-glutamate, the conductance increase associated with non-NMDA receptors becomes the dominant component of the response on cultured spinal cord neurons (Mayer and Westbrook, 1984). This mixed agonist behavior of L-glutamate raises the interesting possibility that a postsynaptic response to presynaptic release of L-glutamate could show markedly different properties depending on the receptor types and their distribution on the postsynaptic neuron.

We have compared the membrane action of several excitatory amino acids on the basis of their voltage sensitivity under voltage clamp. Responses to agonists were recorded at membrane potentials from -100 mV to +30 mV. The steady state chord (ionic) conductances were derived from the relationship,

$$G_x = (E_m - E_x)/I_x,$$

where I_x is the current flow through the agonist-activated channels, E_x is the reversal potential for the response and E_m is the membrane potential. Since the conductance activated by the voltage-dependent agonists decreases with hyperpolarization (approximately an e-fold decrease per 25 mV hyperpolarization; Mayer and Westbrook, 1985), a convenient way to compare agonists is a ratio of chord conductances at a depolarized voltage (+20 mV) and a hyperpolarized voltage (-70 mV). This rectification ratio demonstrates the three categories of voltage sensitivity we have observed (Fig. 4). NMDA, L-aspartate and L-homocysteate have a high rectification ratio due to their strongly voltage-dependent action whereas quisqualate and kainate are nearly voltage-insensitive and thus have a low rectification ratio. The mixed agonists L-glutamate and D-homocysteate show intermediate ratios as does the simultaneous application of a mixture of kainate and NMDA ('mix').

Ionic Selectivity

The reversal potential of currents activated by NMDA is similar to that of kainate and quisqualate currents and close to 0 mV (Mayer and Westbrook, 1984). This suggests a mixed ionic mechanism with an increase in permeability to both Na^+ and K^+; and is not compatible with either a selective Na^+ or Ca^{2+} NMDA channel. Substitution of Cs^+ for K^+ in both voltage clamp (Mayer and Westbrook, 1984) and single channel (Nowak et al., 1984) recordings did not shift the reversal potentials for agonist-evoked responses. Thus, it is likely that both NMDA and non-NMDA channels show little discrimination between monovalent cations. This is similar to acetylcholine-activated channels at the frog endplate (Adams et al., 1980); but, of note, Cs^+ apparently does not permeate glutamate-activated channels at the locust neuromuscular junction (Anwyl, 1977).

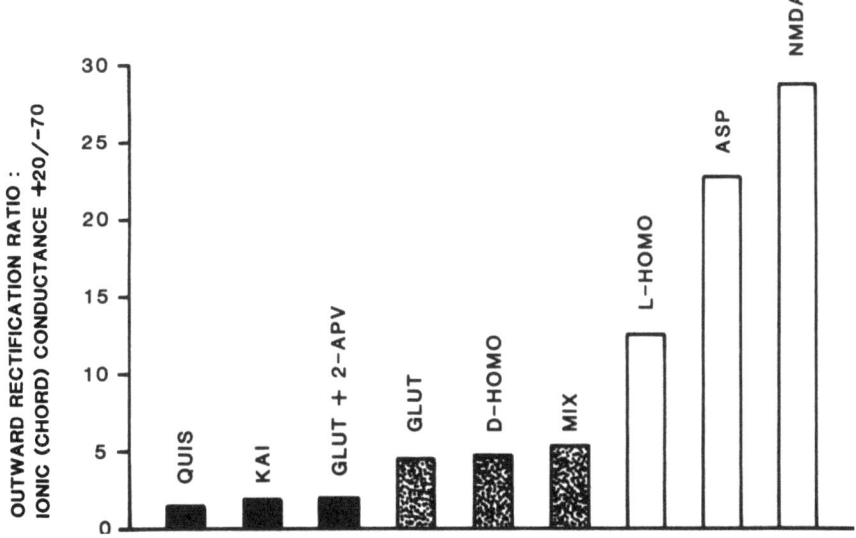

Fig. 4. Excitatory amino acids can be divided into three categories on the basis of their voltage sensitivity. See text for details.

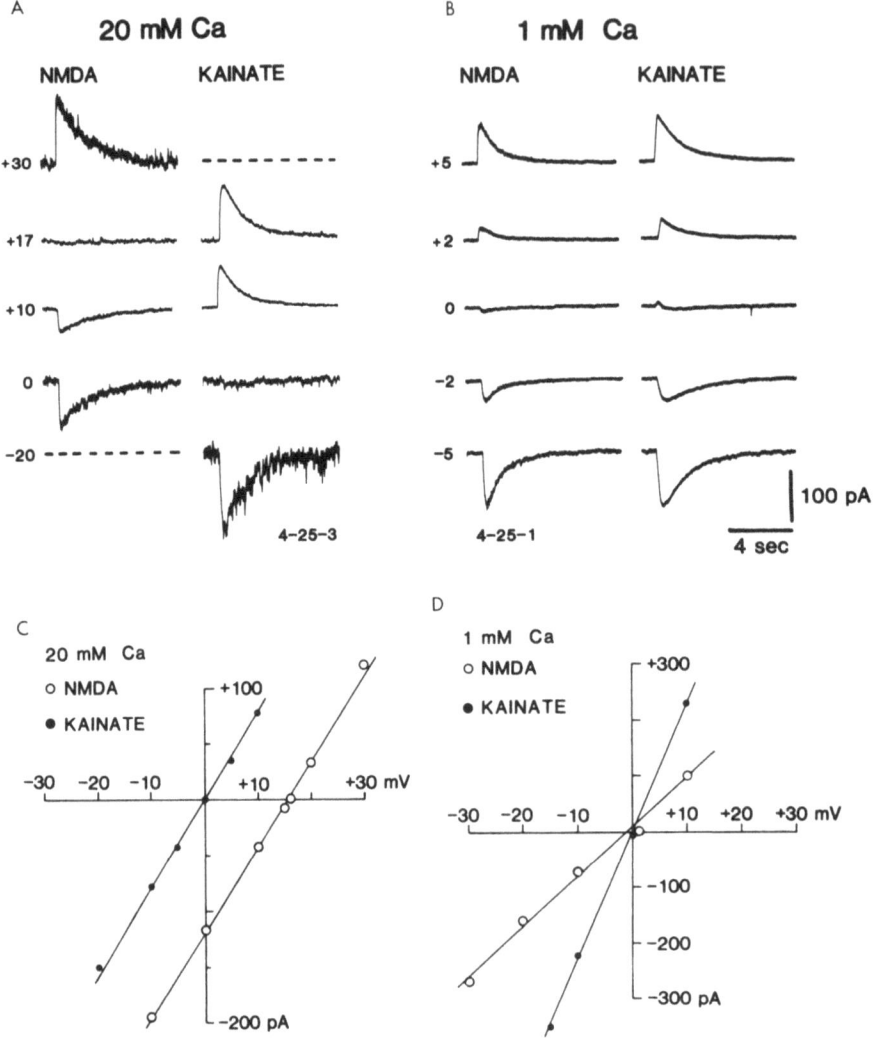

Fig. 5. The reversal potentials for responses to NMDA and kainate in 20 mM $[Ca^{2+}]_o$ (A) were compared with those in 1 mM $[Ca^{2+}]_o$ (B). I-V plots of the peak responses are shown in C and D. Note that the reversals are coincident in 1 mM Ca^{2+} but there is a positive shift of the NMDA reversal in 20 mM Ca^{2+}. Whole-cell recording using discontinuous one-electrode voltage clamp and Cs^+-containing patch electrode.

In a series of experiments performed in recording medium containing 10 mM Na^+, the reversal potential for responses evoked by kainate was shifted to a more negative potential (mean = -16 mV) consistent with a sodium-permeable channel. However, the reversal potential for NMDA responses on the same neurons did not shift, but rather remained near 0 mV (see Fig. 9, Mayer and Westbrook, 1985). In addition, NMDA responses were prolonged and 'faded' or desensitized with continued exposure to low-sodium (see also MacDonald, 1984). This made it difficult to obtain reliable values for the NMDA reversal potential in this medium, but did suggest a difference in permeability between non-NMDA and NMDA channels.

Calcium entry has been suggested to play a role in excitatory amino acid-induced neurotoxicity (Berdichevsky et al., 1983; but see Rothman, 1985). A glutamate-triggered increase in calcium influx has also been demonstrated in frog motoneurons (Bührle and Sonnhof, 1983). In order to determine the Ca^{2+} permeability of the responses evoked by excitatory amino acids, we have compared the reversal potential of kainate and NMDA on neurons in 1 mM $[Ca^{2+}]_o$, and then after raising $[Ca^{2+}]_o$ to 20 mM. Results from one such experiment are shown in Fig. 5. In 1 mM Ca^{2+} ($[Na^+]_o = 140$), both responses reverse near 0 mV; however, in 20 mM Ca^{2+} ($[Na^+]_o = 112$), the NMDA reversal potential shifted to +17 mV while the kainate reversal remained near 0 mV. These results suggest that the channels linked to kainate and NMDA receptors differ in their ionic selectivity. Channels linked to kainate receptors do not appear to have a significant calcium permeability (P_{Ca}) whereas NMDA channels have a higher P_{Ca} at least at depolarized membrane potentials where the reversal potential is measured. At physiological ionic conditions, the calcium flux through the NMDA channels is likely to be small compared to the monovalent cation flux. Whether this calcium flux plays a significant functional role in physiological or pathological processes remains to be determined.

REFERENCES

Adams, D.J., Dwyer, T.M., and Hille, B., 1980, The permeability of end-plate channels to monovalent and divalent metal cations, J. Gen. Physiol., 75:493.

Almers, W., McCleskey, E.W., and Palade, P.T., 1984, A non-selective cation conductance in frog muscle membrane blocked by micromolar external calcium ions, J. Physiol., 353:565.

Altmann, H., Ten Bruggencate, G., Pickelmann, P., and Steinberg, R., 1976, Effects of glutamate, aspartate and two presumed antagonists on feline rubrospinal neurones, Pflüg. Arch., 364:249.

Anwyl, R., 1977, The effect of foreign cations, pH and pharmacological agents on the ionic permeability of an excitatory glutamate synapse, J. Physiol., 273:389.

Ault, B., Evans, R.H., Francis, A.S., Oakes, D.J., and Watkins, J.C., 1980, Selective depression of excitatory amino acid induced depolarizations by magnesium ions in isolated spinal cord preparations, J. Physiol. 307:413.

Berdichevsky, E., Riveros, N., Sanchez-Armass, S., and Orrego, F., 1983, Kainate, N-methylaspartate and other excitatory amino acids increase calcium flux into rat brain cortex cells in vitro, Neurosci. Lett., 36:75.

Bernardi, G., Zieglgänsberger, W., Herz, A., and Puil, E., 1972, Intracellular studies on the action of L-glutamic acid on spinal neurones of the cat, Brain Res., 39:523.

Bührle, Ch. Ph., and Sonnhof, U., 1983, The ionic mechanism of the excitatory action of glutamate upon the membrane motoneurones of the frog, Pflüg. Arch., 396:154.

Dingledine, R., 1983, N-methyl aspartate activates voltage-dependent calcium conductance in rat hippocampal pyramidal cells, J. Physiol., 343:385.

Engberg, I., Flatman, J.A., and Lambert, J.D.C., 1979, The actions of excitatory amino acids on motoneurones in the feline spinal cord, J. Physiol., 228:227.

Evans, R.H., Francis, A.A., and Watkins, J.C., 1977, Selective antagonism by Mg^{2+} of amino acid-induced depolarizations of spinal neurones, Experientia, 33:489.

Evans, R.H., and Watkins, J.C., 1978, Dual sites for antagonism of excitatory amino acid actions on central neurones, J. Physiol., 277:57P.

Flatman, J.A., Schwindt, P.C., Crill, W.E., and Stafström, C.E., 1983, Multiple actions of N-methyl-D-aspartate on cat neocortical neurones in vitro, Brain Res., 266:169.

Hablitz, J.J., and Langmoen, I.A., 1982, Excitation of hippocampal pyramidal cells by glutamate in the guinea-pig and rat, J. Physiol., 325:317.

MacDonald, J.F., and Porietis, A.V., 1982, DL-Quisqualic and L-aspartic acids activate separate excitatory conductances in cultured spinal cord neurones, Brain Res., 245:175.

MacDonald, J.F., Porietis, A.V., and Wojtowicz, J.M., 1982, L-aspartic acid induces a region of negative slope conductance in the current voltage relationship of cultured spinal cord neurones, Brain Res., 237:248.

MacDonald, J.F., and Wojtowicz, J.M., 1982, The effects of glutamate and its analogues upon the membrane conductances of central murine neurones in culture, Can. J. Physiol. Pharmacol., 60:282.

MacDonald, J.F., 1984, Substitution of extracellular sodium ions blocks voltage-dependent decrease of input conductance evoked by L-aspartic acid, Can. J. Physiol. Pharmacol., 62:109.

Mayer, M.L., and Westbrook, G.L., 1984, Mixed-agonist action of excitatory amino acids on mouse spinal cord neurones under voltage clamp, J. Physiol. 354:29.

Mayer, M.L., Westbrook, G.L., and Guthrie, P.B., 1984, Voltage-dependent block by Mg^{2+} of NMDA responses in spinal cord neurones, Nature, 309:261.

Mayer, M.L., and Westbrook, G.L., 1985, The action of N-methyl-D-aspartic acid on mouse spinal cord neurones in culture, J. Physiol., 361:65.

Nowak, L., Bregestovski, P., Ascher, P., Herbet, A., and Prochiantz, A., 1984, Magnesium gates glutamate-activated channels in mouse central neurones, Nature, 307:462.

Olverman, H.J., Jones, W.S., and Watkins, J.C., 1984, L-glutamate has higher affinity than other amino acids for [^{3}H]-D-AP5 binding sites in rat brain membranes, Nature, 307:460.

Rothman, S.M., 1985, The neurotoxicity of excitatory amino acids is produced by passive chloride influx, J. Neurosci., 5:1483.

Segal, M., 1981, The actions of glutamic acid on neurones in the rat hippocampal slice, in: Glutamate as a Neurotransmitter, G. DiChiara and G.L. Gessa, eds., Raven Press, New York, p. 217.

Watkins, J.C., 1981, Pharmacology of excitatory amino acid transmitters, in: Amino Acid Neurotransmitters, F.V. DeFeudis and P. Mandel, eds., Raven Press, New York, p. 205.

A PATCH-CLAMP STUDY OF EXCITATORY AMINO ACID ACTIVATED CHANNELS

P. Ascher and L. Nowak[1]

Laboratoire de Neurobiologie
Ecole Normale Supérieure
46 rue d'Ulm - 75230 Paris Cedex 05
[1]Present Address: Department of Pharmacology
Cornell University Veterinary College
Ithaca, New York 14853

Although the introduction of the patch-clamp techniques (Neher and Sakmann, 1976; Hamill et al., 1981) has been universally recognized as a major breakthrough in the biophysical study of ion channels, these techniques have not yet become a standard tool for pharmacological studies. We shall briefly describe below some of the results that we have obtained, using 'outside-out' patches of mouse central neurons (Nowak et al., 1984), in the characterization of excitatory amino acid (EAA) receptors and channels. In this description, we shall not develop the biophysical results (and in particular the description of kinetics of the channels) but shall concentrate on the results of direct pharmacological and physiological relevance.

COUNTING THE EAA RECEPTOR TYPES

Part of the difficulty encountered in characterizing the EAA receptors has been that these receptors often coexist on a given cell, and that the available agonists and antagonists are not perfectly selective. Single channel studies offer, in principle, a possible answer to the problem of 'counting the number of receptor types' in such a complex situation. Even in the absence of specific agonists or antagonists, and even when different channels coexist on a given membrane patch, the observation of n classes of single channel currents may indicate the presence of n types of channels, and possibly of n types of receptors. This reasoning must be qualified, however, by the possibility that a given receptor may activate different conductance substates (Hamill and Sakmann, 1981; Hamill et al., 1983).

We have tried to apply this approach to the channels activated by N-methyl-D-aspartate (NMDA), quisqualate (QUIS), kainate (KAI) and L-glutamate (L-GLU) in outside-out patches.

The most readily observed channels were those activated by NMDA. In many patches, this agonist activated a single class of channels which showed a linear I-V relations with a slope of about 50 pS in Mg-free solutions (Nowak et al., 1984). More recently, we have observed QUIS channels of 8 pS conductance (resolved directly) and KAI channels of 2 pS conductance (this value was calculated from the noise of the response observed on outside-out patches).

These results fit quite well with the classical notion that there are three main types of EAA receptors: NMDA, QUIS and KAI (McLennan, 1981; Watkins and Evans, 1981), A1, A2, A3 (Foster and Fagg, 1984). However, additional observations indicate that the situation is in fact more complex:

1) We have observed in some patches 'substates' of the NMDA channels having conductances of 35 pS and 18 pS, and differing from the 50 pS state by their mean open time.

2) QUIS and KAI can activate NMDA channels, and therefore are not perfectly selective agonists.

3) QUIS and KAI induced, in some patches, single channel currents which were neither those expected for the 8 pS or 2 pS channels, nor those corresponding to the NMDA channels. These currents were fast and difficult to characterize. In the case of KAI, however, it was possible to determine a conductance of 20 pS.

The complexity of this description is further increased by the observation (Cull-Candy and Ogden, 1985 and Cull-Candy, this volume) of a glutamate activated channel of very low conductance (0.15 pS). Clearly three receptor types are not enough to describe even the 'direct' effects of EAA, not to mention those which may depend on internal messengers (Foster and Roberts, 1981; Garthwaite, 1982; Sladeczek et al., 1985; Nicoletti et al., 1986). The description of the basic conductance states made above should, however, provide a starting point for a new pharmacological characterization of the various receptor types.

PHYSIOLOGICAL FUNCTION OF THE NMDA CHANNELS

The voltage dependence of the I-V relation of the NMDA system in physiological (Mg containing) solutions (Mayer et al., 1984; Nowak et al., 1984) has a number of functional implications. We have discussed some of them in a recent paper (Ascher et al., 1986) and stressed in particular the fact that the 'negative resistance' region of the I-V relation, despite the fact that it is not due to a K-conductance decrease as originally thought (Engberg et al., 1979), has the functional advantages of a K conductance decrease: it depolarizes the cell without increasing its input conductance, and therefore without shunting superimposed EPSPs.

In many cases, however, the role of the negative resistance region will be to facilitate oscillations. The amplitude of these oscillations will depend on the glutamate concentration, as indicated in Fig. 1. The cell is likely to oscillate between E_1 (independent of the glutamate concentration) and E_3, a value which depends on the glutamate concentration. Note, however, that in principle the membrane potential may stabilize for short periods at the intermediate level E_2, from which a hyperpolarization would bring it to E_1, whereas a depolarization would bring it to E_3.

On the bottom part of Fig. 1 is illustrated a possible consequence of the relatively high level of extracellular glutamate. This level has not been measured directly, but may be quite close to that found in the CSF. There, early estimates of up to 30 μM (Curtis and Johnston, 1974) have been revised down, but values around 1 μM can be considered as quite plausible (Ferraro and Hare, 1985). We have observed that this concentration

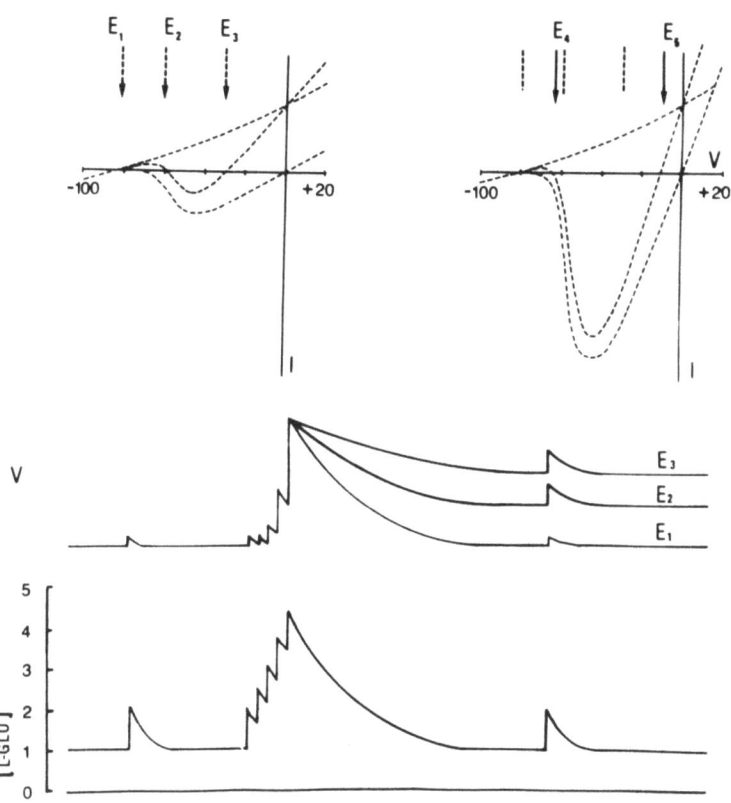

Fig. 1. Upper drawings (I-V curves): An increase in glutamate concentration from 1 to 4 µM will neither alter the reversal potentials of the 'resting' membrane conductance (set near -80 mV) nor that of the glutamate response 0 mV) but will alter the points (E2, E3, E4, E5) at which the compound I-V curve crosses the voltage axis, and which may represent the limits between which the membrane potential will oscillate (see text). Lower traces: If one assumes that at rest the extracellular glutamate concentration is 1 µM, a brief synaptic increase in glutamate concentration (from 1 to 2 µM) will have little effect on membrane potential if it occurs at low frequency, but will be greatly potentiated if a previous depolarization of the cell has triggered a 'regenerative' depolarization.

already produces <u>in vitro</u> a substantial activation of the NMDA conductance. If we assume that such an interstitial concentration exists <u>in vivo</u>, in the presence of Mg and around resting potential, Mg blockade would prevent any substantial depolarization of the cell. However, any increase in the glutamate concentration (or any extraneous depolarization) may relieve the Mg block and lead to a prolonged depolarization. Even if the extracellular glutamate is not high enough to maintain the membrane potential at the plateau value for a very long period, it will create a long lasting 'tail' in response to a brief synaptic input. Note that, in addition, a single EPSP will be greatly potentiated during this tail as compared with the control situation.

CONCLUSION

Outside-out patches are not the universal answer to all the questions concerning the EAA receptors and channels; they are poorly adapted to the study of dendritic or 'subsynaptic' receptors, or to the characterization of responses activated by, or depending upon, internal messengers (Foster and Roberts, 1981; Garthwaite, 1982; Sladeczek et al., 1985; Nicoletti et al., 1986). On the other hand, we think that outside-out patches, beyond their 'biophysical' interest, have a lot to offer for pharmacologists and for physiologists. In some favorable cases it appears indeed possible to go 'from the patch to the neuron'.

REFERENCES

Ascher, P., Nowak, L., and Kehoe, J.S., 1986, Glutamate-activated channels in molluscan and vertebrate neurones, in: Ion Channels in Neural Membranes, J.M. Ritchie and R.D. Keynes, eds., Alan R. Liss, in press.
Cull-Candy, S.G., and Ogden, D.C., 1985, Ion channels activated by L-glutamate and GABA in cultured cerebellar neurones of the rat, Proc. R. Soc. B., 224:367.
Curtis, D.R., and Johnston, G.A.R., 1974, Amino acid transmitters in the mammalian central nervous system, Ergebn. Physiol., 69:97.
Engberg, I., Flatman, J.A., and Lambert, J.D.C., 1979, The actions of excitatory amino acids on motoneurones in the feline spinal cord, J. Physiol., 288:227.
Ferraro, T.N., and Hare, T.A., 1985, Free and conjugated amino acids in human CSF: influence of age and sex, Brain Res., 338:53.
Foster, A.C., and Fagg, G.E., 1984, Acidic amino acid binding sites in mammalian neuronal membranes: their characteristics and relationship to synaptic receptors, Brain Res. Rev., 7:101.
Foster, G.A., and Roberts, P.J., 1981, Kainic acid stimulation of cerebellar cyclic GMP levels: potentiation by glutamate and related amino acids, Neurosci. Lett., 23:67.
Garthwaite, J., 1982, Excitatory amino acid receptors and guanosine 3', 5' cyclic monophosphate in incubated slices of immature and adult rat cerebellum, Neuroscience, 7:2491.
Hamill, O.P., Marty, A., Neher, E., Sakmann, B., and Sigworth, F., 1981, Improved patch clamp techniques for high resolution current recording from cells and cell-free membrane patches, Pflügers Arch., 391:85.
Hamill, O.P., and Sakmann, B., 1981, Multiple conductance states of single acetylcholine receptor channels in embryonic muscle cells, Nature, 294:462.
Hamill, O.P., Bormann, J., and Sakmann, B., 1983, Activation of multiple-conductance state chloride channels in spinal neurones by glycine and GABA, Nature, 309:160.
Mayer, M.L., Westbrook, G., and Guthrie, P.B., 1984, Voltage dependent block by Mg^{2+} of NMDA responses in spinal cord neurones, Nature, 309:261.
McLennan, H., 1981, On the nature of receptors for various excitatory amino acids in the mammalian central nervous system, Adv. Biochem. Psychopharmacology, 27:253.
Neher, E., and Sakmann, B., 1976, Single channel currents recorded from membrane of denervated frog muscle fibers, Nature, 260:799.

Nicoletti, F., Meek, J.L., Iadarola, M.J., Chuang, D.M., Roth, B.L., and Costa, E., 1986, Coupling of inositol phospholipid metabolism with excitatory amino acid recognition sites in rat hippocampus, J. Neurochem., 46:40.

Nowak, L., Bregestovski, P., Ascher, P., Herbet, A., and Prochiantz, A., 1984, Magnesium gates glutamate-activated channels in mouse central neurones, Nature, 307:462.

Sladeczek, F., Pin, J.P., Recasens, M., Bockaert, J., and Weiss, S., 1985, Glutamate stimulates inositol phosphate formation in striatal neurones, Nature, 317:717.

Watkins, J.C., and Evans, R., 1981, Excitatory amino acid transmitters, Ann. Rev. Pharmacol. Toxicol., 21:165.

AMINO ACID ACTIVATED RECEPTOR-CHANNELS AT PERIPHERAL AND CENTRAL SYNAPSES

S.G. Cull-Candy

MRC Receptor Mechanisms Research Group
Department of Pharmacology
University College London

Synaptic channels opened by glutamate and GABA have been examined in locust muscle fibers and cultured cerebellar neurons of the rat. In both systems, the receptors are readily accessible to techniques such as noise analysis (Katz and Miledi, 1972; Anderson and Stevens, 1973) and single channel recording (Neher et al., 1978; Hamill et al., 1981), allowing basic information to be obtained about amino acid activated receptor-channels.

MEMBRANE CHANNELS IN LOCUST MUSCLE FIBERS

Glutamate Receptor-Channels

Locust muscle receives both inhibitory and excitatory innervation; the inhibitory transmitter is thought to be GABA and the excitatory transmitter to be glutamate, although this has not been shown conclusively (see Usherwood and Cull-Candy, 1975). There is, however, fairly compelling evidence that GABA and glutamate, respectively, are the inhibitory and excitatory nerve-muscle transmitters in closely related species such as crayfish (Otsuka et al., 1967; Kawagoe et al., 1984).

Three main techniques have been used to obtain information about the synaptic channels in locust muscle. The nerve has been stimulated to release transmitter and inhibitory or excitatory synaptic currents recorded (Cull-Candy and Miledi, 1982; Cull-Candy, 1984). In other experiments, glutamate or GABA was applied ionophoretically to junctional sites and the membrane noise, associated with the activation of a large number of channels, has been analyzed (Anderson et al., 1978; Cull-Candy and Miledi, 1981). In addition, glutamate (or agonists) have been applied at low concentration in a pipette that can be used for patch-clamp recording, to allow the detection of single channel currents directly (Cull-Candy et al., 1981; Cull-Candy and Parker, 1982).

In order to obtain information about the single channel from glutamate noise, it is necessary to make certain assumptions about the origin of the noise (Katz and Miledi, 1972; Anderson and Stevens, 1973). Firstly, when glutamate - or any similar fast neurotransmitter - acts on the postsynaptic membrane, it produces a response by opening a certain number of channels. The actual number fluctuates around a mean value giving rise

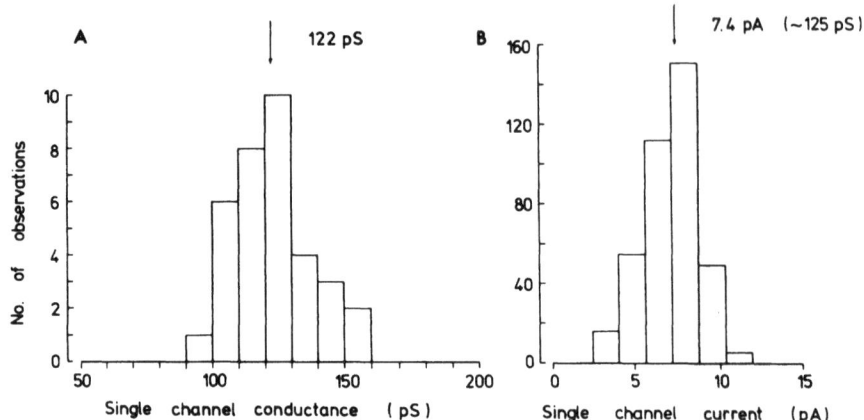

Fig. 1. A: Amplitude distribution of conductances of single channels opened by glutamate. Values are from analysis of glutamate noise obtained at junctions in five muscles; $\gamma = 122 \pm 0.4$ (S.E.) pS indicated by arrow. B: Amplitude distributions of single channel currents obtained with patch-clamp recording from an extrajunctional site. The average single channel current, indicated by arrow, is 7.4 pA; hence for a clamp holding potential of $V_m = -60$ mV, $\gamma \sim 125$ pS.

to fluctuations in membrane current. Secondly, glutamate molecules bind to the postsynaptic membrane causing the transient opening of single channels which produce rectangular current (or conductance) pulses. As has been shown with patch-clamp recording, the conductance pulse may be characterized by a short burst of openings (Colquhoun and Sakmann, 1981; Cull-Candy and Parker, 1982). If a population of channels are simultaneously activated, the envelope of their open times measured with low frequency resolution is described by an exponential distribution where the time constant of the exponential equals the mean burst-length of the channel. With higher frequency resolution, the open-times are usually fitted by the sum of two or more exponentials.

Fig. 1A shows a histogram of the estimates of single channel conductance obtained with noise analysis from five preparations where the recording conditions were particularly good. The average value for the single channel conductance is 122 pS (Anderson et al., 1978), which is roughly five times larger than the conductance of the ACh-activated channel at the vertebrate endplate (Anderson and Stevens, 1973). Extrajunctional glutamate activated D-receptor channels which occur in normal innervated locust muscle fibers (Cull-Candy, 1976), have been examined by means of patch-clamp recording and similar values obtained (Patlak et al., 1979; Cull-Candy et al., 1981). The histogram in Fig. 1B shows the distribution of currents obtained from a single membrane patch and the conductance of this channel is in good agreement with noise estimates. However, values closer to 150 pS have been obtained in many of our experiments (Cull-Candy and Parker, 1983). It is difficult to know if this is a real (although small) difference between junctional and extrajunctional channels or, as seems more likely, simply reflects differences in the techniques used.

Fig. 2 shows examples of single channel currents recorded with a patch-clamp electrode containing 100 µM glutamate. The records are from two muscle fibers voltage-clamped at -110 mV. The single channel current is relatively uniform in size whereas the channel open times can be roughly fitted by an exponential distribution as expected from the noise spectra.

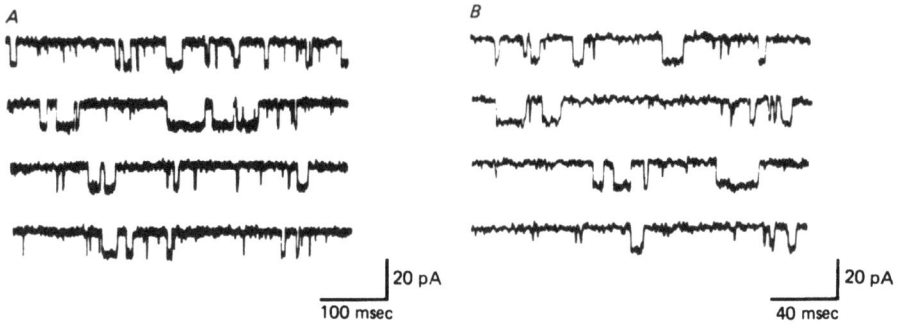

Fig. 2. A, B: records of glutamate-induced single-channel activity from two concanavalin A treated muscle fibers; note the difference in sweep speeds. Cl⁻-free solution was used for the bathing medium and in the patch pipette; in both experiments, the pipette contained 100 μM glutamate. Fibers were voltage-clamped to a potential of -110 mV with a conventional two-micro-electrode clamp, and the patch pipette was placed between the recording and current-passing electrodes. Recording band width was DC to 1 kHz. Inward current is shown as a downward deflection. T = 22°C, calibration 100 or 40 msec and 20 pA. (From Cull-Candy et al., 1981).

There are no double sized events in these recordings, indicating that only one single channel is present under the tip of the patch pipette. This reflects the rather low density of extrajunctional receptors.

GABA-Receptor Channels

GABA-noise spectra can usually be fitted to a single exponential but some of the GABA-spectra show a clear higher frequency component. The burst-length of the GABA channel is decreased by hyperpolarization, as signified by the shift along the frequency axis, and the potential dependence is greater than for the glutamate channel. The mean burst-length of the GABA-channel at -80 mV is between 4-6 msec, markedly longer than the burst length of the glutamate channel (~2.5 msec) and its conductance is only about 20-25 pS, about one fifth the size of the excitatory channel.

In the past, it has proved difficult to record miniature inhibitory currents at many inhibitory synapses as a way of obtaining information about the GABA-channel. In recent experiments, we have examined fibers in which the glutamate receptors have been desensitized. In this situation, with the excitatory currents abolished, m.i.j.c.s. or i.j.c.s. composed of one or more packets (Fig. 3) can be readily detected. The size of the m.i.j.c.s. quantal event can be predicted from the variance of the nerve-evoked inhibitory current and m.i.j.c.s. reverse at the Cl⁻ equilibrium potential. Furthermore, they are unaffected by tetrodotoxin so they do not appear to be multi-quantal events resulting from spontaneous nerve-terminal action potential.

Number of Channels Per Transmitter Quantum

From the size of the miniature inhibitory junction currents and miniature excitatory junctional currents and the properties of the individual synaptic channels (from noise analysis), we have estimated the number of channels opened by a single transmitter packet. A transmitter packet opens about 200 channels at the excitatory synapse and about 1000 channels

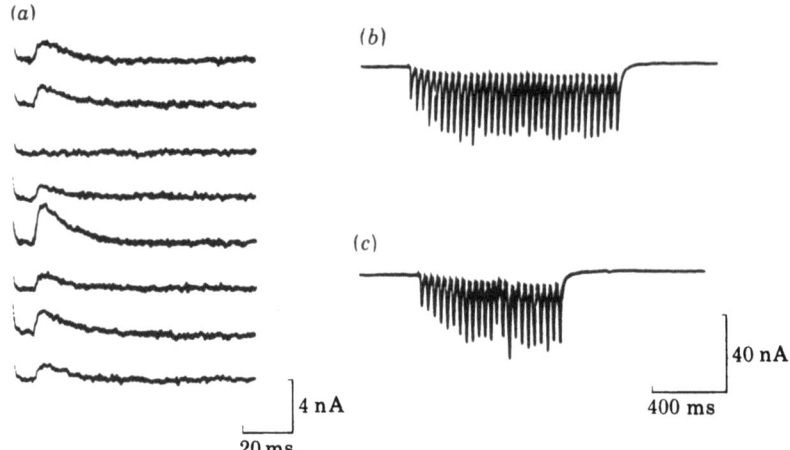

Fig. 3. (a) I.j.c.s. evoked by nerve-stimulation (in normal level of Ca^{2+}). The probability of release of transmitter packets was low. Some impulses fail to release transmitter (third trace down); when present, i.j.c.s. appear to be composed of one, two (seventh trace down) or three (fifth trace down) packets; V_m = -30 mV; calibration 4 nA and 20 ms. (b,c) Trains of i.j.c.s. evoked by high frequency (35 Hz) stimulation (different muscle from (a)). Fluctuations in i.j.c. amplitude can be seen accompanying facilitation of release. V_m = -80 mV; calibration 40 nA and 400 ms. (From Cull-Candy, 1984).

at the inhibitory synapse. So the packet of excitatory transmitter opens a relatively small number of channels when compared with the vertebrate end-plate where the ACh-packet opens 1000-2000 channels (Katz and Miledi, 1972; Anderson and Stevens, 1973). There are several possible explanations for the small number of excitatory channels opened; it could be that the size of the transmitter packet is small or the density of receptors low. However, experimental evidence indicates that glutamate is not very effective at opening postsynaptic channels (see below).

Response of the Individual Receptor to Various Glutamate Concentrations

It was of interest to look at the response of the individual glutamate-receptors to various glutamate-concentrations using patch-clamp recording since in many of our earlier experiments it had been possible to locate membrane patches where only a single channel was active under the tip of the electrode. However, one of the limitations of conventional patch-clamp recording (from cell attached patches) is that the concentration of drug contained within the pipette cannot be readily altered during recording. To overcome this problem, we used a simple system which allowed rapid internal perfusion of the tip of the patch electrode (Cull-Candy, et al., 1981; Cull-Candy and Parker, 1983). The experimental arrangement involves inserting a perfusion pipette down the inside of a patch electrode. The perfusion pipette usually consists of six separate barrels each of which is pre-filled with a different solution. When air pressure is applied to the back of one of the hypodermic needles, the drug solution is injected continuously from one of the barrels into the tip of the electrode, and becomes the predominant concentration in the electrode tip.

Fig. 4 shows an experiment where different glutamate concentrations

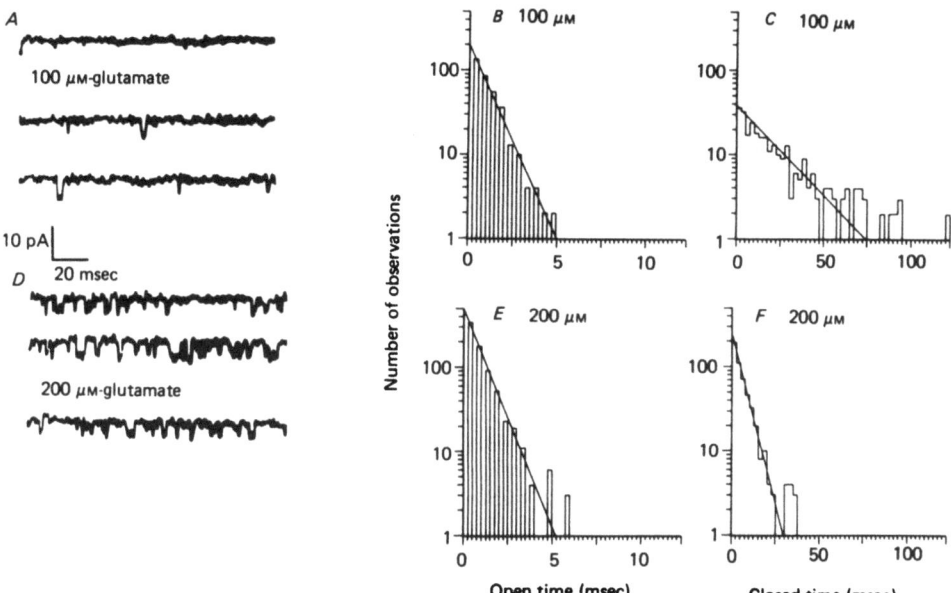

Fig. 4. Examples of channel openings recorded in the presence of two different glutamate concentrations applied to the same membrane patch. A, B and C: traces and corresponding histograms of events obtained with a patch electrode perfused with 100 μM glutamate. D, E and F: patch electrode perfused with 200 μM glutamate. At each concentration, the traces are samples from recordings lasting several minutes (analog tape playback). At this patch, only one channel appeared active: no double sized events were seen even with high glutamate concentrations (600 μM; not shown). B and E: distribution of channel open times for two concentrations can be well fitted with straight lines (on semilogarithmic coordinates), indicating an exponential distribution. Lines fitted to the data correspond to a mean open time of 1 msec in 100 μM glutamate and 0.9 msec in 200 μM glutamate. C and F, distribution of channel closed times. Straight lines fitted to the data correspond to a mean channel closed time of 21 msec in 100 μM glutamate and 5.6 msec in 200 μM glutamate. Calibration: 10 pA, 20 msec; T = 22°C. (From Cull-Candy et al., 1981).

were applied to the membrane patch. The upper traces and histograms are data obtained with 100 μM glutamate, the lower records with 200 μM glutamate. As expected the frequency of opening increased with concentration. The left-hand histograms show the distribution of open times (burst-lengths) on log-linear plots, which can be roughly fitted to a single exponential, and the time-constant does not change with concentration. On the other hand, the closed-time distributions (right-hand histograms) show a decrease in time constant with increasing glutamate concentration.

By measuring the open and closed times, it is possible to get a rough idea of the _effectiveness_ of glutamate at opening channels by measuring the glutamate concentration which holds the channel open for 50% of the time. This occurs at 500 μM glutamate, a concentration which is about ten times larger than the ACh concentration required to hold the ACh channel open for 50% of the time at the vertebrate endplate (Sakmann et al., 1981). If the junctional glutamate receptor behaves in the same way as the extrajunctional receptor, this could clearly account for the small number of channels opened by a transmitter packet.

High Frequency Resolution of Single Glutamate Channels

From patch-clamp experiments with improved time-resolution it was apparent that many channel openings were interrupted by one or more brief closings (Cull-Candy and Parker, 1982, 1983). Since these brief closings were not previously detected with patch-clamp recording or noise analysis using less good time-resolution, the time constant underlying the noise and previous estimates of channel 'life-time' from single channel recordings corresponded to the 'burst length' of a burst of openings, as described for ACh-channels at the vertebrate endplate (Colquhoun and Sakmann, 1981).

It could be that this situation results from several receptor activations occurring in rapid succession. In which case the closed-time distribution should be fitted by a single exponential function. In fact, it is fitted by the sum of two exponentials. By measuring closed and open-time distributions, we found that a glutamate-activated channel normally consists of about 1.6-2.0 openings per burst. The simplest situation to visualize is that a single receptor activation produces a burst of two openings each of which lasts for 1 msec, interrupted by a 100 μsec brief gap. So the 2 msec burst is equivalent to our earlier estimates of channel 'life-time'.

In earlier experiments with less good time resolution, we had compared various glutamate agonists and found that fluoroglutamate produced a channel with a brief 'life-time', whereas quisqualate produced a substantially longer 'life-time'. It seemed that different burst lengths could have two possible origins. All the agonists could produce about two openings per burst and the duration of the individual opening varies. Alternatively, if the duration of the individual opening is fixed, the number of openings per burst varies. It transpires that, on average, fluoroglutamate produces about _one_ opening per burst, glutamate produces about _two_ and quisqualate about _three_ (Fig. 5). However, there is also a slight increase in the duration of each individual opening, with increasing agonist potency, although the increase in the number of openings/burst appears the predominant effect (Cull-Candy and Parker, 1983).

CEREBELLAR NEURONS IN EXPLANT CULTURE

There is a good deal of evidence that glutamate and GABA are widespread neurotransmitters in the mammalian brain (Krnjević, 1974; Nistri and Constanti, 1979). With the advent of patch-clamp recording and its application to tissue cultured neurons, it became possible to examine these receptor-channels directly in central neurons (Hamill et al., 1983; Nowak et al., 1984; Cull-Candy and Ogden, 1985). There are a number of considerable technical advantages in patch-clamping small cultured cells such as neurons since they are amenable to various sorts of experimental manipulation (described by Hamill et al., 1981). For example, it is possible to form a high resistance seal (i.e., 10-100 GΩ) between the pipette and the cell membrane by applying suction to the inside of the patch pipette while pressing it against the cell, and the high resistance seal is maintained when suction ceases (see Hamill et al., 1981). A high resistance seal has two advantages: it reduces thermal noise due to ions moving under the electrode tip and it reduces attenuation of the signal, part of which flows under the rim of the pipette. When the pipette is sealed tightly against a cell, the membrane patch can be punctuated, usually by further suction, without damaging the seal around the pipette rim (Fig. 6). If the pipette contains a medium compatible with the intracellular medium

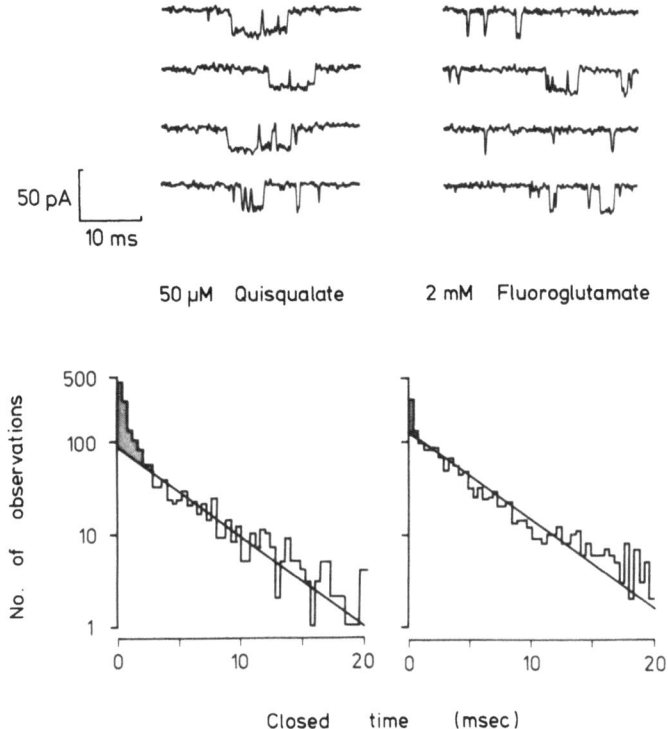

Fig. 5. Recordings of single-channel currents activated by 50 μM quisqualate and 2 mM fluoroglutamate made with a bandwidth of DC-3 kHz. For both agonists, some channel openings can be seen to be interrupted by brief closings, but the mean number of brief closings per burst is higher in the case of quisqualate. Histograms: Distributions of channel-closed times for channels activated by 50 μM quisqualate and 2 mM fluoroglutamate, plotted on semilogarithmic scales. Over much of the time scale, the distributions can be fitted well by single exponential functions (straight lines), except that for both agonists there is a clear excess of very brief closings (shaded areas). The relative number of excess brief closings and hence the number of openings per burst is greater for quisqualate than for fluoroglutamate. (From Cull-Candy and Parker, 1983).

(usually 140 mM CsCl, 5 mM EGTA, 10 mM K HEPES, 0.5 mM $CaCl_2$, 4 mM NaCl), this arrangement can be used very effectively for voltage-clamping the whole cell. The pathway between the pipette and the cell interior is low resistance, so the system allows efficient voltage-clamp of whole cell currents, provided the cell under examination is small (<30 μM diameter). Because in this configuration recording is from the 'whole cell' (with a large membrane area), the background noise is less good. However, it is an effective way of looking at whole cell membrane currents and transmitter activated noise produced by the opening and closing of large numbers of ion channels. In addition, when the pipette is pulled gently away from the cell, a 'neck' of membrane is formed which breaks

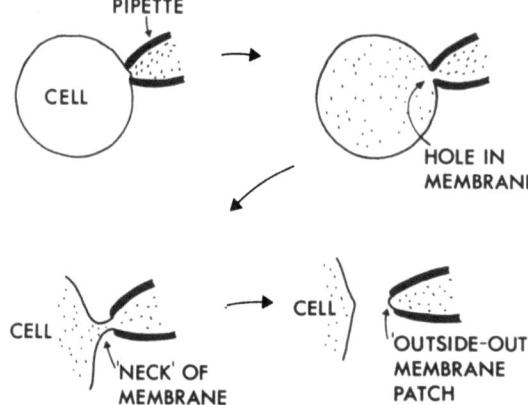

Fig. 6. Whole-cell recording and patch-clamp recording. Diagram of the formation of various configurations of membrane patch (see Hamill et al., 1981) which have been examined in cerebellar neurons. Top left: suction applied to the inside of the pipette allows the formation of a high resistance seal for recording from cell attached patches. Top right: Disruption of the membrane under the electrode tip allows 'whole cell voltage clamp'. If the pipette is then drawn away from the cell, an outside out patch is formed (lower right).

and seals over to produce an 'outside-out' patch on the pipette tip, with the receptors facing the bathing solution. Since the currents are now being recorded from a small patch of membrane, the membrane noise is considerably reduced and it is possible to resolve single channel currents. Alteration of the 'extracellular' medium allows different concentrations of drugs to be applied to the receptor-channels, while the patch-pipette records from the inner surface of the membrane. These are the various approaches that we use for examining cerebellar cells.

One of the initial problems encountered in experiments on cultured neurons is to show the presence of neurons in the culture and to identify neurons and glia from amongst a mixed population of cells. Neurons have a number of distinct morphological characteristics, but more clear cut evidence is necessary to be sure neurons are correctly identified, since glia may possess amino acid receptors and electrically excitable channels. We have used specific monoclonal antibodies to identify neurons and glia. Cultures were reacted with neurofilament selective monoclonal antibody which binds to the protein present in the neurites (Wood and Anderton, 1981). This was then cross-reacted with an antibody conjugated to rhodamine. Most of the neurofilament is contained in the neurites and these parts of the neuron show most clearly. There is also some faint non-specific staining of cell nuclei (in both glial cells and neurons). Cultures were also reacted with antibodies to glial fibrillary acidic protein (GFAP), and glia were identified with indirect immunofluorescence using fluorescein. This approach has allowed us to confirm that neurons could be visually

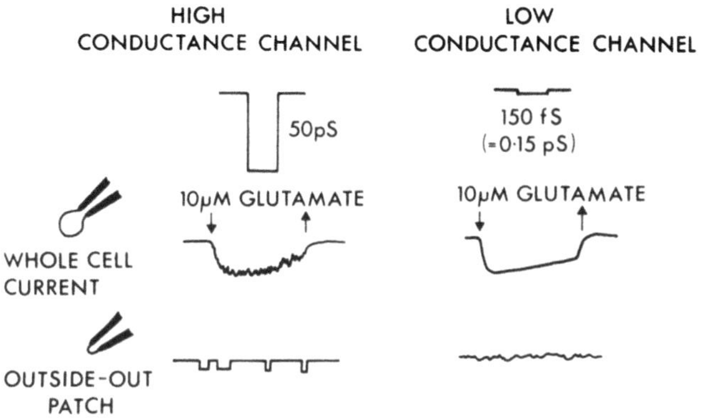

Fig. 7. Cerebellar neurons. Schematic representation of the two main types of response obtained from neurons in explant cultures. Cells with 'high conductance' channels produced a noise response to glutamate in whole cell clamp and single channels were resolved in outside-out patches. Cells with 'low conductance' channels produced little noise increase to glutamate in the whole cell configuration, and outside-out patches gave a noise increase but individual channels were not resolved.

identified with phase contrast or Nomarski interference optics for subsequent patch clamp recording (Cull-Candy et al., 1985).

Most of the cerebellar cells used in our experiments were initially examined in the 'whole cell' configuration, and glutamate or GABA was perfused over the surface of the cell to detect any sensitivity. Glutamate usually produced a sustained inward current as expected if it was opening cation channels. However, it produced two distinct types of response (Cull-Candy and Ogden, 1985) illustrated schematically in Fig. 7. In some cells, the inward current was accompanied by a large noise increase. In others, although the responses were large, there was little obvious change in the noise. In small cells examined with whole cell clamp, the 'large noise' increase was clearly composed of discrete steps, corresponding to the opening of single glutamate channels. The glutamate-noise spectrum can be reasonably well fitted by a single Lorentzian component indicating a mean channel burst-length of about 6 msec, and a single channel conductance of about 50 pS. We further examined these channels by looking directly at single channel currents in outside-out membrane patches. In this situation, the background noise is very low so there is good resolution of individual events. As expected, the size of the current decreased with depolarization, and the extrapolated equilibrium potential for the transmitter was close to zero mV. The conductance of the single channel obtained from the slope of the plot was 48 pS, in good agreement with the noise measurements for the same cells.

In cells where the inward current produced by glutamate was accompanied by very little apparent noise increase, the situation became clearer when outside-out patches were examined from 'high-noise' and 'low-noise' cells. Because the signal to noise ratio was good in this configuration, the single channels were clearly visible in 'high-noise' cells but in 'low-noise' cells only a small noise increase was detected indicating that the conductance of the channel is too small for single channels to be resolved. Analysis of this noise suggested the presence of a channel with a conductance of 140-150 fS.

A number of general conclusions can be drawn at this stage about the glutamate receptors in cerebellar neurons. The current passing through the large conductance channel is in the same range as that which we previously saw in locust muscle and is equivalent to the movement of about 20,000 univalent cations per msec. On the other hand the current passing through the low conductance channel is equivalent to the movement of only 60 ions per msec, which is unusually small for a receptor gated channel. Interestingly, the low conductance of these channels appears to be compensated for in some way. So if you look at the whole cell current produced by a given concentration of glutamate, then the cells with low conductance channels produce a larger total membrane current. This could have two possible explanations. Either, cells with low conductance channels have a higher receptor density or, alternatively, the individual low conductance receptor-channel may have greater sensitivity to glutamate.

ACKNOWLEDGEMENTS

S.G.C.-C. is a Wellcome Trust Senior Lecturer. I am grateful to the MRC and the Wellcome Trust for supporting this work.

REFERENCES

Anderson, C.R., and Stevens, C.F., 1973, Voltage clamp analysis of acetylcholine produced end-plate current fluctuations at frog neuromuscular junction, J. Physiol., 235:655.

Anderson, C.R., Cull-Candy, S.G., and Miledi, R., 1978, Glutamate current noise: post-synaptic channel kinetics investigated under voltage clamp, J. Physiol., 282:219.

Colquhoun, D., and Sakmann, B., 1981, Fluctuations in the microsecond time range of the current through single acetylcholine receptor ion channels, Nature, 294:464.

Cull-Candy, S.G., 1976, Two types of extrajunctional L-glutamate receptors in locust muscle, J. Physiol., 255:449.

Cull-Candy, S.G., and Miledi, R., 1981, Junctional and extrajunctional membrane channels activated by GABA in locust muscle fibers, Proc. R. Soc. Lond. B., 211:527.

Cull-Candy, S.G., Miledi, R., and Parker, I., 1981, Single glutamate-activated channels recorded from locust muscle fibers with perfused patch-clamp electrodes, J. Physiol., 321:195.

Cull-Candy, S.G., and Miledi, R., 1982, Properties of miniature excitatory junctional currents at the locust nerve-muscle junction, J. Physiol., 326:527.

Cull-Candy, S.G., and Parker, I., 1982, Rapid kinetics of single glutamate-receptor channels, Nature, 295:410.

Cull-Candy, S.G., and Parker, I., 1983, Experimental approaches used to examine single glutamate-receptor ion channels in locust muscle fibers, in: Single Channel Recording, B. Sakmann and E. Neher, eds., Plenum Press, New York, p. 389.

Cull-Candy, S.G., 1984, Inhibitory synaptic currents in voltage-clamped locust muscle fibers desensitized to their excitatory transmitter, Proc. R. Soc. Lond. B., 221:375.

Cull-Candy, S.G., and Ogden, D.C., 1985, Ion channels activated by L-glutamate and GABA in cultured cerebellar neurons of the rat, Proc R. Soc. Lond. B., 224:367.

Cull-Candy, S.G., Dilger, P., Ogden, D.C., and Temple, S., 1985, Patch-clamp of rat cerebellar neurones in tissue culture, J. Physiol., 362:45P.

Hamill, O.P., Marty, A., Neher, E., Sakmann, B., and Sigworth, F.J., 1981, Improved patch-clamp techniques for high-resolution current recording from cells and cell-free membrane patches, Pflügers Arch., 391:85.

Hamill, O.P., Bormann, J., and Sakmann, B., 1983, Activation of multiple-conductance state chloride channels in spinal neurones by glycine and GABA, Nature, 305:805.

Katz, B., and Miledi, R., 1972, The statistical nature of the acetylcholine potential and its molecular components, J. Physiol., 665:655.

Kawagoe, R., Onodera, K., and Takeuchi, A., 1984, The uptake and release of glutamate at the crayfish neuromuscular junction, J. Physiol., 354:69.

Krnjević, K., 1974, Chemical nature of synaptic transmission in vertebrates, Physiol. Rev., 54:418.

Neher, E., Sakmann, B., and Steinbach, J.H., 1978, The extracellular patch clamp: a method for resolving currents through individual open channels in biological membranes, Pflügers Arch., 375:219.

Nistri, A., and Constanti, A., 1979, Pharmacological characterization of different types of GABA and glutamate receptors in vertebrates and invertebrates, Prog. Neurobiol., 13:117.

Nowak, L., Bregestovski, P., Ascher, P., Herbert, A., and Prochiantz, A., 1984, Magnesium gates glutamate-activated channels in mouse central neurones, Nature, 307:462.

Otsuka, M., Kravitz, E.A., and Potter, D.D., 1967, Physiological and chemical architecture of a lobster ganglion with particular reference to gamma-aminobutyrate and glutamate, J. Neurophysiol., 30:725.

Patlak, J.B. Gration, K.A.F., and Usherwood, P.N.R., 1979, Single glutmate-activated channels in locust muscle, Nature, 278:643.

Sakmann, B., Patlak, J.B., and Neher, E., 1980, Single acetylcholine-activated channels show burst kinetics in the presence of the desensitizing concentrations of agonist, Nature, 286:71.

Usherwood, P.N.R., and Cull-Candy, S.G., 1975, Pharmacology of somatic nerve-muscle synapses, in: Insect Muscle, P.N.R. Usherwood, ed., Academic Press, London, p. 207.

Wood, J.M., and Anderton, B.H., 1981, Monoclonal antibodies to mammalian neurofilaments, Biosci. Rep., 1:263.

EXPRESSION OF VERTEBRATE AMINO ACID RECEPTORS IN XENOPUS OOCYTES

T.G. Smart[1], A. Constanti[1], K. Houamed[1], G. Bilbe[2,3], D.A. Brown[1], E.A. Barnard[2], and C. VanRenterghem[1,2]

[1]MRC Neuropharmacology Research Group, Department of Pharmacology School of Pharmacy, 29/39 Brunswick Square London, WC1N 1AX
[2]MRC Molecular Neurobiology Research Group, Department of Biochemistry, Imperial College of Science and Technology London, SW7 2AZ
[3]Molecular Genetics Department, Searle Research/Development Lane End Road, High Wycombe, Bucks, HR12 4HL

The apparent ubiquity of both excitatory and inhibitory amino acid receptors in the mammalian central nervous system is now well established. Despite the passage of approximately 23 years since the inception of γ-aminobutyric acid (GABA) and glutamate receptors (Curtis and Watkins, 1963; Krnjević and Phillis, 1963), these receptor sites and their associated 'ion channels' have proved difficult to study quantitatively, mainly because of limited pharmacological accessibility and complicated tissue topography.

These problems have been partly solved by the use of _in vitro_ brain slice preparations and also by the adoption of dissociated neuronal tissue cultures. A different strategy, which we have used, is to extract messenger mRNA from mammalian brain and inject it into an amphibian oocyte, which can then express functional drug receptors in the plasma membrane (Barnard et al., 1982; Smart et al., 1983). The membrane conductance responses to receptor activation can then be conveniently studied by using standard intracellular electrophysiological methods.

mRNA was extracted from day 14 rat or day 19 embryonic chick whole brains using a guanidinium thiocyanate extraction method (Chirgwin et al., 1979; Barnard et al., 1984). Poly (A) -mRNA was isolated by oligo (dT)-cellulose chromatography (Aviv and Leder, 1972) and injected into immature _Xenopus laevis_ oocytes (stage V or VI; Dumont, 1972). After microinjection, the oocytes were incubated in Barth's medium at $21^\circ C$ for 24 hours before recording. All cells were superfused with frog Ringer containing (mM) NaCl 118; KCl 1.9; $CaCl_2$ 2.0; $NaHCO_3$ 2.4; pH 7.4 at room temperature ($20-23^\circ C$).

GABA/GLYCINE-RECEPTORS

The bath-application of either GABA (2-640 μM) or glycine (20 μM-1.6 mM) to oocytes injected with rat brain mRNA, evoked a reversible membrane conductance increase coupled with a depolarization (Fig. 1a, b). Cells

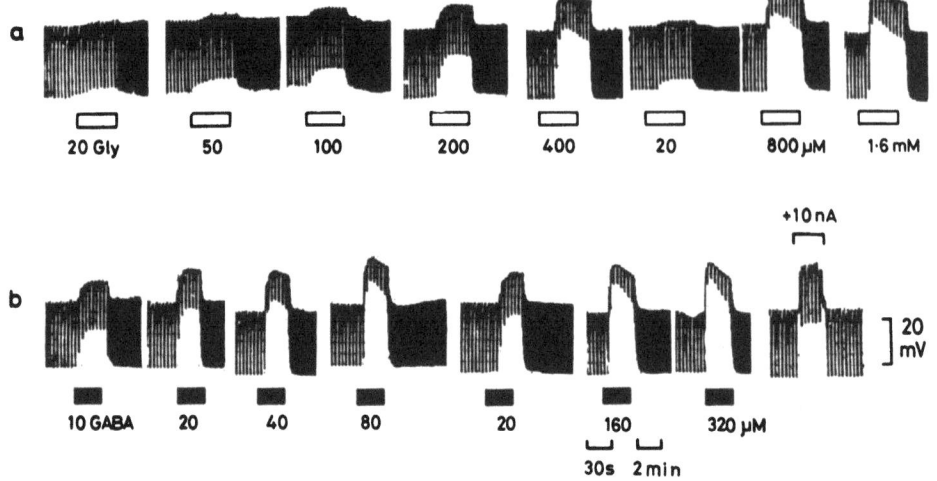

Fig. 1. Membrane conductance responses to bath applied glycine (a) and GABA (b) recorded from a single oocyte after injection with rat brain mRNA. Downward deflections are hyperpolarizing electrotonic potentials (ETPs) in response to current injection. Upward deflections of baseline indicate membrane depolarizations during drug application, indicated by the bars. The responses to either amino acid exhibited desensitization; GABA achieves a higher maximal conductance than glycine in this cell. The conductance increase was due mainly to receptor activation and not to rectification as shown by injecting dc current (+10 nA) to artificially depolarize the cell. The resting potential (rp) was -70 mV.

injected with chick brain mRNA responded similarly, but only to GABA, and non-injected cells showed no sensitivity to either amino acid.

Using rat mRNA, GABA was invariably more active at lower concentrations, although occasionally glycine would produce a greater maximal conductance change compared to GABA in the same cell (cf. Figs. 2 and 7).

The responses to both GABA and glycine were very reproducible and stable, enabling the construction of dose-conductance curves under voltage or current clamp conditions (Fig. 2). The membrane depolarization elicited by GABA or glycine decreased on depolarizing the cell, reversing at ~ -30 mV, which is close to the equilibrium potential for Cl^- ions in these cells (cf. Miledi et al., 1982; Smart et al., 1983). Responses to large doses of GABA and glycine both exhibited some fade or desensitization following microinjection with rat brain mRNA (Fig. 1a, b; Fig. 3B). This fade did not appear to have a common mechanism for these agonists, since cross-desensitization could not be demonstrated (Fig. 3A).

Interestingly, when the peak membrane currents (Ip) elicited under voltage clamp in the presence of 100 μM GABA or 150 μM glycine were analyzed at different holding potentials (V_h), non-linearity was observed at potentials more negative than -50 mV for glycine and -75 mV for GABA. In fact, the glycine curve displayed a distinct negative slope conductance between -70 and -130 mV (Fig. 3C). This rectification may have many causes: one possibility is that glycine may be somehow blocking its own ion channels or that the activated chloride ionophore may show inherent voltage-dependent rectification or inactivation (cf. rectification of iontophoretic GABA responses in the hippocampus, Ashwood et al., 1984). A similar type of current rectification has been observed for the human glycine receptor/

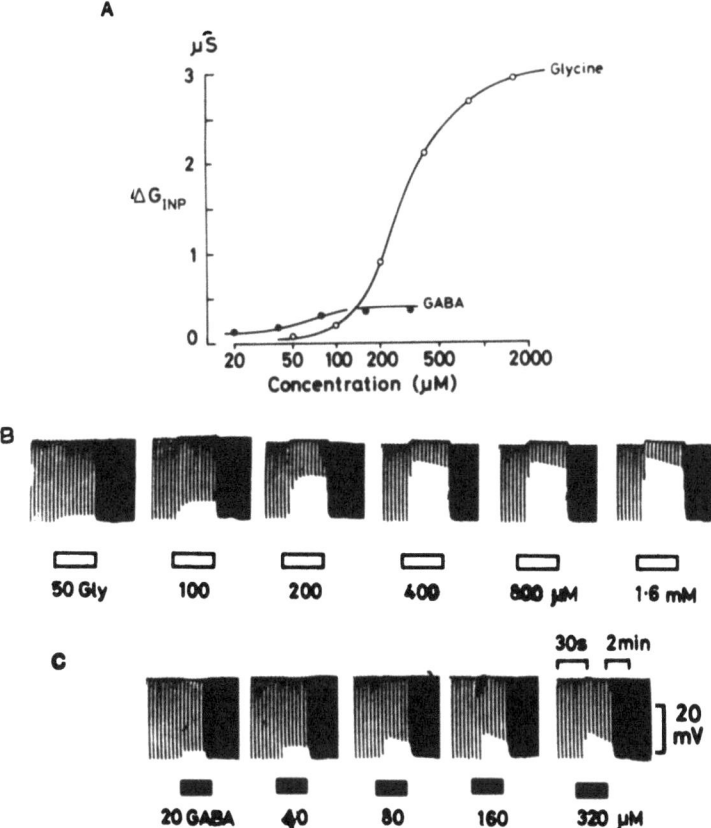

Fig. 2. Variable sensitivity of a rat mRNA injected oocyte to GABA and glycine. (A): log dose-input conductance (ΔG inp) curves for GABA and glycine. Note the lower threshold for GABA but greater maximal conductance for glycine. (B) and (C): corresponding responses to bath applied GABA and glycine; note the fade in the ETPs at high doses, and the greater conductance change achieved by glycine. Cell rp was held at -30 mV by dc current injection: (normal rp = -55 mV).

ionophore incorporated into Xenopus oocytes (Gundersen et al., 1984).

A third possibility is that the synthesized glycine receptor may be coupling to an endogenous Cl^- channel in the oocyte which possesses rectifying properties (Miledi and Parker, 1984) unlike the 'native' neuronal Cl^- ionophore normally associated with the CNS glycine receptor. Interestingly, the differential behavior illustrated in Fig. 3C may suggest that the ion channels activated by GABA and by glycine are not identical.

Both GABA and glycine responses were sensitive to the antagonists picrotoxinin and strychnine. On oocytes injected with chick or rat brain mRNA, GABA responses were also antagonized in an apparently 'competitive' manner by bicuculline (Figs. 4 and 5). The corresponding Schild plot, for the data in Fig. 5A, produced a slope of 0.98 and pA_2 of 5.93 which is similar to the value obtained for the mammalian CNS GABA receptor (5.98; cf. Simmonds, 1982).

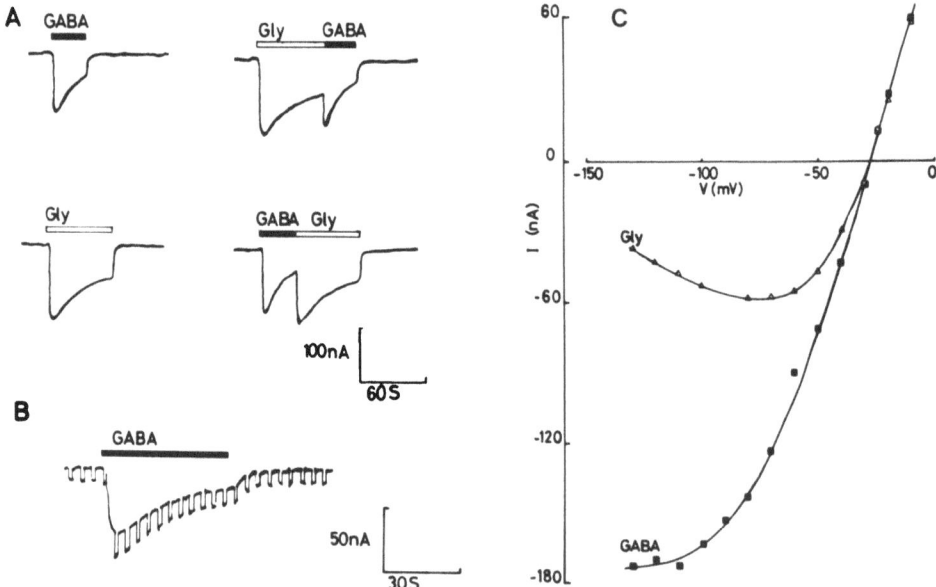

Fig. 3. (A) Inward membrane current evoked by 40 μM GABA and 400 μM glycine in a rat mRNA injected oocyte, under voltage clamp (V_{hold} -60 mV). Note that both GABA and glycine responses faded during drug exposure, but no cross-desensitization was observed when doses were superimposed. (B) Fade of inward current, evoked by 100 μM GABA in a different oocyte clamped at -100 mV. Small (10 mV) hyperpolarizing voltage commands were applied to monitor the membrane input conductance. Note that the current fade was associated with a reduced conductance. (C) Peak membrane currents generated by 100 μM GABA or 150 μM glycine, in a single oocyte under voltage clamp. Note the negative slope conductance region in the glycine curve and non-linearity in the GABA curve at very negative membrane potentials.

Picrotoxinin (1-5 μM) produced a large depression in the GABA dose-conductance curve and a smaller reduction in the glycine curve; both these effects were slow to recover (Fig. 6) on washout. Strychnine (1 μM) in contrast, was a more specific antagonist of glycine responses (Fig. 7A); however, larger doses of strychnine (10 μM) markedly depressed the glycine curve in a clear non-competitive manner and induced a small lateral shift in the GABA curve (Fig. 7B). GABA antagonism at high strychnine doses has been previously reported in the mammalian CNS (Curtis et al., 1971). The expressed GABA receptors resulting from rat and chick mRNA translation retained functional binding subunits for the benzodiazepine and barbiturate potentiators (Smart et al., 1983). Both pentobarbitone (25 μM) and chlorazepate (1 μM) induced a reversible enhancement of GABA but not glycine responses (Fig. 8). Under voltage clamp, pentobarbitone enhanced the GABA-evoked membrane currents and elevated the GABA dose-response relation.

(2) GLUTAMATE/KAINATE RECEPTORS

Rat mRNA-injected cells also acquired a sensitivity to excitatory amino acids such as glutamate, kainate and also quisqualate (Houamed et al.,

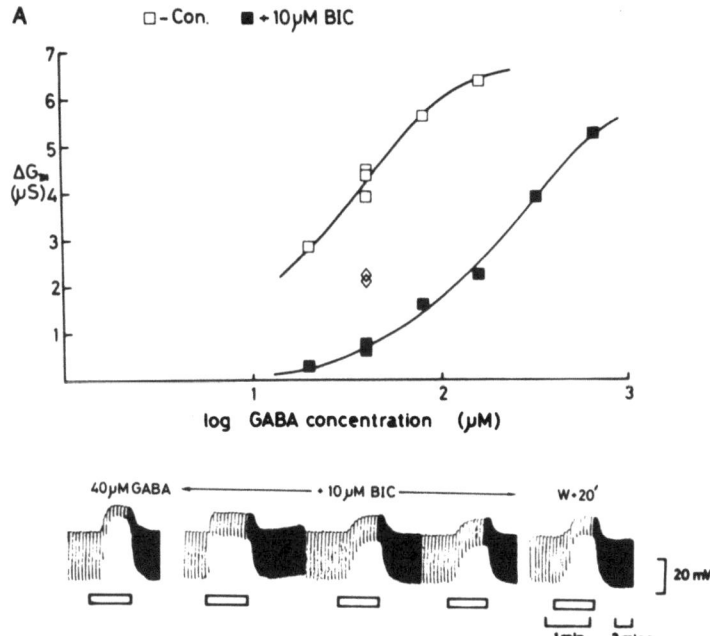

Fig. 4. Effect of bicuculline on GABA responses recorded in a chick mRNA injected oocyte. (A) Dose-conductance curve for GABA in the absence (□) and presence (■) of 10 μM (+)-bicuculline. Note the apparent competitive shift in the curve caused by bicuculline. (B) Single responses to bath applied GABA (40 μM) showing the onset of block by bicuculline and partial recovery.   ◊ = Wash

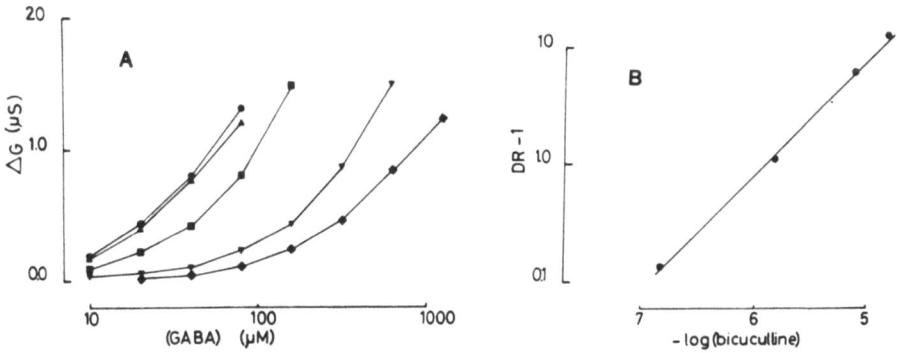

Fig. 5. Effect of bicuculline on GABA responses recorded in a rat mRNA injected oocyte. (A) Log dose-conductance curves for GABA in the absence (●) and presence of (▲), 0.15; (■), 1.5; (▼),7.5; and (◆) 15 μM bicuculline, obtained under voltage-clamp at -100 mV. (B) Corresponding Schild plot for the data in (A). The slope of the line = 0.98; pA_2 = 5.93.

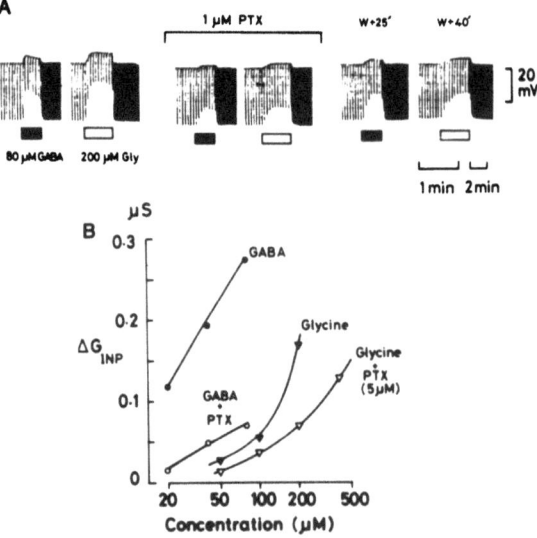

Fig. 6. Antagonism of GABA and glycine responses by picrotoxinin in a rat mRNA injected cell. (A) Conductance changes evoked by GABA and glycine are both reduced by 1 μM picrotoxinin. The GABA response recovers faster than the glycine response rp -51 mV. (B) GABA (●) and glycine (▼) conductance curves are both reduced in the presence of 5 μM picrotoxinin. The GABA curve (○) is depressed more than the glycine curve (▽).

1984). However, responses to aspartate or NMDA were not observed despite the absence of Mg^{2+} in the bathing fluid (Ault et al., 1980; Nowak et al., 1984). This may either reflect a lack of specific NMDA recognition sites or a failure of these sites to couple to functional ion channels. More recently, glutamate and kainate responses were also recorded from oocytes injected with chick brain mRNA (Fig. 9). In general, the Xenopus oocyte appeared to translate a greater variety of receptor proteins when utilizing rat rather than avian mRNA.

Unlike glutamate responses which were often oscillatory in nature (comparable to the muscarinic responses obtained in uninjected cells; Kusano et al., 1982), those to kainate were 'smooth' and showed little tendency to fade during a continuous application (Figs. 9, 10). It is likely that the glutamate response consists of two membrane current components: an oscillatory current inverting at E_{Cl} (Gundersen et al., 1984b) and a smooth current inverting at 0 to -10 mV (Gundersen et al., 1984a,b). The kainate response also reversed at about -10 mV (Fig. 10; cf. also Gundersen et al., 1984a) suggesting some common ionic mediation with the glutamate response.

In summary, the powerful technique of receptor mRNA transcription in the Xenopus oocyte allows not only a quantitative study of receptor/ion channel function, but should also enable elucidation of the requirements for the biosynthesis of receptor polypeptides.

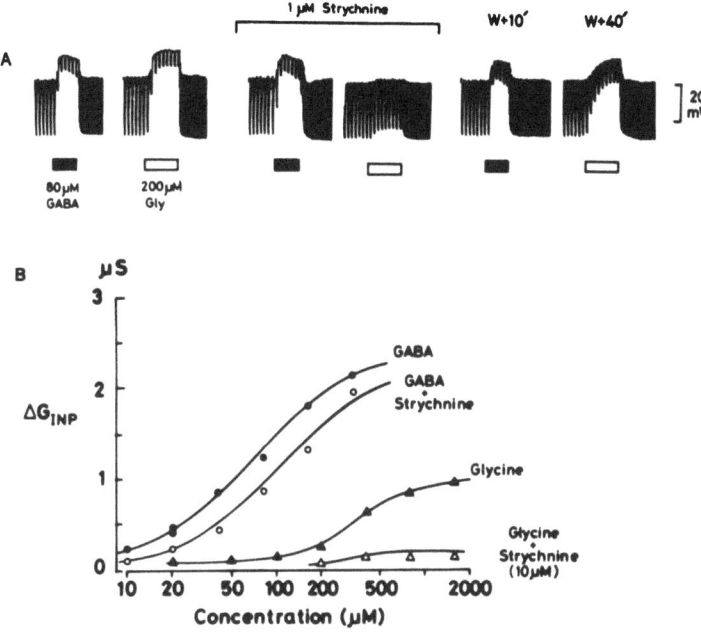

Fig. 7. Effect of strychnine on the responses to bath-applied GABA and glycine in a rat mRNA injected oocyte. (A) 1 μM strychnine preferentially antagonizes the glycine response which almost completely recovers following 40 minutes washing. (B) GABA and glycine dose-conductance curves in the absence (●,▲) and presence (○,△) of 10 μM strychnine. Note the small lateral shift in the GABA curve and the non-competitive depression of the glycine curve in antagonist solution. Reprinted by permission from Figs. 1 and 3 of Houamed et al., 1984.

Our results, and the concurring observations by Miledi and co-workers, clearly show that it is possible to synthesize and insert a CNS drug receptor in a 'foreign cell' membrane, whilst retaining the important binding sites for subsequent modulation of receptor function (see Table 1). The induced GABA receptor in the oocyte exhibits a pharmacological spectrum which is very comparable to that found in the intact CNS. It is also interesting that glycine and excitatory amino acid receptors can be assembled in the same cell, although the pharmacological spectrum of these receptors is yet to be studied in detail. The Xenopus oocyte has been known to translate 'foreign' mRNA from a wide variety of eukaryotic sources (Gurdon et al., 1971) including chick, rat, cat, rabbit and now also human mRNA (Gundersen et al., 1984a; see Table 1).

Currently, we are fractionating the crude mRNA preparations in order to purify specific mRNAs for the translation of only specific types of drug receptor, when injected into single oocytes. This approach has been partly successful, enabling the separation of rat brain mRNAs coding for GABA and glycine receptors, using a sucrose density gradient. The specific mRNA activity is usually spread over more than one fraction, but invariably the fractions containing peak mRNA activities are distinct.

Table 1. Functional Amino-Acid Receptors and Ion-Channel Proteins Expressed in the _Xenopus_ Oocyte

mRNA Source	Receptor	Ion Channel	E_{REV} (mV)	Pharmacological Sensitivity	Notes	Reference
CHICK	GABA	Cl^-	-25	Picrotoxinin - Bicuculline - Pentobarbitone + Chlorazepate +	τ noise 25ms γ noise 4 pS τ chan. 16ms γ chan. 28.5 pS	Miledi et al. 1982 Smart et al. 1983 Miledi et al. 1983
BRAIN	Glutamate			-	oscillatory & smooth components in response	(This study) Fig. 9a/b
	Kainate			-	smooth response	Fig. 9a/b
RAT BRAIN	GABA	Cl^-	-25	Picrotoxinin - Bicuculline - Pentobarbitone + Chlorazepate + Alphaxolone +	sensitive to β-alanine (100 μM - 4mM)	Houamed et al. 1984
	Glycine	Cl^-	-25	Strychnine - (Picrotoxinin -)	i/v shows negative slope region	Houamed et al. 1984
	Glutamate	Cl^-/Na^+? K^+	-24/0	-	(Sensitive to L-aspartate). Oscillatory & smooth response components.	Gundersen et al. 1984b Houamed et al. 1984
	Kainate	Na^+, K^+?	-3.2, -11	-	smooth response linear i/v	Gundersen et al. 1984a, b Houamed et al. 1984
	Quisqualate	-		-	oscillatory response	Gundersen et al. 1984a, b Houamed et al. 1984

HUMAN BRAIN	GABA	Cl⁻	-23	Pentobarbitone + Strychnine -	mRNA from cerebral cortex, 15 wk old foetus	Gundersen et al. 1984c
	Glycine	Cl⁻	-23		i/v shows negative slope region	Gundersen et al. 1984c.
	Kainate	-	-10	-	No glutamate response linear i/v	Gundersen et al. 1984a

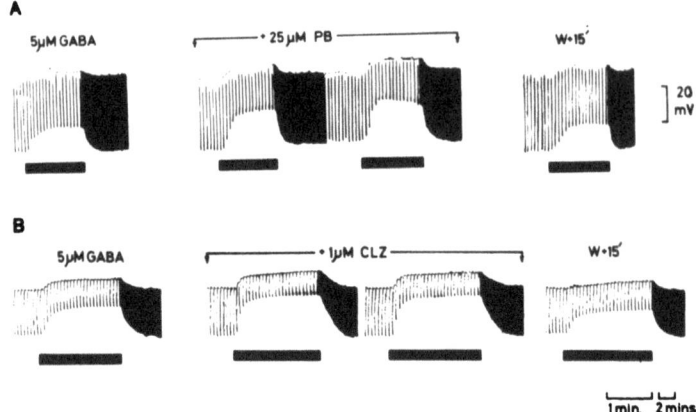

Fig. 8. Enhancement of the GABA-evoked conductance change by 25 μM pentobarbitone (PB, (A)) and by 1 μM chlorazepate (CLZ, (B)) using two different oocytes injected with chick brain mRNA. Note the onset of desensitization in the second successive response in PB. Recovery was rapid after only 15 minutes washing. The enhancement observed with CLZ was smaller compared to that with PB. Recovery was attained after 15 minutes.

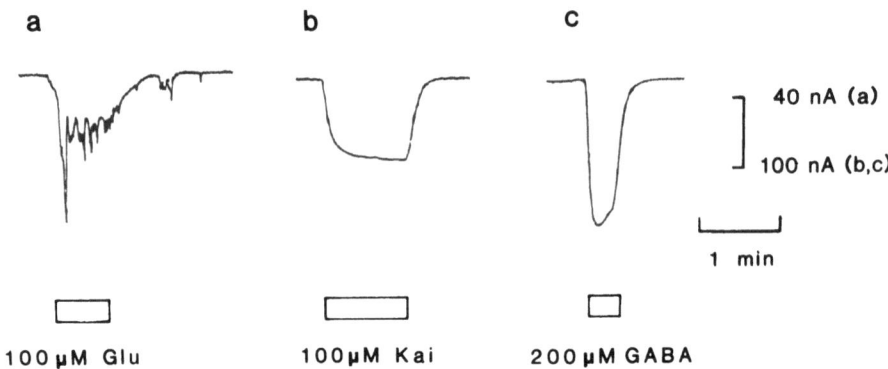

Fig. 9. Inward membrane currents recorded from a single oocyte injected with chick mRNA. (a) Response to 100 μM L-glutamate, note the oscillatory nature of the response; (b) 100 μM kainate, exhibiting only a smooth response; and (c) 200 μM GABA, producing a smooth inward, desensitizing current. V_{hold} -70 mV. Note the different current calibrations.

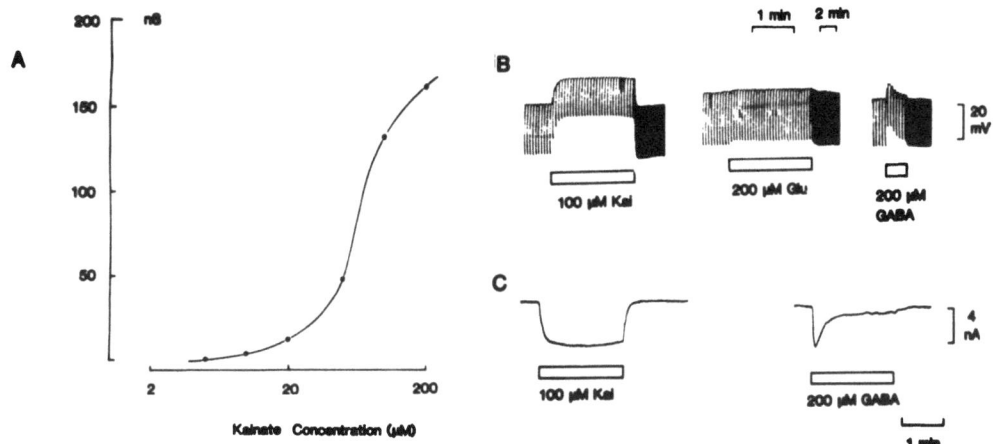

Fig. 10. (A) Log dose-conductance curve to kainate obtained under voltage-clamp in a single oocyte injected with rat mRNA. V_{hold} −100 mV. The reversal potential for the kainate response was −11 mV. (B) Membrane conductance changes to 100 μM kainate (Kai), 200 μM L-glutamate (GLU) and 200 μM GABA recorded under current clamp. Note the smooth maintained response to kainate, the very small response to glutamate and the fading GABA response rp −72 mV. (C) Inward membrane currents to kainate and GABA recorded under voltage clamp. Note the fade in the GABA-induced current. V_{hold} −100 mV. The mRNA used in this experiment was extracted by I. Adcock, Department of Pharmacology, St. Thomas' Hospital Medical School, University of London.

This refined approach will broaden the applications of the Xenopus oocyte to the quantitative study of drug receptors, some of which include:

(i) possible separation of mRNAs, coding for synaptic and extrasynaptic receptor forms in vivo, e.g., GABA;

(ii) testing the mRNAs which direct the synthesis of receptors with overlapping pharmacological spectra, e.g., excitatory amino acids;

(iii) the identification of mRNAs corresponding to the individual polypeptide subunits of the drug receptor, as recently achieved for the nicotinic acetylcholine receptor (Mishima et al., 1984);

(iv) the ability to initiate site directed mutagenesis in the receptor polypeptides, perhaps altering the function of the receptor per se, or of one, or more, modulatory subunits;

(v) and finally, an enriched mRNA source, following identification of the mRNA, would allow the formation and cloning of the corresponding cDNA that codes for the receptor protein, e.g., GABA. Ultimately, the amino acid sequence and subunit structure of the receptor should become available, particularly since the $GABA_A$ receptor complex has been recently isolated and purified (Sigel et al., 1983).

It appears that the Xenopus laevis oocyte will continue to prove an almost ideal vehicle for the application of molecular biology and electrophysiology to the study of CNS drug receptor proteins.

ACKNOWLEDGEMENT

Supported by the MRC.

REFERENCES

Ashwood, T.J., Collingridge, G.L., Herron, C.E., and Wheal, H.V., 1984, Rectification of somatic γ-aminobutyric acid (GABA) responses in rat hippocampal slices, J. Physiol., 357:15.

Ault, B., Evans, R.H., Francis, A.A., Oakes, D.J., and Watkins, J.C., 1980, Selective depression of excitatory amino acid induced depolarizations by magnesium ions in isolated spinal cord preparations, J. Physiol., 307:413.

Aviv, H., and Leder, P., 1972, Purification of biologically active globin messenger RNA by chromatography on oligothymidylic acid-cellulose, Proc. Natl. Acad. Sci. USA, 69:1408.

Barnard, E.A., Miledi, R., and Sumikawa, K., 1982, Translation of exogenous messenger RNA coding for nicotinic acetylcholine receptors produces functional receptors in Xenopus oocytes, Proc. Roy. Soc. Lond. B., 215:241.

Barnard, E.A., Beeson, D., Bilbe, G., Brown, D.A., Constanti, A., Houamed, K., and Smart, T.G., 1984, A system for the translation of receptor messenger RNA and the study of the assembly of functional receptors, J. Receptor Res., 4:681.

Chirgwin, J.M., Przybyla, A.E., McDonald, R.J., and Rutter, W.J., 1979, Isolation of biologically active ribonucleic acid from sources enriched in ribonuclease, Biochemistry, 18:5294.

Curtis, D.R., and Watkins, J.C., 1963, Acidic amino acids with strong excitatory actions on mammalian neurones, J. Physiol., 166:1.

Curtis, D.R., Duggan, A.W., and Johnston, G.A.R., 1971, The specificity of strychnine as a glycine antagonist in the mammalian spinal cord, Exp. Brain Res., 12:547.

Dumont, J.N., 1972, Oogenesis in Xenopus laevis (Daudin) I, J. Morphol., 136:153.

Gundersen, C.B., Miledi, R., and Parker, I., 1984a, Messenger RNA from human brain induces drug- and voltage-operated channels in Xenopus oocytes, Nature, 308:421.

Gundersen, C.B., Miledi, R., and Parker, I., 1984b, Glutamate and kainate receptors induced by rat brain messenger RNA in Xenopus oocytes, Proc. Roy. Soc. Lond. B., 221:127.

Gundersen, C.B., Miledi, R., and Parker, I., 1984c, Properties of human brain glycine receptors expressed in Xenopus oocytes, Proc. Roy. Soc. Lond. B., 221:235.

Gurdon, J.B., Lane, C.D., Woodland, H.R., and Marbaix, G., 1971, Use of frog eggs and oocytes for the study of messenger RNA and its translation in living cells, Nature, 233:177.

Houamed, K.M., Bilbe, G., Smart, T.G., Constanti, A., Brown, D.A., Barnard, E.A., and Richards, B.M., 1984, Expression of functional GABA, glycine and glutamate receptors in Xenopus oocytes injected with rat brain mRNA, Nature, 310:318.

Krnjević, K., and Phillis, J.W., 1963, Iontophoretic studies of neurones in the mammalian cerebral cortex, J. Physiol., 165:274.

Kusano, K., Miledi, R., and Stinnakre, J., 1982, Cholinergic and catecholaminergic receptors in the Xenopus oocyte membrane, J. Physiol., 328:143.

Miledi, R., Parker, I., and Sumikawa, K., 1982, Synthesis of chick brain GABA receptors by frog oocytes, Proc. Roy. Soc. Lond. B., 216:509.

Miledi, R., and Parker, I., 1984, Chloride current induced by injection of calcium into Xenopus oocytes, J. Physiol., 357:173.

Mishina, M., Kurosaki, T., Tobimatsu, T., Morimoto, Y., Noda, M., Yamamoto, T., Terao, M., Lindstrom, J., Takahashi, T., Kuno, M., and Numa, S., 1984, Expression of functional acetylcholine receptor from cloned cDNAs, Nature, 307:604.

Nowak, L., Bregestovski, P., Ascher, P., Herbet, A., and Prochiantz, A., 1984, Magnesium gates glutamate-activated channels in mouse central neurones, Nature, 307:462.

Sigel, E., Stephenson, F.A., Mamalaki, C., and Barnard, E.A., 1983, A γ-aminobutyric acid/benzodiazepine receptor complex of bovine cerebral cortex. Purification and partial characterization, J. Biol. Chem., 258:6965.

Simmonds, M.A., 1982, Classification of some GABA antagonists with regard to site of action and potency in slices of rat cuneate nucleus, Eur. J. Pharmacol., 80:347.

Smart, T.G., Constanti, A., Bilbe, G., Brown, D.A., and Barnard, E.A., 1983, Synthesis of the functional chick brain benzodiazepine barbiturate-GABA receptor complex in mRNA injected Xenopus oocytes, Neurosci. Lett., 40:55.

COMMENTARY - EXCITATORY AMINO ACIDS: PHYSIOLOGICAL STUDIES

D.A. Brown and R. Dingledine

This session was marked by lively discussion on the ionic conductances and physiological roles of excitatory amino acids (EAA). Much of the discussion was centered on the NMDA receptor.

Mg-BLOCK

One of the most important and interesting recent advances in EAA pharmacology has been the demonstration by Ascher and his colleagues (Nowak et al., 1984), followed shortly by Mayer and Westbrook (1985, Mayer et al., 1984), that the NMDA-activated ionic channel is uniquely subject to voltage-dependent block by Mg^{2+} ions, resulting in a region of negative slope conductance in the current-voltage curve negative to about -30 mV. This explains two peculiar features of the NMDA response: the apparent increase in input resistance during the NMDA-depolarization (e.g. Engberg et al., 1979; MacDonald et al., 1982; Dingledine, 1983) and the ability of NMDA to produce membrane potential oscillations and burst-firing. The regenerative inward current necessary for the latter is not produced by the activation of a voltage-dependent inward current as originally thought (MacDonald et al., 1982; Dingledine, 1983), but by the depolarization-induced unblocking of an essentially voltage-independent cation channel. As pointed out by Ascher, the combination of a Mg-blocked NMDA inward current and a linear outward current can give rise to two metastable membrane potentials at the peak and trough of the potential oscillations (about -30 and -70 mV, respectively).

Ca IONS

The NMDA channel does not appear to be a Ca-channel, even though it can be blocked by Mg^{2+} and Co^{2+}. The NMDA current is insensitive to Cd^{2+} (Westbrook), and NMDA channels do not show the 'flickering' characteristics of open-channel block in Cd^{2+} solution (Ascher). The previously reported potentiation of inward Ca-currents by aspartate (MacDonald and Schneiderman, 1984) is probably the result of inadequate space-clamp control, as recognized by MacDonald. Cherubini reported that NMDA increased the amplitude of the Ca-spike and Ca-dependent after-hyperpolarization in hippocampal cells, but this presumably results from the effective increase in input resistance rather than from a direct effect on the Ca-current. Of course, in unclamped cells, the depolarization produced by NMDA, coupled with an effective rise in input resistance, may well activate additional voltage-dependent Ca-currents, which presumably accounts for the observation of Dingledine (1983) that the NMDA-depolarization was reduced by Ca-blockers.

A quite separate question is whether the NMDA channel conducts Ca^{2+} in any significant quantity. Westbrook has provided some evidence that it might, in that the reversal potential for the macroscopic NMDA current showed a clear positive shift of about 20 mV on raising external Ca^{2+} from 1 to 20 mM, whereas no such shift occurred in the reversal potential for the kainate current. However, Ascher's group have not been able to detect a comparable shift in the reversal potential for the single channel current although (as Ascher acknowledged) their data do not preclude a smaller shift. The question is an interesting one because of the possible longer-term effects on cell function of an increased Ca^{2+} influx during activation of NMDA receptors, e.g., in long term potentiation.

EAA CHANNELS

NMDA channels in the unblocked state have a mean open time of 4.7 msec and conductance of about 50 pS in cultured mouse mesencephalic cells (Nowak et al., 1984). Ascher now reports similar open times with several other agonists, including NMDA itself and ibotenate, so that channel lifetime cannot be used as a unique fingerprint to identify the transmitter at NMDA-sites (cf. Crawford and McBurney, 1977). Ascher described at least two quisqualate channels, one of which (with a conductance of about 8 pS) was also activated by glutamate. Since the elementary current deduced from the quisqualate noise spectrum was about half that predicted from the most abundant quisqualate channel, he raised the possibility of an additional, low conductance channel. This set the stage for Cull-Candy's description of a tiny (0.15 pS) glutamate channel in rat cerebellar neurons - comparable to that of the channels carrying the dark current in rod outer segments (Fesenko et al., 1985), and capable of conducting only some 60 ions/msec. Overall, available evidence suggests that glutamate can activate at least three kinetically different types of channel, quisqualate two (possibly three), kainate two and NMDA at least one. The multiplicity of channels activated by the EAA would clearly accord with the multiple receptors revealed by ligand binding and other electrophysiological studies. Further studies may allow a more precise channel/receptor typing.

EAA AND K CHANNELS

Prior to the Mg^{2+}-block interpretation, one of the explanations proffered for the apparent increase in input resistance produced by NMDA was that it reduced K^+-conductance (Engberg et al., 1979). Effects of EAA on K^+-conductances have been resurrected by Cherubini, this time in relation to kainate: he reports that, during the first minutes of superfusion, low (50-200 nM) concentrations of kainate reduce the Ca-dependent after-hyperpolarization in hippocampal pyramidal neurons, and suggests this to result from an effect on the K-conductance rather than on the priming Ca-entry since at this stage Ca-spikes were not depressed, or even increased. With longer applications of kainate, Ca-spikes and Ba-spikes were reduced or completely blocked, this could be attributed to an increased leak current although there are several other possible mechanisms.

This Ca-dependent after-hyperpolarization has been subjected to a great deal of study in recent years and appears to be inhibited by a variety of putative transmitter agents, including acetylcholine (Cole and Nicoll, 1983), noradrenaline (Madison and Nicoll, 1982; Haas and Konnerth, 1983), histamine (Haas and Konnerth, 1983) and CRF (Aldenhoff et al., 1983). Before adding kainate to this list, it will be necessary to confirm such an inhibition under voltage-clamp conditions: the underlying current is very small (Lancaster and Adams, 1984) and only produces a large hyperpolarization because it occurs at the region of highest input resistance, in the potential range between the outward and inward rectifying M and Q-currents (Halliwell and Adams, 1982). Hence it may be as susceptible as the Ba-spike to small changes in leak conductance.

PHYSIOLOGICAL ROLE OF NMDA RECEPTORS

Although evidence has been recently obtained to suggest a participation of NMDA-receptors in some cortical excitatory synaptic potentials (Thomson et al., 1985), on-going Mg^{2+}-block seems to limit the extent to which NMDA receptor-activation drives the primary excitatory synaptic current. Even in the frog spinal cord bathed in Mg^{2+}-free solution, monosynaptic EPSPs appear resistant to the NMDA-antagonist, D-APV (Nistri). However, based on studies with D-APV, it appears that currents generated via NMDA-receptors contribute to polysynaptic excitation in the spinal cord (Nistri, Davidoff) and to synaptically-activated epileptiform paroxysmal depolarizations and after-discharges in the hippocampus (Collingridge, Dingledine, Hablitz, and Swann). As pointed out by Ascher and others, this is to be expected from the dual action of glutamate on NMDA and non-NMDA receptors: activation of the latter will provide a priming inward current, driving the membrane potential toward a region where NMDA-channels are less sensitive to Mg^{2+}-block and therefore susceptible to a second or subsequent charge of glutamate. With cumulative synaptic activation, the NMDA-current might further increase – an interesting form of modulation whereby activation of one ionic channel might potentiate the effect of activating a second ionic channel by the same transmitter. Ascher also raised the possibility that synaptic depolarization might sensitize the NMDA-system to a sufficient extent to be activated by interstitial glutamate. Further information on the precise interstitial concentrations of glutamate and Mg^{2+} ions would assist in evaluating the role of NMDA receptors. The data presented in this symposium certainly support the view that activation of NMDA receptors by endogenous transmitters may contribute strongly to epileptiform discharges.

EAA RECEPTORS

There is little sign of progress yet toward the isolation and structural resolution of the EAA receptors. Radiation inactivation suggests different-sized complexes for quisqualate (52 KD), kainate (77 KD) and NMDA (140 KD) receptors (Honoré), but how far this reflects subunit arrangements is not clear. Analogy with the nicotinic acetylcholine receptor, which behaves rather similarly, suggests a multi-subunit macromolecule probably incorporating the 'channel' in the receptor macromolecule itself. Expression of the glutamate receptor in frog oocytes injected with mRNA extracted from rat brain (Brown; Houamed et al., 1984) therefore would occasion no surprise, and gives no direct clue regarding subunit structure. One peculiarity of this expression system is that glutamate appears to induce two ionic currents – the expected cation current plus another chloride current (Gunderson et al., 1984). The oscillatory nature of this latter current suggests that it is the same current activated via the oocyte's own muscarinic acetylcholine receptors (Kusano et al., 1982). If so, this might imply a more indirect coupling of the receptor to an endogenous Cl-channel.

Although glutamate-activated Cl^--currents have been reported in crustacea (Lingle and Marder, 1981), an alternative, and more attractive, explanation would be that the Cl^- current results from Ca-entry (Barisch, 1983) through the glutamate-activated channels (see above). More refined pharmacological analysis should resolve this question.

REFERENCES

Aldenhoff, J.B., Gruol, D.L., Rivier, J., Vale, W., and Siggins, G.R., 1983, Corticotropin releasing factor decreases postburst hyperpolarizations and excites hippocampal neurons, Science, 221:875.
Barisch, M.E., 1983, A transient calcium-dependent chloride current in the immature Xenopus oocyte, J. Physiol., 342:309.
Cole, A.E., and Nicoll, R.A., 1983, Acetylcholine mediates a slow synaptic potential in hippocampal pyramidal cells, Science, 221:1299.
Crawford, A.C., and McBurney, R.N., 1977, The termination of transmitter action at the crustacean excitatory neuromuscular junction, J. Physiol., 268:711.
Dingledine, R., 1983, N-methyl-aspartate activates voltage-dependent calcium conductance in rat hippocampal pyramidal cells, J. Physiol., 343:385.
Engberg, I., Flatman, J.A., and Lambert, J.D.C., 1979, The actions of excitatory amino acids on motoneurones in the feline spinal cord, J. Physiol., 288:227.
Fesenko, E.E., Kolesnikov, S.S., and Lyubarsky, A.L., 1985, Induction by cyclic AMP of cationic conductance in plasma membrane of retinal outer rod segment, Nature, 313:310.
Gunderson, C.B., Miledi, R., and Parker, I., 1984, Glutamate and kainate receptors induced by rat brain messenger RNA in Xenopus oocytes, Proc. Roy. Soc. Lond. B, 221:127.
Haas, H.L., and Konnerth, A., 1983, Histamine and noradrenaline decrease calcium-activated potassium conductance in hippocampal pyramidal cells, Nature, 302:432.
Halliwell, J.V., and Adams, P.R., 1982, Voltage clamp analysis of muscarinic excitation in hippocampal neurons, Brain Res., 250:71.
Houamed, K.M., Bilbe, G., Smart, T.G., Constanti, A., Brown, D.A., Barnard, E.A., and Richards, B.M., 1984, Expression of functional GABA, glycine and glutamate receptors in Xenopus oocytes injected with rat brain mRNA, Nature, 310:318.
Kusano, K., Miledi, R., and Stinnakre, J., 1982, Cholinergic and catecholaminergic receptors in the Xenopus oocyte membrane, J. Physiol., 328:143.
Lancaster, D., and Adams, P.R., 1984, Single electrode voltage clamp of the slow AHP current in rat hippocampal pyramidal cells, Soc. Neurosci. Abstr., 10:872.
Lingle, C., and Marder, E., 1981, A glutamate-activated chloride conductance on a crustacean muscle, Brain Res., 212:481.
MacDonald, J.F., Porietis, A.V., and Wojtowicz, J.M., 1982, L-aspartic acid induces a region of negative slope conductance in the current-voltage relationship of cultured spinal cord neurons, Brain Res., 237:248.
MacDonald, J.F., and Schneiderman, J.H., 1984, L-aspartic acid potentiates 'slow' inward current in cultured spinal cord neurons, Brain Res., 296:350.
Madison, D.V., and Nicoll, R.A., 1982, Noradrenaline blocks accommodation of pyramidal cell discharge in the hippocampus, Nature, 299:636.
Mayer, M.L., and Westbrook, G.L., 1984, Mixed agonist action of excitatory amino acids on mouse spinal cord neurones under voltage clamp, J. Physiol., 354:29.
Mayer, M.L., Westbrook, G.L., and Guthrie, P.B., 1984, Voltage-dependent block by Mg^{2+} of NMDA responses in spinal cord neurones, Nature, 309:261.
Nowak, L., Bregostovski, P., Ascher, P., Herbet, A., and Prochiantz, A., 1984, Magnesium gates glutamate-activated channels in mouse central neurones, Nature, 307:462.
Thomson, A.M., West, D.C., and Lodge, D., 1985, An N-methylaspartate receptor-mediated synapse in rat cerebral cortex: a site of action of ketamine?, Nature, 313:479.

SESSION VIII
METAL IONS AND EPILEPSY

TRANSITION METAL IONS IN EPILEPSY: AN OVERVIEW

S.H. Chung, B. Gabrielsson and D.K. Norris

National Institute for Medical Research
Mill Hill
London NW7 1AA

Several species of the transition elements of the first series are present throughout the central nervous system in trace quantities. The most reliable estimates of their concentrations in the brain are 27 ppm wet mass for Fe^{2+}, 16 ppm for Zn^{2+}, 2.4 ppm for Cu^{2+}, 0.05 ppm for Co^{2+} and 0.3 ppm Mn^{2+}. Little is known of the role of these transition metals in normal brain function although there are indications that they may participate in some aspects of synaptic transmission. Here we consider how abnormal brain levels of trace metals may be epileptogenic in that they enhance excitatory synaptic mechanisms and reduce inhibitory processes.

PROPERTIES OF TRANSITION METALS

The first transition series begins after scandium when electrons occupy all s and p orbitals of the first (K), second (L) and third (M) shells and the 3d orbitals become filled for the first time. An additional electron successively occupies the set of five d orbitals as we proceed from Ti to Cu in the Periodic Table, and the orbitals become completely filled with 10 electrons in Zn. The unique chemistry of transition elements, and the essence of crystal field theory, derives from the properties of the d orbitals. For detailed mathematical exposition on the chemistry of transition elements, the reader is referred to Orgel (1960), Griffith (1961), Ballhausen (1962), and Figgis (1966).

One of the prominent characteristic features of transition elements is their ability to form complexes, in which the d orbitals of metal atoms coordinate to electron-donating ligand molecules. The number of ligands, either monatomic or polyatomic negative ions or neutral polar molecules, that attach to the metal is its coordination number, which, for the elements of the first transition series, is usually 4 and 6, forming tetrahedral or octahedral complexes. The complexes formed by the first transition elements with more than 80 ligands, coordinating generally through either oxygen or nitrogen, were shown to follow the sequence of stabilities:

Mn < Fe < Co < Ni < Cu > Zn.

This, after the original investigators, is usually referred to as the Iryin-William order. There are a small group of ions, however, including Pd^{2+}, Hg^{2+} and Cu^{2+}, which form their most stable complexes with elements of the second or succeeding periods, such as P, S and As.

Because of their ability to form interstitial complexes and withdraw electrons from the reacting substances, nearly all transition elements have catalytic power. Among the great variety of natural metal-protein complexes are metalloenzymes, in which the metal ion forms a part of the catalytic site. There are several possible ways metal ions can contribute to catalysis. For example, they can accept or donate electrons from ligands via π or σ bonds, thereby promoting acid-base or covalent catalysis. Alternatively, they can serve as a bridge through common coordination to bring the enzyme and substrate into proximity or they may enhance reactivity by cyclizing substrates. A broad generalization one can make about metallic catalysis is that Mg^{2+} is usually the best activating ion for enzymes involving phosphate groups whereas Mn^{2+} and Zn^{2+} are somewhat less active, and with peptidases Co^{2+} or Mn^{2+} are more active than Zn^{2+}. Metal ions high in the order of complex stabilities (e.g., Pb^{2+} and Cu^{2+}) are poor activators and may act as inhibitors. A great many enzymes which depend on thiol groups for their activity are inhibited by heavy metals, and their inhibitory potency roughly follows the order of solubility products of metal sulfides, namely,

$Hg \gg Cu > Pb,Cd > Zn > Co,Ni > Mn$.

Unfortunately, a detailed knowledge of the chemistry of transition metals does not permit us to deduce their roles in the nervous system. Metals in the nervous system may not necessarily saturate all -SH binding sites, as one would expect from the stability constants, since hydrogen ions will compete with the metal for the ligand and anions from the medium will compete with ligand metal. The most general statement we can make is that a metal ion, with its positive charge, is effectively an acid in the Lewis sense and will hydrolyze and precipitate at biological pH if it is not carried as metal complex. It is therefore safe to assume that any metallic cations which find their way into the brain will be complexed to multidentate ligands until they are excreted but the groups which bind the metal are difficult to identify.

METAL IONS, NEUROACTIVE AMINO ACIDS AND EPILEPSY

The capacity of metal ions to induce epilepsy has long been recognized. Focal intracerebral applications of metallic substances, such as cobalt powder (Ross and Craig, 1981; Colasanti et al., 1982; Trottier et al., 1982), alumina cream (Heinemann et al., 1981; Reynolds et al., 1981; Oakley and Ojemann, 1982), iron or haem (Willmore et al., 1978; Willmore and Rubin, 1982), zinc chloride (Itoh and Ebadi, 1982; Chung and Johnson, 1983; Pei et al., 1983), and copper chloride (Chung and Johnson, 1983) induce epileptiform discharges and sometimes spontaneous seizures in animals. For a review of early studies, see Ward (1972). It has been reported that chronic oral treatment with D-penicillamine, an effective metal chelating agent, protects seizure-prone baboons against photic-induced convulsions (Alley et al., 1981), and mice maintained on zinc-deficient diets display protection against electrically induced seizures (Tokuoka et al., 1967).

Given the complexing behavior and catalytic potency of divalent metal ions, we can identify possible sites at which metal ions may regulate neuronal function and, in excess, interfere with the normal synaptic transmission and render the brain hyperexcitable. As illustrated schematically in Fig. 1, an excess of divalent metal ions, firstly, can inhibit the activities of some of the enzymes that catalyze the synthesis and degradation of neurotransmitters, thereby altering the size of pools of releasable transmitters. Moreover, divalent metal ions contained in nerve terminals

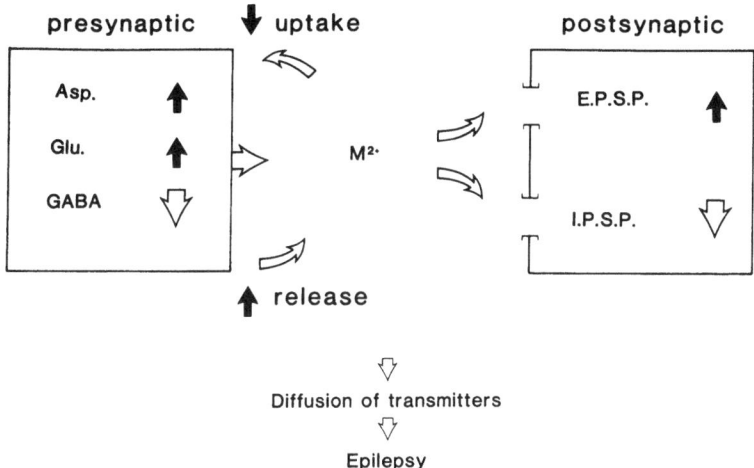

Fig. 1. A schematic diagram linking metal ions, neuroactive amino acid metabolism and epilepsy. Abbreviations: M^{2+}, metal ions; E.P.S.P., excitatory postsynaptic potential, I.P.S.P., inhibitory postsynaptic potential.

are extruded into the extracellular space following intense neuronal activity (Assaf and Chung, 1984; Ben-Ari et al., 1984; Howell et al., 1984). Metal ions released in the extracellular space may in turn potentiate the release of neuroactive amino acids or inhibit the rapid uptake of released transmitter molecules. Finally, metal ions in the extracellular space can interfere with synaptic transmission, either by prolonging the action of excitatory neurotransmitters (Smart and Constanti, 1983) or by rendering the action of released GABA less effective (Smart and Constanti, 1982). Such pleiotropic effects of metal ions, if they occur, will cause an accumulation in the synaptic cleft of charged transmitter molecules which may slowly disseminate to adjacent brain regions.

METAL IONS IN ENZYME CATALYSIS

The presence of metal ions is an essential constituent of some enzymes, whereas for other enzymes maximal reaction occurs when specific metal ions are present at optimum concentrations. In the metalloenzymes, the first group of enzymes, metal ions are held firmly by specialized bonds and dissociation does not occur at all, or is slow under normal physiological conditions. By contrast, the binding of the metal ion in metal-sensitive enzymes is weak, there is no stoichiometry in the metal-protein ratio, and the complex is labile. For these enzymes, metal ions may be construed as playing a regulatory role, since their activities are accelerated or retarded by different ambient concentrations of the metals. Gurd and Wilcox (1956) have presented an excellent review on the binding of metal ions by proteins, peptides and amino acids in which they also discuss in detail the chemistry of mercaptide formation and possible mechanisms of metallic activations of enzymes.

The relevance of metals to epilepsy stems from the fact that two of the essential enzymes involved in amino acid metabolism, glutamate decarboxylase (EC 4.1.1.15) and glutamine synthetase (EC 6.3.1.2), are known to be inhibited by metal ions. In Fig. 2, the activity of glutamate decarboxylase is plotted against the concentration of Zn^{2+} in the incubating medium. In agreement with the findings reported by Wu and Roberts (1974) and Ebadi et al. (1981), we find that the activity of glutamate decarboxylase, the rate-limiting enzyme that determines the level of GABA in the nervous system, is critically dependent on the concentration of Zn^{2+}. The activity of glutamine synthetase, a Mg^{2+} or Mn^{2+} activated enzyme, has long been known to be affected by the relative concentrations of metal ions and it has been suggested that metal ion interaction may be of physiological significance in the control of this enzyme (Meister, 1972). Using crude synaptosomal preparations, it has been shown that a significant inhibition of glutamine synthetase activity occurs in the presence of Zn^{2+} (R. Murphy, personal communication), although the degree of metallic inhibition is less pronounced than that for glutamate decarboxylase. Both of these enzymes contain free sulfhydryl groups in their active sites and are very sensitive to sulfhydryl reagents such as 5,5'-dithiobis-2-nitrobenzoic acid (Wu and Roberts, 1974) and p-hydroxymercuribenzoate (Meister, 1962). The regulation of these enzymes by metal ions appears to be complex. Certain inhibitors of glutamine synthetase, such as carbamyl phosphate, are effective only in the presence of Mn^{2+} (Tate and Meister, 1974).

It is also likely that the activities of glutaminase (EC 3.5.1.2) and glutamate dehydrogenase (EC 1.4.1.3) are modulated by divalent metals, although no quantitative data are available. The equilibrium for both of the enzymes lies in the direction of glutamate formation. Glutamate dehydrogenase was originally thought to be a zinc metalloenzyme but subsequent reports have thrown doubt on the belief that Zn^{2+} is essential for its activity, although it has a high affinity for Zn^{2+} (Smith et al., 1975). The regulatory function of metal ions on bacterial glutamate dehydrogenate is reported by LeJohn et al. (1969). Similarly, bacterial glutaminase is stimulated by divalent cations and the degree of activation

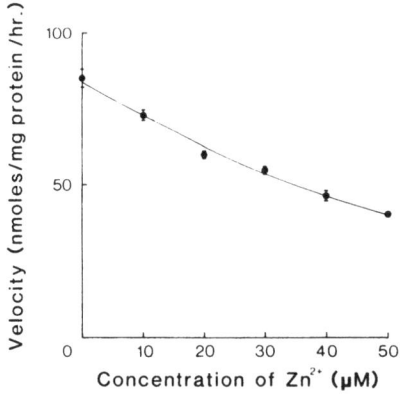

Fig. 2. Effects of Zn^{2+} on the activity of glutamate decarboxylase. The data were obtained from crude synaptosomal fractions from the brains of mice, DBA/2J. The velocity of the enzyme is plotted as a function of the concentration of Zn^{2+} in the incubating medium, and points are fitted with an exponential function. Error bars represent one S.E.M.

was shown to depend on the levels of divalent metals and adenine nucleotides (Prusiner, 1974). Detailed studies on the interaction between metal ions and enzymes involved in neuroactive amino acid metabolism are clearly needed.

METAL IONS AND PYRIDOXAL PHOSPHATE

Pyridoxal-5'-phosphate (PLP) holds an exceptional place among the coenzymes with regard both to the diversity of its catalytic functions and to their significance in biochemical transformations of amino acids. Among the enzymes which use PLP as their prosthetic group are aspartate transaminase (EC 2.6.1.1), GABA-transaminase (EC 2.6.1.19) and glutamate decarboxylase, all of which play crucial roles in anabolic and catabolic pathways of aspartate, glutamate and GABA. The relevance of metal ions to PLP stems from the fact that pyridoxal phosphokinase (EC 2.7.1.35), the enzyme which acts on all three non-phosphorylated vitamers (pyridoxine, pyridoxal and pyridoxamine), is activated by Zn^{2+}, as well as Mn^{2+}, Co^{2+}, Ni^{2+} and Fe^{2+} (McCormick et al., 1961). Ebadi and his colleagues (1981) demonstrated that the addition of 12 µM Zn^{2+} enhances the activity of the enzyme by 50%.

It is known that some PLP dependent enzymes are unsaturated, in that not all are holoenzymes, and the overall activities are increased by the addition of PLP. GABA-transaminase *in vivo* is believed to be normally unsaturated and its activity is critically regulated by the labile pool of PLP, whereas the activity of glutamate decarboxylase is somewhat insensitive to a small increase in PLP (Norris et al., 1985). One would thus expect that an increase in brain levels of PLP following either intraperitoneal injection or by stimulating the activity of pyridoxal kinase, would decrease the level of GABA and enhance glutamate, if only the involvement of the two enzymatic pathways are considered. As activities of numerous other PLP-dependent, metal-sensitive enzymes interact, the situation *in vivo* will undoubtedly prove to be complicated. Nevertheless, evidence suggests that the balance between GABA and glutamate is influenced by metal ions and PLP, excesses of these substances shifting the balance towards the accumulation of glutamate (Chung and Johnson, 1983; Norris et al., 1985).

There are other ways in which metal ions, together with PLP, can influence the excitability of the nervous system. Pyridoxal phosphate and metal ions can catalyze non-enzymatic reactions, including decarboxylation and transamination. A metal ion in the non-enzymatic reaction performs, in an inferior fashion, the same function as the protein in the enzyme. The initial reaction is the formation of an aldimine Schiff base by condensation between the amino group and the 4'-aldehyde group of pyridoxal phosphate. In these azomethines, the metal ion plays a dual role. It stabilizes the coplanar geometry of the intermediate Schiff bases by coordinating the phenolic hydroxyl and imino N atom, as shown below.

Then, the ion polarizes the bonds between this nitrogen and two adjacent carbon atoms by bridging the gap existing in the π-bond between the pyridoxal-substrate imine, so enhancing the electron withdrawing effect of the pyridine ring. Such non-enzymatic reactions are known to occur in model systems, but they occur at rates that are orders of magnitude lower than the corresponding enzymatic reactions, and require concentrations of metal ions, amino acids, and PLP much higher than the overall concentrations found in the brain to proceed at reasonable rates. Nevertheless, the possible occurrence of such reactions in vivo, as well as the role of a similar adduct, namely, PLP-drug-metal ion, in anticonvulsant action (see Chung et al., 1984), should be borne in mind.

RELEASE OF METAL IONS INTO THE EXTRACELLULAR SPACE

To postulate that metal ions play physiological roles in the nervous system, it is necessary to demonstrate that they are not permanently sequestered in a neuronal compartment but, under certain circumstances, are extruded into the extracellular space and transported back into the cell. Hints that metal ions contained in nerve terminals, especially in dense-core vesicles, are extruded in the extracellular space have existed in the literature. Nitsch and Rinne (1981) report that dense-core vesicles undergo exocytosis during epileptic seizures. Also, there is a selective loss of Timm stain in the mossy fiber pathway stimulated electrically for 24 hours (Sloviter, 1985).

Several recent studies have demonstrated that zinc ions contained in hippocampal tissues are released upon electrical stimulation (Howell et al., 1984), depolarization with high K^+ (Assaf and Chung, 1984) and during kainate-induced epileptic seizures (Assaf and Chung, 1984; Ben-Ari et al., 1984). As shown in Fig. 3, the concentration of Zn^{2+} in perfusate collected during the application of kainic acid was higher than that collected during control periods. The largest extrusion of Zn^{2+} occurred during the first 10 minutes of stimulation, when electrophysiological recordings indicated that the neurones were displaying abnormal electrical activity characterized by a transient enlargement of evoked responses followed by a period during which responses could not be evoked. When the tissues were perfused with the normal solution again, the electrical activity slowly returned to normal. There was a small, but statistically significant, decrease of the Zn^{2+} content in the samples collected during the recovery period, suggesting that tissues were accumulating Zn^{2+} contained in the perfusing media. Assaf and Chung (1984) estimated that about 18% of the total Zn^{2+} contained in the tissue was extruded into the perfusing medium when the perfusing solution contained 23.8 mM KCl. If we assume that the extracellular space occupies 15% of the brain, the local concentration of Zn^{2+} at the peak of convulsive activity could reach almost 300 μM. It remains to be investigated if other trace metals existing in the nervous system are also released by similar manipulations. Such a transient increase in the extracellular concentration of metal ions could have a profound effect on neuronal function. For example, they can interfere with the uptake or release of neurotransmitters as well as with the action of transmitter molecules on postsynaptic cells. Moreover, a massive efflux of metal ions into a confined extracellular space may alter the pH of the microenvironment, leading to cellular damage. The CA3 hippocampal neurones, which are innervated by the mossy fiber bundle containing the highest concentration of Zn^{2+}, are among the first group of cells destroyed by kainic acid. The possibility that cellular damage caused by this neurotoxin is due to the release of metal ions deserves careful investigation.

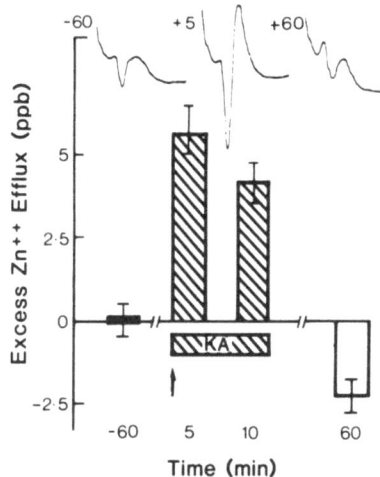

Fig. 3. Release of zinc from hippocampal slices <u>in vitro</u>. The bars represent the average differences in the zinc concentration between the control solution and the perfusates collected during and after the application of 100 μM kainic acid. Each collection period lasted for 5 minutes. The insets indicate that the evoked response was enhanced during kainate application.

INFLUENCE OF METAL IONS ON RELEASE AND UPTAKE OF NEUROACTIVE AMINO ACIDS

Metal ions are known to inhibit the uptake of catecholamines by adrenal medulla (Kirshner, 1962) and retard the receptor binding of acetylcholine (Hulme et al., 1983) and GABA (Baraldi et al., 1984). Similar inhibitory effects were observed with neuroactive amino acids, as shown in Fig. 4. The high affinity uptake mechanisms of both glutamate and GABA were impaired by the presence of Cu^{2+}, Zn^{2+} and Fe^{2+}. As summarized in Table 1, addition of 30 μM Fe^{2+} to the medium inhibited maximum uptake of glutamate by 39%, whereas the same concentration of Zn^{2+} led to an inhibition by 19%. The most effective ion in this respect was Cu^{2+}, causing 41% inhibition of uptake at 10 μM. Similarly, the presence of metal ions inhibited the maximum uptake of GABA. Iron caused an inhibition of GABA uptake with a 33% decrease at 50 μM. Copper was again found to be the most potent inhibitor of uptake, causing 43% inhibition at 20 μM.

Release of neuroactive amino acids, on the other hand, is potentiated by divalent metal ions. Both spontaneous and veratrine-induced glutamate release from purified synaptosomal preparations was considerably enhanced in the presence of 30 μM of Cu^{2+}, Fe^{2+} and, to a lesser extent, Zn^{2+}, as shown in Fig. 5. Potassium-stimulated release was similarly influenced by the divalent ions.

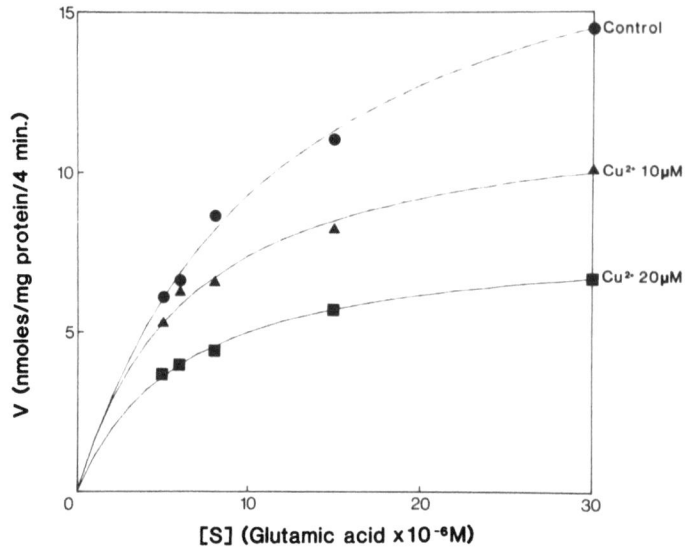

Fig. 4. High affinity glutamate uptake as a function of concentration of glutamate. The Michaelis-Menten equation was fitted through the data points, and the two constants together with their standard errors were calculated. The presence of Cu^{2+} at the concentrations shown inhibited V_{max}. The data were obtained from crude synaptosomal fractions from DBA/2J brains.

POSTSYNAPTIC RECEPTORS

The interaction between metal ions and postsynaptic receptors in the central nervous system has so far received little attention. Two recent reports suggest that zinc ions may modulate postsynaptic responses through mechanisms which are yet to be elucidated. Using in vitro slice preparations, Smart and Constanti (1983) showed that bath-application of Zn^{2+} caused a prolongation of the excitatory postsynaptic potential and irregular oscillation in the membrane potential. Wright (1984) noted that there was an increase in spontaneous firing rate following iontophoretic application of Zn^{2+}. It is not clear, however, if the observed effects in these studies stem from the action of metal ions on the postsynaptic receptors or presynaptic terminals. This problem undoubtedly will be resolved, as systematic investigations are carried out with the aid of a broad array of modern experimental techniques.

If metal ions are indeed released into the extracellular space, and if the released ions in turn potentiate further release of neuroactive amino acids and hinder the uptake mechanisms, then the time course of transmitter removal from the synaptic cleft will be prolonged. The presence of transmitter molecules in the synaptic cleft may, under certain circumstances, influence the action of physiologically released GABA or glutamate.

CONCLUDING REMARKS

We have here briefly outlined the ways in which metal ions in the nervous system can modulate neuronal excitability. Although the crucial roles they play in protein chemistry and enzymology have been long recog-

Table 1. Kinetic parameters of high affinity uptake of glutamate and GABA. The values of V_{max} are given in nmol/mg of protein/4 min. The percentage differences between the control values and experimental values are tabulated. *$p < 0.05$, **$p < 0.01$.

	Glutamate		GABA	
Control	18.3 ± 0.9		15.2 ± 0.6	
Cu^{2+} (10 µM)	10.9 ± 0.9	−41%**	12.0 ± 0.4	−21%**
Cu^{2+} (20 µM)	6.0 ± 0.3	−67%**	8.7 ± 0.4	−43%**
Zn^{2+} (30 µM)	14.7 ± 1.1	−19%**	12.0 ± 1.2	−21%*
Zn^{2+} (60 µM)	11.5 ± 1.2	−37%**		
Fe^{2+} (30 µM)	11.2 ± 1.1	−39%**		
Fe^{2+} (50 µM)			10.2 ± 1.0	−33%**
Co^{2+} (100 µM)	17.5 ± 1.5	− 5%	14.2 ± 1.1	− 6%

nized trace element research in neurobiology is still in its infancy. We hope that the overview we present here may stimulate further research, aimed at elucidating the role of trace metals in brain function. The solution of this challenging research problem should increase our understanding of brain function and of the pathological mechanisms underlying disordered brain function.

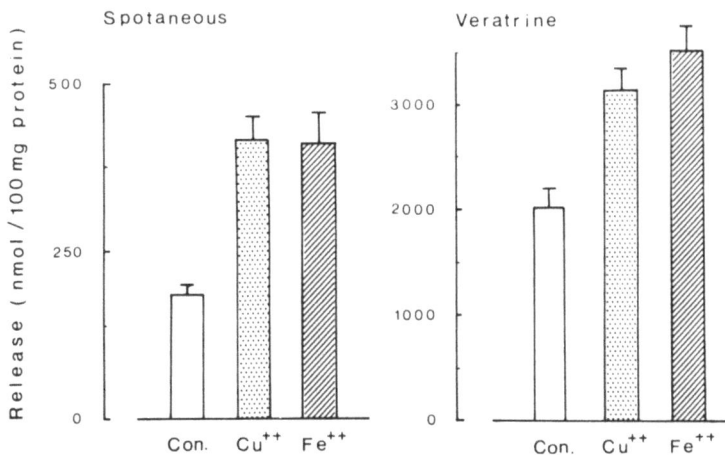

Fig. 5. Influence of metal ions on the release of endogenous glutamate from the nerve terminals. Purified synaptosomes were incubated in the presence of 30 µM of metal ions for 15 minutes. Spontaneous release (left) and veratrine-stimulated release (right) of glutamate under different experimental conditions are shown. Error bars represent one S.E.M.

REFERENCES

Alley, M.C., Killam, E.K., and Fisher, G.L., 1981, The influence of d-penicillamine treatment upon seizure activity and trace metal status in Senegalese baboon, Papio papio, J. Pharmacol. Exp. Ther., 217:138.

Assaf, S.Y., and Chung, S.H., 1984, Release of endogenous Zn^{2+} from brain tissue during activity, Nature, 308:734.

Ballhausen, C.J., 1962, Introduction to Ligand Field Theory, McGraw-Hill Book Co., New York.

Baraldi, M., Caselgrandi, E., and Santi, M., 1984, Effect of zinc on specific binding of GABA to rat brain membranes, in: The Neurobiology of Zinc, C.J. Frederickson, G.A. Howell, and E.J. Kasarskis, eds., Alan R. Liss, Inc., New York, p. 59.

Ben-Ari, Y., Charton, G., Leviel, V., and Rovira, C., 1984, Spontaneous and evoked release of Zn^{2+} in the Ammon's horn of the anesthetized rat, J. Physiol., 357:40P.

Chung, S.H., and Johnson, M.S., 1983, Experimentally induced susceptibility to audiogenic seizure, Exp. Neurol., 82:89.

Chung, S.H., Johnson, M.S., and Gronenborn, A.M., 1984, L-cycloserine: a potent anticonvulsant, Epilepsia, 25:353.

Colasanti, B.K., Lindamood, C., and Craig, C.R., 1982, Effects of marihuana cannabinoids on seizure activity of cobalt-epileptic rats, Pharmacol. Biochem. Behav., 16:573.

Ebadi, M., Itoh, M., Bifano, J., Wendt, K., and Earle, A., 1981, The role of Zn^{2+} in pyridoxal phosphate mediated regulation of glutamic acid decarboxylase in brain, Int. J. Biochem., 12:1107.

Figgis, B.N., 1966, Introduction to Ligand Fields, Interscience Publishers, New York.

Griffith, J.S., 1961, The Theory of Transition-Metal Ions, Cambridge Univ. Press, Cambridge.

Gurd, F.R.N., and Wilcox, P.E., 1956, Complex formation between metallic cations and proteins, peptides, and amino acids, Adv. Protein Chem., XI:311.

Heinemann, U., Konnerth, A., and Lux, H.D., 1981, Stimulation induced changes in extracellular free calcium in normal cortex and chronic alumina cream foci of cats, Brain Res., 213:246.

Howell, G.A., Welch, M.G., and Frederickson, C.J., 1984, Stimulation-induced uptake and release of zinc in hippocampal slices, Nature, 308:736.

Hulme, E.C., Berrie, C.P., Birdsall, N.J.M., Jameson, M., and Stockton, J.M., 1983, Regulation of muscarinic agonist binding by cations and guanine nucleotides, Eur. J. Pharmacol., 94:59.

Itoh, M., and Ebadi, M., 1982, Selective inhibition of hippocampal glutamic acid decarboxylase in zinc-induced epileptic seizures, Neurochem. Res., 7:1287.

Kirshner, N., 1962, Uptake of catecholamines by a particulate fraction of the adrenal medulla, J. Biol. Chem., 237:2311.

LeJohn, H.B., Jackson, S.G., Klassen, G.R., and Sawula, R.V., 1969, Regulation of mitochondrial glutamic dehydrogenase by divalent metals, nucleotides, and α-ketoglutarate, J. Biol. Chem., 244:5346.

McCormick, D.B., Gregory, M.E., and Snell, E.E., 1961, Pyridoxal phosphokinases. I. Assay, distribution, purification, and properties. J. Biol. Chem., 236:2076.

Meister, A., 1962, Glutamine synthesis, in: The Enzymes, Vol. 6, P.D. Boyer, H. Lardy and K. Myrback, eds., Academic Press, New York, p. 443.

Meister, A., 1974, Glutamine synthetase of mammals, in: The Enzymes, Vol. 10, P.D. Boyer, ed., Academic Press, New York, p. 699.

Nitsch, C., and Rinne, U., 1981, Large dense-core exocytosis and membrane recycling in the mossy fiber synapses of the rabbit hippocampus during epileptiform seizures, J. Neurocytol., 10:201.

Norris, D.K., Murphy, R.A., and Chung, S.H., 1985, Alteration of amino acid metabolism in epileptogenic mice by elevation of brain pyridoxal phosphate, J. Neurochem., 44:1403.

Oakely, J.C., and Ojemann, G.A., 1982, Effects of chronic stimulation of caudate nucleus on a preexisting alumina seizure focus, Exp. Neurol., 75:360.

Orgel, L.E., 1960, An Introduction to Transition-Metal Chemistry: Ligand-Field Theory, Methuen and Co., Ltd., New York.

Pei, Y., Zhao, D., Huang, J., and Cao, L., 1983, Zinc-induced seizures: a new experimental model of epilepsy, Epilepsia, 24:169.

Prusiner, S., 1974, Glutaminases of Escherichia coli: properties, regulation and evolution, in: The Enzymes of Glutamine Metabolism, S. Prusiner and E.R. Stadtman, eds., Academic Press, London, p. 293.

Reynolds, A.F., Ojemann, G.A., and Ward, A.A., Jr., 1981, Intracellular records from chronic alumina epileptogenic foci in the monkey, Epilepsia, 22:147.

Ross, S.M., and Craig, C.R., 1981, γ-Aminobytyric acid concentration, L-glutamate-1-decarboxylase activity, and properties of the γ-aminobutyric acid postsynaptic receptor in cobalt epilepsy in the rat, J. Neurosci., 1:1388.

Sloviter, R.S., 1985, A selective loss of hippocampal mossy fiber Timm stain accompanies granule cell seizure activity induced by perforant path stimulation, Brain Res., 330:150.

Smart, T.G., and Constanti, A., 1982, A novel effect of zinc on the lobster muscle GABA receptor, Proc. R. Soc. Lond. B., 215:327.

Smart, T.G., and Constanti, A., 1983, Pre-and postsynaptic effects of zinc on in vitro prepyriform neurones, Neurosci. Lett., 40:205.

Smith, E.L., Austin, B.M., Blumental, K.M., and Nye, J.F., 1975, Glutamate dehydrogenase, in: The Enzymes, Vol. 11, P.D. Boyer, ed., Academic Press, New York, p. 293.

Tate, S.S., and Meister, A., 1974, Glutamine synthetases of mammalian liver and brain, in: The Enzymes of Glutamine Metabolism, S. Prusiner and E.R. Stadtman, eds., Academic Press, London, p. 77.

Tokuoka, S., Takeshi, F., Hiraoka, H., Takashima, M., Fuji, M., and Watanabe, M., 1967, Neurochemical consideration on the alleviating effect of caudal resection of pancreas on epileptic seizures; relationship of zinc metabolism to brain excitability, Bull. Yamaguchi Med. School, 14:1.

Trottier, S., Truchet, M., and Laroudie, C., 1982, Secondary ion microanalysis in the study of cobalt-induced epilepsy in the rat, Exp. Neurol., 76:231.

Tuckwell, H.C., and Miura, R.M., 1978, A mathematical model for spreading cortical depression, Biophys. J., 23:257.

Ward, A.A., Jr., 1972, Topical convulsant metal, in: Experimental Models of Epilepsy, D.P. Purpura, J.K., Penry, D. Tower, D.M. Woodbury, and R. Walter, eds., Raven Press, New York, p. 13.

Willmore, L.J., Hurd, R.W., and Sypert, G.W., 1978, Epileptiform activity initiated by pial iontophoresis of ferrous and ferric chloride on rat cerebral cortex. Brain Res.,152:406.

Willmore, L.J., and Rubin, J.J., 1982, Formation of malonaldehyde and focal brain edema induced by subpial injection of $FeCl_2$ into rat isocortex, Brain Res., 246:113.

Wright, D.M., 1984, Zinc: effect and interaction with other cations in the cortex of the rat, Brain Res., 311:343.

Wu, J.-Y., and Roberts, E., 1974, Properties of brain-glutamate decarboxylase: inhibition studies, J. Neurochem., 23:795.

ZINC-BINDING PROTEINS IN THE BRAIN

M. Ebadi and Y. Hama

Department of Pharmacology
University of Nebraska College of Medicine
42nd Street and Dewey Avenue
Omaha, Nebraska 68105

ABSTRACT

As an essential substance, zinc is involved in maintaining the functions and/or the structures of at least 200 metalloenzymes that participate in numerous biochemical reactions, including the metabolism of proteins and nucleic acids. The steady-state concentration of zinc in the brain must be regulated firmly since both an excess and a deficiency of zinc have been implicated in neurological disorders including epilepsy. Zinc-binding proteins have been detected in the bovine hippocampus, cerebellum, and pineal gland. A metallothionein-like protein has been identified recently in the rat brain which resembles in some but not all aspects a hepatic metallothionein. The synthesis of this protein is stimulated following the administration of zinc and copper but not of cadmium. The zinc-stimulated protein incorporates ^{35}S cysteine 24-fold higher than the native, unstimulated protein; is blocked by actinomycin D; produces two isoforms by ion exchange chromatography on DEAE Sephadex A 25 columns; and by high performance liquid chromatography, depicts a similar but not identical profile to zinc-stimulated hepatic metallothionein. Since the synthesis of this protein is stimulated following the administration of zinc and is depressed in the brains of zinc-deficient rats, it is postulated that the unbound pool of zinc may serve as one of the factors involved in regulating the synthesis of this protein. Since zinc in physiological concentrations stimulates a number of pyridoxal phosphate-dependent reactions and in pharmacological doses inhibits an extensive number of SH-containing enzymes and receptor sites for neurotransmitters, we postulate that the metallothionein-like protein in the brain may have function(s) associated with zinc homeostasis and perhaps events related to synaptic functions.

INTRODUCTION

Zinc, which is the fourth most abundant ion in the brain, is involved in an impressive array of physiological states, including maintaining the functions of enzymes and nucleic acids. The importance of these metabolic processes in central nervous maturation suggests that zinc deficiency may affect neural development.

Role of Zinc in Enzymes

The roles of zinc in more than 200 metalloenzymes can be divided into four major categories (Vallee, 1983).

Catalytic role. Zinc participates directly and is involved in catalysis by enzymes such as carbonic anhydrase, carboxypeptidase, and aldolase.

Structural role. Zinc participates mostly but not exclusively in maintaining the quaternary structure of oligomeric holoenzymes such as aspartate transcarbamylase.

Regulatory (modulatory) role. Zinc regulates enzymatic activity by either activating enzymes (e.g., bovine lens leucine aminopeptidase) or inhibiting them (e.g., porcine kidney aminopeptidase).

Noncatalytic and nonstructural role. Zinc is involved in an obscure fashion in certain metalloenzymes such as human alcohol dehydrogenase or E. coli alkaline phosphatase, but the involvement is neither of a catalytic nor of a structural nature.

Role of Zinc in Nucleic Acid Metabolism

Zinc, which is present in the nucleus, nucleolus, and chromosomes (Fujii, 1954), is required for the synthesis of ribonucleic acid (RNA), deoxyribonucleic acid (DNA), and protein. Thymidine kinase, DNA-dependent RNA polymerase, DNA polymerase, and RNA-dependent DNA polymerase are zinc-dependent enzymes (see O'Dell, 1974; Prasad, 1976; Underwood, 1977; Vallee, 1983; and Dreosti, 1984 for reviews).

Neurobiological Aspects of Zinc Deficiency

Zinc is essential for the growth and development of the fetus (see Hurley et al., 1971; Hurley, 1981; and Ebadi, 1985 for reviews). In addition to congenital malformation, zinc deficiency produces behavioral abnormalities such as impaired avoidance, lower explorative activity, a lower level of physical activity, enhanced maternal contacts (Sandstead et al., 1977, 1978), enhanced aggression (Halas et al., 1975, 1977; Peters, 1978), and learning and memory deficits (Halas and Sandstead, 1975; Halas et al., 1979, 1983; Peters, 1979; Golub et al., 1983). The learning and memory deficits occur faster and are more pronounced in younger animals (Gordon, 1984).

Studies by Halas and Kawamoto (1984) have shown that zinc deficiency impairs memory and causes persistent injury across all regions of the hippocampus, an area of the brain which is essential for the registration and retention of information (Olton et al., 1977, 1978; Buell et al., 1977).

Recent studies have shown that zinc deficiency in rats impairs acquisition of cerebellar granule cells (Dvergsten et al., 1983), impairs maturation of cerebellar purkinje cells (Dvergsten et al., 1984a), and impairs dendritic differentiation of cerebellar basket and stellate cells (Dvergsten et al., 1984b).

The Status of Zinc in Disorders of the CNS

The concentration of zinc is altered in an extensive number of pathological states. Those which are associated with the diseases of the CNS are enumerated in Table 1.

Table 1. Alteration in the Concentration of Zinc in Disorders of CNS

Pathological States	Zinc Status	References
Alcoholism	Reduced in Serum	Wu et al., 1984
Alzheimer-type Dementia	Elevated in CSF	Hershey et al., 1983
Down's Syndrome	Elevated in RBC	Neve et al., 1983
Epilepsy	Reduced in Serum	Palm and Hallmans, 1982
Friedreich's Ataxia	Reduced in Hair	Shapcott et al., 1984
Guillain-Barre Syndrome	Elevated in CSF	Palm et al., 1982
Hepatic Encephalopathy	Reduced in Brain	Baraldi et al., 1983
Multiple Sclerosis	Elevated in RBC	Dore-Duffy et al., 1983
	Elevated in Blood	Rieder et al., 1983
Pick's Disease	Elevated in Hippocampus	Constantinidis and Tissot, 1981
Retinitis Pigmentosa	Elevated in Blood	Mozha, 1974; Eckhert, 1981
	Elevated in Plasma	Bastek et al., 1976
Schizophrenia	Reduced in Brain	Kimura and Kumura, 1965; Pfeiffer and LaMola, 1983
Wernicke-Korsakoff Syndrome	Reduced in Serum	Majumdar et al., 1983

The mechanisms involved in the alteration of the concentration of zinc are not known and deserve investigation.

The Status of Zinc in Epileptiform Seizures

The involvement of metallic salts such as magnesium (Borges and Gucer, 1978), iron (Sypert, 1982), manganese (Papavasiliou et al., 1979), and zinc (Ebadi and Pfeiffer, 1984) in epileptiform seizures has been reviewed. The status of zinc in human epilepsy and in experimentally-induced seizures is summarized in Table 2.

ZINC-BINDING PROTEIN IN THE BRAIN

Proteins that bind zinc may be classified into at least three major

Table 2. Zinc and Epileptiform Seizures

Subjects/Condition	Comments	References
Neonatal Seizures	Seizures occur 4-6 days after birth; low zinc level in CSF; innocous.	Pryor et al., 1981; Goldberg and Sheehy, 1982
Human Epilepsy	Low initial serum concentration of zinc; anticonvulsants increase plasma zinc levels and may cause 'masked zinc deficiency state'.	Pippenger et al., 1980; Palm and Hallmans, 1982
Photically-Induced Seizures, *Papio Papio*	Elevated zinc but not copper serum level	Alley et al., 1981
Rats	ICV adm. of zinc produced epileptic seizures; it was postulated that zinc inhibited Na^+K^+ATPase.	Donaldson et al., 1971
Rats	ICV adm. of zinc produced epileptic seizures and inhibited only the hippocampal GAD by interfering with action of pyridoxal phosphate.	Itoh and Ebadi, 1982b
Audiogenic Mouse	Higher levels of Zn^{2+} and Cu^{2+} in brain; postulated to inhibit GAD.	Chung and Johnson, 1983
Rabbit	IC adm. of Zn produced epileptic seizures associated with 'spike and wave activities' which were often present in the absence of overt manifestations of seizures.	Pei et al., 1983

groups consisting of metalloenzymes, metallothioneins, and metalloproteins other than metalloenzymes or metallothioneins.

<u>Metalloenzyme in the Brain</u>

Carbonic anhydrase, an example of a zinc enzyme, in addition to being a very efficient catalyst of the reversible hydration of CO_2, also catalyzes the hydration of aldehydes and the hydrolysis of certain esters (see Lindskog 1983 for a review). In the brain, no consistent association between the concentration of zinc and the activities of metalloenzymes is apparent. For example, in the rat brain, zinc is present in highest concentration in the hippocampus (Ebadi et al., 1981), whereas the highest activity of carbonic anhydrase is found not in the hippocampus, but in the medulla oblongata (Nair and Bau, 1971). Similarly, in the rabbit brain, the concentration of zinc does not parallel the activity of carbonic anhydrase (Klee and Lieflander, 1965).

<u>Metalloprotein in the Brain Unrelated to Metalloenzyme or Metallothionein</u>

Studies by Baudier et al. (1983) and Baudier and Gerard (1983) using brain have shown that the S100 protein fraction, which is a mixture of S100a (αβ) and S100b (ββ) components, should no longer be considered only as a 'calcium-binding protein' but also as a zinc-binding protein. The S100b protein exhibits two sets of zinc-binding sites with KD values of 10^{-7} to 10^{-6} M. Studies by Hesketh (1983) have shown that at least one zinc-binding site exists on tubulin. Furthermore, a zinc- dependent interaction of bovine S100b protein with the tubulin-microtubule system has been reported (Baudier et al., 1983). Finally, Baudier et al. (1985) have shown that zinc binding on rat S100b protein regulates calcium binding by increasing the calcium affinity of the protein and reducing the antagonistic effects of K^+ on calcium binding. The implication of these findings in the neurobiology of zinc, which may be far reaching, awaits clarification.

Metallothionein-Like Proteins in the Brain

Metallothioneins (Kagi and Vallee, 1960) bind a variety of metals, including cadmium, copper and zinc. A metallothionein has a Mr of 6500 daltons, consists of a single polypeptide chain of 61 amino acids, 25 to 30% of whose residues are cysteine, has a metal-binding capacity of between 5 and 7 g atoms/mol, and contains no disulfide bonds or aromatic amino acids.

Nuclear magnetic resonance (113cd NMR) studies indicate that both cadium and zinc bound to metallothioneins are arranged in two clusters of A and B, binding four and three atoms of metal, respectively. Metal binding appears to occur first in cluster A and then in cluster B, and the metal release takes place from cluster B and then from cluster A (Nielson and Winge, 1983). Metallothioneins contain two isoforms, designated as metallothioneins I and II (Pulido et al., 1966; Kagi and Nordberg, 1979; Klauser et al., 1983). The DNA sequence data indicate that there may be more than two metallothionein genes (Karin and Richards, 1982).

Although metallothioneins have been isolated and characterized primarily from liver, kidney, and intestine (see Kagi and Nordberg, 1979; and Lucier and Hook, 1984 for review), metal-binding proteins (ligands) have not until recently been characterized in brains (Chen and Ganther, 1975; Record et al., 1982; Itoh et al., 1983; Sato et al., 1984a, b; Ebadi et al., 1984; Awad and Ebadi, 1985). The discovery of a metallothionein-like protein in the rat brain by Itoh et al. (1983) was a fortuitous observation which took place while studying the interactions among zinc, glutamic acid decarboxylase, and convulsive seizures in rats (Itoh and Ebadi, 1982b). The intracerebroventricular administration of zinc sulfate (0.3 µmol/10 µl) caused seizures characterized by running fits, vocalizations, fasciculation of facial muscles, myoclonic movements of the limbs, and tonic-clonic convulsion. These seizures were prevented by γ-aminobutyric acid (GABA) but not by other putative transmitters. During the course of our investigation, some interesting observations were made.

- The administration of substantially larger doses of zinc sulfate (up to 100 mg/kg), either intravenously or intraperitoneally, did not cause convulsive seizures in rats.

- Since the convulsive dose of 0.3 µmol/10 µl of zinc sulfate was considerably lower than the concentrations of zinc in 14 regions of rat brain (Ebadi et al., 1981), we postulated that most of the zinc in the brain was bound and did not exist in free form. The fact that the rat did not exhibit spontaneous convulsions supported this contention. Furthermore, we concluded that the intravenously or intraperitoneally administered zinc in large toxicological doses became bound rapidly to albumin, α-macroglobulin, transferrin, ceruloplasmin, haptoglobulin, γ-globulins, metallothioneins, and perhaps other peripheral proteins (Prasad, 1979) becoming unavailable to be transported to the CNS and, therefore, unable to perturb the free concentration of zinc. Furthermore, since the zinc-induced convulsive seizures ceased following a duration of only 15 to 30 minutes, it was concluded that the intracerebroventricularly administered zinc also became bound rapidly to protein in the CNS, restoring the pre-zinc treated state. Based on these observations and conclusions, we searched for and identified zinc-binding proteins in the rat brain.

By using Sephadex G-75 column chromatography calibrated with proteins of known molecular weights and by using other techniques, we detected three separate zinc-binding proteins with apparent estimated molecular weights of 13,000, 15,000 and 210,000 daltons (Itoh and Ebadi, 1982a; Itoh et al., 1983). The two higher molecular weight proteins have not been characterized further, and their properties remain totally unknown. The chromatofocusing of the lowest molecular weight protein isolated from Sephadex G-75 column depicted three separate zinc-binding proteins, which focused at pHs of 6.8, 6.2, and 5.3, respectively (Ebadi, 1984a). The low molecular weight zinc-binding protein resembles in some but not all aspects a hepatic metallothionein for the following reasons. The synthesis of this protein is increased in a dose-dependent fashion by intracerebroventricular (Ebadi et al., 1984; Ebadi and Wallwork, 1985) but not intraperitoneal administration of zinc sulfate (Itoh et al., 1983; Heilmaier and Summer, 1985). Copper, which exists in large concentration in the brain, stimulates the synthesis of metallothionein-like protein, but does not bind to it. Cadmium, which is undetectable in the brain, does not stimulate the synthesis of this protein (Ebadi, 1984b). The zinc-stimulated metallothionein-like protein is attenuated by actinomycin D (Ebadi et al., 1984). The zinc-stimulated metallothionein-like protein incorporates ^{35}S cysteine 24-fold higher than the native, unstimulated protein. In addition, when the partially purified protein is first incubated with ^{65}Zn and then subjected to ion exchange chromatography on DEAE Sephadex A-25 columns, the protein peak that binds ^{65}Zn and the one that incorporates ^{35}S cysteine depict identical elution profiles (Ebadi et al., 1984). The zinc-stimulated metallothionein-like protein, like hepatic metallothionein, produces two isoforms on ion exchange chromatography of DEAE Sephadex A-25 columns (Ebadi et al., 1985).

Comparative high performance liquid chromatographic profiles of zinc treated rat brain metallothionein-like protein and zinc-treated hepatic metallothionein produced similar but not identical patterns (Ebadi et al., 1985). Preliminary studies have shown that metallothionein-like protein isoform I possesses a Mr of 6200 and consists of 60 residues of 12 cysteine and no histidine, arginine, leucine, tyrosine, or phenylalanine (Ebadi et al., 1985).

<u>The Proposed Function of Metallothionein-Like Proteins in the Brain</u>

The mechanisms influencing the synthesis of metallothioneins are complex and may be controlled by numerous physiological and pathological factors including dietary zinc, growth and development, epinephrine, glucagon, glucocorticoids, infection, starvation, stress, diabetes, and arthritis (see Cousins, 1985 for an excellent review). Furthermore, the proposed functions of metallothioneins are the maintenance of zinc homeostasis, the activation of apometalloproteins, and involvement in cellular defense mechanisms and in the detoxification of toxins (see Webb and Cain, 1982 and Cousins, 1985 for reviews).

The factor(s) regulating the synthesis of the metallothionein-like protein in the brain is not for certain and is under investigation. However, since the synthesis of this protein is stimulated following administration of zinc and is depressed in zinc-deficient rats (Ebadi and Wallwork, 1985), one may entertain the possibility that its synthesis may be influenced by the 'free' concentration of zinc.

Furthermore, the function(s) of the metallothionein-like protein in the brain may be distinct from that of the hepatic metallothionein. Although the synthesis of the metallothionein-like protein is stimulated following administration of zinc, this protein is, nevertheless, always present in the brains of zinc-untreated rats (Itoh et al., 1983; Heilmaier and Summer, 1985). Furthermore, this protein can be differentiated from the hepatic metallothionein in that it binds neither copper nor cadmium (Ebadi, 1984b); hence, it may not participate in a metal detoxification mechanism. Finally, there is no reason to believe that the mechanisms involved in maintaining zinc homeostasis in the peripheral organs and in the central nervous system are identical.

We postulate not only that the metallothionein-like protein functions as a donor of zinc to various zinc-requiring systems, but also that its inducibility, binding more and hence preventing the rise of free zinc, may play a decisive role in avoiding CNS toxicity.

Metallothionein-Like Protein and B6-Dependent Enzymes

Zinc may regulate the activity of glutamic acid decarboxylase in a dose-dependent fashion, stimulating it at physiological levels and inhibiting it at pharmacological levels. For example, studies from our laboratory have shown that zinc at 1.7×10^{-7} M stimulates the hippocampal pyridoxal kinase, enhancing the formation of pyridoxal phosphate and, in turn, enhancing the activity of glutamic acid decarboxylase, a pyridoxal phosphate (PLP)-requiring enzyme. On the other hand, zinc at 6.5×10^{-4} M inhibited glutamic acid decarboxylase (GAD) without inhibiting pyridoxal kinase (Ebadi et al., 1981). Itoh and Ebadi (1982b) showed that zinc sulfate administered intracerebroventricularly caused epileptiform seizures, including tonic-clonic convulsions. Since the zinc-induced convulsive seizures were blocked by the administration of GABA, the activity of GAD in zinc-treated animals was studied. When GAD was assayed under conditions in which 0.2 mM PLP were added, no differences were noted in tissues obtained from saline-treated or zinc sulfate-treated animals. In addition, the lack of effects was apparent from 5 to 30 minutes after the administration of zinc. Furthermore, the lack of effects was uniformly shown in hippocampus, thalamus, hypothalamus, amygdala, and caudate nucleus. On the other hand, when the status of GAD in these areas was ascertained in the absence of PLP, the activity was significantly reduced in hippocampus, with a more pronounced effect seen 30 minutes after the administration of zinc. The effect was specific for hippocampus but not for other brain areas studied (Itoh and Ebadi, 1982b).

The relation between the action of GABA as the main inhibitory transmitter in the hippocampal neurons and the genesis of epileptic seizures is now well established (see Ben-Ari and Krnjević, 1981). We interpret these results to indicate that zinc may inhibit the activity of GAD by interfering with the binding of pyridoxal phosphate to APOGAD.

Recent studies by Denner and Wu (1985) have shown that two forms of GAD are found in the rat brain. One form (GAD A) does not require exogenous pyridoxal phosphate for activity whereas another form (GAD B) does. Furthermore, the ratio between GAD A and GAD B is nonuniform throughout the brain area, and the hippocampus contains twice as much GAD B than GAD A. The modification by zinc and perhaps metallothionein-like protein of the activity of other pyridoxal phosphate-requiring enzymes remains to be illuminated.

Table 3. The Effects of Zinc on Select Synaptic Events

Parameters	Studied Effects	Reference
[^{3}H]Aspartate Binding	Zn^{2+} (IC_{50}: 50 µM) inhibited aspartate binding to the hippocampus.	Slevin and Kasaraskis, 1985
[^{3}H]β-Carboline -3-Carboxylate	Zn^{2+} (1 mM) inhibited β-carboline binding to the crude synaptic membrane of rat brain.	Mizuno et al., 1983
Cholecystokinin Concentration	Diethyldithiocarbonate chelating Zn^{2+} increased 3-fold the hippocampal concentration of cholecystokinin.	Stengaard-Pedersen et al., 1984
[^{3}H]Diazepam Binding	Zn^{2+} (1 mM) enhanced [^{3}H]diazepam binding to the crude synaptic membrane of rat brain.	Mizuno et al., 1983
GABA Receptor	Zinc in 10 µM–1 mM depressed the GABA-evoked conductance increase in the lobster muscle by inhibiting Cl^- channel.	Smart and Constanti, 1982
[^{3}H]Glutamate Binding	Zn^{2+} (IC_{50}: 131 µM) inhibited glutamate binding in the hippocampus.	Slevin and Kasarskis, 1985
[^{3}H]Glutamate Binding	Zn^{2+} (IC_{50}: 8 µM) inhibited glutamate binding to the pineal gland.	Govitrapong et al. 1985
[^{3}H]Met-Enkephalin Binding	Zn^{2+} (IC_{50}: 0.7 µM) inhibited enkephalin binding to the hippocampal opiate receptor.	Stengaard-Pederson, 1982
Prolactin Secretion	Zinc (5–10 µM) caused an acute, sustained, and reversible inhibition of prolactin secretion.	Login et al., 1983 Judd et al., 1984
[^{3}H]Spiroperidol Binding	Zn^{2+} (IC_{50}: .7 mM) inhibited D_2-dopamine binding to the pineal gland.	Govitrapong et al., 1985
[^{3}H]Spiperone Binding	Zn^{2+} (IC_{50}: 3.9 µM) inhibited D_2-dopamine binding to the striatum.	De Vries and Beart, 1985
[^{3}H]Substance P	Zn^{2+} (IC_{50}: 0.5 mM) inhibited substance P binding to the salivary gland.	Lee et al., 1983

Metallothionein-Like Protein and Calcium-Mediated Events

Zinc inhibits calcium-mediated activation of phosphodiesterase, Ca^{2+} ATPase, phosphorylase kinase, and adenylate cyclase. Furthermore, zinc inhibits calcium-stimulated histamine release, phagocytosis, chemotaxis, platelet aggregation, and hemoglobin binding to RBC (for reviews, see Brewer et al., 1979 and 1983). The possible involvement of metallothionein-like protein in protecting the calcium-mediated events in the brain is suggested. Indeed, studies by Baudier et al. (1985) have shown that in the absence of KCl, rat brain S100b protein is characterized by two high affinity Ca^{2+} binding sites with K_ds of 2×10^{-5} M and by four lower affinity Ca^{2+} binding sites with K_d values of 10^{-4} M. In the presence of 120 mM KCl, rat brain S100b protein binds two Zn^{2+} ions/mol of protein with K_d values of 10^{-7} and four others with lower affinity and K_d values of 10^{-6} M. Furthermore, these investigators have reported that zinc binding on rat S100b protein regulates calcium binding by increasing the calcium affinity of the protein and reducing the antagonistic effect of K^+ on calcium binding. Studies by Wright (1984) have shown that the application of zinc ions to cortical neurons causes excitation and blocks the depressant effects of calcium ions but not those of manganese or GABA.

Metallothionein-Like Protein and Synaptic Events

Zinc ions and in turn metallothionein-like protein may modulate not only the synthesis but also the release and binding of transmitters (Table 3).

The inhibitory actions of zinc on numerous receptor sites occur in physiological concentrations of zinc, which in 14 regions of the rat brain have been shown to range from 0.5 - 1.2 µmol/g tissue (Ebadi et al., 1981). Furthermore, the concentration of zinc in certain brain areas such as the giant boutons of the hippocampal mossy fibers may be even greater (Haug, 1967; Crawford and Harris, 1984).

REFERENCES

Alley, M.C., Killam, E.K., and Fisher, G.L., 1983, The influence of D-penicillamine treatment upon seizure activity and trace metal status in the Senegalese baboon, Papio papio, J. Pharmacol. Exp. Ther., 217:138.

Awad, A., and Ebadi, M., 1985, The characteristics of metallothioneins in bovine pineal gland, Fed. Proc., 44:3053.

Baraldi, M., Caselgrandi, E., Borella, P., and Zeneroli, M.L., 1983, Decrease of brain zinc in experimental hepatic encephalopathy, Brain Res., 258:170.

Bastek, J., Bogden, A., Cinotti, W., Tenhove, G., Stephans, M., Markopoulos, M., and Charles, J., 1976, Trace metals in a family with sex-linked retinitis pigmentosa, in: Retinitis Pigmentosa, M.B. Landers, III, M.L. Wolbarsht, J.E. Dowling and A.M. Latios, eds., Plenum Press, New York, p. 43.

Baudier, J., and Gerard, D., 1983, Ions binding to S100 proteins: structural changes induced by calcium and zinc on S100a and S100b proteins, Biochemistry, 22:3360.

Baudier, J., Haglid, K., Haiech, J., and Gerard, D., 1983, Zinc ion binding to human brain calcium binding proteins, calmodulin and S100b protein, Biochem. Biophys. Res. Comm., 114:1138.

Baudier, J., Labourdette, G., and Gerard, D., 1985, Rat brain S100b protein: purification, characterization, and ion binding properties. A comparison with bovine S100b protein, J. Neurochem., 44:76.

Ben-Ari, Y., and Krnjević, K., 1981, Actions of GABA on hippocampal neurons with special reference to the aetiology of epilepsy, in: Neurotransmitters, Seizures, and Epilepsy, P.L. Morselli, K.G. Lloyd, W. Löscher, B. Meldrum and E.H. Reynolds, eds., Raven Press, New York, p. 63.

Borges, L.F., and Gucer, G., 1978, Effect of magnesium on epileptic foci, Epilepsia, 19:81.

Brewer, G.J., Aster, J.C., Knutsen, C.A., and Kruckeberg, W.C., 1979, Zinc inhibition of calmodulin: a proposed molecular mechanism of zinc action on cellular functions, Am. J. Hematol., 7:53.

Brewer, G.J., Hill, G.M., Prasad, A.S., and Cossack, Z.T., 1983, Biological roles of ionic zinc, in: Zinc Deficiency in Human Subjects, A.S. Prasad, A.O. Cavdar, G.J. Brewer, and P.J. Aggett, eds., Alan R. Liss, Inc., New York, p. 35.

Buell, S.J., Fosmire, G.J., Ollerich, D.A., and Sandstead, H.H., 1977, Effects of postnatal zinc deficiency on cerebellar and hippocampal development in the rat, Exp. Neurol., 55:199.

Chen, R.W., and Ganther, H.E., 1975, Relative cadmium binding capacity of metallothionein and other cytosolic fraction in various tissues of the rat, Environ. Physiol. Biochem., 5:235.

Chung, S.H., and Johnson, M.S., 1983, Divalent transition-metal ions (Cu^{2+} and Zn^{2+}) in the brains of epileptogenic and normal mice, Brain Res., 280:323.

Constantinidis, J., and Tissot, R., 1981, Role of glutamate and zinc in the hippocampal lesions of Pick's disease, in: Glutamate as a Neurotransmitter, G. DiChiara and G.L. Gessa, eds., Raven Press, New York, p. 413.

Cousins, R.J., 1985, Absorption, transport, and hepatic metabolism of copper and zinc: special reference to metallothionein and ceruloplasmin, Physiol. Rev., 65:238.

Crawford, I.L., and Harris, N.F., 1984, Distribution and accumulation of zinc in whole brain and subcellular fractions of hippocampal homogenates, in: The Neurobiology of Zinc. Part A. Physiochemistry, Anatomy and Techniques, C.J. Frederickson, G.A. Howell, and E.J. Kasarskis, eds., Alan R. Liss, Inc., New York, p. 157.

Denner, L.A., and Wu, J.-Y., 1985, Two forms of rat brain glutamic acid decarboxylase differ in their dependence on free pyridoxal phosphate, J. Neurochem., 44:957.

De Vries, D.J., and Beart, P.M., 1985, Competitive inhibition of [^{3}H] spiperone binding to D-2 dopamine receptors in striatal homogenates by organic calcium channel antagonists and polyvalent cations, Eur. J. Pharmacol., 106:133.

Donaldson, J., St. Pierre, T., Minnich, J.L., and Barbeau, A., 1971, Seizures in rats associated with divalent cation inhibiton of Na$^+$K$^+$ATPase, Can.J. Biochem., 49:1217.

Dore-Duffy, P., Catalanotto, F., Donaldson, J.O., Ostrom, K.M., and Testa, M.A., 1983, Zinc in multiple sclerosis, Ann. Neurol., 14:450.

Dreosti, I.E., 1984, Zinc in the central nervous system: the emerging interactions, in: The Neurobiology of Zinc. Part A. Physiochemistry, Anatomy and Techniques, A.J. Frederickson, G.A. Howell, and E.J. Kasarskis, eds., Alan R. Liss, Inc., New York, p. 1.

Dvergsten, C.L., Fosmire, G.J., Ollerich, D.A., and Sandstead, H.H., 1983, Alterations in the postnatal development of the cerebellar cortex due to zinc deficiency. I. Impaired acquisition of granule cells, Brain Res., 271:217.

Dvergsten, C.L., Fosmire, G.J., Ollerich, D.A., and Sandstead, H.H., 1984a, Alterations in the postnatal development of the cerebellar cortex due to zinc deficiency. II. Impaired maturation of Purkinje cells, Dev. Brain Res., 16:11.

Dvergsten, C.L., Johnson, L.A., and Sandstead, H.H., 1984b, Alterations in the postnatal development of the cerebellar cortex due to zinc deficiency. III. Impaired dendritic differentiation of basket and stellate cells, Dev. Brain Res., 16:21.

Ebadi, M., Itoh, M., Bifano, J., Wendt, K., and Earle, A., 1981, The role of Zn^{2+} in pyridoxal phosphate-mediated regulation of glutamic acid decarboxylase in brain, Int. J. Biochem., 13:1107.

Ebadi, M., 1984a, Characterization of zinc binding ligand in rat brain, Trans. Soc. Neurosci., 14:1062.

Ebadi, M., 1984b, The presence of metallothionein-like protein in rat brain, Fed. Proc., 43, 3317.

Ebadi, M., and Pfeiffer, R.F., 1984, Zinc in neurological disorders and in experimentally induced epileptiform seizures, in: The Neurobiology of Zinc. Part B. Deficiency, Toxicity, and Pathology, C.J. Frederickson, G.A. Howell, and E.J. Kasarskis, eds., Alan R. Liss, Inc., New York, p. 307.

Ebadi, M., White, R.J., and Swanson, S., 1984, The presence and functions of zinc binding proteins in developing and mature brains, in: Neurobiology of Zinc (Part A), C.J. Frederickson, G.A. Howell, and E.J. Kasarkis, eds., Alan R. Liss, Inc., New York, p. 39.

Ebadi, M., 1985, The role of zinc in growth and development, J. Nutr. Growth and Cancer, 2:181.

Ebadi, M., and Wallwork, J.C., 1985, Zinc binding proteins (ligands) in brains of severely zinc deficient rats, Biol. Trace Element Res., 7:129.

Ebadi, M., Babin, D., and Swanson, S., 1985, Amino acid analysis and HPLC characterization of the isoforms of the metallothionein-like protein in rat brain, Trans. Soc. Neurosci., 15:155.

Eckhert, C.D., 1981, Elevated body zinc in rats with inherited dystrophy, J. Heredity, 72:130.

Fujii, T., 1954, Presence of zinc in nucleoli and its possible role in mitosis, Nature, 174:1108.

Goldberg, H.J., and Sheehy, E.M., 1982, Fifth day fits: an acute zinc deficiency syndrome?, Arch. Dis. Childh., 57:632.

Golub, M.S., Gershwin, M.E., and Vijayan, V.K., 1983, Passive avoidance performance of mice fed marginally or severely zinc deficient diets during post-embryonic brain development, Physiol. Behav., 30:409.

Gordon, E.F., 1984, Behavioral correlates of experimental zinc deficiency, in: The Neurobiology of Zinc. Part B. Deficiency, Toxicity, and Pathology, C.J. Frederickson, G.A. Howell, and E.J. Kasarskis, eds., Alan R. Liss, Inc., New York, p. 77.

Govitrapong, P., Hama, Y., Awad, A., and Ebadi, M., 1985, The inhibitory action of zinc and cadmium on D_2 dopamine and glutamate receptors in bovine pineal gland, Trans. Endocr. Soc., 67:1272.

Halas, E.S., and Sandstead, H.H., 1975, Some effects of prenatal zinc deficiency on behavior of the adult rat, Pediatric. Res., 9:94.

Halas, E.S., Hanlon, M.J., and Sandstead, H.H., 1975, Intrauterine nutrition and aggression, Nature, 257:221.

Halas, E.S., Reynolds, G., Rowe, M., Heinrich, M., and Pirc, M., 1977, Comparison of frequency, intensity and duration of aggressive responses in rats, Physiol. Behav., 18:975.

Halas, E.S., Heinrich, M.D., and Sandstead, H.H., 1979, Long term memory deficits in adult rats due to postnatal malnutrition, Physiol. Behav., 22:991.

Halas, E.S., Eberhardt, J.J., Diers, M.A., and Sandstead, H.H., 1983, Learning and memory impairment in adult rats due to severe zinc deficiency during lactation, Physiol. Behav., 30:371.

Halas, E.S., and Kawamoto, J.C., 1984, Correlated behavioral and hippocampal effects due to perinatal zinc deprivation, in: The Neurobiology of Zinc. Part B. Deficiency, Toxicity, and Pathology, C.J. Frederickson, G.A. Howell, and E.J. Kasarskis, eds., Alan R. Liss, Inc., New York, p. 91.

Haug, F.-M.S., 1967, Electron microscopical localization of the zinc in hippocampal mossy fiber synapses by a modified sulphide silver procedure, Histochemistry, 8:355.

Heilmaier, H.E., Summer, K.H., 1985, Metallothionein content and zinc status in various tissues of rats treated with iodoacetic acid and zinc, Arch. Toxicol., 56:247.

Hershey, C.O., Hershey, L.A., Varnes, A., Vibhakar, S.D., Lavin, P., and Strain, W.H., 1983, Cerebrospinal fluid trace element content in dementia: clinical, radiologic, and pathologic correlations, Neurology, 33:1350.

Hesketh, J.E., 1983, Zinc binding to tubulin, Int. J. Biochem., 15:743.

Hurley, L.S., Gowan, J., and Swenerton, H., 1971, Teratogenic effects of short-term and transitory zinc deficiency in rats, Teratology, 4:199.

Hurley, L.S., 1981, Teratogenic aspects of manganese, zinc, and copper nutrition, Physiol. Rev., 61:249.

Itoh, M., and Ebadi, M., 1982a, Zinc binding proteins in rat brain and their interactions with glutamic acid decarboxylase, Fed. Proc., 41:9736.

Itoh, M., and Ebadi, M., 1982b, The selective inhibition of hippocampal glutamic acid decarboxylase in zinc-induced epileptic seizures, Neurochem. Res., 7:1287.

Itoh, M., Ebadi, M., and Swanson, S., 1983, The presence of zinc binding proteins in brain, J. Neurochem., 41:823.

Judd, A.M., MacLeod, R.M., and Login, I.S., 1984, Zinc acutely, selectively and reversibly inhibits pituitary prolactin secretion, Brain Res., 294:190.

Kagi, J.H., and Vallee, B.L., 1960, Metallothionein: a cadmium- and zinc containing protein from equine renal cortex, J. Biol. Chem., 235:3460.

Kagi, J.H., and Nordberg, M., 1979, Metallothionein, Birkhäuser Verlag, Basel, Switzerland.

Karin, M., and Richards, R.I., 1982, Human metallothionein genes-primary structure of the metallothionein-II gene and a related processed gene, Nature, 299:797.

Kimura, K. and Kumura, J., 1965, Polarigraphic determination of zinc levels in the brains of schizophrenics and control patients, Proc. Jap. Acad., 41:943.

Klauser, S., Kagi, J.H.R., and Wilson, K.J., 1983, Characterization of isoprotein patterns in tissue extracts and isolated samples of metallothioneins by reverse-phase high pressure liquid chromatography, Biochem. J., 209:71.

Klee, M.R., and Lieflander, M., 1965, Über das Vorkommen von Zink und Carbonat-hydro-lyase im Kaninchenhirn, Hoppe Seyler's Z. Physiol. Chem., 341:143.

Lee, C.-M., Javitch, J.A., and Snyder, S.H., 1983, ^{3}H-Substance P binding to salivary gland membranes regulation by guanyl nucleotides and divalent cations, Mol. Pharmacol., 23:563.

Lindskog, S., 1983, Carbonic anhydrase, in: Zinc Enzymes, T.G. Spiro, ed., John Wiley and Sons, New York, p. 79.

Login, I.S., Thorner, M.O., and MacLeod, R.M., 1983, Zinc may have a physiological role in regulating pituitary prolactin secretion, Neuroendocrinology, 37:317.

Lucier, G.W., and Hook, G.E., 1984, Metallothionein and cadmium nephrotoxicity, in: Environmental Health Perspectives, 54, U.S. Government Printing Office, Washington, D.C.

Majumdar, S.K., Shaw, G.K., and Thomson, A.D., 1983, Serum zinc, magnesium and calcium status in the Wernicke-Korsakoff syndrome, Drug and Alcohol Dependence, 12:403.

Mizuno, S., Ogawa, N., and Mori, A., 1983, Differential effects of some transition metal cations on the binding of β-carboline-3-carboxylate and diazepam, Neurochem. Res., 8:873.

Mozha, I.B., 1974, Levels of various trace elements in the blood of patients with various types of retinal pathology, Vestn. Oftalmol., 5:59.

Nair, V., and Bau, D., 1971, Studies on the functional significance of carbonic anhydrase in central nervous system, Brain Res., 31:185.

Neve, J., Sinet, P.M., Molle, L., and Nicole, A., 1983, Selenium, zinc and copper in Down's syndrome (trisomy 21): blood levels and relations with glutathione peroxidase and superoxide dismutase, Clin. Chim. Acta, 133:209.

Nielson, K.B., and Winge, D.R., 1983, Order of metal binding in metallothionein, J. Biol. Chem., 258:13063.

O'Dell, B.L., 1974, Role of zinc in protein synthesis, in: Clinical Applications of Zinc Metabolism, W.J. Pories, W.H. Strain, J.M. Hsu, and R.L. Woosley, eds., Charles C. Thomas, Springfield, p. 5.

Olton, D.S., Collison, C., and Werz, M., 1977, Spatial memory and radial arm maze performance of rats, Learning and Motivation, 8:289.

Olton, D.S., Walker, J.A., and Gage, F.H., 1978, Hippocampal connection and spatial discrimination, Brain Res., 139:295.

Palm, R., and Hallmans, G., 1982, Zinc and copper metabolism in phenytoin therapy, Epilepsia, 23:453.

Palm, R., Hallmans, G., and Sjostrom, R., 1982, Zinc concentrations in normal and pathological cerebrospinal fluid, Acta Neurol. Scand., 90:184.

Papavasiliou, P.S., Kutt, H., Miller, S.T., Rosal, V., Wang, Y.Y., and Aronson, R.B., 1979, Seizure disorders and trace metals: manganese tissue levels in treated epileptics, Neurology, 29:1466.

Pei, Y., Zhao, D., Huang, J., and Cao, L., 1983, Zinc-induced seizures: a new experimental model of epilepsy, Epilepsia, 24:169.

Peters, D.P., 1978, Effects of prenatal nutritional deficiency on affiliation and aggression in rats, Physiol. Behav., 20:359.

Peters, D.P., 1979, Effects of prenatal nutrition on learning and motivation in rats, Physiol. Behav., 22:1067.

Pfeiffer, C.C., and LaMola, S., 1983, Zinc and manganese in the schizophrenia, J. Orthomolec. Psych., 12:215.

Pippenger, C.E., Garlock, C., Fernandez, F., Slavin, W., and Iannarone, J., 1980, Effect of antiepileptic drugs on manganese, zinc and copper concentrations in whole blood, RBC, and plasma of epileptic, in: Advances in Epileptology: XIth Epilepsy International Symposium, R. Canger, ed., Raven Press, New York, p. 435.

Prasad, A.S., 1976, Deficiency of zinc in man and its toxicity, in: Trace Elements in Human Health and Disease, A.S. Prasad, ed., Academic Press, New York, p. 1.

Prasad, A.S., 1979, Clinical, biochemical and pharmacological role of zinc, *Ann. Rev. Pharmacol. Toxicol.*, 20:393.

Pryor, D.S., Don, N., and Macourt, D.C., 1981, Fifth day fits: a syndrome of neonatal convulsions, *Arch. Dis. Childh.*, 56:753.

Pulido, P., Kagi, J.H.R., and Vallee, B.L., 1966, Isolation and some properties of human metallothionein, *Biochemistry*, 5:1768.

Record, I.R., Dreosti, I.E., Tulsi, R.S., Fraser, R.S., Fraser, F.J., Buckley, R.A., and Manuel, S.J., 1982, Postnatal accumulation of zinc by the rat hippocampus, *Biol. Trace Element Res.*, 4:279.

Rieder, H.P., Schoettli, G., and Seiler, H., 1983, Trace elements in whole blood of multiple sclerosis, *Eur. Neurol.*, 22:85.

Sandstead, H.H., Fosmire, G.J., Halas, E.S., Jacob, R.A., Strobel., D.S., and Marks, E.O., 1977, Zinc deficiency: effects on brain and behavior of infants, *Am. J. Clin. Nutr.*, 31:844.

Sandstead, H.H., Strobel, D.A., Logan, G.M., Marks, E.O., and Jacob, R.A., 1978, Zinc deficiency in pregnant rhesus monkeys: effects on behavior of infants, *Am. J. Clin. Nutr.*, 31:844.

Shapcott, D., Giguere, R., and Lemieux, B., 1984, Zinc and taurine in Friedreich's ataxia, *Can. J. Neurol. Sci.*, 11:623.

Sato, S.M., Frazier, J.M., and Goldberg, A.M., 1984a, The distribution and binding of zinc in the hippocampus, *J. Neurosci.*, 4:1662.

Sato, S.M., Frazier, J.M., and Goldberg, A.M., 1984b, A kinetic study of the *in vivo* incorporation of ^{65}Zn into the rat hippocampus, *J. Neurosci.*, 4:1671.

Slevin, J.T., and Kasarkis, E.J., 1985, Effects of zinc on markers of glutamate and aspartate neurotransmission in rat hippocampus, *Brain Res.*, 334:281.

Smart, T.G., and Constanti, A., 1982, A novel effect of zinc on the lobster muscle GABA receptor, *Proc. Roy. Soc. Lond.*, B215:327.

Stengaard-Pedersen, K., 1982, Inhibition of enkephalin binding to opiate receptors by zinc ions: possible physiological importance in the brain, *Acta Pharmacol. Toxicol.*, 50:213.

Stengaard-Pedersen, K., Larson, L.-I., Fredens, K., and Rehfeld, J.F., 1984, Modulation of cholecystokinin concentrations in the rat hippocampus by chelation of heavy metals, *Proc. Natl. Acad. Sci. USA*, 81:5876.

Sypert, G.W., 1982, Metallic salts and epileptogenesis, in: *Physiology and Pharmacology of Epileptogenic Phenomena*, M.R. Klee, H.D. Lux, and E.J. Speckman, eds., Raven Press, New York, p. 81.

Underwood, E.J., 1977, Zinc, in: *Trace Elements in Human and Animal Nutrition*, E.J. Underwood, ed., Academic Press, New York, p. 196.

Vallee, B.L., 1983, Zinc in biology and biochemistry, in: *Zinc Enzymes, Vol. 5*, T.G. Spiro, ed., Metal Ion in Biology Series, John Wiley and Sons, New York, p. 3.

Webb, M., and Cain, K., 1982, Functions of metallothionein, *Biochem. Pharmacol.*, 31:137.

Wright, D.M., 1984, Zinc: effect and interaction with other cations in the cortex of the rat, *Brain Res.*, 311:343.

Wu, C.-T., Lee, J.-N., Shen, W.W., and Lee, S.-L., 1984, Serum zinc, copper, and ceruloplasmin levels in male alcoholics, *Biol. Psych.*, 19:1333.

NEUROBEHAVIORAL, NEUROENDOCRINE AND NEUROCHEMICAL EFFECTS OF ZINC

SUPPLEMENTATION IN RATS

M. Baraldi, P. Zanoli, A. Benelli, M. Sandrini, A. Giberti,
E. Caselgrandi[1], G. Tosi[2] and C. Preti[2]

Institute of Pharmacology
Chair of Hygiene[1], Department of Chemistry[2]
Modena University
Modena, Italy

INTRODUCTION

Extensive morphological and biochemical investigations have been devoted to the neurological effects induced by zinc deficiency since this trace metal has been recognized as essential during neurogenesis for a normal development of the central nervous system (for a review see Dreosti, 1984; Sandstead, 1984). Less attention, however, has been addressed to the neurological consequences of an increased supplementation of zinc. In this context it must be mentioned that the peripheral acute administration of zinc intravenously or intraperitoneally in doses up to 100 mg/kg has not been associated with the production of convulsive seizures or other behavioral abnormalities (Ebadi et al., 1984). From these results, it has been concluded that after parenteral acute administration, zinc does not induce any effect since: a) by binding to circulating proteins such as albumin, gamma-globulins and probably other proteins (Prasad, 1979; Disilvestro and Cousins, 1983; Ebadi et al., 1984) it is unavailable to the CNS; b) it exists mostly in bound form in the brain (Ebadi et al., 1984). In an attempt to gain more information on the behavioral effects induced by peripheral supplementation of zinc, we obtained the same results with regard to the lack of convulsions. In our experience, however, acute intraperitoneal injections of zinc sulphate or zinc acetate induced, in a dose-related fashion starting from 50 mg/kg, a sedative effect which caused a 50% mortality rate within six hours at 200 mg/kg. Based on physiological considerations and on our present observations, it seems more likely to suggest that zinc enters the brain but does not reach concentrations sufficient to induce epileptic seizures which can be elicited by its intracerebroventricular administration (Itoh and Ebadi, 1982; Ebadi et al., 1984). The levels of zinc tested in six brain areas of these rats did not show significant variation in comparison with controls, whereas zinc plasma levels were found to be increased. Time-course determinations of zinc in blood and brain areas are needed in order to explain these data. On the other hand, there are several question marks concerning the machinery which regulates zinc homeostasis in the brain. Thus, it is difficult to explain cerebral zinc uptake and turnover (Kasarskis, 1984) and the unaltered zinc levels in the brain of zinc deficient rats (Wallwork et al., 1983; Kasarskis, 1984) as well as the mechanisms which regulate the bound/free zinc ratio.

From these observations and considerations we decided to carry out further studies on:

1) the effects induced by zinc in vitro on the properties of GABA-benzodiazepine and opiate receptors;

2) the neurobehavioral, neuroendocrine and neurochemical effects induced by a subchronic administration of zinc in order to gain more information on its availability to the CNS and on its potential epileptic action;

3) the functional status of the GABA-benzodiazepine receptor unit and of glutamate, opiate and adenosine recognition sites after repeated episodes of epileptic seizures induced by the intraventricular administration of zinc;

4) the ability of diazepam to counteract the zinc-induced epileptic seizures by forming coordinative compounds with zinc in the brain.

EFFECT OF ZINC IN VITRO ON THE PROPERTIES OF GABA-BENZODIAZEPINE AND OPIATE RECEPTORS

We have recently provided evidence (Baraldi et al., 1984b) that in zinc-deprived synaptic brain membranes the readdition of increasing amounts of zinc plays a modulatory role on Na^+-independent GABA binding. While zinc seems to optimize and to increase GABA binding at physiological concentrations ($1-2 \times 10^{-6}$ M), its deficiency or its excess negatively affects the number of GABA binding sites labeled by ^{3}H-GABA. Since by using ^{3}H-GABA as a ligand GABA can bind in presence of calcium ions both $GABA_A$ and $GABA_B$ receptors (Hill and Bowery, 1981), one could surmise that the zinc-induced increase of GABA binding could be due to the ability of zinc to mimic the essential role of calcium in the detection of $GABA_B$ receptors. As shown in Table 1, this is not the case since, although the addition in the incubation medium of zinc or calcium increases ^{3}H-GABA binding, in competition experiments the presence of baclofen decreases the specific binding of ^{3}H-GABA, an effect which is favored by the addition of calcium but inhibited by the presence of zinc. This finding seems to indicate that at physiological concentrations zinc could be essential to optimize the binding of GABA to $GABA_A$ receptors while calcium could be essential for the binding of GABA to $GABA_B$ receptors. Since it is well established that zinc and calcium compete in several biological systems (Daniel et al.,

Table 1. Effect of Zinc and Calcium on Na^+-Independent ^{3}H-GABA (10 nM) Binding to Brain Synaptic Membranes Both in Absence and Presence of Baclofen

Addition in vitro	^{3}H-GABA, Specific Binding (fmol/mg protein)	%
Tris-Buffer	236.2 ± 4.1	100
$ZnCl_2$ 10^{-6} M	272.8 ± 6.2	115
Baclofen 10^{-3} M	187.6 ± 7.4[a]	79
$ZnCl_2$ + Baclofen	214.7 ± 8.1[b]	90
$CaCl_2$ 2.5×10^{-3} M	266.4 ± 6.2	113
$CaCl_2$ + Baclofen	158.3 ± 10.2[b]	67

Values are the mean ± S.E.M. of three separate experiments done in triplicate where GABA (10^{-5} M) was routinely used as displacer. Student's t-test: [a]p < 0.05 vs Tris-buffer; [b]p < 0.05 vs Baclofen.

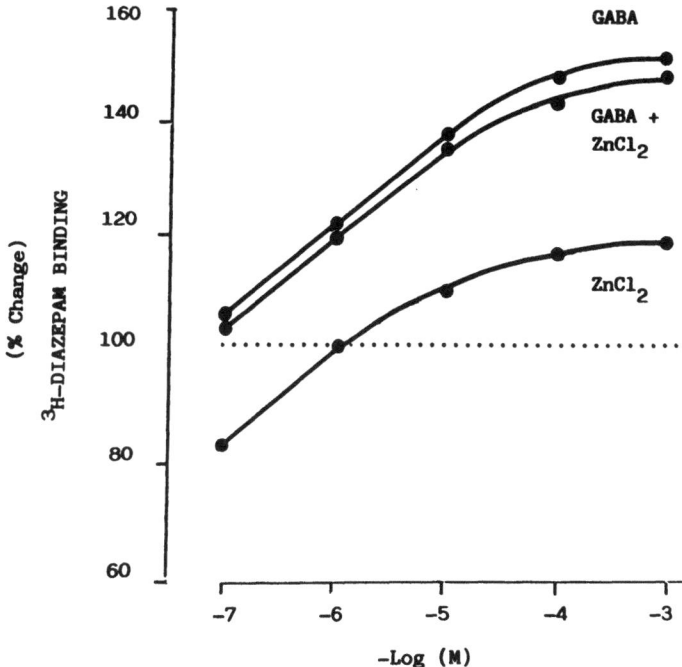

Fig. 1. Effects of zinc, GABA and its combination on ^{3}H-diazepam binding to brain synaptic membranes. ^{3}H-Diazepam binding (1.5 nM) was performed on partially zinc-deprived brain membranes using diazepam (10^{-6} M) as displacer.

1970; Brewer and Oelshlegal, 1974; Bettger and O'Dell, 1981), it could well be that an alteration of the physiological ratio between the two cations at the synaptic level in the brain might affect the functional activity of $GABA_A$ or $GABA_B$ receptors. Interestingly, Mackerer and Kochman (1978) have reported that zinc, among other divalent cations, enhances ^{3}H-diazepam binding when added in the incubation medium in the range of 0.1 to 10 mM. As shown in Fig. 1, we have confirmed these data by adding to partially zinc-deprived synaptic brain membranes, as previously described (Baraldi et al., 1984b), $ZnCl_2$ in the range between 10^{-7} and 10^{-3} M to the incubation medium of ^{3}H-diazepam binding, which was performed using an established method (Baraldi et al., 1984c; Santi et al., 1985). Furthermore, to gain information on the mechanism of this effect, we studied the influence of zinc on GABA-stimulated ^{3}H-diazepam binding.

It is established that GABA enhances the affinity constant of benzodiazepine receptor agonists in vitro (Tallman et al., 1978) and in vivo (Gallager and Tallman, 1983; Baraldi, 1985) by stimulating high affinity GABA receptors. Under our experimental conditions, high zinc concentrations failed to affect the enhancing action of GABA on ^{3}H-diazepam binding (Fig. 1).

From these data, we can tentatively suggest that zinc enhances ^{3}H-diazepam binding directly by affecting the binding properties of the proteins which express the diazepam binding sites or indirectly by affecting environmental biomembrane factors which normally negatively affect ^{3}H-diazepam binding. Whatever the mechanism by which zinc enhances ^{3}H-diazepam binding, it is clear that this cation seems to be essential at physiological concentrations to optimize diazepam binding as it does for GABA binding. At pharmacological-toxicological ($10^{-5} - 10^{-3}$ M) doses, however, zinc still slightly favors ^{3}H-diazepam binding while inhibiting ^{3}H-GABA binding. The finding that in the presence of an excess of diazepam high doses of zinc, which normally block the coupling of GABA to its receptors, did not affect GABA-stimulated ^{3}H-diazepam binding, seems to corroborate our present working hypothesis. Thus diazepam, by forming coordinative complexes with zinc, could prevent the inhibitory action exerted by an excess of this cation on GABA binding to its receptors. Evidence has been provided that diazepam can form _in vitro_ coordination compounds with zinc (Preti and Tosi, 1978a) and other bivalent transition metals (Preti and Tosi, 1978b) such as copper or nickel which have been reported (Mackerer and Kochman, 1978) to enhance ^{3}H-diazepam binding. Furthermore, it must be mentioned that more recently Mizumo et al. (1983) have reported that zinc inhibits the binding of ^{3}H-β-carboline-3-carboxylate, a compound which has proconvulsant activity and which has been classified as a benzodiazepine inversive-agonist (Braestrup et al., 1980).

Despite this interesting complex action of zinc on the GABA-benzodiazepine receptor machinery, it seems necessary to bear in mind (in order to explain the behavioral effects induced by supplementation of zinc _in vivo_) that zinc can also affect the properties of other neurotransmitter systems. Zinc has been described to be a potent inhibitor of adenosine receptor agonist ligands but only a partial inhibitor of antagonists (Marangos et al., 1983). Since adenosine and its analogs induce sedation while adenosine antagonists such as theophylline are stimulants and convulsants, the inhibitory effects of zinc on adenosine receptors could be implicated in the epileptogenic activity of intracerebroventricularly injected zinc. Furthermore, zinc has been shown to reduce, in a dose-related fashion, both the affinity (10^{-5} M) and the number (10^{-3} M) of binding sites of opiate receptors labeled by ^{3}H-D-ALA^2-met^5-enkephalin (Stengaard-Pedersen, 1982) which mostly binds δ-opiate receptor subtypes. This effect _in vitro_ has been attributed to oxidation of opioid receptor SH-groups by zinc ions since the presence of thiol-reducing agents can restore the binding capacity of $ZnCl_2$-treated membranes. This interpretation seems to be in line with the finding that alteration of the SH-group present in opioid receptors suffices to block opioid binding (Simon and Groth, 1975). More recently, we have reported (Baraldi et al., 1984a) that zinc added _in vitro_ inhibits not only ^{3}H-enkephalin binding but also, as shown in Table 2, ^{3}H-naloxone binding (performed as previously described, Baraldi et al., 1983b) to brain membranes, thus confirming and extending the above mentioned finding.

The inhibitory effect of zinc on the opioid system has been proved to be operative also _in vivo_ since a subcutaneous depot injection of zinc strongly reduced the naloxone precipitated withdrawal syndrome in morphine-dependent rats (Baraldi et al., 1984).

Table 2. _In vitro_ Inhibition by Zinc (10^{-5} M) of ^{3}H-Naloxone Binding to High Affinity Opiate Receptors of Whole Brain Rat Membranes

Addition	Binding Characteristics	
	K_D (nM)	Bmax (fmol/mg protein)
Tris buffer	2.6 ± 0.08	400 ± 14
$ZnCl_2$	3.8 ± 0.05*	390 ± 8

Values are the mean ± S.E.M. of three different experiments done in triplicate and were calculated by Scatchard plot analysis. Student's t-test: *$p < 0.01$.

All of these data seem to stress the important role of a deficiency or an excess of zinc in the functional activity and responsiveness of some receptor systems in the CNS.

BEHAVIORAL AND BIOCHEMICAL EFFECTS INDUCED BY PERIPHERAL ZINC ADMINISTRATION

As mentioned in the Introduction, the peripheral supplementation of zinc in animals has not yet been associated with any behavioral abnormality. Since this lack of effect of zinc was due to the fact that the observation was focused mostly on the precipitation of convulsive activity, we decided to reinvestigate this issue by closely scrutinizing the dose-dependent behavioral effects induced by zinc. As reported in Table 3, the oral administration of low amounts of zinc sulphate (0.5-5 mg/kg) or zinc acetate (data not shown) to normal male rats (180-200 g body weight) induces repeated episodes of stretching and yawning (SYS) associated with penile erections (PE). It is noteworthy that SYS associated with PE was originally described by us after the peripheral administration to normal rats of small non stereotypy-inducing doses of dopaminomimetic drugs such as apomorphine (Baraldi and Benassi-Benelli, 1975a,b; 1977), apocodeine (Baraldi et al., 1979), amantadine (Baraldi and Bertolini, 1974), amphetamine (Baraldi and Benassi-Benelli, 1977) and N-n-propyl-norapomorphine (Benassi-Benelli et al., 1978). These effects induced by small doses of apomorphine and other dopaminomimetics were tentatively ascribed to the activation of a subtype of dopamine receptors different from those mediating the appearance of stereotypies (Baraldi and Benassi-Benelli, 1975a) or alternatively to a first step activation of the same type of dopamine receptors which mediate stereotypies. These first observations and suggestions were confirmed and extended by others who attributed SYS and PE induced by apomorphine to the activation of self inhibitory dopamine autoreceptors (Mogilnicka and Klimek, 1977; DiChiara et al., 1978), which also mediate apomorphine-induced hypomotility. However, the recent observation (Serra et al., 1983a) that hypophysectomy prevents SYS and PE but not the hypomotility induced by low doses of apomorphine seems to rule out an implication of dopamine autoreceptors. This hypothesis seems to be further supported by the finding that the selective stimulant of dopamine autoreceptors, (-)3-PPP (Hjorth et al., 1983), fails to induce yawning (Serra et al., 1984) and sexual stimulation (Ahlenius and Larson, 1985), while (+)3-PPP, which stimulates both autoreceptors and postsynaptic dopamine receptors (Hjorth et al., 1983), induces repeated episodes of yawning (Serra et al., 1984). It is worth noting, irrespective of the type of dopamine receptor implicated, that SYS and PE are elicited not only by

Table 3. Penile Erections (PE) and Stretching-Yawning Syndrome (SYS) Induced by the Administration of $ZnSO_4$ to Normal Male Rats

$ZnSO_4$ mg/kg/os	Rats with PE/ Treated Rats	Total* Episodes of PE	Rats with SYS/ Treated Rats	Total* Episodes of SYS
0	2/10	2	2/10	6
0.5	10/10	18	8/10	18
5	6/10	12	4/10	6
50	4/10	4	2/10	2
100	2/10	2	1/10	2

*Within 90 minutes of observation which began one hour after treatment.

small doses of dopaminomimetics but also by adrenocorticotropic (ACTH) and melanocyte stimulating (MSH) hormones (Bertolini and Gessa, 1981). Since, as mentioned before, hypophysectomy prevents SYS and PE induced by apomorphine (Serra et al., 1983a), as does the administration of inhibitors of protein synthesis (Serra et al., 1983b), it is likely that the activation of D_1 and/or D_2 dopamine postsynaptic receptors may affect the release of peptides from the pituitary-hypothalamic area. It could well be that a slight activation of dopamine receptors inhibits the release of opiates such as β-endorphin whose accumulation in the pituitary could trigger a compensatory increased excretion of ACTH-MSH hormones which in turn cause SYS and PE (Bertolini and Gessa, 1981). This hypothesis seems to be supported by the notion that, since the release of β-endorphin from the pituitary is under tonic inhibitory dopaminergic control exerted from the hypothalamus (Tilders and Smelik, 1978), dopamine antagonists stimulate (Farah et al., 1982) β-endorphin secretion and block SYS and PE induced by apomorphine and other dopaminomimetics (Baraldi and Bertolini, 1974; Baraldi and Benassi-Benelli, 1975a,b; 1977; Baraldi et al., 1979; Bertolini and Gessa, 1981). Furthermore, β-endorphin induces analgesia (Tseng et al., 1976) and inhibition of sexual behavior (Meyerson and Terenius, 1977) while ACTH administration causes hyperalgesia and stimulates sexual behavior (Bertolini and Gessa, 1981). These opposite effects of β-endorphin and ACTH are suppressed when the two peptides are simultaneously injected (Fratta et al., 1981). From these findings the hypothesis has been put forward that ACTH peptides might be considered as the endogenous antagonists of opiate receptors (for a review see Bertolini and Gessa, 1981).

Though the issue requires more experimental data to be clarified, how can we interpret the ability of low doses of zinc to induce SYS and PE? As shown in Table 4, the administration of zinc sulphate by oral intubation induces not only SYS and PE but also a hyperalgesic state which peaks at the dose of 5 mg/kg. By increasing the amount of zinc sulphate, there is a disappearance of SYS and PE (see Table 3) while the hyperalgesic state is not enhanced (Table 4). It must be mentioned, however, that at 50-100 mg/kg zinc induces sedation and intermittent episodes of stereotypies.

From this finding and in light of the above reviewed data on dopaminomimetic drugs and the ability of ACTH to induce SYS and PE, we can tentatively suggest that the administration of small doses of zinc might inhibit the release of β-endorphin from the pituitary while favoring the excretion of ACTH. This working hypothesis is supported by the finding that a single subcutaneous depot injection of zinc oxide (110 mg/kg), dissolved in a castoroil-benzylbenzoate vehicle (Brewer et al., 1981), progressively induces, as briefly reported elsewhere (Baraldi et al., 1984a), an increased

Table 4. Hyperalgesia Induced by Zinc Sulphate Administration to Rats (Mean ± S.E.M.)

Zinc Sulphate (mg/kg)	Reaction Time (Seconds)
0	19.50 ± 1.66
0.5	14.68 ± 1.17*
5	9.06 ± 1.03**
50	10.08 ± 1.13**
100	9.50 ± 1.66**

Hot plate test (50 ± 0.5°C) was performed four hours after treatment. Student's t-test: *p < 0.01, **p < 0.001.

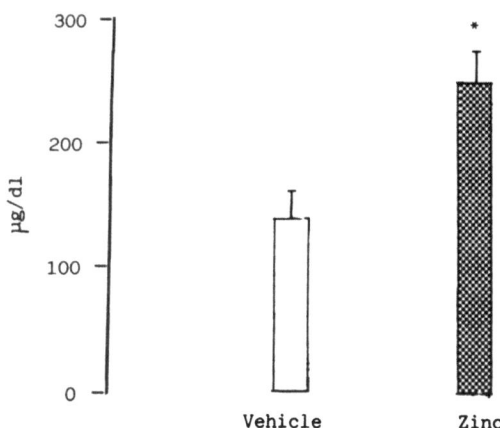

Fig. 2. Zinc plasma concentrations (mean ± SD) in rats 72 hours after a single subcutaneous depot injection of vehicle or ZnO (110 mg/kg). Student's t-test: *p < 0.001.

sensitivity to pain stimuli which becomes significant 24 hours after the injection. When these rats were sacrificed 72 hours after the depot injection, the plasma concentrations of zinc, determined with an atomic absorption spectrometric method as previously described (Baraldi et al., 1983), were found to be 74% higher than those of rats injected with vehicle alone (Fig. 2). Although a similar increased presence of zinc was found in the pituitary (Table 5), only a slight increase of zinc was detected in the hypothalamus and brainstem whereas no changes were found in other brain areas.

This finding together with the provided evidence that zinc increases the release of ACTH from the pituitary in vitro (LaBella et al., 1973), prompted us to test if the increased presence of zinc in the pituitary of our zinc-treated rats could affect the release of β-endorphin-like peptides and ACTH. Indeed, as shown in Table 6, zinc causes a reduction of $ACTH_{1-24}$, tested by using a two-side immunoradiometric assay (IRMA-Holland) following the extraction procedure of Thody et al. (1975), in

Table 5. Concentrations of Zinc in the Pituitary and in Brain Areas of Rats 72 Hours After a Single Subcutaneous Depot Injection of ZnO (110 mg/kg) or Vehicle

Area	ZINC (μg/g dry weight)	
	Vehicle	ZnO
Pituitary	44.3 ± 3.1	77.2 ± 4.6**
Hypothalamus	42.1 ± 3.4	54.2 ± 4.1*
Brainstem	23.3 ± 1.6	32.5 ± 2.0*
Cerebellum	32.7 ± 3.2	34.1 ± 1.9
Hippocampus	55.6 ± 5.3	65.1 ± 4.1
Cortex	38.2 ± 2.8	40.8 ± 3.4

Values are the mean ± S.E.M. of the assay performed in 12 rats/group. Student's t-test: *p < 0.05; **p < 0.01 vs vehicle.

Table 6. Adrenocorticotropic Hormone (ACTH) in Plasma, Anterior Pituitary Lobe (AP) and Medio-Basal Hypothalamus (MBH) of Rats 72 Hours After a Single Subcutaneous Depot Injection of ZnO (110 mg/kg) or Vehicle

Treatment	Plasma (pg/ml)	ACTH$_{1-24}$ AP (ng/mg protein)	MBH (ng/mg protein)
Vehicle	68.2 ± 10.1	91.9 ± 5.4	0.61 ± 0.1
ZnO	119.6 ± 8.1	46.8 ± 3.6*	1.41 ± 0.08*

Reported values are mean ± S.E.M. of RIA performed in duplicate on tissue extracts of 10 rats/group. Student's t-test: *p < 0.001 vs. vehicle.

the anterior pituitary lobe (AP). Since at the same time there is an increase of ACTH in plasma and medio-basal hypothalamus (MBH), we can infer that the reduced presence of ACTH in the AP could be due to an increased release of this hormone. Furthermore, as shown in Table 7, zinc exerts an opposite effect on the β-endorphin-like immunoreactivity (βE-LI), assayed as previously described (Petraglia et al., 1985), in the AP and neurointermediate lobe (NIL). The finding that βE-LI seems to be at steady-state in the MBH but decreased in plasma might be indicative of an accumulation of this peptide in the pituitary because of an inhibited release. Although we have not yet performed the same endocrine studies after the oral administration of the small doses of zinc which induce SYS, PE and hyperalgesia, we can surmise that a similar phenomenon may be operative in inducing these behavioral effects. The hyperalgesic effect induced by the administration of zinc both by depot injection and by oral route could be explained not only by the increased release from AP of ACTH, which seems to exert an anti-opioid effect (for a review see Bertolini and Gessa, 1981), but also by the finding that zinc, as depicted in Fig. 3, reduces $\underline{\text{in vivo}}$ the affinity constant of opiate receptors, as it does $\underline{\text{in vitro}}$ (Stengaard-Pedersen, 1982; Baraldi et al., 1984a). As mentioned before, zinc sulphate and acetate induce, when repeatedly administered per os at doses of 50-100 mg/kg, a sedative effect with intermittent episodes of stereotypies, which seems to prevent the appearance of SYS and PE but not hyperalgesia. To gain some information on this behavioral effect of zinc, we have performed a screening study on the binding properties of opiate, GABA and adenosine receptors after a chronic administration (100 mg/kg/os/day x3) of zinc sulfate. In these experiments, zinc exerted an antiopiate effect by significantly reducing the affinity constant of ^{3}H-naloxone binding, performed

Table 7. β-Endorphin Levels in Anterior (AP), Neurointermediate (NIL) Pituitary Lobes, Medio-Basal Hypothalamus (MBH) and Plasma of Rats 72 Hours after a Single Depot Injection of Vehicle or ZnO (110 mg/kg s.c.)

Treatment	AP	NIL	MBH	Plasma
	(pmol/mg protein)			(fmol/ml)
Vehicle	342 ± 34	1131 ± 122	9.07 ± 0.33	25.3 ± 0.61
ZnO	636 ± 17*	2695 ± 146*	9.98 ± 0.29	12.4 ± 0.38*

Each value represents the mean ± S.E.M. of RIA assay on 15 rats/group. Student's t-test: *p < 0.001.

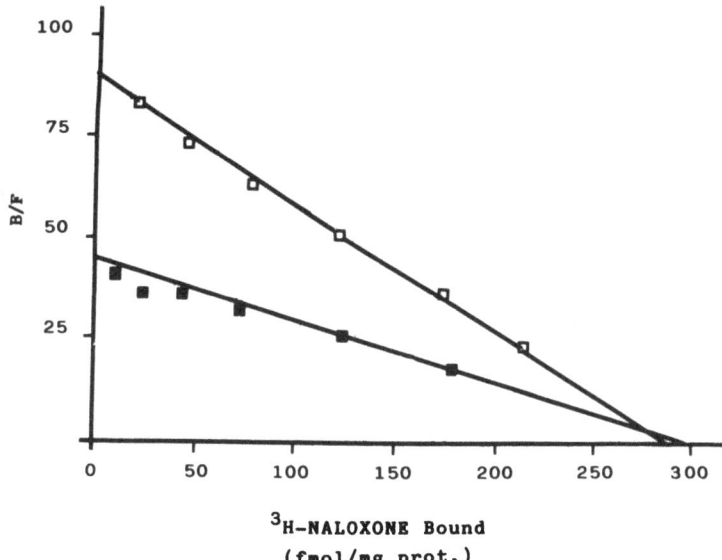

Fig. 3. Scatchard plot analysis of ^{3}H-naloxone binding to hypothalamic membranes prepared from brain of rats sacrificed 72 hours after a single injection of ZnO (110 mg/kg) or vehicle. The affinity constant (K_D) and the maximum number of binding sites (B_{max}), given in nM and fmol/mg protein, are: control: (□) K_D = 3.1 ± 0.2; B_{max} = 290 ± 18; ZnO (■) K_D = 8.1 ± 0.1 (p < 0.001 vs. controls; B_{max} = 300 ± 12).

on brainstem (data not shown). This finding could explain the persistence of the hyperalgesic effect after the chronic administration of high doses of zinc. Furthermore, we found in whole brain membranes of these rats a slight decrease of the affinity constant of high affinity GABA receptors (data not shown), a finding which could be indicative of a slight increase of endogenous GABA. Interestingly, when we studied the binding characteristics of adenosine receptors labeled by the agonist ligand ^{3}H-PIA using the method of Marangos et al. (1983), we found in the hippocampus an increase of both high and low affinity receptors without changes in the affinity constants (data not shown). Since zinc has been described to inhibit adenosine receptors in vitro, we can infer that chronic exposure to high doses of zinc might induce a reactive supersensitivity phenomenon of adenosine receptors, hence contributing to the sedative effect of zinc.

Taken together, these findings indicate that acute and chronic peripheral injection of zinc, although not inducing epileptic seizures even when injected at high doses, can induce in a dose-related fashion behavioral effects which seem to be mediated by its biochemical effects on the CNS. It remains to be clarified, however, why after three days of zinc treatment (100 mg/kg of zinc sulfate or acetate) the levels of zinc in the brain are not enhanced or, as shown in Fig. 4, which depicts the levels of zinc in the hippocampus, slightly decreased. We can tentatively suggest that an excess of zinc in the brain allows the activation of a protective mechanism which eliminates zinc via CSF pathways or cerebral capillaries.

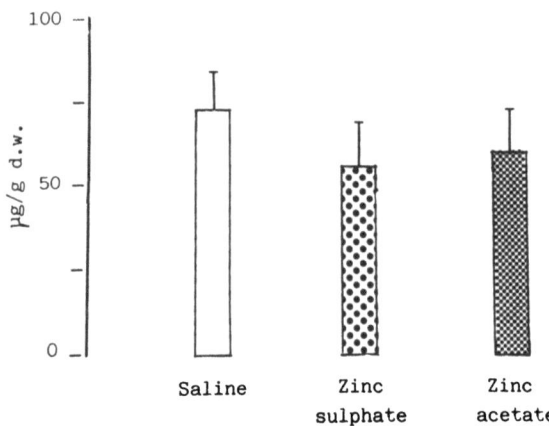

Fig. 4. Zinc levels in hippocampus of control rats and rats chronically treated (100 mg/kg/os/day x3) with zinc (mean ± SD).

CONVULSIVE EFFECT OF INTRACEREBROVENTRICULARLY INJECTED ZINC: ANTAGONISTIC EFFECT OF DIAZEPAM

It has been reported (Izumi et al., 1973) that the intraventricular administration of low doses of zinc induces in female rats SYS similar to that induced by the i.c.v. injection of ACTH (Bertolini and Gessa, 1981) while high doses of zinc produce convulsions (for review see Ebadi et al., 1984). As shown in Table 8, we have confirmed and extended these data since low doses of zinc sulphate i.c.v. injected in male rats induce not only SYS but also PE and since the epileptic seizures induced by higher doses of zinc are inhibited by pretreatment with diazepam. Cannulated rats i.c.v. injected with saline and rats which developed tonic-clonic seizures after the injection of $ZnSO_4$ (10 µl of a 36 mM solution), were sacrificed 15 minutes later. Crude synaptic membranes for binding studies on different areas were prepared after dissection of the brain. The results of these screening studies on the changes of the binding characteristics of brain neurotransmitter systems in i.c.v. zinc injected rats are summarized as qualitative variations in Table 9. Briefly, acute zinc induced seizures are associated with a slightly decreased affinity of high affinity adenosine receptors tested in the hippocampus and of high affinity GABA receptors studied in the cortex. ^{3}H-Naloxone binding performed on brainstem membranes showed a dramatic reduction in the high affinity (K_D 6.1 vs. 1.8 nM)

Table 8. Convulsive Effect of Intraventricular Injection (10 µl) of $ZnSO_4$ and Antagonism by Diazepam of Epileptic Seizures

Pretreatment mg/kg i.p.	Treatment mM/rat i.c.v.	Epileptic Seizures affected/treated rats	% Affected SYS	PE
Saline	Saline	0/10	20	10
Saline	4.5	0/10	100	80
Saline	9	2/10	0	0
Saline	18	6/10	0	0
Saline	36	10/10	0	0
Diazepam	36	4/10	0	0

Diazepam was injected 30 minutes before $ZnSO_4$.

Table 9. Qualitative Changes in Brain Neurotransmitter Systems After Tonic-Clonic Epileptic Seizures Induced by the Intracerebroventricular Injection of $ZnSO_4$ (36 mM)

Neurotransmitter System	^{3}H-Ligand	Brain Area	Changes Affinity	Binding Sites
Adenosine	^{3}H-PIA	Hippocampus	↓	None
GABA	^{3}H-Muscimol	Cortex	None	↓ (high affinity)
Benzodiazepine	^{3}H-Ro15-1788	Cortex	None	None
	^{3}H-Diazepam	Cortex	↑	None
Glutamate	^{3}H-Glutamate	Cortex	↓	↓
Opiate	^{3}H-Naloxone	Brainstem	↓	None

Rats were sacrificed 15 minutes after $ZnSO_4$ injection when repeated episodes of tonic-clonic convulsions have occurred.

without changes in the K_D or B_{max} of the low affinity opiate receptors. The most striking change associated with zinc-induced seizures was found in the binding of ^{3}H-glutamate which undergoes a considerable reduction of both the affinity constant and of the number of recognition sites (K_D 2181 nM, B_{max} 11.8 pmol/mg protein) in comparison with controls (K_D 484, B_{max} 16.4). This effect could be indicative of a massive release of glutamate which might be the major factor responsible for the epileptogenic action of zinc. Finally, in accordance with the action exerted by high doses of zinc in vitro, ^{3}H-diazepam binding showed an increased affinity in zinc-treated rats (K_D 4.8 vs 9.4 nM). This latter finding, together with the evidence that diazepam counteracts zinc-induced seizures and forms coordinative complexes with zinc in in vitro chemical experiments (Preti and Tosi, 1978b), prompted us to examine whether or not diazepam exerts such a property when injected in vivo. For this purpose, diazepam (5 mg/kg i.p.) was injected in normal rats which were sacrificed 30 minutes later by decapitation. The brain was removed, homogenized in Tris-HCl buffer (pH 7.4) and centrifuged twice at 48,000 x g for 20 minutes. The collected supernatants were centrifuged (100,000 x g x 60 minutes) and the new supernatant, after liophylisation, resuspended in 1/10 of the original volume. As shown in Fig. 5, the U.V. spectrophotometric analysis of this extracted material showed three peaks at 200, 280 and 370 nm, which are qualitatively similar to the spectra obtained in the U.V. analysis of the zinc-diazepam complex obtained by laboratory reaction. It is important to note that zinc and diazepam alone or in combination did not give any peak in this area. This first demonstration that diazepam complexes with zinc in vivo seems to be important in explaining the anticonvulsant activity of this drug.

CONCLUSIONS

Although we cannot draw definitive conclusions on the mechanisms by which zinc induces all the above mentioned behavioral, neuroendocrine and neurochemical effects, we have provided some evidence that a peripheral administration of zinc induces, in a dose related fashion, SYS, PE, hyperalgesia and sedation, which seem to be mediated through a central effect, but never epileptic seizures. On the other hand, zinc acutely applied to the CNS causes at low doses, as it does after a peripheral administration, SYS and PE but at high doses, instead of inducing sedation, it provokes seizures which are antagonized by a pretreatment with diazepam. Finally,

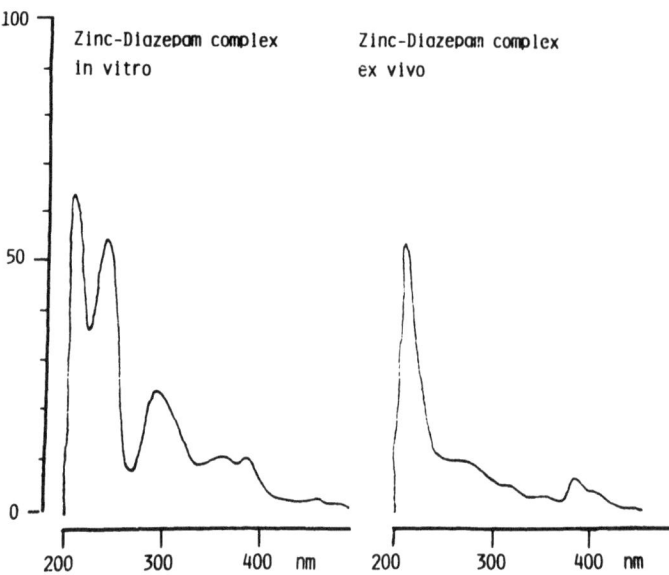

Fig. 5. U.V. spectra of material extracted from brain of rats treated with diazepam. Comparison with the U.V. spectra of the zinc-diazepam complex obtained by laboratory reaction.

we have provided the first demonstration that this antagonistic effect of diazepam could be due to the property of this drug to form a complex with zinc.

ACKNOWLEDGEMENTS

Supported by grants of MPI (60%) and CNR No. 8402182,56.

REFERENCES

Ahlenius, S., and Larson, K., 1985, Apomorphine and haloperidol-induced effects on male rat sexual behavior: no evidence for actions due to stimulation of central dopamine autoreceptors, Neurosci. Lett. Suppl., 18:117.
Baraldi, M., and Bertolini, A., 1974, Penile erections induced by amantadine in male rats, Life Sci., 14:1231.
Baraldi, M., and Benassi-Benelli, A., 1975a, Iduzione di erezioni ripetute nel ratto adulto mediante apomorfina, Riv. Farmacol. Ter., 6:147.
Baraldi, M., and Benassi-Benelli, A., 1975b, Dissociation of the capacity of apomorphine to evoke penile erection and stereotypy following intragastric administration to adult rats, Riv. Farmacol. Ter., 6:361.
Baraldi, M., and Benassi-Benelli, A., 1977, Sexual excitement induced in the adult male rat by low doses of d-amphetamine or apomorphine: suppression by severe stereotyped behavior, Riv. Farmacol. Ter., 8:49.
Baraldi, M., Benassi-Benelli, A., Bernabei, M.T., Cameroni, A., Ferrari, F., and Ferrari, P., 1979, Apocodeine induced stereotypies and penile erections in rats, Neuropharmacology, 18:57.

Baraldi, M., Caselgrandi, E., Borella, P., and Zeneroli, M.L., 1983a, Decrease of brain zinc in experimental hepatic encephalophathy, Brain Res., 258:170.

Baraldi, M., Poggioli, R., Santi, M., Vergoni, A.V., and Bertolini, A., 1983b, Antidepressant and opiates interactions: pharmacological and biochemical evidence, Pharmacol. Res. Comm., 15:843.

Baraldi, M., Caselgrandi, E., and Santi, M., 1984a, Production of withdrawal symptoms in morphine-dependent rats by zinc: behavioral and biochemical studies, Neurosci. Lett., 18:5401.

Baraldi, M., Caselgrandi, E., and Santi, M., 1984b, Effect of zinc on specific binding of GABA to rat brain membranes, in: The Neurobiology of Zinc (Part A), C.J. Frederickson, G.A. Howell, and E.J. Kasarskis, eds., Alan R. Liss, Inc., New York, p. 59.

Baraldi, M., Zeneroli, M.L., Ventura, E., Penne, A., Pinelli, G., Ricci, P., and Santi, M., 1984c, Supersensitivity of benzodiazepine receptors in hepatic encephalopathy due to fulminant hepatic failure in the rat: reversal by a benzodiazepine antagonist, Clin. Sci., 62:167.

Baraldi, M., 1985, Chronic increase of GABA in vivo as a tool to evidentiate the partial agonist property of RO15-1788 by using an in vitro binding assay, Neurosci. Lett. Suppl., 18:31.

Benassi-Benelli, A., Ferrari, F., and Pellegrini-Quarantotti, B., 1978, Penile erection induced by N-n-propyl-norapomorphine in rats, Arch. Int. Pharmacodyn. Ther., 241:128.

Bertolini, A., and Gessa, G.L., 1981, Behavioral effects of ACTH and MSH peptides, J. Endocrinol. Invest., 4:421.

Bettger, W.J., and O'Dell, B.L., 1981, A critical physiological role of zinc in the structure and function of biomembranes, Life Sci., 28:1425.

Braestrup, C., Nielsen, M., and Olsen, C.E., 1980, Urinary and brain β-carboline-3-carboxylates as a potent inhibitor of brain benzodiazepine receptors, Proc. Natl. Acad. Sci. USA, 77:2288.

Brewer, G.S., Ellis, F.B., and Bjork, L., 1981, Parenteral depot method for zinc administration, Pharmacology, 23:254.

Brewer, J., and Oelshlegal, F.J., 1974, Antisuckling effect of zinc, Biochem. Biophys. Res. Comm., 58:854.

Daniel, E.E., Massingham, R., and Nasmyth, P.A., 1970, The mechanism of contractile effects of ouabain and zinc on the rat uterus, J. Pharmacol. Exp. Therap., 173:293.

Di Chiara, G., Corsini, G.U., Mereu, G.P., Tissari, A., and Gessa, G.L., 1978, Self-inhibitory dopamine receptors: their role in the biochemical and behavioral effects of low doses of apomorphine, in: Adv. Biochem. Psychopharmacol., P.J. Roberts, ed., Raven Press, New York, p. 275.

Disilvestro, R.A., and Cousins, R.S., 1983, Physiological ligands for copper and zinc, Ann. Rev. Nutr., 3:261.

Dreosti, I.E., 1984, Zinc in the central nervous system: the emerging interactions, in: The Neurobiology of Zinc (Part A), C.J. Frederickson, G.A. Howell and E.J. Kasarskis, eds., Alan R. Liss, Inc., New York, p. 1.

Ebadi, M., White, R.S., and Swanson, S., 1984, The presence and function of zinc-binding proteins in developing and mature brain, in: The Neurobiology of Zinc (Part A), C.J. Frederickson, G.A. Howell, and E.J. Kasarskis, eds., Alan R. Liss, New York, p. 39.

Farah, J.M., Sapum Malcolm, J.R.D., and Mueller, J.P., 1982, Dopaminergic inhibition of pituitary β-endorphin-like immunoreactivity secretion in the rat, Endocrinology, 110:657.

Fratta, W., Rossetti, Z.L., Poggioli, R., and Gessa, G.L., 1981, Reciprocal antagonism between $ACTH_{1-24}$ and β-endorphin in rats, Neurosci. Lett., 24:71.

Gallager, D.W., and Tallman, J.F., 1983, Consequences of benzodiazepine receptor occupancy, Neuropharmacology, 22:1493.

Hill, D.R., and Bowery, N.C., 1981, ^{3}H-Baclofen and ^{3}H-GABA binding to bicuculline insensitive $GABA_B$ sites in rat brain, Nature, 290:149.

Hjorth, S., Carlsson, A., Clark, D., Swanson, K., Winkinström, H., Sanchez, P., Lindberg, P., Hacksell, U., Arvidsson, L.E., Johansson, A., and Nilsson, J.L.G., 1983, Central dopamine receptor agonist and antagonist actions of the enantiomers of 3PPP, Psychopharmacology, 81:89.

Itoh, M., and Ebadi, M., 1982, The selective inhibition of hippocampal glutamic acid decarboxylase in zinc-induced epileptic seizures, Neurochem. Res., 7:1287.

Izumi, K., Donaldson, J., and Barbeau, A., 1973, Yawning and stretching in rats induced by intraventricularly administered zinc, Life Sci., 12:203.

Kasarskis, E.J., 1984, Regulation of zinc homeostasis in rat brain, in: The Neurobiology of Zinc (Part A), C.J. Frederickson, G.A. Howell, and E.J. Kasarskis, eds., Alan R. Liss, New York, p. 27.

La Bella, F., Dular, R., Stanley, V., and Queen, G., 1973, Pituitary hormone releasing of inhibiting activity of metal ions present in hypothalamic extracts, Biochem. Biophys. Res. Comm., 52:786.

Mackerer, C.R., and Kochman, R.L., 1978, Effects of cations and anions on the binding of ^{3}H-diazepam to rat brain, Proc. Soc. Exper. Biol. Med., 158:393.

Marangos, P.J., Patel, J., Martino, A.M., Dilli, M., and Boulenger, J.P., 1983, Differential binding properties of adenosine receptor agonists and antagonists in brain, J. Neurochem., 41:367.

Meyerson, B., and Terenius, L., 1977, β-endorphin and male sexual behavior, Eur. J. Pharmacol., 42:191.

Mizumo, S., Ogawa, N., and Mori, A., 1983, Differential effects of some transition metal cations on the binding of β-carboline-3-carboxylate and diazepam, Neurochem. Res., 8:873.

Mogilnicka, E., and Klimek, V., 1977, Drugs affecting dopamine neurons and yawning behaviour, Pharmacol. Biochem. Behav., 7:305.

Prasad, A.S., 1979, Clinical, biochemical and pharmacological role of zinc, Ann. Rev. Pharmacol. Toxicol., 20:393.

Preti, C., and Tosi, G., 1978a, Synthesis and spectroscopic studies on group IIB metal 1-4-benzodiazepine-complexes, Trans. Met. Chem., 3:246.

Preti, C., and Tosi, G., 1978b, The complexing behaviour of diazepam towards some bivalent first row transition metals, J. Inorg. Nucl. Chem., 41:263.

Petraglia, F., Baraldi, M., Giarre, G., Facchinetti, F., Santi, M., Volpe, A., and Genazzani, A.R., 1985, Opioid peptides of the pituitary and hypothalmus: changes in pregnant and lactating rats, J. Endocrin., 105:239.

Sandstead, H.H., 1984, Neurobiology of zinc, in: The Neurobiology of Zinc (Part B), C.J. Frederickson, G.A. Howell, and E.J. Kasarskis, eds., Alan R. Liss, New York, p. 1.

Santi, M., Pinelli, G., Ricci, P., Penne, A., Zeneroli, M.L., and Santi, M., 1985, Evidence that 2-phenylpyrazolo|4,3-c|-quinolin-3(5H)-one antagonises pharmacological, electrophysiological and biochemical effects of diazepam in rats, Neuropharmacology, 24:99.

Serra, G., Collu, M., Loddu, S., Celasco, G., and Gessa, G.L., 1983a, Hypophysectomy prevents yawning and penile erection but not hypomotility induced by apomorphine, Pharmacol. Biochem. Behav., 19:917.

Serra, G., Fratta, W., Collu, M., Napoli-Farris, L., and Gessa, G.L., 1983b, Cycloheximide prevents apomorphine-induced yawning, penile erection and grooming in rats, Eur. J. Pharmacol., 86:279.

Serra, G., Collu, M., Serra, A., and Gessa, G.L., 1984, Estrogens antagonize apomorphine-induced yawning in rats, Europ. J. Pharmacol., 104:383.

Simon, E., and Groth, J., 1975, Kinetics of opiate receptor inactivation by sulfhydryl reagents: evidence for conformational change in presence of sodium ions, Proc. Natl. Acad. Sci. USA, 72:2404.

Stengaard-Pedersen, R., 1982, Inhibition of enkephalin binding to opiate receptors by zinc ions: possible physiological importance in the brain, Acta Pharmacol. Toxicol., 50:213.

Tallman, S.F., Thomas, J.W., and Gallager, D., 1978, GABAergic modulation of benzodiazepine binding site sensitivity, *Nature*, 272:383.

Thody, A.J., Penny, R.J., Taylor, D.C., and Taylor, C., 1975, Development of a radioimmunoassay for α-melanocyte-stimulating hormone, *J. Endocrin.*, 67:385.

Tilders, F.J., and Smelik, P.G., 1978, Effects of hypothalamic lesions and drugs interfering with dopaminergic transmission on pituitary MSH content of rats, *Neuroendocrinology*, 25:275.

Tseng, L.F., Loh, H., and Li, C.H., 1976, β-Endorphin as a potent analgesic by intravenous injection, *Nature*, 263:239.

Wallwork, J.C., Milue, D.B., Sims, R.L., and Sandstead, H.H., 1983, Severe zinc deficiency: effects on the distribution of nine elements (potassium, phosphorous, sodium, magnesium, calcium, iron, zinc, copper and manganese) in regions of rat brain, *J. Nutr.*, 112:1895.

EXCITATORY AMINO ACIDS AND DIVALENT CATIONS IN THE KINDLING

MODEL OF EPILEPSY

J.T. Slevin, E.J. Kasarskis, T.C. Vanaman and M. Zurini

V. A. Medical Center and Departments of Neurology
Pharmacology and Biochemistry
University of Kentucky

The epilepsies are chronic disorders of the central nervous system characterized by recurrent convulsive or non-convulsive seizures. Recent estimates suggest between 20 and 40 million people may be affected worldwide. One research strategy for investigating the neurobiological mechanisms underlying the induction and maintenance of the epileptic state has been the development of experimental animal models. Recently, one major model in particular has emerged as a focus of intense research effort: the kindling phenomenon first described by Goddard et al. (1969). The partial seizures, which constitute the behavioral manifestations of the amygdala-kindled model, are considered to represent the best available analogy to human complex partial (temporal lobe, 'psychomotor', limbic) epilepsy (McNamara, 1984a).

Typically, animals are 'kindled' by passage of low levels of electric current via an electrode stereotaxically placed in an appropriate brain structure, particularly areas of the 'limbic' system (e.g., amygdala, entorhinal cortex). The initial response to such stimulation is a strictly localized electrical afterdischarge and, possibly, a behavioral response characteristic of the stimulated brain region. Repeated stimulations, at appropriate intervals, induce changes in the brain which permanently alter its responsiveness to subsequent stimuli of the identical magnitude and duration. The cumulative effect of this heightened neuronal excitability is a major generalized convulsion, evoked by the same stimulus which initially elicited only a focal electrographic disturbance.

The mechanism of kindling appears to involve synaptic transfer of information (Messenheimer et al., 1979). However, contrary to initial expectations, specific pathologic changes in synaptic morphology have not been observed (Slevin and DeKosky, 1986). Consequently, a pharmacologic approach to the study of various neurotransmitter systems in kindled animals has been employed with the anticipation that any abnormalities demonstrated in this model may also play a role in human focal epilepsy, particularly complex partial epilepsy. Current evidence (McNamara, 1984b) favors an attenuating or modulating role for norepinephrine, the GABA/benzodiazepine system, and possibly dopamine. Alteration of these systems, as well as of the excitatory neurotransmitter acetylcholine (Savage and McNamara, 1982), appears to evolve temporally in response to attainment of the kindled state, rather than to serve as an induction mechanism for the kindling phenomenon itself. This association is most evident in the case of the inhibitory neurotransmitters. Both electrophysiologic (Tuff et al., 1983) and neurochemical (McNamara et al., 1980b) studies in the hippocampus indicate that inhibition is not reduced by kindling but, in fact, may be modestly and transiently increased. Hence, attention has shifted to evaluate the contribution of altered excitatory neurotransmission in the induction and maintenance of kindling. The interaction of the excitatory amino acid (EAA) neurotransmitters, L-glutamic and L-aspartic acids, with the total cationic milieu in the hippocampus may play a role in the kindling process and may indicate an important new direction for future research in the epilepsies.

EXCITATORY AMINO ACIDS AND EPILEPSY

In the CNS, excitatory synapses far outnumber inhibitory ones and play an obvious role, particularly in projection pathways. Many (DiChiara and Gessa, 1981) have argued that L-glutamic and L-aspartic acids are major excitatory neurotransmitters in mammalian CNS, however it has been difficult to unequivocally establish this role for them. Nevertheless, several methods have been applied to define and characterize specific pathways, including several within the limbic system, which appear to use either one or the other of these excitatory amino acids for neurotransmission. Fiber connections of the limbic system for which there is strong evidence that an EAA may serve as a neurotransmitter include: the perforant path, hippocampal projections to lateral septal nuclei and mamillary bodies, granule cell mossy fiber projections, pyriform and entorhinal cortical efferents to ipsilateral amygdala, and the lateral olfactory tract (Storm-Mathisen, 1981; Fonnum et al., 1983). Many of these may be activated during the development of kindling; the perforant path, mossy fibers and Schaffer collateral/commissural projections are considered crucial (McNamara, 1984a,b; Dasheiff and McNamara, 1982).

Systematic neuropharmacologic studies, coupled with the synthesis of new radioligands of high specific activity, have indicated the presence of at least three distinct classes of receptors with which the putative neurotransmitters, L-glutamic and L-aspartic acids, and other hypothesized endogenous excitants may interact (Schwarcz and Meldrum, 1985). Glutamate has a high affinity for all receptor subtypes; however, the three classes can be distinguished by their preferential activation by either N-methyl-D-aspartate (NMDA), α-kainic acid (KA), or quisqualic acid (QA). Of these three receptors, only the NMDA site has been well characterized physiologically and it is still not entirely clear which endogenous agonist normally acts at each site.

Excitatory amino acids appear to be important in the pathophysiology of epilepsy. For example, seizure activity can be reliably produced by the systemic or focal intracerebral administration of EAA's, e.g., NMDA, KA, or QA (Zaczek and Coyle, 1982). In experimental status epilepticus, the initial cytopathological changes are observed postsynaptically in pathways subserved by EAAs (Evans et al., 1983) implying that 'overactivity' in these pathways may be involved in the seizures. A third and most convincing line of evidence supporting a role for EAAs in epilepsy is the observation that EAA antagonists are potent anticonvulsants in may animal models, such as audiogenic seizure-susceptible mice (Croucher et al., 1982), photic-sensitive Papio papio (Meldrum et al., 1983), and amygdala-kindled rats (Peterson et al., 1983). These antagonists, including various phosphono-substituted dicarboxylic acids, act selectively on EAA receptors, and most particularly at the NMDA receptor (Schwarcz and Meldrum, 1985).

Despite a surfeit of indirect evidence, demonstration that a neurochemical alteration at an EAA synapse is, of necessity, involved in the etiology and/or perpetuation of the kindling phenomenon has not been forthcoming. However, if kindling is a behavioral reflection of incremental facilitation of EAA neurotransmission, then a clearly progressive modification of any of several synaptic functions (e.g., neurotransmitter release, receptor binding, receptor-effector coupling, reuptake and sequestration) could provide a biochemical basis for this phenomenon.

Of presynaptic processes, both Ca-dependent EAA transmitter release and Na-dependent reuptake have been scrutinized. Liebowitz et al. (1978) could not detect a difference in K-stimulated, Ca-dependent release of [^{3}H] L-glutamic acid from hippocampal slices of entorhinally kindled and control animals. Preliminary data from our group indicate that K-stimulated release of endogenous L-glutamate and L-aspartate from hippocampal slices is similar in amygdala-kindled, electroshocked and sham-stimulated animals, suggesting that EAA release is unaltered in this kindling paradigm. In agreement with the general observation that kindling is not associated with alteration of sodium-dependent reuptake systems, entorhinal kindling does not appear to affect L-glutamate reuptake (Slevin and Ferrara, 1985). We are not aware of published data addressing the possibility of KA-stimulated release, alteration of glutamate catabolic enzymes, or perturbations of the putative neuron-glial glutamate (neurotransmitter) cycle.

Considerably more investigation has been directed toward EAA receptors and their possible role in kindling and a closely related phenomenon, long term potentiation (LTP). When certain synapses in the hippocampus are subjected to brief, high frequency electrical stimulation in a range similar to that used in kindling, the postsynaptic potentials evoked by subsequent stimulation are significantly increased. This effect may last for days, hence the descriptive term LTP. The similarity of the stimulus parameters that effectively induce both kindling and LTP has prompted the speculation that LTP may be the electrophysiologic basis of kindling (Goddard, 1981). Lynch and coworkers (Siman et al., 1985) have presented evidence that the induction of LTP in hippocampal CA1 pyramidal neurons by stimulating Schaffer collateral/commissural fibers is associated with an increase in postsynaptic calcium concentration. This in turn activates the calcium-dependent protease, calpain I, which degrades fodrin, a cytoskeletal structural protein. The resulting alteration of membrane configuration exposes occult postsynaptic glutamate receptors, thereby increasing the postsynaptic potentials subsequently evoked by exposure to glutamate.

Recognizing the electrophysiological similarities between kindling and LTP, Savage and coworkers examined [^{3}H] L-glutamate binding in rats kindled by angular bundle (i.e., perforant pathway) stimulation. In a series of studies (Savage et al., 1982; Savage et al., 1984), they reported a transient increase (one day after completion of kindling) of QA-sensitive [^{3}H] L-glutamic acid binding to an in vitro preparation of hippocampal synaptic membranes depleted of Ca. Because the observed changes were not present one month following completion of kindling, they concluded that the increase in glutamate receptors may be functionally related to LTP and to the induction, but not the maintenance, of the kindled state. Indirect support for this formulation comes from the data of Maru et al. (1982), which indicate that LTP is associated only with the initial stages of kindling and that the robustness of LTP diminishes as kindling evolves to completion.

Using a similar kindling paradigm, Slevin and Ferrara (1985) failed to detect an alteration of [^{3}H] L-glutamate binding to hippocampal membranes at any time during the kindling process evoked by entorhinal cortical stimulation (Fig. 1). Similarly, whether a causal association between LTP and increased postsynaptic glutamate receptor density exists has been

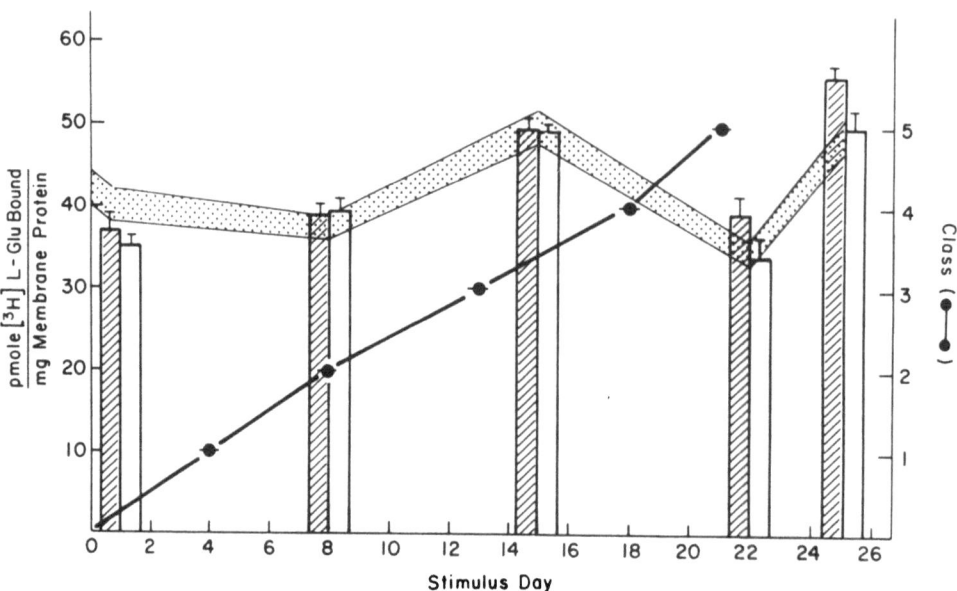

Fig. 1. Kindling time course. Rats receiving the kindling stimulation were killed at stimulation days 1, 8, 15, or 22 (date of second consecutive daily class 5 seizure) and 3 days after the last stimulus-induced seizure (day 25). Each time point represents binding data from six kindled or six control (sham-operated) rats. Stipled band is mean ± S.E. of L-[^{3}H] glutamic acid binding to hippocampal membranes of controls. Hatched bars are mean ± S.E. of binding to hippocampus ipsilateral to the stimulus and open bars are mean ± S.E. of binding to hippocampus contralateral to the stimulus. Solid circles with error bars indicate progression of the animals through kindling stages. There was no difference among groups of L-[^{3}H] glutamic acid binding to hippocampus on any of the days sampled (ANOVA, $p > 0.05$). (From Slevin and Ferrara, 1985).

questioned by Sastry and Goh (1984) who reported that [^{3}H] L-glutamate binding to hippocampal membranes is correlated temporally not with an LTP but with an immediate post-tetanic depression of the population spike. Lastly, recent electrophysiologic data (Dolphin, 1983) indicate that γ-glutamyl-glycine blockade of postsynaptic excitatory amino acid receptors masks but does not prevent LTP of the perforant path, which suggests there is a presynaptic component that involves increased transmitter release from perforant path terminals. In the future, binding studies using more specific ligands, for example [^{3}H] AMPA at QA receptors and [^{3}H] 2-amino-7-phosphonoheptanoic acid at NMDA sites, may clarify these apparent discrepancies.

There have been some pilot studies on the relation of the kainic acid class of EAA receptors to kindling. Binding sites for KA in rat brain are most abundant in the terminal fields of hippocampal mossy fibers and, to a lesser extent, of commissural/association pathways (Monaghan and Cotman, 1982). Both of these neuronal projections are critical for the kindling phenomenon. Recently it has been shown by autoradiography that [^{3}H] KA binding to hippocampal slices from rats kindled by either multiple daily amygdalar or angular bundle stimulations is reduced in

Table 1. [^{3}H] Kainic Acid Binding to the 'High Affinity' Kainate Site in Hippocampus of Kindled Rats

	Kindled	Shock	Control
Right	142 ± 13 n = 15	123 ± 13 n = 13	109 ± 15 n = 14
Left	110 ± 12 n = 12	119 ± 10 n = 15	111 ± 14 n = 14

Kindled animals received once-daily right entorhinal electrical stimulations consisting of a 1-s train of biphasic square wave pulses of 1 ms duration delivered at 60 Hz; current was in the range of 500 μA. 'Shock' animals received 50-mA direct current for 0.8 seconds through ear clips on two consecutive days. Controls received sham operations including insertion of electrodes. Binding was measured to hippocampal homogenates (London and Coyle, 1979) within 24 hours prior to the last seizure.

these subfields bilaterally (Savage et al., 1984). Using the 'high-low' technique of London and Coyle (1979), our group observed an increase of high affinity binding (2 nM site) of [vinylidene - ^{3}H] KA (60 Ci/mmol) but failed to detect a change in total KA binding (high and low affinity sites) to hippocampi ipsilateral to the entorhinal stimulus (Table 1). Whether the same or different sites were measured using these radically different techniques remains to be determined. Moreover, any effect the temporal pattern of stimulations might have had on the biochemical/molecular level is unknown.

In summary, to date there have been no systematic autoradiographic or homogenate binding assays to study the kindling phenomenon which have made use of the recently developed radioligands of agonists and antagonists with relative specificity for the putative classes of excitatory amino acid receptors. Hence, although there are impressive electrophysiologic and behavioral data indicating a role of excitatory amino acid neurotransmitters in the kindling phenomenon, a direct biochemical alteration at the synapse has yet to be demonstrated.

Divalent Cations and EAA Neurotransmission

It has long been recognized that the ionic environment is critical to many synaptic processes related to neurotransmission, for example Ca-dependent synthesis and release and Na-dependent reuptake of transmitter. In addition to these fundamental processes, it has been determined that EAA receptor characteristics may also be critically dependent upon the in vivo concentration of certain cations. For example, external Ca and transmembranal Ca-fluxes may causally relate to glutamate receptor density changes (Baudry et al., 1983) and the ionic channels opened by NMDA are blocked by Mg at concentrations within the range found in vivo (Mayer et al., 1984). Indeed, attempts have been made to classify the heterogeneous EAA binding sites by their activity in specified ionic environments (Mena et al., 1985). Hence, a Na-dependent binding has been characterized with parameters which suggest it represents binding to an uptake site. Na-independent [^{3}H] L-glutamate binding can be resolved into Cl-independent and Cl/Ca-dependent sites. It has been suggested (Schwarcz and Meldrum, 1985) that this latter site represents the quisqualate receptor and the 2-amino-

4-phosphonobutyrate receptor (as defined in electrophysiologic studies). However, the identity of these ionically-defined sites, their relation to [^{3}H] L-glutamate sites defined by agonist/antagonist activity, and to EAA receptors defined electrophysiologically has yet to be clearly established. Recent studies (Mena et al., 1985; Slevin and Kasarskis, 1985) have indicated that both Na-independent L-glutamate binding sites can be inhibited by Zn, thus this transition metal displaces both [^{3}H] L-glutamate (IC_{50} = 130 µM) and [^{3}H] L-aspartate (IC_{50} = 50 µM) from hippocampal membrane homogenates. This inhibition may have physiological importance because it is observed at endogenous Zn concentrations in the central nervous system.

ASSOCIATION OF ZINC WITH EPILEPSY AND KINDLING

Although a single report (Porsche, 1983) suggests that Zn may prevent seizures induced in rats by systemic administration of KA, most work to date indicates that Zn acts as a convulsant. In one study (Wright, 1984), iontophoretically applied zinc excited 45% of rat cortical neurons tested and, when applied concurrently, blocked the depressant effects of Ca. The photosensitive baboon, Papio papio, has a higher serum zinc concentration compared to seizure-resistant primates and Alley et al. (1981) have demonstrated a protective effect against photic-induced seizures by administration of the zinc chelator, D-penicillamine. Taken together, these studies are consistent with the hypothesis that excess zinc may facilitate development of epileptiform activity.

Despite its high level in hippocampus (particularly in association with mossy fibers; Frederickson et al., 1983), its potential regulatory effect on EAA receptors (Slevin and Kasarskis, 1985), and its stimulus-evoked release from mossy fibers (Howell et al., 1984; Charton et al., 1985), little attention has been given to the role of Zn as it may relate to kindling. Recently, Mody and Miller (1985) reported that commissurally kindled rats have a significantly higher concentration of hippocampal zinc than sham stimulated animals. We have examined hippocampal zinc levels in amygdala-kindled animals and find changes similar to those described by Mody and Miller.

In our study, Zn increased approximately two-fold in the hippocampus ipsilateral to the stimulating electrode within 24 hours after kindled rats experienced the second of two daily class 5 seizures, our operating definition for the end point of the kindling process (Fig. 2). Notably, such an elevation did not occur in either cerebellum or neocortex of the same animals and was increased only 1.25-fold in the hippocampus contralateral to the stimulus. Moreover, we observed a delayed elevation of Zn throughout the brain. Thus, animals kindled by amygdalar stimulation to two class 5 seizures and then given no further stimulations for two weeks had up to a 1.6-fold increase in Zn in such disparate brain regions as cerebellum, hippocampus and cortex. Similar changes were not observed following a single electroconvulsive seizure.

These two studies represent the initial attempts to ascribe a role for Zn in kindling. An elevated level of Zn in the brain has been correlated with increased susceptability to seizures (Ebadi and Pfeiffer, 1984) in at least two seizure-prone genetic models (Alley et al., 1981; Chung and Johnson, 1983), and now is apparently linked to kindling (Mody and Miller, 1985). Further detailed work, especially upon the unique pool of Zn associated with mossy fibers, needs to be performed in order to elucidate the temporal relationship of Zn alteration to the induction and perpetuation of kindling. Any of several established effects of Zn on the central nervous system may be operant in kindling. Among these, Zn influences

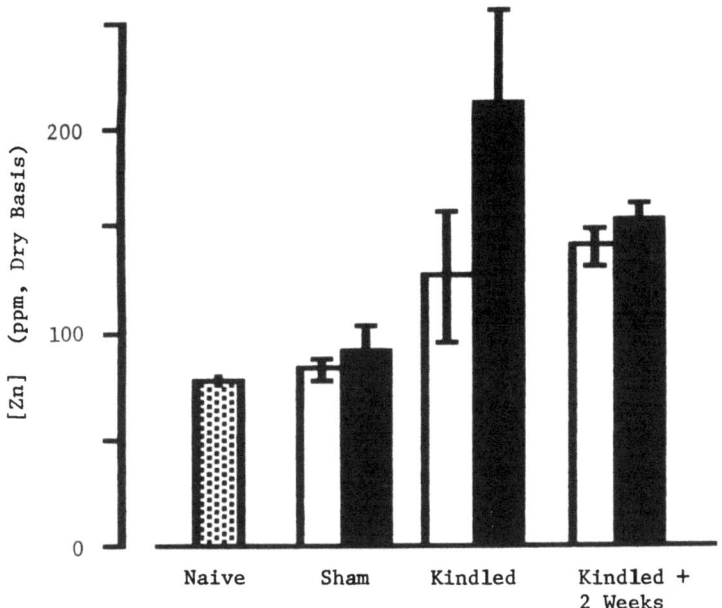

Fig. 2. Alterations in the level of total hippocampal zinc in rats after kindling induced by stimulation of the right amygdala. (Open bars designate left and solid bars designate right hippocampal zinc levels.)

the activity of several enzyme systems involved in electrolyte homeostasis (e.g., carbonic anhydrase, Na/K ATPase; Donaldson et al., 1971) and in metabolism of neurotransmitters (e.g., glutamic acid decarboxylase; Wolf and Schmidt, 1982). Zn modulates the affinity of many neurotransmitter receptors including the EAAs, GABA/benzodiazepine, muscarinic cholinergic, and opiate and has been shown to prolong the EPSPs of cortical neurons (Smart and Constanti, 1983). Furthermore, Zn exerts powerful effects on membrane Na, K and Ca ionic channels by either traversing the channel itself or by modifying the gating membrane proteins (Meves, 1976; Kawa, 1979; Gilly and Armstrong, 1982a,b). Lastly, excess or maldistributed Zn affects the metabolism of other divalent cations including Ca, Mn, Mg and Cu (Donaldson et al., 1971; DeLorenzo, 1984), all of which have been implicated in epileptogenesis and, indirectly, to kindling. Precisely which effect Zn is exerting during kindling remains the subject for future investigation.

CALCIUM DEPENDENT REGULATION IN KINDLED EPILEPSY

Transient or permanent alterations of ongoing neuronal discharge, as occurs in kindling, may be associated with changes related to Ca and the various processes that are regulated by this cellular messenger (DeLorenzo, 1984). As previously noted, a subpopulation of EAA receptors is Ca-dependent and it has been postulated these receptors may have a role in kindling. Wasterlain and Farber (1984) have demonstrated an association between septal kindling and a post hoc inhibition of in vitro Ca-calmodulin-induced phosphorylation of synaptic plasma membrane proteins from rat hippocampus and amygdala-entorhinal area. Substantial evidence suggests that Ca-dependent phosphorylation systems are involved in neurotransmitter synthesis, turnover and release and that they serve as a potential site of action for anticonvulsant drugs (DeLorenzo, 1984).

We have recently initiated studies to test directly whether specific changes occur in the calmodulin (CaM) regulation system or other Ca-regulated enzymes during the process of kindling. Preliminary studies have concentrated on the Ca-pumping membrane ATPase in amygdala-kindled rat brain. Using photoaffinity labeling with [^{125}I]-azido calmodulin, it is possible to detect alterations in CaM target proteins in kindled brain tissues collected by dissection on ice immediately following decapitation of the animals. Membrane fractions prepared from homogenates of these tissues were resolved by SDS-PAGE without further treatment or following photoaffinity labeling with [^{125}I]-azido CaM (Fig. 3). First, there are substantial differences in the patterns obtained with the control and experimental cerebellar membrane preparations when compared to those of hippocampal membranes. More importantly, a moderately strong band of radioactivity, migrating with exactly the same mobility as that expected for the azido-CaM photo adduct with purified erythrocyte membrane Ca-ATPase, is present in hippocampi of controls and in the contralateral hippocampus of amygdala-kindled animals. However, this band appears to be absent or greatly diminished in the ipsilateral hippocampus of amygdala-kindled animals while the overall protein pattern and all other CaM-adducts appear to be largely unchanged (see, for example, the dark band in the autoradiogram at 65 kDa).

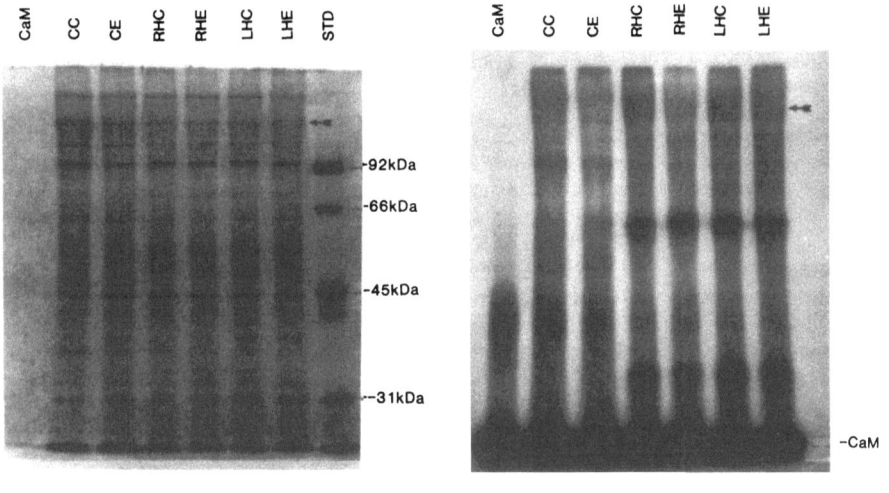

Fig. 3. CaM binding proteins by photoaffinity labeling. Photolabeling of cerebellar and hippocampal membranes with [^{125}I]-azido calmodulin. The left panel shows Coomasie blue stained 10% SDS-PAGE analysis of membrane samples (50 μg total protein) disrupted directly in 1% SDS-1% 2-mercaptoethanol. The right panel shows an autoradiogram obtained from a parallel 10% SDS gel run on the same amount of sample first mixed and photolyzed with [^{125}I]-azido CaM as described by McCartney et al. 1983. Samples: CC = Cerebellum control; CE = Cerebellum experimental; RHC = Right hippocampus control; RHE = Right hippocampal experimental; LHC = Left hippocampal control; LHE = Left hippocampal experimental; CaM = azido CaM control; STD = molecular weight standards of the chain weights noted. Electrode implantation was in the right amygdala in both control and fully-kindled rats.

Table 2. Ca^{2+}-Dependent ATPase Activities of Enriched Synaptosomal Membrane Prepared from the Hippocampus of Control and Amygdala-Kindled Rats

	Basal Activity	CaM-Stimulated Activity	Stimulation Factor
Control Animal	6.84	9.93	1.45
Kindled Animal	3.85	6.02	1.56

Activities are expressed in mol P_i liberated/mg protein/hour. The results are the average of six independent measurements on the two membrane preparations; CaM = calmodulin.

Enzyme assays also indicate an alteration in the level of CaM-activated ATPase in the amygdala-stimulated hippocampus. The amount of detectable Ca^{2+}-ATPase activity, both basal and CaM-activated, is approximately 1.5- to 2-fold lower in synaptosomal membranes prepared from pooled (R+L) hippocampi from kindled rat brain than from surgical controls. This would be expected for an almost complete loss of the enzyme from the ipsilateral kindled hippocampus as suggested by the data shown in Fig. 3.

Conversely, calmodulin-regulated 3', 5'-cyclic AMP phosphodiesterase (PDE) activity appears to be substantially (50%) elevated in hippocampus ipsilateral to the stimulus of amygdala-kindled rats (data not shown). Thus, the apparent decrease in CaM-sensitive membrane Ca-ATPase in the kindled ipsilateral hippocampus is not simply a loss of all CaM-activated enzymes. It should be noted that no differences were detected between the calmodulin contents of individual kindled and control hippocampi as judged by both PDE activator and ELISA assay (data not shown).

CONCLUSIONS

Several excellent reviews of the kindling phenomenon which have appeared in recent years (McNamara et al., 1980a; Kalichman, 1982) stress that, despite intense study, the neurochemical/molecular basis of this electrophysiologically-defined process remains obscure. The simple (but requisite) *in vitro* measure of synaptic markers of neurotransmission for a variety of transmitters has as yet not provided a mechanistic understanding of this phenomenon. Therefore it is reasonable to scrutinize the process on a more complex and possibly more physiological level. We have reviewed and attempted to synthesize observations from three diverse origins - excitatory amino acids, trace metals, and Ca-signaling systems. At first glance, parallel changes in these systems during kindling are intriguing and may provide suitable avenues to formulate new hypotheses regarding the establishment of the epileptic state.

REFERENCES

Alley, M.C., Killam, E.K., and Fischer, G.L., 1981, The influence of D-penicillamine treatment upon seizure activity and trace metal status in the Senegalese baboon, Papio papio, J. Pharmacol. Exp. Therap., 217:138.

Baudry, M., Siman, R., Smith, E.K., and Lynch, G., 1983, Regulation by calcium ions of glutamate receptor binding in hippocampal slices, Europ. J. Pharmacol., 90:161.

Charton, G., Rovira, C., Ben-Ari, Y., and Leviel, V., 1985, Spontaneous and evoked release of endogenous Zn^{2+} in the hippocampal mossy fiber zone of the rat in situ, Exp. Brain Res., 58:202.

Chung, S.H., and Johnson, M.S., 1983, Divalent transition-metal ions (Cu^{2+} and Zn^{2+}) in the brains of epileptogenic and normal mice, Brain Res., 280:323.

Croucher, M.J., Collins, J.F., and Meldrum, B.S., 1982, Anticonvulsant action of excitatory amino acid antagonists, Science, 216:899.

Dasheiff, R.M., and McNamara, J.O., 1982, Intradentate colchicine retards the development of amygdala kindling, Ann. Neurol., 11:347.

DeLorenzo, R.J., 1984, Calmodulin systems in neuronal excitability: a molecular approach to epilepsy, Ann. Neurol., 16:S104.

DiChiara, G., and Gessa, G.L., 1981, Glutamate as a Neurotransmitter, Raven Press, New York.

Dolphin, A.C., 1983, The excitatory amino acid antagonist γ-D-glutamylglycine masks rather than prevents long term potentiation of the perforant path, Neuroscience, 10:377.

Donaldson, J., St.-Pierre, T., Minnich, J., and Barbeau, A., 1971, Seizures in rats associated with divalent cation inhibition of Na^+/K^+ ATPase, Can. J. Biochem., 51:87.

Ebadi, M., and Pfeiffer, R.F., 1984, Zinc in neurological disorders and in experimentally induced epileptiform seizures, in: The Neurobiology of Zinc. Part B: Physiochemistry, Anatomy and Techniques, C.J. Frederickson, G.A. Howell and E.J. Kasarskis, eds., Alan R. Liss, New York, p. 307.

Evans, M.C., Griffiths, T., and Meldrum, B.S., 1983, Early hippocampal changes in the rat following bicuculline and L-allyl lycine-induced seizures: a light and electron microscope study, Neuropathol. Appl. Neurobiol., 9:39.

Fonnum, F., Fosse, V.M., and Allen, C.N., 1983, Identification of excitatory amino acid pathways in the mammalian nervous system, in: Excitotoxins, K. Fuxe, P.J. Roberts, and R. Schwarcz, eds., Macmillan Press, London, p. 3.

Frederickson, C.J., Klitenick, M.A., Manton, W.I., and Kirkpatrick, J.B., 1983, Cytoarchitectonic distribution of zinc in the hippocampus of man and the rat, Brain Res., 273:335.

Gilly, W.F., and Armstrong, C.M., 1982a, Slowing of sodium channel opening kinetics in squid axon by extracellular zinc, J. Gen. Physiol., 79:935.

Gilly, W.F., and Armstrong, C.M., 1982b, Divalent cations and the activation kinetics of potassium channels in squid giant axons, J. Gen. Physiol., 79:965.

Goddard, G.V., McIntyre, D.C., and Leech, C.K., 1969, A permanent change in brain function resulting from daily electrical stimulation, Exp. Neurol., 25:295.

Goddard, G.V., 1981, The continuing search for mechanism, in: Kindling 2, J.A. Wada, ed., Raven Press, New York, p. 1.

Howell, G.A., Welch, M.G., and Frederickson, C.J., 1984, Stimulation-induced uptake and release of zinc in hippocampal slices, Nature, 308:736.

Kalichman, M.W., 1982, Neurochemical correlates of the kindling model of epilepsy, Neurosci. Behav. Rev., 6:165.

Liebowitz, N.R., Pedley, T.A., and Cutler, R.W., 1978, Release of γ-aminobutyric acid from hippocampal slices of the rat following generalized seizures induced by daily electrical stimulation of entorhinal cortex, Brain Res., 138:369.

London, E.D., and Coyle, J.T., 1979, Specific binding of [^{3}H] kainic acid to receptor sites in rat brain, Mol. Pharmacol., 15:492.

Maru, E., Tatsuno, J., Okamoto, J., and Ashida, H., 1982, Development and reduction of synaptic potentiation induced by perforant path kindling, Exp. Neurol., 38:409.

Mayer, M.L., Westbrook, G.L., and Guthrie, P.B., 1984, Voltage-dependent block by Mg^{2+} of NMDA responses in spinal cord neurones, Nature, 309:261.

McCartney, J.E., Klevit, R.E., Blum, J.J., and Vanaman, T.C., 1983, Chemical studies of calmodulin and the regulation of motile systems, in: Calcium Binding Proteins 1983, B. deBernard, G.L. Sottorasa, G. Sandri, E. Carafoli, A.N. Taylor, T.C. Vanaman, and R.J.P. Williams, eds., Elsevier, North Holland Biomedical Press, Amsterdam, p. 273.

McNamara, J.O., Byrne, M.C., Dasheiff, R.M., and Fitz, J.G., 1980a, The kindling model of epilepsy: a review, Prog. Neurobiol., 15:139.

McNamara, J.O., Peper, A.M., and Patrone, V., 1980b, Repeated seizures induce long-term elevation of hippocampal benzodiazepine receptors, Proc. Natl. Acad. Sci. USA, 77:3029.

McNamara, J.O., 1984a, Kindling: an animal model of complex partial epilepsy, Ann. Neurol., 16:S72.

McNamara, J.O., 1984b, Role of neurotransmitters in seizure mechanisms in the kindling model of epilepsy, Fed. Proc., 43:2516.

Meldrum, B.S., Croucher, M.J., Badman, G., and Collins, J.F., 1983, Anti-epileptic action of excitatory amino acid antagonists in the photosensitive baboon, Papio papio, Neurosci. Lett., 39:101.

Mena, E.E., Monaghan, D.T., Whitmore, S.R., and Cotman, C.W., 1985, Cations differentially affect subpopulations of L-glutamate receptors in rat synaptic plasma membranes, Brain Res., 329:319.

Messenheimer, J.A., Harris, E.W., and Steward, O., 1979, Sprouting fibers gain access to circuitry transynaptically altered by kindling, Exp. Neurol., 65:469.

Meves, H., 1976, The effect of zinc on the late displacement current in squid giant axons, J. Physiol., 254:787.

Mody, I., and Miller, J.J., 1985, Levels of hippocampal calcium and zinc following kindling-induced epilepsy, Can. J. Physiol. Pharmacol., 65:159.

Monaghan, D.T., and Cotman, C.W., 1982, The distribution of [^{3}H] kainic acid binding sites in rat CNS as determined by autoradiography, Brain Res., 252:91.

Peterson, D.W., Collins, J.F., and Bradford, H.F., 1983, The kindled amygdala model of epilepsy: anticonvulsant action of amino acid antagonists, Brain Res., 275:169.

Porsche, E., 1983, Zinc prevents kainic acid induced seizures in rats, IRCS Med. Sci., 11:599.

Sastry, B.R., and Goh, J.W., 1984, Long-lasting potentiation in hippocampus is not due to an increase in glutamate receptors, Life Sci., 34:1497.

Savage, D.D., and McNamara, J.O., 1982, Kindled seizures reduce a select subpopulation of ^{3}H-QNB binding sites in rat dentate gyrus, J. Pharmacol. Exp. Ther., 222:670.

Savage, D.D., Nadler, V.J., and McNamara, J.O., 1984, Reduced kainic acid binding in rat hippocampal formation after limbic kindling, Brain Res., 323:128.

Savage, D.D., Werling, L.L., Nadler, V.J., and McNamara, J.O., 1982, Selective increase in L-[^{3}H] glutamate binding to a quisqualate-sensitive site on hippocampal synaptic membranes after angular bundle kindling, Europ. J. Pharmacol., 85:255.

Savage, D.D., Werling, L.L., Nadler, V.J., and McNamara, J.O., 1984, Selective and reversible increase in the number of quisqualate-sensitive glutamate binding sites on hippocampal synaptic membranes after angular bundle kindling, Brain Res., 307:332.

Schwarcz, R., and Meldrum, B.S., 1985, Excitatory amino acid antagonists provide a therapeutic approach to neurological disorders, Lancet, 2:140.

Siman, R., Baudry, M., and Lynch, G., 1985, Regulation of glutamate receptor binding by the cytoskeletal protein fodrin, Nature, 313:225.

Slevin, J.T., and Ferrara, L.P., 1985, Lack of effect of entorhinal kindling on L[^{3}H] glutamic acid presynaptic uptake and postsynaptic binding in hippocampus, Exp. Neurol., 89:48.

Slevin, J.T., and Kasarskis, E.J., 1985, Effects of zinc on markers of glutamate and aspartate neurotransmission in rat hippocampus, Brain Res., 334:281.

Slevin, J.T., and DeKosky, S.T., 1986, Stability of sialogangliosides in kindled hippocampus, Exp. Neurol., 91:208.

Smart, T.G., and Constanti, A., 1983, Pre- and postsynaptic effects of zinc on in vitro prepyriform neurones, Neurosci. Lett., 40:205.

Storm-Mathisen, J., 1981, Glutamate in hippocampal pathways, in: Glutamate as a Neurotransmitter, G. DiChiara and G.L. Gessa, eds., Raven Press, New York, p. 43.

Tuff, L.P., Racine, R.J., and Adamac, R., 1983, The effects of kindling on GABA-mediated inhibition in the dentate gyrus of the rat. I. Paired pulse depression, Brain Res., 277:79.

Wasterlain, C.G., and Farber, D.B., 1984, Kindling alters the calcium/calmodulin-dependent phosphorylation of synaptic plasma membrane proteins in rat hippocampus, Proc. Natl. Acad. Sci. USA, 81:1253.

Wolf, G., and Schmidt, W., 1982, Zinc as a putative regulatory factor of glutamate dehydrogenase activity in glutamergic systems, in: Neuronal Plasticity and Memory Formation, C. Ajmone-Marsan and H. Matthies, eds., Raven Press, New York, p. 437.

Wright, D.M., 1984, Zinc: effect and interaction with other cations in the cortex of the rat, Brain Res., 311:343.

Zaczek, R., and Coyle, J.T., 1982, Excitatory amino acid analogues: neurotoxicity and seizures, Neuropharmacology, 21:15.

EFFECT OF ZINC ON NEURONAL ACTIVITY IN THE RAT FOREBRAIN

D.M. Wright

Department of Anatomy
The Medical School
University of Bristol, U.K.

INTRODUCTION

Zinc is an important trace element. Manifestations of deficiency include: impaired growth and maturation, depressed healing and immunity, dermatitis, hypogonadism and impotence and impaired neuropsychological function (see Stanstead et al., 1983). There are about 150 metalloenzymes which are activated by zinc. It is often thought that it is in this manner that zinc ions exert their biological action. Pathological changes however can occur rapidly after dietary zinc deprivation, and a recent review (Bettger and O'Dell, 1981) suggests that the activity of only a few of these metalloenzymes is significantly depressed. The authors argue that zinc exerts physiological roles independent of those of the zinc metalloenzymes, and suggest an important role in the maintenance of membrane structure and function such that deficiency would result in destabilization and altered membrane properties. Conversely, elevated extracellular concentrations of zinc would have a stabilizing or protective effect.

Zinc ions therefore appear to have a ubiquitous role in biomembranes. Within the central nervous system, however, an uneven distribution of zinc occurs. The highest concentrations are found in the hippocampus, specifically in the giant boutons of mossy fibers (Haug, 1967), hypothalamus, and cerebral cortex (Donaldson et al., 1973). The functional significance of these findings is unclear. A few studies have looked at the effect of a decrease in zinc levels. For example, Hesse (1979) reported that in rats made chronically zinc deficient, normal frequency potentiation of evoked responses in hippocampal subfield CA3 did not occur. Similarly, chelation of zinc with H_2S caused marked decrements and irreversible blockade of transmission in the mossy fiber system (von Euler, 1962). In contrast, Danscher and co-workers (1975), who used diphenylthiocarbazone to chelate zinc, observed an initial excitatory effect but found no evidence of a reduction in mossy fiber transmission or in a changed frequency potentiation or post-tetanic responses. Another study, however, (Crawford et al., 1973) showed both an initial excitatory response and a decrement in frequency potentiation after zinc chelation. Few studies have looked at the effect of an increase in zinc levels. These have shown, however, that zinc administration to the nervous system, rather than having a stabilizing effect, increases neuronal excitability and causes convulsions. Furthermore, higher levels of zinc have been found in the brains of epileptogenic mice than in normal mice (Chung and Johnson, 1983). What then is the nature

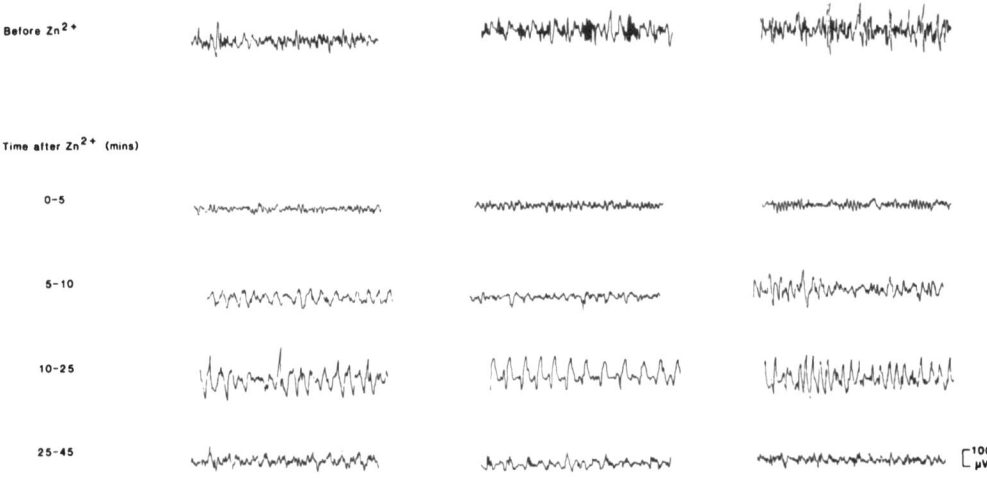

Fig. 1. Effect of intraventricular Zn^{2+} on rat ECOG. Effect of intraventricular injection of zinc chloride (500 μg, 2 μl) on the ECOG of urethane-anesthetized rats. Results are from three different rats (A, B, and C) immediately before and at various times after the injection. Each trace shows the typical ECOG from each time epoch. Bursts of repetitive spikes were particularly prevalent between 10-25 minutes. Recordings are bipolar and contralateral to the injection site.

of the effects of zinc and has this relevance for the etiology of epilepsy?

ZINC-INDUCED SEIZURES

Injection of zinc chloride at doses of 200-500 μg into the lateral cerebral ventricle of urethane-anesthetized Wistar rats caused epileptogenic ECOG activity. Typical examples from such experiments are shown in Fig. 1. The immediate response to the injection was a period of low amplitude, high frequency activity lasting several minutes but with episodes of spike-slow wave complexes. Subsequently, the amplitude of the ECOG increased and was characterized by longer periods of spike-slow wave activity. Clonic movements were often seen, usually consisting of the raising and extension of the forepaws and ipsilateral 'waving' movements. In most animals, excessive salivation was noted together with hyperventilation and whisker twitching. The ECOG pattern after 30 minutes consisted of spike-wave discharges and periods of smaller amplitude, higher frequency activity during which convulsive activity usually occurred. Two animals died within 15 minutes of receiving a cumulative dose of 1 mg. Zinc-induced seizures observed in these urethane-anesthetized rats appear to be similar to those reported in the unanesthetized rabbit (Pei et al., 1983). In rabbits, seizures usually occurred after a mean latent period of 5.8 hours and most died following the onset of continuous tonic seizures between 24-36 hours after injection. The antiepileptic drugs diphenylhydantoin, sodium valproate, phenobarbital and nitrazepam were reported to afford some protection against the zinc-induced seizures.

It is well known that a number of metal ions can cause convulsions, indeed topical administration of cobalt or aluminium oxide to the cortex is a method often used to provide chronic experimental models of epilepsy (Purpura, 1972). It is possible that post-traumatic epilepsy may be the consequence of iron released from hemoglobin since intracortical injection of hemoglobin to rats produces chronic focal spike activity (Rosen and Frumin, 1979), as does ferric chloride injection (Lange et al., 1980). Intracortical injection of copper ions in the pigeon also causes convulsions. It was suggested this was due to a disturbance of the membrane pump (Peters et al., 1966). This hypothesis was also taken up by Donaldson and co-workers (1981) who reported a correlation between the in vivo convulsant effects of various divalent cations and inhibition of microsomal Na^+-K^+-ATPase. Zinc and copper ions were the most potent whilst manganese and ferrous ions were less effective. Intraventricular injection of zinc ions (10-1000 µg) was found to cause stereotyped behavior, jumping, running, leaping and tonic and clonic seizures (which were invariably fatal); these effects were similar to those evoked by ouabain. Subsequently, it was reported that both ouabain and zinc preferentially inhibited Na^+-K^+-ATPase in the hippocampus and that both caused marked elevations in the potassium ion concentration of cerebrospinal fluid (Donaldson et al., 1972). A number of investigators have shown that increasing potassium ion levels causes convulsions. Zuckerman and Glaser (1968), for example, have reported that not only did increasing the concentration of potassium ions produce paroxysmal activity but that administration of single shocks to the hippocampus during a period of high potassium application also caused convulsions, suggesting that stimuli which ordinarily did not cause epileptiform activity might do so if potassium levels were disturbed. The ability of zinc to increase extracellular potassium levels is thus an attractive explanation for its convulsant effects.

Neuronal activity is profoundly influenced by the activity not only of potassium ions but also by calcium ions. Zinc may also act on calcium systems. A decrease in extracellular levels of calcium has long been known to enhance neuronal excitability (Frankenhaeuser and Hodgkin, 1957). Current theories on the mechanism underlying epilepsy suggest alterations in calcium-dependent processes may be involved. Heinemann and co-workers (1977), for example, have shown that a decrease in extracellular calcium ions may accompany and precede the onset of ictal activity and changes in extracellular potassium. Recently, a decrease in calcium binding protein during kindling-induced epilepsy was demonstrated by Miller and Bainbridge (1983), who suggested that an increase in neuronal excitability could occur as a result of decreased buffering of intracellular calcium. It is of interest therefore that zinc can antagonize a number of calcium-mediated processes (Bettger and O'Dell, 1981), particularly those involving the calcium binding protein calmodulin (Brewer et al., 1979).

Whilst zinc-induced seizures may be the result of an alteration in potassium and/or calcium ion concentrations, a further possibility should also be considered, namely that they may occur as a consequence of changes in the activity of a number of enzymes important in neurotransmitter metabolism. In particular, zinc may enhance glutamatergic and reduce GABAergic function through inhibition of glutamate decarboxylase (Wu and Roberts, 1974). GABA fulfills an important role as an inhibitory neurotransmitter. There is evidence to suggest that decreased inhibition is of importance in epileptogenesis (see Ward et al., 1969). During tetanic stimulation, for example, there is a marked decrement of hippocampal IPSPs and conductance increases evoked by GABA. Ben-Ari and co-workers (1979) suggested therefore that 'an important element in the mechanism of onset of seizures is a failure of the normal synaptic inhibitory control, at least partly owing to the fading of the action of GABA'.

STUDIES OF ZINC ACTION AT A CELLULAR LEVEL

Although evidence from a number of quite diverse studies has implicated zinc as a causative agent in epilepsy, a role for zinc in neuronal function, which is central to an understanding of this problem, has yet to be established. Recently, it has been shown that zinc ions can be released during excitation (Assaf and Chung, 1984; Howell et al., 1984) and could thus participate directly in synaptic transmission. Accordingly, the effect of zinc ions on neuronal activity was examined following iontophoretic administration.

Experiments were performed on male albino rats which were anesthetized with urethane (1.25 g/kg, i.p.) and placed in a stereotaxic frame. Extracellular action potentials of single neurons in the fronto-parietal cortex were recorded via a sodium chloride (3 M) filled barrel of a five-barrelled glass micropipette. The neuronal activity was fed to a pulse- shape window discriminator and the resulting pulses were quantified by a microcomputer based ratemeter system (Leendertz and Wright, 1983) and expressed on a polygraph as frequency-histograms. Other barrels of the micropipette contained various combinations of the following solutions: zinc chloride (0.2 M), calcium chloride (0.1 M), manganese chloride (0.1 M), barium chloride (0.2 M), magnesium chloride (0.1 M), cerium chloride (0.1 M in 0.1 M NaCl), GABA (0.2 M), glutamate (0.2 M), D-Ala^2-Met^5-enkephalinamide (10 mM in 0.1 M NaCl), morphine sulphate (0.07 M), sodium chloride (3 M, for testing for current artifacts) and were used for iontophoretic drug application. The criterion for effect (i.e., excitation or inhibition) was a > 30% change in firing frequency following iontophoresis.

The effects of zinc and other cations on the firing rate of cortical neurons are summarized in Table 1. Most commonly, zinc caused excitations and these were often prolonged in duration and associated with high-frequency bursts of activity (Fig. 2). Excitatory effects were also seen after application of barium and in an even larger percentage of neurons (72.7%). These excitations were similar in nature to those evoked by zinc; inhibitions were never seen. In contrast, calcium, magnesium, manganese and cerium had depressant effects. These inhibitions were rapid in onset and short in duration; excitations were never seen. Calcium depressed the activity of a large number of neurons (67/75) and reduced glutamate-evoked activity of 4/5 quiescent neurons. Manganese had similar potent depressant effects affecting 22/28 (78.6%) neurons, and magnesium depressed the activity of 2/6 neurons. Cerium appeared to be the least potent, depressing the activity of only 7/32 (21.9%) neurons, but the resistance of the cerium-containing barrel was often high so that only small currents could be applied during the tests.

Table 1. Effect of Zinc and Other Cations on the Firing Frequency of Cortical Neurons

	Excitation	Depression	No Effect	n
Zinc	45.2	9.6	45.2	31
Barium	72.7	0	27.3	11
Calcium	0	89.3	0.7	75
Cerium	0	21.9	78.1	32
Magnesium	0	33.3	66.7	6
Manganese	0	78.6	21.4	28

Numbers are percentages of cells in each category.
Adapted from Wright, 1984.

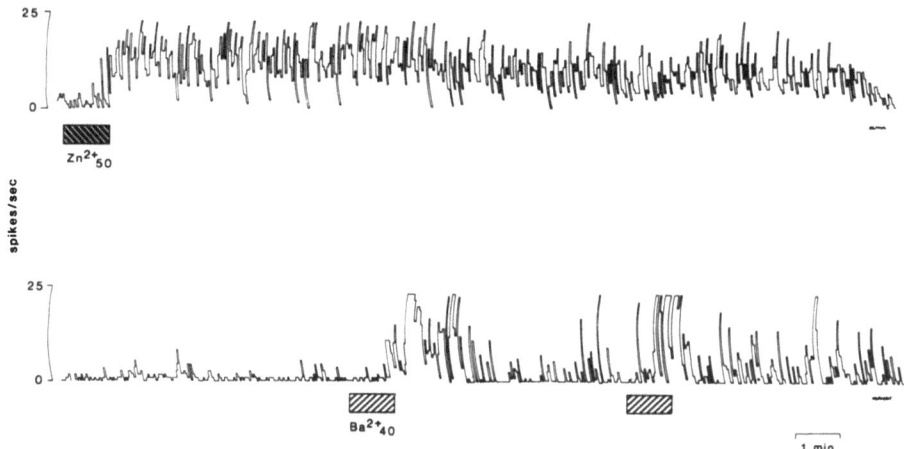

Fig. 2. Effect of zinc and barium on cortical activity. Excitatory effects of iontophoretically applied zinc and barium ions to cortical neurons. Upper and lower traces are polygraph records of the firing frequency of two different neurons. Bars indicate the iontophoretic application of the cations and numbers the currents (nA) used.

Whereas in these studies zinc was found to cause excitations, Rozear and co-workers (1971) reported only depressant effects of zinc applied iontophoretically to cortical neurons of the rat. This may be an example of a species difference since excitatory effects were observed in different studies in the rat. For example, in the rat hippocampal slice preparation bath-application of zinc caused an increase in the amplitude of evoked potentials in subfield CA3 following stimulation of the mossy fibers and occasionally multiple population spikes were observed (Brown et al., 1985). Similarly, in the rat olfactory slice preparation zinc caused a prolongation of the EPSP accompanied by irregular oscillations in membrane potential, which was usually sufficient to initiate repetitive neuronal discharge (Smart and Constanti, 1983). It was also observed that shortly after zinc ion application there was an increase in both the frequency and amplitude of spontaneous post-synaptic potentials. Interestingly, Smart and Constanti also noted that after some period of EPSP prolongation following zinc administration transmission was blocked, often irreversibly at higher zinc concentrations. This finding may provide an explanation for the results of Rozear et al. (1971). (In this regard it may be noted that in the cat barium caused excitation (Krnjević et al., 1971) similar to that observed in the rat). Both barium (Werman and Grundfest, 1961) and zinc reduce potassium permeability and a slow inward leak of sodium ions may be a common mechanism underlying the excitations they cause in anesthetized rats.

Zinc and Calcium Antagonism

As has been discussed earlier, there is evidence to suggest zinc may disrupt calcium-dependent systems and this may in turn predispose neuronal tissue to aberrant forms of activity. Further evidence for this was obtained from studies in which zinc and calcium ions were simultaneously iontophoresed onto cortical neurons in anesthetized Wistar rats. It was found that the depressant effects of calcium were blocked in 23/29 (79.3%) neurons by zinc (Table 2 and Fig. 3). Manganese and cerium on the other hand failed to antagonize the effects of calcium (Fig. 4). Indeed, when applied with higher currents, they too had a depressant effect. Although it has been suggested that this may occur indirectly as a consequence

Table 2. Effect of Zinc on Calcium Induced Depression of Cortical Firing and Comparison With Other Cations

	Block	No Effect	n
Zinc	79.3	20.7	29
Barium	20.0	80.0	10
Cerium	0	100	6
Manganese	0	100	13

Numbers are percentages of cells in each category.
Adapted from Wright, 1984.

of the release of calcium (Phillis and Limacher, 1974), in the present study it could not be blocked by zinc. This also shows that it is unlikely that the interaction between calcium and zinc is simply the consequence of a simple summing of the excitatory and inhibitory influences of these ions. Furthermore, excitatory responses to zinc only occurred in half the studies and only rarely at the currents required to block the effects of calcium. Also, barium, which elicited excitatory responses from a greater number of cortical neurons than did zinc, was ineffective at preventing calcium-induced depression of neuronal firing. An action of zinc to block a neuronal calcium system is also inferred from recent studies in the hippocampal slice preparation, which showed that paired-pulse potentiation in subfield CA3 was attenuated by zinc (Brown et al., 1985) since this is commonly believed to be a calcium-dependent phenomenon indicative of a short-term increase in synaptic efficacy.

A depressant effect of calcium on neuronal firing rate has been reported previously (Krnjević, 1965; Kato and Somjen, 1969) and is thought to occur as a consequence of an interference with sodium permeability. It has been postulated that extracellular calcium ions are drawn into the sodium channel and plug the conducting pore (Woodhull, 1973). Zinc also decreases sodium permeability and it has been suggested that this occurs because of an electrostatic attraction to gating charges and the resultant stabilization of the sodium channel in the closed confirmation. Thus, whilst an explanation for the apparently specific blockade of calcium action by zinc can as yet only be a matter for conjecture, it appears that zinc and calcium ions may compete for binding sites in the vicinity of the sodium channel. However, potassium channels have been found to be more sensitive to zinc than sodium channels where the interaction between zinc ions and gating charges is weak and readily overcome by sufficient depolarization. The balance of these opposing influences may be such that no change in neuronal firing would be detected.

<u>Zinc and GABAergic Function</u>

Whilst zinc may indrectly affect GABAergic systems by inhibiting enzymes involved in GABA metabolism, a more direct action can also occur. Smart and Constanti (1982) found that zinc (and copper) could depress GABA-evoked conductance change in the lobster muscle, possibly by binding with imidazole groups thought to be present at the GABA receptor-ionophore complex. They tentatively suggested that zinc (and copper) could induce epilepsy by antagonizing endogenous GABA-mediated inhibition. As can be seen in Fig. 4, this now seems unlikely as there is no evidence for such an action in the mammalian central nervous system. Indeed, in a subsequent study in the rat olfactory cortex, Smart and Constanti (1983) found that zinc ions actually enhanced GABA-evoked depolarization and

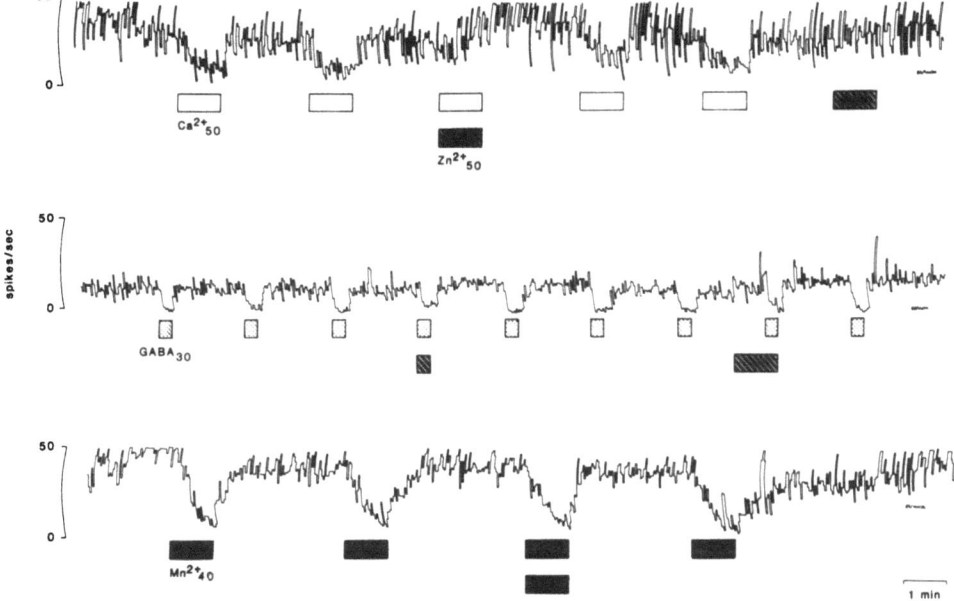

Fig. 3. Zinc: interaction studies with other cations in the cortex. Comparison of the effect of simultaneous iontophoretic application of zinc ions with calcium ions (top trace), GABA (middle trace) and manganese (lower trace). Each trace is a polygraph record of the firing frequency of a different cortical neuron. Zinc ions, which had no effect on the basal firing frequency of the neuron illustrated in the top trace, blocked the depressant effect of calcium ions. The lower trace shows that the depressant effect of GABA or manganese ions were not blocked. (Bars indicate iontophoretic application of cations and numbers the currents (nA) used).

conductance increase. Zinc also had no effect on GABA responses of rat ganglionic neurons. In light of these results and those showing opposite effects of zinc on cortical neural activity in the rat (Wright, 1984) and cat (Rozear et al., 1971), it seems that extrapolation of zinc effects between species should be treated with caution.

Zinc and Opiate Action

The action of opiates may be affected by the presence of zinc ions. A number of reports have linked opiate action and divalent ions in various ways (for review see Chapman and Way, 1980). In the present study, it was found that the inhibitory effect of the enkephalin analog D-Ala^2-Met^5-enkephalinamide (DAME) on the activity of cortical neurons could be blocked by simultaneous iontophoresis of zinc ions. An example of this is shown in Fig. 5. This observation is in keeping with an earlier report by Stengaard-Pedersen (1982) that the stereospecific binding of tritiated DAME in synaptosomal preparations of rat hippocampus, cerebral cortex and basal ganglia could be completely blocked by zinc ions. This effect appears to involve not only a decrease in binding affinity but also a decrease in the number of binding sites, and may involve oxidation of essential opiate receptor SH groups. The data also suggest that endogenous zinc concentrations in the aforementioned brain areas are compatible with

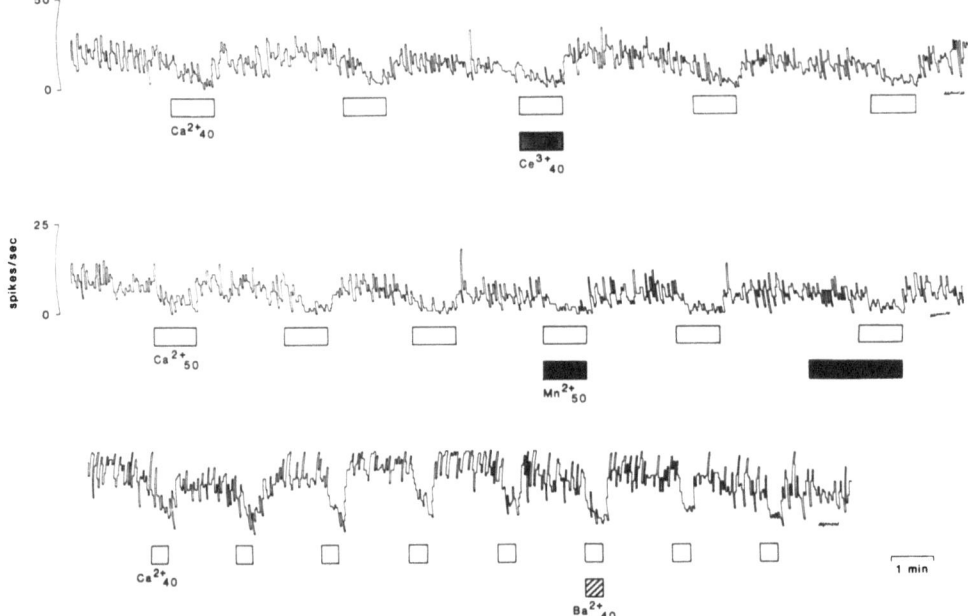

Fig. 4. Effect of cations on cortical activity. Comparison of the effect of simultaneous application of cerium (top trace), manganese (middle trace) and barium ions (lower trace) with calcium ions on neuronal activity. Each trace is a polygraph record of the firing frequency of a different cortical neuron. No interactions occurred between these ions (cf. effect of zinc ions on calcium-induced depressions shown in Fig. 3). Bars indicate iontophoretic application of cations and numbers the currents (nA) used.

those necessary for *in vitro* blockade of opiate binding. It is of interest that the distribution of zinc ions and enkephalin-like immunoreactivity in the hippocampus is identical, both being confined to the mossy fibers region (Stengaard-Pedersen et al., 1981), since this also suggests that the interaction between zinc ions and enkephalins may have physiological significance. Enkephalins are known to have epileptogenic properties (Frenk et al., 1978), perhaps as a consequence of disinhibition (Zieglgänsberger et al., 1978). Paradoxically, zinc ions appear to be able to block the convulsive actions of these opioids yet can also cause convulsions.

SUMMARY

Zinc ions, which are unevenly distributed in the CNS and can be released from nerve terminals, have been implicated as causative agents in epileptogenesis. The present study has shown that intraventricular administration to anesthetized rats causes seizure activity of the ECOG and convulsions. Since the manner in which zinc influences neuronal activity and triggers convulsions is unclear, studies were also made of its effect on spontaneous and evoked activity in the rat forebrain. It was found that iontophoretic application of zinc to cortical neurons causes slow and often prolonged increases in firing rate, usually accompanied by bursts of high frequency discharge in just under half the studies. Another cation, barium, evoked excitatory responses of a similar type and a reduction in potassium permeability may underlie the effects of both cations. In contrast, calcium,

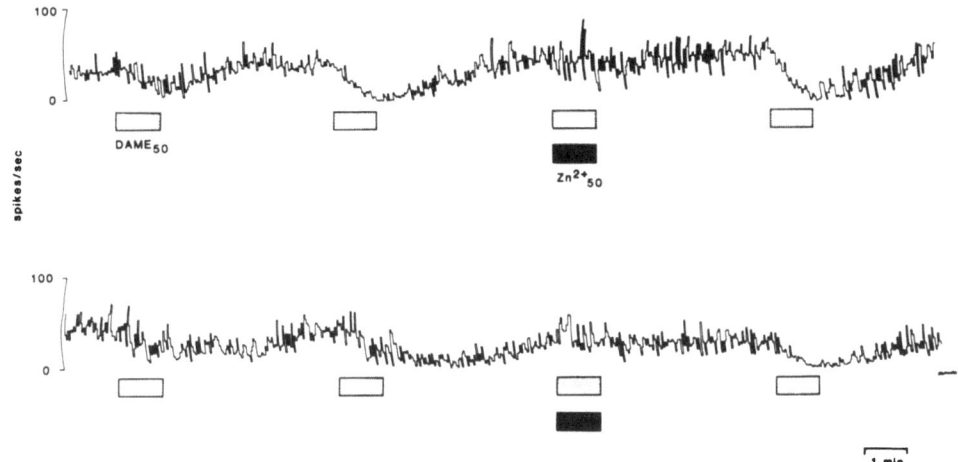

Fig. 5. Zn^{2+}/D-Ala^2-Met^5-enkephalinamide interaction in rat cortex. Continuous polygraph record of the electrical activity of a cortical neuron to show that the depressant response to the iontophoretic application of D-Ala^2-Met^5-enkephalinamide (DAME) could be blocked by simultaneous iontophoresis of zinc ions. (Bars indicate iontophoretic application and numbers the currents (nA) used).

magnesium, manganese and cerium caused short duration depressant effects. The depression induced by calcium, but not by the other cations, could be blocked by zinc. Similarly, in the hippocampus zinc depressed calcium-dependent potentiation in subfield CA3 evoked by paired-pulse stimulation of mossy fibers; excitatory effects (namely an increase in spike amplitude and appearance of multiple population spikes) were seen at higher zinc concentrations. The depressant effects of an enkephalin analog on cortical firing rate were also blocked by zinc, consistent with studies from another laboratory suggesting enkephalin/zinc interactions. In contrast, the depressant effect of GABA could not be blocked by zinc, although an antagonism has been reported in the lobster muscle.

Firm conclusions regarding the mechanism(s) underlying the triggering of seizure activity by zinc cannot yet be drawn, but the results of these studies would be consistent with an interference with calcium and/or potassium ion activity rather than with GABA binding sites.

ACKNOWLEDGEMENTS

The author is grateful to Mrs. J. Readman for excellent technical assistance and to the A.F.R.C. for financial support.

REFERENCES

Assaf, S.Y., and Chung, S.H., 1984, Release of endogenous Zn^{2+} from brain tissue during activity, Nature, 308:734.
Ben-Ari, Y., Krnjević, K., and Reinhardt, W., 1979, Hippocampal seizures and failure of inhibition, Can. J. Physiol. Pharmacol., 57:1462.
Bettger, W.J., and O'Dell, B.L., 1981, A critical physiological role of zinc in the structure and function of biomembranes, Life Sci., 28:1425.

Brewer, G.J., Aster, J.C., Knutsen, C.A., and Kruckeberg, W.C., 1979, Zinc inhibition of calmodulin: a proposed molecular mechanism of zinc action on cellular functions, Amer. J. Hematol., 7:53.

Brown, M.W., Khulusi, S.S., and Wright, D.M., 1985, Effects of zinc ions on paired-pulse potentiation in the rat hippocampal slice, J. Physiol., 360:35P

Chapman, D.B., and Way, E.L., 1980, Metal ion interactions with opiates, Ann. Rev. Pharmacol. Toxicol., 20:553.

Chung, S.H., and Johnson, M.S., 1983, Divalent transition-metal ions (Cu^{2+} and Zn^{2+}) in the brains of epileptogenic and normal mice, Brain Res., 280:323.

Crawford, I.L., Doller, H.J., and Connor, J.D., 1973, Diphenylthiocarbazone effects on evoked waves and zinc in the rat hippocampus, Pharmacologist, 15:2.

Danscher, G., Shipley, M.T., and Andersen, P., 1975, Persistent function of mossy fibre synapses after metal chelation with DEDTC (Antabuse), Brain Res., 85:522.

Donaldson, J., St.-Pierre, T., Minnich, J.L., and Barbeau, A., 1971, Seizures in rats associated with divalent cation inhibition of Na^+-K^+-ATPase, Can J. Biochem., 49:1217.

Donaldson, J., Minnich, J.L., and Barbeau, A., 1972, Ouabain-induced seizures in rats: regional and subcellular localization of ^{3}H-ouabain associated with Na^+-K^+-ATPase in brain, Can. J. Biochem., 50:888.

Donaldson, J., St.-Pierre, T., Minnich, J. L., and Barbeau, A., 1973, Determination of Na^+, K^+, Mg^{2+}, Cu^{2+}, Zn^{2+}, and Mn^{2+} in rat brain regions, Can. J. Biochem., 51:87.

Euler, C. von, 1962, On the significance of the high zinc content in the hippocampal formation, in: Physiologie de l'Hippocampe, P. Passouant, ed., Editions du Centre National de la Recherche Scientifique, Paris, p. 135.

Frankenhaeuser, B., and Hodgkin, A.L., 1957, The action of calcium on the electrical properties of squid axons, J. Physiol., 137:218.

Frenk, H.G., Urea, G., and Liebeskind, J.C., 1978, Epileptic properties of leucine- and methionine-enkephalin. Comparison with morphine and reversibility by naloxone, Brain Res., 147:327.

Haug, F.M.S., 1967, Electron microscopical localization of the zinc in hippocampal mossy fibre synapses modified by sulphide silver procedure, Histochemie, 8:355.

Heinemann, U., Lux, H.D., and Gutnick, M.J., 1977, Extracellular free calcium and potassium during paroxysmal activity in the cerebral cortex of the cat, Exp. Brain Res., 27:237.

Hesse, G.W., 1979, Chronic zinc deficiency alters neuronal function of hippocampal mossy fibres, Science, 205:1005.

Howell, G.A., Welch, M.G., and Frederickson, C.J., 1984, Stimulation-induced uptake and release of zinc in hippocampal slices, Nature, 308:736.

Kato, G., and Somjen, G.G., 1969, Effects of micro-iontophoretic administration of magnesium and calcium on neurones in the central nervous system of cats, J. Neurobiol., 2:181.

Krnjević, K., 1965, Actions of drugs on single neurones in the cerebral cortex, Brit. Med. Bull., 21:10.

Krnjević, K., Pumain, R., and Renaud, L., 1971, Effects of Ba^{2+} and tetraethylammonium on cortical neurones, J. Physiol., 215:223.

Lange, S.C., Neafsey, E.J., and Wyler, A.R., 1980, Neuronal activity in chronic ferric chloride epileptic foci in cats and monkey, Epilepsia, 21:251.

Leendertz, J., and Wright, D.M., 1983, A microcomputer based system for spike processing at low cost, J. Neurosci. Meth., 9:1.

Miller, J.J., and Baimbridge, K.G., 1983, Biochemical and immunohistochemical correlates of kindling-induced epilepsy: role of calcium binding protein, Brain Res., 278:322.

Pei, Y., Zhao, D., Huang, J., and Cao, L., 1983, Zinc-induced seizures: a new experimental model of epilepsy, Epilepsia, 24:169.

Peters, R.A., Shorthouse, M., and Walshe, J.M., 1966, Studies on the toxicity of copper 2: the behaviour of microsomal membrane ATPase on the pigeon's brain tissue to copper and some other metallic substances, Proc. Roy. Soc. Lond., 166:285.

Phillis, J.W., and Limacher, J.J., 1974, Effects of some metallic cations on cerebral cortical neurones and their interactions with biogenic amines, Can. J. Physiol. Pharmacol., 52:566.

Purpura, D.P., ed., 1972, Experimental Models of Epilepsy, Raven Press, New York.

Rosen, A.D., and Frumin, N.V., 1979, Focal epileptogenesis after intracortical hemoglobin injection, Exp. Neurol., 66:277.

Rozear, M., DeGroof, R., and Somjen, G., 1971, Effects of micro-iontophoretic administration of divalent metal ions on neurones of the central nervous system of cats, J. Pharmacol. Exp. Ther., 176:109.

Smart, T.G., and Constanti, A., 1982, A novel effect of zinc on the lobster muscle GABA receptor, Proc. Roy. Soc. B., 215:327.

Smart, T.G., and Constanti, A., 1983, Pre- and post-synaptic effects of zinc on in vitro prepyriform neurones, Neurosci. Lett., 40:205.

Stengaard-Pedersen, K., Fredens, K., and Larsson, L.I., 1981, Enkephalin and zinc in the hippocampal mossy fibre system, Brain Res., 212:230.

Stengaard-Pedersen, K., 1982, Inhibition of enkephalin binding to opiate receptors by zinc ions: possible physiological importance in the brain, Acta Pharmacol. Toxicol., 50:213.

Triggle, D.J., and Swamy, V.C., 1980, Pharmacology of agents that effect calcium: agonists and antagonists, Chest. Suppl. 1, 78:174.

Ward, A.A., Jasper, H.H., and Pope, A., 1969, Clinical and experimental challenges of the epilepsies, in: Basic Mechanisms of the Epilepsies, H.H. Jasper, A.A. Ward and A. Pope, eds., Little, Brown and Co., Inc., Boston, p. 1.

Werman, R., and Grundfest, H., 1961, Graded and all-or-none electrogenesis in arthropod muscle. II. The effects of alkali-earth and onium ions on lobster muscle fibres, J. Gen. Physiol., 44:997.

Woodhull, A.M., 1973, Ionic blockade of sodium channels, J. Gen. Physiol., 61:687.

Wright, D.M., 1984, Zinc - effect and interaction with other cations in the cortex of the rat, Brain Res., 311:343.

Wu, J.Y., and Roberts, E., 1974, Properties of brain l-glutamate decarboxylase: inhibition studies, J. Neurochem., 23:759.

Zieglgänsberger, W., Siggins, G., French, E., and Bloom, F., 1978, Effects of opioids on single unit activity, in: Characteristic and Function of Opioids, Developments in Neuroscience, Vol. 4, J. M. Van Ree and L. Terenius, eds., Elsevier/North-Holland, Amsterdam, p. 75.

Zuckerman, E.C., and Glaser, G.H., 1968, Hippocampal epileptic activity induced by local ventricular perfusion with high-potassium cerebrospinal fluid, Exp. Neurol., 20:87.

RELATIONSHIP OF GLUTAMIC ACID AND ZINC TO KINDLING OF THE RAT AMYGDALA:

AFFERENT TRANSMITTER SYSTEMS AND EXCITABILITY IN A MODEL OF EPILEPSY

I.L. Crawford

University of Texas and Veterans Administration
Department of Neurology and Epilepsy Center
Dallas, Texas 75216

INTRODUCTION

The seizure threshold in limbic structures is low, most notably in the amygdaloid nuclear complex. The susceptibility of the amygdala to epileptiform activity can be enhanced by several manipulations including repeated electrical stimulation (kindling) or repeated injections of certain chemical agents. The progressive change in electrical activity is characterized by gradual increasing afterdischarge, spread of seizure activity to remote brain regions and, eventually, generalized convulsions. A goal of this research project is to identify neurochemical mechanisms involved in the cause and termination of electrographic seizures and the spread of abnormal activity.

Higher serum levels of copper have been measured in epileptic patients and beagles than in healthy control persons or dogs matched for age and sex (Brunia and Buyze, 1972). Barbeau and Donaldson (1974) reported significantly lower serum zinc concentrations in treated epileptics than in control subjects. Chronic oral treatment of seizure-prone baboons with D-penicillamine reduced induction of seizures and lowered abnormally elevated plasma zinc (Alley et al., 1981). Chung and Johnson (1983) studied higher levels of zinc and copper in the brain of audiogenic mice. Physiologic (Smart and Constanti, 1983; Wright, 1984) and neurochemical (Slevin and Kasarskis, 1985) studies suggest zinc may act directly on neuronal membranes. These observations raise the possibility that concomitant changes in plasma and brain concentrations of zinc may contribute to hyperexcitability in temporal lobe structures; especially when zinc in the brain appears focally concentrated in regions known to have a low threshold for epileptic seizures (Isaacson, 1982), both for man and experimental animals.

Maske (1955) suggested that a metal was localized endogenously in high concentrations within the limbic system when he noted that parts of the hippocampal formation had a highly selective affinity for dithizone. The _in vivo_ deposition of a zinc-dithizonate complex was supported by subsequent observations that ^{65}Zn accumulated in the same hippocampal area as that stained by dithizone (Hassler and Soremark, 1968; Crawford and Connor, 1975). Heavy metal staining was shown by Timm (1958) to be most intense in a zone subjacent to the pyramidal cell body layer. The location of the stained area corresponds with the hippocampal mossy fiber system. Glutamic acid is a putative transmitter released by granule cells

(Crawford and Connor, 1973); identified immunohistochemically in mossy fiber terminals (Storm-Mathisen et al., 1983); and actively taken up into mossy boutons (Storm-Mathisen, 1981; Szutowicz et al., 1983).

The amygdaloid complex, a nuclear mass in the temporal lobe, is another brain area which stains heavily for metals (Haug, 1974). The reaction intensity varies in parts of the basal, cortical, lateral and central nuclei. Hall (1975) speculated that at least part of the zinc content of the amygdala is located in synaptic endings of fibers arising in the pyriform cortex. Axons from pyriform neurons are thought to ascend in the external capsule and provide short collaterals which form boutons de passage on dendrites of cells in the lateral and other strongly stained amygdalar nuclei. The transition metal stain is located in synaptic vesicles within asymmetric synaptic contacts in the amygdala and hippocampus (Haug, 1967; Perez-Clausell and Danscher, 1985).

The histochemical evidence from in vivo and in vitro staining patterns suggests zinc is differentially distributed in brain. Although dithizone is not strictly specific for any metal, its stability constant with zinc is high and zinc-dithizonate can be identified by a characteristic absorption spectrum. In all mammalian brains thus far examined, including human, quantitative chemical analyses have consistently shown the concentration of zinc in the amygdala and hippocampus is one of the most distinguishing neurochemical properties of the limbic system structures that invariably contain relatively more zinc compared to other brain areas (Danscher et al., 1975; Crawford and Harris, 1984). By way of contrast, other transitional metals such as copper, manganese, and iron are more concentrated elsewhere in the nervous system (Hui et al., 1979). The chemical composition of the hippocampal formation is not remarkably different from that of other CNS regions with regard to protein, carbohydrate, lipid and water content (Crawford, 1983).

The hypothesis tested by the experiments in our current studies is that glutamic acid and zinc are involved in the kindling process. Central administration of zinc (Donaldson et al., 1971; Pei et al., 1983), or analogs of excitatory amino acids (Meldrum, 1984) can induce electrographic seizures. One of the major mechanisms that terminates the action of glutamate is a high affinity uptake system. The epileptogenic actions of glutamate may be mediated by activation of specific subtypes of amino acid receptors. Frequent stimulation of receptors by repeated release of endogenous excitatory transmitters may occur in the kindled amygdala. We have made selective lesions to determine whether kindling involves the release and uptake of endogenous dicarboxylic acids that are epileptogenic. Another series of studies was designed to determine whether the cholinergic excitatory transmitter system afferent to the amygdala was involved, since transitional metal ions affect the uptake of norepinephrine and choline by brain synaptosomes (Prakash et al., 1973) and the metals are present in cholinergic synaptic vesicles (Schmidt et al., 1980).

In our studies, which assess the effects of selective lesions in afferents from the pyriform cortex to the amygdala, we reasoned that, if these terminal afferents are glutamergic and zinc-rich, the high affinity glutamate uptake and zinc content of amygdalar tissue should be reduced by the lesions, and if the afferents are involved in kindling, the development and spread of seizures would be affected. The results gained from these investigations provide information on the epileptogenic properties of the excitatory amino acids, and the role zinc and the amino acid transmitters may have in experimental seizure models as well as in clinical epilepsy.

The electrical stimulation of zinc-rich limbic structures in unanesthetized, freely moving animals provides an opportunity to examine trace metals in a model of epileptogenesis which closely approximates the human condition of focal temporal lobe (complex partial) epilepsy. These studies should also identify neurochemical mechanisms involved in the cause and termination of electrographic seizures.

KINDLING

Of the various animal models of epilepsy, the kindling model (Goddard et al., 1969) offers the best advantages for studying the role of neurotransmitters in the development and maintenance of convulsive activity. In this model, an initially subconvulsant electrical stimulus is applied for a brief period at regular intervals (usually once daily). When this is done, there is a progressive change in electrical acitivty, with gradually increasing afterdischarge, and then spread of afterdischarge to remote brain regions. Eventually, clinically apparent seizures develop, with a pattern of development and spread specific for the brain region stimulated. Once a fully developed seizure occurs, it will occur with each successive stimulation, even if the successive stimuli are not administered for a period of several months, indicating that there has been a persistent change in the convulsive threshold to this type of electrical stimulation. The area of the brain which may be kindled most rapidly is the amygdalar complex. Daily electrical stimulation of the basolateral amygdalar nucleus results in the development of gradually increasing afterdischarge, and is associated with a series of progressively more complicated behavioral seizures (Racine, 1972). The fully developed seizure looks very much like certain types of temporal lobe seizures in man, with vigorous running movements of the forelimbs, lip-smacking, chewing, and rearing and falling backwards. Goddard found that the kindling effect could be prevented by a destructive lesion, restricted to the basolateral nucleus, suggesting that the chemical change underlying the altered electrical threshold was probably restricted to the area stimulated.

This model offers many advantages to the investigator interested in studying the biochemistry and pharmacology of epilepsy. In contrast to chemical models, no drug is injected, and therefore one does not have to distinguish the relevant from the non-relevant effects of the drug. Thus far, few morphological alterations have been found in the kindled focus, in contrast to injection of heavy metals or thermal injury to produce convulsions, procedures which result in neuronal loss and glial proliferation. Perhaps most important, the model allows one to explore the basis of the underlying change in excitability in the absence of biochemical and electrical events caused by the seizure activity. In all these features, kindling more closely resembles the situation in human epilepsy than do most other animal models of epilepsy.

Much evidence of a pharmacological nature has accumulated, indicating that neurotransmitters are involved in the epileptic process and that manipulation of neurotransmitters by drugs results in alteration of seizure thresholds (Crawford, 1979). Since transmission of information between mammalian nerve cells is almost exclusively via chemical transmitters, the probability is high that alterations in the synthesis, uptake, or release of one of these substances is involved in the development of kindling. Furthermore, it is quite possible that currently available antiepileptic drugs exert their actions via their effects on neurotransmitter systems. For example, valproic acid elevates concentrations of brain GABA, and benzodiazepines may exert their effects by enhancing the effectiveness of GABA at central synapses (Meldrum, 1984).

If one could interfere with excitatory neurotransmission to the amygdala, it seems a reasonable possibility that this would result in decreased amygdalar excitability and therefore in elevated seizure threshold and longer time to kindling. If such experiments were successful in thus altering this animal model of epilepsy, they could ultimately lead to new approaches to control of epilepsy in man. We therefore designed experiments to test whether interference with excitatory transmission to the amygdala would result in retardation or abolition of the kindling process.

EXCITATORY TRANSMITTERS AND AMYGDALAR KINDLING

Intra-amygdalar injection of nanomolar glutamate produces prolonged afterdischarge and behavioral convulsions (Crawford and Wooten, 1979). Kainic acid, a rigid analog of glutamic acid and a glutamergic agonist, causes prolonged focal status epilepticus when injected into the amygdala (Ben-Ari and Lagowska, 1978) but kindling with glutamate has not yet been reported.

Grossman (1963) first noted that injection of acetylcholine or cholinergic agonists into the basolateral amygdaloid complex caused epileptiform spike discharges and behavioral psychomotor seizures. Vosu and Wise (1975) produced kindling by amygdalar injection of cholinergic agonists. Arnold et al. (1973) found that atropine retarded amygdaloid kindling and that reserpine, which depletes brain catecholamines and serotonin, facilitated seizure development. However, when drug administration ceased, the rats required additional stimulations before they re-developed motor seizures. GABA antagonists also produce kindling. Opioids and naloxone, an antagonist of the enkephalins (endogenous CNS opiates) have effects on kindling (Caldecott-Hazard et al., 1982; Le Gal La Salle et al., 1977) but it is not yet clear how neurotransmitter mechanisms are involved in the effects. The content of enkephalin (Vindrola et al., 1981) and zinc (Mody and Miller, 1985) are increased in brain tissue from kindled rats.

BIOCHEMICAL CHANGES ASSOCIATED WITH KINDLING

Just as anatomical studies have revealed no remarkable morphological changes associated with kindling, remarkably few biochemical changes have been found. The few persistent changes noted have been quantitatively minor. Various authors have noted alterations involving dopamine, with decreased levels, increased turnover, increased number of receptors, and decreased tyrosine hydroxylase activity (Kalichman, 1982). Since kindling is not affected by lesions of the substantia nigra, the relevance of these findings to the process of kindling is unclear, and they probably represent effects rather than an underlying pathophysiological phenomenon. McNamara (1978) has reported a transient decrease in muscarinic binding sites and a later transient decrease in noradrenergic binding sites, but the relevance of these effects is unclear, since the kindling phenomenon is persistent. No long-lasting change was found in norepinephrine levels, glutamic acid decarboxylase, choline acetylase, glutaminase, dopamine beta-hydroxylase, or taurine synthetase (Kalichman, 1982). A selective increase in the number of quisqualate-sensitive glutamate binding sites was measured by Savage et al. (1984) when they studied hippocampal synaptic membranes after angular bundle kindling. The reversibility of the effect suggested that glutamate may be involved in the induction rather than in the maintenance of hippocampal kindling.

LESION AND STIMULATION STUDIES IN AMYGDALAR KINDLING

Goddard and his coworkers (1969), in their initial studies, found that the kindled convulsive response could be abolished by making a small electrolytic lesion (less than 0.5 mm in diameter) at the electrode tip. This indicates that the effect is probably linked to a local change in the amygdala, rather than resulting from remote effects of current spread or distant synaptic effects. Furthermore, a rough correlation could be made between the ease of kindling of various brain regions and the density of their connections to the amygdala.

Only a few studies have been reported with regard to the effect of brain lesions on amygdalar kindling, and even fewer brain lesions have been made with the objective of selectively modifying neurotransmitter input. Corcoran et al. (1974) found that intraventricular injection of 6-hydroxydopamine, which destroys catecholamine terminals in the brain, resulted in prolongation of afterdischarge and reduced the duration of time until behavioral kindling. Lesions of substantia nigra or median raphe, which presumably decrease dopamine or serotonin selectively, have not had consistent effects on time to kindling. Stimulation of the median raphe, which presumably release serotonin in the amygdala, was found to shorten time to kindling in one study (Racine and Cosina, 1979), and to prolong time to kindling in another (Kovacs and Zoll, 1974). Medial forebrain bundle stimulation, which would be expected to release norepinephrine, dopamine, and serotonin in the amygdala, prolonged time to kindling (Dubicka et al., 1978). Wada and Wake (1977) have found that lesions of the corpus callosum and anterodorsal cortex shorten time to kindling, and that lesions of the midbrain reticular formation block propagation of the kindled seizures, while lesions of anteromesial cortex, orbital cortex, and nonspecific thalamus have no effect on time to kindling. Lesions restricted to ventrobasal thalamus and interpeduncular nucleus (Ackerman and Engel, 1978) are of particular interest, since they are associated with an increased time to kindling. Little is known of the neurotransmitter(s) involved in these latter lesions.

COMBINED LESION AND BIOCHEMICAL STUDIES OF KINDLING

In our studies, we have combined electrophysiological and biochemical approaches to examine changes following lesions of specific excitatory neurotransmitter inputs to the amygdala. We feel this approach is particularly important because we can assess marked variation in the effectiveness of lesions in producing biochemical changes.

Our initial studies were concerned with looking at changes in neurotransmitter and cyclic nucleotide levels in the amygdala associated with the kindling process and excitatory transmitters (Walker et al., 1981). Such studies formed an essential background for looking at the effect of lesions on the physiology and biochemistry of kindling, and stimulated subsequent studies. We measured the content of cyclic AMP and cyclic GMP in control and kindled cortex and amygdala: no significant differences were observed. This suggests that elevated cyclic AMP levels reported to accompany experimental seizures of various kinds (Ferrendelli and Kinscherf, 1977) are a consequence of epileptiform activity rather than being fundamental to the genesis of seizure states. This finding illustrates an advantage of the kindling model over other models of epilepsy in distinguishing cause from effect.

We then explored the possibility that lesions of neurotransmitter afferent systems may affect the kindling process and increase understanding of it. The left medial forebrain bundle (MFB) was lesioned and amygdalar electrodes were placed for kindling on the same side (Wooten et al., 1978). Control animals were implanted with electrodes but not lesioned. One day after the third successive stage 5 seizure, significant reductions in ipsilateral dopamine (90%), norepinephrine (68%), and serotonin (54%) were noted in the striatum, with similar but less marked reductions in cortex. The animals with MFB lesions progressed through the usual 5 stages of behavior kindling without any change in days to complete kindling. However, stereotypical seizure behavior was attenuated, particularly in the early stages of development. Interestingly, the average duration of afterdischarge was significantly longer for rats with MFB lesions than for non-lesioned control animals. The MFB lesions evidently dissociated kindling and afterdischarge, which would appear to contradict previous suggestions that the two processes are intimately linked (Racine, 1972). It also suggests that one of the monoamines is involved in the feedback inhibition of amygdalar afterdischarge. The approach we initially took has promise of yielding some intriguing results. However, it provided few clues to suggest a favorable way to significantly modify the kindling process. The MFB lesions did not prolong kindling and selective lesions of the serotonergic or noradrenergic input to the amygdala appear to facilitate seizure development rather than retard it.

CHOLINERGIC AFFERENTS AND AMYGDALAR KINDLING

Acetylcholine (ACh) is widely distributed in the brain, with marked regional variations (Cheney et al., 1975). Iontophoresis of ACh onto brain cells may cause excitation with rapid onset (nicotinic), or excitation with slow onset (muscarinic; Burchfiel et al., 1979). Physiological evidence suggests that ACh is a transmitter in several central pathways including the septo-hippocampal tract, the habenulo-interpeduncular tract, and in striatal interneurons. With regard to the amygdala, lesions of the magnocellular preoptic nuclei reduce choline acetylase activity (ChAT) in the basolateral amygdala by 65%, indicating that this nuclear mass sends a major cholinergic projection to the amygdala, probably via the ventral amygdalofugal pathway (Emson, 1979). We tested the hypothesis that cholinergic mechanisms underlie the kindling phenomenon by lesioning the lateral preoptic area, reasoning that if the hypothesis were true, such lesions would delay or abolish the kindling process (Walker et al., 1985). Kindling in unlesioned animals increased acetylcholine levels about two-fold in the amygdala and hippocampus, without altering the activity of ChAT.

In rats with preoptic lesions, no alterations of the kindling process were detected despite a marked reduction (50%) in acetylcholine concentration and ChAT activity in the amygdala. Although the lesion caused a significant reduction in cholinergic input to the amygdala, no change was produced in the duration of time to kindling. The animals with lesions of the lateral preoptic area, which reduced ChAT activity by 30% or more kindled in 9 ± 2 days, whereas animals with a sham lesion kindled in 11 ± 2 days; not a statistically significant difference.

Our data agree with other studies (Kalichman, 1982) indicating that alterations in the cholinergic system do accompany the kindling process. Ours was the first study in which acetylcholine itself was measured and its concentration was increased in both amygdala and hippocampus. However, the lack of effect of lesions in a major excitatory input on the kindling process suggests that the cholinergic system is affected, but it may not be the primary system involved in kindling.

GLUTAMERGIC AFFERENTS AND KINDLING

The amino acids, rather than the monoamines or acetylcholine, are emerging as transmitter substances of special interest in relation to epilepsy (Crawford, 1979; Meldrum, 1984). Experimental evidence raises the possibility that the excitatory and inhibitory amino acids, through their synaptic actions, exert powerful effects on neuronal excitability and might participate in neuronal paroxysms. Epilepsy may be a disease of neurotransmitter imbalance between excitatory and inhibitory amino acid systems.

Glutamate satisfies many criteria as an excitatory neurotransmitter. It is unevenly distributed in nervous tissue (Perry, 1971), and there exists a relatively specific, high affinity, sodium-dependent uptake system for glutamate in nerve endings (aspartate is accumulated via the same system; Wofsey et al., 1971). The most definitive evidence comes from experiments which indicate that lesions of specific neuronal pathways result in a decrease of glutamate (but not of other amino acids) as well as a decrease in high affinity uptake of glutamate (Fonnum, 1978). Iontophoresis of glutamate powerfully excites virtually all neurons thus far studied (Sawada et al., 1983), and accumulating evidence suggests that it is an excitatory physiological transmitter in certain pathways (Storm-Mathisen, 1981). Another criterion for transmitter function is calcium-dependent release, which is present in the intact hippocampus, and absent in the hippocampus following a lesion of the perforant path (Storm-Mathisen, 1977). In the cerebral cortex, electron microscopic studies suggest that a very high proportion (14-15%) of the nerve terminals accumulate glutamate by a high affinity mechanism.

Previous work has provided neurochemical evidence that aspartate and glutamate may be neurotransmitters released by terminals in several archicortical and paleocortical pathways. The pyriform cortex is a seizure sensitive area with considerable involvement in clinical complex partial seizures (Penfield and Kristiansen, 1951) and in the development of subcortical kindled convulsions (McIntyre and Wong, 1985). Interictal discharges usually begin first in the amygdala-pyriform region regardless of the location of the initial kindling site. In man, the uncinate gyrus, a homology of the rat pyriform cortex, has been repeatedly implicated in temporal lobe epilepsy (Falconer et al., 1964).

Since removal of pyriform cortex markedly decreased high affinity glutamate uptake in amygdala (Walker and Fonnum, 1983), we designed experiments to selectively lesion glutamergic input to the amygdala and then study the resultant changes in kindling and its associated phenomena (Crawford et al., 1983). Surgical details for the removal of the pyriform cortex and dissection of the amygdala have been reported previously (Walker et al., 1981). High affinity glutamate uptake was determined according to Divac et al. (1977) and is based on the premise that terminals of glutamergic neurons have a system for high affinity uptake of glutamic acid that is sodium dependent. This uptake can be easily distinguished from the sodium independent glutamate binding process at this concentration range. Briefly, the uptake media was added in 0.5 ml aliquots to 5 ml glass tubes. Homogenates (10 µl) were added to each tube, and preincubated for 20 minutes. Tritium labeled glutamate (40.3 Ci/mmol) was added to each tube (final glutamate concentration: 0.1 µM). Uptake was stopped after 3 minutes by addition of ice cold buffer (0.9% NaCl; 500 mg/l bovine serum albumin). Contents of each tube were then filtered by suction on 0.45 µm nitrocellulose membranes and were washed twice with 4 ml aliquots of the same buffer system.

Our studies indicated that removal of pyriform cortex markedly decreases high affinity glutamate uptake in amygdala while amygdalar uptake is not affected by other cortical lesions. Thus, we were able to selectively lesion glutamergic input to the amygdala and then look at the resultant changes in kindling and its associated phenomena. Pyriform ablation markedly prolonged the kindling time to about five weeks compared to less than two weeks for controls (p < 0.001). In the amygdala of kindled sham operated rats, high affinity glutamate uptake was not significantly different from that in the contralateral amygdala of the same animals (27.5 ± 3.4 vs 21.7 ± 2.1 μmol/hr/g wet weight). In contrast, removal of the pyriform cortex reduced high affinity glutamate uptake to less than half of that in the amygdala of the unoperated side.

In a second series of experiments, rats were kindled and then the pyriform cortex was ablated. Sham operated controls rekindled within 2 days, while experimental rats required 7 ± 2 days to regain stage 5 seizures. These findings indicate the potential importance of a glutamergic input in the development of the kindling process and its less critical role in the persistence of kindling. The results also suggest that the influence of glutamergic afferents may be largely postsynaptic.

RELATIONSHIP OF AMYGDALAR EXCITABILITY TO ZINC AND GLUTAMERGIC AFFERENTS

Reasoning that the amygdala is a zinc rich limbic structure and that the transition metals may be demonstrated histochemically, we used a modification (Sloviter, 1985) of the Timm method to stain the brains from a series of rats in which the pyriform cortex had been ablated unilaterally. The silver sulfide reactivity in the ipsilateral amygdalar nuclei was markedly less intense than that on the contralateral side. Since other metals are visualized by the Timm's stain, we injected a similar series of experimental rats with diphenylthiocarbazone (dithizone) which is a more specific intravital stain that forms zinc-dithizonate *in vivo*. The red chromogen was much less obvious in the ipsilateral amygdala than in the unoperated side or in the hippocampus.

It seemed probable that any drug with strong metal binding properties has the potential for profoundly influencing function in limbic structures containing zinc rich neuronal elements. The electrophysiologic effects of dithizone was studied in awake, unrestrained rats (Crawford et al., 1973; Drust and Crawford, 1984) and *in vitro* (Doller and Crawford, 1984).

Intraperitoneal injections of dithizone (70 mg/kg) in control rats caused overt motor responses similar to those elicited by amygdaloid stimulation, primarily very stereotyped movements of the face and neck muscles such as sniffing, chewing, head turning and a suspiciously high correlation with symptoms of seizure activity. All of the mentioned responses are characterized as rhythmic and as lacking strict localization within the amygdala: the area with the lowest threshold for face and jaw movements, the basolateral nucleus is also the area with the lowest seizure threshold. Some motor responses were accompanied by either clonic twitches or partial seizures when the dose was increased to 100 mg/kg. Electrographic records showed that changes occurred in the amygdala prior to similar abnormal activity in the hippocampus or frontal cortex. Epileptiform 'burst' discharges recorded from the amygdala in awake rats occurred concomitant with intravital staining of amygdalar sites. Dithizone caused 'Stage 4' seizures in kindled rats that had received no electrical stimulation within one week prior to injection. We have reported similar results on studies in the hippocampus. The action of zinc chelation may regulate or affect the excitability of the amygdala and hippocampus. Our results indicate that dithizone can produce time and dose dependent electrographic seizures that seem to originate in the amygdala and hippocampus.

The physical chemistry of the group IIB metals may provide clues to predict function and pharmacology in neural tissues. Zinc has the capacity to form coordinate bonds with functional groups in biologically active molecules present in neural tissues, and is usually complexed to organic ligands rather than free in the cytosol or interstitial fluids. Coordination enables zinc to bind reversibly and with a particular stereochemistry with bioactive ligands and chelating drugs which are electron donors.

If zinc functions in neurons, such as in synaptic transmission, and is a locus for drug action, then it would be important to identify agents that affect the morphology of zinc-rich fibers. Such chemicals could be tools for probing physiologic and neurochemical mechanisms. A number of studies have dealt with the effects of systemic administration of drugs on metallohistochemistry. Periodic treatment with dithizone prevented mossy fiber staining for histochemically reactive metals (Danscher and Haug, 1971). This was interpreted as metal depletion, although zinc concentration in brain tissues was not quantitatively analyzed. However, this conclusion has not been confirmed in subsequent quantitative studies from other laboratories (Fjerdingstad et al., 1977), including ours. Single intraperitoneal injections of either alloxan or oxine (both are diabetogenic and are structural analogs of diphenylhydantoin) intensified the Timm's stain in the hippocampus of rats. Disulfiram or its metabolite, diethyldithiocarbamate, diminished the Timm's reaction, while calcium disodium edetate had no effect (Danscher et al., 1973). Trace metal content of the whole brain is markedly increased by several chelating agents (Lakomaa et al., 1982).

All of these compounds as metal chelators could presumably alter the metallochemistry of the metal rich pathways by removing, masking or sequestration of zinc and may, in turn, affect neurotransmission. The histological patterns revealed by the Timm method in the hippocampus and the amygdala-pyriform cortical areas have a remarkable correspondence to patterns of termination of putative glutamergic pathways in which a zinc-glutamate complex may be a site for drug action (Crawford, 1983). These observations raise the possibility that concomitant changes in plasma and brain concentrations of zinc may lead to hyperexcitability in temporal lobe structures, especially when zinc in the brain appears largely concentrated in regions known to have a low threshold for epileptic seizures for both humans and experimental animals.

POTENTIAL SIGNIFICANCE

If our hypothesis is proven correct, these experiments could ultimately prove of great significance to the study of neurochemical mechanisms of epilepsy and perhaps lead to new approaches to the therapy of epilepsy in man. If it is found that the kindling process leads to an increase in levels of one or another excitatory transmitters in the amygdala, it would be the first biochemical change to suggest a logical mechanism for the increased excitability caused by kindling. If lesions of specific excitatory afferents result in prolongation or abolition of the kindling response, it might ultimately be possible to use a similar approach in patients with intractable seizures. If pharmacologic blockade of the excitatory afferents proves effective in prolonging or blocking kindling, new drug approaches to prevention or treatment of seizures in man might result in time.

REFERENCES

Ackerman, R.F., and Engel, J., 1978, Lesions of the interpeduncular nucleus retard development of amygdaloid-kindled seizures in rats, Soc. Neurosci. Abstr., 4:139.

Alley, M.C., Killam, E.K., and Fisher, G.L., 1981, The influence of D-penicillamine treatment upon seizure activity and trace metal status in the sengalese baboon, Papio Papio, J. Pharmacol. Exp. Ther., 271:138.

Arnold, P.S., Racine, R.J., and Wise, R.A., 1973, Effects of atropine, reserpine, 6-hydroxydopamine, and handling on seizure development in the rat, Exp. Neurol., 40:457.

Barbeau, A., and Donaldson, J., 1974, Zinc, taurine and epilepsy, Arch. Neurol., 30:52.

Ben-Ari, Y., and Lagowska, J., 1978, Action epileptogene induite par des injections intra-amygdaliennes d'acide kainique, C.R. Acad. Sc. Paris, 287:813.

Brunia, C.H.M., and Buyze, G., 1972, Serum copper levels and epilepsy, Epilepsia, 13:621.

Burchfiel, J.L., Duchowny, M.S., and Duffy, F.H., 1979, Neuronal supersensitivity to acetylcholine by kindling in the rat hippocampus, Science, 204:1096.

Caldecott-Hazard, S., Shavit, Y., Ackerman, R.F., Engel, J. Jr., Frederickson, R.C.A., and Liebeskind, J.C., 1982, Behavioral and electrographic effects of opioids on kindled seizures in rats, Brain Res., 251:327.

Cheney, D.L., LeFevre, H.F., and Racagni, G., 1975, Choline acetyltransferase activity and mass fragmentographic measurement of acetylcholine in specific nuclei and tracts of rat brain, Neuropharmacology, 14:801.

Chung, S.H., and Johnson, M.S., 1983, Divalent transition-metal ions (Cu^{2+} and Zn^{2+}) in the brains of epileptogenic and normal mice, Brain Res., 280:323.

Corcoran, M.E., Fibiger, H.C., McCaughran, J.A., and Wade, J.A., 1974, Potentiation of amygdaloid kindling and metrazol-induced seizures by 6-hydroxydopamine in rats, Exp. Neurol., 40:471.

Crawford, I.L., and Connor, J.D., 1973, Localization and release of glutamic acid in relation to the hippocampal mossy fibre pathway, Nature, 244:442.

Crawford, I.L., Doller, H.J., and Connor, J.D., 1973, Diphenylthiocarbazone effects on evoked waves and zinc in the rat hippocampus, Pharmacologist, 15:197.

Crawford, I.L., and Connor, J.D., 1975, Zinc and hippocampal function, J. Orthomol. Psychiat., 4:39.

Crawford, I.L., 1979, Neurotransmitters: relevance to neurologic disease and therapy, in: The Treatment of Neurological Diseases, R. Rosenberg ed., Spectrum, New York, p. 525.

Crawford, I.L., and Wooten, W.C., 1979, Kainic acid elicits electrographic epileptiform activity after central and parenteral administration to awake rats, Soc. Neurosci. Abstr., 5:191.

Crawford, I.L., 1983, Zinc and the hippocampus: histology, neurochemistry, pharmacology and putative functional relevance, in: Neurobiology of the Trace Elements, I.E. Dreosti, and R.M. Smith, eds., Humana Press, New Jersey, p. 163.

Crawford, I.L., Walker, J.E., Homan, R.W., and Barletta, Maria, 1983, Glutamergic afferents from pyriform cortex to amygdala are involved in electrical kindling of the rat, Soc. Neurosci. Abstr., 9:762.

Crawford, I.L., and Harris, N.F., 1984, Distribution and accumulation of zinc in whole brain and subcellular fractions of hippocampal homogenates, in: The Neurobiology of Zinc. Part A: Physiochemistry, Anatomy, and Techniques, C.J. Frederickson, G.A. Howell, and E.J. Kasarskis, eds., Alan R. Liss, New York, p. 157.

Danscher, G., and Haug, F.M.S., 1971, Depletion of metal in the rat hippocampal mossy fibre system by intravital chelation with dithizone, Histochemie, 28:211.

Danscher, G., Haug, F.M.S., and Fredens, K., 1973, Effect of diethyldithiocarbamate (DEDTC) on sulphide silver stained boutons, Exp. Brain Res., 16:521.

Danscher, G., Hall, E., and Fredens, K., 1975, Heavy metals in the amygdala of the rat: zinc, lead, and copper, Brain Res., 94:167.

Divac, I., Fonnum, F., and Storm-Mathisen, J., 1977, High affinity uptake of glutamate in the terminals of corticostriatal axons, *Nature*, 266:377.

Doller, H.J., and Crawford, I.L., 1984, Effect of dithizone on transmission in the hippocampal slice preparation, in: *The Neurobiology of Zinc. Part B: Deficiency, Toxicity, and Pathology*, C.J. Frederickson, G.A. Howell, and E.J. Kasarskis, eds., Alan R. Liss, New York, p. 163.

Donaldson, J., St. Pierre, T., Minnich, J.L., and Barbeau, A., 1971, Seizures in rats associated with divalent cation inhibition of Na^+-K^+ ATPase, *Can. J. Biochem.*, 49:1217.

Drust, E.G., and Crawford, I.L., 1984, Effects of dithizone on spontaneous electrical activity recorded from the rat hippocampus in vivo, in: *The Neurobiology of Zinc. Part B: Deficiency, Toxicity, and Pathology*, C.J. Frederickson, G.A. Howell, and E.J. Kasarskis, eds., Alan R. Liss, New York, p. 155.

Dubicka, I., Frank, J.M., and McCutcheon, B., 1978, Attenuation of a convulsive syndrome in the rat by lateral hypothalamic stimulation, *Physiol. Behav.*, 20:31.

Emson, P.C., Paxinos, G., Le Gal La Salle, G., Ben-Ari, Y., and Silver, A., 1979, Choline acetyltransferase and acetylcholinesterase containing projections from the basal forebrain to the amygdaloid complex of the rat, *Brain Res.*, 165:271.

Falconer, M.G., Serafetinides, E.A., and Corsellis, J.A.N., 1964, Etiology and pathogenesis of temporal lobe epilepsy, *Arch. Neurol.*, 10:233.

Ferrendelli, J.A., and Kinscherf, D.A., 1977, Cyclic nucleotides in epileptic brain: effects of pentylenetetrazol on regional cyclic AMP and cyclic GMP levels in vivo, *Epilepsia*, 18:525.

Fjerdingstad, E., Danscher, G., and Fjerdingstad, E.J., 1977, Changes in zinc and lead content of rat hippocampus and whole brain following intravital dithizone treatment as determined by flameless atomic absorption spectrophotometry, *Brain Res.*, 130:369.

Fonnum, F., ed., 1978, *Amino Acids as Chemical Transmitters*, Plenum Press, New York.

Goddard, G.V., McIntyre, D.C., and Leech, C.K., 1969, A permanent change in brain function resulting from daily electrical stimulation, *Exp. Neurol.*, 25:295.

Grossman, S.P., 1963, Chemically induced epileptiform seizures in the cat, *Science*, 142:409.

Hall, E., 1975, The anatomy of the limbic system, in: *Neural Integration of Physiological Mechanisms and Behavior*, G.J. Morgenson and F.R. Calaresu, eds., Univ. of Toronto Press, Toronto, p. 68.

Hassler, O., and Soremark, R., 1968, Accumulation of zinc in mouse brain, *Arch. Neurol.*, 19:117.

Haug, F.-M.S., 1967, Electron microscopical localization of zinc in hippocampal mossy fibre synapses by a modified sulfide silver procedure, *Histochemie*, 8:355.

Haug, F.-M.S., 1974, Light microscopical mapping of the hippocampal region, pyriform cortex, and cortico-medial amygdaloid nuclei of the rat with Timm's sulphide silver method. Area dentata, hippocampus and subiculum, *Z. Anat. Entwickl. Gesch.*, 145:1.

Hui, K.-S., Davis, B.A., and Boulton, A.A., 1979, Mass spectrometric identification of Cu, Zn, Fe, Co, Mn, Mg, and Pb in mammalian brain, *J. Neurosci. Res.*, 4:169.

Isaacson, R.L., 1982, *The Limbic System*, Plenum Press, New York.

Kalichman, M.W., 1982, Neurochemical correlates of the kindling model of epilepsy, *Neurosci. Biobehav. Rev.*, 6:165.

Kovacs, D.A., and Zoll, J.G., 1974, Seizure inhibition by median raphe nucleus stimulation in the rat, *Brain Res.*, 70:165.

Lakomaa, E.L., Sato, S., Goldberg, A.M., and Frazier, J.M., 1982, The effect of sodium diethyldithiocarbamate treatment on copper and zinc concentrations in rat brain, *Toxicol. Appl. Pharmacol.*, 65:286.

Le Gal La Salle, G., Calvino, B., and Ben-Ari, Y., 1977, Morphine enhances amygdaloid seizures and increases inter-ictal spike frequency in kindled rats, Neurosci. Lett., 6:255.

Maske, H., 1955, Über den topochemischen Nachweis von Zink im Ammonshorn verschiedener Säugetiere, Naturwissenschaften, 42:424.

McIntyre, D.C., and Wong, R.K.S., 1985, Modification of local neuronal interactions by amygdala kindling examined in vitro, Exp. Neurol., 88:529.

McNamara, J.O., 1978, Muscarinic cholinergic receptors participate in the kindling model of epilepsy, Brain Res., 154:415.

Meldrum, B., 1984, Amino acid neurotransmitters and new approaches to anticonvulsant drug action, Epilepsia, 25(Suppl 2):140.

Mody, I., and Miller, J.J., 1985, Levels of hippocampal calcium and zinc following kindling-induced epilepsy, Can J. Physiol. Pharmacol., 63:159.

Pei, Y., Zhao, D., Huang, J., and Cao, L., 1983, Zinc-induced seizures: a new experimental model of epilepsy, Epilepsia, 24:169.

Penfield, W., and Kristiansen, K., 1951, Epileptic Seizure Patterns, Thomas, Springfield, Ill.

Perez-Clausell, J., and Danscher, G., 1985, Intravesicular localization of zinc in rat telencephalic boutons. A histochemical study, Brain Res., 337:91.

Perry, T.L., Berry, K., Diamond, S., and Mok, C., 1971, Regional distribution of amino acids in human brain obtained at autopsy, J. Neurochem., 18:513.

Prakash, N.J., Fontana, J., and Henkin, R.I., 1973, Effect of transitional metal ions on (Na^+ and K^+) ATPase activity and the uptake of norepinephrine and choline by brain synaptosomes, Life Sci., 12:249.

Racine, R.J., 1972, Modification of seizure activity by electrical stimulation. II. Motor seizures, Electroenceph. Clin. Neurophysiol., 32:281.

Racine, R.J., and Cosina, D.V., 1979, Effects of midbrain raphe lesions or systemic p-chlorophenylalanine on the development of kindled seizures in rats, Brain Res. Bull., 4:1.

Sawada, S., Takada, S., and Yamamoto, C., 1983, Selective activation of synapses near the tip of drug-ejecting microelectrode, and effects of antagonists of excitatory amino acids in the hippocampus, Brain Res., 267:156.

Schmidt, R., Zimmerman, H., and Whittaker, V.P., 1980, Metal ion content of cholinergic synaptic vesicles isolated from the electric organ of torpedo: effect of stimulation-induced transmitter release, Neuroscience, 5:625.

Slevin, J.T., and Kasarskis, E.J., 1985, Effects of zinc on markers of glutamate and aspartate neurotransmission in rat hippocampus, Brain Res., 334:281.

Sloviter, R.S., 1985, A selective loss of hippocampal mossy fiber Timm stain accompanies granule cell seizure activity induced by perforant path stimulation, Brain Res., 330:150.

Smart, T.G., and Constanti, A., 1983, Pre- and postsynaptic effects of zinc on in vitro prepyriform neurones, Neurosci. Lett., 40:205.

Storm-Mathisen, J., 1977, Glutamic acid and excitatory nerve endings: reduction of glutamic acid uptake after axotomy, Brain Res., 120:379.

Storm-Mathisen, J., 1981, Glutamate in hippocampal pathways, in: Glutamate as a Neurotransmitter, G. Di Chiara and G.L. Gessa, eds., Raven Press, New York, p. 43.

Storm-Mathisen, J., Leknes, A.K., Bore, A.T., Vaaland, J.L., Edminson, P., Haug, F.-M.S., and Ottersen, O.P., 1983, First visualization of glutamate and GABA in neurones by immunocytochemistry, Nature, 301:517.

Szutowicz, A., Harris, N.F., Srere, P.A., and Crawford, I.L., 1983, ATP-citrate lyase and other enzymes of acetyl-CoA metabolism in fractions of small and large synaptosomes from rat brain hippocampus and cerebellum, J. Neurochem., 41:1502.

Timm, F., 1958, Zur Histochemie des Ammonshorngebietes, Z. Zellforsch., 48:548.

Vindrola, O., Briones, R., Asai, M., and Dernandez-Guardiola, A., 1981, Amygdaloid kindling enhances the enkephalin content in the rat brain, Neurosci. Lett., 21:39.

Vosu, H., and Wise, R.A., 1975, Cholinergic seizure kindling in the rat: comparison of caudate, amygdala, and hippocampus, Behav. Biol., 13:442.

Wada, J.A., and Wake, A., 1977, Dorsal frontal, orbital, and mesial frontal cortical lesions and kindling in cats, Can. J. Neurol. Sci., 4:107.

Walker, J.E., Mikeska, J.A., and Crawford, I.L., 1981, Cyclic nucleotides in the kindled amygdala of the rat, Brain Res. Bull., 6:1.

Walker, J.E., and Fonnum, F., 1983, Regional cortical glutamergic and aspartergic projections to the amygdala and thalamus of the rat, Brain Res., 267:371.

Walker, J.E., Hirsch, S., and Crawford, I.L., 1985, Interruption of cholinergic afferent pathways to the amygdala failed to alter electrical kindling, Exp. Neurol., 88:742.

Wofsey, A.R., Kuhar, M.J., and Synder, S.H., 1971, A unique synaptosomal fraction which accumulates glutamic and aspartic acids in brain tissue, Proc. Natl. Acad. Sci. USA, 68:1102.

Wooten, W.C., Walker, J.E., and Crawford, I.L., 1978, Medial forebrain bundle lesions prolong amygdalar afterdischarge without changing the onset of behavioral seizures in the kindled rat, Soc. Neurosci. Abstr., 4:148.

Wright, M.D., 1984, Zinc: effect and interaction with other cations in the cortex of the rat, Brain Res., 311:343.

COMMENTARY - METAL IONS AND EPILEPSY

S.H. Chung and I.L. Crawford

The session on Metal Ions and Epilepsy brought together scientists who presented a series of papers that explored the putative interactions between excitatory transmitters and the epileptogenic actions of transition metals. The origins of this emerging area of research may be traced, in part, to the acute and chronic seizure activity caused by transition metals, the presence of high focal concentrations of certain metals in specific pathways, and the increasingly apparent importance of amino acid transmitters in the pathophysiology of epilepsy. To foster this new and exciting field, the session included investigators who used interdisciplinary approaches to bridge the gap between neuroscience and bioinorganic chemistry. Advances in this new approach to understanding epilepsy may be easier to achieve because it is less likely that obvious directions for investigation have already been exploited. The rate of future progress will depend on the ability of researchers to think about trace element physiology, epileptic disease processes and cellular mechanisms at a molecular level.

In an overview, Chung outlined the basic chemistry and biophysical characteristics of the 'first row' transition metals. In a broader perspective it should be recalled that only 25 of the 103 elements in the periodic table are essential for functions in biologic systems. The individual concentrations of fourteen of the essential metals is less than 0.01% of total body weight and are considered to be 'trace' elements. This contrasts markedly with the abundant, almost ubiquitous, presence of excitatory amino acids in the nervous system. Among the transition and group II elements the abundance of zinc in the brain is second only to that of iron.

Chung stressed the importance of metal ions in the extracellular space and the potential for interaction with release or uptake of transmitters as well as action on postsynaptic receptors. This viewpoint was expanded to include homeostasis within neurons by Ebadi who presented evidence for a group of zinc binding metallothionein-like proteins in brain. The synthesis of the metalloprotein was stimulated remarkably following central injections of zinc, and was blocked by actinomycin D. Evidence from studies in other systems suggests that metal ions are not incorporated into the growing ribosome-bound polypeptide chain until the protein is fully formed. According to this view, a metal does not induce its own coordination site, but its interaction with the metalloprotein awaits the expression of the genetic message. These metal-binding proteins may establish an equilibrium between bound and free zinc in nerve terminals. Control of the physiologic concentration of 'free' metal may, in turn, modulate the activity of transmitters at release, uptake and receptor site that are metal sensitive. The molecular characterization, regulation and synthesis of the proteins remain to be elucidated.

Baraldi presented results based on his studies of the behavioral and neurochemical effects of large parenteral doses of zinc. He emphasized the general concept that it is very difficult to perturb zinc homeostasis in the brain, although zinc plasma levels may have extremely large fluctuations. The behavioral syndrome caused by peripherally administered zinc resembled the effects of dopamine and its agonists, although other data showed an important endocrine component related to ACTH and endorphins. Central administration of zinc caused seizures which were associated with reduction of glutamate binding affinity and number of recognition sites, perhaps indicative of a massive release of glutamate. These results taken with evidence for zinc interaction with the GABA-benzodiazepine receptor complex may eventually be used to explain, in part, the mechanisms underlying zinc induced seizures. The observations may also be relevant to epilepsy since excitatory and inhibitory transmitters constantly interact on the membrane of a neuron. Any of these factors may determine whether the summated effects become unbalanced and the subsequent incoordination causes reverberations in the nervous system that are finally expressed as seizures.

Slevin reported on studies of the kindling model of epilepsy and its usefulness to examine the relationship of excitatory amino acids to divalent cations. Although the results of these studies failed to show a change in glutamate binding to hippocampal membranes from kindled rats, he found that sodium-independent glutamate binding sites are inhibited by zinc and the concentrations of the metal are increased in the kindled hippocampus. How these changes are related to kindling phenomena remains to be answered. It is pertinent to note the dramatic changes in calcium binding protein, calmodulin and other calcium regulated enzymes that have been reported from other studies on models of epilepsy, since zinc interactions with other divalent cations are well established in many neurologic functions.

Wright addressed the issue of transition metal actions on neuronal membranes. In contrast to previous reports in the literature, he showed that iontophoretically applied zinc caused slow, prolonged increases in the firing rate of half the cortical neurons studied. Although dose-effect studies are difficult with iontophoretic methods, Wright made similar observations using an *in vitro* slice technique. He interpreted these results as an interference with calcium mediated mechanisms or an action on sodium-potassium ATPase rather than an effect on GABA binding sites. He based this conclusion largely on his important observation of a relatively selective block of calcium depressant actions when zinc was administered simultaneously.

Crawford presented preliminary results from recent experiments designed to explore mechanisms linking glutamic acid and zinc to epileptogenesis. He showed that lesions of zinc-rich, glutamergic afferents from the pyriform cortex to amygdala retarded the kindling process. It was noted that another zinc containing pathway, the mossy fibers of hippocampal granule cells, may also be glutamergic and participate in the development of kindled seizures. Although ablation of neurons in the pyriform cortex and their axons is technically very difficult, anatomical studies are providing knowledge that is needed to refine the necessary surgical procedures. If zinc-rich glutamergic afferents influence epileptogenesis, and if lesions of the afferents block kindling, it may be possible to use a similar approach in patients with intractable seizures.

Knowledge emerging from all the phases of investigation presented in the session provide new incentives to experiments relevant to the cause and mechanisms that contribute to epilepsy. It is clear that epilepsy is a multifactorial process, and that multidiscipline interrelated research is a productive approach. Studies of metal ions and excitatory amino acids as they relate to seizure disorders have great promise, but to integrate results and realize their full potential, researchers will need to encompass a wide variety of disciplines.

SESSION IX
SEIZURES AND BRAIN DAMAGE: THE EXCITOTOXIC LINK

INCITING EXCITOTOXIC CYTOCIDE AMONG CENTRAL NEURONS

J.W. Olney

Department of Psychiatry and McDonnell Center for the Study
of Higher Brain Function
Washington University
St. Louis, Missouri

INTRODUCTION

One of my major research goals in recent years has been to answer a simple question: 'Can one CNS neuron excite another CNS neuron to death?' I suspect that the answer is yes, although admittedly the evidence is not all in. As a prime example, it seems likely that for millions of years, neurons in the brains of temporal lobe epileptics have been hyper-exciting their fellow neurons to death. In a series of animal studies, we have shown that regardless of the method used to stimulate limbic seizure activity, if such activity persists long enough (>1 hour), it results in a type of acute brain damage which ultrastructurally resembles that induced by the excitotoxic transmitters, glutamate (Glu) and aspartate (Asp). Moreover, in some of our experiments the seizure activity was induced specifically in pathways thought to use Glu or Asp as transmitter and the brain damage observed was selectively localized to postsynaptic neurons innervated by such pathways. Thus, when I refer to 'inciting excitotoxic cytocide among central neurons', I am suggesting that one can, by appropriately stimulating seizure circuits in animal brain, cause certain neurons—those that use Glu or Asp as transmitter—to excite other central neurons to death. Here I will review evidence from animal research favoring an excitotoxic explanation for seizure-related brain damage (SRBD) and will also discuss recent evidence shedding new light on the mechanism(s) underlying excitotoxic phenomena.

SEVERAL APPROACHES FOR INDUCING SRBD

Kainic Acid: A Promising but Problematic SRBD Model

Arousal of interest in the possible complicity of excitatory amino acids in SRBD was initially stimulated by the discovery that kainic acid (KA), a rigid Glu analog, is both a powerful convulsant and brain damaging agent (Olney et al., 1974). As a convulsant, KA is rather unique; instead of producing isolated grand mal or generalized seizures, as many convulsants do, it induces 'limbic status epilepticus' (Olney, 1981) lasting for hours and featuring frequently recurring episodes of rearing on hind limbs with peroral frothing and head and forepaw clonus. Predictably, after approximately one hour of such seizure activity, acute neurodegenerative changes resembling those associated with Glu neurotoxicity appear in various limbic and related brain regions (Olney et al., 1979; Fig. 1). Since a similar pattern of brain damage is known to occur in human temporal lobe epilepsy (Corsellis and Meldrum, 1976), and since the KA SRBD syndrome is highly reproducible and very readily induced by either parenteral or intracranial administration of KA, it provides a useful animal model of temporal lobe epilepsy (Nadler et al., 1978; Ben-Ari et al., 1979). The KA model, however, is not ideal for studying the mechanism(s) by which seizures beget brain damage because KA produces powerful Glu-like excitotoxin activity which conceivably could explain disseminated Glu-like lesions independent of a seizure mechanism. Evidence that diazepam prevents KA-induced seizures and disseminated brain damage without interfering with the direct toxic action of KA at its site of local injection in brain (Ben-Ari et al., 1979; Fuller and Olney, 1981) strongly supports a seizure mechanism rather than direct toxic action as basis for the disseminated lesions. To explain the Glu-like appearance of disseminated lesions induced by KA, we have proposed that they are, in fact, caused by Glu, i.e., excessive seizure-mediated release of endogenous Glu (Olney, 1981). It can always be argued, however, that KA contributes to the toxic process, e.g., by stimulating release of Glu from Glu terminals, by blocking reuptake of Glu, by acting together with endogenous Glu at Glu receptors, etc. To further clarify this issue, we have pursued alternate approaches for inducing SRBD.

Folate-Induced SRBD Syndrome

Several years ago we observed (Olney et al., 1981a,b) that folic acid (FA) or its reduced derivative, folinic acid, when injected directly into the amygdala or striatum, induces KA-like repetitive rearing seizures and distant brain damage without causing local tissue damage at the injection site. Ultrastructurally, the cytopathology induced at distant sites by FA appears identical to the disseminated damage induced by KA. Moreover, like the KA syndrome, both the seizure activity and distant brain damage are easily blocked by diazepam pretreatment (Fuller et al., 1981). The ability of FA to induce seizure-related distant lesions without damage at the local injection site has been confirmed in rats by McGeer et al. (1983) and Tremblay et al. (1983) and in cats by Kaijima et al. (1984). FA thus mimics the KA property of inducing sustained seizures and distant brain damage but lacks the confounding KA property of destroying neurons by a Glu-like direct toxic action; it follows that the distant toxicity of FA cannot readily be explained by its diffusion to distant sites, since once having diffused there, it lacks direct toxic action. Our FA findings support the conclusion that prolonged limbic seizure activity per se can result in acute brain damage having Glu/Asp cytopathological characteristics.

Dipiperidinoethane-Induced SRBD Syndrome

Prompted by the observation of Levine and Sowinski (1980) that systemic injection of dipiperidinoethane (DPE) results in a limbic pattern of brain damage, we administered DPE systemically to adult rats and found that it causes KA-like repetitive rearing seizures and a KA-like pattern of acute brain damage (Olney et al., 1980). Since injecting DPE directly into the amygdala did not cause either seizures or brain damage, we proposed that DPE may owe its convulsant and brain damaging potential to an active metabolite generated _in vivo_. We noted that the DPE molecule bears no resemblance to KA but does resemble the cholinergic agent tremorine (Olney et al., 1980).

Cholinergic Mechanism Underlies DPE SRBD Syndrome

Since the cholinergic activity of tremorine is thought to arise from its _in vivo_ conversion to oxotremorine, we injected DPE-di-N-oxide, an

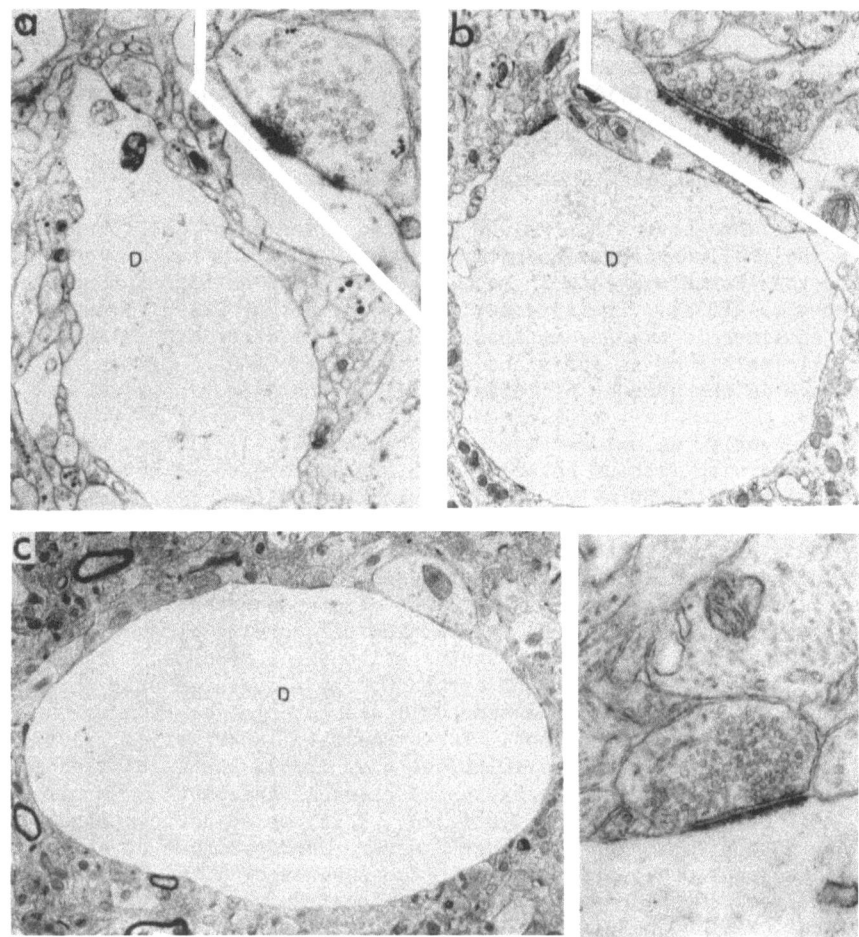

Fig. 1. The cytopathology shown in these three panels appears identical; each panel displays an axodendritic synaptic scene in which the postsynaptic dendrite (D) is undergoing acute edematous degeneration while the presynaptic axon terminal retains a normal healthy appearance. The scene in a is from the arcuate nucleus of the mouse hypothalamus following subcutaneous administration of Glu. The scene in b is from the CA1 region of the rat hippocampus following persistent limbic seizure activity induced by subcutaneous administration of kainic acid. Panel c depicts the reaction in the CA3 apical dendritic field of the rat hippocampus following two hours of electrical stimulation of the perforant path. The cytopathology in a is clearly induced by exogenous Glu; we propose that the cytopathology in b and c is induced by endogenous Glu (a and b x10,000; c x6,000; enlarged views x36,000).

oxidized DPE derivative synthesized by J.F. Collins, into the rat amygdala and found (Olney et al., 1982) that it produces repetitive rearing seizures and a FA-like (distant but not local) pattern of limbic brain damage. This supported our hypothesis that the DPE molecule is converted *in vivo* to an active oxidized derivative capable of inducing a SRBD syndrome, apparently by interacting with the central cholinergic transmitter system. In further support of this interpretation, Baron et al. (1985) subsequently showed that DPE is both a cholinesterase inhibitor and cholinergic antagonist whose dual opposing actions negate any toxic effects in vivo, whereas DPE-di-N-oxide is a pure cholinesterase inhibitor which, by promoting synaptic build-up of acetylcholine, could conceivably cause prolonged seizures.

Known Cholinergic Agents Produce an SRBD Syndrome

Pursuing the above findings, we injected a series of cholinergic agonists and cholinesterase inhibitors directly into the rat amygdala and found that certain agents in each category cause an SRBD syndrome (Olney et al., 1983a). The syndrome induced by either DPE-di-N-oxide or known cholinergic agents can most accurately be described as FA-like in that it consists of seizures and a distant disseminated pattern of brain damage in the absence of local damage at the injection site.

More recently, we demonstrated (Honchar et al., 1983) that when rats are pretreated with lithium chloride (2-3 mEq/kg sc) and 24 hours later given pilocarpine (20-30 mg/kg sc) or physostigmine (0.4 mg/kg sc), a severe KA-like SRBD syndrome develops which, if not interrupted by diazepam treatment, is uniformly lethal after 5-10 hours of status epilepticus. The same dose of lithium or cholinergic agent alone causes neither seizures nor brain damage. Using surface and depth electrode recordings, Zorumski et al. (1984) have characterized the lithium/pilocarpine syndrome electrographically as a rapidly evolving status epilepticus syndrome. By electronmicroscopy, we have found that the cytopathology associated with this syndrome has the same Glu-like excitotoxic characteristics that we have described in other SRBD syndromes. Very recently, Turski et al. (1983) reported, and we have confirmed (Olney et al., 1986), that a similar SRBD syndrome can be induced in rats by intraperitoneal treatment with pilocarpine alone but it requires a dose of 400 mg/kg. Thus, by an unknown mechanism lithium permits pilocarpine to induce a severe SRBD syndrome at approximately 1/20 the dose that would be required in the absence of lithium pretreatment. The lithium/cholinotoxic syndrome is prevented by either atropine or diazepam pretreatment. After the seizure activity is well established, atropine does not influence either the seizures or brain damage process but diazepam arrests the seizures and attenuates the brain damage (Olney et al., 1983c). Although a cholinergic mechanism underlies the induction of seizures in this SRBD model, we propose that secondary activation of Glu/Asp pathways and excessive release of these excitotoxic transmitters is the mechanism giving rise to brain damage.

INDUCING SRBD BY STIMULATING SPECIFIC EXCITOTOXIN TRACTS

Focal Motor Seizures and Distant Thalamic Lesions

In other experiments, we have shown (Collins and Olney, 1982) that persistant focal motor seizure activity induced by topical (supradural) application of various convulsants to the rat sensorimotor cortex results in Glu-like local lesions in specific thalamic nuclei that receive Glu innervation from cortical neurons involved in the seizure process. We chose this particular preparation as an alternative to the limbic seizure model in an effort to facilitate interpretation of results, i.e., the system is anatomically simple and it is relatively well established that a specific neuronal pathway (corticothalamic) being primarily activated uses Glu as transmitter (Fonnum et al., 1981). Our finding that persistent discharge activity in a Glu pathway causes acute Glu-type damage, selectively affecting distant neurons upon which Glu is being released, strongly supports the hypothesis that endogenous Glu may play a role in the cellular damage associated with seizures.

Sustained Perforant Path Stimulation and Hippocampal Damage

Sloviter (1983) has elegantly shown that persistent electrical stimulation of the perforant path (putative Glu excitatory input to the hippocampus) causes KA-like electrophysiological and light-microscopic histopathological changes in the rat hippocampus. In rats prepared by Sloviter, we have demonstrated (Olney et al., 1983b) that the acute hippocampal cytopathology induced by perforant path stimulation is ultrastructurally indistinguishable from the seizure-linked cytopathology induced in the hippocampus by KA, FA, DPE or cholinergic agents, which in turn has the dendrosomatotoxic/axon-sparing characteristics of the lesion that any excitotoxin, including Glu or Asp, induces locally when injected into brain. We propose that an explanation for these several correlations lies in the fact that persistent hippocampal discharge activity, a common denominator linking perforant path stimulation with KA, FA or cholinergic drug treatment, entails excessive release of endogenous excitotoxins (Glu or Asp) at many hippocampal synapses. The pattern of dendrosomal damage in each case follows a laminar distribution corresponding closely with putative Glu/Asp innervation patterns in the hippocampus.

ULTRASTRUCTURAL SIMILARITY BETWEEN SEIZURE-INDUCED AND GLU-INDUCED LESIONS

Although the several SRBD syndromes we have studied vary with respect to the mechanism of seizure induction, we have been impressed that in each case, after approximately one hour of sustained seizure activity, a type of acute cytopathology emerges which has all of the ultrastructural characteristics of Glu-induced brain damage. To document this observation in prior publications (Olney et al., 1979; Olney, 1981; Collins and Olney, 1982; Honchar et al., 1983; Olney et al., 1983a,b; Olney et al., 1986) we have emphasized a single dramatic feature of Glu neurotoxicity which is consistently reproduced in SRBD syndromes - the acute dendrotoxic/axon-sparing type of cytopathology (Fig. 1). However, a systematic comparison (see below) reveals that every other major feature of Glu neurotoxicity also occurs in SRBD syndromes. Conversely, every major aspect of the cytopathology associated with SRBD syndromes can be explained in terms of Glu neurotoxicity.

Massive edematous swelling is the most characteristic pathological change observed in neuronal somata of the retina or hypothalamus (Fig. 2) following sc administration of Glu (Olney, 1969, 1971). This type of Glu-like cytopathology is reproduced in every SRBD syndrome we have examined, but it is the predominant response only in certain brain regions and/or cell types. For example, it is the characteristic response of primary neurons in the amygdala (Fig. 3a), thalamus and lateral septum and of interneurons in the olfactory cortex, neocortex and hippocampus. Specific changes in intracellular organelles and intranuclear chromatin, which previously have been described as a feature of edematous degeneration of nerve cells in Glu neurotoxicity (Olney, 1971, 1981), are all faithfully reproduced as a feature of neuronal edematous degeneration in SRBD syndromes. This is an extremely acute type of neuronal degeneration which proceeds to end-stage neuronal necrosis within hours and culminates in early phagocytosis of the dead cell.

Vacuolar condensation (also known as dark cell degeneration) of neuronal somata was described many years ago (Olney, 1969) as a consistent, albeit not the most prominent, feature of the neurotoxic reaction seen in the retina following sc administration of Glu. This is also a consistent feature of the cytopathology induced in the hypothalamus by sc Glu (Fig. 2). In SRBD syndromes, vacuolar condensation is the characteristic response of pyramidal neurons in the olfactory and hippocampal cortices and (less

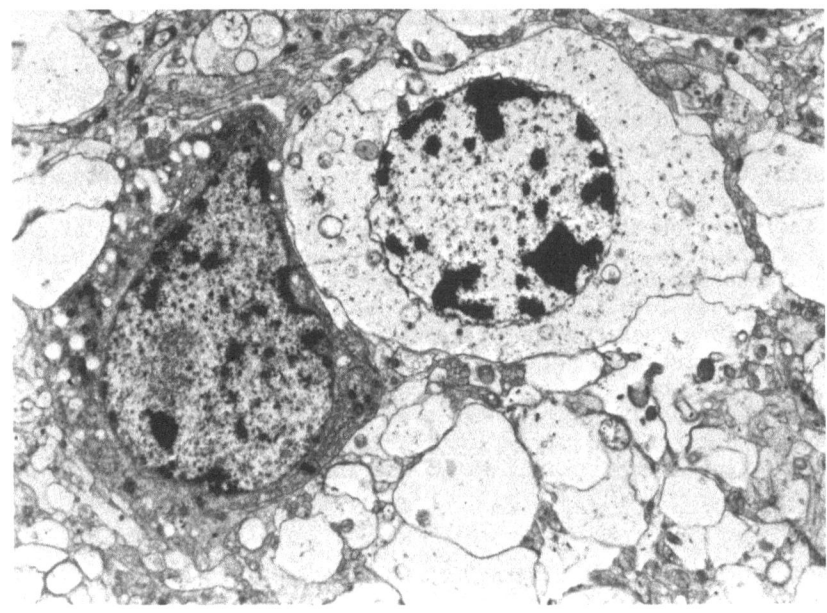

Fig. 2. This electron micrograph depicts two neurons undergoing degeneration in the infant monkey hypothalamus four hours after subcutaneous administration of Glu. The neuron on the right is exhibiting signs of acute edematous degeneration, including pathological changes in cytoplasmic organelles and nuclear chromatin. The neuron on the left is undergoing dark cell degeneration, including condensation and vacuolar changes in the cytoplasm and clumping of nuclear chromatin. Some of the swollen processes in the surrounding neuropil are dendritic, others are glial (x9,000).

conspicuously) in neocortex; however, this type of neuronal degeneration not infrequently affects scattered neurons in every brain region involved in the SRBD reaction (Fig. 3b). This is a less acute type of cytopathology since neurons in various stages of vacuolar condensation can be seen as unphagocytized dark cell profiles in the damaged brain region for days following the toxic event.

Edematous changes in glia have been described as a prominent feature of the toxic reaction in circumventricular brain regions following sc administration of Glu or in any brain region following direct injection of an excitotoxin (Olney, 1971; McGeer et al., 1978). These changes are barely perceptable following low neurotoxic doses of an excitotoxin, but are a conspicuous feature of the more severe reaction to higher doses (Fig. 4a). Edematous swelling of glia is a prominent feature of the cytopathology associated with persistent seizure activity, especially in pyramidal layers of the hippocampus (Fig. 4b), olfactory cortex and portions of neocortex. While the mechanistic basis of this gliotoxic response remains to be clarified, its characteristic presence both in Glu-induced and seizure-induced lesions supports the hypothesis that an excitotoxic mechanism underlies the latter as well as the former.

To the best of my knowledge, all available ultrastructural evidence supports the view that SRBD syndromes have in common a certain type of cytopathology which is indistinguishable from Glu-induced cytopathology.

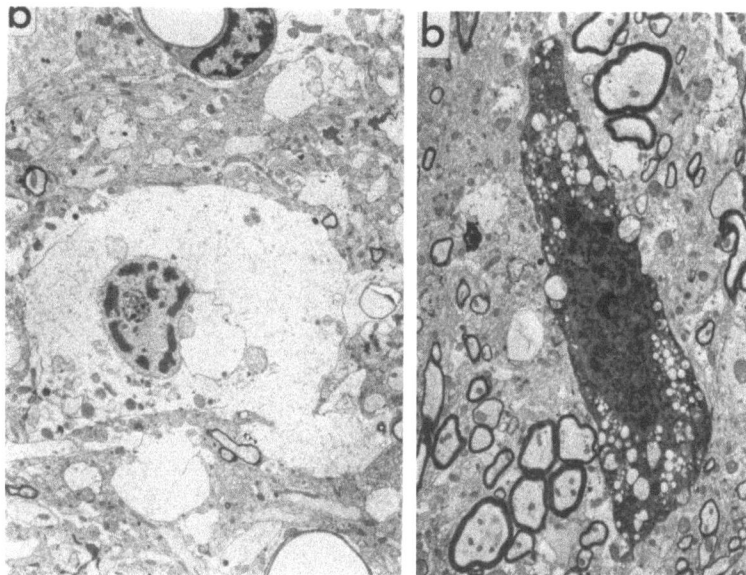

Fig. 3. In a, a neuron is undergoing acute edematous degeneration in the cortical nucleus of the rat amygdala following continuous limbic seizure activity induced by subcutaneous administration of pilocarpine (400 mg/kg). The pathological changes in this neuron are indistinguishable from those resulting from the direct toxic action of exogenous Glu, as illustrated in Fig. 2. In b, a neuron is undergoing dark cell degeneration in the ventromedial nucleus of the thalamus after two hours of continuous focal motor seizure activity induced by topical application of bicuculline to the dura over the motor cortex. We propose that seizure-mediated release of endogenous Glu from corticothalamic axons ending on this neuron is responsible for this toxic reaction. The vacuolated condensed appearance of the cytoplasm with clumping of nuclear chromatin gives this cell an appearance identical to the cell in Fig. 2 undergoing dark cell degeneration following exposure to exogenous Glu administered subcutaneously (a = x5,000; b = x7,000).

Tremblay et al. (1983), however, on the basis of light microscopic observations, have argued that the disseminated lesions induced by intraamygdaloid FA and KA differ from one another. According to these authors, KA damages the CA3 hippocampal region but FA does not and FA damages the pyriform cortex, whereas KA does not. Moreover, they maintain that the distant lesions induced by FA differ from those induced by KA in that the FA lesions have an 'anoxic-ischemic' appearance. Thus, they consider the KA and FA syndromes to be different both in pattern of lesion distribution and in type of resulting cytopathology. The former issue is addressed in the next paragraph and the latter in a separate section below.

First, it should be pointed out that the type of cytopathology induced by KA and FA may be identical by ultrastructural analysis even if differences in lesion pattern are discernable by light microscopy. However, I am not convinced that there are reproducible differences in the KA and FA lesion patterns. Certainly in some of our FA-treated rats the CA3 hippo-

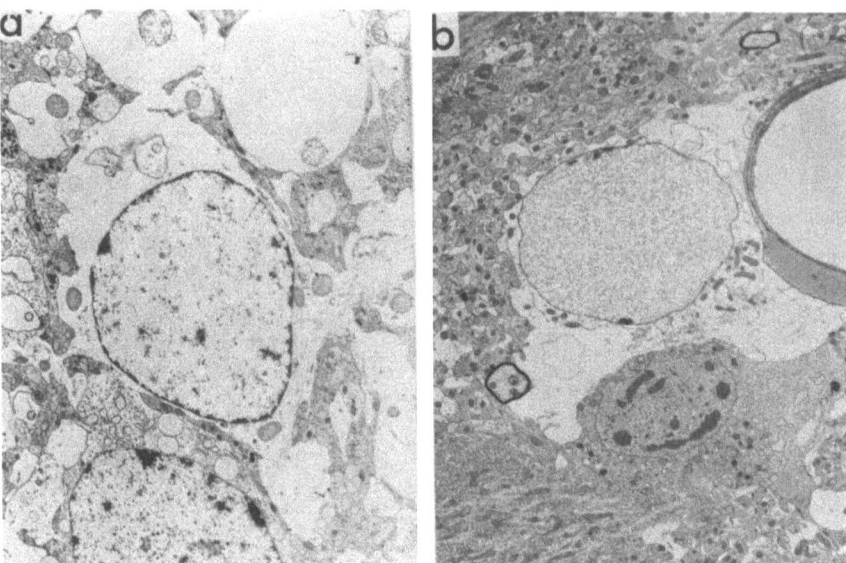

Fig. 4. Panel a depicts a swollen edematous glial cell in the infant monkey hypothalamus four hours after subcutaneous administration of Glu. Glial swelling characteristically accompanies Glu neurotoxicity whenever a high enough dose of Glu has been employed to cause a severe toxic reaction. In b, a swollen glial cell is present in the CA1 region of the hippocampus following continuous electrical stimulation of the perforant path. The edematous glial cell is in contact with a blood vessel at right, and there is an acutely necrotic neuron immediately subjacent to the swollen glial cell. This degenerating neuron displays pathological changes in nuclear chromatin and cytoplasmic organelles characteristic of a neuron undergoing Glu-induced edematous degeneration; however, the cytoplasm and nucleoplasm are relatively condensed as is typically seen in dark cell degeneration. Thus, perforant path stimulation has caused this neuron to display a mixed picture comprising some features of both types of degeneration that neurons are known to undergo following excessive exposure to Glu (a = x5,000; b = x7,000).

campal area was as badly damaged as we typically find following KA treatment (Fig. 5) and in KA-treated rats the pyriform cortex was as badly damaged as we typically find in FA-treated rats. Moreover, it has been reported by numerous laboratories that the pyriform cortex is one of the brain regions most frequently damaged by any mode of KA administration. Kaijima et al. (1984), who studied the electrographic, neuropathological and clinical responses of cats injected with FA into the amygdala, comment on this issue as follows, 'In contrast to Tremblay's work, our anatomic results clearly show pathologic damage in hippocampal area CA3.' These authors did find that the initial electrographic changes induced locally in the amygdala by FA injection differed from those induced by KA, but as soon as recurrent limbic seizure activity was recordable, all EEG signs of the FA- and KA-induced seizure states were identical. Kaijima et al. (1984) also found the behavioral seizure picture following FA in cats to be identical to that following KA injection.

We have compared the pattern of disseminated brain damage induced by systemic lithium/pilocarpine treatment with that induced by systemic

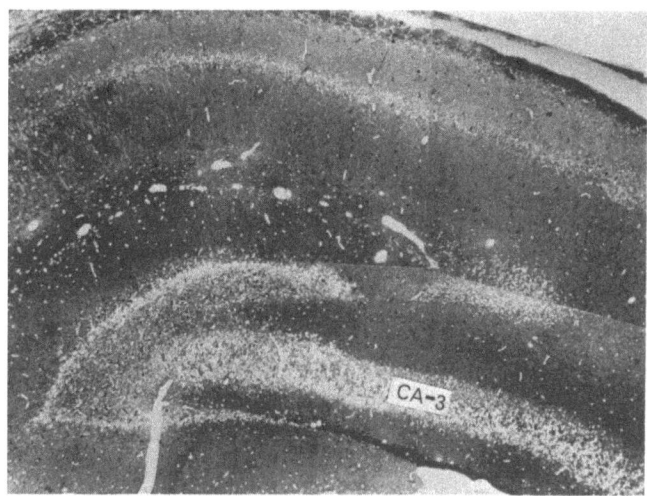

Fig. 5. A light micrograph depicting the acute pattern of edematous degeneration occurring in the rat hippocampus four hours following intraamygdaloid injection of folic acid (50 nmol). This pattern of degenerative change is indistinguishable from that seen in association with prolonged seizure activity induced by systemic or intraamygdaloid administration of kainic acid (for example, see Olney et al., 1979). The pattern of degeneration in both the kainic acid and folic acid syndromes clearly includes the CA3 region (x120).

KA and conclude that the patterns are remarkably similar, although minor differences can be discerned. For example, although systemic KA damages the frontoparietal cerebral cortex and substantia nigra (Schwob et al., 1980), cellular degeneration in these regions is more consistently and more dramatically present following lithium/pilocarpine treatment. Conversely, although the lithium/pilocarpine lesion pattern characteristically includes hippocampal damage, this is possibly a more dramatic feature of KA treatment. It is our impression that these differences are most clearly evident in animals that have been examined after relatively brief or mild seizure activity, whereas after more prolonged or more severe activity the patterns become indistinguishable. Thus, it is possible that some of the circuits activated initially in one syndrome are different from those activated initially in the other, but as either syndrome progresses to an advanced stage of status epilepticus, many if not all of the same limbic and related circuits become involved and the resulting patterns of brain damage become increasingly similar. It bears reiterating, however, that the type of cytopathology in all SRBD syndromes may be identical even if there are minor differences in the pattern of lesion distribution among syndromes.

EXCITOTOXIC ANTAGONISTS MAY PROTECT AGAINST SRBD

Currently we are employing an *in vitro* retinal preparation (discussed further below) that is useful for screening agents for excitotoxin agonist or antagonist activity. Pursuant to the report by Anis et al. (1983)

that the dissociative anesthetics, ketamine and phencyclidine, specifically antagonize the excitatory activity of NMA (but not KA) in the spinal cord, we tested these agents and found (Olney et al., 1986a) that they are exceptionally powerful in blocking the neurotoxic action of NMA (but not KA) on retinal neurons (Fig. 6). Because blood brain barriers are permeable to these agents, we explored their ability to alter the SRBD action of KA. Since they do not block the excitatory action of KA, we reasoned that they should not prevent KA from inducing seizures, yet their potent blockade of NMA receptors might prevent brain damage from being mediated by endogenous Glu or Asp at NMA-type receptors. We found that these agents do protect against the seizure-mediated cytopathology associated with KA treatment (Olney and Fuller, unpublished observations). Confounding the interpretation, however, is the observation that the seizure activity itself was substantially suppressed, presumably due to blockade of NMA receptors which are an integral part of circuits involved in the propogation of seizure activity. Thus, it is possible that either an anticonvulsant or antitoxin principle - or both - may underlie the protection conferred against brain damage.

ANOXIC AND EPILEPTIC BRAIN DAMAGE LINKED BY EXCITOTOXIN MECHANISM

An association has long been recognized between epilepsy, especially temporal lobe epilepsy, and a certain pattern of neuronal loss primarily affecting the cerebral cortex, hippocampus and thalamus (Corsellis and Meldrum, 1976). A resemblance between this pattern of brain damage and that associated with anoxia, and the fact that convulsions sometimes temporarily interrupt respiration and interfere with cerebral oxygenation, has led to the view that oxygen deficiency may be the primary underlying cause of epilepsy-related brain damage. Accordingly, the cytopathology associated with either anoxia or epilepsy is often referred to as 'anoxic-ischemic' brain damage. Recent evidence suggests that a common mechanism may indeed underlie these two forms of brain damage; however, it appears that the common denominator may be an abnormal accumulation of endogenous excitotoxic transmitter rather than anoxia. As Rothman (1984 and this volume) has shown, pharmacological antagonism of Glu excitatory receptors protects dissociated cultured hippocampal neurons from degeneration under in vitro anoxic conditions that would otherwise result in acute neuronal necrosis. Benveniste et al. (1984) have shown that in vivo brain ischemia produces a massive acute increase in the extracellular concentration of Glu and Asp. Simon et al. (1984) found that the injection of an amino acid antagonist into the in vivo rat hippocampus during experimental brain ischemia protects against neuronal degeneration which would otherwise predictably occur under such anoxic conditions. Collectively, these data suggest that there may be no meaningful distinction between anoxic and epileptic brain damage, since in either case an excitotoxic mechanism may underlie the toxic event and the resulting tissue pathology may have an excitotoxic character. If no meaningful distinction can be made between anoxic and seizure-mediated brain damage, this weakens the Tremblay et al. (1983) argument that seizure-related cytopathology induced by FA can be distinguished from that induced by KA by its anoxic-ischemic appearance.

PROGRESS IN UNDERSTANDING EXCITOTOXIC MECHANISM

It is clear that the excitotoxic action of Glu and related analogs is a depolarization-mediated process and it was proposed years ago that abnormal permeability of the plasma membrane with loss of control over ion homeostasis might play a prominent role in the pathophysiology of the neuronal necrosis that ensues (Olney, 1978). However, specific evidence for this hypothesis was not generated until quite recently. Using a modifi-

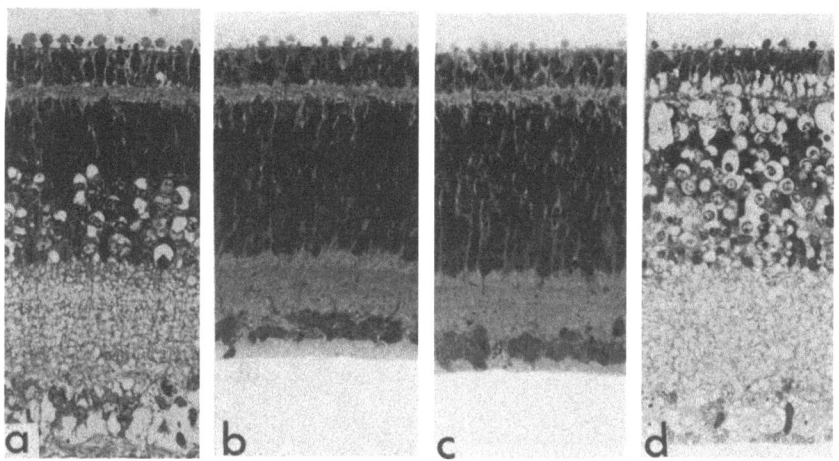

Fig. 6. When the 15 day chick embryo is incubated for 30 minutes in modified Hank's balanced salt solution containing NM(DL)A (200 µM), it results in a highly reproducible edematous neurodegenerative reaction affecting many neural elements in the inner half of the retina (panel a). Elsewhere (Samson et al., 1984; Olney et al., 1986a) we have shown that specific antagonists of the NMDA receptor effectively block this toxic reaction; for example, it is totally prevented by D-aminophosphonovalerate and D-aminophosphonoheptanoate at concentrations of 25 and 75 µM or higher, respectively. The retina in panel b was exposed to NM(DL)A (200 µM) in the presence of phencyclidine (PCP; 0.5 µM) and total protection against the neurotoxic action is shown. In panel c, ketamine (5 µM) afforded total protection against NM(DL)A (200 µM). The concentrations of PCP and ketamine cited here are minimal effective blocking concentrations. Higher concentrations provide total protection and lower concentrations provide partial or no protection. In panel d, the retina was incubated with a high concentration of PCP (3 mM) together with KA at its lowest effective toxic concentration (25 µM). The toxic reaction shown is that typically induced by KA (25 µM); thus, PCP fails to block KA toxicity just as Anis et al. (1983) have shown that it fails to block the excitatory action of KA. Ketamine at 3 mM also fails to block KA toxicity (not shown) (x150).

cation of methods described by Reif-Lehrer et al. (1975), we have used the in vitro chick embryo retina to address this mechanistic issue. When the chick embryo retina is incubated in balanced salt solution containing Glu or any of its excitotoxic analogs, a typical excitotoxin lesion develops within 30 minutes (Olney et al., 1986b). The neurodegenerative process appears by electron microscopy to be the same as that induced in the retina of immature mice by sc administration of Glu, and excitatory amino acid receptor antagonists effectively protect the chick embryo retina from the toxic action of excitatory amino acid agonists (Samson et al., 1984). To determine whether Glu neurotoxicity is dependent on Ca^{2+} influx, as some have proposed, we incubated the chick retina in medium containing toxic concentrations of Glu either with or without Ca^{2+}. The neurotoxic reaction was at least as severe in the absence as in the presence of Ca^{2+}

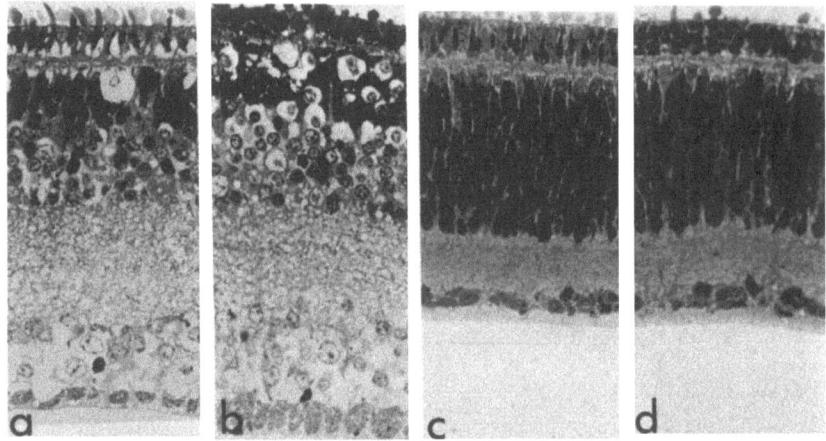

Fig. 7. Incubating the chick embryo retina for 30 minutes in modified Hank's balanced salt solution containing 1 mM Glu results in an acute toxic reaction as depicted in panel a. When Ca^{2+} is excluded from the incubation medium, Glu (1 mM) is just as effective in causing a neurotoxic reaction (panel b). The toxic reaction remains as severe if Ca^{2+} is excluded and EGTA (2 mM) is added to chelate residual Ca^{2+} (not shown). When benzoyl choline, an impermeant cation, is substituted for Na^+ in the incubation medium (panel c), or isethionate, an impermeant anion, is substituted for Cl^- (panel d), Glu (1 mM) does not cause a toxic reaction (x150).

(Price et al., 1985). In a systematic evaluation of the role of other ions in Glu neurotoxicity we found that Na^+ and Cl^- are essential for this neurotoxic reaction (Olney et al., 1986b), i.e, substituting the impermeant cation benzoyl choline or impermeant anion isethionate for Na^+ or Cl^- respectively prevented Glu neurotoxicity (Fig. 7). These experiments were conducted with NMA and KA also with similar results - NMA and KA neurotoxicity, like Glu neurotoxicity, requires Na^+ and Cl^- but not Ca^{2+} in the incubation medium (Olney et al., 1986b). Rothman (1985 and this volume) conducted similar ion substitution experiments and found that the neurotoxic action of Glu on cultured hippocampal neurons is Na^+ and Cl^- but not Ca^{2+} dependent.

Although the above findings suggest that Ca^{2+} plays no role in excitotoxicity, Choi (1985) recently reported that exposure of cultured cerebrocortical neurons to Glu for only a few minutes, then removing Glu from the medium, results in delayed neuronal death in 24 hours, provided Ca^{2+} is present when the culture is initially exposed to Glu. Choi's findings do not necessarily contradict ours; rather, they raise the intriguing possibility that there are two mechanisms by which excitotoxins can induce neuronal death, one being the acute fulminating reaction studied by us and by Rothman, which is not Ca^{2+} dependent, and the other being a slowly evolving neurotoxic process which is Ca^{2+} dependent. Since ultrastructurally one can distinguish two types of neurodegenerative reactions to Glu, one being acutely fulminating (edematous degeneration) and the other a slowly evolving process (vacuolar condensation), it is an interesting possibility that an acute Ca^{2+} independent mechanism underlies the former and a delayed Ca^{2+} dependent mechanism the latter. This dual mechanistic hypothesis warrants serious consideration as it may have important implications regarding the role of excitotoxic transmitters in anoxic and epileptic brain damage, i.e., it would imply that excitotoxin-induced cell death can occur either abruptly or more gradually, depending on the duration of excitotoxin exposure, and that even brief exposure (e.g., resulting from brief anoxia or seizure activity) may be enough to cause the slow and subtle demise of central neurons. This interpretation also has potentially important therapeutic implications since a slowly evolving cell death process may provide greater opportunity for therapeutic intervention.

SUMMARY

Here I have reviewed evidence from electron microscopic studies showing that each of several sustained limbic seizure syndromes is associated with a type of acute brain damage which is ultrastructurally indistinguishable from the brain damage induced by Glu and other excitotoxins. In addition, I have presented evidence that persistent stimulation of specific axonal tracts that use Glu as transmitter results in Glu-like excitotoxic degeneration of postsynaptic neurons innervated by such tracts. Phencyclidine and ketamine, which powerfully block the neurotoxicity of the Glu analog NMA, protect against seizure-related brain damage. This may be explained by either an anticonvulsant or antiexcitotoxic mechanism, or both. Recent evidence suggests that an excitotoxic mechanism (excessive activation of Glu/Asp receptors) may underlie both seizure-mediated and anoxic brain damage. The acute fulminating type of neuronal degeneration induced by Glu is a Na^+ and Cl^- but not Ca^{2+} dependent phenomenon. According to a recent study, however, Glu may induce neuronal necrosis not only by an acute Ca^{2+} independent process but by a more slowly evolving Ca^{2+} dependent process. If, as these data suggest, an excitotoxic mechanism underlies brain damage associated with anoxia and epilepsy, a better understanding of excitotoxic mechanisms may lead eventually to prophylactic approaches for preventing such forms of brain damage.

ACKNOWLEDGEMENTS

Supported in part by NIH grant MH37967 and Research Scientist Award MH38894.

REFERENCES

Anis, N.A., Berry, S.C., Burton, N.R., and Lodge, D., 1983, The dissociative anaesthetics, ketamine and phencyclidine, selectively reduce excitation of central mammalian neurones by N-methylaspartate, Br. J. Pharmacol., 79:565.

Baron, B.M., Kashman, Y., and Sokolovsky, M., 1985, Neurotoxicity of dipiperidinoethane due to in vivo conversion to a selective cholinesterase inhibitor, Brain Res., 331:164.

Ben-Ari, Y., Tremblay, E., Ottersen, O.P., and Naquet, R., 1979, Evidence suggesting secondary epileptogenic lesions after kainic acid: pretreatment with diazepam reduces distant but not local brain damage, Brain Res., 165:362.

Benveniste, H., Drejer, J., Schousboe, A., and Diemer, N.H., 1984, Elevation of the extracellular concentrations of glutamate and aspartate in rat hippocampus during transient cerebral ischemia monitored by intracerebral microdialysis, J. Neurochem., 43:1369.

Choi, D.W., 1985, Two mechanisms underlying glutamate neurotoxicity in cortical cell culture, Soc. Neurosci. Abstr., 11:153.

Collins, R.C., and Olney, J.W., 1982, Focal cortical seizures cause distant thalamic lesions, Science, 218:177.

Corsellis, J.A.N., and Meldrum, B.S., 1976, in: Greenfield's Neuropathology, 3rd Edition, W. Blackwood and J.A.N. Corsellis, eds., Arnold, London, p. 771.
Fonnum, F., Storm-Mathisen, J., and Divac, I., 1981, Biochemical evidence for glutamate as neurotransmitter in corticostriatal and corticothalamic fibres in rat brain, Neuroscience, 6:863.
Fuller, T.A., and Olney, J.W., 1981, Only certain anticonvulsants protect against kainate neurotoxicity, Neurobehav. Toxicol. Teratol., 3:355.
Fuller, T.A., Olney, J.W., and Conboy, V.T., 1981, Diazepam markedly attenuates the neurotoxicity of folic acid, Soc. Neurosci. Abstr., 7:811.
Honchar, M.P., Olney, J.W., and Sherman, W.R., 1983, Systemic cholinergic agents induce seizures and brain damage in lithium-treated rats, Science, 220:323.
Kaijima, M., Riche, D., Rousseva, S., Moyanova, S., Dimov, S., and Le Gal La Salle, G., 1984, Electroencephalographic, behavioral, and histopathologic features of seizures induced by intraamygdala application of folic acid in cats, Exp. Neurol., 86:313.
Levine, S., and Sowinski, R., 1980, Lesions of amygdala, piriform cortex and other brain structures due to dipiperidinoethane intoxication, J. Neuropathol. Exp. Neurol., 39:56.
McGeer, E.G., McGeer, P.L., and Singh, K., 1978, Kainic acid-induced degeneration of neostriatal neurons: dependency upon corticostriatal tract, Brain Res., 139:381.
McGeer, P.L., McGeer, E.G., and Nagai, T., 1983, GABAergic and cholinergic indices in various regions of rat brain after intracerebral injections of folic acid, Brain Res., 260:107.
Nadler, J.V., Perry, B.W., and Cotman, C.W., 1978, Preferential vulnerability of hippocampus to intraventricular kainic acid, in: Kainic Acid as a Tool in Neurobiology, E.G. McGeer, J.W. Olney and P.L. McGeer, eds., Raven Press, New York, p. 219.
Olney, J.W., 1969, Glutamate-induced retinal degeneration in neonatal mice. Electron microscopy of the acutely evolving lesion, J. Neuropathol. Exp. Neurol., 28:455.
Olney, J.W., 1971, Glutamate-induced neuronal necrosis in the infant mouse hypothalamus: an electron microscopic study, J. Neuropathol. Exp. Neurol., 30:75.
Olney, J.W., Rhee, V., and Ho, O.L., 1974, Kainic acid: a powerful neurotoxic analogue of glutamate, Brain Res., 77:507.
Olney, J.W., 1978, Neurotoxicity of excitatory amino acids, in: Kainic Acid as a Tool in Neurobiology, E. McGeer, J.W. Olney, P. McGeer, eds., Raven Press, New York, p. 95.
Olney, J.W., Fuller, T., and DeGubareff, T., 1979, Acute dendrotoxic changes in the hippocampus of kainate treated rats, Brain Res., 176:91.
Olney, J.W., Fuller, T.A., Collins, R.C., and DeGubareff, T., 1980, Systemic dipiperidinoethane mimics the convulsant and neurotoxic actions of kainic acid, Brain Res., 200:231.
Olney, J.W., 1981, Kainic acid and other excitotoxins: a comparative analysis, in: Glutamate as a Neurotransmitter, G. DiChiara and G.L. Gessa, eds., Raven Press, New York, p. 375.
Olney, J.W., Fuller, T.A., and deGubareff, T., 1981a, Kainate-like neurotoxicity of folates, Nature, 292:165.
Olney, J.W., Fuller, T.A., deGubareff, T., and Labruyere, J., 1981b, Intrastriatal folic acid mimics the distant but not local brain damaging properties of kainic acid, Neurosci. Lett., 25:207.
Olney, J.W., Collins, J.F., and deGubareff, T., 1982, Dipiperidinoethane neurotoxicity clarified, Brain Res., 249:195.
Olney, J.W., deGubareff, T., and Labruyere, J., 1983a, Seizure-related brain damage induced by cholinergic agents, Nature, 301:520.

Olney, J.W., deGubareff, T., and Sloviter, R.S., 1983b, 'Epileptic' brain damage in rats induced by sustained electrical stimulation of the perforant path. II. Ultrastructural analysis of acute hippocampal pathology, Brain Res. Bull., 10:699.

Olney, J.W., Honchar, M.P., and Sherman, W.R., 1983c, Diazepam prevents lithium-pilocarpine neurotoxicity in rats, Soc. Neurosci. Abstr., 9:401.

Olney, J.W., Collins, R.C., and Sloviter, R.S., 1986, Excitotoxic mechanisms of epileptic brain damage, in: Basic Mechanisms of Epilepsy, A.V. Delgado-Escueta, A.A. Ward, and D.M. Woodbury, eds., Raven Press, New York, in press.

Olney, J.W., Price, M.T., Fuller, T.A., Labruyere, J., Samson, L., Carpenter, M., and Mahan, K., 1986a, The anti-excitotoxic effects of certain anesthetics, analgesics and sedative-hypnotics, Neurosci. Lett., in press.

Olney, J.W., Price, M.T., Samson, L., and Labruyere, J., 1986b, The ionic basis of excitotoxin-induced neuronal necrosis, Neurosci. Lett., in press.

Price, M.T., Olney, J.W., Samson, L., and Labruyere, J., 1985, Calcium influx accompanies but does not cause excitotoxin-induced neuronal necrosis, Brain Res. Bull., 14:369.

Reif-Lehrer, L., Bergenthal, J., and Hanninen, L., 1975, Effects of monosodium glutamate on chick embryo retina in culture, Invest. Ophthalmol., 14:114.

Rothman, S.M., 1984, Synaptic release of excitatory amino acid neurotransmitter mediates anoxic neuronal death, J. Neurosci., 4:1884.

Rothman, S.M., 1985, The neurotoxicity of excitatory amino acids is produced by passive chloride influx, J. Neurosci., 5:1483.

Samson, L., Olney, J.W., Price, M.T., and Labruyere, J., 1984, Kynurenate protects against excitotoxin-induced neuronal necrosis in chick retina, Soc. Neurosci. Abstr., 10:24.

Schwob, J.E., Fuller, T., Price, J.L., and Olney, J.W., 1980, Widespread patterns of neuronal damage following systemic or intracerebral injections of kainic acid: a histological study, Neuroscience, 5:991.

Simon, R.P., Swan, J.H., Griffiths, T., and Meldrum, B.S., 1984, Blockade of N-methyl-D-aspartate receptors may protect against ischemic damage in the brain, Science, 226:850.

Sloviter, R.S., 1983, 'Epileptic' brain damage in rats induced by sustained electrical stimulations of the perforant path. I. Acute electrophysiological and light microscopic studies, Brain Res. Bull., 10:675.

Tremblay, E., Cavalheiro, E., and Ben-Ari, Y., 1983, Are convulsant and toxic properties of folates of the kainate type?, Eur. J. Pharmacol., 93:283.

Turski, W.A., Cavalheiro, E.A., Schwarz, M., Czuczwar, S.J., Kleinrok, Z., and Turski, L., 1983, Limbic seizures produced by pilocarpine in rats: behavioural, electroencephalographic and neuropathological study, Behav. Brain Res., 9:315.

Zorumski, C.F., Collins, R.C., Olney, J.W., and Clifford, D.B., 1984, Lithium-pilocarpine seizures: in vivo studies, Soc. Neurosci. Abstr., 10:552.

SELECTIVE AND NON-SELECTIVE SEIZURE RELATED BRAIN DAMAGE PRODUCED BY

KAINIC ACID

Y. Ben-Ari, A. Repressa, E. Tremblay, and L. Nitecka

LPN, CNRS, Gif-sur-Yvette, F-91190 France
and INSERM-U29, 123 Bd de Port-Royal
Hôpital de Port-Royal, 75014 Paris, France

Parenteral or intracerebral administration of the potent excitatory amino acid kainic acid (KA) produces a seizure and brain damage syndrome in which limbic structures play a central role. This has been investigated extensively, notably since it constitutes a useful animal model of human temporal lobe epilepsy. In the present report, I shall briefly discuss a number of aspects concerning these actions of KA — in particular the effects produced by parenteral injections of the toxin (e.g., Ben-Ari, 1985 for a recent review).

LIMBIC STATUS EPILEPTICUS PRODUCED BY PARENTERAL ADMINISTRATION OF KAINATE

Electrographic records made in chronically implanted rats indicate that i.v. or i.p. injections of kainate produce first paroxysmal discharges localized to the hippocampal formation, these involve after various delays other limbic structures — in particular the amygdala — and at a third stage secondarily generalized seizures are observed in other cortical and extracortical structures (e.g., Ben-Ari et al., 1981; Lothman and Collins, 1981). Corresponding to these stages, the animal first displays staring, then individual limbic motor seizures and at a later stage more complex signs including various types of stereotyped behavior. 2-deoxyglucose (2DG) studies have also illustrated with remarkable details the anatomical substrates of these seizures, notably the preferential involvement of the hippocampus and amygdala (Fig. 1; Ben-Ari et al., 1981; Lothman and Collins, 1981). The seizures evoked by kainate are similar to those induced by daily repetitive stimulation of limbic structures, notably the amygdala (i.e. kindling; Ben-Ari et al., 1981; Lothman and Collins, 1981). In keeping with observations made in human epileptics and animal experiments, there is good evidence that the amygdala in particular plays a crucial role in the motor expression of the seizures (e.g., Ben-Ari, 1985). Thus, it is only once the amygdala and associated structures are involved (2nd stage) that the typical limbic signs are observed.

There is little doubt that the rise in metabolism seen with the 2DG method occurs in relation to the propagation of paroxysmal discharges to neuronal populations which are axonally interconnected (Ben-Ari et al., 1981; Lothman and Collins, 1981). Starting from the hippocampal formation as also indicated from maturation studies (Tremblay et al.,

RAT

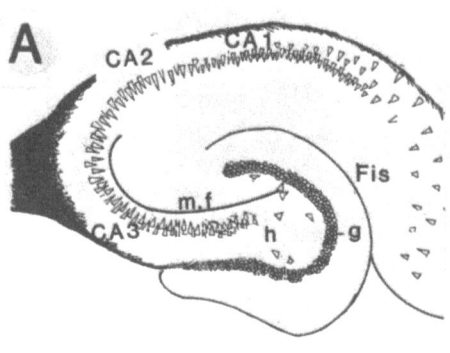 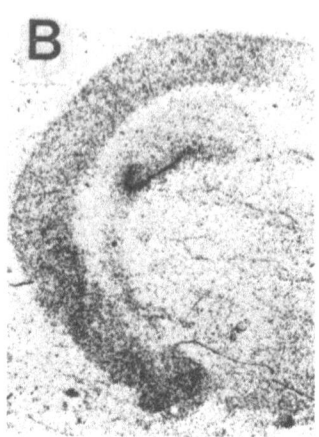

HUMAN

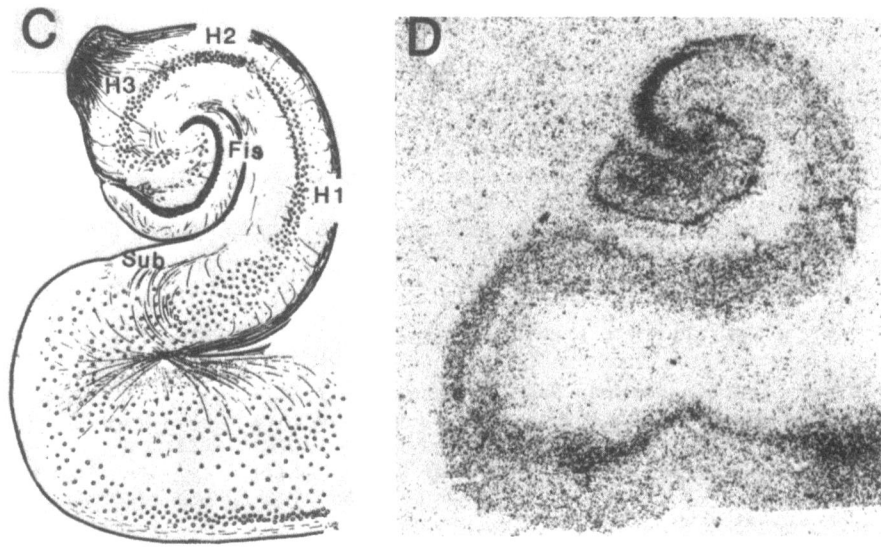

Fig. 1. Autoradiographs to depict the distribution of kainate binding sites in the rat (top) and human (bottom) hippocampi. Left sides: schematic diagrams of the hippocampus. Right sides: autoradiographs obtained as described elsewhere (Berger and Ben-Ari, 1984). Note the high density of sites in the CA3-H3 mossy fiber zone. 48 hour post-mortem delays (Tremblay et al., 1985).

1984), the seizure activity will involve the lateral septum and other targets of the pre- and post-commissural efferent projections of the hippocampal formation. With almost no exceptions, the structures which are labeled during severe limbic motor seizures, are directly connected with the hippocampal formation, the amygdala and the medio-dorsal nucleus of the thalamus (e.g., Tremblay et al., 1984). These observations reflect the importance of synaptic activity in mediating the brain damage; lesion experiments have also provided direct evidence in this respect (see below).

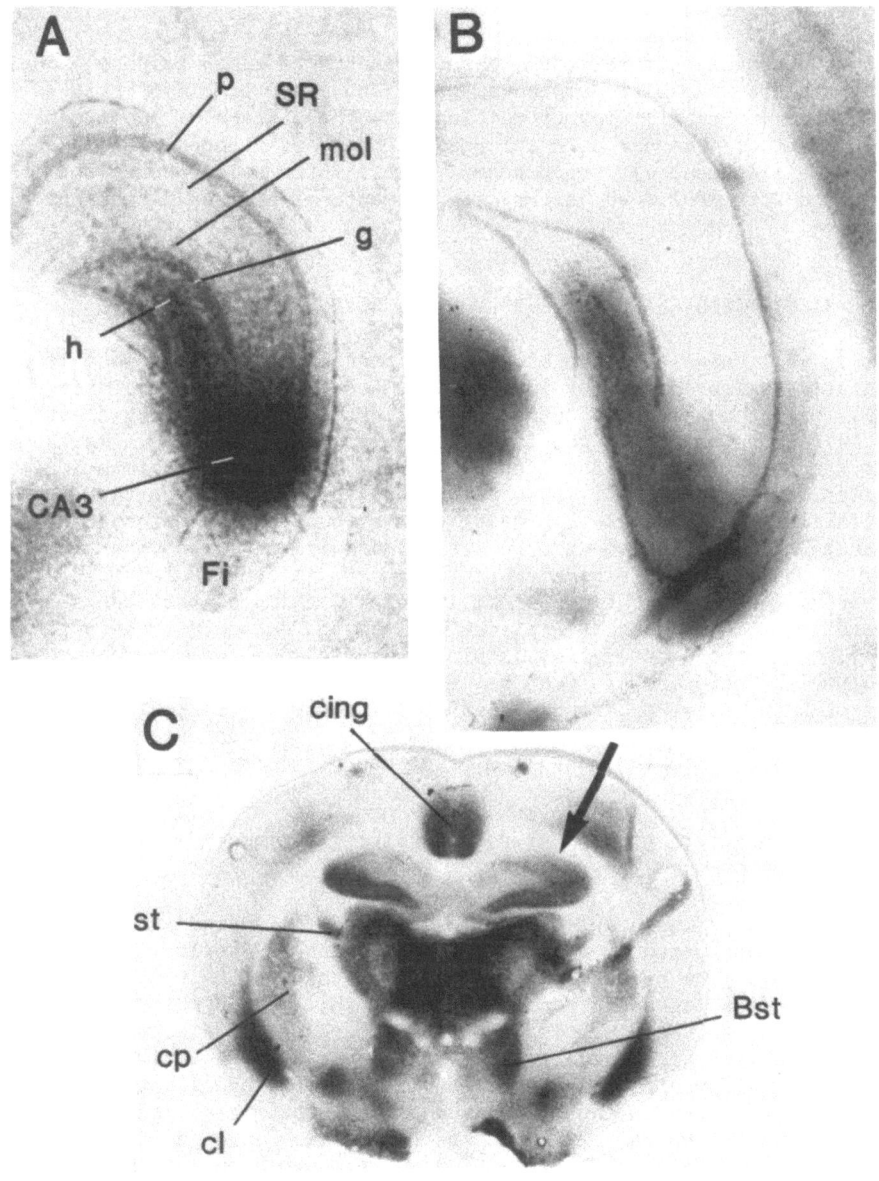

Fig. 2. 2-deoxyglucose autoradiographs to depict the metabolic rise in the hippocampus following parenteral administrations of KA. In A-B from immature (12 days) and adult rats, respectively, the 2-DG films were superimposed upon the Nissl stained sections from which they were obtained; note that the rise is restricted to the CA3 mossy fiber region: C. From an adult rat to depict the rise in other limbic structures following repetitive limbic motor seizures (Ben-Ari et al., 1981; Tremblay et al., 1985).

Following various survival periods, histopathological alterations are observed in limbic structures; there is excellent parallelism between the 2DG map and the subsequent lesions (Olney el al., 1979; Schwob et al., 1980; Ben-Ari et al., 1981).

The parenteral kainate model bears relevance to human temporal lobe epilepsy since a) the hippocampus, amygdala and other limbic structures play a central role in its symptomatology; b) the pattern of brain damage is reminiscent of 'Ammon's horn sclerosis', the most frequent lesion seen in chronic epileptics; c) 'spontaneous' limbic motor seizures are consistently noted with prolonged delays after parenteral KA (see Nitecka et al., this volume).

HOW MUCH KAINIC ACID REACHES THE VULNERABLE ZONES?

It is well known that the blood brain barrier is poorly permeable to excitatory amino acids, notably glutamic acid (Le Fauconnier, this volume). Direct measures of the concentrations of KA traversing the barrier (e.g., Le Fauconnier and Berger et al., this volume) give values in the hundred nanomolar range in the hippocampus and other brain regions (one hour after administration). These values are actually smaller than the concentrations of classical impermeant agents (such as saccharose) used to evaluate the blood volume in the dissected samples. Even when the opening of the blood brain barrier is taken into account (Ruth, this volume) by measuring the concentrations of KA in seizing animals, the change is not dramatic; this is readily explicable since the concentrations of KA in the blood one hour after the administration is very low (Le Fauconnier, this volume). Therefore, it must be kept in mind that the initial direct effects of kainate are produced by rather low concentrations of kainate. These are several orders of magnitude less than the doses used in various studies to investigate the mechanism of KA neurotoxicity (see below).

HIGH AFFINITY SPECIFIC BINDING SITES FOR KAINATE: DISTRIBUTION AND ROLE IN THE SYNDROME

The presence of saturable specific, high affinity binding sites for KA has been demonstrated by several groups (London and Coyle, 1979; Schwarcz and Fuxe, 1979). The regional distribution of these sites has been mapped by means of autoradiography (Monaghan and Cotman, 1982; Berger and Ben-Ari, 1983; Unnerstall and Wamsley, 1983) notably in recent studies using a radioactive KA with high specific activity (Tremblay et al., 1985). The highest levels of specific binding sites in the brain are located in the stratum lucidum of CA3 in the hippocampus, i.e., the region innervated by the mossy fibers which originates in the granular layer of the fascia dentata (Fig. 2). This is also conspicuous in human hippocampi; the post mortem delays (up to 48 hours) do not reduce the labeling (Tremblay et al. 1985). A recent study also indicates that this labeling is due to slowly displaceable sites (Berger, Charton and Ben-Ari, submitted). The CA3 pyramidal neurons are also the most vulnerable in the brain to the excitotoxic action of KA (Nadler et al., 1980) and bursting activity is produced by KA already with concentrations of 20-50 nM in CA3 pyramidal cells recorded in slice preparations (Robinson and Deadwyler, 1981; Cherubini et al., this volume). Also, the amygdala is highly enriched in high affinity slowly displaceable sites; this and other observations suggest that these sites are preferentially located on neurons which undergo selective degeneration after severe KA induced seizures (Berger, Charton and Ben-Ari, submitted).

All of these observations, therefore, suggest that in spite of the poor permeability of the blood brain barrier to KA, the concentrations of toxin which eventually reach the CA3 zone will produce paroxysmal discharge, which in turn will propagate to progressively activate the entire limbic system (see below). It is, however, not clear whether the primary site of action is on the mossy fibers terminals or the CA3 pyramidal cell dendrites, i.e., whether the KA sites are located on the pre- or postsynaptic elements. Both kainate administration in the amygdala and intrahippocampal injections of colchicine, which respectively destroy preferentially the pyramidal neurons of CA3 and the mossy fibers, will reduce or abolish the binding sites in this zone (Monaghan and Cotman, 1981; Tremblay, Repressa and Ben-Ari, unpublished observations). Whatever the exact localization of these sites, paroxysmal discharge will be generated in the CA3 field leading to orthodromic (and presumably antidromic) activation of the entire hippocampal formation in a first stage. This <u>in situ</u> sequence of events (i.e., genesis of paroxysmal discharge in CA3 and spread to other hippocampal fields) has been directly shown to occur in the slice preparation with a variety of convulsant drugs, notably KA (Robinson and Deadwyler, 1981).

VULNERABLILTY OF GABAERGIC INTERNEURONS TO KAINATE

Electrophysiological observations suggest that in the hippocampus the recurrent GABAergic innervation is labile and is removed during epileptogenesis (Ben-Ari et al., 1979). Administration of kainate also produces a reduction of GABAergic inhibition. Thus, parenteral administration of the toxin causes a reduction of the paired-pulse inhibition in the fascia dentata (Sloviter and Damiano, 1981) and in slice preparations KA reduces the IPSP perhaps by a presynaptic mechanism (Fisher and Alger, 1984); since KA also reduces the afterhyperpolarization triggered by calcium entry (Cherubini et al., this volume), this would also tend to favor repetitive discharge (Madison and Nicoll, 1983). Recent studies also indicate that in kainate treated animals there is a reduction in the IPSPs and afterhyperpolarizations in the hippocampus weeks after the treatment suggesting a permanent effect (Ashwood et al., 1983; Franck and Schwartzkroin, 1985).

We have recently obtained anatomical evidence in keeping with this using antibodies directed against GABA. In preliminary studies, in spite of some variability, hilar interneurons were frequently destroyed in KA treated animals with long survival periods (Figs. 3 and 4; Nitecka, Woodson and Ben-Ari, unpublished observations). This will tend to disinhibit the granular neurons and enhance the spread of seizure discharge along the mossy fibers (see below). Along these lines, it is important to note that following parenteral (but not intra-amygdaloid or intracerebroventricular) KA, hilar neurons are often selectively destroyed (Ben-Ari et al., 1981; Nitecka et al., 1985). Local coagulative infarcts are often conspicuous in this region (Nitecka and Tremblay, this volume). It is possible that the destruction of hilar interneurons is due to these infarcts.

MECHANISMS OF SEIZURE RELATED BRAIN DAMAGE

Following severe seizures produced by KA, the damage in several limbic structures is characterized by a widespread necrosis. Thus, in the pyriform cortex, the necrosis involves the entire frontobasal and temporobasal portions of the brain under the rhinal fissure. The necrosis includes perivenous hemorrhages, astroglial scarring and vascular sprouting. This <u>non selective seizure related brain damage</u> is of an anoxic-ischemic type, presumably due to overactivity, release of water and metabolites and subsequent edema. This, by compressing drainage vessels against the skull in the fronto- basal region, results in local disturbance of blood flow

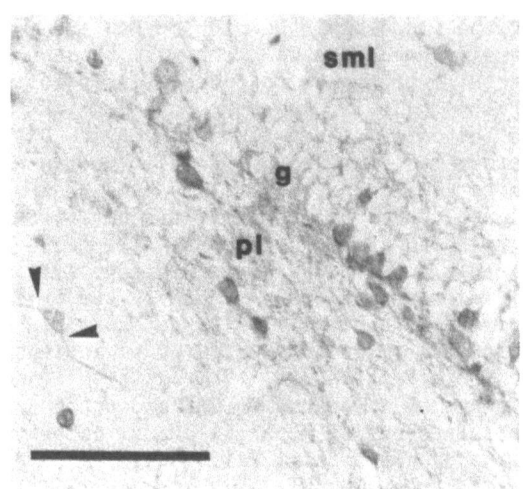

Fig. 3. Distribution of presumed GABA-ergic neurons in the hilus of the fascia dentata. GABA antibodies coupled with an HRP procedure. Note the numerous GABA containing cells in the polymorph zone, the processes of the neurons are ocassionally visible (arrow). Similar observations were made in rats following brief seizures (and survival times) after administration of kainate.

and anoxic ischemic changes (Sperk et al., 1983). This type of damage is produced in a number of regions, notably the CA1 hippocampal field which is more vulnerable to hypoxic-ischemic procedures than CA3 (Wieloch, this volume). Interestingly, these regions (CA1, but also septum) contain only the low affinity component of KA binding sites (Berger, Charton and Ben-Ari, submitted).

In contrast, the <u>selective seizure related brain damage</u> is causally related to the propagation of seizure discharge through a well characterized synaptic connection and without the involvement of more global deleterious effects associated with the convulsions. The evidence in favor of this type of damage is compelling for the CA3 region. Thus, direct local measures of the pO_2, pCO_2 and blood flow in this region indicate that the damage in CA3 is not due to a mismatch between local metabolic demands and blood flow (Fig. 5; Pinard et al., 1984). Several lines of evidence (including lesion or electrical stimulation experiments) suggest that the mossy fibers play a crucial role in this damage (Nadler et al., 1980,1981; Sloviter, 1983; see also Sloviter and Nadler, this volume). Developmental studies show that the damage in this zone will only be produced once the mossy fibers synapses are mature (Nitecka et al., 1984).

Two types of mechanisms have been proposed to underline this selective seizure related damage. The first one implies excessive entrance of calcium associated with the seizures which may cause neuronal damage by means of proteolysis. In support of this view, histochemical data suggest a limited Ca^{2+} binding capacity (Jande et al., 1981), and a particularly important decrease in extracellular Ca^{2+} during repetitive electrical

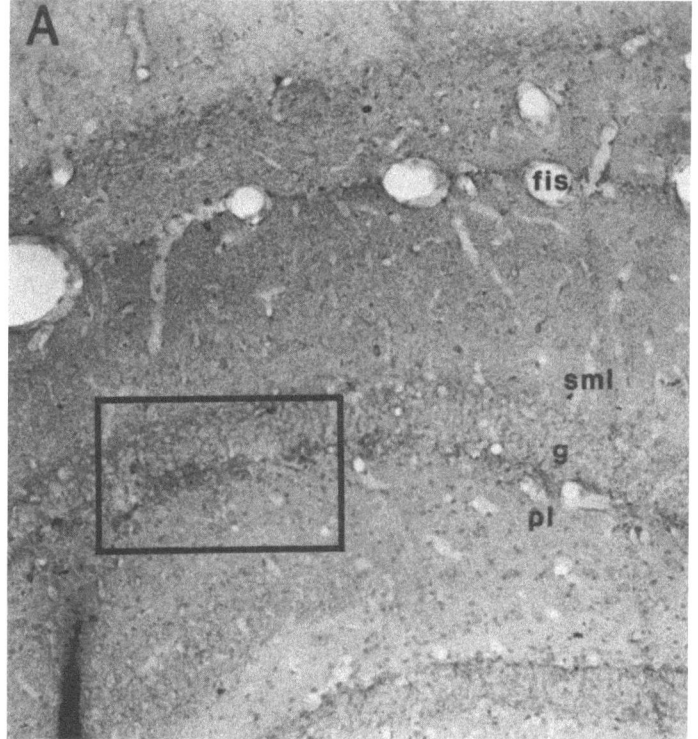

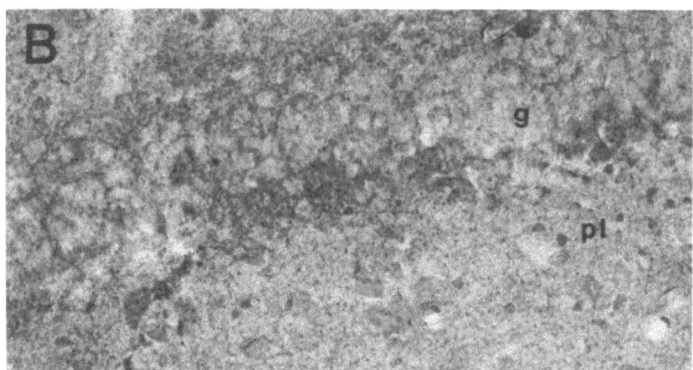

Fig. 4. Vulnerability of hilar GABAergic neurons to parenteral administration of KA. GABA positive neurons are surrounded by glia (arrows).

stimulation (Krnjević et al., 1980). However, measures of extracellular Ca^{2+} following microiontophoretic ejection of KA are not readily compatible with this hypothesis (Heinemann and Pumain, this volume), also in slice preparations Ca^{2+} spikes are reduced or blocked by KA (Cherubini et al., this volume). Rothman (1985 and this volume) has suggested that the neurotoxicity of KA in hippocampal cultures is due to a steady depolarization which leads to an influx of chloride. This leads to cation entry, which results in water entry and cell lysis. However, exceedingly high concentrations of KA were used in this study (100 μM). It is thus possible that this mechanism is more relevant to the non-selective seizure related type of damage, notably in view of the data suggesting that excitatory amino acids play a role in its etiology (Wieloch, this volume).

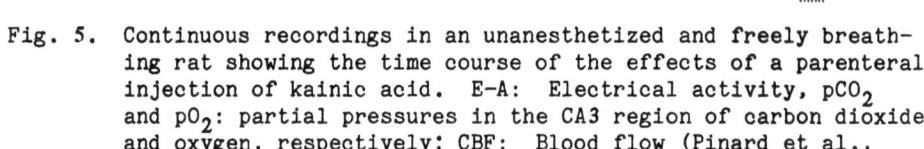

Fig. 5. Continuous recordings in an unanesthetized and freely breathing rat showing the time course of the effects of a parenteral injection of kainic acid. E-A: Electrical activity, pCO_2 and pO_2: partial pressures in the CA3 region of carbon dioxide and oxygen, respectively; CBF: Blood flow (Pinard et al., 1985).

Other observations also indicate that factor(s) perhaps exclusively localized in the mossy fibers may be released and produce the damage during the seizures. Since this region is particularly enriched in high affinity KA binding sites (see above), it is possible that the mossy fibers release an endogenous kainate agent which may be glutamic acid although the evidence is not compelling. A number of endokain candidates have been suggested, including folates (Olney et al., 1981). A detailed analysis of the effect of intra-amygdaloid injections of folates has, however, not confirmed this suggestion; this procedure failed to produce the selective seizure related damage seen in CA3 (Tremblay et al., 1984). Further studies should be made, notably by identifying the transmitter and modulators released from the mossy fibers.

It is also possible that the toxic factor is not the transmitter itself but another agent released with the transmitter during severe seizure discharge. Zn^{2+} is a likely agent. Zn^{2+} is particularly concentrated in the mossy fibers in particular in the mossy fiber terminals (Haug, 1967,1973). The concentrations of Zn^{2+} in this region is amongst the highest in the brain (Chung and Johnson, 1983). Zn^{2+} is also accumulated by the mossy fibers and released upon their activation (Assaf and Chung, 1984; Charton et al., 1985). Thus, in a push-pull study made in situ, it was found that when the cannula is located in the vicinity of the mossy fibers, a brief K^+ pulse produced an evoked release of Zn^{2+} (assayed by atomic absorption) which can reach levels two orders of magnitude higher than the blank; there is no evoked release when the cannula is in the fimbria, distant zones in the hippocampus or adjacent extrahippocampal regions (Charton et al., 1985; Fig. 6). The mechanisms through which Zn^{2+} might produce the damage are not clear notably in view of the large spectrum of cellular activities in which Zn^{2+} is involved. Interestingly, the hippocampal levels of zinc are increased in kindled animals (Slevin et al., this volume).

Whatever the exact mechanism underlying both the selective and nonselective damage and the contribution of excitatory amino acids in its etiology, it is important to stress the likely contribution of hippocampal

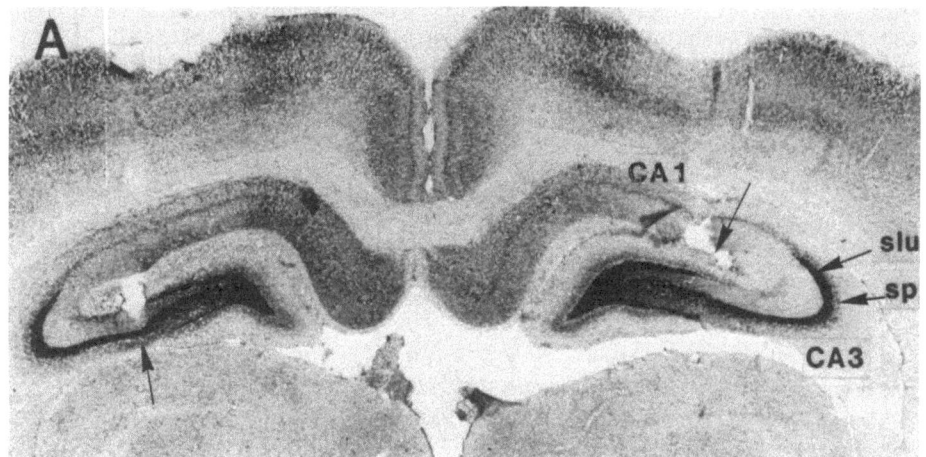
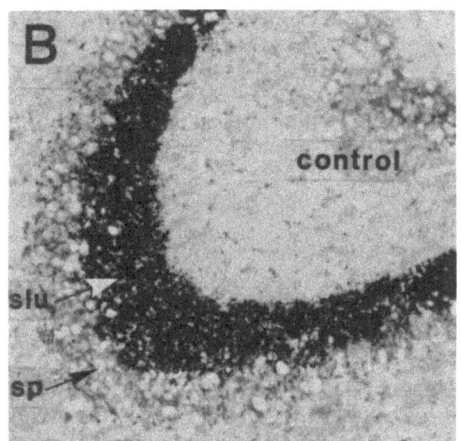
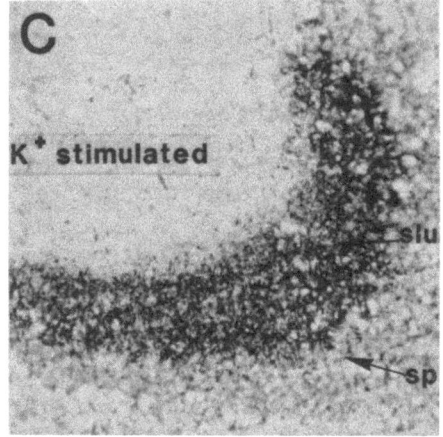

Fig. 6. Depolarization of the mossy fiber region produces a reduction of histochemically demonstrable zinc. Timm stain. Bilateral push-pull cannulae were stereotaxically introduced in situ in the hippocampus (A). When the cannula was adequately located (in the vicinity of the mossy fibers), a K^+ pulse produced a reduction of the Timm positive material in comparison to the control side.

plasticity in epileptogenesis. Thus, sprouting of mossy fibers deprived of their post-synaptic targets may create epileptogenic neuronal circuits in otherwise quiescent zones (Tauck and Nadler, 1985 and in this volume). This may underline the progressive character of pathological alterations associated with the epilepsies in humans (Scheibel and Scheibel, 1973).

REFERENCES

Assaf, S.Y., and Chung S., 1984, Release of endogenous Zn^{2+} from brain tissue during activity, Nature, 308:734.
Ashwood, T.J., Lancaster, B., and Wheal, H.V., 1983, Bursting activity in the kainic acid (KA) lesioned rat hippocampus is associated with a reduction in GABA mediated inhibition, J. Physiol., 336:59P.

Ben-Ari, Y., Krnjević, K., and Reinhardt, W., 1979, Hippocampal seizures and failure of inhibition, Can J. Physiol. Pharmacol., 57:1462.

Ben-Ari, Y., Tremblay, E., Riche, D., Ghilini, G., and Naquet, R., 1981, Electrographic, clinical and pathological alterations following systemic administration of kainic acid, bicuculline and pentetrazole: metabolic mapping using the deoxyglucose method with special reference to the pathology of epilepsy, Neuroscience, 6:1361.

Ben-Ari, Y., 1985, Limbic seizure and brain damage produced by kainic acid: mechanisms and relevance to human temporal lobe epilepsy, Neuroscience, 14:375.

Berger, M., and Ben-Ari, Y., 1983, Autoradiographic visualization of ^{3}H kainic acid receptor subtypes in the rat hippocampus, Neurosci. Lett., 39:237.

Charton, G., Rovira, C., Ben-Ari, Y., and Leviel, V., 1985, Spontaneous and evoked release of endogenous Zn^{2+} in the hippocampal mossy fiber zone of the rat in situ, Exp. Brain Res., 58:202.

Chung, S.H., and Johnson, M.S., 1983, Divalent transition-metal ions (Ca^{2+} and Zn^{2+}) in the brains of epileptogenic and normal mice, Brain Res., 280:323.

Fisher, R.A., and Alger, B.E., 1984, Electrophysiological action of kainic acid (KA) induced epileptiform activity in the rat hippocampal slice, J. Neurosci., 4:1312.

Franck, J.E., and Schwartzkroin, P.A., 1983, Kainate lesioned hippocampi become epileptogenic, Soc. Neurosci. Abstr., 9:908.

Haug, F.M.S., 1967, Electron microscopical localization of the zinc in hippocampal mossy fiber synapses by a modified sulphide silver procedure, Histochemie, 8:355.

Haug, F.M.S., 1973, Heavy metals in the brain. A light microscopic study of the rat with Timm's sulphide silver method. Methodological considerations and cytological and regional staining patterns, Adv. Anat. Embryol. Cell Biol., 47:1.

Jande, S.S., Maler, L., and Lawson, E.M., 1981, Immunohistochemical mapping of vitamin D dependent calcium binding protein in brain, Nature, 294:765.

Krnjević, K., Morris, M.E., and Reiffenstein, R.J., 1980, Changes in extracellular Ca^{2+} and K^+ activity accompanying hippocampal discharges, Can. J. Physiol. Pharmacol., 58:579.

London, E.D., and Coyle, J.T., 1979, Specific binding of ^{3}H kainic acid to receptor sites in rat brain, Molec. Pharmacol., 15:492.

Lothman, E.W., and Collins, C.R., 1981, Kainic acid-induced limbic motor seizures: metabolic, behavioral, electroencephalographic and neuropathological correlates, Brain Res., 218:299.

Madison, D.V., and Nicoll, R.A., 1984, Control of the repetitive discharges of rat CA1 pyramidal neurons in vitro, J. Physiol., 354:319.

Monaghan, D.T., and Cotman, C.W., 1982, The distribution of ^{3}H kainic acid binding sites in rat CNS as determined by autoradiography, Brain Res., 252:91.

Nadler, J.V., and Cuthbertson, G.J., 1980, Kainic acid neurotoxicity toward hippocampal formation: dependence on specific excitatory pathways, Brain Res., 195:47.

Nadler, J.V., Evenson, D.A., and Cuthbertson, G.J. 1981, Comparative toxicity of kainic acid and other acidic amino acids toward rat hippocampal neurons, Neuroscience, 6:2505.

Nadler, J.V., Evenson, D.A., and Smith, E.M., 1981, Evidence from lesion studies for epileptogenic and non-epileptogenic neurotoxic interactions between kainic acid and excitatory innervation, Brain Res., 205:405.

Nitecka, L., Tremblay, E., Charton, G., Bouillot, J.P., Berger, M., and Ben-Ari, Y., 1984, Maturation of kainic acid seizure-brain damage syndrome in the rat. II. Histopathological sequelae, Neuroscience, 13:1073.

Olney, J.W., Fuller, T.A., and De Gubareff, T., 1981, Folates have kainate like neurotoxicity, Nature, 292:165.

Pinard, E., Tremblay, E., and Seylaz, J., 1984, Blood flow compensates oxygen demand in the vulnerable CA3 region of the hippocampus during kainate-induced seizures, Neuroscience, 13:1039.

Robinson, J.H., and Deadwyler, S.A., 1981, Kainic acid produces depolarization of CA3 pyramidal cells in the in vitro hippocampal slice, Brain Res., 221:117.

Rothman, S.M., 1985, The neurotoxicity of excitatory amino acids is produced by passive chloride influx, J. Neurosci., 6:1483.

Scheibel, M.E., and Scheibel, A.B., 1973, Hippocampal pathology in temporal lobe epilepsy. A Golgi survey, in: Epilepsy. Its Phenomenon in Man, M.A.B. Brazier, ed., Academic Press, New York, p. 311.

Schwarcz, R., and Fuxe, K., 1979, ^{3}H kainic acid binding: relevance for evaluating the neurotoxicity of kainic acid, Life Sci., 24:1471.

Schwob, J.E., Fuller, T., Price, J., and Olney, J.W., 1980, Widespread patterns of neuronal damage following systemic or intracerebral injections of kainic acid: a histological study, Neuroscience, 5:991.

Sloviter, R.S., and Damiano, B.P., 1981, On the relationship between kainic acid-induced epileptiform activity and hippocampal neuronal damage, Neuropharmacology, 20:1003.

Sloviter, R.S., 1983, Epileptic brain damage in rats induced by sustained electrical stimulation of the perforant path. I. Acute electrophysiological and light microscopic studies, Brain Res. Bull., 10:675.

Sperk, G., Lassmann, H., Baran, H., Kish, S.J., Seitelberger, F., and Hornykiewicz, O., 1983, Kainic acid induced seizures: neurochemical and histopathological changes, Neuroscience, 10:1301.

Tauck, D.L., and Nadler, J.V., 1985, Evidence of functional mossy fiber sprouting in hippocampal formation of kainic acid treated rats, J. Neurosci.,5:1016.

Tremblay, E., Berger, M., Nitecka, L., Cavalheiro, E., and Ben-Ari, Y., 1984, A multidisciplinary study of folic acid neurotoxicity: interactions with kainate binding sites and relevance to the aetiology of epilepsy, Neuroscience, 12:569.

Tremblay, E., Nitecka, L., Berger, M.L., and Ben-Ari, Y., 1984, Maturation of kainic acid seizure-brain damage syndrome in the rat. I. Clinical, electrographic and metabolic observations, Neuroscience, 13:1051.

Tremblay, E., Repressa, A., and Ben-Ari, Y., 1985, Autoradiographic localization of KA binding sites in human hippocampi, Brain Res., 343:378.

Unnerstall, J.R., and Wamsley, J.K., 1983, Autoradiographic localization of high-affinity ^{3}H kainic acid binding sites in the rat forebrain, Eur. J. Pharmacol., 86:361.

ON THE ROLE OF SEIZURE ACTIVITY AND ENDOGENOUS EXCITATORY AMINO ACIDS

IN MEDIATING SEIZURE-ASSOCIATED HIPPOCAMPAL DAMAGE

R.S. Sloviter

Neurology Center, Helen Hayes Hospital
New York State Department of Health
West Haverstraw, New York 10993 and the
Departments of Pharmacology and Neurology
Columbia University, New York, New York 10032

INTRODUCTION

For many years, the brain damage associated with epilepsy was thought to be the result of anoxia believed to occur in brain regions susceptible to the effects of seizures (Meldrum and Corsellis, 1984). This was a reasonable supposition since anoxia causes neuron loss in some of the same brain regions affected in epilepsy (Meldrum and Corsellis, 1984). Two concurrent and related lines of research during the past fifteen years have significantly changed this view of epileptic brain damage. Meldrum and others have shown that although anoxia causes cell death, neither anoxia nor decreased blood flow occur in the affected brain regions during seizure activity (Meldrum and Corsellis, 1984). The second derives from the discovery by Olney (1978) that a number of endogenous excitatory amino acids are directly neurotoxic. This finding led to the 'excitotoxic' hypothesis which states that excitation and cell death are causally related and that, therefore, abnormal concentrations of endogenous excitatory amino acids could be responsible for the neuropathologic lesions seen in a number of neurologic disorders (Olney and de Gubareff, 1978). The subsequent discovery by Olney et al. (1974) of the neurotoxic and convulsant properties of the glutamate analog kainic acid served as a stimulus to research in experimental epilepsy in particular, since this compound rapidly and reliably produces a seizure state in normal animals that is associated with a pattern of brain damage (Nadler et al., 1978) similar to that seen in the brains of many chronic human epileptics (Meldrum and Corsellis, 1984). Our interest in this area of research began with the desire to understand how kainate caused hippocampal seizure activity and whether a direct neurotoxic effect of kainate or seizure activity per se causes hippocampal cell death. Olney et al. (1974) originally suggested that kainate produces excitation and cell death by an agonist action at glutamate receptors. Accordingly, the sensitivity of different hippocampal cell types to kainate (Nadler et al., 1978) was suggested to be due to differences in glutamate receptor density (Olney et al., 1979). On the basis of studies showing that transection of the mossy fiber pathway (Nadler and Cuthbertson, 1980) or pretreatment with diazepam (Ben-Ari et al., 1979) prevented kainate-induced hippocampal damage, it was alternately suggested that the hippocampal damage caused by kainate is due to seizure activity induced in the hippocampal granule cells by kainate (Ben-Ari et al., 1979; Nadler and Cuthbertson,

1980). According to this view, seizure activity in the mossy fiber pathway damages CA3 pyramidal cells by an unspecified mechanism that does not involve glutamate or glutamate receptors (Ben-Ari et al., 1979; Nadler and Cuthbertson, 1980). The results of studies by Olney et al., (1974), Nadler and colleagues (1980) and Crawford and Connor (1973) led us to form an hypothesis that accomodated both theories. Nadler et al. (1980) showed that among the hippocampal cells most sensitive to the neurotoxic effects of kainate were the cells of the dentate hilus. Some of these interneurons receive dense innervation from the granule cells (Amaral, 1978) and are believed to mediate recurrent inhibition in the granule cell layer (Andersen et al., 1966). We hypothesized at the time that if, as had been suggested by Crawford and Connor (1973), the granule cells use glutamate as a transmitter, then the inhibitory interneurons that receive dense innervation from the granule cells might possess the highest density of glutamate receptors. If, as suggested by Olney and colleagues (1974), kainate acts via glutamate receptors, then these inhibitory interneurons might be preferentially damaged by kainate. Since the loss of inhibition is associated with the onset of seizure activity (Roberts, 1980), we predicted that kainate injection might decrease inhibition first and cause granule cell seizure activity as a result. According to this scenario, granule cell seizure activity would release glutamate from the mossy fibers and cause damage to the CA3 pyramidal cells as a result. This hypothesis would explain how kainate initiates granule cell seizure activity and why transection of the mossy fiber pathway or diazepam protects the CA3 pyramidal cells. Since it was not known at the time if kainate affected inhibition or even if it caused hippocampal granule cell seizure activity, this seemed a worthwhile starting point. Our initial study with kainic acid, and experiments that were the logical extension of it, are reviewed in this chapter. They provide evidence that seizure activity <u>per se</u> causes neuronal damage and that the release of endogenous excitatory amino acids in high concentrations during seizures may mediate epileptic brain damage.

METHODS

Details of the electrophysiological and surgical methods used in these studies have been described previously (Sloviter and Damiano, 1981a,b; Sloviter, 1983). Neuroanatomical techniques utilized include light microscopy using Nissl stains on paraffin- and plastic-embedded tissues (Sloviter, 1983; Sloviter and Dempster, 1985), the sulphide/silver stain for transitional metals (Sloviter, 1985), the Rapid Golgi method of silver dichromate impregnation (Sloviter, 1983), immunocytochemistry to visualize glia (Sloviter and Dempster, 1985) and electron microscopy (Olney et al., 1983; Sloviter and Dempster, 1985).

RESULTS AND DISCUSSION

<u>Kainate-Induced Effects</u>

Our initial study (Sloviter and Damiano, 1981a) was designed to determine if kainate affects granule cell recurrent inhibition before it produces seizures and whether or not kainate produces granule cell seizure activity during the period in which CA3 pyramidal cell damage ensues. Rats anesthetized with urethane were evaluated for recurrent inhibition (using the twin pulse technique) and granule cell responsiveness to perforant path stimuli (Andersen et al., 1966). They were then given kainate (10 mg/kg i.v.) and were evaluated continuously for four hours after injection. Fig. 1 shows that kainate first caused a decrease in inhibition (the ability of the first of two granule cell spikes to inhibit the second) and, later,

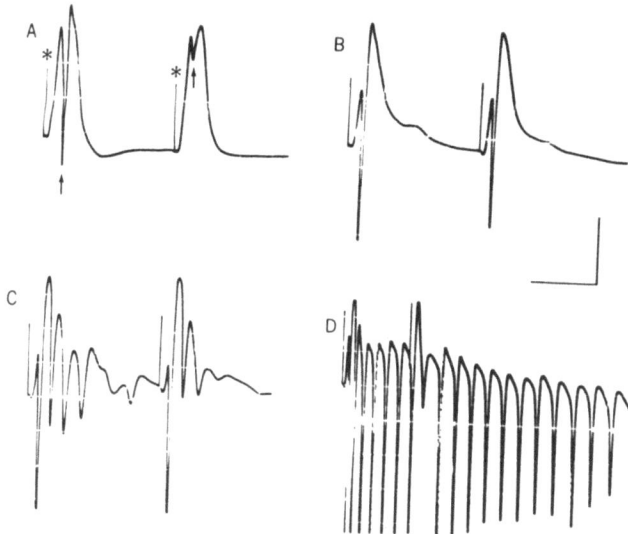

Fig. 1. Effect of kainic acid on hippocampal granule cell activity. A: Pre-drug control. Recurrent inhibition is reflected as the ability of the first granule cell spike (arrow) to prevent the second. B: 50 minutes after kainate (10 mg/kg i.v.). C: 58 minutes. D: 66 minutes. Bars: 20 msec in A,B and C; 40 msec in D; 10 mV all panels (from Sloviter and Damiano, 1981a).

multiple granule cell spiking and epileptiform discharges in response to the same stimuli. Kainate did not increase the amplitude of the extracellularly recorded dendritic negative potential (corresponding to the intracellular EPSP) at the time that granule cell spikes increased and inhibition decreased (unpublished). This experiment showed that kainate: 1) decreased inhibition *in vivo* before granule cell seizure activity was produced and 2) evoked epileptiform discharges in the granule cell layer at a time coincident with the damage to CA3 pyramidal cells. The acute hippocampal damage caused by kainate is shown in Fig. 2. Our conclusion that inhibition is decreased *in vivo* by kainate has been addressed recently by four groups using the *in vitro* hippocampal slice preparation. Westbrook and Lothman (1983) failed to find evidence of decreased inhibition whereas three other groups (Fisher and Alger, 1984; Kehl et al., 1984; Lancaster and Wheal, 1984) concluded that kainate does decrease inhibition. Although our first experiment clearly showed that kainate caused granule cell seizure activity, it did not provide evidence that the seizures in the granule cell layer were causally related to the damage of the CA3 pyramidal cells. This experimental approach highlighted a major problem in the interpretation of results from experiments in which convulsants are used to evoke seizure activity and brain damage. That is, the inability to separate effects that may be due to seizure activity *per se* from those that are the result of a direct neurotoxic effect of the convulsant or other conditions that may develop as a result of widespread seizures, widespread damage and motor convulsions. Therefore, we reasoned that if seizure activity in the hippocampal mossy fiber pathway was the sole cause of hippocampal pyramidal cell loss, then perforant path stimulation that evokes granule cell seizure activity should produce the same pattern of damage as kainate without involving the interpretational problems inherent in the use of kainate or other chemical convulsants.

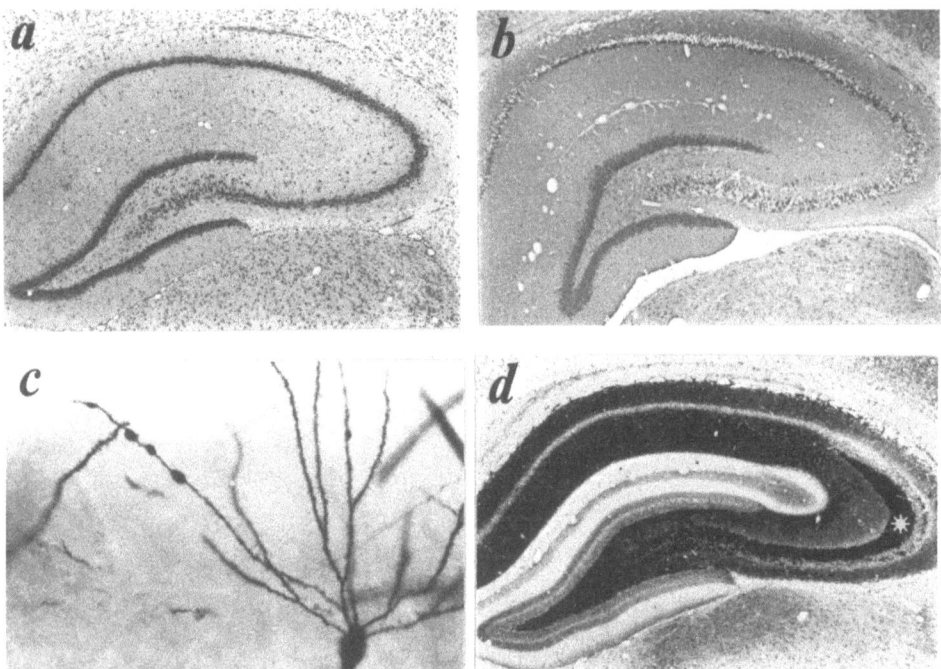

Fig. 2. Acute morphologic effects of kainate. A: control. B: 4 hours after kainate (12 mg/kg i.p.). C: Rapid Golgi stained CA1 pyramidal cell showing focal dendritic swellings. D: normal Timm stained hippocampus showing location of mossy fiber pathway (*) for reference to acute damage pattern in B (from Sloviter and Dempster, 1985).

Perforant Path Stimulation

Figs. 3 and 4 show the effects of 24 hours of intermittent perforant path stimulation on granule cell activity, recurrent inhibition and hippocampal morphology. Unlike kainate, perforant path stimulation produced damage primarily on the side of stimulation. Stimulation-induced damage showed the same hierarchy of sensitivity of hippocampal cell types as caused by kainate (Nadler et al., 1978). This consisted of damage to the interneurons of the hilus and to CA3a and CA3c pyramidal cells.

The active seizure period is characterized by a variety of morphologic changes that are no longer present hours or days after the seizures occur. These effects include acute glial swelling and dendritic dilatation in regions innervated by excitatory pathways. These effects, shown to occur after kainate (Olney et al., 1979; Sloviter and Damiano, 1981a), were reproduced by perforant path stimulation (Sloviter and Damiano, 1981a,b; Olney et al., 1983). Figs. 5a and 5b show the CA1 region of the hippocampus from animals either given kainic acid (12 mg/kg i.p.) four hours before perfusion (a) or stimulated electrically for two hours at 20 Hz (b). Light microscopy of Rapid Golgi- and Nissl-stained thin sections clearly show these acute dendritic swellings (Sloviter and Damiano, 1981b; Sloviter, 1983). Ultrastructural analysis showed that the dendritic swellings induced by electrical stimulation were of the 'axon-sparing' type and therefore virtually identical to those caused by kainate (Olney et al., 1983). In summary, stimulation of the perforant path that evoked granule cell

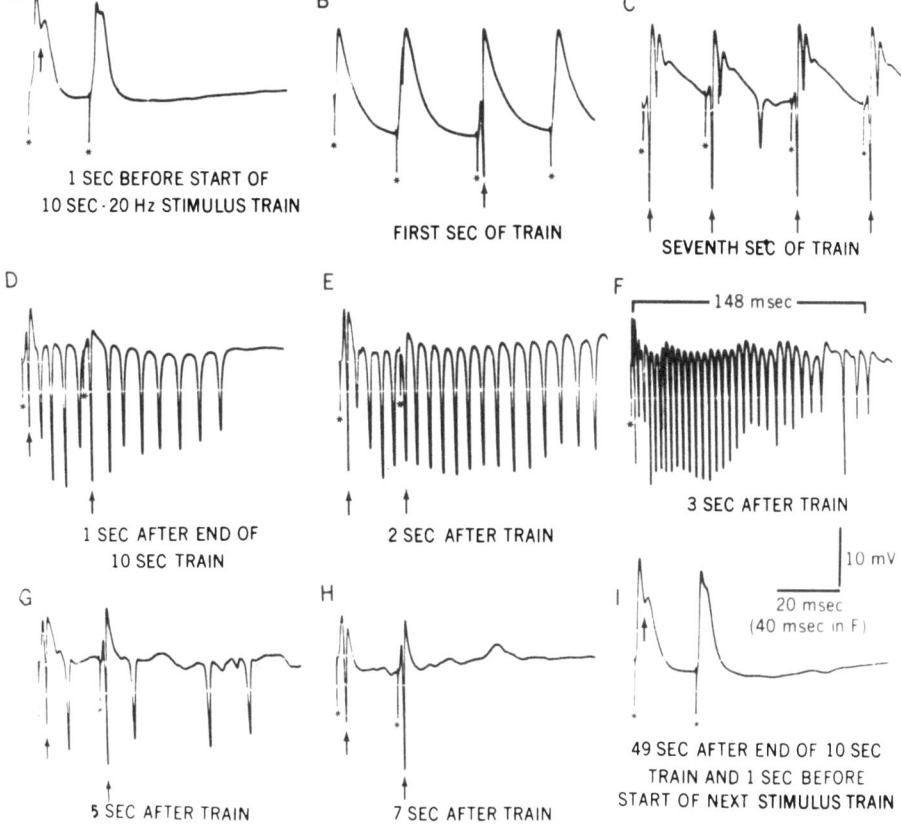

Fig. 3. Responses of ipsilateral granule cells to perforant path stimulation. This figure shows the neuronal events in a single 60-second cycle that was repeated 1,440 times in a 24-hour period (from Sloviter, 1983).

seizure activity, but which did not cause motor convulsions, reproduced all of the effects we observed in kainate-treated hippocampi. These effects include the loss of recurrent inhibition in the granule cell layer, acute glial and dendritic swelling and irreversible neuronal damage showing the same hierarchy of sensitivity as after kainate. These results with electrical stimulation demonstrate that seizure activity per se is neurotoxic to the cells that receive it, that the hippocampal effects of systemically- or intraventricularly-administered kainate are primarily the result of the seizures kainate produces and that the similar damage seen in human epileptic brains (Meldrum and Corsellis, 1984) is most likely the result of excessive excitatory activity from the temporal neocortex to the hippocampus.

Possible Endogenous Mediators of Epileptic Brain Damage

Since the experiments described above using perforant path stimulation showed that increased presynaptic activity in excitatory pathways caused irreversible postsynaptic damage, the next question that arose concerned the identity of the transmitter or transmitters that, by their sustained release, might initiate or mediate the observed seizure-induced damage. On the basis of his studies with excitatory amino acids and a variety of convulsant agents, Olney (1983) suggested that seizure activity causes damage by releasing 'endogenous excitotoxin', probably glutamate or aspar-

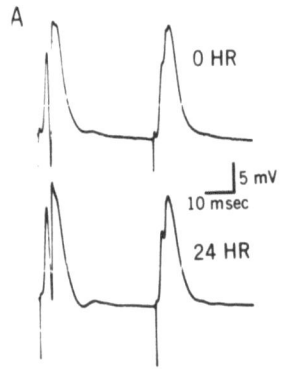 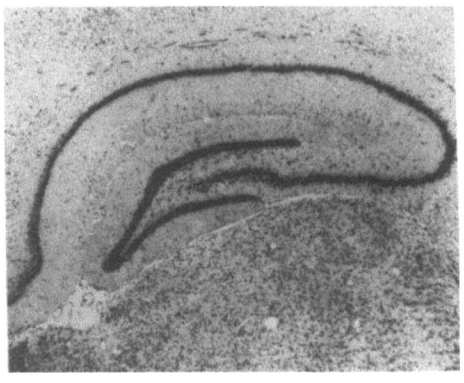

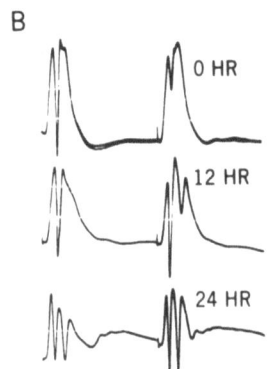 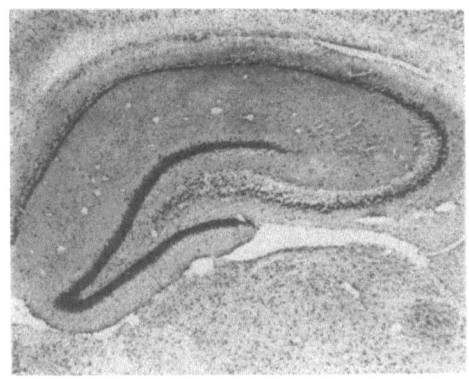

Fig. 4. Effect of 24-hour perforant path stimulation on granule cell recurrent inhibition and hippocampal morphology. A and A': Control. B and B': Stimulated. Note the decreased inhibition (ability of first spike to inhibit next) and kainate-like damage (from Sloviter, 1983).

tate, in concentrations that cause irreversible neuronal death. This has been doubted by several investigators (Nadler et al., 1978; Nadler and Cuthbertson, 1980; Mangano and Schwarcz, 1983; Ben-Ari, 1985), primarily on the basis of a number of studies in which glutamate, when injected directly into the hippocampus, has been either ineffective in reproducing kainate's effects (Nadler et al., 1978) or very weak by comparison, and then only in the immediate region of the injection site (Nadler et al., 1981; Mangano and Schwarcz, 1983; McBean and Roberts, 1984). Ben-Ari (1985) has recently noted that since glutamate has only been shown to cause damage long after injection, it is difficult to understand how glutamate could mediate seizure-associated damage when that damage is caused so rapidly by either kainate or electrical stimulation. In fact, neither glutamate nor aspartate had ever been evaluated for their ability to reproduce immediately the acute hippocampal changes characteristic of seizure-associated hippocampal damage. Previous studies have evaluated morphology days or weeks after a single injection or infusion, a period when the characteristic seizure-associated changes are no longer present. Therefore,

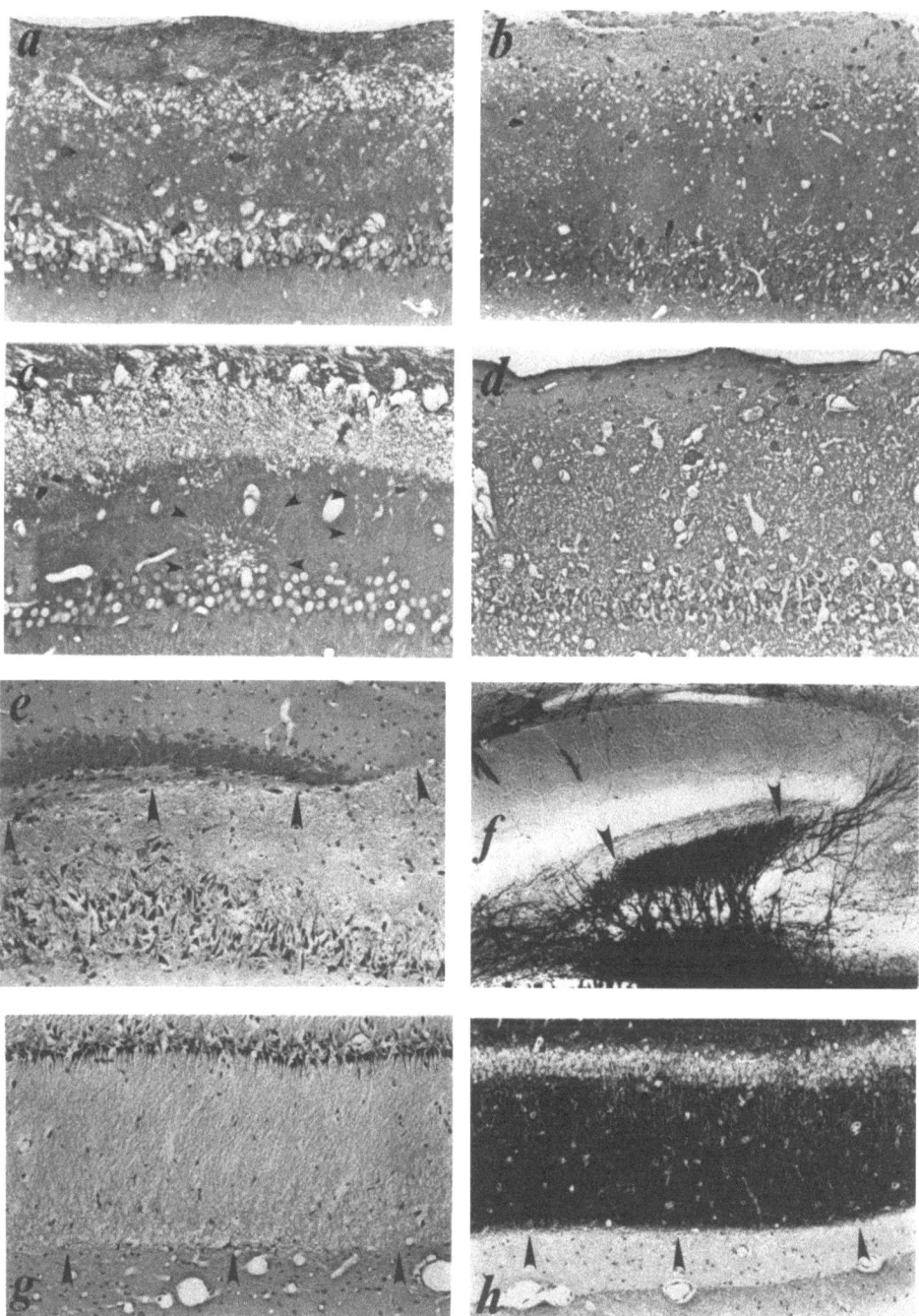

Fig. 5. Acute hippocampal morphology. A: CA1 region four hours after kainate. B: Two-hour perforant path stimulation. C: after one hour of glutamate injections (3 μmoles/5 min/1 hr). D: after same dose GABA. E: CA3 region after glutamate. F: Rapid Golgi-stained hippocampus showing that the location of acute damage in F corresponds to the CA3 pyramidal cells. G: CA1 region after glutamate. H: Timm-stained hippocampus showing that the damage to CA1 stops where the commissural innervation of CA1 (G) ends (arrows) (from Sloviter and Dempster, 1985).

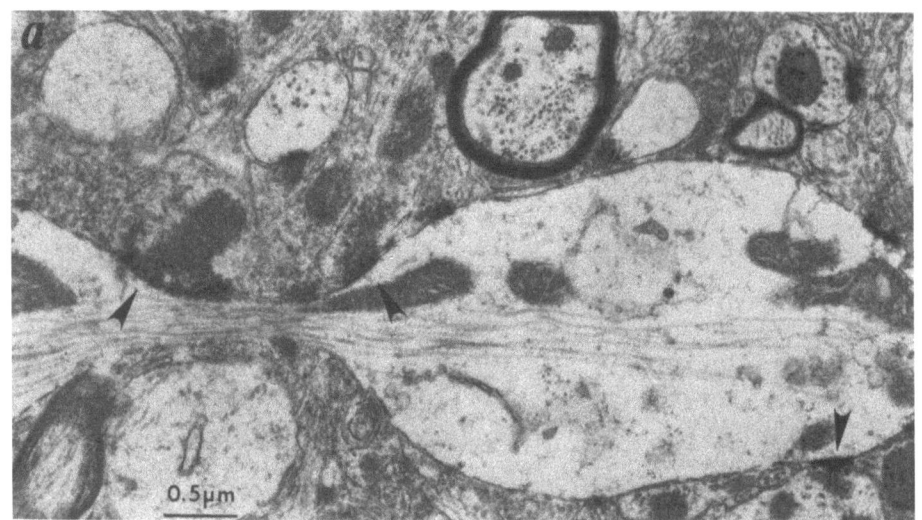
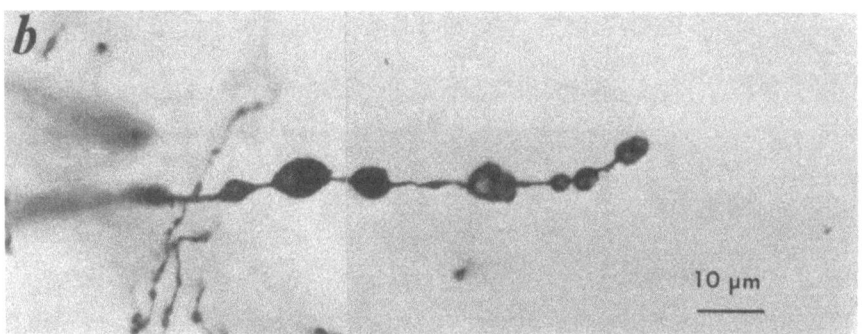
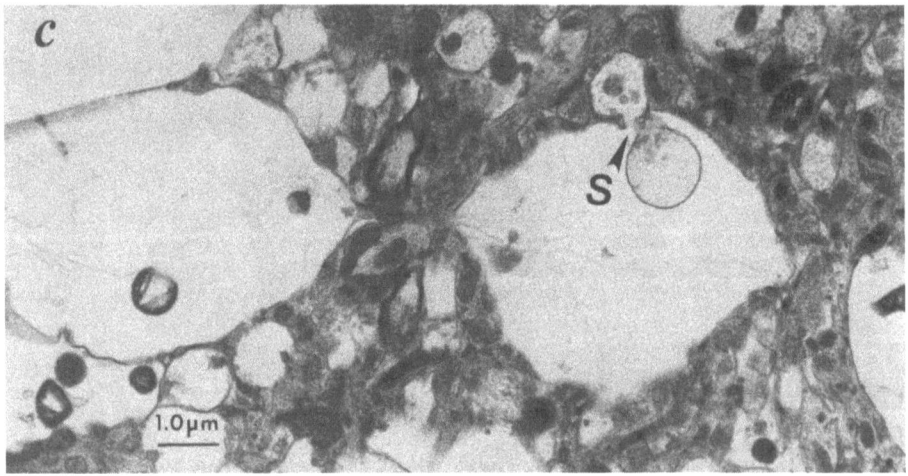

Fig. 6. Acute dendritic swelling after kainate or glutamate injection. A: four hours after kainate (12 mg/kg i.p.). Arrows denote synapses. B: Rapid Golgi-stained CA1 dendrite after one hour of glutamate injections (3 μmoles/5 min/1 hr). C: glutamate (same dose). S denotes dendritic spine surrounded by relatively normal terminals. Note similarity between acute swellings caused by kainate or glutamate (from Sloviter and Dempster, 1985).

it had not been determined if high tissue concentrations of these compounds are capable of rapidly producing the acute morphologic features caused by kainate or perforant path stimulation.

To address this question, we recently injected glutamate, aspartate, GABA, acetylcholine, NaCl or CSF into the lateral ventricle every five minutes for one hour. This route of administration was chosen over intrahippocampal injection to avoid the interpretative problem of differentiating between mechanical and chemical damage. Injections were given every five minutes in order to maintain high tissue concentrations that might otherwise be counteracted by the protective processes of uptake, diffusion and metabolism. Morphology was evaluated immediately after the one hour of injections as well as four weeks later in order to determine if neuron loss and gliosis occurred (Sloviter and Dempster, 1985). Fig. 5 shows the acute effects of glutamate and GABA on hippocampal morphology. Glutamate and aspartate (3 µmoles/5 min for one hour) produced acute glial and dendritic swelling as well as neuronal necrosis in the regions adjacent to the ventricle. GABA, closely related structurally to glutamate but inhibitory in nature, produced only acute glial swelling. Acetylcholine produced no morphologic changes in the tissue adjacent to the ventricle but did cause dendritic swellings in the regions innervated by the temporo-ammonic portion of the perforant path, presumably as a result of seizure activity induced in this pathway. Rapid Golgi light microscopy and electron microscopy of glutamate and aspartate-treated hippocampi revealed that, like kainate and perforant path stimulation, both excitatory amino acids caused focal dendritic swellings of the 'axon-sparing' type (Fig. 6). Four weeks after injection, glutamate and aspartate, but not GABA or acetylcholine, caused irreversible neuron loss and reactive gliosis (Sloviter and Dempster, 1985). These data show that high doses of glutamate or aspartate are capable of reproducing all of the acute morphologic effects caused by kainate or perforant path stimulation and that they do so in the same short period during which seizure-induced damage occurs (Sloviter and Dempster, 1985).

The hypothesis that glutamate is responsible for the seizure-induced damage is supported by two recent findings. One is the immunocytochemical localization of glutamate in the terminals of the mossy fibers (Storm-Mathisen et al., 1983). The other is the autoradiographic localization of kainate-displaceable glutamate receptors in the hippocampal region innervated by the mossy fiber terminals (Monaghan and Cotman, 1982; Monaghan et al., 1983; Greenamyre et al., 1985). This suggests that glutamate is the natural ligand for these 'kainate' receptors. The evidence supporting the view that an endogenous amino acid receptor agonist, possibly glutamate, mediates epileptic brain damage, can now be summarized as follows: 1) a variety of convulsants, some of which are not directly neurotoxic, cause seizure activity, convulsions and a similar pattern of brain damage (Olney, 1983); 2) electrical stimulation of the perforant path, which causes seizure activity in hippocampal pathways but not motor convulsions, causes focal hippocampal damage identical to that caused by convulsants (Sloviter and Damiano, 1981b; Olney et al., 1983; Sloviter, 1983); 3) the hippocampal mossy fiber pathway, which projects from the granule cells of area dentata to the CA3 pyramidal cells of regio inferior, contains glutamate (Crawford and Connor, 1973; Storm-Mathisen et al., 1983), has the capacity to take up glutamate from the extracellular space (Storm-Mathisen and Iversen, 1979) and probably releases glutamate upon electrical stimulation (Crawford and Connor, 1973); 4) the region of the hippocampus that receives the mossy fiber innervation contains a high density of high affinity glutamate binding sites (Monaghan et al., 1983; Greenamyre et al., 1985); and 5) glutamate and aspartate, when injected directly into the brain, produce morphologic changes in the hippocampus that are qualitatively similar to those caused by convulsants or perforant path stimulation (Sloviter and Dempster, 1985).

Although the case for glutamate as the mediator of epileptic brain damage seems quite strong, it has not been proved. Glutamate is probably contained in all cells in its role in intermediary metabolism and therefore the presence of glutamate does not, in itself, imply a transmitter function. Nor have all the criteria (Werman, 1966) been met for identifying glutamate as the transmitter of any pathway, let alone the mossy fibers in particular. Glutamate-like immunoreactivity (Storm-Mathisen et al., 1983) could signify a glutamate-containing moiety distinct from glutamic acid or any antigenic determinant with immunologic reactivity. Similarly, glutamate uptake in vitro does not necessarily imply physiologic function. The availability of relatively specific and selective amino acid antagonists would clarify these issues to a considerable degree. The possibility that other endogenous ligands could mediate seizure-related damage warrants study. It remains to be determined if endogenous excitatory compounds other than glutamate and aspartate, e.g., quinolinic acid (Schwarcz et al., 1984), are capable of reproducing all of the acute morphologic effects described above.

Another substance known to be present in the mossy fiber pathway has been suggested to be involved in some way in the seizure-induced damage to the CA3 pyramidal cells. This substance is zinc. Although a functional role for synaptic metals in general, and zinc in particular, has been suggested for the past quarter century (Crawford, 1983), the reason for zinc's presence in neural pathways is obscure. Since the mossy fibers are rich in zinc (Haug, 1967, 1973) and because mossy fiber seizure activity is lethal to cells that receive it, we addressed the role of mossy fiber zinc in the seizure-induced phenomena described above by evaluating the pattern of histochemically stainable metal in both the stimulated- and unstimulated hippocampi of the same animal (Sloviter, 1985). The results of that study showed that after 24 hours of intermittent stimulation, there was a selective and nearly total loss of histochemically stainable metal in the mossy fiber pathway of the stimulated hippocampus. The basis of the sulphide/silver stain is the precipitation of reactable metal by sulphide. Therefore, the selective loss of staining could be due to either a release and loss of synaptic zinc or a change in its presynaptic chemical disposition that reduces the amount of metal available to react with the sulphide. In view of recent reports that treatments that cause hippocampal depolarization release synaptic zinc (Assaf and Chung, 1984; Howell et al., 1984; Charton et al., 1985), zinc release during seizures could play a role in the postsynaptic damage. A presynaptic role for zinc distinct from a release phenomenon also seems possible. It may be significant that glutamate and zinc, both of which are believed to be present in the mossy fibers (Haug, 1967; Crawford and Connor, 1973; Storm-Mathisen et al., 1983), interact chemically in vitro to form glutamatozinc (II) dihydrate (Freeman, 1967). There is also a striking parallelism between hippocampal regions that contain glutamate-like immunoreactivity (Storm-Mathisen et al., 1983) and high affinity glutamate binding sites (Monaghan et al., 1983; Greenamyre et al., 1985) and those regions that contain synaptic metal (Haug, 1967, 1973). If excessive activity in excitatory amino acid-containing pathways is potentially lethal to the cells they innervate, and therefore undesirable biologically, perhaps synaptic zinc, capable of forming a polymeric complex with synaptic glutamate, plays a physiologically protective role by binding presynaptic glutamate available for release. The additional possibility that zinc, if released, may potentiate or inhibit the binding of amino acids to their receptors by a similar chemical interaction also deserves consideration. If zinc does play either a pre- or postsynaptic role in seizure-related hippocampal damage, the evidence analogous to that described above for glutamate, remains to be obtained. In particular, zinc has not yet been shown to produce the acute seizure-associated morphologic changes induced by perforant path stimulation or by glutamate or aspartate.

Relevance to Clinical Epilepsy

The results described above demonstrate that electrical stimulation of a single excitatory pathway evokes focal seizure activity and irreversibly damages the cells that receive synaptic input from the discharging neurons. This effect is produced in normal adult animals without causing motor convulsions or a cessation of breathing. Continuous recording of electrical activity in the cell layer in which the seizure activity originates shows that granule cell spiking and epileptiform activity occurs throughout the period during which the damage to the immediately adjacent hilus and hippocampus occurs. This eliminates the possibility that there is general anoxia in the hippocampus during this period. Regardless of the final mechanism responsible for seizure-associated cell death, the results presented above provide conclusive evidence that excessive activity in at least some excitatory pathways damages the cells they innervate. If seizure activity *per se*, distinct from metabolic disturbances that may accompany seizures, is the cause of epileptic brain damage occurring after status epilepticus or a lifetime of periodic seizure episodes (Meldrum and Corsellis, 1984), then each seizure must be regarded as potentially brain damaging. Therefore, an important question to be answered is: how much seizure activity can be endured before irreversible brain damage results? This question is presently under investigation in our laboratory but preliminary results show that irreversible hippocampal damage is produced in normal anesthetized adult rats in less than one hour by perforant path stimulation at 20 Hz. Although it is impossible to determine how many minutes of brain-damaging seizures in adult rats are equivalent to a given amount of seizure activity in pediatric or adult patients, it is clear that, regardless of the state of oxygenation, sustained neuronal firing rapidly causes significant permanent damage. Given the strikingly similar patterns of hippocampal damage in experimental animals and epileptic humans (Meldrum and Corsellis, 1984), this factor should clearly be taken into account when considering how soon to begin pharmacological or surgical intervention.

ACKNOWLEDGEMENTS

This research was supported by grants from the Epilepsy Foundation of America and the National Institute of Neurological and Communicative Disorders and Stroke (R01 NS 18201).

REFERENCES

Amaral, D.G., 1978, A Golgi study of cell types in the hilar region of the hippocampus in the rat, *J. Comp. Neurol.*, 182:851.

Andersen, P., Holmqvist, B., and Voorhoeve, P.E., 1966, Entorhinal activation of dentate granule cells, *Acta Physiol. Scand.*, 66:448.

Assaf, S.Y., and Chung, S.-H., 1984, Release of endogenous zinc from brain tissue during activity, *Nature*, 308:734.

Ben-Ari, Y., Tremblay, E., Ottersen, O.P., and Naquet, R., Evidence suggesting secondary epileptogenic lesions after kainic acid: pre-treatment with diazepam reduces distant but not local brain damage, *Brain Res.*, 165:362.

Ben-Ari, Y., 1985, Limbic seizure and brain damage produced by kainic acid: mechanisms and relevance to human temporal lobe epilepsy, *Neuroscience*, 14:375.

Charton, G., Rovira, C., Ben-Ari, Y., and Leviel, V., 1985, Spontaneous and evoked release of zinc in the hippocampal mossy fiber zone of the rat in situ, Exp. Brain Res., 58:202.

Crawford, I.L., and Connor, J.D., 1973, Localization and release of glutamic acid in relation to the hippocampal mossy fiber pathway, Nature, 244:442.

Crawford, I.L., 1983, Zinc and the hippocampus, in: Neurobiology of the Trace Metals, I.E. Dreosti and R.M. Smith, eds., Humana, Clifton, New Jersey, p. 163.

Fisher, R.S., and Alger, B.E., 1984, Electrophysiological mechanisms of kainic acid-induced epileptiform activity in the rat hippocampal slice, J. Neurosci., 4:1312.

Freeman, H., 1967, Crystal structure of metal-peptide complexes, Adv. Protein Chem., 22:257.

Greenamyre, J.T., Olson, J.M.M., Penney, J.B., and Young, A.B., 1985, Autoradiographic characterization of N-methyl-D-aspartate-, quisqualate- and kainate-sensitive glutamate binding sites, J. Pharmacol. Exp. Ther., 233:254.

Haug, F.-M.S., 1967, Electron microscopical localization of the zinc in hippocampal mossy fiber synapses by a modified sulphide silver procedure, Histochemie, 8:355.

Haug, F.-M.S., 1973, Heavy metals in the brain, Adv. Anat. Embryol. Cell Biol., 47:1.

Howell, G.A., Welch, M.G., and Frederickson, C.J., 1984, Stimulation-induced uptake and release of zinc in hippocampal slices, Nature, 308:736.

Kehl, S.J., McLennan, H., and Collingridge, G.L., 1984, Effects of folic and kainic acids on synaptic responses of hippocampal neurones, Neuroscience, 11:111.

Lancaster, B., and Wheal, H.V., 1984, Chronic failure of inhibition of the CA1 area of the hippocampus following kainic acid lesions of the CA3/4 area, Brain Res., 295:317.

Mangano, R.M., and Schwarcz, R., 1983, Chronic infusion of endogenous excitatory amino acids into rat striatum and hippocampus, Brain Res. Bull., 10:47.

McBean G.J., and Roberts, P.J., 1984, Chronic infusion of L-glutamate causes neurotoxicity in rat striatum, Brain Res., 290:372.

Meldrum, B.S., and Corsellis, J.A.N., 1984, Epilepsy, in: Greenfield's Neuropathology, J.H. Adams, J.A.N. Corsellis and L.W. Duchen, eds., Wiley, New York, p. 921.

Monaghan, D.T., and Cotman, C.W., 1982, The distribution of 3-H-kainic acid binding sites in rat CNS as determined by autoradiography, Brain Res., 252:91.

Monaghan, D.T., Holets, V.R., Toy, D.W., and Cotman, C.W., 1983, Anatomical distributions of four pharmacologically distinct 3-H-L-glutamate binding sites, Nature, 306:176.

Nadler, J.V., Perry, B.W., and Cotman, C.W., 1978, Intraventricular kainic acid preferentially destroys hippocampal pyramidal cells, Nature, 271:676.

Nadler, J.V., and Cuthbertson, G.J., 1980, Kainic acid neurotoxicity toward hippocampus: dependence on specific excitatory pathways, Brain Res., 195:47.

Nadler, J.V., Evenson, D.A., and Cuthbertson, G.J., 1981, Comparative toxicity of kainic acid and other acidic amino acids toward rat hippocampal neurons, Neuroscience, 6:2505.

Olney, J.W., Rhee, V., and Ho, O.L., 1974, Kainic acid: a powerful neurotoxic analogue of glutamate, Brain Res., 77:507.

Olney, J.W., and de Gubareff, T., 1978, Glutamate neurotoxicity and Huntingtons chorea, Nature, 271:557.

Olney, J.W., Fuller, T., and de Gubareff, T., 1979, Acute dendrotoxic changes in the hippocampus of kainate treated rats, Brain Res., 76:91.

Olney, J.W., 1983, Excitotoxins, in: Excitotoxins, K. Fuxe, P. Roberts, and R. Schwarcz, eds., Macmillan, London, p. 82.

Olney, J.W., de Gubareff, T., and Sloviter, R.S., 1983, 'Epileptic' brain damage in rats induced by sustained electrical stimulation of the perforant path. II. Ultrastructural analysis of acute hippocampal pathology, Brain Res. Bull., 10:699.

Roberts, E., 1980, Epilepsy and antiepileptic drugs: a speculative synthesis, in: Antiepileptic Drugs: Mechanisms of Action, G.H. Glaser, J.K. Penry and D.M. Woodbury, eds., Raven Press, New York, p. 667.

Schwarcz, R., Brush, G.S., Foster, A.C., and French, E.D., 1984, Seizure activity and lesions after intrahippocampal quinolinic acid injection, Exp. Neurol., 84:1.

Sloviter, R.S., and Damiano, B.P., 1981a, On the relationship between kainic acid-induced epileptiform activity and hippocampal neuronal damage, Neuropharmacology, 20:1003.

Sloviter, R.S., and Damiano, B.P., 1981b, Sustained electrical stimulation of the perforant path duplicates kainate-induced electrophysiological effects and hippocampal damage in rats, Neurosci. Lett., 24:279.

Sloviter, R.S., 1983, 'Epileptic' brain damage in rats induced by sustained electrical stimulation of the perforant path. I. Acute electrophysiological and light microscopic studies, Brain Res. Bull., 10:675.

Sloviter, R.S., 1985, A selective loss of hippocampal mossy fiber Timm stain accompanies granule cell seizure activity induced by perforant path stimulation, Brain Res., 330:150.

Sloviter, R.S., and Dempster, D.W., 1985, 'Epileptic' brain damage is replicated qualitatively in the rat hippocampus by central injection of glutamate or aspartate but not by GABA or acetylcholine, Brain Res. Bull., 15:39.

Storm-Mathisen, J., and Iversen, L.L., 1979, Uptake of 3-H glutamic acid in excitatory nerve endings: light and electron microscopic observations in the hippocampal formation of the rat, Neuroscience, 4:1237.

Storm-Mathisen, J., Leknes, A.K., Bore, A.T., Vaaland, J.L., Edminson, P., Haug, F.-M.S., and Ottersen, O.P., 1983, First visualization of glutamate and GABA in neurones by immunocytochemistry, Nature, 301:517.

Werman, R.A., 1966, Criteria for identification of a central nervous system transmitter, Comp. Biochem. Physiol., 18:745.

Westbrook, G.L., and Lothman, E.W., 1983, Cellular and synaptic basis of kainic acid-induced hippocampal epileptiform activity, Brain Res., 273:97.

KAINIC ACID SEIZURES AND NEURONAL CELL DEATH: INSIGHTS FROM STUDIES OF

SELECTIVE LESIONS AND DRUGS

J.V. Nadler, M.M. Okazaki, M. Gruenthal, B. Ault[1], and
D.R. Armstrong

Department of Pharmacology, Duke University Medical Center
Durham, North Carolina 27710, USA
[1]Department of Pharmacology, Burroughs-Wellcome Co., Research
Triangle Park, North Carolina 27709, USA

INTRODUCTION

Hippocampal pathology (Ammon's horn sclerosis (AHS)) is a well-recognized finding in the brains of temporal lobe epileptics (Meldrum and Corsellis, 1984). Classical AHS involves extensive loss of neurons from hippocampal area CA1 (h_1, Sommer sector), a less extensive neuronal deficit in the CA3-CA4 area (h_3-h_5, endblade, endfolium), and relative sparing of neurons in the h_2 area ('resistant zone'; area CA2 and the adjacent portion of area CA3a which contains the mossy fiber endbulb) and in the fascia dentata. Most commonly, some neuronal loss in other brain regions, particularly the amygdala, thalamus and cerebral neocortex, accompanies the hippocampal lesion. Although the near-total loss of neurons from area CA1 is the most striking feature of AHS in many patients, there is reason to believe that the most vulnerable neurons are the CA3 hippocampal pyramidal cells and the morphologically diverse neurons of area CA4 (Margerison and Corsellis, 1966).

AHS was first described more than a century ago (Sommer, 1880), yet even today the question of its relation to limbic or complex partial seizures remains unresolved. AHS, regardless of etiology, might contribute to the development of temporal lobe epilepsy in a number of ways, such as by deletion of inhibitory interneurons or the formation of abnormal excitatory circuits (Tauck and Nadler, 1985). Conversely, numerous studies have established that seizures, or at least status epilepticus, can produce lesions similar to AHS in experimental animals (Meldrum, 1983). In these cases, neuronal cell death evidently results from excessive seizure activity per se and not from secondary effects, such as hypoxia or hypotension. We have employed the potent convulsant kainic acid (KA) as a tool with which to investigate the conditions under which limbic seizures lead to neuronal cell death (Nadler, 1981; Ben-Ari, 1985).

KA is especially useful for this purpose, because it provokes in experimental animals seizures that appear to originate in and remain largely confined to the limbic system (Ben-Ari et al., 1980, 1981; Menini et al., 1980; Lothman and Collins, 1981). Moreover, when sufficient doses are administered, the seizures are accompanied by brain lesions that closely resemble AHS. Although most workers have administered KA parenterally, we find that administration by the intracerebroventricular (i.c.v.) route has several important advantages. First, the lesions produced by any given dose of KA are much more consistent from one animal to the next. Second, i.c.v. KA preferentially destroys hippocampal pyramidal cells over neurons in other regions of the brain, whereas this distinction is much less obvious in animals given intravenous KA. Third, by varying the dose of KA one can readily produce a graded series of lesions similar to those identified in temporal lobe epileptics. Finally, the edema-related degeneration of CA1 pyramidal cells often found in animals given KA parenterally (Lassmann et al., 1984) occurs only in those animals given i.c.v. KA that experience the most intense electrographic and behavioral seizures. Considering the somewhat variable neuropathological findings in temporal lobe epileptics, all of which are generally classed as AHS, it seems a moot point whether i.c.v. or parenteral administration of KA more accurately replicates AHS in experimental animals.

Several lines of evidence suggest that lesions made by parenterally-administered KA and by KA injected into a brain region distant from the site of neuronal degeneration result from excessive seizure activity, rather than from a direct toxic action of the convulsant (Nadler, 1981; Ben-Ari, 1985). Less extensive evidence supports a similar mechanism for the neuronal degeneration that follows i.c.v. administration of KA. Thus administration of diazepam (Ben-Ari et al., 1980) or acute interruption of excitatory pathways within the limbic system (Nadler and Cuthbertson, 1980; Nadler et al., 1981) reduces the extent of neuronal degeneration. The studies described here strongly reinforce the conclusion that brain lesions made by i.c.v. KA are predominantly seizure-mediated.

INSIGHTS FROM STUDIES OF CONVULSANTS AND ANTICONVULSANTS

The notion that brain lesions made by i.c.v. KA result from excessive limbic seizure activity implies that other convulsants administered by the same route will produce similar lesions, provided that they elicit limbic seizure activity of the same duration. Furthermore, anticonvulsants would be expected to reduce the total seizure duration and extent of neuronal degeneration in parallel. Studies designed to test these predictions have also allowed us to define the relationship between limbic seizure duration and neuronal cell death.

These studies were performed on adult male Sprague-Dawley rats chronically implanted with EEG recording electrodes bilaterally in area CA3 of the rostral hippocampus and in the basolateral amygdala. Neocortical surface EEG, when desired, was recorded from skull screws inserted in the parietal bones. Convulsants were infused through a cannula placed in one of the lateral ventricles at the level of the septum. They were dissolved in 2.5 µl of artificial CSF (Elliott, 1969) and delivered at a rate of 0.2 µl/min, while the animal was temporarily restrained. Continuous EEG recordings were obtained before, during and for four hours after the end of the infusion. One day later, the animal was perfused with buffered paraformaldehyde and the extent of neuronal degeneration was determined histologically by use of silver impregnation (Nadler and Evenson, 1983) as well as cresyl violet.

Effects of KA and Bicuculline Methiodide

Animals treated with KA alone were given a dose (0.94–1.41 nmol) sufficient to produce a prolonged status epilepticus and extensive brain lesions in every case. EEG recordings initially demonstrated an increase in frequency and/or amplitude. Animals first exhibited evidence of limbic seizure activity 5–15 minutes after the beginning of the infusion (Fig. 1).

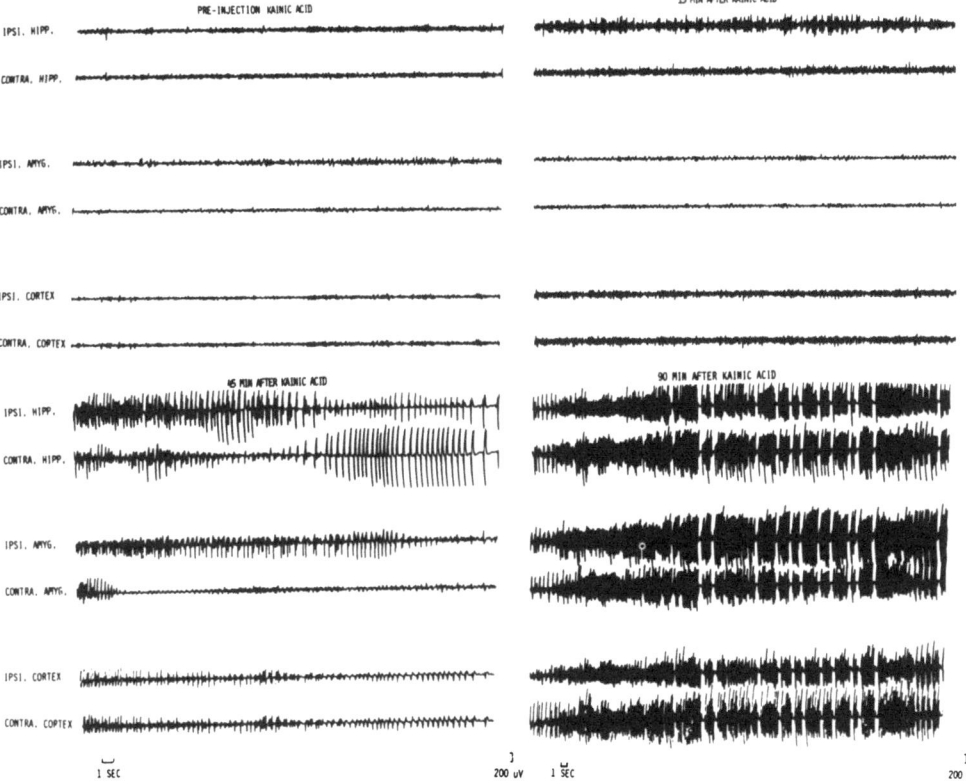

Fig. 1. EEG recordings taken before and at various times after i.c.v. administration of KA. Note the different time calibration for the 90-minute recording.

Each discrete ictal episode (defined as a five-second or longer period of synchronized sharp waves at least twice the preictal amplitude and separated by no more than two seconds recorded simultaneously from at least two limbic regions) detected during the first 20-30 minutes was first recorded from either the ipsilateral hippocampal lead only or from the ipsilateral hippocampal lead and one other (normally the ipsilateral amygdala lead). The duration of each ictal episode progressively increased and the duration of interictal periods progressively decreased during approximately the first hour until interictal periods essentially disappeared. This condition, amounting to a status epilepticus, persisted well beyond the end of the recording period.

Behavioral seizures correlated temporally with EEG events. During discrete electrographic seizures, animals remained immobile with fixed gaze and exhibited facial automatisms and some head nodding (equivalent to class 1-2 limbic kindled seizures (Racine, 1972)). Status epilepticus was associated with essentially continuous head nodding; forelimb clonus and rearing (equivalent to class 3-4 limbic kindled seizures) were only occasionally observed. Class 3-4 seizures were associated with activation of the neocortical EEG. At all other times, neocortical records appeared to reflect activity volume-conducted from underlying limbic regions.

Hippocampal lesions involved nearly all CA3 pyramidal cells and some CA4 neurons throughout the rostrocaudal extent of this region ipsilateral

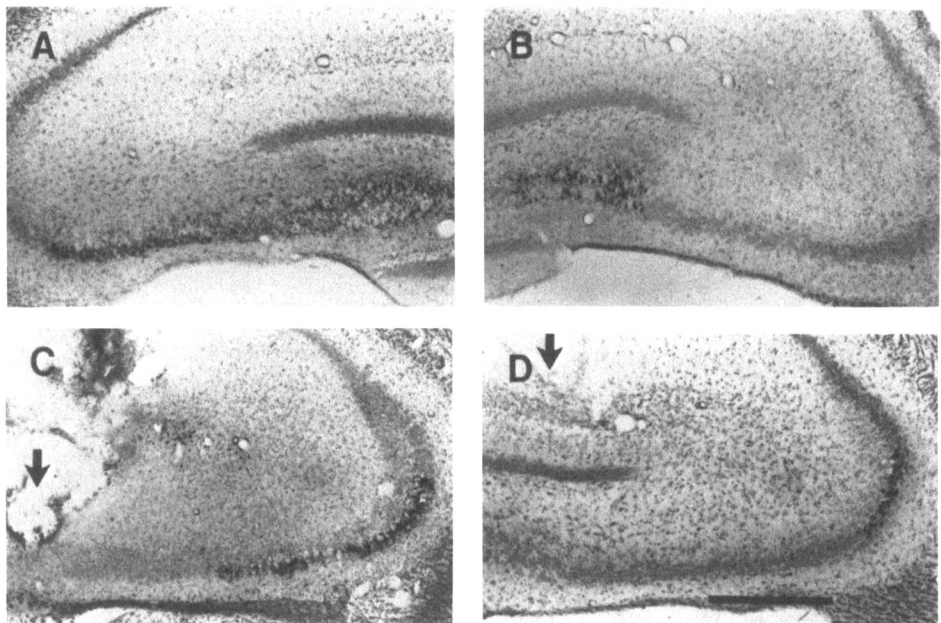

Fig. 2. Degenerating neurons in hippocampal area CA3 after i.c.v. infusion of convulsants with or without an anticonvulsant. Silver impregnation. (A) KA alone, (B) BMI alone, (C) KA + baclofen, (D) KA + phenobarbital. Filled arrows indicate electrode tracks. Note degeneration after i.c.v. BMI and attenuation of the CA3 lesion by baclofen and phenobarbital. Scale bar = 0.5 mm.

to the i.c.v. cannula (Fig. 2A), as well as some rostral CA3 neurons contralateral to the cannula. Sparse bilateral degeneration of CA1 pyramidal cells was noted in several cases, but damage to the h_2 area and fascia dentata occurred only in those rats that suffered the most extreme seizures. Other regions of neuronal degeneration included the basolateral and medial posterior amygdala, pyriform cortex, claustrum-insula region, several of the thalamic nuclei, zona incerta and layers III, V and VI of some portions of the cerebral neocortex.

I.c.v. administration of 5-6 nmol of bicuculline methiodide (BMI) produced electrographic and behavioral seizure patterns quite distinct from those produced by i.c.v. KA. Electrographic seizure activity began abruptly and simultaneously in all regions and consisted of alternating periods of slow (0.5-1 Hz) and fast (>3 Hz) activity (Fig. 3). Total seizure duration in six rats ranged from 2 to 123 minutes. There was no evidence of the rhythmic 'cycling' characteristic of KA seizures. Behavioral seizures consisted initially of a few brief tonic-clonic convulsions followed, after the aminal was released from the restrainer, by wild running. These behaviors did not correlate with electrographic events and ceased within 15 minutes after the end of the infusion. Electrographic seizures persisted beyond this point in three animals and were then associated with kindled seizure-like behaviors identical to those observed after KA infusion.

Despite the different seizure patterns elicited by i.c.v. KA and BMI, they produced similar patterns of brain damage (Fig. 2A, B). Degenerating CA3 pyramidal cells (in either area CA3a or CA3c) and a few degenerating

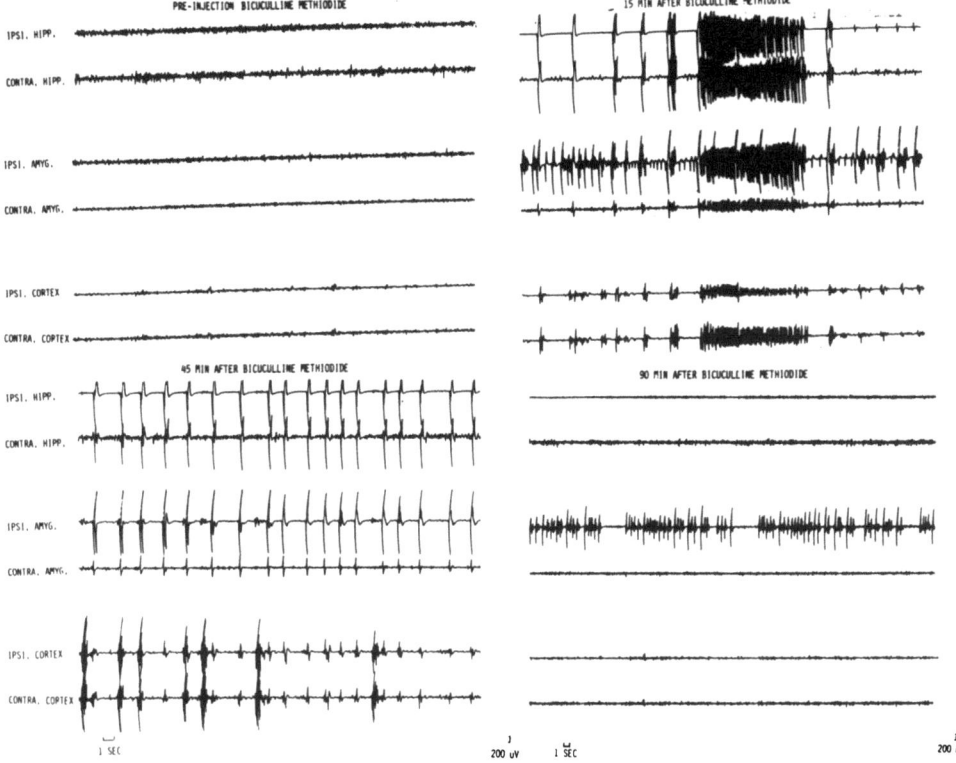

Fig. 3. EEG recordings taken before and at various times after i.c.v. administration of BMI. Note the different time calibration for the 90-minute recording.

neurons in other vulnerable regions were identified in the three BMI-treated rats that experienced 57-123 minutes of limbic seizure. No degenerating neurons were identified in three other rats that experienced 18 minutes of limbic seizure or less.

Anticonvulsant Actions of Phenobarbital and Baclofen

Phenobarbital and baclofen were tested for their ability to inhibit KA-induced electrographic seizures, on the one hand, and neuronal degeneration, on the other. Phenobarbital is employed clinically to suppress many types of epileptic seizure, including complex partial seizures, whereas baclofen is virtually untested as an anticonvulsant in man. Nevertheless, both drugs reduce the rate of KA-induced bursting in the hippocampal CA3 area in vitro at concentrations within the range of therapeutic CSF levels (Clifford et al., 1982; Gruenthal et al., 1984). Phenobarbital (40 mg/kg, i.p.) was administered 15 minutes before the beginning of KA infusion and baclofen (5 mg/kg, i.p.) was administered at the end of the infusion. At these doses, both drugs were mildly sedating.

Phenobarbital and baclofen each prevented status epilepticus in five of six rats given 0.94 nmol of KA. In the ten successful trials total residual limbic seizure activity ranged from 5 to 94 minutes. The anticonvulsants reduced the abundance of neuronal degeneration in parallel with the reduction in duration of electrographic limbic seizures (Fig. 2C,D). In animals protected from status epilepticus, residual neuronal degeneration, if present, was essentially confined to the rostral CA3 area ipsilateral to the i.c.v. cannula, whereas the two rats that experienced status epilepticus developed widespread lesions characteristic of animals treated with KA alone. Even in the ipsilateral CA3 area anticonvulsants considerably reduced the extent of the damage. In three animals, no degenerating neurons could be found. It should be noted that 40 mg/kg of phenobarbital does not protect hippocampal neurons from focally-injected KA (Zaczek et al., 1981) and neither does 5 mg/kg of baclofen, at least in anesthetized rats (unpublished observations).

Six additional rats were treated with baclofen 15 minutes before i.c.v. infusion of BMI. Five rats experienced from 0 to 24 minutes of limbic seizure and neuronal degeneration was absent. The sixth rat experienced a prolonged status epilepticus and suffered brain damage equivalent to that found in rats whose status epilepticus was provoked by i.c.v. KA.

Relation Between Limbic Seizure Duration and Neuronal Cell Death

By pooling the data from our pharmacological studies, we have derived a relationship between the occurrence of neuronal cell death and the total duration of limbic seizure activity (Fig. 4). In the hippocampus ipsilateral to the i.c.v. cannula, neuronal degeneration invariably appeared after at least 40 minutes of electrographically recorded limbic seizure. In some cases, we detected seizure-related hippocampal degeneration after as little as 15-23 minutes.

INSIGHTS FROM STUDIES OF SELECTIVE LESIONS

By making selective lesions of various hippocampal pathways singly and in combination, we found that only mossy fiber lesions protected CA3 pyramidal cells from the lethal effect of i.c.v. KA (Table 1; Nadler and

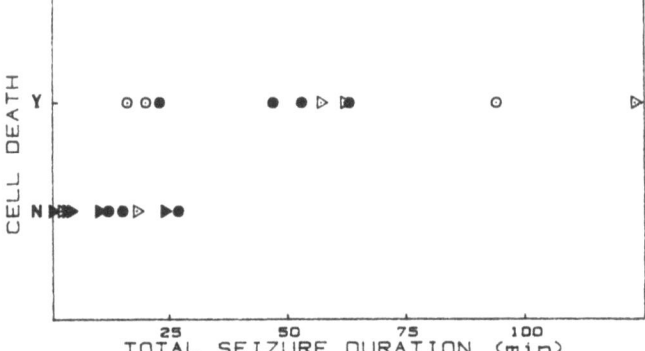

Fig. 4. Neuronal degeneration (Y) or lack of neuronal degeneration (N) in the ipsilateral hippocampus related to total limbic seizure duration after i.c.v. administration of BMI alone (open triangles), BMI + baclofen (solid triangles), KA + baclofen (solid circles) or KA + phenobarbital (open circles). Subjects that developed status epilepticus have been excluded.

Table 1. Prevention of Hippocampal Neuronal Cell Death by Prior Lesion

Lesion of	I.C.V. Kainic Acid	Focal Kainic Acid
Mossy fibers	+	−
Entorhinal cortex	−	±
Septum	−	±
Entorhinal cortex + Septum	−	+
Contralateral hippocampus	−	−

The indicated lesions were made 2.5-4 days before administration of either 3.75 (i.c.v.) or 2.34 nmol (focal) of KA under pentobarbital anesthesia. (+) All vulnerable neurons protected, (±) protection of CA1 pyramidal cells and dentate granule cells only, (−) no protection.

Cuthbertson, 1980; Nadler et al., 1981). Mossy fiber lesions were effective immediately, suggesting that the destruction of CA3 pyramidal cells depends on impulse flow within the mossy fiber pathway. In contrast, mossy fiber lesions did not protect CA3 pyramidal cells from focally-injected KA; only ablation of septal and entorhinal innervation prevented cell death in this instance and the protective effect appeared only after enough time had elapsed for terminal degeneration. Thus, the critical excitatory pathways for destruction of neurons by i.c.v. and focally-injected KA are completely different. This was the first clear evidence that KA injected by the two routes destroys hippocampal neurons by different mechanisms.

We initially interpreted our lesion data to suggest that KA-induced degeneration of CA3 pyramidal cells requires a functional mossy fiber pathway simply because the mossy fibers serve as an obligatory link in the generation of hippocampal seizures. This hypothesis predicts that a mossy fiber lesion should protect CA3 pyramidal cells from destruction by i.c.v. KA only if it also attenuates the hippocampal seizure. We have tested this hypothesis in chronically-implanted unanesthetized rats subjected to a mossy fiber lesion immediately before implantation of the i.c.v. cannula and EEG electrodes.

Mossy fiber lesions were made in either of two ways: by transecting this pathway unilaterally with a Scouten wire knife over a 4-mm length of the rostral hippocampal formation (N = 18) or by destroying the dentate granule cells with colchicine (Nadler and Cuthbertson, 1980) unilaterally (N = 4) or bilaterally (N = 5). Unilateral lesions were always made on the side of the brain where the cannula was subsequently placed. Animals injected with colchicine were allowed to recover for three weeks before administration of KA, in order to avoid non-specific effects of colchicine. Animals whose mossy fibers had been transected were allowed to recover for either one (N = 15) or three (N = 3) weeks. In a separate series of animals, Timm's staining confirmed that these procedures drastically reduced the mossy fiber innervation of area CA3 and revealed little or no sprouting of the residual mossy fibers within three weeks.

In 19 control animals (sham-operated or vehicle-injected), the electrographic (Fig. 5), behavioral and neuropathological (Fig. 6A) effects of i.c.v. KA were the same as those described above. Mossy fiber lesions protected 8 of 27 subjects from the development of status epilepticus. Also, in half the animals with unilateral mossy fiber lesions, ictal events were recorded in the hippocampus contralateral to the i.c.v. cannula before the ipsilateral hippocampus (Fig. 5). In all animals, the total duration of seizure activity recorded from the denervated hippocampus approximated

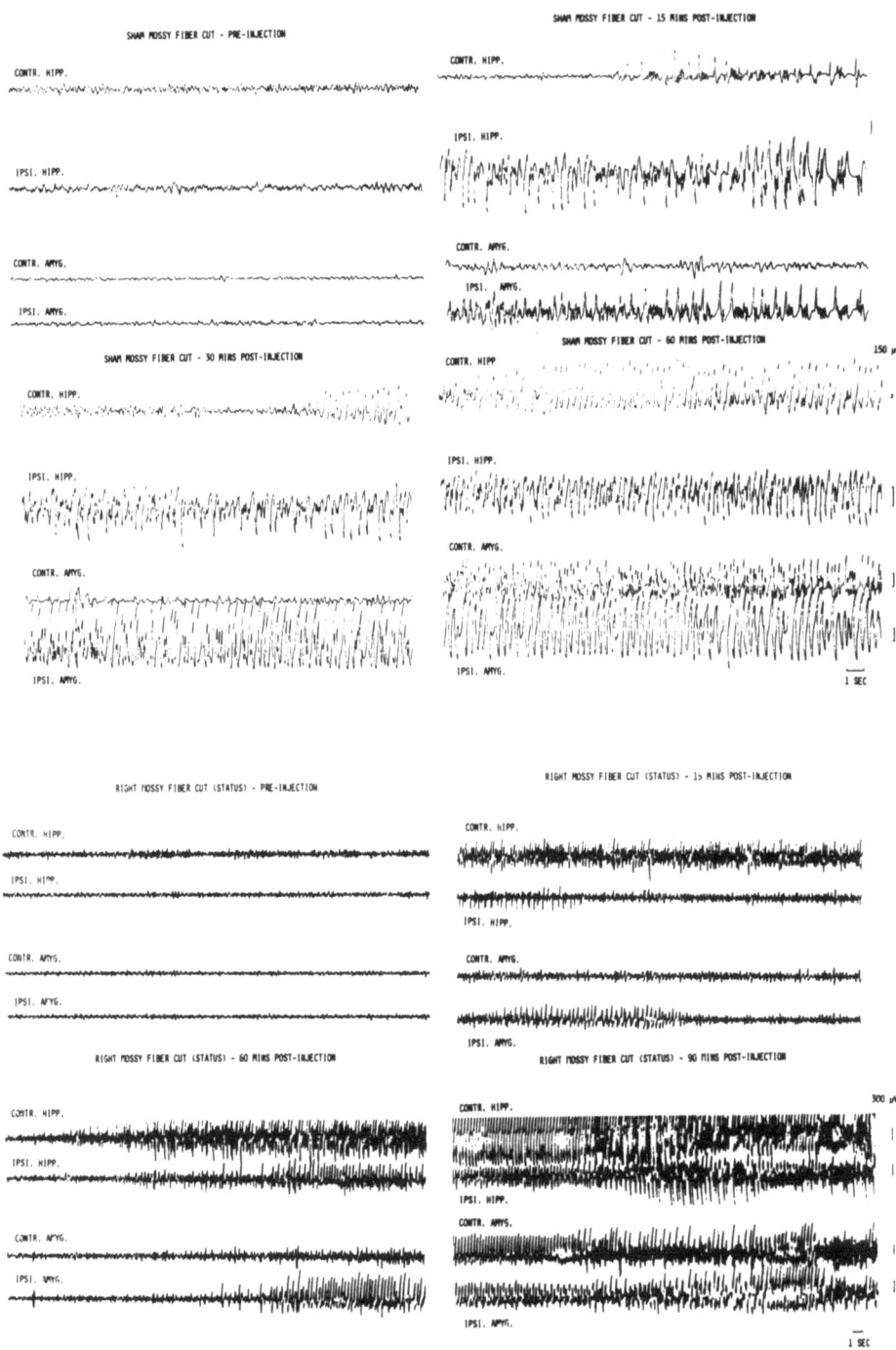

that recorded from the other limbic regions. However, mossy fiber lesions always substantially reduced or abolished neuronal cell death in the ipsilateral CA3 area, even when they failed to reduce the duration or severity of limbic seizures (Fig. 6B-D). The protective effect of a transection was limited to those CA3 pyramidal cells deprived of mossy fiber innervation; caudal to the transection, neuronal degeneration was as extensive as in controls. Colchicine injections destroyed nearly all dentate granule cells in both the rostral and caudal hippocampal formation, except at the caudal pole. Only at the caudal pole, where most granule cells remained intact, did KA destroy an appreciable number of CA3 pyramidal cells. The specific protective effect of mossy fiber lesions was confined to the denervated hippocampus. Neurons in other brain regions were protected from degeneration only when the mossy fiber lesion prevented the development of status epilepticus.

These results refute our initial hypothesis that mossy fiber lesions protect CA3 hippocampal pyramidal cells by blocking the hippocampal seizure. Destruction of these neurons by i.c.v. KA requires an intact mossy fiber pathway, but the seizures do not.

DISCUSSION

These results reinforce our conclusion that i.c.v. KA destroys CNS neurons predominantly by inducing limbic seizures, not through a direct excitotoxic action. The lesion could be reproduced with a very different convulsant administered in the same fashion and attenuated by anticonvulsants effective against the seizures *in vivo* and against KA-induced bursting in the rat hippocampal slice. Our results also suggest an explanation for the extreme vulnerability of CA3 hippocampal pyramidal cells to seizure-induced cell death: namely, a specific neurotoxic interaction with a substance or substances contained within the mossy fibers.

Significance of Total Seizure Duration

The relative vulnerability of CA3 hippocampal pyramidal cells to seizure-mediated neuronal cell death can be clearly discerned when one compares lesions from animals that experienced different durations of limbic seizure activity. Degeneration of these neurons can be demonstrated in some animals after as little as 15 minutes of limbic seizure within a four-hour recording period, whereas other vulnerable neurons consistently degenerate only after hours of seizure. It may appear surprising that such brief seizure activity can be so damaging, but our results agree fairly well with the report of Nevander et al. (1984), who induced seizures with fluorothyl. We do not know whether the same total seizure duration would be so damaging if the individual ictal episodes were spread over a longer period of time. Nevertheless, it can be easily understood from our results how repeated complex partial seizures could lead to the progressive development of Ammon's horn sclerosis. The seizure-mediated degeneration of CA3-CA4 neurons may then lead to the insidious development by axon sprouting of abnormal excitatory circuits, whose effect could be to exacerbate the epileptic condition (Tauck and Nadler, 1985).

Fig. 5. EEG recordings taken before and at various times after i.c.v. administration of KA to a sham-operated rat and a rat with an ipsilateral mossy fiber transection. Note that the transection initially reduced seizure activity in the ipsilateral hippocampus compared to the contralateral hippocampus, but all regions attained status epilepticus in 90 minutes. Time and amplitude calibrations differ for the two sets of recordings.

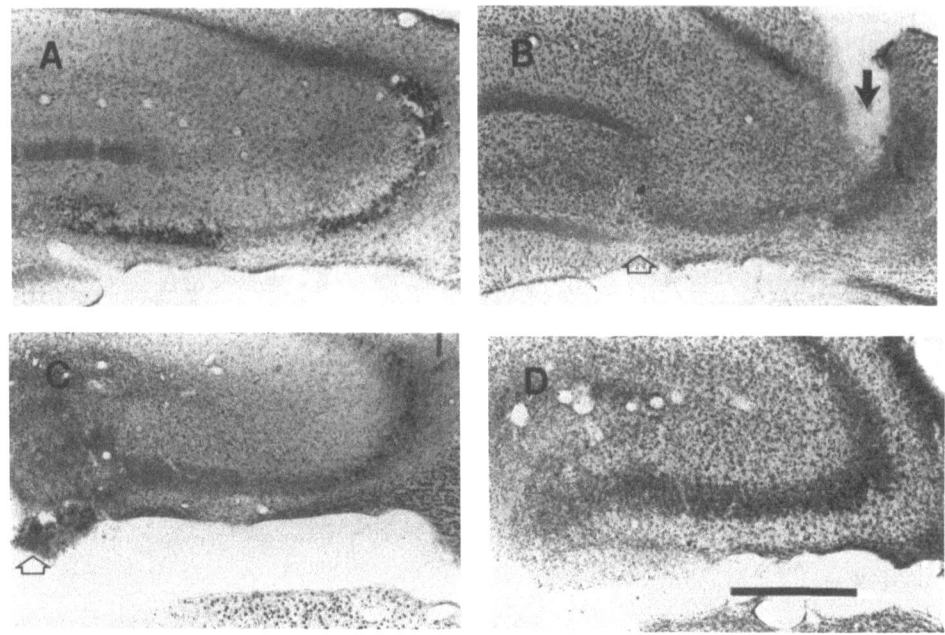

Fig. 6. Degenerating neurons in the ipsilateral hippocampal area CA3 after i.c.v. infusion of KA in animals previously injected unilaterally with artificial CSF alone (A), subjected to a unilateral mossy fiber transection (B,C) or injected unilaterally with colchicine (D). Silver impregnation. B was protected from status epilepticus, but C and D were not. Note the lack of neuronal degeneration in B and D and minimal degeneration in C. Filled arrows indicate electrode tracks. Open arrows indicate mossy fiber transection. Scale bar = 0.5 mm.

Significance of Interaction with the Mossy Fiber Pathway

The work of Sloviter (1983) demonstrated that hippocampal CA3-CA4 neurons can be destroyed by overdriving their mossy fiber innervation. The effect of stimulating other pathways was not studied, however. Perhaps the most striking result of our recent investigations has been the very different incidences of CA3 pyramidal cell degeneration according to the presence or absence of an intact mossy fiber pathway, regardless of total seizure duration. Evidently, hippocampal seizures can be sustained through the activation of excitatory circuits that bypass the mossy fibers (involving, for example, commissural, longitudinal associational and/or septohippocampal pathways), but these seizures are less damaging than those which involve the mossy fibers. Nitecka et al. (1984) arrived at a similar conclusion on the basis of developmental studies.

One possible explanation for the central role of the mossy fiber pathway lies in the dependence of KA receptors on the integrity of this pathway. The mossy fiber terminal zone of area CA3 contains a high density of KA receptors (Monaghan and Cotman, 1982; Unnerstall and Wamsley, 1983), most of which are of the high affinity subtype (Berger and Ben-Ari, 1983). This distribution must at least partially account for the extraordinary sensitivity of CA3 pyramidal cells to the convulsant (Westbrook and Lothman,

1983) and depolarizing (Robinson and Deadwyler, 1981) actions of KA. Transection of the mossy fibers (unpublished observations) or their destruction with colchicine (Monaghan and Cotman, 1982) reduces the density of KA receptors in the CA3 area. Thus, the protective effect of a mossy fiber lesion may result, in part, from the loss of these receptors. Against this notion, however, is our observation that mossy fiber lesions confer protection against KA-induced degeneration within 10 minutes (Nadler et al. 1981). It seems unlikely that KA receptors could turn over or downregulate in so short a time.

Another possibility is that KA releases a neurotoxin from the mossy fiber boutons which then destroys the postsynaptic cells. Mossy fiber boutons contain a number of neuroactive substances, including glutamate, dynorphin-related opioid peptides, Zn^{2+}, enkephalins and cholecystokinin. KA can directly release glutamate from synaptic terminals (Ferkany and Coyle, 1983; Pastuszko et al. 1984) and dynorphin-related opioid peptides from mossy fiber boutons (Chavkin et al., 1983). Baclofen can block the releasing action of KA (Potashner and Gerard, 1983). However, the concentrations of KA needed to achieve measurable directly-evoked release in those systems studied to date are several orders of magnitude above that which is adequate to induce epileptiform activity in area CA3. KA might, of course, release a neurotoxic substance from the mossy fiber boutons indirectly by eliciting seizures that activate this pathway. Of the neuroactive compounds contained within the mossy fiber boutons, glutamate would seem the least promising candidate for a specific neurotoxin, since it is also released by other excitatory afferent pathways in area CA3.

A third possible neurotoxic mechanism would be a synergistic postsynaptic interaction between KA and one or more of the neuroactive substances released by the mossy fibers. Such an interaction between KA and glutamate has been described (Shinozaki and Konishi, 1970). However, a study of CA3 hippocampal pyramidal cells revealed no synergism between KA and glutamate (Sawada and Yamamoto, 1984). Postsynaptic interactions between KA and other neuroactive substances remain to be investigated.

Finally, one must consider the possible role of disinhibition. Exposure of the hippocampal formation to KA acutely impairs synaptic inhibition in pyramidal cells (Fisher and Alger, 1984), and both synaptic and intrinsic types of inhibition remain subnormal for days or weeks (Lancaster and Wheal, 1984; Franck and Schwartzkroin, 1985). Disinhibition in the CA3 area leads to epileptiform discharge and would thus presumably contribute to pyramidal cell death. However, it is difficult to explain the necessity for a functional mossy fiber pathway on the basis of disinhibition. It should be noted that the distribution of KA receptors in the CA3-CA4 area corresponds to the distribution of mossy fibers, not to that of inhibitory interneurons and their projections. Furthermore, the destruction of hippocampal GABAergic neurons demonstrated after parenteral administration of KA (Heggli and Malthe-Sørenssen, 1982; Sperk et al., 1983) seems to be less prominent after i.c.v. administration. Fewer degenerating interneurons can be identified in histological sections and bilateral i.c.v. infusions of KA that destroy most of the CA3 pyramidal cells do not significantly reduce Na^+-dependent GABA uptake in any hippocampal region (unpublished observations).

The foregoing discussion emphasizes the need for more detailed information on the physiology of the mossy fiber pathway under both normal and epileptic conditions. An important, but unanswered, question is whether the mossy fibers are required only for KA-induced AHS or whether a destructive effect of excessive mossy fiber activity also plays an obligatory role in clinical AHS. If the latter is the case, then inhibitors of mossy fiber transmission, and possibly also KA receptor antagonists, might be useful therapeutic agents in temporal lobe epilepsy.

ACKNOWLEDGEMENTS

We thank Ciba-Geigy Corp. (Ardsley, N.Y.) for supplies of baclofen. These studies were supported by NIH grant NS 17771 and an NIH postdoctoral fellowship (M.G.)

REFERENCES

Ben-Ari, Y., Tremblay, E., Ottersen, O.P., and Meldrum, B.S., 1980, The role of epileptic activity in hippocampal and 'remote' cerebral lesions induced by kainic acid, Brain Res., 191:79.

Ben-Ari, Y., Tremblay, E., Riche, D., Ghilini, G., and Naquet, R., 1981, Electrographic, clinical and pathological alterations following systemic administration of kainic acid, bicuculline or pentetrazole: metabolic mapping using the deoxyglucose method with special reference to the pathology of epilepsy, Neuroscience, 7:1361.

Ben-Ari, Y., 1985, Limbic seizure and brain damage produced by kainic acid: mechanisms and relevance to human temporal lobe epilepsy, Neuroscience, 14:375.

Berger, M., and Ben-Ari, Y., 1983, Autoradiographic visualization of [^{3}H]-kainic acid receptor subtypes in the rat hippocampus, Neurosci. Lett., 39:237.

Chavkin, C., Bakhit, C., Weber, E., and Bloom, F.E., 1983, Relative contents and concomitant release of prodynorphin/neoendorphin-derived peptides in rat hippocampus, Proc. Natl. Acad. Sci. USA, 80:7669.

Clifford, D.B., Lothman, E.W., Dodson, W.E., and Ferrendelli, J.A., 1982, Effect of anticonvulsant drugs on kainic acid-induced epileptiform activity, Exp. Neurol., 76:156.

Elliott, K.A.C., 1969, The use of brain slices, in: Handbook of Neurochemistry, Volume 2, A. Lajtha, ed., Plenum Press, New York, p. 103.

Ferkany, J.W., and Coyle, J.T., 1983, Kainic acid selectively stimulates the release of endogenous excitatory acidic amino acids, J. Pharmacol. Exp. Ther., 225:399.

Fisher, R.S., and Alger, B.E., 1984, Electrophysiological mechanisms of kainic acid-induced epileptiform activity in the rat hippocampal slice, J. Neurosci., 4:1312.

Franck, J.E., and Schwartzkroin, P.A., 1985, Do kainate-lesioned hippocampi become epileptogenic?, Brain Res., 329:309.

Gruenthal, M., Ault, B., Armstrong, D.R., and Nadler, J.V., 1984, Baclofen blocks kainic acid-induced epileptiform activity, Soc. Neurosci. Abstr., 10:184.

Heggli, D.E., and Malthe-Sørenssen, D., 1982, Systemic injection of kainic acid: effect on neurotransmitter markers in piriform cortex, amygdaloid complex and hippocampus and protection by cortical lesioning and anticonvulsants, Neuroscience, 7:1257.

Lancaster, B., and Wheal, H.V., 1984, Chronic failure of inhibition of the CA1 area of the hippocampus following kainic acid lesions of the CA3/4 area, Brain Res., 295:317.

Lassmann, H., Petsche, U., Kitz, K., Baran, H., Sperk, G., Seitelberger, F., and Hornykiewicz, O., 1984, The role of brain edema in epileptic brain damage induced by systemic kainic acid injection, Neuroscience, 13:691.

Lothman, E.W., and Collins, R.C., 1981, Kainic acid induced limbic seizures: metabolic, behavioral, electroencephalographic and neuropathological correlates, Brain Res., 218:299.

Margerison, J.H., and Corsellis, J.A.N., 1966, Epilepsy and the temporal lobes, Brain, 89:499.

Meldrum, B.S., 1983, Metabolic factors during prolonged seizures and their relation to nerve cell death, in: Status Epilepticus: Mechanisms of Brain Damage and Treatment, Advances in Neurology, Volume 34, A.V. Delgado-Escueta, C.G. Wasterlain, D.M. Treiman, and R.J. Porter, eds., Raven Press, New York, p. 261.

Meldrum, B.S., and Corsellis, J.A.N., 1984, Epilepsy, in: Greenfield's Neuropathology, J.H. Adams, J.A.N. Corsellis, and L.W. Duchen, eds., John Wiley and Sons, New York, p. 921.

Menini, C., Meldrum, B.S., Riche, D., Silva-Comte, C., and Stutzmann, J.M., 1980, Sustained limbic seizures induced by intraamygdaloid kainic acid in the baboon: symptomatology and neuropathological consequences, Ann. Neurol., 8:501.

Monaghan, D.T., and Cotman, C.W., 1982, The distribution of [^{3}H]kainic acid binding sites in rat CNS as determined by autoradiography, Brain Res., 252:91.

Nadler, J.V., and Cuthbertson, G.J., 1980, Kainic acid neurotoxicity toward hippocampal formation: dependence on specific excitatory pathways, Brain Res., 195:47.

Nadler, J.V., 1981, Kainic acid as a tool for the study of temporal lobe epilepsy, Life Sci., 29:2031.

Nadler, J.V., Evenson, D.A., and Smith, E.M., 1981, Evidence from lesion studies for epileptogenic and non-epileptogenic neurotoxic interactions between kainic acid and excitatory innervation, Brain Res., 201:405.

Nadler, J.V., and Evenson, D.A., 1983, Use of excitatory amino acids to make axon-sparing lesions of hypothalamus, in: Hormone Action, Part H: Neuroendocrine Peptides, Methods in Enzymology, Volume 103, P.M. Conn, ed., Academic Press, New York, p. 393.

Nevander, G., Ingvar, M., Auer, R., and Siesjö, B., 1984, Irreversible neuronal damage after short periods of status epilepticus, Acta Physiol. Scand., 120:155.

Nitecka, L., Tremblay, E., Charton, G., Bouillot, J.P., Berger, M.L., and Ben-Ari, Y., 1984, Maturation of kainic acid seizure-brain damage syndrome in the rat. II. Histopathological sequelae, Neuroscience, 13:1073.

Pastuszko, A., Wilson, D.F., and Erecińska, M., 1984, Effects of kainic acid in rat brain synaptosomes: the involvement of calcium, J. Neurochem., 43:747.

Potashner, S.J., and Gerard, D., 1983, Kainate-enhanced release of D-[^{3}H]-aspartate from cerebral cortex and striatum: reversal by baclofen and pentobarbital, J. Neurochem., 40:1548.

Racine, R.J., 1972, Modification of seizure activity by electrical stimulation. II. Motor seizure, Electroenceph. Clin. Neurophysiol., 32:281.

Robinson, J.H., and Deadwyler, S.A., 1981, Kainic acid produces depolarization of CA3 pyramidal cells in the in vitro hippocampal slice, Brain Res., 221:117.

Sawada, S., and Yamamoto, C., 1984, Fast and slow depolarizing potentials induced by short pulses of kainic acid in hippocampal neurons, Brain Res., 324:279.

Shinozaki, H., and Konishi, S., 1970, Actions of several anthelmintics and insecticides on rat cortical neurones, Brain Res., 24:368.

Sloviter, R.S., 1983, Epileptic brain damage in rats induced by sustained electrical stimulation of the perforant path. I. Acute electrophysiological and light microscopic studies, Brain Res. Bull., 10:675.

Sommer, W., 1880, Erkrankungen des Ammonshorns als aetiologisches Moment der Epilepsie, Arch. Psychiatr. Nervenkrkh., 10:631.

Sperk, G., Lassmann, H., Baran, H., Kish, S.J., Seitelberger, F., and Hornykiewicz, O., 1983, Kainic acid induced seizures: neurochemical and histopathological changes, Neuroscience, 10:1301.

Tauck, D.L., and Nadler, J.V., 1985, Evidence of functional mossy fiber sprouting in hippocampal formation of kainic acid-treated rats, J. Neurosci., 5:1016.

Unnerstall, J.R., and Wamsley, J.K., 1983, Autoradiographic localization of high-affinity [^{3}H]kainic acid binding sites in the rat forebrain, Eur. J. Pharmacol., 86:361.

Westbrook, G.L., and Lothman, E.W., 1983, Cellular and synaptic basis of kainic acid-induced hippocampal epileptiform activity, Brain Res., 273:97.

GLUTAMATE AND ANOXIC NEURONAL DEATH IN VITRO

S.M. Rothman

Departments of Pediatrics, Anatomy and Neurobiology, and
Neurology
Washington University School of Medicine
St. Louis, Missouri

INTRODUCTION

Identifying the factors responsible for the exquisite sensitivity
of the brain to anoxia and ischemia has been very difficult with currently
available preparations. During the past three years, our laboratory has
utilized two fairly well characterized in vitro preparations to study
the effects of anoxia on central mammalian neurons. The experiments described below, which employed cultures of dissociated hippocampal neurons
and slices of rat hippocampus, implicate the synaptic release of glutamate
(Glu) in the pathophysiology of anoxic neuronal death.

NEUROBIOLOGY OF HIPPOCAMPUS IN CULTURE

Our initial experiments were performed with cultures of dissociated
rat hippocampal neurons, obtained from fetuses at 17-18 days gestation
(Banker and Cowan, 1977; Peacock et al., 1979; Rothman, 1984). Neurons
in these cultures develop extensive processes, excitatory and inhibitory
synaptic activity, and physiological properties similar to those observed
in pyramidal neurons in the intact hippocampus (Peacock, 1979; Rothman
and Cowan, 1981; Bartlett and Banker, 1984; Rothman and Samaie, 1985).
These cultures are composed of many neurons which would probably have
become pyramidal cells in the mature animal.

Neurons in these cultures demonstrate clear-cut depolarizing responses
to the putative amino acid transmitters Glu and aspartate (Asp), as well
as to the 'pure' excitatory amino acid agonists kainate (KA), N-methyl-
D-aspartate (NMDA), and quisqualate (Quis; Rothman and Samaie, 1985).
In an attempt to characterize the receptor and transmitter responsible
for excitatory synaptic transmission in these cultures, we compared the
EPSP's evoked by monosynaptic intracellular stimulation with amino acid
responses produced by microperfusion of known concentrations of agonist.
We specifically looked at antagonism of both synaptic and amino acid responses by known blockers of excitatory amino acids and also at differences
in the voltage sensitivity of various responses (Mayer and Westbrook,
1984).

Depolarizations produced by Glu were effectively blocked by γ-D-glutamylglycine (DGG) and cis-2,3-piperidine dicarboxylic acid (PDA). Asp responses were reduced by both these chemicals and also aminophosphonovaleric acid (APV) and D-α-aminoadipic acid (DAA). The latter two substances had little effect on Glu responses, measured at resting potential (Table 1).

KA responses were very effectively reduced by DGG but not APV, while the latter completely eliminated NMDA depolarizations. Quis responses were only partly blocked by DGG. DGG, and to a lesser extent PDA, antagonized the excitatory synaptic potentials.

The voltage dependency of the different amino acid depolarizations was quite interesting. Both Glu and Asp responses diminished as the neuron was hyperpolarized to potentials more negative than -100 mV. NMDA produced a similar effect. However, the EPSP's and KA and Quis depolarizations were all linear, increasing with membrane hyperpolarization (Rothman and Samaie, 1985). Glu responses became much more linear in the presence of APV, an observation made by others as well (Westbrook and Mayer, 1984).

These results all suggested that Glu was the synaptic transmitter, acting at KA and/or Quis receptors when it was released by presynaptic terminals, but binding NMDA receptors in addition when it was applied exogenously. Since all our neurons demonstrated spontaneous synaptic activity, and only about two-thirds responded to Quis, it seemed more likely that KA rather than Quis receptors were actually present at post-synaptic membranes across from nerve terminals. Our results also indicate that cultured hippocampal neurons possess NMDA receptors, as do hippocampal pyramidal neurons in the slice (Dingledine, 1983). They do not illuminate the role these receptors play in normal synaptic physiology, however.

Inhibitory synapses have also been physiologically identified in the hippocampal cultures (Peacock, 1979; Rothman and Samaie, 1985). They almost certainly utilize γ-aminobutyric acid as their transmitter, because they are blocked by concentrations of bicuculline which have little effect on glycine responses and almost completely antagonize GABA induced hyperpolarizations. A number of other putative neurotransmitters have no direct effects on neuronal resting membrane potentials in these cultures. These include: acetylcholine, adenosine, enkephalin, norepinephrine, serotonin, and somatostatin. These chemicals may all modulate synaptic transmission but cannot be conventional transmitters in culture (Segal, 1983; Rosenberg et al. 1984).

Table 1. Reduction of Amino Acid Depolarizations and EPSP's by Antagonists

	APV	DAA	DGG	PDA
Glu	+/-	-	++	+
ASP	++	++	++	++
KA	-		++	
NMDA	++			
Quis	-		+/-	
EPSP	-	-	++	+

++Depolarization reduced by at least 80%
+Depolarization reduced by 65-70%
+/-Depolarization reduced by 25-50%
-Depolarization reduced by less than 15%
(Adapted from Rothman and Samaie, J. Neurophysiol, 1985)

ANOXIA IN CULTURE

In view of the extreme sensitivity of the intact hippocampus to anoxia and ischemia, it seemed reasonable to determine whether these cultures were also dependent on a supply of oxygen. In early experiments, we found that cultures continued to develop normally if exposed to cyanide, an inhibitor of oxidative phosphorylation (Fig. 1B). Placement in 95% nitrogen/ 5% carbon dioxide (95% N_2/5% CO_2) within 48 hours of plating also had no detectable effect on the cultures (Fig. 1C).

Older cultures (at least two weeks in vitro) were extremely vulnerable to either treatment. The addition of cyanide produced neuronal swelling within 30-60 minutes, which was followed by neuronal disintegration (Fig. 2A). Placing these cultures in 95% N_2/5% CO_2 produced similar morphologic effects, but on a different time scale. Neurons took 12-24 hours to degenerate, although the end result of anoxia was identical to cyanide treatment (Fig. 2C). The marked difference in oxygen dependence between our young and old cultures was initially a puzzle. One possible explanation for this observation was that the development of synaptic interactions, almost universal in the mature cultures but absent from the younger ones (Jackson et al., 1982), was responsible for sensitivity to anoxia.

In order to investigate this hypothesis, high concentrations of magnesium (10 mM) were added to cultures immediately prior to cyanide or nitrogen exposure to eliminate synaptic activity. In both situations the magnesium was protective (Fig. 2B, 2D). The cultures were virtually unchanged after a prolonged anoxic challenge, sufficient to completely eliminate neurons in control cultures (Rothman, 1983). Intracellular recording from neurons in magnesium treated cultures demonstrated normal resting membrane potentials and action potentials, further indicating that the preparations were healthy.

The relationship between synaptic activity and anoxic neuronal death was initially a puzzle. One possible explanation was that synaptic activity resulted in the release of Glu, the most likely excitatory transmitter in our cultures and also a potent toxin for central mammalian neurons (Olney, 1978). Preliminary experiments designed to test this hypothesis by blocking synaptic activity with a post synaptic antagonist of Glu were unsuccessful. The Glu antagonist then available, PDA, was toxic when placed in the cultures for more than a few minutes. In more recent experiments, the antagonist DGG (10 mM), which blocks depolarization caused by Glu, Asp, KA and NMDA, and the EPSP, was capable of protecting the cultures from prolonged anoxia (Rothman, 1984). As DGG also blocked the toxicity of exogenous Glu in these cultures (Rothman, 1984) and has no

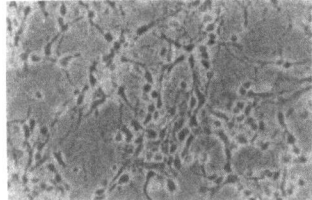 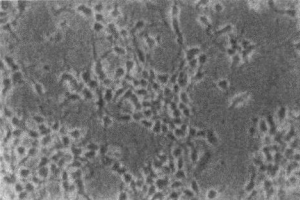 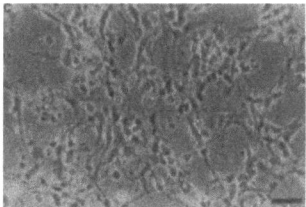

Fig. 1. Anoxia in young cultures. (A) Typical field from a 48 hour old culture demonstrating early development of neurons in vitro. Fields in (B) and (C) show neurons the same age plated into medium containing cyanide (1 mM) or lacking oxygen. The cultures were indistinguishable. Scale: 50 µm. (Reproduced from Rothman, 1983).

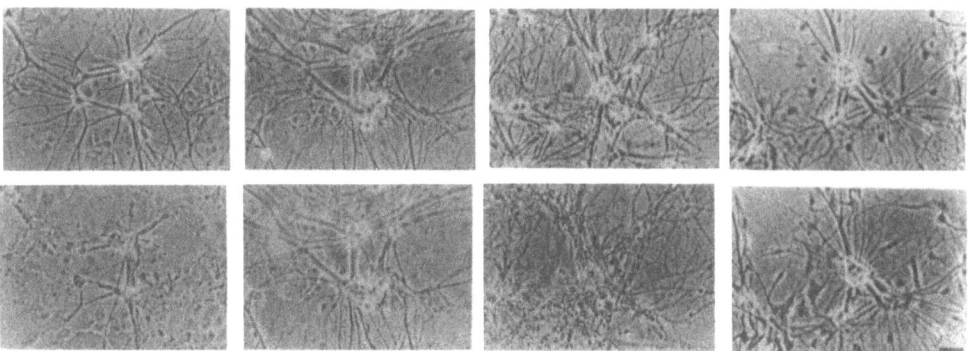

Fig. 2. Anoxia in mature cultures. (A) Top shows typical field of neurons in a culture prior to cyanide treatment, while bottom shows the dramatic loss of cells in the same field 21 hours after cyanide. (C) These two photomicrographs show the same field before (top) and 14 hours after (bottom) the culture was put into a 95% N_2/5% CO_2 environment. Again, there is a total disintegration of the neurons. (B) and (D) These two sequences show that 10 mM magnesium eliminates the dramatic effects of cyanide (B) and anoxia (D). Fields photographed after both insults appeared virtually unchanged. Scale: 50 μm. (Reproduced from Rothman, 1983).

known presynaptic effects (Crunelli et al., 1983), it is almost certain that DGG protects by preventing the Glu which accumulates during anoxia from binding to postsynaptic receptors. It is unclear whether Glu accumulates because release increases during anoxia (Bosley et al., 1983), reuptake decreases (Silverstein et al., 1985), or some combination of the two exists.

When these experiments were performed, we assumed that DGG protected cultures from anoxia because it markedly reduced receptor binding. At that time we did not investigate the protective effects of APV, because there was no evidence that it reduced synaptic potentials in our cultures. Its major site of action was the NMDA receptor. In recent experiments, APV (100-200 μM) has been just as effective as DGG in protecting cultured neurons from anoxia. This indicates that NMDA receptors must play a role in anoxic neuronal death even though it is unclear how they participate in conventional synaptic transmission.

The success of APV in protecting the cultures from anoxia led us to try the dissociative anesthetic ketamine as well. A number of recent experiments have shown that ketamine selectively antagonizes NMDA produced depolarizations, so ketamine might also protect the cultures from anoxia (Thomson et al., 1985). We found that cultures pretreated with ketamine (25 or 100 μM) were not obviously altered by a period of oxygen deprivation (one day) sufficient to produce disintegration of control cultures. Intracellular recording also verified that neurons in the ketamine treated cultures were healthy. As a further assay of neuronal viability, we also measured levels of ATP in ketamine protected cultures, as well as control cultures, APV treated cultures, and pentobarbital treated cultures (Table 2).

ATP/protein ratios were approximately 70% of control (preanoxia) levels in the cultures treated with ketamine, either 25 or 100 μM. Pentobarbital, even at 1 mM, did little to maintain ATP levels during anoxia and cultures treated with pentobarbital were visually indistinguishable from controls after anoxia. Ketamine also prevented the neuronal disintegration seen after NMDA is added to cultures and blocked the depolarization

Table 2. ATP Levels (μmoles/g protein) in Anoxic Cultures
(average ± 1 SD for three cultures)

1.	Control - preanoxia	2.34 ± 0.1
2.	Control - postanoxia	.29 ± .09
3.	APV (200 μM) - postanoxia	1.56 ± .19
4.	Ketamine (100 μM) - postanoxia	1.65 ± .29
5.	Ketamine (25 μM) - postanoxia	1.65 ± .43
6.	Pentobarbital (1 mM) - postanoxia	.63 ± .04

produced by NMDA application during intracellular recording. The effects of ketamine in anoxia, therefore, are unlikely to be due to some nonspecific anesthetic action. Rather, its ability to prevent activation of NMDA type receptors probably explains its protective effects in anoxic cultures.

MECHANISM OF EXCITATORY AMINO ACID NEUROTOXICITY

Neurons in the hippocampal cultures die within an hour when exposed to Glu, Asp, KA, and NMDA. As the neurotoxicity of excitatory amino acids (EAA) appeared to be directly related to anoxic neuronal death in these cultures, it seemed relevant to dissect out the mechanism of EAA neurotoxicity. The cultures provide an ideal preparation for these sorts of experiments, as the ionic composition of the bathing media can be easily and quickly changed.

We first established that electrical activity was unnecessary for EAA-induced neuronal death. When tetrodotoxin (1 μg/ml) was added to medium to suppress action potentials, and extracellular fluid containing no added calcium and high concentrations of magnesium was used to eliminate synaptic activity, cell death still occurred a short time after the addition of Glu, KA or NMDA (Rothman, 1985). These experiments also suggested that calcium influx was not important for rapid EAA induced neuronal death. The cultured neurons were not immediately affected by the calcium ionophore A23187, another indication that calcium influx was not important for rapid 'excitotoxicity'.

EAA neurotoxicity could be blocked, however, if the cultures were exposed to any of the amino acids after the extracellular sodium was removed and replaced with the impermeable cation benzoylcholine. This suggested that either sodium influx, depolarization, and/or passive chloride influx as a consequence of depolarization could be responsible for EAA neurotoxicity (Kuffler et al., 1984). The latter explanation was supported by another set of experiments. Cultures were exposed to EAA in extracellular solution from which the bulk of the chloride was removed and replaced with a sulfate, a relatively impermeable anion, and sucrose, to maintain osmotic tension. Under these conditions, the rapid neuronal disintegration produced by Glu, KA or NMDA was no longer evident. Further support for the hypothesis that passive chloride influx was responsible for excitotoxicity came from experiments in which cultures were depolarized with extracellular solution containing up to 140 mM potassium. When chloride was present as the balancing anion neurons rapidly swelled and died. Substitution of sulfate for chloride prevented this phenomenon.

These results, taken together, suggest a straightforward mechanism for excitotoxicity. The depolarization produced by EAA results in chloride influx down its electrochemical gradient into the neuron. This is coupled to more cation influx to maintain charge balance and results in movement of water into the neuron in an attempt to maintain osmotic pressure. Eventually the water influx produces lysis of the neuron (Fig. 3A; Rothman, 1985).

Choi (1985) has recently described a delayed form of Glu-neurotoxicity which may be related to calcium influx (Fig. 3B). If cultured cortical neurons are briefly exposed to Glu, so that they do not rapidly disintegrate, they slowly deteriorate over the next day and eventually die. However, if calcium is not present during the period of exposure to Glu, the amount of neuronal death is dramatically reduced. We have recently confirmed Choi's observations in our hippocampal cultures. In addition, we have found that APV or DGG, added to the medium after the initial Glu exposure, blocks this delayed form of excitotoxicity. Therefore, ongoing activation of EAA receptors may be necessary for delayed excitotoxicity. Calcium may be toxic because it greatly increases presynaptic Glu release and/or the number of receptors for EAA (Lynch and Baudry, 1984), and not because it irreversibly damages the biochemical machinery of the neuron.

ANOXIA IN HIPPOCAMPAL SLICE

Our observations on anoxia in hippocampal cultures have been extended

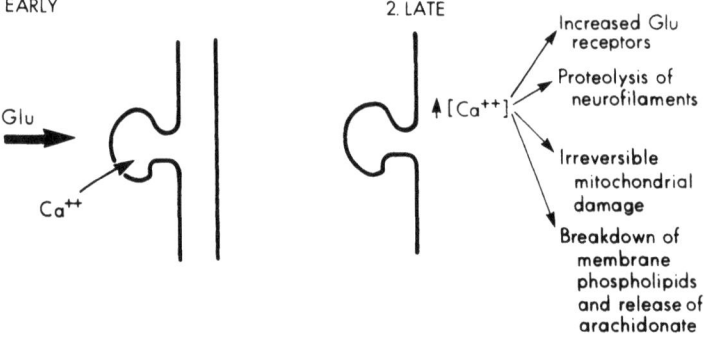

Fig. 3. The mechanism of EAA neurotoxicity. (A) Rapid excitotoxicity is produced by sodium influx leading to depolarization, chloride influx, and finally osmotic lysis of the neuron. (B) Delayed EAA toxicity is likely related to calcium influx and may be caused by calcium activating a variety of biochemical pathways within the neuron. (Adapted from Rothman and Olney, 1986).

to the hippocampal slice preparation. When slices are exposed for 40-50 minutes to buffer solution equilibrated with 95% N_2/5% CO_2, the orthodromic CA1 field potential disappears and is markedly attenuated after reoxygenation (Table 3; Fig. 4A; Clark et al., 1985). Field potentials could sometimes be found closer to the stimulating electrode or after the stimulating voltage was increased; but when potentials were compared at the same site and stimulus strength, they were almost always dramatically reduced. If magnesium (10 mM) was added to the perfusate prior to anoxia, there was almost a complete recovery of field potentials, and it was possible to find neurons with normal properties with intracellular recording (Table 3; Fig. 4B). The EAA antagonists kynurenate (Robinson et al., 1984), APV and ketamine all preserved the field potential as well.

Our results confirm previous observations on anoxia in the hippocampal slice (Kass and Lipton, 1982) and provide evidence that our conclusions from tissue culture can be extended to a more intact preparation.

CLINICAL IMPLICATIONS

These experiments indicate that synaptic release of EAA, most likely Glu, is responsible for anoxic neuronal death in cultures and slices of hippocampus. Recent reports also support the hypothesis that ischemic neuronal death in vivo is directly related to activation of NMDA type receptors (Simon et al., 1984). Therapies designed to reduce the release of EAAs or block their binding to receptors might, therefore, be expected to reduce the morbidity and mortality of stroke, perinatal asphyxia and cerebral hypoxia. While continued trials of specific EAA antagonists seem important, a more attractive immediate strategy might be to investigate the use of ketamine and related drugs (cyclazocine) in whole animal models of cerebral hypoxic/ischemic injury. Ketamine has been used clinically

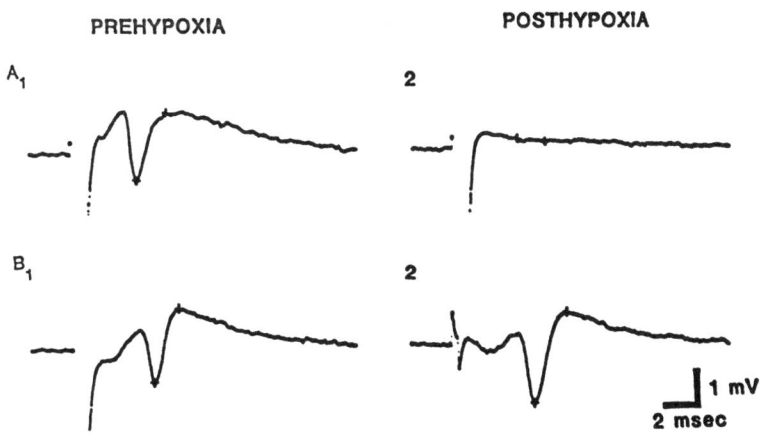

Fig. 4. (A) Typical CA1 population spike before (A1) after (A2) 40 minutes of anoxia. (B) Population spike before (B1) and after (B2) anoxia in a slice protected with 10 mM magnesium.

Table 3. Recovery of CA1 Population Spike in Anoxic Hippocampal Slices (n = number of slices)

1. Control	1.9% (20)
2. Magnesium (10 mM)	97% (17)*
3. Kynurenate (5-10 mM)	89% (12)*
4. APV (10-100 μM)	89% (10)*
5. Ketamine (25 μM)	99% (7)

*Kynurenate and APV data averaged together

for 15 years, easily crosses the blood brain barrier, and is readily available. It may, therefore, be possible to rapidly determine whether blockade of EAA receptors will be a rational therapy for a group of devastating human neurological diseases.

ACKNOWLEDGEMENTS

R. Hauhart and Dr. J. Thurston performed all the ATP determinations. G. Clark, M. Samaie and J. Solomon did the slice experiments. P. Mallott typed the manuscript. This work was supported by the NINCDS, the Epilepsy Foundation of America and the American Heart Association.

REFERENCES

Banker, G., and Cowan, W.M., 1977, Rat hippocampal neurons in dispersed cell culture, Brain Res., 126:397.

Bartlett, W.P., and Banker, G.A., 1984, An electron microscopic study of the development of axons and dendrites by hippocampal neurons in culture. II. Synaptic relationships, J. Neurosci., 4:1954.

Bosley, T.M., Woodhams, P.L., Gordon, R.D., and Balázs, R., 1983, Effects of anoxia on the stimulated release of amino acid neurotransmitters in the cerebellum in vitro, J. Neurochem., 40:189.

Choi, D.W., 1985, Glutamate neurotoxicity in cortical cell culture is calcium-dependent, Neurosci. Lett., 58:293.

Clark, G., Samaie, M., and Rothman, S., 1985, Blockade of synaptic transmission protects rat hippocampal slice from hypoxia, Ann. Neurol., 18:385.

Crunelli, V., Forda, S., and Kelly, J.S., 1983, Blockade of amino acid-induced depolarizations and inhibition of excitatory post-synaptic potentials in rat dentate gyrus, J. Physiol., 341:627.

Dingledine, R., 1983, N-methylaspartate activates voltage dependent calcium conductance in rat hippocampal pyramidal cells, J. Physiol., 343:385.

Jackson, M.D., Lecar, H., Brenneman, D.E., Fitzgerald, S., and Nelson, P.G., 1982, Electrical development in spinal cord cell culture, J. Neurosci., 2:1052.

Kass, I.S., and Lipton, P., 1982, Mechanisms involved in irreversible anoxic damage to the in vitro rat hippocampal slice, J. Physiol., 332:459.

Kuffler, S.W., Nicholls, J.G., and Martin, A.R., 1984, From Neuron to Brain, Sinauer, Sunderland, MA.

Lynch, G., and Baudry, M., 1984, The biochemistry of memory: a new and specific hypothesis, Science, 274:1057.

Mayer, M.L., and Westbrook, G.L., 1984, Mixed-agonist action of excitatory amino acids on mouse spinal cord neurons under voltage clamp, J. Physiol., 354:29.

Olney, J.W., 1978, Neurotoxicity of excitatory amino acids, in: Kainic Acid as a Tool in Neurobiology, E.G. McGeer, J.W. Olney, and P.L. McGeer, eds., Raven, New York, p. 95.

Peacock, J.H., 1979, Electrophysiology of dissociated hippocampal cultures from fetal mice, Brain Res., 169:247.

Peacock, J.H., Rush, D.F., and Mathers, L.H., 1979, Morphology of dissociated hippocampal cultures from fetal mice, Brain Res., 169:231.

Robinson, M.B., Anderson, K.D., Koerner, J.F., 1984, Kynurenic acid as an antagonist of hippocampal excitatory transmission, Brain Res., 309:119.

Rosenberg, P.A., Schweitzer, J.S., and Dichter, M.A., 1984, Norepinephrine increases synaptic activity in rat cerebral cortex in dissociated cell culture, Soc. Neurosci. Abstr., 10:70.

Rothman, S.M., and Cowan, W.M., 1981, A scanning electron microscope study of the in vitro development of dissociated hippocampal cells, J. Comp. Neurol., 195:141.

Rothman, S.M., 1983, Synaptic activity mediates death of hypoxic neurons, Science, 270:536.

Rothman, S., 1984, Synaptic release of excitatory amino acid neurotransmitter mediates anoxic neuronal death, J. Neurosci., 4:1884.

Rothman, S.M., 1985, The neurotoxicity of excitatory amino acids is produced by passive chloride influx, J. Neurosci., 5:1483.

Rothman, S.M., and Samaie, M., 1985, The physiology of excitatory synaptic transmission in cultures of dissociated rat hippocampus, J. Neurophysiol. 54:701.

Rothman, S.M., and Olney, J.W., 1986, Glutamate and the pathophysiology of hypoxic/ischemic brain damage, Ann. Neurol., 19:105.

Segal, M., 1983, Rat hippocampal neurons in culture: responses to electrical and chemical stimuli, J. Neurophysiol., 50:1249.

Silverstein, F.S., Buchanan, F., and Johnston, M.V., 1985, Hypoxia-ischemia causes severe but reversible depression of striatal synaptosomal ^{3}H-glutamate uptake, Ann. Neurol., 18:122.

Simon, R.P., Swan, J.H., Griffiths, T., and Meldrum, B.S., 1984, Blockade of N-methyl-d-aspartate receptors may protect against ischemic damage in the brain, Science, 226:850.

Thomson, A.M., West, D.C., and Lodge, D., 1985, An N-methylaspartate receptor-mediated synapse in rat cerebral cortex: a site of action of ketamine?, Nature, 313:479.

Westbrook, G.C., and Mayer, M.L., 1984, Glutamate currents in mammalian spinal neurons: resolution of a paradox, Brain Res., 301:375.

QUINOLINIC ACID: A PATHOGEN IN SEIZURE DISORDERS?

R. Schwarcz, C. Speciale, E. Okuno, E.D. French and
C. Köhler[1]

Maryland Psychiatric Research Center, P.O. Box 21247
Baltimore, Maryland 21228 U.S.A.
[1]Department of Pharmacology, ASTRA-Läkemedel
Södertälje, Sweden

Multiple lines of experimental evidence, exemplified in several chapters of this volume, point to a prominent if not necessarily causative involvement of brain N-methyl-D-aspartate (NMDA) receptors in seizure phenomena. At the present time, the situation is reminiscent of that encountered in schizophrenia research where it appears that dopamine receptors are 'somehow' involved in psychiatric symptomatology (Snyder et al., 1974). While the theoretical framework for studies in those two seemingly quite unrelated areas is thus remarkably similar, one considerable difference exists: there is no question as to the identity of the endogenous agonist of the dopamine receptor. In fact, the catecholamine has lent its name to the 'dopamine hypothesis' of schizophrenia. In contrast, no endogenous NMDA-agonist of similar prominence has emerged, which could be generally accepted as a candidate for a pathogen in human epileptic disorders. Yet the characterization of such an endogenous compound is eminently relevant, given the non-endogenous nature of NMDA and the apparent non-selectivity of glutamate for the NMDA-subtype of excitatory amino acid receptors (Foster and Fagg, 1984). Here we will review the growing body of evidence suggesting that quinolinic acid (pyridine 2,3-dicarboxylic acid; QUIN) may fulfill such a role.

QUINOLINIC ACID AS AN ENDOGENOUS CONVULSANT

For several decades, QUIN has been known to exist in mammalian urine and peripheral tissues (Henderson and Hirsch, 1949; Henderson and Ramasarma, 1949). By the mid 1960's, its metabolism had been elucidated and its role as an intermediate in the kynurenine pathway, converting tryptophan to nicotinamide adenine dinucleotide (NAD) had been documented (Nishizuka and Hayaishi, 1963; Gholson et al., 1964). Not until 1983 (Wolfensberger et al., 1983), however, was QUIN's presence in the brain established, where it occurs at a concentration of approximately 1 micromolar. Work on brain QUIN had been stimulated greatly by the observation that it can, upon microiontophoretic application, excite cortical neurons and that its physiological properties are apparently mediated by NMDA-receptors (Stone and Perkins, 1981; Perkins and Stone, 1983). Subsequent studies revealed that intracerebrally applied QUIN can cause lesions of the axon-sparing type, thus qualifying it as an endogenous 'excitotoxin' (Schwarcz

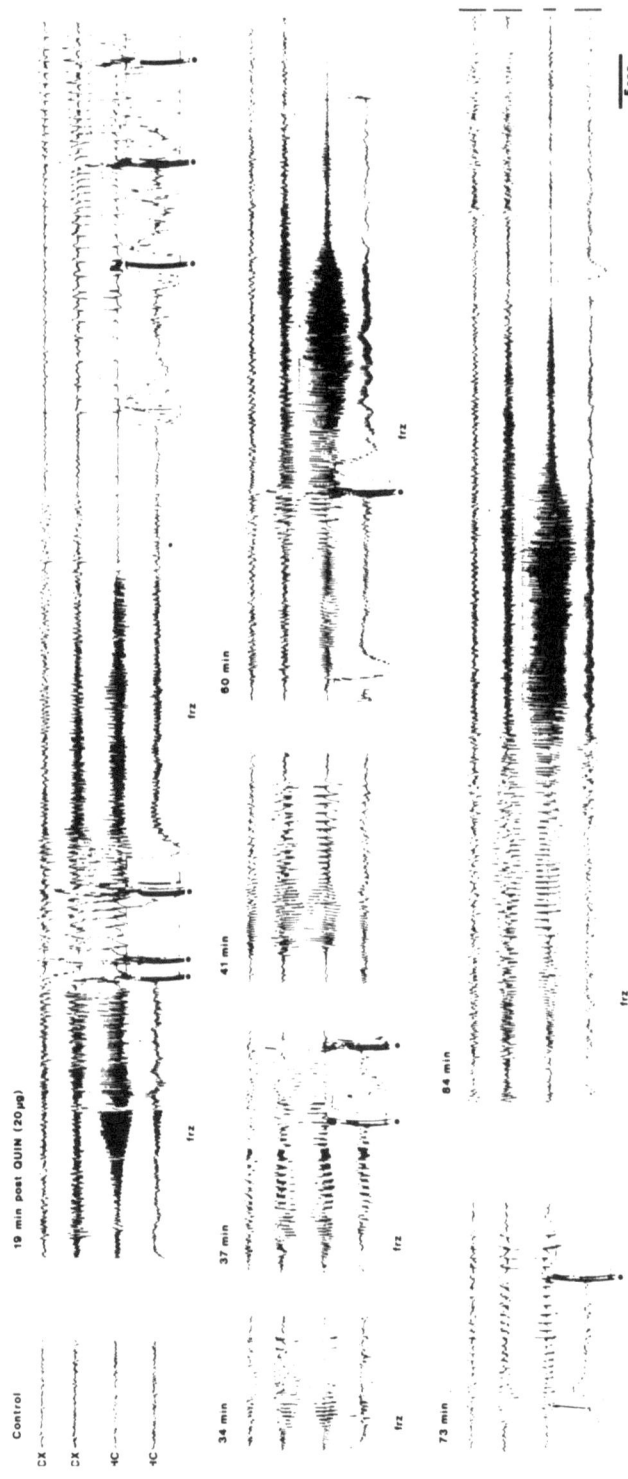

Fig. 1. Characteristic EEG changes associated with QUIN-induced seizures. Records are from an animal injected with 120 nmol QUIN. Postinjection times are indicated. The second record of the top and bottom row show long multicomponent seizure episodes composed of high- and low-voltage fast activity, high-voltage synchronized spiking, and a period of postictal depression (more prominent in the lHC lead) followed by high-voltage synchronized spiking. The latency to onset of the first seizure was 19 minutes. The second row and first record of the third row show the more characteristic short-duration, repetitive ictal episodes caused by QUIN. These events predominantly consisted of large-amplitude spikes and occasional interspike bursts of 10 Hz (see 37-minute record). The dots indicate wet-dog shakes; frz, frozen appearance of the animal during seizure episodes. (From Schwarcz et al., 1984a).

et al., 1983). In confirmation of the neurophysiological data, the toxic properties of QUIN, too, could be demonstrated to be mediated by NMDA-receptors since specific NMDA-antagonists were able to prevent QUIN-induced nerve cell loss (Schwarcz et al., 1984a,b).

Notably, QUIN's excitotoxic effects show pronounced regional selectivity in the rat brain. Of the brain areas tested so far, the hippocampus appears to be the most susceptible to both QUIN's excitatory (Perkins and Stone, 1983) and degenerative (Schwarcz and Köhler, 1983) effects. Furthermore, as originally reported by Lapin and his collaborators (Lapin, 1981; Lapin et al., 1982), QUIN is a potent convulsant in a variety of species. It is, in fact, one of the most potent endogenous seizure provoking agents discovered so far. When applied in nmol quantities to the hippocampus of an unanesthetized rat, QUIN causes a specific pattern of seizure activity, best characterized by repetitive periods of high-voltage spiking typically lasting approximately 20 seconds (Fig. 1) interspersed with occasional longer multicomponent episodes of approximately 60 seconds duration (Schwarcz et al., 1984a). Thus, the seizure pattern resembles, but is not identical to, that elicited by the intrahippocampal administration of such exogenous excitatory amino acids as kainate (French et al., 1982) and ibotenate (Aldinio et al., 1983). Although QUIN's excitotoxic properties are shared by kainate and ibotenate (Schwarcz et al., 1984b), QUIN causes a unique array of sequelae following its intracerebral application. A point in case involves the neuropathological consequences of intrahippocampal QUIN injections. Here QUIN, unlike kainate, produces highly localized lesions which in no case have been noted to affect areas distant from the injection site (e.g., the extremely kainate-sensitive pyriform cortex, cf. Ben-Ari, 1985). On the other hand, QUIN, at low doses, exerts preferential toxicity against pyramidal cells (with the exception of CA2), while granule cells degenerate only after exposure to larger amounts of the toxin (Fig. 2). This is in marked contrast to the neurodegenerative properties of ibotenate or NMDA, which are equally toxic to all hippocampal neuronal cell types (Köhler et al., 1979; Nadler et al., 1981).

QUINOLINIC ACID METABOLISM IN THE BRAIN

As an integral part of the kynurenine pathway (leading from tryptophan to NAD), the enzymes responsible for the immediate biosynthesis (3-hydroxyanthranilic acid oxygenase; 3HAO) and degradation (quinolinic acid phosphoribosyltransferase; QPRT) of QUIN were identified and characterized in peripheral tissues by the mid 1960's (Nishizuka and Hayaishi, 1963;

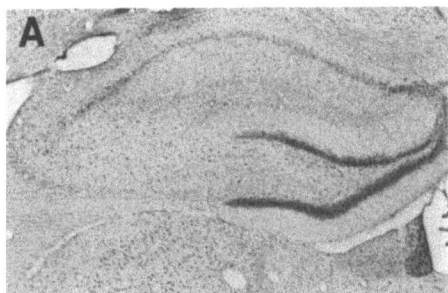
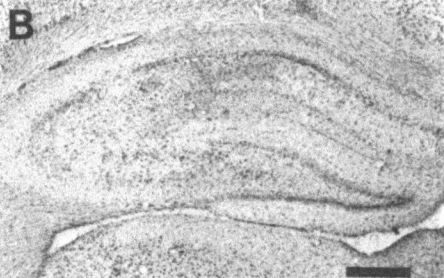

Fig. 2. Photomicrographs of QUIN-induced lesions in the rat hippocampus. Note the pattern of degeneration caused by the injection of 30 nmol QUIN (A) and the total neuronal loss resulting from 120 nmol QUIN (B). Bar: 500 μm.

Gholson et al., 1964). Using novel radiochemical methodologies we recently were able to measure the activities of both enzymes in rat and human brain tissue (Foster et al., 1985a, b; Schwarcz et al., 1985; Foster et al., 1986). Thus, QUIN, which does not penetrate the blood-brain barrier under normal physiological conditions (Foster et al., 1984), can be synthesized and catabolized by the brain. Moreover, there exists a pronounced variation of both 3HAO and QPRT between brain areas. Interestingly, the hippocampal formation appears to contain the two enzymes in a ratio which favors the accumulation of QUIN, i.e., high levels of 3HAO and rather low levels of QPRT. The absolute values for both enzymes in rat and human hippocampus (Table 1) indicate that powerful regulatory mechanisms must exist, which serve to prevent the accumulation of QUIN in the cell – and possibly its expulsion into the extracellular space. It appears that the brain is indeed equipped with an array of endogenous factors (such as Fe^{2+}, Mg^{2+}, oxygen and some proteins), which can directly or indirectly influence and modulate QUIN metabolism. The levels of brain QUIN are also dependent on the availability of its bio-precursor 3-hydroxyanthranilic acid, which can probably be synthesized from tryptophan in cerebral tissue (Gál and Sherman, 1980; Moroni et al., 1984). Studies of the <u>in vivo</u> disposition of QUIN and its relationship to QUIN's enzymatic machinery are still in their infancy. Work with QPRT, however, has led to the development of novel concepts concerning a possible role for QUIN in normal and abnormal brain function.

CELLULAR LOCALIZATION OF QPRT IN THE RAT HIPPOCAMPUS

Immunohistochemical techniques, utilizing antibodies prepared against purified rat liver QPRT (Okuno and Schwarcz, 1985), proved instrumental in defining the cell types associated with the QUIN-catabolizing enzyme and therefore, by inference, with QUIN itself. In a semiquantitative manner, a good correlation was obtained between the regional distribution of QPRT in the brain (determined biochemically; Foster et al., 1985b) and that of small (5-7 μm diameter) QPRT-containing glial cells, which were sometimes noted to be in close association with nerve cell bodies (Fig. 3). However, the nature of QPRT-positive glial cells has not been sufficiently investigated. While unequivocal identification will require analysis at the ultrastructural level, many of these QPRT-containing cells appear to be oligodendroglia. Notably, QPRT-antibodies and those recognizing glial fibrillary acidic protein, an established marker for astroglia, show different staining patterns in the rat brain (unpublished observation).

QPRT-containing glial cells are scattered throughout the entire hippocampal formation of the rat (Figs. 3A-C). Some of them possess short processes without clearcut orientation within the hippocampus, while others

Table 1. 3HAO and QPRT Activities in Rat and Human Hippocampus

	3HAO (pmol/hr/mg protein)	QPRT (fmol/hr/mg protein)
Rat	1420 ± 130 (5)	111 ± 12 (8)
Human	727 ± 202 (21)	131 ± 11 (7)

Enzymes were determined in crude preparations of fresh rat and stored ($-80°C$) and thawed human tissue according to the standard conditions established by Foster et al. (1985b, 1986). Data represent the mean ± S.E.M. of the number of individual specimens given in parentheses.

are often arranged in the form of 'caps' on neuronal somata. This is in marked contrast to QPRT-positive cells of the pyriform cortex, which often possess long processes perpendicular to the pial surface (micrographs not shown). In some instances (Fig. 3D), more than one QPRT-containing cell may be associated with a hippocampal neuron. Moreover, small QPRT-positive elements, possibly representing end feet, can frequently be observed on the soma surface. Thus, the hippocampal QUIN system is clearly highly organized yet its distribution is not obviously correlated to any known neuroactive compound. The functional significance of these specific anatomical arrangements and, in particular, their possible relevance for understanding QUIN-related seizure phenomena are not immediately obvious. QUIN can not be actively taken up by brain cells (Foster et al., 1984) but may enter via passive diffusion once extracellular levels exceed a critical limit. Under those circumstances, the close physical proximity of highly epileptogenic neurons to cells containing a high concentration of the specific QUIN-catabolizing enzyme, QPRT, may practically serve as a protective device against QUIN-induced convulsions or neuronal degeneration.

EFFECTS OF HIPPOCAMPAL LESIONS ON 3HAO AND QPRT

If QUIN and its metabolic enzyme(s) are indeed localized in glial cells, they should not be lost from the tissue following the ablation of neuronal elements. In fact, a unilateral injection of the excitotoxin ibotenic acid, sufficient to cause the degeneration of all neurons throughout the dorsal hippocampus (Schwarcz et al., 1979), resulted in a delayed but dramatic local _increase_ in the activities of both 3HAO and QPRT (Fig. 4). In contrast, no significant change in the two enzymes was noted in the hippocampus following cholinergic deafferentation by fornix-fimbria transection (Table 2). Thus, it appears that the loss of neuronal cell bodies but not of afferent fibers is capable of initiating the enzymatic reactions. It remains to be seen if the biochemically determined increases in the activity of both 3HAO and QPRT correspond to _de novo_ synthesis or merely represent an 'up-regulation' of the enzymes. The time course of the observed enzymatic changes favors the former interpretation. Furthermore, immunohistochemical data obtained with anti-QPRT antibodies should reveal if reactive glia, known to proliferate following excitotoxic insults (Coyle et al., 1978), contain QPRT and thus account for the increase in QPRT activity measured in whole tissue homogenates.

Complementary data were obtained when brain 3HAO and QPRT activities

Table 2. Effects of Cholinergic Deafferentation on Hippocampal Enzyme Activities

	3HAO	QPRT	GAD	CAT
Dorsal hippocampus	109 ± 14	118 ± 4	101 ± 15	31 ± 6**
Ventral hippocampus	92 ± 8	130 ± 12	85 ± 8	15 ± 2**

Unilateral fornix-fimbria transections were performed in rats seven days prior to sacrifice as described by Schwarcz and Köhler (1980). Enzyme activities (cf. legend to Fig. 4) were determined in deafferentated and non-transected contralateral hippocampus of six rats and calculated on a mg protein basis. Data for all enzymes are expressed as a percentage ± S.E.M. of the non-transected contralateral hippocampus. **$p < 0.01$ (paired t-test) as compared to controls.

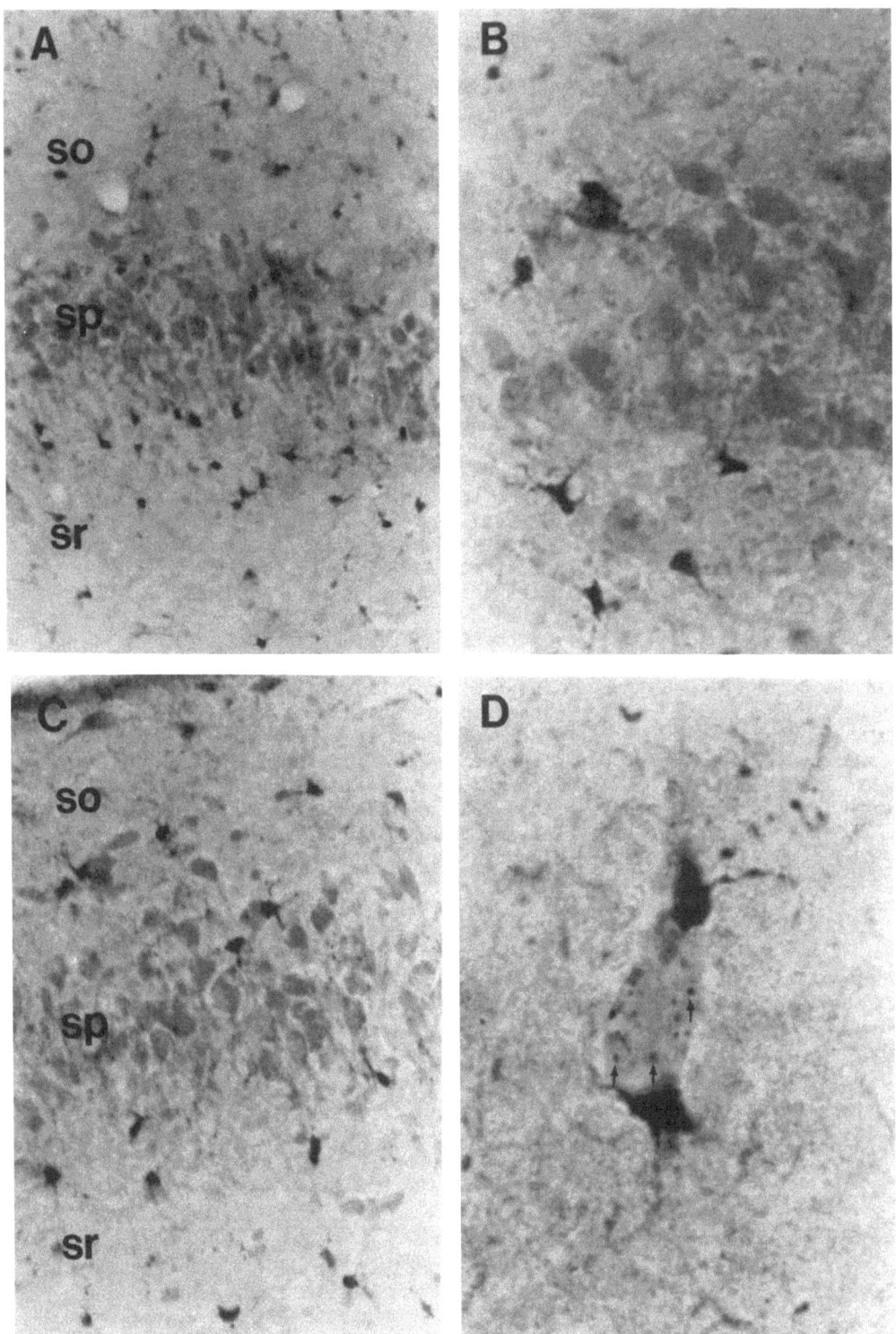

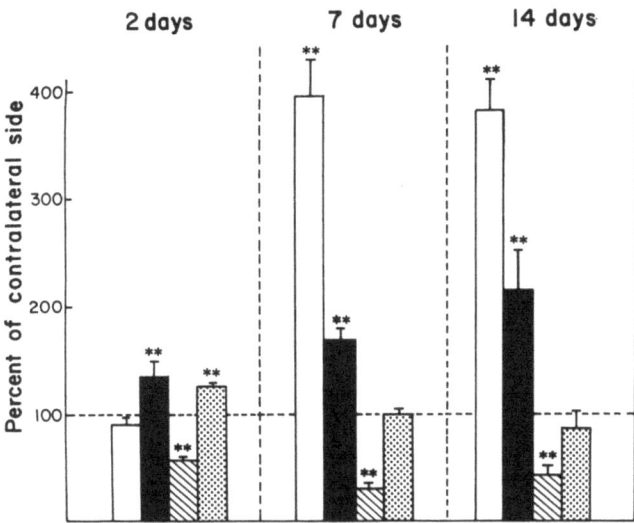

Fig. 4. Activities of 3HAO (open bars), QPRT (solid bars), glutamic acid decarboxylase (GAD; hatched bars) and choline acetyltransferase (CAT; stippled bars) in the rat hippocampus at various timepoints following the local injection of 120 nmol ibotenic acid (1 μl, pH 7.4). Enzyme activities were determined according to the methods of Foster et al. (1985b and 1986 for QPRT and 3HAO, respectively), Wilson et al. (1972; GAD) and Bull and Oderfeld-Novak (1971; CAT). Data, calculated on a mg protein basis, are expressed as a percentage of the uninjected contralateral hippocampus and are the mean ± S.E.M. of six animals in each group. GAD decreases indicate the success of the ibotenate lesion while unchanged CAT activity reflects the integrity of septo-hippocampal cholinergic afferents. **$p < 0.01$ as compared to the contralateral hippocampus (paired t-test).

Fig. 3. Photomicrographs showing QPRT-immunoreactive glial cells in different layers of hippocampal subfields CA1 (A, B), CA3 (C) and hilus of the area dentata (D). Fixed brains were cut on a freezing microtome and 30 μm thick sections incubated floating free in rabbit anti-QPRT antiserum (Okuno and Schwarcz, 1985; diluted 1:6,000 in phosphate buffered saline, pH 7.4, containing 0.2% Triton X-100 and 1% normal goat serum) for 3-7 days. The antigen-antibody complex was made visible by the method of Hsu et al. (1981). Scattered QPRT-positive cells can be seen in all laminae (A-C) and can sometimes be found in close contact with neuronal cell bodies (B, D). Sporadically, more than one QPRT-containing glial cell embraces a neuron and QPRT-positive structures resembling end feet can be seen on the soma surface (arrows in D). Abbreviations: so: stratum oriens; sr: stratum radiatum; sp: stratum pyramidale. Magnifications: A: x160; B: x400, C: x250, D: x1000.

Table 3. Effects of Systemic Kainate Administration on Enzyme Activities in the Rat Brain

	3HAO	QPRT	GAD
Hippocampus	397 ± 107**	258 ± 72*	68 ± 13
Pyriform Cortex	339 ± 63**	129 ± 34	34 ± 7*

Activities of 3HAO, QPRT and GAD (the latter as a marker of neuronal degeneration) were measured in kainic acid (12 mg/kg, i.p.; N = 5) treated and control (saline treated; N = 6) rats seven days after the injection. For methodological references, see legend to Fig. 4. Tissue from both hemispheres was pooled for every animal to yield a single assay point. Data were calculated on a mg protein basis and represent the mean ± S.E.M. of the percentage change in the kainate group as compared to the control group. **$p < 0.01$; *$p < 0.05$ vs. controls (t-test).

were measured following the systemic administration of kainic acid (12 mg/kg) to rats. This treatment regimen causes a well-established, highly reproducible limbic seizure syndrome, which has been suggested to provide an excellent animal model for temporal lobe epilepsy (Ben-Ari, 1985). Notably, the nerve cell loss which occurs throughout the limbic system is not entirely due to the direct action of the convulsant excitotoxin, kainic acid, but also to seizure activity per se. In particular, lesions in the pyriform cortex and the amygdala, in contrast to those occurring in the hippocampus, may be due to the secondary action of locally released excitatory amino acids such as glutamate (cf. Collins and Olney, 1982). However, increases in the activities of the QUIN-metabolizing enzymes were observed in both hippocampus and pyriform cortex (Table 3) while unlesioned brain regions such as the striatum and the olfactory bulb did not have such a reaction to kainate administration (data not shown). As with ibotenate lesions of the hippocampus, these effects could not be observed prior to the occurrence of neuronal degeneration, e.g., during the acute phases of the kainate-induced limbic syndrome (data not shown). Thus, it appears that nerve cell loss, possibly irrespective of the chemical nature of the toxic agent causing the degeneration, results in a substantial increase in 3HAO and QPRT activity.

Regardless of the cellular localization of the 'reactive' QUIN system and the mechanism(s) by which it is triggered, it is tempting to speculate about possible functional consequences of an increase in QUIN metabolism. In tissue, which is partially or totally depleted of neuronal cell bodies, (e.g., the lesioned hippocampus), the ratio between 3HAO and QPRT (see above) is tilted in favor of QUIN production at the expense of QUIN degradation. Unfortunately, due to the lack of a method sufficiently sensitive to determine QUIN levels (or, more appropriately, QUIN turnover) in brain tissue, it is unknown at present if the effects described here can indeed be translated into a large increase in the level or turnover of QUIN itself. As discussed earlier, regulatory mechanisms may well counteract the ability of lesioned hippocampal tissue to generate hyperphysiological quantities of this endogenous excitotoxin. If, however, increased amounts of QUIN are produced locally following the ablation of nerve cell bodies, and if the toxin is expelled from the QUIN-producing cellular entity (e.g., by release or diffusion) into the extracellular space, it is plausible that surrounding intact nerve cells may be detrimentally affected. Extrapolation of this concept to the situation in temporal lobe epilepsy is certainly premature and must take into consideration differences between the effects of a single intrahippocampal excitotoxin injection in animals and seizure disorders in humans. However, since several forms of epilepsy can reasonably be termed a progressive brain disorder (Hauser, 1983), QUIN could be envisioned to constitute a trigger factor in a cascade of events, originating with the (seizure-related?) death of just a few neurons. Fortunately, this idea can be tested experimentally so that an evaluation can be expected in the near future.

SYNOPSIS

The evidence for an involvement of QUIN in human seizure disorders is clearly circumstantial. Importantly, QUIN is not a classical neurotransmitter and may thus play only a negligible or no role at all in normal brain function (Foster et al., 1984). We have yet to understand if and how such a possibly inert metabolite may turn into a pathogen. Several crucial questions remain to be addressed before a case can be made for a 'quinolinic acid hypothesis' of temporal lobe epilepsy. Among the most prominent ones figure the extracellular concentration of QUIN in the human brain under normal and pathological ('epileptic') conditions, the relationship between QUIN metabolism in the brain and its extracellular concentration and, a related issue, the regulation of cerebral QUIN metabolism (i.e., turnover). It is of equal importance to assess if NMDA-receptors, particularly those in the hippocampus and other parts of the limbic system, can exert a modulatory function upon brain QUIN. Unquestionably, future experiments with selective NMDA-antagonists will prove useful for the elucidation of such possible (feedback) interactions.

ACKNOWLEDGEMENTS

We thank Ms. H. O'Brien for expert secretarial assistance. This work was supported by USPHS grant NS-16102.

REFERENCES

Aldinio, C., French, E.D., and Schwarcz, R., 1983, Effects of intrahippocampal ibotenic acid and their blockade by (-)2-amino-7-phosphonoheptanoic acid: morphological and electroencephalographical analysis, Exp. Brain Res., 51:36.
Ben-Ari, Y., 1985, Limbic seizure and brain damage produced by kainic acid: mechanisms and relevance to human temporal lobe epilepsy, Neuroscience, 14:375.
Bull, G., and Oderfeld-Novak, B., 1971, Standardization of a radiochemical assay of choline acetyltransferase and a study of the activation of the enzyme in rabbit brain, J. Neurochem., 19:935.
Collins, R.C. and Olney, J.W., 1982, Focal cortical seizures cause distant thalamic lesions, Science, 218:177.
Coyle, J.T., Molliver, M.E., and Kuhar, M.J., 1978, In situ injection of kainic acid: a new method for selectively lesioning neuronal cell bodies while sparing axons of passage, J. Comp. Neurol., 180:301.
Foster, A.C., and Fagg, G.E., 1984, Acidic amino acid binding sites in mammalian neuronal membranes: their characteristics and relationship to synaptic receptors, Brain Res. Rev., 7:103.
Foster, A.C., Miller, L.P., Oldendorf, W.H., and Schwarcz, R., 1984, Studies on the disposition of quinolinic acid after intracerebral or systemic administration in the rat, Exp. Neurol., 84:428.
Foster, A.C., Whetsell, W.O., Jr., Bird, E.D., and Schwarcz, R., 1985a, Quinolinic acid phosphoribosyltransferase in human and rat brain: activity in Huntington's disease and in quinolinate-lesioned rat striatum, Brain Res., 336:207.

Foster, A.C., Zinkand, W.C., and Schwarcz, R., 1985b, Quinolinic acid phosphoribosyltransferase in rat brain, J. Neurochem., 44:446.

Foster, A.E., White, R.J., and Schwarcz, R., 1986, Synthesis of quinolinic acid by 3-hydroxyanthranilic acid oxygenase in rat brain tissue in vitro, J. Neurochem., in press.

French, E.D., Aldinio, C., and Schwarcz, R., 1982, Intrahippocampal kainic acid, seizures and local neuronal degeneration: relationships assessed in unanesthetized rats, Neuroscience, 7:2525.

Gál, E.M., and Sherman, A.D., 1980, L-Kynurenine: its synthesis and possible regulatory function in brain, Neurochem. Res., 5:223.

Gholson, R.K., Ueda, I., Ogasawara, N., and Henderson, L.M., 1964, The enzymatic conversion of quinolinate to nicotinic acid mononucleotide in mammalian liver, J. Biol. Chem., 239:1208.

Hauser, W.A., 1983, Status epilepticus: frequency, etiology, and neurological sequelae, in: Status Epilepticus: Mechanisms of Brain Damage and Treatment, A.V. Delgado-Escueta, C.G. Wasterlain, D.M. Treiman and R.J. Porter, eds., Raven Press, New York, p. 3.

Henderson, L.M., and Hirsch, H.M., 1949, Quinolinic acid metabolism. I. Urinary excretion by the rat following tryptophan and 3-hydroxyanthranilic acid administration, J. Biol. Chem., 181:667.

Henderson, L.M., and Ramasarma, G.B., 1949, Quinolinic acid metabolism. III. Formation from 3-hydroxyanthranilic acid by rat liver preparations, J. Biol. Chem., 181:687.

Hsu, S.M., Raine, L., and Fanger, H., 1981, Use of avidin-biotin-peroxidase complex (ABC) in immunoperoxidase techniques: a comparison between ABC and unlabelled antibody (PAP) procedures, J. Histochem. Cytochem., 29:557.

Köhler, C., Schwarcz, R., and Fuxe, K., 1979, Intrahippocampal injection of ibotenic acid provide histological evidence for a neurotoxic mechanism different from kainic acid, Neurosci. Lett., 15:223.

Lapin, I.P., 1981, Kynurenines and seizures, Epilepsia, 22:257.

Lapin, I.P., Prakhie, I.B., and Kiseleva, I.P., 1982, Excitatory effects of kynurenine and its metabolites, amino acids and convulsants administered into brain ventricles: differences between rats and mice, J. Neural Transm., 54:229.

Moroni, F., Lombardi, G., Carla, V., and Moneti, G., 1984, Studies on the content, synthesis and disposition of quinolinic acid in physiology and pathology, Clin. Neuropharmacol., 7(Suppl):448.

Nadler, J.V., Evenson, D.A., and Cuthbertson, G.J., 1981, Comparative toxicity of kainic acid and other acidic amino acids toward rat hippocampal neurons, Neuroscience, 6:2505.

Nishizuka, Y., and Hayaishi, O., 1963, Studies on the biosynthesis of nicotinamide adenine dinucleotide. I. Enzymic synthesis of niacin ribonucleotides from 3-hydroxyanthranilic acid in mammalian tissues, J. Biol. Chem., 238:3369.

Okuno, E., and Schwarcz, R., 1985, Purification of quinolinic acid phosphoribosyltransferase from rat liver and brain, Biochim. Biophys. Acta, 841:112.

Perkins, M.N., and Stone, T.W., 1983, Pharmacology and regional variations of quinolinic acid-evoked excitation in the rat central nervous system, J. Pharmacol. Exp. Therap., 226:551.

Schwarcz, R., Hökfelt, T., Fuxe, K., Jonsson, G., Goldstein, M., and Terenius, L., 1979, Ibotenic acid induced neuronal degeneration: a morphological and neurochemical study, Exp. Brain Res., 37:199.

Schwarcz, R., and Köhler, C., 1980, Evidence against an exclusive role of glutamate in kainic acid neurotoxicity, Neurosci. Lett., 19:243.

Schwarcz, R., and Köhler, C., 1983, Differential vulnerability of central neurons of the rat to quinolinic acid, Neurosci. Lett., 38:85.

Schwarcz, R., Whetsell, W.O., Jr., and Mangano, R.M., 1983, Quinolinic acid: an endogenous metabolism that causes axon-sparing lesions in rat brain, Science, 219:316.

Schwarcz, R., Brush, G.S., Foster, A.C., and French, E.D., 1984a, Seizure activity and lesions following intrahippocampal injection of quinolinic acid, Exp. Neurol., 84:1.

Schwarcz, R., Foster, A.C., French, E.D., Whetsell, W.O., Jr., and Köhler, C., 1984b, Excitotoxic models for neurodegenerative disorders, Life Sci., 35:19.

Schwarcz, R., White, R.J., and Whetsell, W.O., Jr., 1985, 3-hydroxyanthranilic acid oxygenase: activity in lesioned rat striatum and in human brain tissue, Soc. Neurosci. Abstr., 11:240.6.

Snyder, S.H., Banerjee, S.P., Yamamura, H.I., and Greenberg, D., 1974, Drugs, neurotransmitters and schizophrenia, Science, 184:1243.

Stone, T.W., and Perkins, M.N., 1981, Quinolinic acid: a potent endogenous excitant at amino acid receptors in rat CNS, Eur. J. Pharmacol., 72:411.

Wilson, S.H., Schrier, B.K., Farber, J.L., Tompson, E.J., Rosenberg, R.N., Blume, A.J., and Nirenberg, M.W., 1972, Markers for gene expression in cultured cells from nervous system, J. Biol. Chem., 247:3159.

Wolfensberger, M., Amsler, U., Cuénod, M., Foster, A.C., Whetsell, W.O., Jr., and Schwarcz, R., 1983, Identification of quinolinic acid in rat and human brain tissue, Neurosci. Lett., 41:247.

COMMENTARY - SEIZURES AND BRAIN DAMAGE: ARE EXCITATORY AMINO ACIDS INVOLVED?

Y. Ben-Ari and R. Schwarcz

The program of this session, which was deliberately chosen as the final part of the symposium, was designed to critically assess the evidence for a participation of excitatory amino acids in seizure phenomena and associated selective nerve cell death. The speakers agreed that excitatory transmission mediated by acidic amino acids is likely to play an important role in both the initiation and propagation of epileptic phenomena. However, substantial discussion centered around the following questions: 1) which mechanisms (if any) link paroxysmal discharges and excitotoxic brain damage?, 2) is there a primary locus in the brain for the initiation of excitotoxin-induced seizures?, 3) is glutamate a viable candidate as an endogenous trigger of seizure phenomena and associated lesions and 4) which ionic movements are primarily responsible for excitotoxin-induced neuropathological changes?

THE EXCITOTOXIC LINK

In its most simplistic form, Olney's original excitotoxic hypothesis states nothing more than the fact that a quantitative parallelism exists between the neuroexcitatory and neurotoxic potencies of acidic amino acids. Excitotoxic lesions are ultrastructurally identified by an 'axon-sparing' appearance, i.e., the degeneration of neuronal somata and dendrites and the preservation of axon terminals of neurons originating outside the lesion area. Probably one of the most elementary questions that has to be asked in the context of this symposium is: can the excitotoxic hypothesis be extended to embrace seizure-related neuronal loss, or, in other words, do 'convulsotoxins' exist? If they do, to follow the argument further, is there any evidence for the neuropathological equivalence of lesions caused by convulsive excitatory amino acids and other convulsants (such as bicuculline) in experimental animals? If yes, are these lesions identical to those observed in humans? Unfortunately, no information is presently available on the exact nature - ultrastructurally speaking - of seizure-related brain damage in humans. It is therefore important to realize that the extrapolation of all data presented in this session has to be exercised with caution until the excitotoxic quality of lesions in human epileptics can be unequivocally demonstrated.

Substantial evidence was presented in favor of a role of seizures in the precipitation of excitotoxin-mediated neurodegeneration. It is equally important to note, however, that excitotoxic damage can occur without concomitant seizure phenomena (e.g., under hypoglycemic or hypoxic conditions or after local injection of only modestly convulsive excitotoxins such as ibotenic acid or quisqualic acid into the brain of experimental animals). The issue of the special nature of seizure-related excitotoxic cell death was discussed by several speakers, who focused much of their attention on the vulnerability of hippocampal pyramidal cells.

THE ISSUE OF CA3 PYRAMIDS

There can be no doubt that hippocampal CA3 cells play a special role in epilepsy. They have been shown to function as pacemaker neurons and also appear to be among the most susceptible cells in the brain with regard to neurodegenerative events in human seizure disorders. The interest of excitotoxicologists in this group of neurons was greatly sparked by the finding that a) they are extremely vulnerable to the toxic effects of kainic acid (KA) and b) they contain (pre- or postsynaptically) the highest density of KA-binding sites in the brain of most species investigated so far, including man. Indeed, some experimental evidence argues in favor of an implication of these cells in KA-induced seizures: 1) A quantitative correlation exists between seizure severity and subsequent damage to CA3 cells; 2) Deoxyglucose mapping shows a virtual overlap between KA-activated areas and those undergoing degeneration; 3) Diazepam pretreatment abolishes hippocampal damage following an intraamygdaloid KA injection; 4) Acute transection of the mossy fiber system protects against hippocampal nerve cell loss when KA is applied intracerebroventricularly. Similarly, KA-induced paroxysmal discharges are substantially diminished by this treatment; 5) In the course of postnatal development, there is a close temporal association between the first precipitation of seizures and hippocampal lesions caused by KA.

These findings, taken together, clearly establish a close relationship between CA3 pyramids, KA-induced lesions and seizures. They also, however, raise pertinent questions regarding the pathogenesis of epilepsy. For instance, why can several of the above phenomena also be observed after the administration of bicuculline or cholinergic agonists? What is the reason for the failure of acute mossy fiber transections to protect CA3 cells from locally applied KA? Which aspects of the 'limbic circuit' are essential for KA-induced discharges and brain damage to occur? And most importantly, since KA is an exogenous excitotoxin, does KA mimic the actions of a 'real' convulsant, possibly an endogenous excitotoxin such as glutamate?

GLUTAMIC ACID AND QUINOLINIC ACID

Since much of the attention has been focused on the mossy fiber system and its postsynaptic element, it is tempting to speculate about the mediation of seizures and cell damage by neuroactive substances released directly onto the susceptible (i.e., vulnerable) CA3 neurons. As pointed out, paroxysmal discharges can be mediated by well-defined anatomical routes and the hippocampal mossy fibers may be just one of several important synaptic pathways in this respect. However, Sloviter's work puts special emphasis on the role of glutamate, a putative transmitter of both the perforant path and the mossy fiber system. KA, according to his view, mimics the actions of glutamate, which can be released from the mossy fiber system following perforant path stimulation. In other words, glutamate itself, under pathological conditions in humans, could produce the same effects as KA and may thus constitute the missing link in the excitotoxic hypothesis of epilepsy. If this interpretation is correct (after all, glutamate does interact powerfully with the KA-receptor), several important questions remain to be addressed: what is the trigger mechanism leading to an overactive glutamate function and subsequent paroxysmal discharges and brain damage? What are the particular neurochemical and neuroanatomical features of the mossy fiber system with regard to glutamate, which could

account for the preferential damage of CA3 cells? For example, could
there be as yet undiscovered synergistic actions between glutamate and
other endogenous excitants such as acetylcholine, aspartate, etc.? Similarly, can axon-sparing 'epileptic' damage be caused by other than classic
excitotoxins?

Work with quinolinic acid indicates that there exist non-neurotransmitter substances in the brain, which may, in epilepsy, accumulate intrasynaptically at a concentration sufficient to precipitate seizures as well as
selective excitotoxic brain damage. Here, too, it will be necessary not
only to critically assess if a compound like quinolinic acid fulfills
the stringent criteria to be considered a candidate for an endogenous
convulsotoxin. Work on novel substances must also aim at unraveling basic
aspects of their neurobiological fate, including mechanisms which govern
their expulsion into the synapse. Finally, novel concepts regarding epileptogenicity may arise from new knowledge on the cellular localization of
quinolinic acid and, possibly, related convulsive metabolites: e.g.,
is it possible, as implied by Schwarcz, that glial cells have a role in
the precipitation of seizure phenomena (cf. also Ward, this volume)?

SELECTIVE NEURONAL DEGENERATION IN EPILEPSY: SYNOPSIS

It appears that much of the confusion and, in part, controversy concerning 'epileptic' nerve cell death is due to the fact that the question
has not been uniformly defined by all investigators participating in the
discussion. The issue, when raised by Ben-Ari and Nadler, focuses greatly,
often exclusively, on hippocampal subfields known to be affected in temporal
lobe epilepsy. Their experiments employ specific pathways and synaptic
circuitry in order to explain selective cell death in the disease. Olney's
position differs by de-emphasizing the issue of selective regional vulnerability and stressing the overall principle that both local and 'distant'
nerve cell loss following the application of a convulsive excitotoxin
is of the axon-sparing type. Therefore, no fundamental difference exists
between, e.g., hippocampal and cortical damage following systemic KA administration. The latter, according to Olney, may simply be due to the release
of an endogenous excitotoxin secondary to the excitation (or beginning
degeneration) occurring at the site of primary action.

Viewed in the context of the above comments on the as yet unresolved
'peculiarity' of the hippocampal mossy fiber system, the two standpoints
are by no means incompatible. It will certainly be important, however,
to carefully assess the contribution of seizures themselves to this anatomically defined area in order to examine if seizure-related ionic movements
could yield the net result of a disproportionately large accumulation
of glutamate or an endogenous congener in, e.g., the CA3 region of the
hippocampal formation. It is this very level of analysis where Rothman's
data on the ionic prerequisites and dependencies of excitotoxic cell death
come into play: it has been sufficiently documented that ionic imbalances
- different ones under different experimental conditions! - can play an
important part in the initiation and propagation of axon-sparing neuronal
degeneration. _In vitro_ systems such as Rothman's can be expected to eventually explain why (and how) CA3 cells die in the course of seizure disorders
while CA1 neurons remain apparently unaffected. Changes in the experimental
conditions (e.g., to hypoxia) may, in the same system, yield an explanation
of the preferential vulnerability, again axon-sparing, of CA1 cells when
seizures are absent.

SYMPOSIUM INTERNATIONAL
INSERM-CNRS

ACIDES AMINES EXCITATEURS
ET EPILEPSIE

EXCITATORY AMINO ACIDS
AND SEIZURE DISORDERS

1-5 Septembre 1985
Chateau de Fillerval
France

CONTRIBUTORS

M. Abreu
Departments of Neuroscience,
Pharmacology, and Psychiatry
The Johns Hopkins University School
of Medicine
600 N. Wolfe Street
Baltimore, Maryland 21205 USA

D.G. Amaral
The Salk Institute for Biological
Studies and the Clayton Foundation
for Research − California Division
P.O. Box 85800
San Diego, California 92138 USA

D.R. Armstrong
Department of Pharmacology
Duke University Medical Center
Durham, North Carolina 27710 USA

P. Ascher
Laboratoire de Neurobiologie
Ecole Normale Supérieure
46 rue d'Ulm
75230 Paris, Cedex 05, France

B. Ault
Department of Pharmacology
Burroughs-Wellcome Co.
Research Triangle Park, North Carolina 27709 USA

T.L. Babb
Department of Neurology
UCLA School of Medicine
Los Angeles, California 90024 USA

M. Baraldi
Institute of Pharmacology
Via Campi 287
Modena University
Modena, Italy

H. Baran
Institute of Biochemical Pharmacology
Borschkegasse 8A
A-1090 Vienna, Austria

E.A. Barnard
MRC Molecular Neurobiology Research Group
Department of Biochemistry
Imperial College of Science and Technology
London, SW7 2AZ England

M. Beaujean
Departments of Neurobiology and Organic
Chemistry
The Weizmann Institute of Science
P.O. Box 26
Rehovot, Israel 76100

Y. Ben-Ari
LPN, CNRS
Gif-sur-Yvette, F-91190 France
Present Address: U29 INSERM
123 Bd de Port-Royal
75014 Paris, France

A. Benelli
Institute of Pharmacology
Via Campi 287
Modena University
Modena, Italy

M.L. Berger
Institute of Biochemical Pharmacology
Borschkegasse 8A
A-1090 Vienna, Austria

G. Bernard
INSERM U26
200 Rue du Faubourg St Denis
75010 Paris, France

G. Bilbe
MRC Molecular Neurobiology Research Group
Department of Biochemistry
Imperial College of Science and Technology
London, SW7 2AZ England and
Molecular Genetics Department
Searle Research/Development
Lane End Road
High Wycombe, Bucks HR12 4HL England

R. Blakely
Departments of Neuroscience, Pharmacology,
and Psychiatry
The Johns Hopkins University School
of Medicine
600 N. Wolfe Street
Baltimore, Maryland 21205 USA

D.W. Bonhaus
Departments of Medicine (Neurology) and
Pharmacology
Duke University Medical Center and
Epilepsy Research Laboratory
Veterans Administration Medical Center
Durham, North Carolina 27705 USA

D.A. Brown
MRC Neuropharmacology Research Group
Department of Pharmacology
School of Pharmacy
29/39 Brunswick Square
London, WC1N 1AX England

D.O. Carpenter
Division of Laboratories and Research
New York State Department of Health
Albany, New York 12201 USA

E. Caselgrandi
Institute of Pharmacology
Via Campi 287
Modena University
Modena, Italy

E. Cherubini
LPN, CNRS
Gif-sur-Yvette, F-91190 France
Present Address: U29 INSERM
123 Bd de Port-Royal
75014 Paris, France

S.H. Chung
National Institute for Medical Research
The Ridgeway, Mill Hill
London, NW7 1AA England

A. Constanti
MRC Neuropharmacology Research Group
Department of Pharmacology
School of Pharmacy
29/39 Brunswick Square
London, WC1N 1AX England

C.W. Cotman
Department of Psychobiology
University of California
Irvine, California 92717 USA

J.T. Coyle
Departments of Neuroscience, Pharmacology,
and Psychiatry
The Johns Hopkins University School
of Medicine
600 N. Wolfe Street
Baltimore, Maryland 21205 USA

I.L. Crawford
University of Texas and Veterans Administration
Department of Neurology and Epilepsy Center
Dallas, Texas 75216 USA

W.E. Crill
Department of Physiology and Biophysics
University of Washington
School of Medicine
Seattle, Washington 98195 USA

M. Cuénod
Brain Research Institute
University of Zürich
August-Forelstr. 1
CH-8029 Zürich, Switzerland

S.G. Cull-Candy
MRC Receptor Mechanisms Research Group
Department of Pharmacology
University College London
London, WC1 England

T. Dalkara
Departments of Anaesthesia Research and
Physiology
McGill University
3655 Drummond St.
Montreal (Quebec), H3G 1Y6 Canada

P. David
Department of Neurobiology
The Weizmann Institute of Science
P.O. Box 26
Rehovot, Israel 76100

R. Dingledine
Department of Pharmacology
University of North Carolina at Chapel Hill
Chapel Hill, North Carolina 27514 USA

K.Q. Do
Brain Research Institute
University of Zürich
August-Forelstr. 1
CH-8029 Zürich, Switzerland

M. Ebadi
Department of Pharmacology
University of Nebraska College of Medicine
42nd Street and Dewey Avenue
Omaha, Nebraska 68105 USA

D. Eisenberg-Tamarin
Department of Neurobiology
The Weizmann Institute of Science
P.O. Box 26
Rehovot, Israel 76100

B. Engelsen
Norwegian Defense Research Establishment
Division for Environmental Toxicology
P.O. Box 25
N-2700 Kjeller, Norway

U. Erez
Department of Neurobiology
The Weizmann Institute of Science
P.O. Box 26
Rehovot, Israel 76100

J.M.H. ffrench-Mullen
Department of Neurology
The Johns Hopkins University School
of Medicine
600 N. Wolfe Street
Baltimore, Maryland 21205 USA

B.O. Fischer
Anatomical Institute
University of Oslo
Karl Johans gate 47
N-0162 Oslo 1, Norway

R. Fisher
Department of Neurology
The Johns Hopkins University School
of Medicine
600 N. Wolfe Street
Baltimore, Maryland 21205 USA

J.A. Flatman
Department of Physiology and Biophysics
University of Washington
Seattle, Washington 98195 USA

F. Fonnum
Norwegian Defense Research Establishment
Division for Environmental Toxicology
P.O. Box 25
N-2700 Kjeller, Norway

V.M. Fosse
Norwegian Defense Research Establishment
Division for Environmental Toxicology
P.O. Box 25
N-2700 Kjeller, Norway

A.C. Foster
Merck, Sharp and Dohme, Ltd.
Neuroscience Research Centre
Terlings Park, Eastwick Road
Harlow, Essex U.K.

J.E. Franck
Department of Physiology and Biophysics
University of Washington
Seattle, Washington 98195 USA

E.D. French
Maryland Psychiatric Research Center
P.O. Box 21247
Baltimore, Maryland 21228 USA

H. Frenk
Department of Psychology
Tel Aviv University
Tel Aviv, Israel

B. Gabrielsson
National Institute for Medical Research
The Ridgeway, Mill Hill
London, NW7 1AA England

M. Gho
Gif-sur-Yvette, F-91190 France
Present Address: U29 INSERM
123 Bd de Port-Royal
75014 Paris, France

A. Giberti
Institute of Pharmacology
Via Campi 287
Modena University
Modena, Italy

O. Goldberg
Department of Organic Chemistry
The Weizmann Institute of Science
P.O. Box 26
Rehovot, Israel 76100

G. Goping
Laboratory of Neuropathology and Neuroanatomical Sciences
NINCDS, NIH
Bethesda, Maryland 20892 USA

M. Gruenthal
Department of Pharmacology
Duke University Medical Center
Durham, North Carolina 27710 USA

H. Hagberg
Institute of Neurobiology
University of Göteborg
P.O. Box 33031
S-400 33 Göteborg, Sweden

Y. Hama
Department of Pharmacology
University of Nebraska College of Medicine
42nd Street and Dewey Avenue
Omaha, Nebraska 68105 USA

A. Hamberger
Institute of Neurobiology
University of Göteborg
P.O. Box 33031
S-400 33 Göteborg, Sweden

U. Heinemann
Max Planck Institute of Psychiatry
Department of Neurophysiology
Am Klopferspitz 18a
D-8033 Planegg, F.R.G.

P.L. Herrling
Wander Research Institute
P.O. Box 2747
CH-3001 Bern, Switzerland

O. Hornykiewicz
Institute of Biochemical Pharmacology
Borschkegasse 8A
A-1090 Vienna, Austria

K. Houamed
MRC Neuropharmacology Research Group
Department of Pharmacology
School of Pharmacy
29/39 Brunswick Square
London, WC1N 1AX England

I. Jacobson
Institute of Neurobiology
Univeristy of Göteborg
P.O. Box 33031
S-400 33 Göteborg, Sweden

D. Johnston
Neuroscience Program, Section of Neurophysiology
Department of Neurology
Baylor College of Medicine
1 Baylor Plaza
Houston, Texas 77030 USA

E.J. Kasarskis
V.A. Medical Center and Departments of Neurology,
Pharmacology and Biochemistry
University of Kentucky
800 Rose Street
Lexington, Kentucky 40506 USA

A.E. King
Department of Pharmacology
St. Bartholomew's Hospital Medical College
University of London
Charterhouse Square
London EC1M 6BQ, England

G.L. King
Department of Pharmacology
University of North Carolina at Chapel Hill
Chapel Hill, North Carolina 27514 USA

K. Kitz
Institute of Biochemical Pharmacology
Borschkegasse 8A
A-1090 Vienna, Austria

I. Klatzo
Laboratory of Neuropathology and Neuroanatomical
Sciences
NINCDS, NIH
Bethesda, Maryland 20892 USA

Ch. Köhler
Astra Läkemedel AB
Department of Neuropharmacology
Södertälje, Sweden

K.J. Koller
Departments of Neuroscience, Pharmacology,
and Psychiatry
The Johns Hopkins University School
of Medicine
600 N. Wolfe Street
Baltimore, Maryland 21205 USA

K. Krnjević
Department of Physiology
McGill University
3655 Drummond Street
Montreal (Quebec) H3G 1Y6, Canada

I. Kurcewicz
Unité de Recherches sur l'épilepsie
INSERM U97
2 ter rue d'Alesia
F-75014 Paris, France

H. Lassmann
Neurological Institute
University of Vienna
Schwarzspanierstr. 17
A-1090 Vienna, Austria

J.W. Lazarewicz
Medical Research Centre
Polish Academy of Sciences
Warsaw, Poland

F.J. Lebeda
Neuroscience Program, Section of Neurophysiology
Department of Neurology
Baylor College of Medicine
1 Baylor Plaza
Houston, Texas 77030 USA

J.M. Lefauconnier
INSERM U26
200 Rue de Faubourg St Denis
75010 Paris, France

A. Lehmann
Institute of Neurobiology
University of Göteborg
P.O. Box 33031
S-400 33 Göteborg, Sweden

J. Louvel
Unité de Recherches sur l'épilepsie
INSERM U97
2 ter rue d'Alesia
F-75014 Paris, France

A. Luini
Department of Neurobiology
The Weizmann Institute of Science
P.O. Box 26
Rehovot, Israel 76100

J.F. MacDonald
Playfair Neuroscience Unit
University of Toronto
Toronto Western and Wellesley Hospitals
Toronto, Ontario M5T 2S8 Canada

C. Matute
Brain Research Institute
University of Zürich
August-Forelstr. 1
Ch-8029 Zürich, Switzerland

M.L. Mayer
Laboratory of Developmental Neurobiology
National Institute of Child Health and
Human Development
National Institutes of Health
Bethesda, Maryland 20892 USA

J.O. McNamara
Departments of Medicine (Neurology) and
Pharmacology
Duke University Medical Center and Epilepsy
Research Laboratory
Veterans Administration Medical Center
Durham, North Carolina 27705 USA

B. Meldrum
Neurology Department
Institute of Psychiatry
Denmark Hill
London SE5 8AF, England

Z. Miljkovic
Playfair Neuroscience Unit
University of Toronto
Toronto Western and Wellesley Hospitals
Toronto, Ontario, Canada

D.T. Monaghan
Department of Psychobiology
University of California
Irvine, California 92717 USA

J.V. Nadler
Department of Pharmacology
Duke University Medical Center
Durham, North Carolina 27710 USA

A. Nistri
Department of Pharmacology
St. Bartholomew's Hospital Medical College
University of London
Charterhouse Square
London EC1M 6BQ, Great Britain

L. Nitecka
LPN, CNRS
Gif-sur-Yvette, F-91190 France
Present Address: U29 INSERM
123 Bd de Port-Royal
75014 Paris, France

C. Nitsch
Anatomical Institute
University of Munich
Pettenkoferstr. 11
D-8000 München 2, F.R.G.

D.K. Norris
National Institute for Medical Research
The Ridgeway, Mill Hill
London NW7 1AA, England

L. Nowak
Department of Pharmacology
Cornell University Veterinary College
Ithaca, New York 14853 USA

M.M. Okazaki
Department of Pharmacology
Duke University Medical Center
Durham, North Carolina 27710 USA

E. Okuno
Maryland Psychiatric Research Center
P.O. Box 21247
Baltimore, Maryland 21228 USA

J.W. Olney
Department of Psychiatry and McDonnell Center
for the Study of Higher Brain Function
Washington University
St. Louis, Missouri 63110 USA

L. Ory-Lavollée
Departments of Neuroscience, Pharmacology,
and Psychiatry
The Johns Hopkins University School
of Medicine
600 N. Wolfe Street
Baltimore, Maryland 21205 USA

O.P. Ottersen
Anatomical Institute
University of Oslo
Karl Johans gate 47
N-0162 Oslo 1, Norway

R.H. Paulsen
Norwegian Defense Research Establishment
Division for Environmental Toxicology
P.O. Box 25
N-2700 Kjeller, Norway

U. Petsche
Institute of Biochemical Pharmacology
Borschkegasse 8A
A-1090 Vienna, Austria

C. Preti
Institute of Pharmacology
Department of Chemistry
Via Campi 287
Modena University
Modena, Italy

J.L. Price
Department of Anatomy and Neurobiology
Washington University School of Medicine
St. Louis, Missouri 63130 USA

R. Pumain
Unité de Recherches sur l'épilepsie
INSERM U97
2 ter rue d'Alesia
F-75014 Paris, France

A. Repressa
LPN, CNRS
Gif-sur-Yvette, F-91190 France
Present Address: U29 INSERM
123 Bd de Port-Royal
75014 Paris, France

E. Rinvik
Anatomical Institute
University of Oslo
Karl Johans gate 47
N-0162 Oslo 1, Norway

S.M. Rothman
Departments of Pediatrics, Anatomy and
Neurobiology, and Neurology
Washington University School of Medicine
St. Louis, Missouri 63178 USA

C. Rovira
LPN, CNRS
Gif-sur-Yvette, F-91190 France
Present Address: U29 INSERM
123 Bd de Port-Royal
75014 Paris, France

F.T. Russchen
Department of Anatomy
Vrije Universiteit
Amsterdam, The Netherlands

P.A. Rutecki
Neuroscience Program, Section of Neurophysiology
Department of Neurology
Baylor College of Medicine
1 Baylor Plaza
Houston, Texas 77030 USA

R.E. Ruth
Institute for the Study of Developmental
Disabilities and Committee on Neuroscience
The University of Illinois at Chicago
Chicago, Illinois 60608 USA

M. Sandrini
Institute of Pharmacology
Via Campi 287
Modena University
Modena, Italy

J.H. Schneiderman
Playfair Neuroscience Unit
University of Toronto
Toronto Western and Wellesley Hospitals
Toronto, Ontario, Canada

R. Schwarcz
Maryland Psychiatric Research Center
P.O. Box 21247
Baltimore, Maryland 21228 USA

P.A. Schwartzkroin
Departments of Neurological Surgery
and Physiology and Biophysics
University of Washington
Seattle, Washington 98195 USA

P.C. Schwindt
Department of Physiology and Biophysics
University of Washington
Seattle, Washington 98195 USA

F. Seitelberger
Neurological Institute
University of Vienna
Schwarzspanierstr. 17
A-1090 Vienna, Austria

C. Shin
Departments of Medicine (Neurology) and
Pharmacology
Duke University Medical Center and Epilepsy
Research Laboratory
Veterans Administration Medical Center
Durham, North Carolina 27705 USA

J.T. Slevin
V.A. Medical Center and Departments of Neurology,
Pharmacology, and Biochemistry
University of Kentucky
800 Rose Street
Lexington, Kentucky 40506 USA

R.S. Sloviter
Neurology Center
Helen Hayes Hospital
New York State Department of Health
West Haverstraw, New York 10993 USA
and The Departments of Pharmacology and Neurology
Columbia University
New York, New York 10032 USA

T.G. Smart
MRC Neuropharmacology Research Group
Department of Pharmacology
School of Pharmacy
29/39 Brunswick Square
London WC1N 1AX, England

W. Spain
Department of Physiology and Biophysics
University of Washington
Seattle, Washington 98195 USA

C. Speciale
Maryland Psychiatric Research Center
P.O. Box 21247
Baltimore, Maryland 21228 USA

G. Sperk
Institute of Biochemical Pharmacology
Borschkegasse 8A
A-1090 Vienna, Austria

C.E. Stafstrom
Department of Physiology and Biophysics
University of Washington
Seattle, Washington 98195 USA

J. Storm-Mathisen
Anatomical Institute
University of Oslo
Karl Johans gate 47
N-0162 Oslo 1, Norway

P. Streit
Brain Research Institute
University of Zürich
August-Forelstr. 1
CH-8029 Zürich, Switzerland

Y. Tayarani
INSERM U26
200 Rue du Faubourg St Denis
75010 Paris, France

V.I. Teichberg
Department of Neurobiology
The Weizmann Institute of Science
P.O. Box 26
Rehovot, Israel 76100

G. Tosi
Institute of Pharmacology
Department of Chemistry
Via Campi 287
Modena University
Modena, Italy

E. Tremblay
LPN, CNRS
Gif-sur-Yvette, F-91190 France
Present Address: U29 INSERM
123 Bd de Port-Royal
75014 Paris, France

W.A. Turski
Wander Research Institute (a Sandoz Research Unit)
P.O. Box 2747
CH-3001 Bern, Switzerland

G. Urca
Departments of Physiology and Pharmacology
Tel Aviv University
Tel Aviv, Israel

T.C. Vanaman
V.A. Medical Center and Departments of Neurology, Pharmacology and Biochemistry
University of Kentucky
800 Rose Street
Lexington, Kentucky 40506 USA

N.M. van Gelder
CRSN/Dép de physiologie
Faculté de médecine
Université de Montréal
C.P. 6128, succursale A
Montréal, Quebéc H3C 3J7 Canada

G.W. Van Hoesen
Department of Anatomy
University of Iowa
Iowa City, Iowa 52242 USA

C. VanRenterghem
MRC Neuropharmacology Research Group
Department of Pharmacology
School of Pharmacy
29/39 Brunswick Square
London WC1N 1AX, England
MRC Molecular Neurobiology Research Group
Department of Biochemistry
Imperial College of Science and Technology
London SW7 2AZ, England

A. Vezzani
Maryland Psychiatric Research Center
P.O. Box 21247
Baltimore, Maryland 21228 USA

A.A. Ward, Jr.
Department of Neurological Surgery
University of Washington School of
Medicine
Seattle, Washington 98195 USA

G.L. Westbrook
Laboratory of Developmental Neurobiology
National Institute of Child Health and
Human Development
National Institutes of Health
Bethesda, Maryland 20892 USA

W.O. Whetsell, Jr.
Department of Pathology
Vanderbilt University Medical Center
Nashville, Tennessee 37232 USA

T. Wieloch
Laboratory for Experimental Brain Research
University Hospital
University of Lund
S-221 85 Lund, Sweden

M.P. Witter
Department of Anatomy
Vrije Universiteit
Amsterdam, The Netherlands

D.M. Wright
Department of Anatomy
The Medical School
University of Bristol
Bristol BS8 1TD, England

C. Yim
Departments of Anaesthesia Research and
Physiology
McGill University
3655 Drummond Street
Montreal (Quebec) H3G 1Y6, Canada

R. Zaczek
Departments of Neuroscience, Pharmacology,
and Psychiatry
The Johns Hopkins University School
of Medicine
600 N. Wolfe Street
Baltimore, Maryland 21205 USA

P. Zanoli
Institute of Pharmacology
Via Campi 287
Modena University
Modena, Italy

M. Zurini
V.A. Medical Center and Departments of
Neurology, Pharmacology, and Biochemistry
University of Kentucky
Lexington, Kentucky 40506 USA

INDEX

Acidic peptides in the brain, 375–382
Afterhyperpolarization (AHP), 407, 476, 481, 488, 540
  blockade by kainate, 477, 478, 651
  enhanced by NMDA, 478, 479
Alumina gel focus, 105, 158, 159, 451, 453
Alzheimer's disease and KA binding sites, 247–249
α-Aminoadipic acid, 238, 487, 688
Aminooxyacetic acid, 340
Ammon's horn sclerosis
  neuropathology, 117–120, 650, 673
  surgery, 115, 116, 163
AMPA (α-amino-3-hydroxy-5-methyl-4-isoxazolepropionic acid)
  autoradiography, 245–247
Amphibian spinal cord in vitro, 255, 485–493
Amygdaloid complex
  amygdalo-cortical projections, 10, 11, 13, 14, 100
  amygdalo-hippocampal connections, 5–10, 74, 100
  excitability, 613–619
  functional considerations, 29, 144
  general inputs to, 44, 45
  intra-amygdaloid connections, 55–63, 614–618
  nerve cell loss following kainate, 149, 214, 637, 649, 676
  neurotransmitter systems in the, 45, 614–618
  sensory inputs to, 36–41, 100
  zinc in the, 612, 618
Animal models of epilepsy, 139, 140, 147, 157, 158, 322, 333, 349, 440, 475, 631, 632, 647, 673
Anomalous rectification, 476
Anoxic neuronal damage, 135, 136,

Anoxic neuronal damage (continued) 637, 640, 651, 659, 687–694
  in cultures, 135, 687–692
  in slices, 692, 693
Antagonists of EAA, 131, 238–240, 242, 243, 253, 296, 303–305, 321–327, 364, 378, 385, 471, 639–641
Anticonvulsant actions
  of antagonists of EAA, 253, 259, 296, 304, 321–327, 349, 385, 425, 471, 588
  of baclofen, 677, 678
  of GABA, 343
  of taurine, 354
APB (Aminophosphonobutyric acid), 245, 246, 378, 379, 385
APH (Aminophosphonoheptanoic acid), 131, 133, 135, 239, 240, 255–257, 298, 304, 308, 322, 323, 325, 378, 385, 425
APV (Aminophosphonovaleric acid; =APP), 239–243, 298, 304, 322, 364, 365, 379, 425, 429, 431, 432, 442, 465–471, 473, 478, 482, 485–487, 492, 493, 501, 502, 641, 688, 690, 692
Aspartate
  action on hippocampal neurons, 429–435, 688
  action on spinal neurons, 426–428, 497, 498
  effect on extracellular ion concentrations, 441–443, 457
  extracellular levels, 128, 129, 132, 286
  release of, 254, 265
Aspartate aminotransferase, 128, 264
Astroglia swelling, 128, 224, 451, 453, 454, 636, 638, 662, 666
Astrogliosis, 107, 108, 111, 149–153, 161, 229, 337, 451, 651, 701
ATP stores and neuronal damage, 131, 132, 136, 690, 691
Autoradiography
  of amino acid uptake sites, 54

Autoradiography (continued)
  of EAA binding sites, 160, 161, 237-249, 306, 309, 310, 317, 378, 648, 649, 666

Ba^{2+} spikes, 403, 480, 540
Bed nucleus of stria terminalis, 25
Bicuculline, 527, 529
  induced damage, 176, 180, 181, 184, 185, 637, 676-678
  induced release of amino acids, 365, 369
  induced seizures, 176, 177, 185, 365, 371, 465, 466, 471, 637, 676-678
Blood brain barrier
  damage of, 175-186, 223, 225, 232
  principles, 175, 176, 191
  and seizures, 175-186, 197, 207, 208
  and transport of EAA, 191, 192, 200-207, 651, 700
  transport of macromolecules, 183-186
Brain damage produced by
  cholinergic agents, 632-635
  electrical stimulation of the perforant path, 635, 662-665
  folates, 632, 635, 638-640
  ICV bicuculline, 676-678
  ICV kainate, 674-684
  intracerebral ibotenate, 701, 703, 704
  intracerebral kainate, 199, 214, 632, 679
  parenteral kainate, 148-154, 214-217, 223-229, 631, 632, 651-654, 661, 662, 704, 711
  quinolinic acid, 308, 309, 350, 354-359, 668, 699
Brain uptake index of kainate, 200-202, 205, 206
Brainstem, 28, 75, 76, 101, 143
Burst generation
  and convulsant drugs, 466, 471-473, 677
  endogenous, 392
  and GABAergic inhibition, 471, 472
  in hippocampal slices, 392, 414, 466
  model, 397, 398, 407-409, 471-473
  and NMDA, 326, 406, 407, 433, 471, 472, 487

CA1
  effects of EAA, 159, 379, 471-473, 476-482, 665
  hypoxic/ischemic damage, 132, 161, 693, 694, 711
  lesions after KA, 149, 151, 152, 161, 214-216, 224-225, 633, 638, 666, 674
  PDS in, 391-393, 466
  synchronized bursting activity in, 414
CA3
  kainate receptors, 244, 648, 650, 651, 682, 683
  lesions after kainate, 149, 151, 160, 214-216, 633, 638, 639, 651-654, 660, 675-682, 710, 711
  PDS in, 392, 393
Ca^{2+}
  blockers, 405, 477
  conductance, 403, 407, 444
  dependent potassium channels (gKCa), 476, 480-482
  and EAA, 245-247, 364-370, 425, 435, 441-445, 465, 467, 476-479, 505, 653
  extracellular concentration of, 109, 129, 363-371, 407, 408, 435, 441-445, 454-457, 497, 500, 504, 505, 652, 653, 691
  influx and neuronal damage, 229, 306, 326, 457, 505, 641, 652, 691, 692
  influx during seizures, 109, 110, 306, 326, 370, 435, 444, 457, 481, 652
  spikes, 403, 426, 435, 465, 467, 477-479, 481, 540, 653
  uptake, 364, 365, 367-369
Calmodulin (CaM) regulation system, 593-595
Caudate nucleus, 132, 256, 286-291
Cerebellar neurons in explant culture, 518-522
Cerebral blood flow and seizures, 180-183, 232, 651
Cerebral cortex
  and actions of EAA, 406, 407, 439-445
  and GABA mediated inhibition, 158
  neuronal damage in, 150
  slice recordings, 406, 407
Cerebral glucose utilization, 119
  and kainate seizures, 649, 650
  and kindling, 140
Cerebrovascular permeability, 211
Changes in extracellular space in epilepsy, 161, 449-457
Chronic cobalt epilepsy, 440, 443, 444

Cl⁻
  and damage produced by EAA, 642, 643, 653, 691
  and EAA receptors, 245-247, 541
  extracellular levels of, 453-457
  and GABA function, 341, 526, 527
Coastline burst index, 467-471
Cortical epileptic focus, 105, 440
Cysteic acid, 254
Cysteine sulfinic acid, 254

Dark cell degeneration, 635-638
Decrease of inhibition and epilepsy, 108, 158, 159, 339-343, 469-473, 475, 651-653, 660, 683
Dihydrokainate, 365, 367, 369
Dipiperidinoethane, 632-635
Dithizone, 612, 618, 619

EAA
  containing pathways, 58-63, 253, 258, 263-275, 286, 588, 634, 635, 637, 669, 674
  effects in cultures, 354-358, 425-436, 497-505, 507-510, 518-522, 687-692
Edematous swelling, 224-227, 232, 337, 341, 451, 453, 635, 636-639, 651, 662
Endoplasmic reticulum
  swelling of, 224
Entorhinal cortex, 10, 71-73, 83-95, 227
Ephaptic currents, 413-422, 461
Epileptogenesis
  and circuitry, 392, 393
  cortical, 105, 158, 391, 403
  and EAA, 159, 161, 331-343, 409, 444, 445, 455-457, 465-473, 475-482, 541, 626, 669
  and ephaptic currents, 162, 413, 416-422
  hippocampal, 116, 120, 391-398, 454-457, 465-473, 480-483, 654, 655
  and interneurons, 159, 651
  and PDS, 165, 391-396, 402, 541
  and release from inhibition, 111, 112, 122, 158, 159, 469-473, 475, 651, 683
Epileptic predisposition, 332, 333
Excitatory postsynaptic current, 395

Excitatory postsynaptic potential (EPSP), 162, 391, 393-398, 461, 467, 472, 487, 488, 492, 493, 508, 510, 547, 603, 687-689
Excitotoxicity
  of cholinergic agents, 632-635, 651-655
  definition, 304, 305, 659
  endogenous excitotoxins, 308, 350, 359, 635, 659, 663, 697, 704, 710, 711
  of folates, 632, 635, 638-640
  of kainate, 307, 631, 651-655, 660, 661
  mechanisms of, 305-312, 326, 358, 359, 642, 643, 667, 668, 691, 692, 709-711
  and neurodegenerative disorders, 127, 131-134, 285, 305, 326, 386, 637, 643, 659, 704, 705
  of other EAA, 307, 308, 354, 699-705
  of perforant path stimulation, 635, 662-665
  role of mossy fibers in, 652, 659, 660, 678-683, 710
  and seizures, 110, 482, 631-636, 638-640, 643
Extracellular space, 109, 229, 352, 421, 440, 444, 449-457, 547, 550, 704
Extravasation of proteins during seizures, 177, 179, 183, 212, 217-219, 225, 229, 232

Fluorocitrate, 287, 288, 291
Fluoroglutamate, 518, 519
Folic acid, 632, 635, 638-640

GABA
  channels in locust muscle, 515
  inhibition in epileptic hippocampus, 108, 111, 121, 158, 159, 651-654
  metabolism and seizures, 334, 337, 339-343, 549, 563, 604, 605
  receptors in oocytes, 525-535
Giant synaptic potential hypothesis, 392, 394-396
Glutamate
  action on neurons in vitro, 486-493, 500-503, 507-509, 514, 515, 688
  binding sites, 237-243, 304, 377, 378
  brain metabolism of, 286-291, 333
  and Ca^{2+}, 366, 641-643
  channels in locust muscle fibers, 514-518, 521, 522

Glutamate (continued)
  by cleavage of NAAG, 380–382
  complex with zinc, 619
  effect on extracellular ion concentrations, 441–443
  extracellular levels, 128, 129, 132, 286, 369, 516, 517, 640
  induced neurotoxicity, 135, 285, 632–638, 641, 642, 663–669, 689, 692, 709–711
  pools in the brain, 129, 285–292, 318, 335
  receptors in oocytes, 528, 530, 532–535
  release of, 254, 265, 288–290, 304, 331–343, 369, 371, 502, 551–553, 612, 689, 692
  transfer constant, 194, 195, 197
  uptake into brain, 200
γ-Glutamic acid diethylester (GDEE), 324, 471
γ-Glutamylaminomethylsulphonate (GAMS), 323–325, 471
γ-Glutamylglycine, 297–299, 471, 590, 688–690, 692
Glycine
  receptors in oocytes, 525–528, 530, 531
GVG (γ-vinyl GABA), 141–144, 287, 291

Hippocampal formation
  anatomy and organization, 6–9, 67–69, 100
  cortical connections, 73, 74
  cultures, 354–357, 687–692
  entorhinal projections, 71–73
  ephaptic actions in, 414
  in situ recordings, 352, 353, 414, 419
  interneurons, 71, 651, 660
  intrinsic connections, 69–71
  plasticity of receptors in, 243, 247–249
  seizure induced neuronal damage in, 115–123, 149, 215–218, 224–226, 633, 639, 651, 659, 660, 669, 674–684, 699, 710, 711
  slice recordings, 379, 392, 414, 454–457, 465, 475, 476, 651
  subcortical connections, 74–76
  and zinc, 550, 551, 560, 565, 579, 580, 592, 593, 612, 618, 619, 626, 654
Homocysteic acid, 254–259, 317, 443, 444, 498, 503

Human cortex
  slice recordings, 162–169
Human (temporal lobe) epilepsy, 105, 110, 115–123, 147, 151, 154, 157, 162–169, 211, 331, 339, 342, 445, 475, 560, 613, 632, 647, 650, 659, 663, 669, 683, 684, 704, 705, 711
3-Hydroxyanthranilic acid oxygenase, 699–701, 703, 704
3-Hydroxy-2-quinoxaline carboxylic acid (HQC), 299, 300
Hypertrophy of capillary vessels, 149, 226
Hypoglycemia
  amino acid metabolism in, 286–290
  neuronal damage after, 127–131, 227, 363, 709
  and release of EAA, 128, 129, 285, 289
Hypothalamus, 23–27, 75, 101, 636

Immunocytochemistry
  aspartate-like, 58, 59
  CCK-like, 88, 89
  CRF-like, 91
  GABA-like, 55, 84, 85, 87, 151, 651–653
  GAD, 84, 85, 87, 121–123
  GFAP, 520
  glutamate-like, 58, 59, 264, 318, 612, 667
  homocysteate-like, 257, 258
  IgG-like, 212
  methodology, 54
  NPY-like, 93
  QPRT, 700–703
  somatostatin-like, 91, 92
  substance P-like, 93
  taurine-like, 57
  VIP-like, 89–91
In vivo brain dialysis, 129, 350–352, 363–371, 386
Inhibitory post synaptic potential (IPSP)
  and burst generation, 465
  and epileptogenesis, 158, 159, 472, 547, 601, 651
Insular cortex, 14
Interictal discharges, 391, 401
Interleukin-1, 111
Ion sensitive electrodes, 440, 450, 451, 493
Ionic changes during epileptic seizures, 439–445, 449–457
Ischemic damage, 115, 127, 132–136, 160, 223, 240, 285, 321, 326, 337, 338, 363, 445, 637, 640, 651, 693
Isoelectricity, 128

K^+
  accumulation, 108, 451-454
  conductances, 407, 476, 478, 479, 481, 488, 508
  effects of EAA on, 455-457, 476-482, 540, 541
  extracellular concentration of, 108, 109, 128, 161, 229, 392, 408, 451-456, 601
Kainate
  and anomalous rectification, 476
  binding sites (receptors), 160, 161, 199, 243, 244, 295, 304, 324, 439, 475, 590, 591, 648, 650-652, 682, 683, 710
  and blood brain barrier, 179, 199-208, 211, 225, 651
  and Ca^{2+} mediated events, 364, 365, 367, 369, 441-445, 455-457, 476-478
  cerebral blood flow, 182, 651
  derivatives of, 296-298
  effect on extracellular ion concentrations, 441-444, 455-457
  entry into the brain, 199-208, 650, 651
  and IgG-like immunoreactivity, 214, 217-219
  neuropathological changes, 147-154, 215-218, 224-227, 232, 307, 349, 482, 631, 632, 641, 651-654, 660-662, 674-684
  parenteral administration of, 147, 148, 179, 200-208, 212-218, 223-229, 365, 632, 647-654, 660, 704, 711
  receptors in oocytes, 528, 530, 532-535
  and reduction of inhibition, 159, 651, 661
  and release of EAA, 199, 307, 365, 366, 369, 370, 632
  and release of zinc, 550
  seizures produced by, 147, 148, 161, 179, 182, 197, 199, 207, 211, 213, 214, 349, 365, 475, 482, 550, 631, 632, 647-655, 660-662, 673-684, 705, 709-711
  spontaneous limbic motor seizures, 147, 148
  transfer constant, 194
β-Kainic acid, 324, 325
Ketamine, 426, 428-434, 436, 461, 640, 641, 690, 691, 693, 694

Kindling
  mechanisms of, 139-144, 161, 587-595, 612-619, 626
  microinjection of muscimol and, 141
  role of EAA in, 325, 588-592, 616-619
  role of substantia nigra in, 141-144
Klüver-Bucy syndrome, 35
Kynurenic acid, 309, 318, 324, 351, 352, 386, 693
Kynurenine pathway, 699

Lateral olfactory tract, 375, 379
Locust muscle fibers, 513-518
Long term potentiation (LTP), 240, 242, 258, 325, 472, 589, 590

Metalloenzymes, 547, 558
Metallothioneins, 560-565, 625
Methionine sulphoximine, 178, 179, 287, 290
Methoxypyridoxine, 177, 178
Mg^{2+}
  antagonism of anoxic damage, 689, 690, 693
  antagonism of NMDA, 407, 425, 436, 465, 485, 487-493, 497-500, 509, 539-541
  extracellular levels of, 461, 691
Microvessels
  EAA uptake into, 195, 196
Mitochondria
  swelling of, 224
Models of epileptiform activity, 397, 398, 401, 407-409, 439, 440, 444, 471, 472, 613
Monoamines, 94, 135, 339, 564, 575, 576, 615, 616, 697
mRNA extraction, 525

Na^+
  current, 405-407
  efflux receptor assay, 296, 317
  extracellular concentration of, 407, 408, 442-444, 453-457, 505
N-acetyl-aspartic acid (NAA), 375
N-acetyl-aspartylglutamate (NAAG), 318, 375-382, 386, 387
Negative resistance region, 508
Network-driven burst, 392, 393, 397, 398
NM(D)A (N-methyl-(D)-aspartate)
  action on spinal neurons, 486-493, 498-505
  antagonists, 131, 238-240, 242, 243, 258, 259, 296, 304, 321-327, 364, 378, 385,

NM(D)A (N-methyl-(D)-aspartate)
  (continued)
 antagonists (continued)
    425, 436, 465-473, 482,
    492, 640, 690, 699, 705
  blockade of neurotoxic effects,
    309, 641
  and burst discharge, 406, 407,
    433, 436, 465, 487, 539
  and Ca^{2+} mediated events, 364-366,
    368-370, 441-445, 455-457,
    465, 478, 479, 505, 539,
    540
  channels, 444, 497-505, 508,
    509
  induced lesions, 141, 308, 641,
    642, 690
  receptors, 131, 237-243, 254-
    259, 295, 304, 317, 321,
    323, 386, 406, 409, 425,
    441, 444, 461, 465, 468,
    470-473, 485, 493, 497,
    498, 508, 530, 539-541,
    588, 640, 688, 690, 693,
    697

β-ODAP (L-3-oxyalylamino-2-amino-
    propionic acid), 425, 427,
    430, 433, 435, 436
Oocytes
  receptor expression in, 525-535
Organotypic cultures, 354-358

Pacemaker neurons, 105, 107, 108,
    710
Paroxysmal current, 395
Paroxysmal depolarizing shift (PDS),
    165, 304, 391-396, 401,
    402, 439, 461, 466, 472
Patch clamp, 507-510, 513-522
Penicillin induced epilepsy, 333,
    334, 393, 401
Pentylenetetrazole, 176, 297, 298
Peptides of kainic acid, 296-298
Phencyclidine (PCP), 426, 428-430,
    433, 436, 640, 641
Phosphoethanolamine, 364-366, 371
Picrotoxin, 297, 465-468, 471,
    528, 530
Piperidinedicarboxylic acid (PDA),
    308, 688, 689
Protection from neuronal damage
    by
  experimental lesion, 678-683,
    710
  ketamine, 640, 690
  NMDA antagonists, 131, 309,
    326, 689, 699
  taurine, 353-358, 371

Pyriform cortex
  and changes after KA, 149, 153,
    224, 228, 638, 651, 676

Quinolinic acid, 307-312, 318,
    349-359, 370, 668, 697-
    705, 711
Quinolinic acid phosphoribosyl-
    transferase (QPRT), 699-
    704
Quisqualate, 255, 256, 379, 425,
    427, 441-444, 455, 456,
    486-493, 498-500, 507,
    508, 518, 519, 528, 688
  receptors, 134, 245, 295, 304, 324,
    441, 485, 493, 497, 508, 591,
    614, 688

Release
  of EAA, 129, 207, 254, 265, 271,
    288-290, 311, 325, 331-343,
    366, 369, 386, 589, 612, 660,
    689, 692, 710, 711
  of endogenous sulfur containing
    amino acids, 254-257, 350-352
  of endogenous zinc, 550, 654
  of metal ions, 550
  of phosphoethanolamine, 364-366,
    371
Retina
  effects of EAA in, 641, 642
Retrograde tracers, 6, 8, 22, 54,
    59-63, 84, 258, 265-271,
    274, 318
Reversal potential, 394, 395, 503-
    505

Safety factor of antidromic
    invasion, 416
Schaffer collateral system, 242,
    466, 467, 475
Seizure duration and sclerosis,
    678-682
Seizures induced by
  cholinergic agents, 323, 614,
    616, 632-634
  EAA, 147, 148, 159, 161, 179, 182,
    197, 199, 200, 213, 214, 331-
    343, 351, 354, 365, 369, 378,
    379, 433-436, 445, 465-473,
    475-482, 588, 631-636, 638-640,
    643, 647-654, 660-662, 673-684,
    698, 699, 701
  metal ions, 105, 158, 440, 451,
    546, 547, 559-563, 571, 580,
    581, 592, 599-601, 612
  other chemical convulsants, 176-179,
    185, 322, 333, 365, 401, 440,
    451, 453, 466, 637, 676-678

Selective seizure related brain damage, 115-123, 147-154, 214-217, 223-229, 449, 482, 631-636, 638-640, 643, 652-654, 660-669, 673-684, 709-711
Septal complex, 74
Single channel currents, 508, 513-515, 519, 521, 522
Sommer sector, 9, 115, 116, 151
Spike accomodation, 467
Spike and wave primary epilepsy, 338, 339
Spontaneous
  limbic motor seizures, 147, 148, 650
  rhythmic synaptic events, 167
  spreading epileptiform activity, 422
Spreading depression of Leao, 333, 337, 338
Status epilepticus, 186, 213, 214, 337, 365, 371, 449, 457, 588, 631, 639, 647, 678, 679, 681
Striatum, 21, 22
Strychnine, 401, 414, 527, 528, 531
Subiculum (Pre-,Para), 8-10, 73-75, 85, 119, 121-123
Substantia innominata, 19-21
Substantia nigra
  burst firing in, 143
  microinjections into, 141, 142, 323
  role in kindling, 143, 144
Sulfur containing amino acids, 253-259, 303
Synchronized neuronal discharge, 123, 401, 402, 408, 414, 422

Taurine, 57, 333-337, 349-359, 364-366, 371, 386, 387
Temporal lobe
  anatomical organization, 11, 13, 14
  epilepsy, 14, 99, 115-117, 163, 475, 617, 640, 647, 673, 674, 684, 704, 705, 711
  and zinc, 612
Tetraethylammonium (TEA), 403, 404, 406, 407, 426, 467, 477-479
Tetramethylammonnium (TMA), 450-452, 455, 456
Tetrodotoxin (TTX), 364, 367, 369, 403, 404, 426, 430, 441, 442, 467, 476-479, 481, 486, 487, 489-493, 498, 515, 691

Thalamus, 22, 23, 75, 149, 150, 634
Transfer constant of EAA, 194, 195, 197
Transition metals, 545-553
  and EAA release and uptake, 551-553
  in enzyme catalysis, 547-550
  and epilepsy, 545-553, 601, 611
  general properties, 546, 547, 602
  release into extracellular space, 547, 550, 625
Transmembrane potential, 414, 415, 417-421, 453
Trifluoroperazin, 287

Uptake of amino acids, 200-208, 265, 271, 338, 380, 381, 551-553, 589, 612, 617, 666
  into brain microvessels, 193-197, 231

Voltage clamp, 391, 403, 426-436, 497-502, 526, 528, 529, 533
Voltage dependency EAA mechanism, 407, 425, 433, 456, 461, 497-505, 507-510

Zinc
  behavioral effects induced by, 575, 576, 626,
  binding protein in the brain, 559-564
  and calcium antagonism, 564, 601, 603, 604
  deficiency and CNS disorders, 557-560, 571, 599
  deficient diet, 546
  and EAA, 550-553, 564, 581, 593, 612, 618, 619, 626, 668
  effects on brain receptors, 565, 572-581, 593, 626
  effects on the firing rate, 602-607
  and enzyme catalysis, 547-549, 558, 563, 599
  and excitotoxicity, 550, 668, 669
  and GABA-benzodiazepine receptors, 572-574, 580, 581, 626
  and GABAergic function, 549, 563, 601, 604, 605
  induced seizures, 546, 559-563, 571, 580, 581, 592, 599-601, 611
  and kindling, 592, 593, 612, 618, 619
  and neuroendocrine function, 576-578, 626
  and opiate function, 564, 574, 578, 579, 605-607
  release of endogenous, 550, 551, 602, 654, 668

MIX
Papier aus verantwortungsvollen Quellen
Paper from responsible sources
FSC® C105338

If you have any concerns about our products,
you can contact us on
ProductSafety@springernature.com

In case Publisher is established outside the EU,
the EU authorized representative is:
Springer Nature Customer Service Center GmbH
Europaplatz 3, 69115 Heidelberg, Germany

Printed by Libri Plureos GmbH
in Hamburg, Germany